VIIIth INTERNATIONAL ASTRONAUTICAL CONGRESS
BARCELONA 1957

VIII. INTERNATIONALER ASTRONAUTISCHER KONGRESS

VIIIe CONGRÈS INTERNATIONAL D'ASTRONAUTIQUE

PROCEEDINGS

BERICHT **COMPTES RENDUS**

HERAUSGEGEBEN VON EDITORIAL BOARD COMITÉ DES RÉDACTEURS

P. J. BERGERON · W. VON BRAUN · K. A. EHRICKE · F. HECHT · J. MARIAL
E. SÄNGER · K. SCHÜTTE · L. I. SEDOV · L. R. SHEPHERD · S. F. SINGER

SCHRIFTLEITUNG EDITOR-IN-CHIEF RÉDACTEUR EN CHEF

F. HECHT

MIT 233 FIGUREN WITH 233 FIGURES AVEC 233 FIGURES

Springer-Verlag Wien GmbH 1958

ISBN 978-3-662-39020-7 ISBN 978-3-662-39990-3 (eBook)
DOI 10.1007/978-3-662-39990-3

Softcover reprint of the hardcover 1st edition 1958

Vorwort

Der vorliegende Sammelband der anläßlich des VIII. Internationalen Astronautischen Kongresses, der vom 6. bis 12. Oktober 1957 in Barcelona stattfand, gehaltenen Vorträge umfaßt alle vom zuständigen Prüfungskomitee der Internationalen Astronautischen Föderation approbierten Vorträge mit Ausnahme eines, dessen Manuskript verspätet einlief.

Der VIII. Kongreß der Internationalen Astronautischen Föderation stand unvorhergesehenerweise im Zeichen des ersten erfolgreichen Erdsatellitenstartes, des „Sputnik I" (4. Oktober 1957). Die kommende Entwicklung zeichnete sich jedoch schon weitgehend in den Vorträgen dieses Kongresses ab, obwohl diese ja bereits vor dem Start abgefaßt worden waren. Sie sind deshalb auch vom Standpunkt der Geschichte der Raumfahrt besonders interessant.

Die Herausgeber des vorliegenden Berichtsbandes sind dem Springer-Verlag in Wien dafür zu Dank verpflichtet, daß dieser keine Mühe und Kosten scheute, um das Sammelwerk mit großer Sorgfalt und technisch einwandfrei auszugestalten. Besonderer Dank gebührt ferner den unermüdlichen Übersetzern der den einzelnen Vorträgen vorausgehenden Kurzfassungen: Herrn Dr. Leslie R. Shepherd, Chilton, Berks., United Kingdom, für die Übersetzungen in die englische Sprache, und Herrn Professor Ing. Baudouin Fraeijs de Veubeke, Université de Liége, Belgique, für die zahlreichen Übersetzungen in die französische Sprache. Ohne die uneigennützige Hilfe dieser beiden Kollegen wäre der Berichtsband nicht in der vorliegenden Form zustande gekommen.

Friedrich Hecht, Wien

Contents — Inhalt — Sommaire

Einige Arbeiten, deren Manuskripte besonders fruh vorlagen, sind schon im Heft 1 des Bandes 4 (1958) der „Astronautica Acta" erschienen. Deshalb sind in dem vorliegenden Berichtsband zunächst in alphabetischer Reihung alle übrigen Vorträge in englischer, deutscher, französischer oder italienischer Sprache abgedruckt; an diese schließen die Arbeiten in russischer Sprache, ebenfalls in alphabetischer Reihenfolge, an. Als letzte folgen dann, wieder in alphabetischer Folge, die obenerwähnten, schon in den „Astronautica Acta" erschienenen Arbeiten. Diese Anordnung erfolgte *ausschließlich* aus drucktechnischen Gründen.

Der Vortrag „Observation of a Satellite near its Culmination" von Professor Dr. Ing. A. Boni, Roma, konnte in den vorliegenden Sammelband nicht mehr aufgenommen werden, da das endgültige Manuskript der Redaktion verspätet eingesandt wurde; er wird noch im Jahre 1958 im Band 4 der „Astronautica Acta" erscheinen.

Essai de contribution à l'autopropulsion nucléaire

Par

J.-J. Barré [1]

(Avec 7 Figures)

Résumé — Zusammenfassung — Abstract

Essai de contribution à l'autopropulsion nucléaire. Il est d'abord rappelé que les réacteurs nucléaire utilisables pour l'autopropulsion devraient fonctionner à des températures très élevées, de l'ordre de celles qui règnent dans les foyers des autopropulseurs classiques, soit 2500 à 3000° K.

Cette remarque amène l'auteur à utiliser la première résonnance de la courbe des sections efficaces du plutonium, résonnance qui se situe vers 3350° K avec une section efficace de plus de 3000 barns. Le plutonium devrait alors être utilisé à l'état fondu, ce qui présenterait d'ailleurs certains avantages; la difficulté serait de franchir à la mise en route la zône de fonctionnement instable qui s'étend de 800° K à 3350° K.

Chauffée à de telles températures, l'ammoniac pourrait atteindre une vitesse de sortie supérieure à 4200 m/s. Pour accroître cette vitesse de sortie, il est proposé ensuite de fonctionner à détente réchauffée. La cession de chaleur par convection ne pouvant être envisagée durant la détente, il est proposé de chauffer le gaz par le rayonnement des parois, le gaz ayant été préalablement opacifié par mise en suspension de particules de carbone. Cette solution permettrait d'atteindre quelques 5200 m/s (toujours avec l'ammoniac).

Pour aller plus loin, l'auteur fait appel à "l'hyperchauffe" c'est-à-dire à la chauffe du fluide par freinage en son sein des fragments de fission échappés au matériau fissible. Deux schémas sont proposés, l'un avec du carbure d'uranium 235 à l'état solide, l'autre avec du plutonium à l'état gazeux. La vitesse de sortie serait alors voisine de 6000 m/s.

En soustrayant à l'hyperchauffe une partie du fluide, par prélèvement à la sortie de la préchauffe convective, la vitesse de sortie pourrait encore être notablement augmentée. (Après refroidissement par rayonnement de la portion de fluide prélevée, cette dernière serait recyclée.)

L'auteur attire particulièrement l'attention du lecteur sur le caractère purement spéculatif de cet exposé qui n'a donné lieu de sa part à aucune recherche expérimentale.

Beitrag zum Problem des nuklearen Triebwerkes. Es sei zunächst daran erinnert, daß fur Triebwerke verwendbare Kernreaktoren bei sehr hoher Temperatur arbeiten müßten, die in die Größenordnung derjenigen zu liegen kämen, die in den Brennkammern der klassischen Triebwerke herrschen, d. h. bei etwa 2500 bis 3000° K.

Diese Bemerkung veranlaßt den Verfasser, die erste Resonanzstelle in der Kurve der Wirkungsquerschnitte des Plutoniums zu benützen. Diese Resonanz liegt bei 3350° K bei einem Wirkungsquerschnitt von mehr als 3000 barn. Infolgedessen müßte das

---

[1] Ingénieur Militaire en Chef des Fabrications d'Armement, 49, rue du Maréchal Foch, Versailles (Seine et Oise), France.

Plutonium in geschmolzenem Zustand angewendet werden, was übrigens gewisse Vorzüge mit sich brächte; die Schwierigkeit bestünde darin, bei der Inbetriebnahme die Zone der instabilen Funktion zu überschreiten, die sich von 800° K bis 3350° K erstreckt.

Bei einer Aufheizung auf solche Temperaturen könnte Ammoniak eine Ausströmungsgeschwindigkeit von mehr als 4200 m/sec erreichen. Um diese Geschwindigkeit noch zu erhöhen, wird vorgeschlagen, bei Entspannung unter Wiedererhitzung zu arbeiten. Da der Wärmeverlust durch Konvektion während der Entspannung nicht in Betracht gezogen werden kann, wird vorgeschlagen, das Gas durch die Strahlung der Wände zu erhitzen, nachdem es vorher durch Suspension von Kohlenstoffpartikeln undurchsichtig gemacht worden ist. Diese Methode würde die Erreichung von etwa 5200 m/sec gestatten (vorausgesetzt ist immer Ammoniak).

Darüber hinaus schlägt der Verfasser eine „Überhitzung" vor, d. h. die Erhitzung des strömenden Mediums durch Bremsung der aus dem spaltbaren Material stammenden Spaltungsfragmente im Strömungsinneren. Es werden zwei Schemata vorgeschlagen: das eine mit 235 U-Carbid in festem Zustand, das andere mit Pu in gasförmigem Zustand. Die Auströmungsgeschwindigkeit läge dann bei ungefahr 6000 m/sec.

Die Ausströmungsgeschwindigkeit könnte aber noch beträchtlich vergrößert werden, wenn ein Teil des strömenden Mediums durch Anzapfung im überhitzten Bereich nach der konvektiven Vorerhitzung abgezweigt würde. (Nach Abkühlung des abgezweigten Strömungsanteils durch Abstrahlung könnte dieser wieder in den Zyklus zurückgeführt werden.)

Der Verfasser betont besonders den rein spekulativen Charakter der Abhandlung, die er nicht experimentell prüfte.

Contribution to the Problem of the Nuclear Rocket Engine. It is recalled that nuclear reactors for propulsion of rockets must operate at very high temperatures, of the same order as those obtaining in the thrust chambers of conventional engines, namely, 2,500 to 3,000° K. This fact suggests to the author that the first capture cross section resonance of plutonium-239 should be utilised. This is situated at a neutron energy corresponding to a temperature of 3,350° K and has an effective cross section of 3,000 barns. At this temperature the plutonium might be used in the molten state, which, in other respects, presents certain advantages. Some difficulty would be encountered in crossing the unstable region of neutron temperature between 800 and 3,350° K.

Heated to such a temperature, ammonia could attain an exhaust velocity of 4,200 m/s. It is suggested that a greater exhaust velocity might be obtained by heating the gas during expansion. This would be done by radiative rather that convective heat transfer, the gas being rendered opaque by means of a suspension of carbon particles. In this manner it would be possible to achieve, with ammonia, an exhaust velocity of 5,200 m/s.

Further increase in performance might be achieved by a process which the author describes as "superheating", that is to say, direct heating of the working fluid by recoiling fission fragments escaping from the fissile material. Two schemes are proposed, the one involving uranium-235 carbide in the solid state, the other involving gaseous plutonium. The exhaust velocity resulting, in the first case, would be in the neighbourhood of 6,000 m/s.

In the superheating process, the working gas passes twice through the reactor core, removing heat by convective transfer during the first transit and by direct fission fragment recoil during the second passage. If, after the first transit, the fluid is divided and only part of it is subjected to superheating, the exhaust velocity may be still further augmented. (The portion of the fluid which is not superheated is cooled by radiation and then recycled.)

The author emphasises that the proposals made in his paper are purely speculated and do not refer to any experimental work carried out on his part.

Dès 1948, notre très honoré Président, Monsieur SHEPHERD, publiait en collaboration avec Monsieur CLEAVER, le schéma d'autopropulseur nucléothermique que voici (Fig. 1).

Dans ce dispositif, le fluide propulsif est chauffé par convection en passant à travers un réacteur nucléaire fonctionnant à très haute température, puis subit une détente isentropique dans une tuyère convergente-divergente que réfrigère le propulsif avant de pénétrer dans la chambre de chauffe.

"Très haute température" avons nous dit; en effet, c'est la température initiale du fluide qui, dans une détente isentropique, conditionne la vitesse d'éjection; la relation est bien connue:

$$w_\sigma{}^2 = \frac{2\gamma}{\gamma - 1} \cdot \frac{\mathfrak{R}}{\mathfrak{M}} \cdot (T_i - T_\sigma). \qquad (1)$$

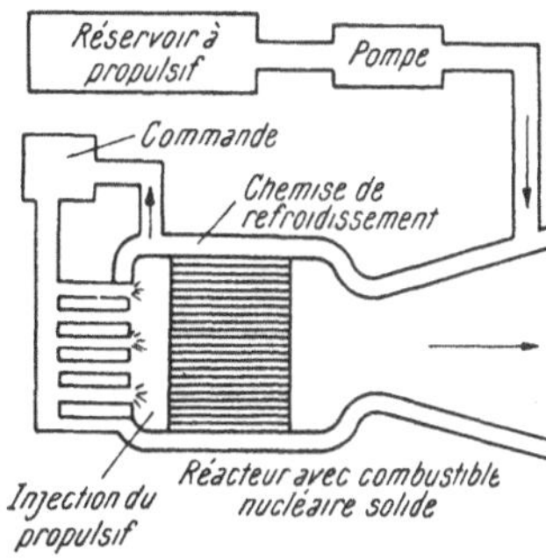

Fig. 1

Or les fusées usuelles à propergols chimiques fonctionnent déjà à des températures avoisinant 3000° K; sous peine d'être surclassé par ces dernières, l'autopropulseur ci-dessus devra donc être équipé d'un réacteur nucléaire fonctionnant à des températures de cet ordre.

De tels réacteurs n'ayant pas encore été réalisés, il convient avant tout d'examiner si, à priori, la chose est possible ou non.

Réacteurs nucléaires pour températures élevées

Bien que l'étude des réacteurs nucléaires ne soit ni de notre ressort, ni de notre compétence, la conception de réacteurs fonctionnant à des températures très élevées ne nous paraît pas utopique à priori.

En effet, tout d'abord, il ne semble pas inconcevable de faire fonctionner un réacteur avec des matériaux fissibles fondus; bien des réacteurs à phase liquide ne sont-ils pas déjà en service, certains utilisant même un alliage fondu de Bismuth et d'Uranium ?

Le fonctionnement en phase liquide élimine d'ailleurs tous les inconvénients dûs aux changements d'états allotropiques des matériaux fissibles, changements d'états qui détériorent rapidement les réacteurs à combustible solide.

Par ailleurs, les tensions de vapeur de ces matériaux fondus sont particulièrement faibles: fondant vers 1400° K, l'Uranium n'atteindrait sa température d'ébullition sous la pression atmosphérique qu'aux environs de 4200° K[1].

D'autre part, dans un bain de métal en fusion, les "poisons" produits par les fissions se décanteraient d'eux-mêmes et il serait relativement facile de les soutirer et de les évacuer en les soufflant dans la tuyère avec le propulsif (à condition, bien entendu, d'opérer dans l'espace ou, au moins, en zône désertique).

Ce bain pourrait être contenu dans un creuset réfractaire, en graphite par exemple, la pression étant supportée par des parois métalliques refroidies par le propulsif comme le sont celles d'un éjecteur par le propergol.

Enfin la décroissance de réactivité due à l'élévation de température du réacteur ne parait pas constituer un obstacle rédhibitoire; de 300° K à 3000° K la section efficace de U^{235} passe en effet de 450 à 130 barns[2].

[1] D'après Rare Metals Handbook.

[2] Le barn vaut 10^{-24} cm^2.

Quant au Pu^{239}, il ne suit pas du tout la loi en 1/v; comme on peut le voir sur la Fig. 2, il a sensiblement la même section efficace à 4500° K qu'à 300° K, mais, dans l'intervalle, à 3350° K, cette section efficace a atteint la valeur considérable de plus de 3000 barns; il faudrait franchir avec précaution cette pointe de réactivité, les variations de température ayant tendance à déstabiliser le régime du réacteur sur la branche ascendante de 800 à 3350° K; mais sur la branche qui descend après le maximum, les variations de température redeviennent autorégulatrices. Comme on le verra tout à l'heure, cette particularité du plutonium paraît susceptible d'être exploitée pour réaliser "l'hyperchauffe" en phase gazeuse, étant donné les hautes températures nécessaires[1].

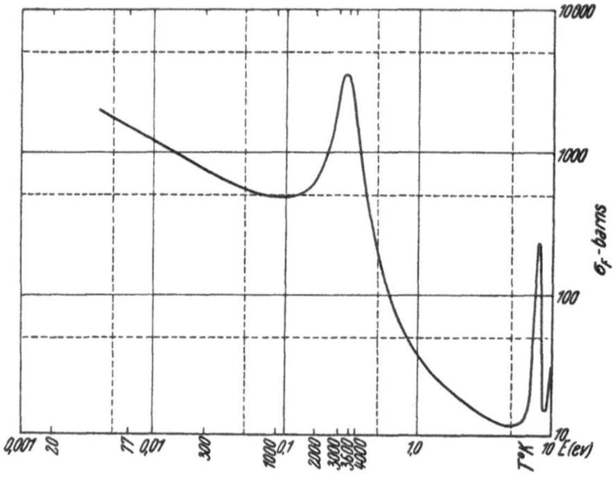

Fig. 2. Sections de fission efficaces du $_{94}Pu^{239}$ (d'après D. J. Hughes et ses collaborateurs)

Admettant donc l'hypothèse optimiste de la possibilité de réaliser un jour ou l'autre des réacteurs fonctionnant aux environs de 3000° K, il parait intéressant tout d'abord de noter que le dispositif simple décrit ci-dessus présenterait déjà deux avantages sur les fusées classiques fonctionnant à des températures analogues.

Progrès réalisables par le dispositif a chauffe convective et détente isentropique

Le premier de ces progrès, très sensible aux Astronautes et futurs découvreurs de mondes inconnus que nous sommes tous ici (n'est-il pas vrai ?), serait la quasi-certitude de pouvoir se ravitailler partout en propulsif de retour: l'eau, la neige carbonique, des minéraux facilement fusibles seraient en effet beaucoup plus faciles à trouver que du pétrole et de l'acide nitrique ou tout autre comburant!

Le second, plus subtil, résiderait dans un choix de propulsifs beaucoup plus étendu; le propulsif n'étant plus nécessairement un produit de combustion, l'on pourrait adopter des corps de masse moléculaire, de chaleur spécifique, d'atomicité et de densité les plus favorables.

C'est ainsi qu'il résulte d'une prospection effectuée, sous la direction du Professeur Serruys, par la Section de Thermodynamique du Centre d'Etudes des Projectiles Autopropulsés[2], que l'ammoniac NH_3, chauffé à 3000° K puis

[1] Pour réaliser l'hyperchauffe en phase solide, il faudrait au contraire utiliser un modérateur à très basse température (et ce serait alors aussi valable pour U^{235}). Pour Pu^{239}, les sections efficaces extrapolées seraient de 3000 b avec des neutrons de 20° K (H_2 liquide) et de 1500 b à 77° K (N_2 liquide). A noter que la coexistence de très hautes et de très basses températures est couramment réalisée dans les fusées utilisant l'oxygène liquide.

[2] Organisme dépendant de la Direction des Etudes et Fabrications d'Armement et placé sous la direction scientifique du Professeur H. Moureu.

détendu isentropiquement de 20 à 0,2 hpz[1], acquièrerait une vitesse de sortie de 4200 m/s correspondant à une vitesse efficace de 4440 m/s (soit à une consommation spécifique de 2,25 Kg de propulsif par tonne de poussée).

Si l'on remarque que la densité de l'ammoniac solide atteint la valeur très honorable de 0,82, on voit qu'une telle performance constituerait un sérieux pas en avant de celles des engins actuels et justifierait le recours à l'énergie nucléaire.

Mais l'on peut espérer beaucoup mieux, et ce sera là l'objet principal de cette causerie.

Voies ouvertes au développement de l'autopropulseur nucléothermique

Dans leur mémoire de 1948 précité, MM. Shepherd et Cleaver indiquaient déjà, entre autres, ces deux voies suivantes:

1° — réchauffe de la détente par réassociation chimique du fluide dissocié en cours de chauffe;

2° — accroissement de la température du propulsif au delà de celle des parois, par freinage des fragments de fission au sein même du fluide.

Ce dernier processus que nous qualifierons "hyperchauffe" dans la suite, impose une division extrême du matériau fissible; cette condition a conduit les Auteurs du mémoire susvisé à la conception d'un réacteur fonctionnant en phase gazeuse.

Ce sont des voies tout à fait analogues dont nous allons essayer maintenant d'entreprendre le défrichement.

Réchauffe de la détente

Nous remarquerons d'abord que les excellentes performances signalés plus haut pour l'ammoniac relèvent précisément de la première proposition précitée: elles sont dues en effet à l'importante dissociation subie par ce corps à 3000° K et à l'énergie dégagée par les réassociations qui se produisent au cours de la détente isentropique (au sens étendu de ce terme).

C'est d'une idée analogue que procède le principe du propulseur nucléothermochimique du Professeur Serruys (Fig. 3).

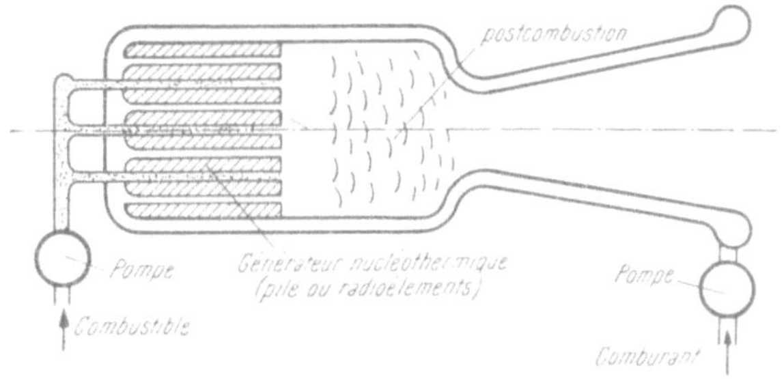

Fig. 3

Prenant pour propulsif un combustible et un comburant, l'auteur du projet envisage de les chauffer séparément dans un réacteur nucléaire à la sortie duquel ils se mélangeraient, subissant alors une "post-combustion" qui se prolongerait au cours de la détente.

Malheureusement, en revenant aux propergols chimiques, les avantages signalés tout à l'heure disparaissent, en particulier en ce qui concerne le choix du fluide à détendre; il s'en suit que les performances calculées sont nettement inférieures à celles que l'on avait cru pouvoir escompter à priori.

---

[1] Il est rappelé que l'hectopièze (1,019 Kg/cm²) se situe entre le Kg/cm² et l'Atmosphère (1,033 Kg/cm²).

Le véritable avantage de ce processus, avantage non négligeable d'ailleurs, serait de pouvoir réaliser encore d'excellentes performances en développant dans le réacteur nucléaire des températures bien moins élevées que celles envisagées jusqu'ici, soit 1500 ou 2000° K.

L'on peut enfin penser à une réchauffe nucléaire de la détente. Pour cela, le recours à la convection est exclu: en effet, sur les bords exceptés, l'écoulement de la veine est laminaire et la vitesse de translation avoisinant les vitesses d'agitation moléculaire, l'on ne peut espérer une "réchauffe à cœur". D'ailleurs les cœfficients de transfert des parois à la couche limite, même supposée turbulente, seraient faibles, de l'ordre de 0,1 w/cm² et degré de différence de température; pour une couche limite laminaire, ce coefficient descendrait à 0,01 w/cm² degré et même en-dessous.

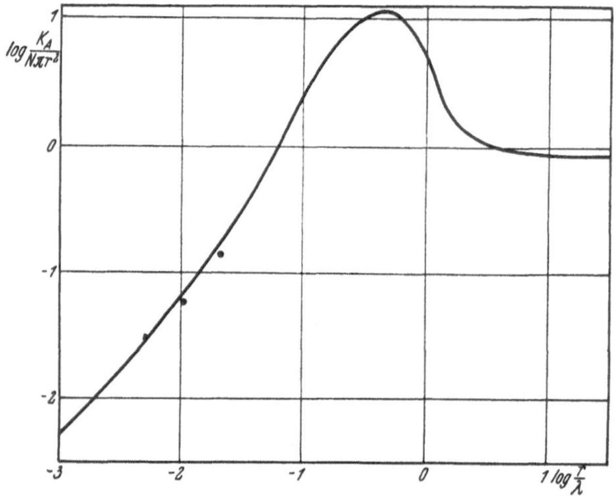

Fig. 4. Coefficient d'absorption du carbone, valeurs relatives (● valeurs empruntées à PEPPERHOFF)

Il faudrait donc recourir au rayonnement. Avec des parois en graphite ou en oxydes de terres rares (ou zircone, par exemple), le pouvoir émissif atteindrait l'unité et le rayonnement des parois suivrait la loi de STEFAN.

Mais pour que ce rayonnement soit absorbé par le fluide il serait nécessaire d'opacifier ce dernier en y mettant en suspension des particules de carbone.

Il convient que ces dernières soient assez nombreuses pour absorber toute l'énergie rayonnée et la recéder au gaz par chocs moléculaires; mais il faut, par contre, que leur débit-masse ne soit qu'une fraction du débit-masse du courant gazeux pour ne pas ralentir ce dernier.

La courbe que voici (Fig. 4), tirée d'un mémoire du Dr. F. ROESSLER du Laboratoire de Recherches de St-Louis, donne les variations de $\log \frac{K_A}{\pi \cdot r^2 \cdot N}$ en fonction de $\log \frac{r}{\lambda}$,

K_A étant le coefficient d'absorption,
r le rayon des particules,
N leur nombre par cm³,
λ la longueur d'onde de la lumière absorbée.

Or, d'après la loi de WIEN, à 3000° K le λ correspondant au maximum de la courbe de répartition d'énergie du spectre du corps noir est très voisin de 1 μ, c'est-à-dire qu'elle se situe dans le proche infra-rouge.

En donnant aux particules un rayon de l'ordre de 0,3 μ l'on voit que le coefficient moyen d'absorption K_A sera approximativement:

$$K_A \cong 4\pi \cdot 9 \cdot 10^{-10} \cdot N \qquad (2)$$
$$= 1{,}13 \cdot 10^{-8} \cdot N.$$

Si R est le rayon du divergent en un point donné de son axe, et que l'on désire en ce point une absorption de 99% il faudra que:

$$K_A \cdot R = \log_e \frac{I_0}{I} = \log_e \frac{100}{1} = 4{,}605 \cong 4{,}6 \qquad (3)$$

$$\text{d'où: } N = \frac{1}{R} \cdot \frac{4{,}6}{1{,}13} \cdot 10^8 \cong 4{,}07 \cdot 10^8 \cdot \frac{1}{R} \cong 4 \cdot 10^8 \cdot \frac{1}{R}.$$

Par exemple, pour un rayon de sortie de 100 cm, cela donnerait:

$$N = 4 \cdot 10^6.$$

La masse spécifique du carbone étant de l'ordre de 2, cela représenterait une densité de particules de $9 \cdot 10^{-7}$ g/cm³. Or sous 0,5 hpz et à 2750° K, la densité de l'ammoniac, dissocié serait de $1{,}9 \cdot 10^{-5}$. La proportion de carbone serait donc de 4,7 g pour 100 g d'ammoniac. En supposant les particules en équilibre de température et de vitesse avec le gaz, la perte de vitesse serait donc de l'ordre de 2,4% mais la densité du propulsif serait légèrement accrue.

Pour des particules de 0,05 μ l'absorption serait sensiblement la même, mais le transfert de particules à gaz serait plus aisé; il serait donc indiqué de chercher à s'approcher de cette dernière dimension.

Avec cette absorption de 99% et en admettant une différence de température de 200° K, la puissance transmise par cm² serait donnée par la formule:

$$P/S = 5{,}71 \cdot 10^{-12} \cdot 0{,}99 \cdot (3000^4 - 2800^4) \qquad (4)$$
$$= 113 \text{ w/cm}^2.$$

Toujours avec l'ammoniac, pour une poussée de 1 tonne, avec une détente de 20 à 0,2 hpz, il faudrait fournir à la détente 25800 kw; les parois de la tuyère devraient donc avoir une superficie de 22,8 m², ce qui semble réalisable.

La vitesse de sortie atteindrait 5200 m/s, et ceci constituerait un progrès notable sur les 4200 m/s de la détente isentropique à partir de 3000° K.

L'on va voir maintenant que l'hyperchauffe permettrait encore une nouvelle amélioration des performances.

Hyperchauffe

Comme il a été dit tout à l'heure, l'hyperchauffe consisterait à transformer en chaleur, au sein même du fluide, l'énergie cinétique des fragments de fission; pour cela on favoriserait l'évasion de ces derniers hors du matériau fissible en réduisant l'épaisseur de ce "combustible" nucléaire à l'équivalent de 1 micron de combustible pur.

Un dégagement résiduel de chaleur dans la lame fissible ne pouvant être évité, il serait nécessaire de la refroidir; par ailleurs, pour assurer la résistance mécanique d'une lame aussi mince, il faudrait l'épauler. L'organisation suivante (Fig. 5) répond à ces deux conditions: la couche fissible (composé réfractaire d'Uranium ou de Plutonium) est étalée sur les parois internes de tubes en graphite disposés en faisceau dans une enceinte résistant à la pression.

Dans ces conditions, environ 50% des fragments de fission sont freinés dans la couche fissible ou le graphite tandis que l'autre moitié s'échappe à l'intérieur des tubes.

Le fluide propulsif arrive dans la jupe de refroidissement de la tuyère puis traverse le faisceau en léchant l'extérieur des tubes qu'il refroidit tandis que lui-même s'échauffe; il ressort par le haut des tubes à une température de l'ordre de 2500 à 3000° K et redescend par l'intérieur de ceux-ci en s'hyperchauffant.

Son enthalpie initiale s'étant accrue de ΔH dans le parcours ascendant, s'accroît sensiblement de la même quantité pendant l'hyperchauffe, ce qui revient à très peu près à multiplier par $\sqrt{2}$ la vitesse limite correspondant à la chauffe simple.

Le diamètre interne des tubes est conditionné par la distance de freinage des particules dans le gaz propulsif. Pour de l'ammoniac sous 50 hpz et à 4000° K, ce diamètre devrait être de l'ordre de 1,4 cm. (Si l'on tient compte du surcroît de freinage dû à l'intense ionisation produite par les fragments de fission, ce diamètre pourrait sans doute être réduit.)

Il convient maintenant de rechercher les conditions de réactivité d'un tel faisceau placé à l'intérieur d'une enveloppe à laquelle serait donnée la forme favorable d'un cylindre de hauteur égale à son diamètre. Se reportant à la Fig. 5, l'on voit que le volume interessé par un élément tubulaire est celui du prisme hexagonal exinscrit à la paroi interne du tube considéré. Avec un matériau fissible de masse spécifique 18,7 et des tubes de 2 mm d'épaisseur en graphite de masse spécifique 1,6 les densités de présence sont égales à $2,9 \cdot 10^{-3}$ g/cm³ pour le matériau fissible, à 0,57 g/cm³ pour le graphite et à 0,022 pour l'ammoniac dissocié à la température moyenne de 3000° K et sous 50 hpz[1].

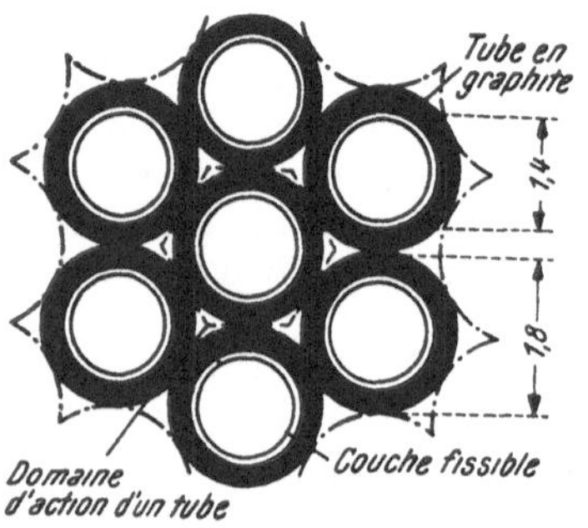

Fig. 5

Partant des données tirées de "The Science and Engineering of Nuclear Power" édité par CLARK GOODMAN en 1947, de "Pile Neutron Research" de DONALD J. HUGUES de 1953 et du mémoire "Notions fondamentales relatives aux réacteurs nucléaires" publié par M. SURDIN du C.E.A. dans le numéro spécial de "l'Onde Electrique" d'Octobre 1955, l'on déduit les caractéristiques de fonctionnement ci-après[2]:

			NH₃ à 50 hpz et 3000°	
	U^{235}	C	N	H
Nombre N d'atomes par cm³ · 10^{-24}	$7 \cdot 10^{-6}$	$2,85 \cdot 10^{-2}$	$7,8 \cdot 10^{-4}$	$2,34 \cdot 10^{-3}$
Section de fission $\sigma_f \cdot 10^{-24}$	170	0	0	0
Section de capture $\sigma_c \cdot 10^{-24}$	33	$1,5 \cdot 10^{-3}$	0,59	0,11
Section de diffusion $\sigma_d \cdot 10^{-24}$	0	4,55	10,4	26
$N \cdot \sigma_f$	$1,19 \cdot 10^{-3}$	0	0	0
$N \cdot \sigma_c$	$2,31 \cdot 10^{-4}$	$4,3 \cdot 10^{-5}$	$4,6 \cdot 10^{-4}$	$2,58 \cdot 10^{-4}$
$N \cdot \sigma_d$	0	$4,3 \cdot 10^{-2}$	$8,1 \cdot 10^{-3}$	$6,1 \cdot 10^{-2}$

Notre propos n'étant pas de faire un projet de réacteur, mais de chercher l'ordre de grandeur des dimensions minima, l'on utilisera la formule très approxi-

[1] Dans le réacteur — bouilleur de la Caroline du Nord, la densité de présence de U^{235}, était de 47,6 mg/cm³, mais il y avait également une densité plus grande de matériaux absorbants: 0,1 g/cm³ d'hydrogène, 43,5 mg/cm³ de Soufre et 278 mg/cm³ de U^{238}.

[2] La solution "à neutrons froids" ne sera pas développée dans le présent exposé.

mative ci-dessous, qui ne s'applique en toute rigueur qu'à des neutrons monocinétiques et à un réacteur sans réflecteur:

$$B^2 \cdot D = A \tag{5}$$

avec:

$$B^2 = \frac{10,7}{R^2} \text{ pour un cylindre du carré de rayon } R \text{ cm} \tag{6}$$

$$D = \frac{1}{3 \sum N \cdot \sigma_d} \tag{7}$$

$$A = (v-1) \cdot \sum N \cdot \sigma_f - \sum N \cdot \sigma_e. \tag{8}$$

v étant égale à 2,50 neutrons créés en moyenne à chaque fission, il vient, tous calculs faits:

$$R \cong 200 \text{ cm.}$$

Cette valeur est très approximative et peut-être optimiste, mais il serait possible de disposer un réflecteur de 30 à 40 cm d'épaisseur tout autour du réacteur, ce qui ramènerait le rayon R aux alentours de la valeur ci-dessus.

Un tel réacteur à l'uranium ne conviendrait donc que pour de très gros engins; son volume serait en effet de 51,2 m³ et le bilan-poids s'établirait comme suit:

U^{235}	149 kg
Graphite	29,2 t
Azote	535 kg
Hydrogène	114 kg

Total de contenu:	30,5 t	
Nombre de tubes:	45 200	
Surface de chauffe externe:	1840 m²	
Puissance approximative:	gaz transparent:	372 Mw
	gaz opaque:	1860 Mw
Poussée ($w = 5880$ m/s):	gaz transparent:	12,4 t
	gaz opaque:	62 t.

Remplaçant U^{235} par Pu^{239}, l'on pourrait réduire le rayon à quelque 40 cm, les données ci-dessus étant alors réduites en conséquence, mais il est bien peu probable qu'il existe des composés du Pu restant à l'état solide au-delà des 3400° K que postulent les fortes sections efficaces signalés plus haut; l'emploi de Pu^{239} n'est donc à retenir que pour le réacteur gazeux dont il sera question plus loin.

Ce qui précède laissant quelqu' espoir quant à la réalisation d'un réacteur nucléaire à hyperchauffe, il parait intéressant de rechercher le mode d'exploitation optimum de cette possibilité: détente isentropique, isotherme ou détente mixte?

Soit donc (Fig. 6) E le point représentatif de l'état du fluide à son entrée dans le foyer sous la pression p_i et T_l l'isotherme limite pour une chauffe convective ou rayonnante du fluide par les parois.

Le diagramme relatif à la détente isentropique comprend:

— la chauffe isobare, EL,

— l'hyperchauffe isobare LI_1,

— la détente isentropique $I_1 \sigma_1$.

Etant donné la disposition du réacteur susvisé, les gains d'enthalpie pendant la chauffe et l'hyperchauffe sont sensiblement équivalents:

$$H_{I_1} - H_L = H_L - H_E = \frac{P}{2\mu}, \tag{9}$$

P étant la puissance développée par le réacteur et μ le débit-masse du propulsif.

Le point de sortie σ_1 est déterminée par la pression de sortie des gaz p_σ que l'on s'est fixé[1]; la vitesse de sortie est alors donnée par:

$$w_{\sigma_1}{}^2 = 2\,(H_{I_1} - H_{\sigma_1}) = 2\cdot[2\,H_L - H_E - H_{\sigma_1}]. \qquad (10)$$

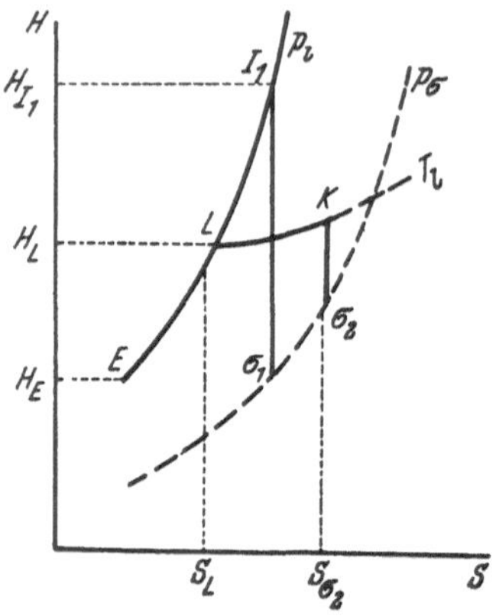

Fig. 6

Le diagramme relatif à la détente mixte comporte:

— la chauffe isobare EL,
— l'hyperchauffe isotherme LK,
— la détente isentropique $K\sigma_2$.

Pour les mêmes raisons que ci-dessus, la Puissance fournie le long de EL est du même ordre que celle fournie le long de LK et l'on peut écrire:

$$\frac{P}{2\,\mu} = H_L - H_E = T_l\cdot(S_{\sigma_2} - S_L); \qquad (11)$$

(l'on voit que ce n'est qu'au cas où L serait sur l'isobare p_σ que la détente mixte se résoudrait en une détente simplement isotherme).

La vitesse de sortie est donnée ici par:

$$w_{\sigma_2}{}^2 = 2\cdot[T_l\,(S_{\sigma_2} - S_L)] + H_L - H_{\sigma_2} = 2\cdot[2\,H_L - H_E - H_{\sigma_2}]; \qquad (12)$$

H_{σ_2} étant supérieur à H_{σ_1}, l'on voit que:

$$w_{\sigma_1} > w_{\sigma_2}.$$

C'est à dire que la détente isentropique est plus favorable que la détente isotherme complétée par une détente isentropique.

Matérialisons tout cela par quelques chiffres, le propulsif étant toujours de l'ammoniac, la température limite étant de 2800° K et la détente de 50 à 0,1 hpz.

Supposons d'abord cette détente isentropique; du diagramme (H, S) l'on tire:

$HE = 2370$ cal/mole
$HL = 54.000$ cal/mole
d'où: $H_{I_1} = 105.630$ cal/mole
et: $T_{I_1} = 4100°$ K;

L'abaque donne alors: $H_{\sigma_1} = 33.000$ cal/mole
d'où: $H_{I_1} - H_{\sigma_1} = 72.630$ cal/mole, soit: 17,9 Kj/T.
et: $w_{\sigma_1} = 5990$ m/s.

Passant au cas de la détente isotherme, il vient:

$$S_{\sigma_2} - S_L = \frac{5{,}16\cdot 10^4}{2{,}8\cdot 10^3} = 18{,}4 \text{ cal/mole}\cdot\text{degré}.$$

[1] Pour des raisons de poids et d'encombrement, il eut été plus rationnel de déterminer les éléments de sortie en se fixant la poussée par unité de surface de la section de sortie:

$$\frac{F}{A_\sigma} = \varrho_\sigma\cdot w^2{}_\sigma + p_\sigma = C^{te};$$

mais cela aurait compliqué inutilement l'exposé.

L'abaque donne alors:

$$H_k = 61.200 \text{ cal/mole}$$
$$H_{\sigma_2} = 39.500 \text{ cal/mole}$$

d'où: $2\,H_L - H_E - H_{\sigma_2} = 66.130$ cal/mole, soit: $16{,}3 \cdot 10^6\ K_j/\text{t}$

d'où: $w_{\sigma_2} = 5710$ m/s.

L'on voit sur cet exemple que la détente adiabatique l'emporte sur la détente isotherme qui serait d'ailleurs beaucoup plus difficile à réaliser.

Possibilité d'accroissement de la vitesse d'éjection

Etant donné que nous sommes riches en énergie, mais toujours indigents en masse éjectable, l'on pourrait accroître les vitesses d'éjection moyennant un gaspillage d'énergie.

Pour cela, après avoir traversé le faisceau de bas en haut, le fluide éjectable serait partagé en deux fractions, l'une subissant l'hyperchauffe et l'autre étant dirigée vers des radiateurs où elle rayonnerait son énergie dans l'espace; une fois refroidie, elle serait réinjectée dans l'éjecteur avec un appoint convenable de propulsif frais. L'hyperchauffe s'appliquant alors à un débit-masse plus faible, la température de ce dernier s'accroît, si l'on opére par dètente isentropique, ou la partie isotherme de la détente augmente, si l'on opère par détente mixte.

Précisons ceci: soit n % la part de débit-masse initial qui subit l'hyperchauffe et considérons d'abord le cas de la détente isentropique. L'équation (9) devient:

$$\frac{P}{2\mu} = H_L - H_E = \frac{n}{100}\,(H_{I_1}{}' - H_L) \tag{13}$$

d'où: $$H_{I_1}{}' = H_L + \frac{100}{n}\,(H_L - H_E) = H_I + \frac{100-n}{n}\,(H_L - H_E). \tag{14}$$

$H_{I_1}{}'$ est donc bien supérieur à H_{I_1}, mais il faut toutefois remarquer que H_σ va croître également, mais à un degré moindre.

La vitesse sera donnée par:

$$w_{\sigma_1}{}'^2 = H_{I_1}{}' - H_{\sigma_1}{}'. \tag{15}$$

Pour fixer les idées, soit 75% la part de flux hyperchauffé, les autres données etant celles de la précédente application numérique; il vient alors:

$$H_{I_1}{}' = \frac{7}{3}\,H_L - \frac{4}{3}\,H_E = 122.840 \text{ cal/mole}$$

d'où: $T_{I_1}{}' = 4340^0\ K$ (au lieu de 4100)

et $w_{\sigma_1}{}' = 6380$ m/s (au lieu de 5990).

La puissance rayonnée serait alors le 1/8 de la puissance totale.

En réduisant n à 50%, il viendrait:

$$H_{I_1}{}'' = 3\,H_L - 2\,H_E = 157.260 \text{ cal/mole}$$
$$T_{I_1}{}'' = 4750^0 \text{ K}$$
$$w_{\sigma_1}{}'' = 7050 \text{ m/s.}$$

La puissance rayonnée atteindrait ici le 1/4 de la puissance totale.

Dans le cas de la détente mixte, la formule (11) deviendrait:

$$\frac{n}{100} \cdot T_L\,(S'\,k - S_L) = H_L - H_E \tag{16}$$

et: $$w_{\sigma_2}{}'^2 = \frac{100}{n} \cdot (H_L - H_E) + H_L - H_{\sigma_2}{}'. \tag{17}$$

Comme application numérique, nous allons rechercher ici la proportion $N/100$ qui correspond à la détente strictement isotherme.

Il faut pour cela que le point K se trouve sur l'isobare 0,1 hpz; l'abaque donne alors: $S_{K'} - S_L = 33{,}0$ cal/mole · degré

d'où: $\frac{n}{100} = 0{,}186.$

Ceci correspond au rayonnement dans l'espace des 40,7% de la puissance totale du réacteur[1].

Mais sans dépasser la température de 2800° K, l'on pourrait obtenir ainsi une vitesse d'éjection considérable: 11.200 m/s.

Malheureusement, la réalisation de l'hyperchauffe pour une détente très poussée nécessiterait, en raison de la raréfaction du fluide propulsif, une "opacisation" de ce fluide vis à vis des fragments nucléaires, par introduction de particules de carbone, ce qui n'irait pas sans diminuer notablement la vitesse de sortie.

A noter qu'en détente isentropique, la proportion ci-dessus de 18,6% conduirait à des températures initiales de l'ordre de 7000° K, ce qui poserait de sérieux problèmes pour la réfrigération des parois.

Bien entendu, des vitesses aussi élevées correspondraient à des poussées modérées qui ne sauraient convenir qu'une fois atteinte la "libération".

Réacteur à gaz fissible

A l'instar de Monsieur Shepherd, l'on peut penser encore à un réacteur gazeux fonctionnant, comme le précédent, avec de la vapeur de plutonium (ou d'un composé de ce corps) à la température optimum de 3600° K et sous une pression totale de 50 hpz. Avec un gaz contenant 1 atome de plutonium par molécule, il suffirait que ce gaz exerce une pression partielle de 3 hpz, environ, pour réaliser la même densité de présence que dans le réacteur solide susvisé, soit $2{,}9 \cdot 10^{-3}$ g/cm³.

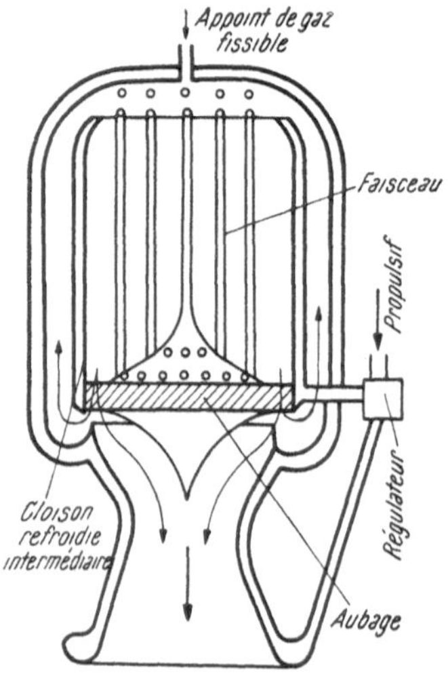

Fig. 7

Le modérateur serait constitué d'une part par le propulsif présent dans le réacteur et, d'autre part, par un faisceau tubulaire de tubes en graphite (Fig. 7) qui seraient réfrigérés par une partie du propulsif injecté de manière à garder la température de 3600° K malgré la température bien plus élevée du mélange gazeux contenu dans le réacteur.

Enfin, il serait probablement possible de séparer avant détente le gaz fissible lourd du gaz propulsif léger. L'on peut concevoir, par exemple, un aubage fixe, réfrigéré par une partie du propulsif injecté et qui imprimerait au fluide un mouvement de rotation d'ensemble à vitesse constante; ceci créerait une distribution radiale des pressions partielles suivant la formule:

$$\frac{P}{p} = e^{A\, \mathfrak{M}\, (R^2 - r^2)\, \omega^2}, \tag{18}$$

[1] A 2800° K, la surface radiante nécessaire pour évacuer dans le vide ces 40,7 % de la puissance du réacteur ne serait que de l'ordre de 1200 cm² par Mégawatt de puissance du réacteur.

A 2000° K, elle ne serait encore que de 4620 cm².

P et p étant les pressions partielles aux distances respectives R et r de l'axe, pour le gaz de masse moléculaire $\mathfrak{M}$.

Entre les deux gaz, l'on aurait alors la relation:

$$\frac{P}{p} = \left(\frac{P'}{p'}\right)^{\frac{\mathfrak{M}}{\mathfrak{M}'}}. \tag{19}$$

Dans le cas du plutonium et de l'ammoniac dissocié, l'on aurait:

$$\frac{\mathfrak{M}}{\mathfrak{M}'} = \frac{239}{8,5} = 28 .$$

Le table ci-dessous donne, pour cette valeur, la correspondance des rapports de pression:

$\frac{P'}{p'}$	1,1	1,5	2
$\frac{P}{p}$	14,43	$8,527 \cdot 10^4$	$6,31 \cdot 10^8$

L'on voit que pour $\frac{P'}{p'} \geqslant 1,5$ il y aura très peu de gaz fissible dans le cercle de rayon r qui marque l'entrée de la tuyère de détente.

Pour $r = \frac{R}{2}$ et $\frac{P'}{p'} = 1,5$ pour l'ammoniac dissocié, la formule (18) donne $w = 5440$, ce qui correspond à une vitesse tangentielle périphérique de 1950m/s pour $R = 36$ cm.

Dans ces conditions, dans le cercle de rayon 18 cm, les pressions partielles moyennes de l'ammoniac et du Pu seront respectivement inférieurs à 31,5 et $3,5 \cdot 10^{-5}$ hpz: pour 8,5 t d'ammoniac, il ne s'échapperait que 30 g de Pu, ce qui paraît admissible.

Pour éviter cette perte, l'on pourrait penser soit à prévoir un deuxième étage de séparation centrifuge, soit à remplacer le gaz fissible par de la poussière fissible formant une sorte d'aérosol; des réacteurs de cette sorte ont été proposés en particulier par les Néerlandais J. Went et H. Bruyn (Nucléonics, Septembre 1954).

Conclusions

Etant donné le caractère purement spéculatif du présent exposé, l'absence de tout contrôle expérimental grève lourdement les déductions qu'on peut être tenté d'en tirer.

Les conclusions qui suivent sont donc très problématiques, ne l'oublions pas; elles présupposent, en particulier:

— la réalisation de réacteurs nucléaires fonctionnant à des températures de 2500 à 3000° K;

— la possibilité de transférer par conductibilité et convection des puissances très élevées;

— enfin, pour certaines d'entre elles, la mise au point de réacteurs nucléaires aptes à l'hyperchauffe.

Nous pensons avoir montré que ces réalisations ne paraissent pas impossibles, à priori, mais c'est tout de même sous réserves, explicitement formulées ou non, que nous nous sommes permis au cours de cette causerie d'avancer les propositions ci-après:

1° — l'autopropulsion nucléaire permettrait d'envisager, avec une quasi-certitude, la possibilité d'un ravitaillement en propulsif de retour;

2° — un réacteur nucléaire fonctionnant entre 2500 et 3000° K permettrait, en faisant choix d'un propulsif convenable, de développer des vitesses d'éjection surclassant nettement celles des autopropulseurs classiques (4000 à 4500 m/s en utilisant l'ammoniac pour propulsif);

3° — l'opacification du gaz propulsif à l'aide d'une faible proportion de noir de fumée permettrait de réchauffer la détente et de porter les vitesses ci-dessus à quelques 5500 m/s;

4° — la mise au point d'un réacteur apte à l'hyperchauffe (à "combustible" solide, gazeux ou à l'état d'aérosol), permettrait de porter le propulsif à des températures de l'ordre de 4000° K et de développer, par détente isentropique, des vitesses d'éjection avoisinant 6000 m/s;

5° — enfin, l'accroissement de la part relative de l'hyperchauffe par rayonnement dans l'espace d'une partie de l'énergie de chauffe porterait les limites ci-dessus aux environs de 4800° K et 7000 m/s, respectivement... et même bien au-delà s'il était encore possible de protéger les parois contre de tels échauffements.

Comme on le voit, l'ampleur des performances escomptées justifie pleinement, semble-t-il, l'intérêt qu'il y aurait à résoudre les problèmes de technique nucléaire dont elles dépendent et nous serions trop heureux si ces spéculations, trop hardies peut-être, pouvaient attirer sur ces problèmes l'attention compétente des nucléoniciens.

Essai de contribution à la propulsion ionique

Par

J.-J. Barré [1]

(Avec 6 Figures)

Résumé — Zusammenfassung — Abstract

Essai de contribution à la propulsion ionique. Dans cet exposé, purement spéculatif, l'auteur, commence par rappeler la théorie des disruptures de CRANBERG et conclut à la nécessité de dématérialiser les électrodes au maximum en les réduisant à des grilles ou à des quinconces d'électrodes tubulaires servant à l'injection des ions dans le champ accélérateur.

Le vide très poussé de l'espace interplanétaire est un facteur favorable pour la mise en œuvre de gradients de potentiel très élevés: 500 000 V/cm.

L'exposé s'attache ensuite aux aspects suivants de la propulsion ionique, le terme "ion" s'étendant aux corpuscules solides ou liquides chargés.

1°. Charge maximum de diverses particules compatible avec la résistance mécanique de ces dernières. Les cas des particules solides, des gouttelettes et des microbulles sont successivement examinés.

2°. Electrisation des particules. Les processus suivants sont envisagés: frottement; dépôt de molécules ionisés; contact avec une électrode; thermo-émission ou photo-émission d'électrons.

3°. Effet de la charge d'espace. En sus du cas classique des ions préformés, l'auteur étudie le cas de la formation des ions par le champ accélérateur, à l'entrée de ce dernier, et celui de leur formation par ce même champ, mais en cours de parcours, par charge progressive.

Ce dernier cas s'avère comme devant être le plus avantageux, du moins théoriquement.

4°. Contrainte mécanique des armatures. Ces contraintes sont importantes, les armatures constituant les organes de poussée, il convient donc de les profiler avec soin.

Beitrag zum Problem des Ionenantriebes. In dieser rein spekulativen Abhandlung erwähnt der Verfasser die Theorie von CRANBERG und schließt auf die Notwendigkeit, die Masse der Elektroden möglichst zu verringern, indem man sie zu Gitterrosten oder versetzten, röhrenförmigen Elektroden verkleinert, die zur Einfuhrung der Ionen in das Beschleunigungsfeld dienen.

Das sehr niedrige Vakuum des interplanetarischen Raumes ist ein Faktor, der die Verwirklichung sehr hoher Potentialgradienten, wie 500000 V/cm, begünstigt.

Die Abhandlung wendet sich hierauf den folgenden Gesichtspunkten des Ionenantriebes zu, wobei der Ausdruck „Ion" sich auf feste oder flüssige, geladene Teilchen bezieht.

1. Die maximale Ladung verschiedener Teilchen, die mit ihrem mechanischen Widerstand noch vereinbar ist. Die Fälle fester Teilchen, von Tröpfchen und Mikrogasblasen werden nacheinander untersucht.

---

[1] Ingénieur Militaire en Chef des Fabrications d'Armement, 49, rue du Maréchal Foch, Versailles (Seine et Oise), France.

2. Elektrische Aufladung der Teilchen. Die folgenden Methoden werden betrachtet: Reibung, Ablagerung ionisierter Moleküle, Beruhrung mit einer Elektrode, Thermo- oder Photoemission von Elektroden.

3. Wirkung der Raumladung. Neben dem klassischen Fall im voraus gebildeter Ionen studiert der Verfasser die Bildung der Ionen beim Eintritt in ein Beschleunigungsfeld und den Fall ihrer Erzeugung durch progressive Aufladung entlang der durchlaufenen Strecke desselben Feldes.

Dieser letztgenannte Fall erweist sich als der vorteilhafteste, zum mindesten theoretisch.

4. Mechanische Beschränkung der Armatur. Diese Beschränkungen sind wesentlich. Da die Armaturen die Organe des Schubes darstellen, müssen sie mit Sorgfalt geplant werden.

Contribution to the Ionic Propulsion. This purely speculative treatment is begun by recalling the CRANBERG Theory and concludes with the necessity of dematerializing to the maximum the electrodes in converting them to grilles or to a quincunx arrangement of tubular electrodes used for the injection of ions in the accelerating field.

The vacuum of interplanetary space is a favorable factor for the employment of very high potential drops of 500,000 V/cm.

The discussion then goes on to the following aspects of ionic propulsion (the term *ion* being extended to charged solid or liquid corpuscles):

(1) Maximum charge of diverse particles compatible with their mechanical resistance (the cases of solid particles, of droplets and of microparticles are successively examined).

(2) Electrification of particles (the following processes are envisioned: friction, deposit of ionized molecules, contact with an electrode, thermo- or photo-emission of electrons).

(3) Effect of the space charge (in addition to the classic case of preformed ions, the case of ion formation by the field accelerator is studied, both at its entrance, and during the course of travel, by progressive charge. This latter case has, at least theoretically, been established as the most advantageous).

(4) Mechanical constraint of the surfaces (important since these surfaces constitute the means of thrust, they should be carefully profiled).

Un propulseur ionique se compose d'un générateur d'énergie électrique à potentiel élevé alimentant un éjecteur de corpuscules[1] électrisés.

Seul l'éjecteur retiendra aujourd'hui notre attention. Remarquons tout de suite que cet éjecteur ne saurait être considéré comme une simple extrapolation de l'accélérateur linéaire de particules d'un générateur VAN DE GRAAF, ou autre.

En effet, les caractéristiques de fonctionnement d'un éjecteur de propulsion ont des ordres de grandeur totalement différents de celles d'un accélérateur, comme le montre le tableau I:

Tableau I

Caractéristiques	Accélérateur	Ejecteur
Puissance	0 à 10 Kw	1000 à 10000 Kw
Intensité du courant particulaire ou corpusculaire	qq m A	qq A
Gradients de potentiel	$5 \cdot 10^3$ V/cm	$5 \cdot 10^5$ V/cm
Charge spécifique des particules ou corpuscules	α: 27000 c b/g e: 10^8 cb/g	1 cb/g
Vitesse des particules ou corpuscules	7000 km/s	de 10 à 50 Km/s
Contrainte des armatures	qq mg/cm^2	qq kg/cm^2
Degré de vide	10^{-3} à 10^{-4} mm Hg	0 (vide spatial)

[1] Le terme de "corpuscule" s'applique ici à de petites quantités de matière pouvant descendre jusqu'aux grosses molécules; le terme de particules est réservé aux constituants élémentaires ou quasi-élémentaires de la matière, c'est-à-dire, de l'électron à l'hélion α.

De telles différences ne sont évidemment pas sans influer profondément sur la structure, voire même sur la conception des appareils.

Les points que nous nous proposons de développer ici concernent:

— l'évaluation de la charge spécifique maximum des différents corpuscules;

— le choix du mode d'ionisation;

— la valeur et l'influence de la charge d'espace;

— les contraintes mécaniques des armatures.

Dans la plupart de ces questions, le gradient de potentiel joue un rôle capital, une valeur élevée de ce gradient étant d'ailleurs toujours à rechercher.

De ce point de vue, le fonctionnement dans le vide spatial est un facteur très favorable; en effet, tout risque de court-circuit par effluves se trouve éliminé et, si les supports d'armatures sont convenablement disposés et suffisamment isolants, les gradients de potentiel ne se trouvent plus limités que par le seuil d'amorçage des "disruptures". Avant de pousser plus avant, il parait donc nécessaire de rappeler les principales caractéristiques de ce phénomène.

Décharges disruptives dans le vide[1]

D'après la loi de Paschen, la rigidité du vide absolu devrait être infinie; or, en fait, dans les vides les plus poussés qui ont pu être réalisés, il se produit des décharges disruptives dont L. Cranberg a donné l'explication suivante[2]:

Sous l'effet de la pression électrostatique qui s'exerce à la surface des armatures, une particule infime se détache de l'une d'elles et se précipite sur l'armature opposée où elle arrive à une vitesse telle que l'absorption de son énergie cinétique provoque la fusion du métal[3] et le départ de plusieurs particules; ces dernières se précipitent à leur tout sur la première électrode en provoquant une nouvelle avalanche, encore plus fournie, et ainsi de suite jusqu'à ce que le développement en chaîne engendre une avalanche suffisamment fournie pour emporter la quasi-totalité de la charge de l'électrode.

Pour loi de cette disrupture, l'auteur propose:

$$\Delta U = K \cdot \sqrt{d}\,, \tag{1}$$

K étant une constante qui caractérise le couple d'électrodes et d représentant l'écartement de ces dernières[4].

On en tire pour l'intensité critique du champ provoquant la disrupture:

$$\frac{\Delta U}{d} = \frac{K}{\sqrt{d}}\,. \tag{2}$$

---

[1] Les unités employées dans cet exposé seront en principe celles du système c g s électrostatique; toutefois les potentiels seront parfois exprimés en volts. Dans ce cas les potentiels seront représentés par la lettre V eu lieu de la lettre U.

[2] Lawrence Granberg, The Initiation of Electrical Breakdown in Vacuum, J. Appl. Physics, pp. 518 et seq. (1952).

[3] Dans certaines circonstances, la température développée avoisinerait le million de degrés.

[4] Pour arriver à cette loi, l'auteur suppose que la chaîne se déclenche quand l'énergie W fournie par unité de surface à la première électrode-cible dépasse une certaine valeur qui caractérise la paire d'électrodes considérée. Or, W est précisément le produit de la différence de potentiel ΔU par la densité de charge du fragment, densité qui est proportionelle à l'intensité du champ à la surface de l'électrode critique; cette intensité étant sensiblement égale au quotient $\Delta U/d$, il vient:

$$W = \frac{\Delta U^2}{d} = C^{te}\,.$$

L'intensité de ce champ critique est donc d'autant plus élevée que les électrodes sont plus rapprochées.

D'après le mémoire précité, le gradient peut atteindre 10^6 volts/cm entre deux sphères de molybdène de 1 cm de diamètre et distantes de 3 mm; ces sphères, polies, avaient au préalable été dégazées à la température du rouge sombre.

L'on peut déduire de la formule (2) ci-dessus que le gradient critique augmente pour une même longueur d'accélération, quand on fractionne également la chute de potentiel à l'aide d'électrodes auxiliaires également réparties.

Le gradient critique, égal à chaque gradient partiel, devient en effet pour un sectionnement en n intervalles:

$$\frac{\Delta U'}{d} = \frac{\Delta U'/n}{d/n} = \frac{K}{\sqrt{d/n}} = \sqrt{n} \cdot \frac{\Delta U}{d}\,; \tag{3}$$

l'on voit que le gradient critique est multiplié par $\sqrt{n}$.

Ce fractionnement est couramment utilisé dans les tubes accélérateurs.

Mais on peut aussi envisager de supprimer la disrupture en en supprimant la cause et, ceci, en "dématérialisant" au moins partiellement, l'une des électrodes; il suffirait pour cela, semble-t-il[1] d'utiliser pour électrodes des grilles et non des surfaces pleines; cette substitution ne change pratiquement la distribution du champ qu'au voisinage immédiat de la grille où règne localement et sur une très petite épaisseur un champ dU/dn extrêmement intense, donc propice à l'ionisation (cf. Fig. 1, extraiie de "*Electricité*" de G. Bruhat).

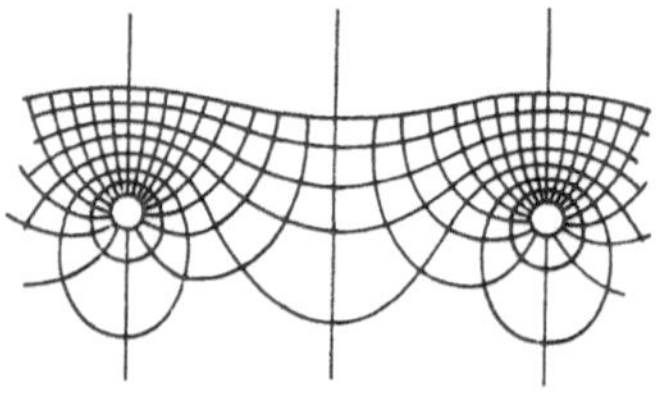

Fig. 1

Par ailleurs, bien que toutes les lignes du champ aboutissent aux fils de la grille, du fait de leur inertie, les fragments générateurs de disrupture ne suivront pas ces lignes de force dans leur région de grande courbure et passeront donc dans les mailles (contrairement à ce que feraient des électrons dénués pratiquement d'inertie).

Dans ce cas, la densité moyenne σ, donc le champ local dU/dn, se trouve multipliée par le rapport de la surface de l'armature à celle des fils qui la composent:

$$\frac{dU}{dn} = \frac{S}{s} \cdot \frac{\Delta U}{d}\,. \tag{4}$$

Ce rapport peut être très élevé, de l'ordre de 100 à 1000, par exemple.

Bien entendu, l'énergie $\overline{\Delta U^2/d}$ prend alors la valeur plus grande $\Delta U \cdot \frac{dU}{dn}$ et des amorces primaires de disrupture partiront sans doute à chaque instant des fils de grilles; elles passeront en quasi-totalité dans les vides de l'électrode opposée et il en sera de même des rares amorces secondaires réémises par cette deuxième armature de sorte que l'amorçage en chaîne devrait, en définitive, se trouver notablement retardé.

Ce risque se trouverait d'ailleurs à peu près totalement éliminé en réduisant l'armature émettrice aux corpuscules qu'il s'agit d'électriser, ces corpuscules étant disposés en quiconce dans un plan. Dans le cas de corpuscules liquides, par exemple, ces dernières perleraient aux orifices d'un faisceau de conduits capillaires. Dans un tel dispositif, ce serait finalement la résistance mécanique

[1] Ceci, n'ayant pas encore reçu la sanction de l'expérience, n'est avancé que sous toutes réserves et ne doit être considéré pour l'instant que comme une vue de l'esprit.

des corpuscules qui limiterait la disrupture, cette dernière correspondant d'ailleurs à la projection des dits corpuscules ou de leurs fragments.

Avec une densité de n particules de rayon r au cm^3, il viendrait ici:

$$\frac{dU}{dn} = \frac{1}{n \cdot \pi \cdot r^2} \cdot \frac{\Delta U}{d}\,. \tag{5}$$

Nous arrêterons là cette nécessaire digression sur les décharges disruptives pour aborder les différents points qui font l'object de cette communication.

I. Charge spécifique maximum d'un corpuscule

Rappelons d'abord que la charge spécifique q/m est liée à la vitesse w imprimée par une chute de potentiel ΔU U.E.S., ou ΔV volts, par la formule:

$$\frac{q}{m} = \frac{1}{2} \cdot \frac{w^2}{\Delta U} = 150 \cdot \frac{w^2}{\Delta V}\,. \tag{6}$$

Le tableau II donne les valeurs de q/m, en unités U.E.S. qui sont nécessaires pour obtenir la vitesse w km/s pour une chute de potentiel ΔV volts.

Tableau II

w km/s / ΔV volts	5	10	20	50	100	200
10^5	$3{,}75 \cdot 10^8$	$1{,}5 \cdot 10^9$	$6 \cdot 10^9$	$3{,}75 \cdot 10^{10}$	$1{,}5 \cdot 10^{11}$	$6 \cdot 10^{11}$
$5 \cdot 10^5$	$7{,}5 \cdot 10^7$	$3 \cdot 10^8$	$1{,}2 \cdot 10^9$	$7{,}5 \cdot 10^9$	$3 \cdot 10^{10}$	$1{,}2 \cdot 10^{11}$
10^6	$3{,}75 \cdot 10^7$	$1{,}5 \cdot 10^8$	$6 \cdot 10^8$	$3{,}75 \cdot 10^9$	$1{,}5 \cdot 10^{10}$	$6 \cdot 10^{10}$
$5 \cdot 10^6$	$7{,}5 \cdot 10^6$	$3 \cdot 10^7$	$1{,}2 \cdot 10^8$	$7{,}5 \cdot 10^8$	$3 \cdot 10^9$	$1{,}2 \cdot 10^{10}$

L'on va considérer sucessivement les cas suivants:

a — corpuscules solides;
b — bulles;
c — gouttelettes liquides;
d — molécules ionisées.

a) Cas des corpuscules solides

Une densité électrostatique σ développe une pression égale à $2\pi\sigma^2$; cette pression tend à faire éclater le corpuscule qui ne peut donc, de ce fait, être chargé indéfiniment. En supposant le corpuscule sphérique, ce qui sera toujours admis dans la suite, et en désignant par r son rayon et par τ sa limite de résistance à la rupture, il vient la condition:

$$2\pi\sigma^2 \cdot \pi \cdot r^2 \leqslant \tau \cdot \pi \cdot r^2\,; \tag{7}$$

par ailleurs, en désignant par ϱ la masse spécifique du corpuscule, il vient aussi:

$$m = \frac{4\pi}{3} r^3 \cdot \varrho \tag{8}$$

et

$$q = 4\pi r^2 \cdot \sigma\,. \tag{9}$$

Eliminant σ entre (7) et (9), il vient:

$$r_{lim} = \frac{3}{\varrho \cdot \frac{q}{m}} \cdot \sqrt{\frac{\tau}{2\pi}}\,. \tag{10}$$

On voit que r limite est d'autant plus petit que q/m est plus grand, ce qui conduit à envisager l'emploi de corpuscules très petits, de l'ordre de dimension des aérosols et pouvant descendre jusqu'au centième de micron.

A titre d'exemple, le tableau III donne en fonction de w et ΔV les rayons limites de particules d'aluminium ($\varrho = 2{,}7$); ce tableau a été établi en prenant pour τ sa valeur macroscopique de 8 kg/mm², soit $8 \cdot 10^8$ cgs, mais il est bien certain que la limite de rupture de très petits fragments est beaucoup plus grande de sorte que les valeurs données par le tableau constituent des limites inférieures des rayons minima.

Tableau III. r_{lim} *(w, ΔV), en μ, pour l'Aluminium*

ΔV volts \ w km/s	5	10	20	50	100	200
10^5	0,33	0,08	0,02			
10^6	3,32	0,83	0,21	0,03	0,01	
$5 \cdot 10^6$	16,60	4,17	1,04	0,17	0,04	0,01

Ce tableau a été limité aux valeurs de r supérieures à 0,01 μ. Il est à remarquer qu'en cas d'éclatement de la particule, chaque fragment garderait la même charge spécifique que la particule primitive, mais la pression électrostatique moyenne serait diminuée, la charge totale restant la même alors que la surface augmenterait (la présence d'arêtes vives sur les fragments pourrait amener à des valeurs locales plus élevées pour la pression électrostatique).

Avec 1 Million de volts, l'on voit que des vitesses de l'ordre de 20 et même 50 km/s pourraient être atteintes assez aisément. Il n'est pas dit, d'ailleurs, que l'Aluminium constitue le matériau idéal dans le cas présent.

Remarquant qu'une sphère creuse offre une grande surface pour un faible poids, l'on pourrait penser à remplacer les sphérules pleins considérés précédemment par des "perles" creuses à parois très minces. Si ε est l'épaisseur de ces parois, l'on trouve facilement qu'à rayon r égal le q/m limite serait accru dans le rapport:

$$\frac{q/m \text{ perle}}{q/m \text{ sphérule pleine}} = \frac{1}{3}\sqrt{\frac{2r}{\varepsilon}}, \tag{11}$$

les vitesses d'éjection étant multipliées par la racine carrée de ce rapport.

Mais autre que de telles perles seraient très fragiles, elles auraient une densité gravimétrique très faible et seraient très encombrante; il faudrait donc les former au fur et à mesure de leur emploi, par exemple en formant des microbulles solidifiées par refroidissement[1].

Ceci nous conduit tout naturellement à l'examen du comportement des bulles à parois liquides.

b) Cas des bulles à parois liquides

Ce n'est plus ici la résistance à la rupture qui entre en jeu, mais la tension superficielle[2]. Faute d'expérience, l'auteur se trouve réduit à des conjectures et il ne faut voir dans ce qui suit que de simples vues de l'esprit.

[1] Un évent ménagé dans les parois permettrait de les dégazer intérieurement pour éliminer la contrainte due à la pression intérieure.

[2] Pour toutes les questions de tension superficielle, l'auteur s'est référé au tome "Capillarité" de la Physique de M. BOUASSE.

Soit donc une bulle de rayon r et d'épaisseur ε; p étant la pression qui règne à l'intérieur de la bulle et A étant la tension superficielle des parois de cette dernière, l'équation d'équilibre s'écrit:

$$p + 2\pi\sigma^2 = \frac{4A}{r}. \tag{12}$$

Or:

$$\sigma = \frac{q}{4\pi \cdot r^2} \tag{9 bis}$$

et[1]:

$$m = \varrho \cdot 4\pi r^2 \varepsilon \tag{13}$$

où ϱ représente la masse spécifique des parois de la bulle. Mettant q/m en évidence, (11) s'écrit:

$$2\pi\left(\varrho \cdot \varepsilon \cdot \frac{q}{m}\right)^2 = \frac{4A}{r} - p; \tag{14}$$

cette relation montre que q/m sera d'autant plus grand que ε, p et r seront plus petits.

Pour l'eau, savonneuse ou non, ε limite peut être pris égal à 0,01 μ (10^{-6} cm) la bulle crevant quant en un de ses points l'épaisseur des parois tombe en-dessous de 0,0012 μ.

Quant à la pression p, elle est au moins égale à la tension de vapeur du liquide des parois, soit de l'ordre de $3 \cdot 10^3$ cgs à la température ordinaire.

A étant, toujours pour l'eau, de l'ordre de 70 (cgs) l'on peut écrire

$$\left(\frac{q}{m}\right)_{lim} = 10^6 \cdot \sqrt{\frac{\frac{280}{r} - 3000}{2\pi}}; \tag{15}$$

ceci montre que r doit être inférieur à 0,093 cm (soit 930 μ); l'on ne peut donc considérer que des microbulles dont la formation poserait d'ailleurs un problème technique qui ne sera pas abordé ici.

D'après (6), la relation ci-dessus peut s'écrire

$$r_{lim} = \frac{280}{3000 + 4{,}5 \cdot 10^{-8}\pi \cdot \frac{w^4}{\Delta V^2}}. \tag{16}$$

Cette formule a permis d'établir le tableau IV qui donne le rayon limite (en μ) en fonction de w et ΔV, dans le cas de l'eau.

Tableau IV. r_{lim} *(en* μ*) en fonction de* w *et* ΔV *(pour l'eau)*

w km/s / ΔV volts	5	10	20	50	100
10^5	3,15	0,197			
$5 \cdot 10^5$	72,7	4,9	0,309		
10^6	235	123	1,23	0,031	
$5 \cdot 10^6$	835	321	29,6	0,795	0,049

L'on voit que les rayons limite sont nettement supérieurs à ceux des sphérules d'Aluminium et cela d'autant plus que ces rayons sont plus grands (ε restant

[1] Les bulles étant très petites (cf. plus loin), la masse de gaz est tout à fait négligeable vis à vis de celle des parois.

constant), c'est-à-dire que les champs accélérateurs sont plus intenses et les vitesses plus faibles.

Bien entendu tout ceci n'est que théorique, la technique des microbulles restant à créer. Quoiqu'il en soit, cette petite étude va permettre d'éclairer le cas plus compliqué des gouttelettes liquides.

c) Cas des gouttelettes liquides

Soit d'abord une gouttelette isolée dans l'espace vide; la pression qui règne à l'intérieur du liquide est:

$$p = \frac{2A}{r} - 2\pi\sigma^2. \tag{17}$$

Quand σ atteint une valeur telle que cette pression interne tombe en-dessous de la tension de vapeur correspondant à la température de la goutte, il semble bien qu'une cavitation doive apparaître au sein de cette dernière qui devient une bulle, ce qui ramène au cas précédent.

Toutefois, à moins d'utiliser des liquides très volatils ou d'opérer à des températures élevées, la tension de vapeur sera négligeable vis à vis de la pression électrostatique et l'on peut écrire que les conditions limites correspondent à:

$$\sigma = \sqrt{\frac{1}{\pi} \cdot \frac{A}{r}}, \tag{18}$$

ce qui, tous calculs faits, conduit à:

$$r_{lim} = \sqrt[3]{\frac{9}{\pi} \cdot \frac{A}{\varrho^2 \cdot \left(\frac{q}{m}\right)^2}}. \tag{19}$$

Le tableau V, établi à partir de cette formule, donne pour l'eau ($\varrho = 1$ et $A = 70$) et le mercure ($\varrho = 13{,}6$ et $A = 471$), les rayons maxima en fonction de w et ΔV.

Tableau V. r_{lim} (w, ΔV) *en* μ, *pour l'eau et le mercure*

ΔV_{volts} \ $u_{km/s}$		5	10	20	50	100	200
eau	10^5	0,11	0,04	0,02			
	10^6	0,52	0,21	0,08	0,02	0,01	
	$5 \cdot 10^6$	1,53	0,61	0,24	0,07	0,03	0,01
Hg	10^5	0,04	0,01				
	10^6	0,17	0,07	0,03	0,01		
	$5 \cdot 10^6$	0,51	0,20	0,08	0,02	0,01	

Considérons maintenant le cas où la goutte n'est plus isolée dans l'espace, mais perle à l'extrémité d'un ajutage à bords tranchants[1] qui l'alimente en liquide sous la pression constante p (Fig. 2).

Le liquide étant supposé "mouillant" et tout phénomène électrostatique étant pour l'instant exclu, le ménisque, initialement hémisphérique, va cheminer vers l'orifice, concavité en avant (position 1); arrivé à l'orifice il s'applatit, puis

[1] Pour éliminer l'effet "angle de raccordement".

se bombe de plus en plus en sens inverse. Si la pression p n'est pas trop forte, il s'arrête en une position telle que 2 pour laquelle, r_a désignant le rayon de l'ajutage:

$$p = \frac{2A}{r} = \frac{2A}{r_a} \cdot \cos\alpha > \frac{2A}{r_a}. \tag{20}$$

Pour $p = \frac{2A}{r_a}$, le rayon du ménisque passe par une valeur minimum égale au rayon de l'ajutage; la goutte est alors hémisphérique et l'équilibre instable.

Si la pression dépasse cette valeur limite $\frac{2A}{r_a}$, il n'y a plus d'équilibre possible et la goutte éclate en gouttelettes plus petites; ceci fait baiser momentanément la pression vers l'orifice de l'ajutage et le cycle recommence.

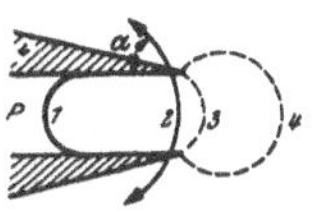

Fig. 2

Si maintenant l'ajutage est porté à un certain potentiel, le ménisque se charge et il convient d'ajouter la pression électrostatique $2\pi\sigma^2$ à la pression p de l'équation (20). Mais que se passe-t-il alors en cas d'éclatement? Les molécules ionisées qui constituent la charge de la goutte s'arrachent-elles seules? Il semblerait plutôt que chacune va entraîner avec elle un agglomérat de molécules neutres se trouvant dans son voisinage; mais, faute de données expérimentales, il parait bien difficile de préciser davantage.

Quoiqu'il en soit, en supposant qu'à l'extrême limite de charge, il se détache une goutte de rayon r_a chargée à la densité limite, c'est encore le tableau V qui donne les rayons limites *d'ajutage* permettant d'obtenir une vitesse w sous une chute de potentiel ΔV.

d) Cas des molécules

Du fait de la petitesse des molécules, les forces mises en jeu ne sont plus de même nature que pour les particules, aussi ce cas diffère-t-il profondément des précédents. Ici, c'est par excès que pèchent les charges spécifiques; en effet, pour une corps de masse moléculaire $\mathfrak{M}$ la charge spécifique d'une molécule ionisée une fois est:

$$\frac{q}{m} = \frac{1{,}64 \cdot 10^{14}}{\mathfrak{M}} \text{ U.E.S./g}, \tag{21}$$

d'où d'après (6):

$$\mathfrak{M} = 1{,}1 \cdot 10^{12} \cdot \frac{\Delta V}{w^2}. \tag{22}$$

Le tableau VI donne $\mathfrak{M}$ en fonction de w et ΔV, en supposant toutes les molécules ionisées une fois. (Pour un taux d'ionisation $1/n$ et en supposant l'équipartition des vitesses maintenue en cours d'accélération, les $\mathfrak{M}$ du tableau devraient être divisés par n; mais cette équipartition suppose une pression relativement élevée, d'où risques de décharge par effluves.)

Tableau VI. $\mathfrak{M}$ $(w, \Delta V)$

w km/s / ΔV volts	25	50	100	500	1000
10^2	17,6	4,4	1,1		
10^3	176	44	11		
10^4	1760	440	110	4,4	1,1
10^5			1100	44	11
10^6				440	110

Comme on le voit, les potentiels mis en jeu sont beaucoup plus faibles et ils s'adaptent mieux à des générateurs du type thermopiles ou photopiles solaires qu'aux générateurs électronucléaires qui débiteraient sous des tensions de l'ordre du mégavolt.

II. Ionisation des corpuscules

A priori, quatre processus peuvent être envisagés pour l'électrisation des corpuscules, à savoir:

— le frottement,
— le dépôt de molécules ionisées,
— le contact avec une électrode,
— l'émission d'électrons par les corpuscules.

a) Electrisation par frottement

L'électrisation par frottement[1] n'est citée ici que pour mémoire; elle n'apparait pas susceptible, en effet, de réaliser une charge régulière et homogène de corpuscules aussi ténus.

L'on pourrait toutefois penser à recourir à une sorte de "meulage" de substances convenables, meulage qui produirait à la fois les corpuscules et la charge de ces derniers; mais un tel procédé aurait sans doute l'inconvénient d'ajouter à la dispersion des charges celle des dimensions.

b) Electrisation par dépôt de molécules ionisées

Ce processus a été spécialement étudié par Monsieur PAUTHENIER qui l'a utilisé d'abord pour le dépoussièrage des gaz de combustion, puis pour constituer l'ioniseur de la machine électrostatique qui porte son nom (cf. "Journal de Physique et le Radium" de Décembre 1932).

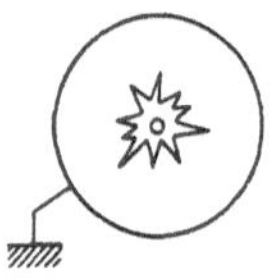
Fig. 3

Le principe est le suivant: une atmosphère ionisée est créee par un fil fin porté à un haut potentiel négatif, ce fil étant disposé selon l'axe d'un cylindre mis à la terre (cf. Fig. 3). Un flux de corpuscules initialement neutres est soufflé dans le cylindre et ceux des corpuscules qui se trouvent en dehors de la gaîne lumineuse de l'effet corona se chargent négativement et se dirigent vers le cylindre où ils se déposent si leur vitesse axiale est suffisamment faible.

Bien que très élégant, ce procédé présente de sérieux inconvénients pour l'application qui nous occupe présentement, soit:

— faible intensité du courant particulaire;

— limitation de la charge spécifique par les dimensions minima imposées aux corpuscules (10μ);

— limitation supplémentaire de cette charge spécifique du fait que la présence obligée d'une atmosphère gazeuse limite le champ inducteur;

— grosses difficultés de transvasement des corpuscules électrisés de l'atmosphère gazeuse au vide très poussé de l'éjecteur;

— enfin, lenteur de la charge (0,2 s dans un champ de 2400 volts/cm).

Pour toutes ces raisons, ce processus d'électrisation ne semble pas applicable ici.

[1] Le mécanisme de cette électrisation est bien connu: sous l'influence des f e m de contact, les faces soumises au frottement se chargent comme les armatures d'un condensateur de très grande capacité; quand on les écarte la capacité diminue considérablement tandis que la différence de potentiel augmente.

c) Electrisation par contact avec une électrode

Soit une sphérule conductrice placée dans un champ $\frac{-dU}{dn}$ (cf. Fig. 4) et dont les charges d'un certain signe (—, pour fixer les idées) peuvent s'écouler sous l'influence du champ. Dans le processus actuel, c'est par le ou les points de contact de la sphérule avec l'armature positive de l'ioniseur que se produirait cet écoulement.

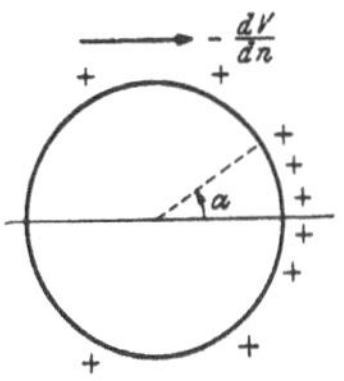

Fig. 4

Dans ces conditions, le champ $\frac{-dU}{dn}$ va provoquer l'apparition de charges positives qui seront réparties sur la sphérule de manière à créer à l'intérieur de cette dernière un champ antagoniste égal et opposé au champ inducteur. Il est facile, mais assez laborieux, de vérifier que cette répartition s'effectue selon la loi:

$$\sigma = -\frac{3}{4\pi}\cdot\frac{dU}{dn}\cdot(1+\cos\alpha), \tag{23}$$

σ étant la densité électrostatique et α l'angle polaire marqué sur la Fig. 4.

La charge totale q est alors donnée par:

$$q = -\int_0^{\pi}\frac{3}{4\pi}\cdot\frac{dU}{dn}(1+\cos\alpha)\cdot 2\pi r^2\cdot\sin\alpha\cdot d\alpha,$$

soit:

$$q = -3r^2\cdot\frac{dU}{dn} = -\frac{r^2}{100}\cdot\frac{dV}{dn}. \tag{24}$$

Pour une sphérule pleine, il vient alors:

$$\frac{q}{m} = -\frac{g}{4\pi\cdot\varrho\cdot r}\cdot\frac{dU}{dn} = -\frac{3}{400\pi\cdot\varrho\cdot r}\cdot\frac{dV}{dn}. \tag{25}$$

S'il s'agit d'un corpuscule solide, il suffit de comparer cette expression à (10) pour en tirer:

$$\left(\frac{dV}{dn}\right)_{lim} = -400\cdot\sqrt{\frac{\pi}{2}\cdot\tau}. \tag{26}$$

Pour $\tau = 10^8$, valeur faible d'ailleurs, comme il a déjà été dit, il viendrait ainsi:

$$\left(\frac{dV}{dn}\right)_{lim} \cong 5\cdot 10^6 \text{ volts/cm.}$$

Le tableau VII donne $\frac{q}{m}$, en U.E.S., en fonction de $\varrho\cdot r$, et de $\frac{dV}{dn}$.

Tableau VII. $\frac{q}{m}\left(\varrho\cdot r, \frac{dV}{dn}\right)$

$-\frac{dV}{dn}$ volts/cm \ ϱ r g/cm^2	10^{-4}	$51\cdot0^{-5}$	10^{-5}	$5\cdot10^{-6}$	10^{-6}
10^5	$3{,}8\cdot10^6$	$7{,}6\cdot10^6$	$3{,}8\cdot10^7$	$7{,}6\cdot10^7$	$3{,}8\cdot10^8$
$5\cdot10^5$	$1{,}9\cdot10^7$	$3{,}8\cdot10^7$	$1{,}9\cdot10^8$	$3{,}8\cdot10^8$	$1{,}9\cdot10^9$
10^6	$3{,}8\cdot10^7$	$7{,}6\cdot10^7$	$3{,}8\cdot10^8$	$7{,}6\cdot10^8$	$3{,}8\cdot10^9$
$5\cdot10^6$	$1{,}9\cdot10^8$	$3{,}8\cdot10^8$	$1{,}9\cdot10^9$	$3{,}8\cdot10^9$	$1{,}9\cdot10^{10}$

Pour obtenir les vitesses d'éjection w correspondantes, il suffit d'éliminer q/m entre (6) et (25), ce qui donne:

$$w = \sqrt{\frac{1}{2000 \cdot \pi} \cdot \frac{\Delta V \cdot \left|\frac{dV}{dn}\right|}{\varrho \cdot r}} \text{ cm/s.} \tag{27}$$

Le nomogramme ci-contre (Fig. 5) qui traduit cette relation, permet de déterminer très rapidement w en fonction de dV/dn, ϱ et r.

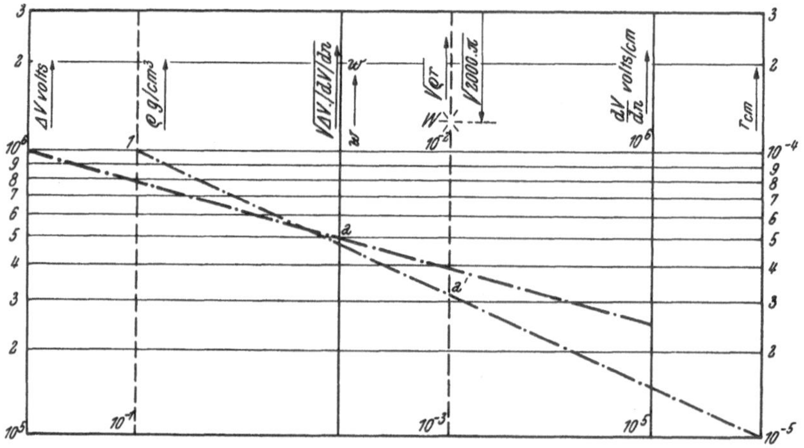

Fig. 5. Nomogramme par points alignés et report, de la relation:

$$w = \sqrt{\frac{1}{2000 \cdot \pi} \cdot \frac{\Delta V \cdot |dV/dn|}{\varrho \cdot r}} \text{ cm/s.}$$

Mode d'emploi: Joignant ΔV à $\left|\frac{dV}{dn}\right|$, on détermine un point a sur l'axe w. Joignant ϱ à r, on détermine un point a' sur l'axe $\sqrt{\varrho \cdot r}$. On porte à partir de a : $\overrightarrow{aw} \equiv \overrightarrow{a'W}$. L'abscisse de w est la valeur cherchée.

Exemple figuré:

$\Delta V = 10^6$ volts

$\left|\frac{dV}{dn}\right| = 2{,}5 \cdot 10^5$ volts/cm

$\varrho = 1$

$r = 10^{-5}$ cm $(0{,}1\mu)$

$w = 1{,}98 \cdot 10^6$ cm/s (19,8 km/s)

S'il s'agit d'une sphérule liquide formée par un ajutage de rayon r_a, c'est à l'expression (19) qu'il faut maintenant comparer (25), ce qui donne:

$$\left(\frac{dV}{dn}\right)_{lim} = 400 \sqrt{\frac{\pi \cdot A}{r}}. \tag{28}$$

Pour l'eau ($A = 70$) et le mercure ($A = 471$), le tableau VIII donne $\left(\frac{q}{m}\right)_{lim}$ en U.E.S., et $\left(\frac{dV}{dn}\right)$ limite en fonction de r[1].

[1] Les expressions de $\frac{q}{m}$ et de w sont naturellement les mêmes que ci-dessus.

Tableau VIII. $\left(\frac{dV}{dn}\right)_{lim}$ *et* $\left(\frac{q}{m}\right)_{lim}$ *en fonction de r pour* H_2O *et Hg (r en μ)*

	r (μ)	5	1	0,5	0,1	0,05	0,01
H_2O	$\left(\frac{dV}{dn}\right)_{lim}$ volts/cm	$2{,}7\cdot 10^5$	$6\cdot 10^5$	$8{,}5\cdot 10^5$	$1{,}9\cdot 10^6$	$2{,}7\cdot 10^6$	$6\cdot 10^6$
	$\left(\frac{q}{m}\right)_{lim}$ U.E.S.	$1{,}25\cdot 10^6$	$1{,}4\cdot 10^7$	$3{,}95\cdot 10^7$	$4{,}45\cdot 10^8$	$1{,}25\cdot 10^9$	$1{,}4\cdot 10^{10}$
Hg	$\left(\frac{dV}{dn}\right)_{lim}$ volts/cm	$7\cdot 10^5$	$1{,}55\cdot 10^6$	$2{,}2\cdot 10^6$	$4{,}9\cdot 10^6$	$7\cdot 10^6$	$1{,}55\cdot 10^7$
	$\left(\frac{q}{m}\right)_{lim}$ U.E.S.	$2{,}4\cdot 10^5$	$2{,}7\cdot 10^6$	$7{,}55\cdot 10^6$	$8{,}5\cdot 10^7$	$2{,}4\cdot 10^8$	$2{,}7\cdot 10^9$

On voit que l'eau permet, pour un même diamètre de gouttes de réaliser des q/m 5,25 fois plus grands que ceux du mercure et par conséquent, à même ΔV, des vitesses 2,3 fois plus élevées.

d) Ionisation par émission d'électrons

Dans ce processus, l'écoulement des charges négatives s'effectuerait par effet thermoélectrique ou photoélectrique, les particules étant placées dans le champ à une température ou avec un éclairement convenables.

Les formules de charge sont identiques à celles du cas précédent et les conclusions relatives aux $\frac{q}{m}$ et aux $\left(\frac{dV}{dn}\right)_{lim}$ sont donc aussi les mêmes.

Il est clair que ce processus ne présente aucun des inconvénients du procédé par dépôt de molécules ionisées; en particulier, les intensités mises en jeu peuvent être beaucoup plus considérables et les durées de charge incomparablement plus réduites.

En effet, l'intensité de saturation, du courant électronique, intensité qui se maintient jusqu'à des différences de potentiel, de quelques volts, peut-être de l'ordre de l'ampère par cm². Le temps de charge dt d'une sphère de rayon r est alors donné, d'après (24), par:

$$-3r^2\cdot\frac{dU}{dn} = 3\cdot 10^9\cdot 1\cdot 2\pi r^2\cdot dt$$

en supposant l'émission répartie uniformément sur le seul hémisphère qui soit convenablement orienté. Ceci donne:

$$dt = -\frac{10^{-9}}{2\pi}\cdot\frac{dU}{dn} = -5{,}3\cdot 10^{-13}\cdot\frac{dV}{dn}. \qquad (29)$$

Pour un champ de 10^6 volts/cm, la durée de charge ne serait donc que de $5{,}3\cdot 10^{-7}$ seconde et elle serait encore moindre pour des champs plus faibles.

Par rapport à l'ionisation par contact, la grosse différence du présent procédé est de permettre aux corpuscules de continuer à se charger durant leur accélération, le champ croissant par suite de l'effet "charge d'espace". Quitte même à ce qu'ils éclatent en route, les fragments continuent à se charger, ce qui permet d'accroître notablement la charge spécifique et, par conséquent la vitesse d'éjection.

Ce point va d'ailleurs être précisé dans le paragraphe consacré à l'influence de la charge d'espace, qui va maintenant être abordé. Ce n'est qu'après l'examen de ce problème que pourront être tirées les conclusions relatives aux différents processus d'ionisation qui viennent d'être envisagés.

III. Limitation de la poussée unitaire par la charge d'espace

Deux cas sont à considérer selon qu'il s'agit d'ions à charge spécifique $\frac{q}{m}$ constante ou de corpuscules émetteurs.

Le premier cas se subdivise lui-même en deux selon que les ions $\frac{q}{m}$ sont préformés et injectés dans le champ accélérateur ou qu'ils se forment sur l'armature amont par l'action même du champ accélérateur.

a) Injection, dans le champ, d'ions préformés

C'est le cas classique auquel correspond la formule bien connue de Langmuir[1]

$$\left(\frac{I}{S}\right)_{max} = \frac{\sqrt{2}}{27 \cdot 10^9 \cdot \pi} \cdot \sqrt{\frac{q}{m}} \cdot \frac{\Delta U^{3/2}}{d^2} \text{ ampères/cm}^2, \tag{30}$$

formule où I/S représente la densité de courant (en ampères/cm²) dans le champ accélérateur $\Delta U/d$ (exprimé en U.E.S.).

Cette limitation de la densité de courant par la charge d'espace entraîne celle de la poussée par unité de section de l'éjecteur: F/s.

En effet, le débit-masse μ/s des corpuscules par unité de section est donnée par:

$$\frac{\mu}{S} = \frac{3 \cdot 10^9 \, (I/S)_{max}}{q} \cdot m \tag{31}$$

et leur vitesse de sortie, par:

$$w_1 = \sqrt{2 \frac{q}{m} \cdot \Delta U}, \tag{32}$$

d'où:

$$\left(\frac{F}{S}\right)_{max} = 3 \cdot 10^9 \cdot \sqrt{2 \frac{\Delta U}{q/m}} \cdot \left(\frac{I}{S}\right)_{max}. \tag{33}$$

Dans le cas présent des ions préformés et injectés, cette poussée s'écrit, d'après (30)[2]:

$$\left(\frac{F}{S}\right)_{max} = \frac{2}{9\pi} \cdot \left(\frac{\Delta U}{d}\right)^2 = 8{,}23 \cdot 10^{-7} \cdot \left(\frac{\Delta V}{d}\right)^2. \tag{34}$$

$\Delta V/d$ étant limité à $5 \cdot 10^5$ volts/cm, la poussée maximum dans ce cas est donc:

$$\left(\frac{F}{S}\right)_{max} = 2{,}06 \cdot 10^5 \text{ cgs},$$

soit: 2 tonnes/m².

[1] La démonstration de cette formule, et des formules analogues correspondant aux deux autres cas, est donnée en annexe.

[2] Lyman Spitzer a donné une formule équivalente dans son mémoire "Trajectoires Interplanétaires entre les orbites des Satellites" (J. Amer. Rocket Soc. de Mars-Avril 1952).

b) Ions formés par le champ à l'entrée de l'éjecteur

La charge des ions est alors fonction du champ accélérateur lui-même. Comme il est démontré en annexe, les formules précédentes deviennent:

$$\left(\frac{\dot{I}}{S}\right)_{b\,max} = 3{,}76 \cdot 10^{-11} \cdot \sqrt{\frac{q}{m}} \cdot \overline{\frac{\Delta U}{d^2}}^{3/2} \tag{35}$$

$$\left(\frac{F}{S}\right)_{b\,max} = \frac{1{,}6}{\pi} \cdot \left(\frac{\Delta U}{d}\right)^2, \tag{36}$$

soit, par rapport au cas a:

$$\left(\frac{\dot{I}}{S}\right)_{b\,max} = 0{,}72 \cdot \left(\frac{\dot{I}}{S}\right)_{a\,max} \tag{35 bis}$$

$$\left(\frac{F}{S}\right)_{b\,max} = 0{,}72 \cdot \left(\frac{F}{S}\right)_{a\,max}. \tag{36 bis}$$

La poussée maximum par m^2 tombe donc ici à 1,44 tonnes/m^2.

Cette infériorité serait cependant largement compensée par une bien plus grande facilité d'organisation, le transfert des ions étant supprimé, ioniseur et éjecteur ne formant plus qu'un seul et même appareil.

Il y a lieu de remarquer par ailleurs que l'ionisation des corpuscules serait renforcée par la surintensité du champ au voisinage immédiat des fils de grille et de l'extrémité tranchante des ajutages d'injection.

c) Corpuscules électrisés dans le champ de l'éjecteur par émission d'électrons

Comme dans le cas précédent, l'électrisation est fonction du champ accélérateur. Mais celui-ci croissant de l'amont à l'aval du fait de l'effet "charge d'espace", la charge moyenne des corpuscules se trouve augmentée.

Les calculs reproduits en annexe conduisent alors aux expressions suivantes:

$$\left(\frac{\dot{I}}{S}\right)_{max} = \frac{8{,}8}{\pi} \cdot 10^{-11} \cdot \sqrt{\frac{3\,r^2}{m \cdot d}} \cdot \left(\frac{\Delta U}{d}\right)^2, \tag{37}$$

d'où le débit-masse:

$$\frac{\mu}{S} = 3 \cdot 10^9 \cdot \left(\frac{\dot{I}}{S}\right)_{max} \cdot \frac{m}{q_{max}}$$

avec:

$$q_{max} = 3 \cdot r^2 \cdot \frac{3}{2} \cdot \frac{\Delta U}{d},$$

d'où:

$$\frac{\mu}{S} = \frac{5{,}87 \cdot 10^{-2}}{\pi} \cdot \frac{m}{r^2} \cdot \sqrt{\frac{3 \cdot r^2}{m \cdot d}} \cdot \frac{\Delta U}{d}. \tag{38}$$

Dans le cas présent, la charge étant variable, la vitesse d'éjection w_1 n'est plus donnée par la formule (32); son expression devient ici:

$$w_1 = 1{,}6 \cdot \sqrt{\frac{3\,r^2}{m} \cdot d} \cdot \frac{\Delta U}{d}, \tag{39}$$

d'où:

$$\left(\frac{F}{S}\right)_{max} = \frac{0{,}281}{\pi} \cdot \left(\frac{\Delta U}{d}\right)^2. \tag{40}$$

Cette dernière expression est la seule comparable aux expressions correspondantes des cas a et b, soit:

$$\left(\frac{F}{S}\right)_{c\ max} = 1{,}265 \cdot \left(\frac{F}{S}\right)_{a\ max} = 1{,}76 \cdot \left(\frac{F}{S}\right)_{b\ max}. \qquad (40\ \text{bis})$$

Du point de vue poussée, ce processus est donc le plus favorable; la poussée maximum par m^2 passe à 2,53 tonnes.

Explicitant $m = \varrho \cdot \frac{4\pi}{3} r^3$ dans les expressions (38) et (39), ces dernières s'écrivent:

$$\frac{\mu}{S} = 0{,}117 \cdot \sqrt{\frac{r}{\pi \cdot d \cdot \varrho}} \cdot \frac{\Delta U}{d} \qquad (38\ \text{bis})$$

et

$$w_1 = 2{,}4 \cdot \sqrt{\frac{d}{\pi \cdot r \cdot \varrho}} \cdot \frac{\Delta U}{d}. \qquad (39\ \text{bis})$$

Ces expressions montrent que la vitesse d'éjection sera d'autant plus grande et le débit-masse d'autant plus petit que r sera lui-même plus petit.

En définitive, ce processus est, théoriquement, le plus avantageux; mais, pour des considérations pratiques, c'est sans doute le deuxième processus qui donnera lieu aux première réalisations en raison de sa simplicité.

IV. Contrainte mécanique des armatures

Ces contraintes s'évaluent par le calcul classique des forces qui agissent sur les armatures des condensateurs; il vient ici, par unité de surface d'armature:

$$\frac{f}{S} = \frac{1}{4\pi} \cdot \left(\frac{\Delta U}{d}\right)^2 = \frac{1}{36 \cdot 10^4 \cdot \pi} \cdot \left(\frac{\Delta V}{d}\right)^2. \qquad (41)$$

Ceci donne, environ 0,01 hpz pour 10^5 volts/cm et 1 hpz pour 10^6 volts/cm.

Cette dernière valeur est assez importante et justifierait peut-être une organisation des grilles où les fils seraient remplacés par des lamelles orientées "dans le lit du vent ionique". Il est à noter d'ailleurs que la charge d'espace diminuerait les contraintes sur l'armature d'entrée, mais les accroîtrait sur l'armature de sortie, ce qui constitue une nouvelle raison pour dématérialiser le plus possible cette dernière.

Dans tous les cas, l'entretoisement des différents organes posera des problèmes de résistance des matériaux assez compliqués, d'autant que les entretoises seront en matériaux isolants de qualités mécaniques assez médiocres et que leurs formes seront sans doute assez compliquées pour accroître leur résistance ohmique et éviter les "claquages". Selon la méthode classique, les différentes sections du bâti devront être protégées contre les surtensions locales par une chaîne de résistances assurant une répartition uniforme de la différence des potentiels des armatures.

V. Conclusions

Les deux principales difficultés rencontrées au cours de la présente étude ont été:

— l'obtention d'ions de charges spécifiques convenables, de l'ordre de 10^8 à 10^{10} U.E.S./g;

— le risque de disruptures consécutives aux valeurs très élevées de l'intensité des champs électrostatiques qui sont nécessaires tant pour ioniser les particules que pour les accélérer.

En ce qui concerne l'ionisation, les deux processus qui se sont montrés les plus prometteurs sont l'électrisation par contact, la plus facile à réaliser, et l'électrisation par émission électronique qui donne les plus fortes poussées par unité de surface.

Les particules solides sont capables, à dimensions égales, de supporter des charges spécifiques plus élevées que les gouttelettes liquides du fait de la plus grande fragilité de ces dernières. Par contre, celles-ci sont plus faciles à produire et à calibrer.

Pour ce qui est de l'accélération des ions, l'effet "charges d'espace" réduit la poussée par cm^2 de section de l'accélérateur à des valeurs relativement faibles. Cette poussée unitaire ne pourrait atteindre des valeurs raisonnables qu'à condition de mettre en œuvre des champs accélérateurs de l'ordre de 100.000 à 500.000 volts/cm.

Enfin, en ce qui concerne l'organisation de l'éjecteur, les règles suivantes semblent pouvoir être formulées:

— fusion de l'ioniseur et de l'accélérateur en un organe unique;

— dématérialisation aussi poussée que possible en constituant les armatures par des grillages formés de lamelles orientées parallèlement au champ local;

— dessin soigné des supports isolants qui devront être faits avec des matériaux de première qualité en raison des fortes tensions mises en jeu.

En terminant ce trop long développement sur les éjecteurs électrostatiques, il convient de rappeler une fois encore que toutes ces vues sont purement théoriques et n'ont reçu aucune sanction de l'expérience. L'expérimentation dans les conditions idéales de l'espace interastral serait d'ailleurs quasi impossible dans les laboratoires terrestres où le degré de vide de cet espace ne peut encore être atteint; il serait tout de même possible, et fort intéressant d'exécuter des essais préliminaires dont les résultats seraient susceptibles de supporter une certaine extrapolation.

Annexe

Effet de la Charge d'espace

a) Ions à $\frac{q}{m}$ constant

Bien que le problème soit classique, il a paru intéressant de le reprendre ici; en vue surtout de la compréhension du cas suivant des particules émettrices.

Soit donc un champ $\frac{-dU}{dz}$ régnant entre deux armatures planes supposées indéfinies (cf. Fig. 6). Pour simplifier les écritures, l'armature positive sera supposée au potentiel U_0 et l'armature négative au potentiel $U_1 = U_0 - \Delta U$. Par raison de symétrie, dans chaque plan parallèle aux armatures, toutes les grandeurs de même espèce (champ, vitesse des particules, etc.) ont même valeur.

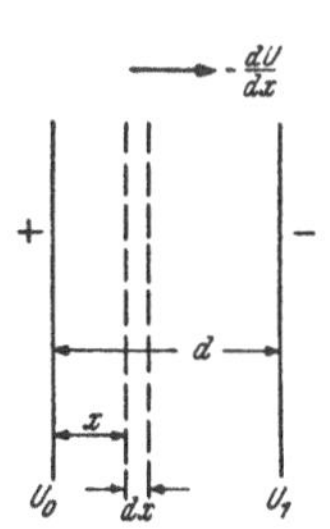

Fig. 6

Tout d'abord, l'écoulement étant supposé cylindrique et en régime permanent si $n(x)$ est le nombre d'ions par cm^3 à la distance x de l'armature positive et si $w(x)$ est leur vitesse, la conservation du flux de particules conduit à la relation:

$$n \cdot w = A = C^{te}, \tag{A-1}$$

I/S étant la densité de courants en ampères/cm^2, il vient:

$$\frac{I}{S} = n \cdot w \cdot \frac{q}{3 \cdot 10^9}. \tag{A-2}$$

d'où:

$$A = \frac{3 \cdot 10^9}{q} \cdot \frac{\dot{I}}{S}. \tag{A-3}$$

Considérons maintenant le feuillet limité par les plans x et $(x + dx)$; les charges positives incluses dans ce feuillet produisent en amont du plan x un champ $-2\pi \cdot \varrho \cdot n \cdot dx$ et, en aval du plan $(x + dx)$, un champ $+2\pi \cdot \varrho \cdot n \cdot dx$. (En effet, le champ d'un plan indéfini chargé uniformement étant indépendant de la distance au plan, pour les régions extérieures au feuillet, toutes les charges de ce dernier peuvent être supposées réparties sur un plan à la densité de n par centimètre *carré.*)

En franchissant le feuillet, le champ accélérateur $\frac{-dU}{dx}$ varie donc de:

$$d\left(-\frac{dU}{dx}\right) = +4\pi \cdot q \cdot n \cdot dx,$$

ce qui s'écrit, d'après (IV-41):

$$U'' = -4\pi \cdot q \cdot n(x) = -4\pi \cdot A \cdot \frac{q}{w(x)}. \tag{A-4}$$

Enfin, le théorème des forces vives donne:

$$m \cdot w \cdot dw = -q \cdot dU,$$

d'où, en supposant w_0 très petit (w_0 nul impliquerait $n_0\ \infty$):

$$w \cong \sqrt{2\frac{q}{m}(U_0 - U)}. \tag{A-5}$$

Portant cette valeur dans (A-4), il vient la relation:

$$U'' = -\frac{4\pi A \cdot q}{\sqrt{2\frac{q}{m}(U_0 - U)}}; \tag{A-6}$$

U' ne figurant pas dans cette équation, cette dernière s'intègre sans difficulté, ce qui donne:

$$U'^2 - U_0'^2 = 8 \cdot \pi \cdot A \cdot q \cdot \sqrt{\frac{2}{q/m}(U_0 - U)}, \tag{A-7}$$

d'où, d'une armature à l'autre:

$$U_1'^2 - U_0'^2 = 8 \cdot \pi \cdot A \cdot q \cdot \sqrt{\frac{2}{q/m} \cdot \Delta U}. \tag{A-8}$$

Pour aller plus avant, il convient de distinguer le cas où les ions q/m, préformés dans un ioniseur, sont injectés dans le champ à travers l'électrode positive, du cas où les ions se forment à même cette électrode sous l'influence du champ $-U_0'$ qui règne à sa surface.

1°. Ions préformés et injectés

$A \cdot q$ étant proportionnel à la densité de courant, (A-7) montre que cette dernière sera d'autant plus élevée que $-U_0$ sera plus petit et, par conséquent, U_1 plus grand, ces deux quantités variant en sens inverses. Or les ions étant supposés injectés à faible vitesse, $-U_0'$ ne peut devenir négatif sous peine de refouler les ions; la densité de courant maximum correspond donc à la nullité de U_0'.

Dans ces conditions, l'intégration de (A-7) donne:

$$(U_0 - U)^3 = \frac{1}{2}(q \cdot \pi \cdot A \cdot q)^2 \cdot \frac{x^4}{q/m},$$

d'où, pour $x = d$:

$$\overline{\Delta U^3} = \frac{1}{2}(q \cdot \pi \cdot A \cdot q)^2 \frac{d^4}{q/m}.$$

Compte-tenu de (A-3), il vient alors, en ampères/cm²:

$$\left(\frac{I}{S}\right)_{max} = \frac{\sqrt{2}}{27 \cdot 10^9 \cdot \pi} \sqrt{\frac{q}{m}} \cdot \frac{\overline{\Delta U}^{3/2}}{d^2} \qquad \text{(A-9)}$$

ce qui n'est autre que la formule classique de LANGMUIR.

On en déduit la poussée $\frac{F}{S}$ par cm² de section de l'accélérateur; on a:

$$\frac{F}{S} = (n \cdot w \cdot m) \cdot w = A \cdot m \cdot w, \qquad \text{(A-10)}$$

d'où, d'après (A-3, 5 et 8):

$$\left(\frac{F}{S}\right)_{max} = \frac{2}{q\pi} \cdot \frac{\overline{\Delta U^2}}{d^2} \cong 8 \cdot 10^{-7} \cdot \left(\frac{\Delta V}{d}\right)^2. \qquad \text{(A-11)}$$

Dans son mémoire "Trajectoires interplanétaires entre les orbites des satellites" paru dans le J. Amer. Rocket Soc., Mars-Avril 1952, LYMAN SPITZER, mettant en évidence la vitesse d'éjection w, donne la formule équivalente ci-dessous, à l'adaptation près des notations:

$$S = 36 \cdot \pi \cdot \left(\frac{q}{m}\right)^2 \cdot \frac{d^2}{w^4} \cdot P; \qquad \text{(A-12)}$$

en y remplaçant la puissance P par l'expression équivalente $\frac{1}{2} Fw$, on retrouvera sans difficulté la formule (A-11).

Le tableau ci-dessous donne la poussée en tonnes par m² en fonction de l'intensité du champ en volts/cm:

V/d volts/cm	10^4	$5 \cdot 10^4$	10^5	$5 \cdot 10^5$	10^6
F/S tonnes/m²	$8 \cdot 10^{-4}$	$2 \cdot 10^{-2}$	$8 \cdot 10^{-2}$	2	8

Comme il a été annoncé au début, on voit que l'obtention de poussées utilisables nécessiterait des champs très intenses.

2°. *Ions formés par le champ initial*

La charge q est alors proportionnelle à $-U'_0$, et par conséquent, la densité de courant devient proportionnelle à $-A \cdot U'_0$.

Dans ce cas, ce n'est donc plus $U'_0 = 0$ qui donne le courant maximum puisque ce dernier est alors nul, les ions ne se formant pas.

(A-7) s'écrit alors, en désignant par K, comme précédemment, le coefficient de proportionnalité de q à $(-U')$:

$$U' = \sqrt{U_0'^2 + 8\pi A \sqrt{2mK(-U_0')(U_0 - U)}}; \qquad \text{(A-13)}$$

d'autre part, en multipliant le 1er membre de (A-7) par dU' et le 2ème membre par la quantité égale $U''\,dx = \dfrac{-4\pi A q}{\sqrt{2\frac{q}{m}(U_0 - U)}}$, il vient en intégrant de 0 à 1:

$$U_1'^3 - U_0'^3 - 3\,U_0'^2 (U_1' - U_0') = +96\pi^2 A^2 m K (-U_0')\,d \qquad \text{(A-14)}$$

avec

$$U_1'^2 = U_0'^2 + 8\pi A \sqrt{2\,m K (-U_0')\,\Delta U}. \qquad \text{(A-15)}$$

Eliminant $A^2\,m\,K$ entre ces relations, il vient, toutes simplifications faites:

$$3\,U_1'^2 - 2\left(2\frac{\Delta U}{d} - 3\,U_0'\right)U_1' - U_0'\left(8\frac{\Delta U}{d} - 3\,U'_0\right) = 0 \qquad \text{(A-16)}$$

d'où l'on tire la condition:

$$-\frac{U_0'}{\Delta U/d} < \frac{1}{3} \qquad \text{(A-17)}$$

et l'expression:

$$U_1' = \frac{\Delta U}{d}\left[\frac{2}{3} - \frac{U_0'}{\Delta U/d} + \frac{2}{3}\sqrt{1 + 3\cdot\frac{U_0'}{\Delta U/d}}\right]; \qquad \text{(A-18)}$$

de cette dernière et de (A-15), l'on tire:

$$A\cdot(-U_0') = \frac{(\Delta U/d)^2}{18\pi\sqrt{2K\cdot m\cdot d}}\cdot\sqrt{\frac{U_0'}{\Delta U/d}}\cdot\left[2 + \left(2 - 3\frac{U'_0}{\Delta U/d}\right)\cdot\sqrt{1 + 3\frac{U_0'}{\Delta U/d}}\right]. \qquad \text{(A-19)}$$

Prenant la dérivée par rapport à $\dfrac{-U_0'}{\Delta U/d}$, l'on trouve que l'optimum correspond à:

$$-U_0' = 0{,}25\cdot\frac{\Delta U}{d}; \qquad \text{(A-20)}$$

on en déduit, sans difficulté:

$$\left(\frac{I}{S}\right)_{max} = 0{,}72\,\frac{\sqrt{2}}{27\cdot 10^9\cdot\pi}\cdot\sqrt{\frac{q}{m}}\cdot\frac{\Delta U^{3/2}}{d^2} \qquad \text{(A-21)}$$

et

$$\left(\frac{F}{S}\right)_{max} = 0{,}72\,\frac{2}{q\pi}\left(\frac{\Delta U}{d}\right)^2, \qquad \text{(A-22)}$$

soit (à même $\frac{q}{m}$ pour I/S), les 72% des valeurs correspondant aux ions préformés et injectés.

b) Particules émettrices

Ici deux remarques préliminaires s'imposent en ce qui concerne les électrons émis par les particules en cours d'accélération:

1) Tout électron capté par une particule sera supposé immédiatement réémis par celle-ci;

2) en raison des vitesses moyennes considérables des électrons, leur participation à la charge d'espace sera négligée.

Ceci posé, les équations fondamentales sont ici:

$$nw = A = C^{te} \qquad \text{(A-23)}$$

sans changement, par rapport à (A-1), les particules continuant à se conserver, nonobstant les variations de leur charge. Mais ici l'intensité varie d'une tranche à l'autre et dans l'expression (A-3) de A, I et y doivent être considérés comme des fonctions de x, soit:

$$A = \frac{3 \cdot 10^9}{q(x)} \cdot \frac{I(x)}{S} \tag{A-24}$$

avec, d'après (24):

$$q(x) = -K\,U'; \tag{A-25}$$

la relation des forces vives devient:

$$m\,w\,dw = -q\,dU = K\,U'\,dU \tag{A-26}$$

et la relation (A-4) demeure, à cela près que q et w sont maintenant des fonctions de x:

$$U'' = -4\pi A \frac{q(x)}{w(x)} = 4\pi A K \frac{U'}{w(x)}. \tag{A-27}$$

Tirant w de cette relation, l'on écrit:

$$w = 4\pi A K \frac{U'}{U''} = 4\pi A K \frac{dU}{dU'}, \tag{A-28}$$

d'où, d'après (A-26):

$$m\,dw = \frac{1}{4\pi A} U'\,dU' \tag{A-29}$$

et, en intégrant:

$$m(w - w_0) = \frac{1}{8\pi A}(U'^2 - U_0'^2). \tag{A-30}$$

Eliminant w entre cette équation et (A-28), il vient:

$$w_0 + \frac{1}{8\pi A m}(U'^2 - U'^2_0) = 4\pi A K \frac{dU}{dU'} = 4\pi A K \frac{U'}{dU'}\,dx. \tag{A-31}$$

Intégrant, comme précédemment, par rapport à dU, d'une part, et à dx d'autre part, de $x = 0$ à $x = d$, l'on obtient les deux relations, en désignant par B la quantité $32\pi^2 mK$:

$$(8\pi A m w_0 - U_0'^2)(U_1' - U_0') + \frac{U_1'^3 - U_0'^3}{3} = -A^2 B \Delta U \tag{A-32}$$

$$(8\pi A m w_0 - U_0'^2)\log\frac{U_1'}{U_0'} + \frac{U_1'^2 - U_0'^2}{2} = A^2 \cdot B \cdot d. \tag{A-33}$$

En raison de la transcendance de la deuxième équation, l'on ne peut poursuivre comme dans le cas précédent.

Revenant donc à l'expression (A-10) de la poussée unitaire, que nous désignerons ici par Φ, il vient d'après (A-30) et en négligeant w_0 vis à vis de w_1:

$$\Phi = \frac{E}{S} = A\,m\,w_1 = \frac{U_1'^2 - U_0'^2}{8\pi}; \tag{A-34}$$

il s'agit de rechercher les conditions qui rendent cette expressions maximum.

Posant alors:

$$U_1' - U_0' = y \text{ et } \frac{U_1'}{U_0'} = z$$

et w_0 étant considéré comme négligeable, les relations (A-32 et 33) s'écrivent alors, en remarquant que:

$$U_0' = \frac{y}{z-1} \text{ et } U_1' = \frac{yz}{z-1}:$$

$$\begin{cases} -\dfrac{y^3}{z-1} + \dfrac{(z^3-1)\,y^3}{3\,(z-1)^3} = -A^2 B \Delta U \\ -\dfrac{y^2}{(z-1)^2} \log_e \dfrac{U_1'}{U_0'} + 4\pi\Phi = A^2 B d. \end{cases}$$

Simplifiant le 1er membre de la première et substituant à y^2, dans la seconde, l'expression tirée de:

$$U_1'^2 - U_0'^2 = 8\pi\Phi = \frac{z+1}{z-1} y^2, \tag{A-35}$$

il vient:

$$\begin{cases} y^3 \dfrac{z+2}{3\,(z-1)} = -A^2 B \Delta U & \text{(A-36)} \\ 4\pi\Phi = \dfrac{A^2 B d}{1 - 2\dfrac{\log_e z}{z^2-1}}. & \text{(A-37)} \end{cases}$$

Eliminant A^2B et Y entre ces trois dernières relations, il vient, tous calculs faits (en remarquant que y est négatif puisque $|U'_1| > |U'_0|$ et que U'_1 et U'_0 sont eux-mêmes négatifs);

$$\sqrt{8\pi\Phi} = \frac{3}{2}\frac{\Delta U}{d} \sqrt{\frac{z+1}{z-1}}\, \frac{z^2 - 1 - 2\log_e z}{(z-1)(z+2)}. \tag{A-38}$$

Cette relation montre que Φ est fonction croissante de z et est donc maximum pour $z\ \infty$, c'est à dire pour $U'_0 = 0$ (ce qui, dans ce cas, n'était pas évident à priori); l'on en conclut:

$$\left(\frac{F}{S}\right)_{max} = \frac{9}{32\cdot\pi}\left(\frac{\Delta U}{d}\right)^2. \tag{A-39}$$

Cette valeur est 1,265 fois plus grande que celle relative au premier cas (ions préformés) et 1,76 fois plus grand que celle relative au second cas (ions formés par le champ initial).

Les autres caractéristiques ont les valeurs suivantes:

$$\begin{cases} U_1'\,max = \dfrac{3}{2}\dfrac{\Delta U}{d} \\ A = \dfrac{0{,}176}{\pi}\dfrac{1}{\sqrt{Kmd}}\dfrac{\Delta U}{d} \\ w_1 = 1{,}6\sqrt{\dfrac{K}{m}d\,\dfrac{\Delta U}{d}} \\ \left(\dfrac{I}{S}\right)_{max} = \dfrac{8{,}8}{\pi}\,10^{-11}\sqrt{\dfrac{K}{md}}\left(\dfrac{\Delta U}{d}\right)^2. \end{cases}$$

Multi-Directional G-Protection in Space Vehicles

By

Harald J. von Beckh[1], SAI

(With 5 Figures)

Abstract — Zusammenfassung — Résumé

Multi-Directional G-Protection in Space Vehicles. It is known that maximum human tolerance to *G*-loads is obtained if the accelerations are acting at right angles to the long axis of the body.

The author describes a device, termed "Anti-*G*-Capsule", which is pivoted about the lateral axis of the craft and assumes automatically a position such that the resultant of all acting accelerations is perpendicular to the heart-head line of the operator.

This *G*-protection seems especially important during the reentry phase where the operator would be subjected to severe accelerations whose directions and intensity vary.

The ejection- and stabilization mechanism of this device would also afford an analogous *G*-protection during and after escape from a disabled space vehicle within the lower layers of the atmosphere.

Schutzvorrichtung gegen mehrseitige Beschleunigung in Raumfahrzeugen. Es ist bekannt, daß Beschleunigungen wesentlich besser vertragen werden, wenn sie senkrecht zur Längsachse des Körpers einwirken.

Der Autor beschreibt eine um eine Querachse drehbare „Anti-*G*-Kapsel", die automatisch den Insassen so lagert, daß die Resultierende aller wirkenden Beschleunigungen senkrecht zur Herz-Kopf-Linie einwirkt.

Dieser Beschleunigungsschutz scheint besonders wesentlich beim Wiedereintritt eines Raumfahrzeuges in dichtere Luftschichten, wobei besonders hohe und in Beziehung auf Intensitat und Einwirkungsrichtung ständig wechselnde Beschleunigungen zu erwarten sind.

Außerdem wurde der Ausstoß- und Stabilisierungsmechanismus in Notfällen ein „Aussteigen" aus dem Raumfahrzeug in erdnahen Schichten ermöglichen.

Protection anti-G multidirectionnelle dans les engins spatiaux. On sait que la tolérance physiologique aux accélérations est la plus grande quand celles-ci sont dirigées perpendiculairement à l'axe longitudinal du corps humain.

L'auteur décrit un dispositif, appelé "capsule anti-*G*", pivotant autour d'un axe transversal à l'engin et prenant automatiquement une position telle que la résultante des accélérations soit perpendiculaire à l'axe cœur-tête du sujet.

Cette protection semble spécialement importante pour la phase de repénétration pendant laquelle le sujet serait soumis à des accélérations importantes de direction et d'intensité variables.

Les mécanismes d'éjection et de stabilisation du dispositif offrent une protection analogue pendant et après l'abandon d'un véhicule désemparé dans les couches basses de l'atmosphère.

---

[1] Space Biology Branch, Aero Medical Field Laboratory, Holloman Air Force Base, New Mexico, U.S.A.

Introduction

Typical escape trajectories of multistaged rocket craft produce linear accelerations that are high, but can be tolerated by the human operator if he is properly positioned. Centrifuge experiments [19] which simulated the accelerations of a three-stage escape trajectory have in addition demonstrated that human subjects in supine position, can perform satisfactorily a dual pursuit task. We can assume then, that selected crew members could assist in the control of such a vehicle.

Radial accelerations would be produced for extended periods, but are of such low intensity that they would not present physiological problems.

However, even small pilot errors or imperfections of the automatic guidance system could produce especially in the re-entry phase excessive G-loads. This would subject the human operator to severe accelerations whose directions, rate of onset, and intensity continuously vary. These accelerations could, however, again be tolerated if the operator were simultaneously positioned transversely to the G-resultant.

In the following report a device is described, which could grant this multidirectional G-protection by automatic positioning. The resultant of all acting accelerations would be presented at right angles to the heart-head line of the operator.

This device, termed the "Anti-G Capsule"[1] would at the same time afford an analogous G-protection during and after escape from a disabled space vehicle within the lower layers of the atmosphere.

Review of Literature

A review of the literature indicates the following studies to be pertinent:

BUEHRLEN [4, 5] reported in 1937 centrifuge runs with humans in supine position with peaks up to 17 G.

WIESEHOEFER [33] recognized that the continuous prone or supine position achieves protection only against radial G-loads, but leaves the pilot unprotected against the linear loads in direction of the flight path, which are especially high in catapult take-off and in arrested landings. He installed a backward tilting seat in the second cockpit of a dive bomber version of a Heinkel He 50 fighter aircraft. This seat allowed the operator the control of the craft in conventional position and tilted him backwards to a supine position when the acceleration reached values of 3 G. In 16 flights, 5 different subjects resisted 7 G during 15 seconds without having reached their limits of tolerance. As usual in acceleration test flights, the control pilot flew in the "crouch" position. Although the backward tilting took place after the acceleratory stress had begun, no vertigo or other labyrinthine troubles were observed.

Since that time intensive investigations of the tolerance to transverse G have been performed. STAUFFER [22, 23] exposed humans for 5 to 8 seconds to 12 G and DUANE, BECKMAN, ZIEGLER and HUNTER [8] exposed themselves for 5 seconds to 15 G, which these authors considered as a "voluntary endpoint". STRUGHOLD [21], RUFF [20], and CLARK et al. [6] considered human tolerance to be 12 G during 60 seconds.

As the transverse position in the plane presents operational and constructional difficulties (poor visibility, necessity of additional controls and instrument panel,

[1] The "Anti-G Capsule" has already been described in a previous paper [34], in terms of atmospheric flight and escape. The present discussion extends this concept to space flight.

complicated escape devices) extensive efforts were made to study the protective effect of partial supination, with and without combination of anti-G garments and straining (GELL and HUNTER [12]; BALLINGER and DEMPSEY [2]; MARTIN and HENRY [17]; VON DIRINGSHOFEN [30]; DORMAN and LAWTON [7]).

It was demonstrated that only a backward tilt beyond 77 degrees exceeded the protection afforded by Anti-G garments and that maximum protection was reached only at 85 degrees.

GELL [11] accomplished remarkable flight experiments with a supinating seat in the second cockpit of an F7F-2N "Tigercat" fighter aircraft. The pilot withstood high radial G-loads during pull outs from steep dives and during spirals for 30 seconds, using an autopilot "P1K" when supinated. GELL also corraborated the experiment of WIESEHOEFER by demonstrating that backward tilting within a G-field does *not* cause vestibular troubles.

Following the principle of the supinating seat, this author describes a conceptual device, which could grant a "*multi-directional*" G-protection by automatic supine positioning. This device is termed the "*Anti-G Ejection Capsule*".

Description of the Anti-G Ejection Capsule

The operator's seat is integrated for escape purposes in an oblate-spheroid, pressurized capsule (Fig. 1), which is pivoted in the lateral axis of the craft allowing free rotation through 360 degrees.

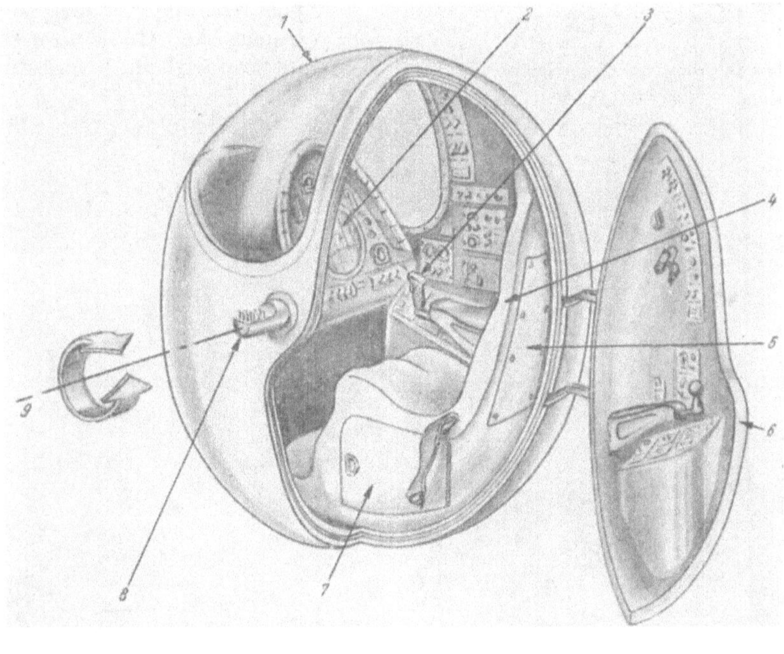

Fig. 1. *1* Removeable panel for escape in water; *2* Plan-position indicator, TV-Screen, or radar scope; *3* Control grip and pivoting armrest for control under high accelerations; *4* Anthropomorphically contoured and cushioned seat to support maximum area under load. Seat provided with lap, shoulder, and ankle straps; *5* Access door for installation of counterweights as needed; *6* Pressure-tight entrance door containing left console and throttle armrest control; *7* Equipment access door; *8* Pivot shaft, bearings, and electrical connections to allow full 360° rotation of capsule while maintaining control via slip-rings behind panel; *9* Lateral axis of the aircraft or space-vehicle

The center of gravity, determined by the eccentric location of pilot and part of the equipment, is located such that within the plane of symmetry, the line from the axis of rotation to the center of gravity is perpendicular to the line between heart and head of the operator (Fig. 2).

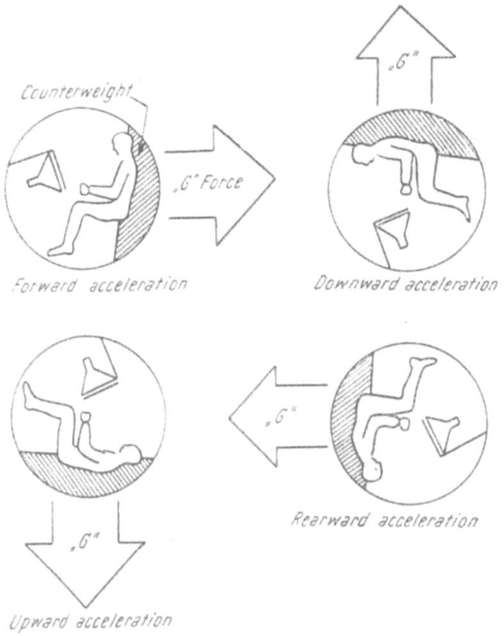

Fig. 2. These pictures show how the capsule moves, as the direction of acceleration changes

If the direction of the resultant of acting accelerations changes, the capsule assumes the new adequate position either directly by the acceleratory stress itself, or indirectly through remote sensing and external power. Thus G-loads act again transversely and press the operator in his anthropomorphologically designed and energy absorbing seat, and seat-back. Adequate damping of the axis avoids oscillatory movements.

Operational controls, e.g. control stick, rudder pedals, throttle and emergency controls are constructed to be handled in supine position under action of high G-loads, i.e., the control stick is integral with the armrest and its

Fig. 3. *1* Capsule extended and locked in position for crew entry; *2* Drogue housing main parachute canopy. Ejection shown downward, but can be effected in any other direction

axis of neutral position coincides with the resultant of G-forces. Displays are integrated in the cabin and turn with the operator. They are connected via slip rings with the pivot shaft of the capsule. Provisions are made for direct vision, or indirectly through displays for control from any position of the capsule.

It is foreseen that the pilot can lock the capsule in the conventional position, which allows him direct vision during that part of the flight path, when high radial accelerations are not expected. This could be. for instance, during take-off, landing, and eventually as an emergency procedure if malfunctioning of indirect vision displays occurs.

During and after ejection, the capsule functions similarly as in flight, because the ejection mechanism acts upon the pivot axis. The pivot shaft is fixed on both ends to a fork-shaped device which in turn is attached to a drogue housing main chute canopy (Fig. 3). This prevents post-ejection tumbling and places the pilot transverse to the decelerative stress. The main chute system and energy absorbing and buoyant qualities of the capsule allow landing on either earth or water (Fig. 4).

Fig. 4. *1* Main chute deployed from drogue container which has aneroid and "Q" sensing elements; *2* Telescoping emergency antenna; *3* Dye-marker dispenser; *4* Retractable air vent with water drain

Survival and comfort devices may be installed as required analogous to existing capsule configurations.

Function of the Capsule in Flight and During Escape

A. In Flight: The capsule will be locked in the conventional position for entry of the operator and take-off. As soon as the capsule is unlocked the pilot's position will be always perpendicular to the resultant G-load, i.e., in supine position.

Lateral accelerations would not cause rotation of the capsule, but, since they already act in a transverse direction (right-left or left-right), protection would be necessary only against traumatic injuries, caused by collision with the inner structure or side walls of the capsule. However, these lateral movements of the operator will be avoided by the anthropomorphic configuration of the seat and an adequate restraint.

In zero gravity conditions, the capsule remains in any position at rest. The restraint prevents separation of the pilot from his seat. Therefore he remains protected if a G-field is again produced.

B. During Escape: Theoretically the capsule can be ejected in any direction, but upward ejection seems the most desirable for facilitating escape at low altitudes and "off the deck".

If experimentation shows that positioning by the ejection stress itself produces too high angular accelerations, the operator could be positioned immediately before ejection by external power (cartridge). Entering in the airstream, the operator would be positioned by the deceleratory stress itself. Another possibility would be to eject with the capsule locked in its telescopically extensible fork. In this case the aerodynamically designed drogue housing would act primarily as stabilizer without adding appreciably to the drag of the capsule as in the former case. Both solutions would position the pilot transversely to the deceleratory stress.

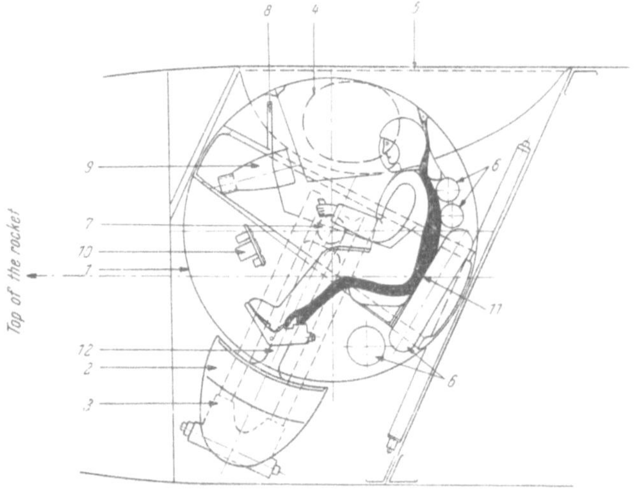

Fig. 5. Installation of the anti-*G* capsule in a space vehicle. *1* Shell of anti-*G* capsule; *2* Housing of main chute system; *3* Catapult mounting; *4* Entrance hatch; *5* Space vehicle canopy (removable); *6* Equipment; *7* Pivot shaft with slip-rings; *8* TV-Screen; *9* Radar scope; *10* Part of instrumentation; *11* Anthropomorphical seat configuration; *12* Telescopic fork-shaped stabilizing arms

In the same position he would confront the opening shock of the main parachute system and would descend in a horizontal position. The energy absorbing qualities of the seat and lower structure of the capsule would allow the pilot to tolerate the landing impact.

After landing on earth he leaves by the entrance door or by the removable panel. After landing in water he has at his disposal a survival kit, dyemarker dispenser, telescoping emergency antenna, retractable air vent with water drain, etc., and leaves, when rescued, by the removable panel.

Discussion

BALLINGER [2] as well as PRESTON-THOMAS et al. [19] reported valuable data concerning the tolerance to accelerations as would be expected in multistage manned rocket crafts. Assuming exhaust velocities between 2.5 and 4 km/sec they used the human centrifuge to demonstrate that men in the supine position and undergoing a typical series of *G*-loads with peaks of 8 *G*, 5.8 *G* and 5.8 *G* in the same run, could perform a dual pursuit task satisfactorily. This would indicate that selected crewmen could assist in the control of such a vehicle, especially if they were helped by adequate computer devices.

Rocket craft, during typical escape trajectories, would produce only insignificant radial loads except in the re-entry, but rocket interceptors could produce high radial loads in their curved trajectories.

For instance, at a velocity of 3000 m/sec (10, 800 km/h) a very small deviation from the linear flight-path, corresponding to a circle of 200 km diameter, would in conventional position produce the intolerable longitudinal load of 9.17 G.

If we assume that the thrust of the vehicle simultaneously produces 5 linear G's, the resultant G-load reaches 10.3 G.

Both accelerations are intolerable in longitudinal direction at longer duration and only "multi-directional" G-protection by continuous positioning transverse to the resultant force could make survival feasible.

The physiological end points and symptoms produced by transverse G have been studied in detail and defined by numerous investigators [4, 8, 17, and 24].

It was reported that stiffness and weariness was noted by all subjects, but no appreciable recovery time was required, and no black-out or impending unconsciousness occurred. However, vertigo which persisted 24 to 48 hours was a common finding and could be interpreted as probably caused by edema within the vestibular apparatus. No permanent damage was reported [8].

A special problem concerning the function of the capsule should be mentioned: The backward-forward rotation within a G-field.

Both GELL [11] and WIESEHOEFER [33] found that backward tilting during the acceleratory stress did *not* cause discomfort or diminution of the operators capability to control the craft. In the anti-G capsule much larger excursions will take place. In transition from positive to negative acceleration the rotatory movement would be through 180 degrees. Therefore the incidence of labyrinthine and other symptoms should be studied with suitable devices.

The valuable results of WEISS et al. [32] obtained from experiments with spin tables are only of partial application for the described device, for this type of movement would be experienced only to a small extent in the capsule.

The human centrifuge could serve for studying prolonged G-conditions during the flight when the capsule is pivoted with its axis perpendicular to the centrifuge arm. However angular, and above all, Coriolis Accelerations have to be considered.

For reproducing the escape conditions of the capsule, a rocket sled, similar to that developed by J. P. STAPP [24] would be suitable, if the capsule were mounted with the pivot shaft in vertical position.

The subject could be positioned before the run with his head-heart line in direction of movement, would swing 90 degrees backwards during the acceleration phase, and finally 180 degrees forwards during the deceleration phase.

The conception of control devices, maneuverable under high G-loads, as it is shortly mentioned in the section "Description of the Capsule", would be studied and will be described in a later publication.

In reference to the escape problem from aircraft at supersonic speeds, we note that with increasing speed, the linear ejection force must be increased in order to clear the tail structure. In conventional ejection systems, this linear stress acts from head to foot, and jeopardizes the structural integrity of the vertebral column.

This induced LOMBARD [16] to consider ejection from a jack-knifed position and MOHRLOCK [18] to experiment with a torso harness-vest. This device consists of a vest which is attached to the upper part of the seat-back and diminishes appreciably the G-load to the lumbar vertebrae. This torso harness is postulated to increase the tolerance by 50%.

A more radical solution utilizes the "B" (Bobsled) seat, developed by Convair and the Stanley [9] Ejection Capsule. Both bring the pilot in a nearly supine position prior to ejection, so that he receives the ejection load transversely.

It is necessary to consider that the problem is twofold in that supine ejection will not protect the pilot against the linear *G*'s encountered during deceleration of the capsule, which will now act in longitudinal direction.

In spite of the advantage of a smaller frontal area and additional forward thrust, this stress could become intolerable at extreme speeds and high altitudes, especially as the latter prolongs the decelerative phase of escape [13].

Conclusions[1]

The author is aware of the numerous technical difficulties involved with the realization of the described device, such as the weight ratio, the transmission of all connections via sliprings, the adequate damping to avoid oscillatory movements and the maintenance of the environment of the sealed capsule.

However, it seems that a simplified modification of the anti-*G* capsule, termed "Anti-*G* Swing", could be used before all these problems are completely solved, for instance, in animal-carrying nosecones.

In this case the complicated slipring-transmission problem would be lessened, because only a reduced number of connections would be necessary for registration of the physiological reactions of the animal.

In recent centrifuge experiments [35], small mammals were located on an "Anti-*G* Swing", which positioned them similarly to the anti-*G* capsule. Since it was found that albino mice could tolerate several hundred *G* for durations up to half a minute, it seems that their survival, even after partly uncontrolled trajectories, would be feasible, if they were adequately positioned.

Another series of experiments recently started, using albino rats located on an anti-*G* swing, which was horizontally fixed on the platform of a rocket-sled, show the advantage of this device, which converts high linear accelerations and decelerations to a comparatively low centrifugal load.

Another application of this device would be in those nosecones, that are spin-stabilized. Although animal compartments could be stabilized by gimbaling devices, it seems to be a simpler solution to place the animals in an anti-*G* swing, located at some distance from the axis of the cone.

In this case the animal could tolerate in continous supine position the resultant of the centrifugal load, produced by the spining of the cone, and the linear accelerations of the trajectory.

In conclusion it is pertinent to mention another aspect of physiologica reactions to weightlessness: In 1954 the author described the effects of post-acceleration weightlessness. In these experiments the pilot abruptly pulled out of a dive, producing a positive acceleration of about 6.5 *G*, which caused gray-out or black-out of the subject [29]. Immediately after the pull-out, the aircraft was flown along the ascending arc of the subgravity-parabola.

In other runs an identical *G*-load was produced, but after pull-out the aircraft was flown in horizontal flight under one-*G*-conditions.

It seemed that the recovery time from black-out and gray-out lasted longer in the first case. The subjects stated also that they "felt better" when pull-out was followed by horizontal flight than by a sub-gravity parabola.

This could be explained by the relaxation of the muscles during weightlessness ("Fallreflex") which may have caused a delay in the return of blood to the right ventricle. Also one could suppose that the complicated reflex mechanisms of the autonomic pressure-receptors and regulators, which are normally "calib-

[1] These conclusions were received April 9, 1958 (Editor).

rated" and used to functioning in a normal G-field, are disturbed by the weightless condition.

On the other hand, a subject exposed to a multiple G-load immediately after a period of weightlessness would experience greater stress than if he went from an one-G-field to a multiple G-load.

Actual experiments in F-94 C fightercrafts, in which high G-loads, produced by pull-outs and diving spirals, are applied immediately before or after the sub-gravity parabola, are being conducted to substantiate the former findings. Cinematographic and adequate instrumentation for physiological data records the effects of pre-weightlessness and post-weightlessness accelerations.

Considering these additional hazards of the combination of acceleration and weightlessness, it seems that an adequate and continous G-protection is specially important in satellite and space projects.

Acknowledgement

The author is indebted to Mr. VICTOR BUTTS and Mr. H. GOTTSCHALK for technical advice, valuable suggestions and for the drawings contained in this paper.

References

1. H. G. ARMSTRONG and J. W. HEIM, The Effect of Acceleration on the Living Organism. J. Aviat. Med. **9**, 209 (1938).
2. E. R. BALLINGER and C. A. DEMPSEY, The Effects of Prolonged Acceleration on the Human Body in the Prone and Supine Positions. WADC 52—250, July, 1952.
3. E. L. BECKMAN, T. D. DUANE, J. E. ZIEGLER and H. N. HUNTER, Human Tolerance to High Positive G Applied at a Rate of 5 to 10 G per second. J. Aviat. Med. **25**, 50 (1954).
4. L. BUEHRLEN, Versuche über die Bedeutung der Richtung beim Einwirken von Fliehkräften auf den menschlichen Körper. Luftfahrtmedizin **1**, 308 (1937).
5. L. BUEHRLEN, Spitzenbeschleunigungen in zwei verschiedenen Lagen. Luftfahrtmedizin **2**, 287 (1938).
6. W. G. CLARK, J. P. HENRY, P. O. GREELEY and D. P. DRURY, Studies on Flying in the Prone Position. Nat. Research Council Aviat. Med. Rep. 466, 1945.
7. P. J. DORMAN and R. W. LAWTON, Effect on G Tolerance of Partial Supination Combined with the Anti-G Suit. J. Aviat. Med. **27**, 490 (1956).
8. M. D. DUANE, E. L. BECKMAN, J. E. ZIEGLER and H. N. HUNTER, Human Studies of 15 Transverse G. J. Aviat. Med. **26**, 298 (1955).
9. R. A. FROST, Engineering Problems in Escape from High Performance Aircraft. J. Aviat. Med. **28**, 74 (1957).
10. O. GAUER, The Physiological Effects of Prolonged Acceleration. German Aviat. Med. WWII, p. 554.
11. C. F. GELL, Modification of F7F, Installation of Supine Seat and Related Components, Inflight Evaluation of the Seat. NADC-MAL 5104 and L 5208 (1952).
12. C. F. GELL and N. H. HUNTER, Physiological Investigation of Increasing Resistance to Black Out by Progressive Backward Tilting to the Supine Position. J. Aviat. Med. **25**, 568 (1954).
13. J. W. GOODRICH, Escape from High Performance Aircraft. WADC Techn. Note 56—7.
14. J. F. HEGENWALD and B. A. BLOCKLEY, Survivable Supersonic Ejection, 27th Annual Meeting, Aero Medical Association, Chicago, 1956.
15. J. P. HENRY, Studies of the Physiology of Negative Acceleration. WADC 5953, Oct. 1950.
16. C. F. LOMBARD, Can Change of Position Effect Ejection Tolerance? 28th Annual Meeting, Aero Medical Association, Denver, May 1957.

17. E. E. MARTIN and J. P. HENRY, The Supine Position as a Means of Increasing Tolerance to Accelerations. WADC No. 6025, August 1950.
18. H. F. MOHRLOCK, Aircraft Performance Systems Related to Escape Systems. J. Aviat. Med. **28**, 59 (1957).
19. H. PRESTON-THOMAS, R. EDELBERG, J. P. HENRY, J. MILLER, W. SALZMAN and G. ZUIDEMA, Human Tolerance to Multistage Rocket Acceleration Curves. J. Aviat. Med. **26**, 390 (1955).
20. S. RUFF, Brief Acceleration: Less than One Second. German Aviat. Med. WWII, p. 584.
21. S. RUFF and H. STRUGHOLD, Compendium of Aviation Medicine. Leipzig: J. A. Barth, 1939.
22. F. R. STAUFFER, The Effect of High Acceleration Forces upon Certain Physiological Functions of Human Subjects Placed in a Modified Supine Position, U.S. Naval School of Aviation Medicine, NM 001 059 02 01, Oct. 1949.
23. F. R. STAUFFER, Certain Physiological Responses in Man Changing from the Sitting Position to Supine Position during Radial Acceleration. U.S. Naval School of Aviation Medicine, NM 001 059 02 02, Feb. 1950.
24. J. P. STAPP and D. C. HUGHES, Supersonic Deceleration and Windblast. J. Aviat. Med. **27**, 407 (1956).
25. J. P. STAPP and W. C. BLOUNT, A Compressed Air Catapult for High Impact Forces. J. Aviat. Med. **28**, 281 (1957).
26. D. G. SIMONS and J. P. HENRY, Electroencephalographic Changes Occurring during Negative Accelerations. Air Force Techn. Rep. No. 5966, May 1950.
27. F. D. VAN WART and J. T. BARTER, A Method for Calculating the Center of Gravity of the Human Body for the Development of Stabilized Escape Systems. 28th Annual Meeting, Aero Med. Association, Denver, May 1957.
28. H. J. VON BECKH, Fisiologia del Vuelo (Spanish), p. 114. Buenos Aires: Ed. Alfa, 1955.
29. H. J. VON BECKH, Experiments with Animals and Human Subjects under Sub and Zero Gravity Conditions during the Dive and Parabolic Flight. J. Aviat. Med. **25**, 235 (1954).
30. H. VON DIRINGSHOFEN, "Long Chair" Position for Fighter Pilots. J. Aviat. Med. **26**, 467 (1955).
31. H. VON DIRINGSHOFEN, "Medical Guide for Flying Units" (English Translation). Toronto Press, 1941.
32. H. S. WEISS, R. EDELBERG, P. V. CHARLAND and J. I. ROSENBAUM, Animal and Human Reaction to Rapid Tumbling. J. Aviat. Med. **25**, 5 (1954).
33. H. WIESEHOEFER, Über Flugversuche zur Frage der Erträglichkeit hoher Beschleunigungen bei liegender Unterbringung der Flugzeuginsassen. Luftfahrtmedizin **4**, 145 (1940).
34. H. J. VON BECKH, Multi-directional G-Protection in Flight and During Escape. Second European Congress of Aviation Medicine, Stockholm, Sept. 16—19, 1957.
35. H. J. VON BECKH and G. J. D. SCHOCK, Centrifuge Experiments on High-G-Loads in Mice and Their Possible Alleviation by Multi-directional Anti-G-Devices. 29th Annual Meeting of the Aero Medical Association, Washington, March 24–26, 1958.

Über die Strömung von Zweiphasengemischen

Von

H. Bednarczyk[1], ÖGfW

(Mit 1 Abbildung)

Zusammenfassung — Abstract — Résumé

Über die Strömung von Zweiphasengemischen. Es wird ein Gleichungssystem für die instationäre Fadenströmung eines Zweiphasengemisches, bestehend aus fein verteiltem Festkörper und Gasphase, hergeleitet. Die Dichte des Festkörpers wird konstant angenommen; von einem Massenaustausch zwischen fester Phase und Gasphase wird abgesehen. Es wird gezeigt, daß die Einführung einer Relativgeschwindigkeit zwischen beiden Phasen notwendig ist. Die sich aus dem Vorhandensein von Reibungskräften zwischen fester und gasförmiger Phase im Verein mit der Relativgeschwindigkeit ergebende Reibungsleistung muß als dem Substanzelement des Zweiphasengemisches zugeführte Reibungswärme in den ersten Hauptsatz eingefügt werden.

On the Flow of Two-Phase Mixtures. A system of equations is derived for the instationary thread flow of a two-phase mixture consisting of a finely distributed solid and a gaseous phase. The density of the solid is supposed to be constant. Mass exchange between the solid and the gaseous phase is disregarded. The necessity of an introduction of a relative velocity between both phases is shown. The existence of frictional forces between the solid and the gaseous phase due to the mentioned relative velocity leads to a frictional effect. It must be inserted into the first main law of thermodynamics, in form of a frictional heat which is fed to the element of substance of the two-phase mixture.

Sur l'écoulement d'un mélange de deux phases. Dérivation d'un système d'équations pour l'écoulement non-stationnaire d'un mélange constitué d'une phase gazeuse et d'une fine suspension solide. La densité de la phase solide est supposée constante; la diffusion massique entre phases étant négligée. La nécessité d'une vitesse d'écoulement relative entre les deux phases est démontrée. L'existence de forces de friction entre la phase solide et la phase gazeuse, dues à la vitesse relative, produit un effet de friction. Celui-ci doit être introduit dans la première loi fondamentale de la thermodynamique, en usant une forme de chaleur de friction qui est communiquée à l'élément substantiel du mélange constitué des deux phases.

In gewissen Fällen wird man in der Raketentechnik vor die Aufgabe gestellt, Strömungsvorgänge zu berechnen, bei denen das strömende Medium nicht mehr bis in die kleinsten Teile homogen ist, sondern aus zwei (oder auch mehreren) Phasen besteht, die so innig miteinander vermengt sind, daß sie makroskopisch als (Quasi-) Kontinuum in Erscheinung treten. Als Beispiele seien genannt: Die Düsenströmung in der wegen ihrer großen Einfachheit noch immer häufig verwendeten Feststoffrakete (strömendes Medium gewöhnlich Gasgemisch +

[1] Technische Hochschule, Wien IV, Karlsplatz 13, Österreich.

+ Festkörper); die Strömung in der neuerdings stark propagierten Wasserdampfrakete (Wasser + Wasserdampf); schließlich die interessanten, wenn auch sehr schwierig zu beschreibenden Vorgänge bei der Brennstoffeinspritzung in die Brennkammer einer Flüssigkeitsrakete (flüssiger Brennstoff + Gasgemisch).

Wenn wir nun im folgenden darangehen, Gleichungen für die Strömung vcn Zweiphasengemischen aufzustellen, so haben wir in diese Gleichungen vor allem eine wesentliche Tatsache einzubauen: die Relativgeschwindigkeit zwischen den beiden Phasen. Daß eine solche existieren muß, leuchtet sofort ein, wenn man sich vor Augen hält, daß die feste, beziehungsweise flüssige Phase jeweils von der Gasphase ,,mitgenommen" werden muß und die ,,mitnehmenden" Reibungskräfte mit verschwindender Relativgeschwindigkeit gleichfalls verschwinden. Die eben erwähnten Reibungskräfte, die durch das Vorhandensein einer zweiten Phase bedingt sind, leisten nun im Zweiphasengemisch eine gewisse innere Arbeit, die als Reibungswärme Einfluß auf die Thermodynamik des Zweiphasengemisches nimmt[1].

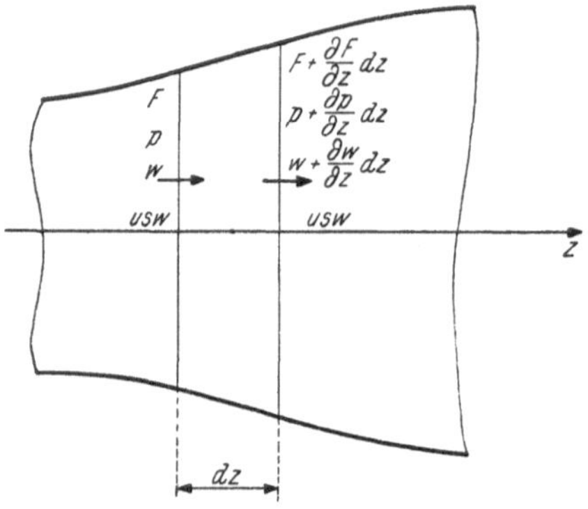

Abb. 1

Um den Weg zur Ableitung der Gleichungen möglichst deutlich werden zu lassen, wollen wir uns auf die instationäre Fadenströmung eines Festkörper-Gas-Gemisches beschränken. Die Dichte ϱ_f des Festkörpers sei konstant und es finde kein Massenaustausch zwischen Festkörper und Gasphase statt. Der Festkörper habe die Temperatur des Gases in seiner nächsten Umgebung. Wenn der Festkörper genügend fein verteilt ist, kann diese Annahme gemacht werden. Wir greifen nun (s. Abb. 1) aus dem Stromfaden ein gewisses makroskopisches Volumselement $dV = Fdz$ heraus. Es soll hinreichend groß sein, so daß in ihm das Zweiphasengemisch als Quasikontinuum aufgefaßt werden kann. Es soll also sinnvoll sein, innerhalb dV von Mittelwerten der Gasdichte, des Druckes, der Strömungsgeschwindigkeit usw. zu sprechen.

Bezeichnen wir mit:
dM_f die Masse des Festkörpers in dV,
dM_G die Gasmasse in dV,
dM die gesamte Masse (Gas + Festkörper) in dV
sowie mit
dV_f das Volumen des Festkörpers in dV,
dV_G das Volumen des Gases in dV,
so bedeuten

$$\varrho_f = \frac{dM_f}{dV_f} \text{ die Dichte des Festkörpers,}$$

$$\varrho_G = \frac{dM_G}{dV_G} \text{ die Dichte des Gases,} \tag{1}$$

[1] Es ist dem Verfasser bisher nicht bekannt geworden, daß die Analyse der Strömungsvorgänge in einem Zweiphasengemisch mit Hilfe der genannten Grundvorstellungen durchgeführt wurde.

$\varrho = \frac{dM}{dV}$ die mittlere Dichte des Zweiphasengemisches im Volumselement dV, und wir erhalten unter Verwendung des spezifischen Gasgehaltes

$$x = \frac{dM_G}{dM} \tag{2}$$

und der Beziehungen

$$\begin{aligned} dV &= dV_G + dV_f \\ dM &= dM_G + dM_f \end{aligned} \tag{3}$$

die bekannte Formel

$$\frac{1}{\varrho} = \frac{1}{\varrho_f} + x\left(\frac{1}{\varrho_G} - \frac{1}{\varrho_f}\right). \tag{4}$$

Weiters finden wir aus (1), (2) und (3)

$$\frac{dV_G}{dV} = x\frac{\varrho}{\varrho_G} \tag{5}$$

$$\text{und} \quad \frac{dV_f}{dV} = (1 - x)\frac{\varrho}{\varrho_f}$$

zwei Gleichungen, die wir noch öfter benötigen werden. Die mittlere Linearabmessung der Festkörperpartikel (in der Technik meist „mittlerer Durchmesser" genannt) wollen wir mit δ, ihre Anzahl in der Volumseinheit des Zweiphasengemisches mit N bezeichnen. Die Kontinuitätsgleichung für die feste Phase können wir dann folgendermaßen ausdrücken: Die Abnahme der Anzahl der in dV vorhandenen Festkörperteilchen in der Zeiteinheit muß gleich sein der Anzahl der Teilchen, die in der Zeiteinheit aus dV abströmen, vermindert um die Anzahl der Teilchen, die in der Zeiteinheit in dV einströmen, das heißt:

$$-\frac{\partial N}{\partial t} \cdot F \cdot dz = \left(F + \frac{\partial F}{\partial z}dz\right)\left(v + \frac{\partial v}{\partial z}dz\right)\left(N + \frac{\partial N}{\partial z}dz\right) - FvN,$$

beziehungsweise ausgerechnet

$$F\frac{\partial N}{\partial t} + \frac{\partial}{\partial z}(FvN) = 0. \tag{6}$$

In (6) bedeutet v die mittlere Geschwindigkeit der Festkörperteilchen. Da die Teilchenzahl eine etwas unhandliche Größe darstellt, wollen wir sie durch bequemere Variablen ausdrücken. Es wird sich zeigen, daß es hierbei im wesentlichen auf den spezifischen Gasgehalt x und die Gemischdichte ϱ ankommt. Wir können nämlich das (mittlere) Volumen eines Festkörperteilchens proportional δ^3 setzen, also

$$\text{Teilchenvolumen} = \alpha\delta^3 \tag{7}$$

schreiben. Mit (7) gewinnen wir die Gleichung

$$\alpha\delta^3 \cdot \varrho_f \cdot N \cdot dV = dM \tag{8}$$

und hieraus unter Verwendung von (1), (3) und (5)

$$N = \frac{(1 - x)}{\alpha\,\delta^3} \cdot \frac{\varrho}{\varrho_f}. \tag{9}$$

(9) in (6) eingesetzt, liefert uns als endgültige Kontinuitätsgleichung für die feste Phase:

$$F\frac{\partial}{\partial t}\Big[(1 - x)\,\varrho\Big] + \frac{\partial}{\partial z}\Big[(1 - x)\,\varrho \cdot vF\Big] = 0. \tag{10}$$

Bei der Aufstellung der Massenbilanz für die Gasphase müssen wir beachten, daß im Ausdruck für den Massenstrom der Gasphase durch die Oberfläche von dV nicht die gesamte Fläche F einzusetzen ist, da die Festkörperströmung im Mittel einen Bruchteil dieser Fläche für sich beansprucht. Die für die Gasströmung zur Verfügung stehende „freie Fläche" wollen wir mit F_G bezeichnen. Hiermit lautet die Massenbilanz für die Gasphase:

$$-\frac{\partial}{\partial t}(\varrho_G dV_G) = \left((F_G + \frac{\partial F_G}{\partial z} dz\right)\left(w + \frac{\partial w}{\partial z} dz\right)\left(\varrho_G + \frac{\partial \varrho_G}{\partial z} dz\right) - F_G w \varrho_G, \qquad (11)$$

wenn w die Strömungsgeschwindigkeit des Gases bedeutet.

Um einen Zusammenhang zwischen der gesamten Querschnittsfläche F und der „freien Fläche" F_G zu finden, berechnen wir zunächst den Querschnitt F_f, den die Festkörperströmung im Mittel benötigt. Wie man sich leicht überlegt, muß nämlich die Beziehung

$$\varrho_f\, vF_f = \varrho_f a\delta^3\, N\, v \cdot F_f \qquad (12)$$

bestehen. Aus (12) und (9) findet man sodann

$$F_f = (1 - x)\frac{\varrho}{\varrho_f} F \qquad (13)$$

und, da

$$F_G = F - F_f \qquad (14)$$

gelten muß, über (4)

$$F_G = x\frac{\varrho}{\varrho_G} F. \qquad (15)$$

Unter Verwendung von (5) und (15) folgt aus (11) nach kurzer Rechnung die Kontinuitätsgleichung für die Gasphase:

$$F\frac{\partial}{\partial t}(x\,\varrho) + \frac{\partial}{\partial z}(x\,\varrho\, wF) = 0. \qquad (16)$$

Als Nächstes wollen wir die beiden Bewegungsgleichungen für die feste Phase und die Gasphase herleiten. Bezeichnen wir mit r_v die je Volumseinheit des Gemisches zwischen fester und gasförmiger Phase ausgetauschte Kraft, so haben wir als Bewegungsgleichung für die feste Phase

$$\varrho_f dV_f \frac{Dv}{dt} = r_v \cdot dV \qquad (17)$$

anzusetzen. Wir haben in (17), da wir schließlich das Geschwindigkeitsfeld der festen Phase und nicht die Bahnkurven der einzelnen Festkörperteilchen berechnen wollen, gleich die substanzielle Beschleunigung

$$\frac{Dv}{dt} = \frac{\partial v}{\partial t} + v\frac{\partial v}{\partial z}$$

eingeführt. Ziehen wir (5) heran, so erhalten wir (17) in der Gestalt

$$(1 - x)\,\varrho\frac{Dv}{dt} = r_v. \qquad (18)$$

Die Bestimmung der Bewegungsgleichung für die Gasphase geht in analoger Weise vor sich. Wir haben nur zu beachten, daß der Druck p auf das in dV eingeschlossene Gasvolumen dV_G auf der gesamten Querschnittsfläche F wirksam ist, und nicht nur auf der durch (15) definierten Fläche F_G. Wir schreiben also

$$\varrho_G dV_G\frac{Dv}{dt} = -r_v\, dV - \left(p + \frac{\partial p}{\partial z} dz\right)\left(F + \frac{\partial F}{\partial z} dz\right) + p\frac{\partial F}{\partial z} dz + pF. \qquad (19)$$

Der Ausdruck $p \frac{\partial F}{\partial z} dz$ stellt die Druckkraft dar, die von der Mantelfläche des Stromfadens längs dz auf das Gasvolumen dV_G ausgeübt wird. Vereinfachen wir (19) mit Hilfe von (5), so verbleibt als Bewegungsgleichung für die Gasphase

$$x\,\varrho \frac{Dw}{dt} = -r_v - \frac{\partial p}{\partial z}. \tag{20}$$

Um einen Ausdruck für die Volumskraft r_v zu finden[1], gehen wir von der Kraft r_i auf ein einzelnes Festkörperteilchen vom „mittleren Durchmesser" δ aus. Diese können wir mit gewisser Berechtigung in Anlehnung an die STOKESsche Formel proportional dem Produkt aus δ und der Relativgeschwindigkeit $(w-v)$ zwischen Festkörpern und Gas ansetzen:

$$r_i = A\delta\,(w - v). \tag{21}$$

Die Volumskraft r_v ergibt sich dann aus (21) einfach durch Multiplikation mit N:

$$r_v = N\cdot A\delta\,(w - v). \tag{22}$$

Eliminieren wir hierin N mit Hilfe von (9), so führt (22) auf

$$r_v = \frac{k}{\delta^2}(1-x)\frac{\varrho}{\varrho_f}(w - v), \tag{23}$$

wenn wir $k = \frac{A}{a}$ als neue Konstante einführen. Mit einem Kraftansatz gemäß (23) lassen sich die Bewegungsgleichungen für Festkörper und Gas in der Gestalt

$$\frac{Dv}{dt} = \frac{k}{\varrho_f \delta^2}(w - v) \tag{24}$$

und

$$\frac{Dw}{dt} = -\frac{k}{\varrho_f \delta^2}\cdot\frac{(1-x)}{x}(w - v) - \frac{1}{x\,\varrho}\frac{\partial p}{\partial z} \tag{25}$$

schreiben.

Es handelt sich noch darum, den ersten Hauptsatz der Thermodynamik für das vorgelegte Zweiphasengemisch zu formulieren. Die Hauptsätze, für die Masseneinheit von Festkörper und Gasphase getrennt angeschrieben, lauten:

$$dq_f = du_f \tag{26 a}$$

$$dq_G = di_G - \frac{dp}{\varrho_G}; \tag{26 b}$$

q_f ... dem Festkörper zugeführte Wärmemenge
u_f ... innere Energie des Festkörpers
q_G ... dem Gas zugeführte Wärmemenge
i_G ... Enthalpie des Gases.

Um aus diesen den Hauptsatz für das Zweiphasengemisch zu gewinnen, überlegen wir uns folgendes: Die Reibungswärme q_r, die zufolge der Relativbewegung von fester und gasförmiger Phase in der Zeiteinheit der Masseneinheit des Gemisches zugeführt wird, muß sich im Verhältnis der Massenanteile auf Festkörper und Gas übertragen:

$$\frac{d\,q_r}{dt} = (1 - x)\frac{d\,q_f}{dt} + x\frac{d\,q_G}{dt}. \tag{27}$$

[1] Zunächst unter Vernachlässigung des Druckintegrals.

Die in der Zeiteinheit der Masseneinheit des Gemisches zugeführte Reibungswärme ist einfach zu ermitteln. Sie ist gleich der Reibungsleistung, die durch das Produkt Reibungskraft je Masseneinheit mal Relativgeschwindigkeit gegeben ist:

$$\frac{d\,q_r}{dt} = \frac{r_v}{\varrho}(w - v) \tag{28}$$

oder, unter Verwendung von (23)

$$\frac{d\,q_r}{dt} = \frac{k}{\varrho_f \delta^2}(1 - x)(w - v)^2. \tag{29}$$

Bei der Bildung der Ausdrücke $\frac{d\,q_f}{dt}$ und $\frac{d\,q_G}{dt}$ haben wir zu beachten, daß die Übertragung der Wärme auf den Festkörper und das Gas nicht lokal, sondern substanziell erfolgt. Wir haben daher zu schreiben:

$$\frac{d\,q_f}{dt} = \frac{\partial u_f}{\partial t} + v\frac{\partial u_f}{\partial z} \tag{30a}$$

und

$$\frac{d\,q_G}{dt} = \frac{\partial i_G}{\partial t} + w\frac{\partial i_G}{\partial z} - \frac{1}{\varrho_G}\left(\frac{\partial p}{\partial t} + w\frac{\partial p}{\partial z}\right). \tag{30b}$$

(29), (30a) und (30b) in (27) eingesetzt, ergibt den gesuchten ersten Hauptsatz für das Zweiphasengemisch:

$$\frac{k}{\varrho_f \delta^2}\frac{(1-x)}{x}(w - v)^2 = \frac{1-x}{x}\left(\frac{\partial u_f}{\partial t} + v\frac{\partial u_f}{\partial z}\right) + \frac{\partial i_G}{\partial t} + w\frac{\partial i_G}{\partial z} - \frac{1}{\varrho_G}\left(\frac{\partial p}{\partial t} + w\frac{\partial p}{\partial z}\right). \tag{31}$$

Die Gleichungen (4), (10), (16), (24), (25) und (31) reichen aus, um im Verein mit einem Ausdruck für die spezifische Wärme des Festkörpers

$$C = C(p, T) \tag{32a}$$

und einer Zustandsgleichung für das Gas

$$T = T(p, \varrho_G) \tag{32b}$$

(T ... absolute Temperatur)

die interessierenden Größen $v, w, \varrho, \varrho_G, x, p, T$ für die instationäre Fadenströmung des vorgelegten Zweiphasengemisches zu berechnen. Nimmt man die spezifische Wärme des Festkörpers konstant an, und nimmt man ferner an, daß sich die Gasphase wie ein ideales Gas verhält, so vereinfacht sich der Hauptsatz (31) zu:

$$\frac{k}{\varrho_f \delta^2}\cdot\frac{1-x}{x}(w - v)^2 =$$
$$= \frac{(1-x)\,C + x\,c_p}{x}\frac{\partial T}{\partial t} + \frac{(1-x)\,C v + x\,c_p\,w}{x}\cdot\frac{\partial T}{\partial z} - \frac{1}{\varrho_G}\left(\frac{\partial p}{\partial t} + w\frac{\partial p}{\partial z}\right) \tag{33}$$

und die Zustandsgleichung (32b) zu

$$T = \frac{p}{R\,\varrho_G}, \tag{34}$$

wobei c_p die spezifische Wärme des Gases bei konstantem Druck und R die individuelle Gaskonstante bedeuten. Eine Diskussion der abgeleiteten Gleichungen an Hand von charakteristischen Spezialfällen soll bei späterer Gelegenheit erfolgen.

The Weight of Minimum Cost Orbital Ferry Vehicles

By

Björn Bergqvist[1], SIS

(With 8 Figures)

Abstract — Zusammenfassung — Résumé

The Weight of Minimum Cost Orbital Ferry Vehicles. Using the general procedure, with continuous weight minimizing by the aid of stress calculations, and earlier described by the S.I.S. project study team, the influence of the burnout load factor and the number of steps on the minimum launching weight has been calculated for a tandem multi-step, constant mass flow orbital ferry vehicle with chemical liquid propellants. A trajectory launching angle of 50 degrees and an inner orbit height of 70 km are postulated and the more economic alternative of no-step recovery accepted. Quite new points of view on the values of the optimum structural factors have been obtained. The optimum burnout load factors are low, being slightly less than 4 and 6 for 4 and 3 steps, respectively. The S.I.S. recommends its procedure as the only one known, that combines all important relevant relationships to give a realistic answer in the project stage to the questions about the choice of variables associated with optimum design in terms of energy, weight and cost.

Das Gewicht billigster Kreisbahn-Lastraketen. Fur eine Lastrakete, die fur die Fahrt auf einer Kreisbahn rings um die Erde gebaut ist, mit chemischen, flussigen Treibstoffen mit konstanter Forderung und mit mehreren Stufen hintereinander arbeitend, hat die S.I.S.-Studiengruppe fur Projektierung den Einfluß des Brennschlußlastfaktors und der Stufenzahl auf das kleinste Startgewicht berechnet. Die Berechnung wurde nach der allgemeinen Methode, die fruher beschrieben worden ist, durchgefuhrt, wobei man das Gewicht mittels Festigkeitsberechnungen fortlaufend auf ein Minimum bringt. Ein Startwinkel von 50 Grad und eine innere Kreisbahnhöhe von 70 km sind dabei vorausgesetzt. Es wird das Prinzip, die Stufen aus okonomischen Ursachen verloren zu geben, verwendet. Ganz neue Gesichtspunkte bei den Werten der optimalen Zellenfaktoren wurden gewonnen. Die optimalen Brennschlußlastfaktoren sind klein, ein wenig kleiner als 4 und 6 fur 4 bzw. 3 Stufen. Die S.I.S. empfiehlt ihre Methode als die einzige bekannte, die alle wichtigen und zugehorigen Zusammenhänge so kombiniert, daß in dem Stadium des Entwurfs eine realistische Lösung des Problems der Wahl derjenigen bestimmenden Faktoren erhalten werden kann, die mit optimaler Konstruktion hinsichtlich Energie, Gewicht und Kosten verknupft sind.

Le poids d'engins orbitaux de transport de coût minimum. Une analyse de l'influence du facteur de charge en fin de combustion et du nombre d'étages en série sur le poids minimum au lancement d'un véhicule de transport orbital à propulsion chimique et débit constant. La recherche du poids minimum utilise un processus de réduction continu, basé sur le calcul des tensions de structure, tel que décrit antérieurement

---

[1] Research Leader of the Swedish Interplanetary Society (SIS), Råsundavagen, 62, III, Solna, Sweden.

par l'équipe d'étude des projets de la Société Interplanétaire Suédoise. La trajectoire envisagée comporte un angle de lancement de 50^0 et une hauteur orbitale de 70 km. Il est admis comme plus avantageux du point de vue économique que les étages ne soient pas récupérés. Les facteurs de structure optimums apparaissent sous un jour nouveau. Les facteurs de charge en fin de combustion sont faibles; ils sont légèrement inférieurs à 4 et 6 pour 4 et 3 étages respectivement. La S.I.S. recommande son procédé de calcul comme le seul connu qui fasse intervenir toutes les relations importantes pour une réponse réaliste, dans le stade avant-projet, au problème du choix des variables intervenant dans une conception optimale du point de vue énergie, poids et coût.

1. Introduction

The approach to a method of designing orbital ferry vehicles to minimum launching weight, which has been proposed by the S.I.S. project study team and was introduced to the International Astronautical Congress in Rome, 1956 [1], has been further developed and refined. It now merits the name of a project procedure and has been numerically carried through in several alternatives.

The main aim of the present investigation was to find the trends, not to determine that particular design in the project stage that implies the absolute minimum weight or cost. Prior to the numerical results shown, those thoughts and arguments are presented, therefore, that are necessary in order to arrive at a satisfactory quantitative mapping of the influence on launching weight of the most important variables. Also, the difference between optimality with respect to energy, weight and cost is explained.

2. Postulates

The main postulates are the same as mentioned in [1], namely:

1. Decision to build manned space platform of some hundred tons of weight.

2. Moderate size of crewless ferry rockets, carrying a payload of 1 ton each. This makes possible the application of series manufacturing methods and cuts production cost. This argument will probably hold for even the considerably fewer crew rockets, these having payloads of the same order of magnitude.

3. Steps should not be recovered. The arguments for this alternative being considerably cheaper are the same as before. Generally spoken, this alternative means also more regular service and thus a valuable improvement in reliability. The result is the absolutely clean hull, without wings or fins.

4. To-days technique is assumed but reliability of navigation and steering mechanisms assumed to be considerably improved. Tandem multi-step design and chemical liquid propellants with constant mass flow are used.

3. Main Factors Varied

The no-step recovery being our first great investigation result, the discovery of the decrease in structural factor ε (step empty weight divided by step full weight) with the increase in mass ratio r_e, other factors held constant, was the other main result and stems from the continuous application of the weight minimizing procedure by the aid of stress and weight calculations. As said in [1], this fact has brought the weight of a three step vehicle to compare not unfavourably with that of a four step one. Therefore, N, i.e. the number of steps, is a natural main factor, even more than under other circumstances.

Another primary variable is n_{cb}, or the burnout load factor, i.e. the ratio between thrust acceleration at burnout and earth surface gravity acceleration. The values of n_{cb} are contemplated in terms of possible danger to men and equipment from the very beginning. Too low n_{cb}-values give great gravity losses for the ascending path (OBERTH's "synergic curve") and a corresponding increase in the required value of v_k, the characteristic velocity. At launching, see Fig. 1, too small a value of n_{cb} does not allow the vehicle to lift at the first moment of burning. Too great a value, on the other hand, brings about too heavy structural elements and motor plants. Consequently, there exists an optimum value of n_{cb}. The determination of this has been the central task of our investigation this time.

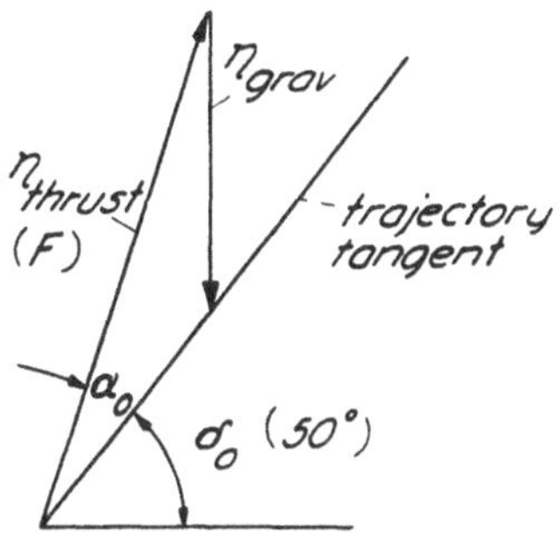

Fig. 1a. Accelerations at launching

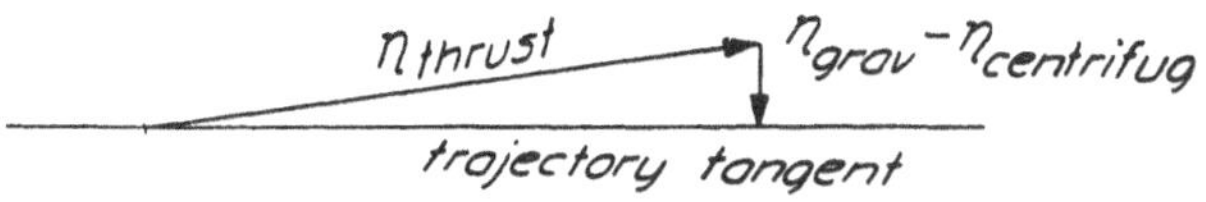

Fig. 1b. Accelerations near inner orbit

4. Factors Not Varied in the Main Investigation

As bi-propellant, fuming nitric acid and hydrazine have been provided.

As was stressed in [1], another operation sequence of the optimum calculations than that one shown there might prove more practical. This point has been cleared up and has given the following results.

The lower launching angle δ_0 (marked a_0 in [1]), the lower gravity losses. This is true without exception but does not mean such a great change in v_k and r_e as expected. δ_0 was therefore held constant at 50 degrees, lower values being disadvantageous of certain reasons.

From preliminary considerations of kinetic and solar radiation temperatures, the inner orbit radius R_i has been held equal to earth radius + 70 km. The smaller R_i, the smaller gravity losses, because it costs less work to lift a diminishing mass a smaller distance. Certain effects appear at decreasing R_i to offset this gain but are quite insignificant.

When the values of n_{cb}, N, δ_0 and R_i have been assumed, the form of the ascending trajectory accepted, the preliminary size, form, center of gravity (CG) and center of air pressure (CP) of the vehicle derived and the combustion chamber pressure (here: 30 kg/cm^2 or 43 psi) and the motor nozzle area—throat area ratio [here: 4,5 for the first (launching) step and 20 for the other ones] settled, there results an univocal r_e-value.

Introducing necessary structural values, see Appendix, paragraphs 2A and 3A, a certain n_{cb} — r_e -combination corresponds to a certain minimum weight sacrifice G_{tmin} (marked R_{tmin} in [1]), by virtue of the condition of minimum weight applied throughout, as well as corresponding values of $\varepsilon_{optimum}$ and $\lambda_{optimum}$, λ being the payload ratio or sub-rocket payload weight divided by the initial subrocket weight.

Of certain practical reasons, the hull wall thickness has been given a constant value for a certain step throughout this preliminary investigation, but different for $N = 3$ and 4. This constantness is a close approximation to reality.

The same ascending trajectory form has been used for all alternatives, with the exception of those with small n_{cb} for which, see Fig. 1, the trajectory angle δ_o must be carefully chosen in order that the "curvature loss" $F\,(1-\cos\alpha)$ should not be exceedingly great because of the great burning time. The shape of the trajectory is characterized, see Fig. 6, by downward curvature along the whole curve and a smooth tangent to the inner orbit in the very moment when the total velocity v reaches the value of the orbital velocity on this height. It has also been found that, apart from alternatives with small n_{cb}, the form of the curve does not have any appreciable influence on v_k when δ_o and R_t are settled.

In Chapter 6 the influence of some secondary variables one by one has been exemplified.

5. Main Results

Using motor plant, propellant and structural basic values given in Chapter 4 and in Appendix paragraphs 2A and 3A, and with the general procedure as described very briefly in Appendix paragraph 4A, the curves in Fig. 2—4 have been obtained. n_{cb} and r_e are equal for all steps each time.

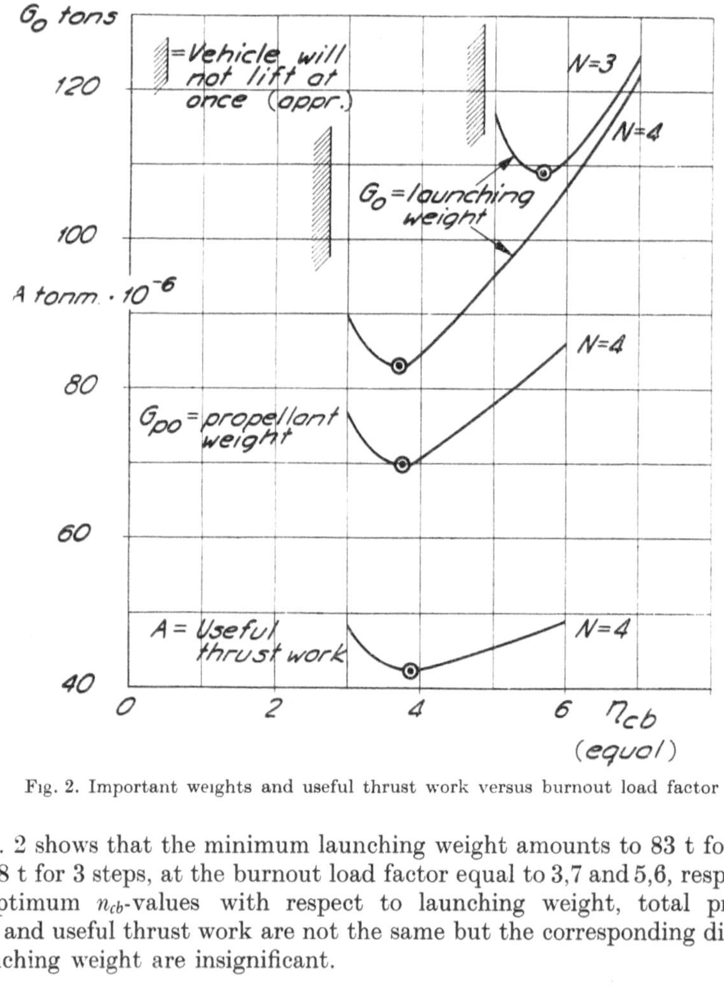

Fig. 2. Important weights and useful thrust work versus burnout load factor

Fig. 2 shows that the minimum launching weight amounts to 83 t for 4 steps and 108 t for 3 steps, at the burnout load factor equal to 3,7 and 5,6, respectively. The optimum n_{cb}-values with respect to launching weight, total propellant weight and useful thrust work are not the same but the corresponding differences in launching weight are insignificant.

Fig. 3 shows that the optimum mass ratio increases slowly with decreasing burnout load factor. At the same time, the optimum structural factor decreases rapidly, being considerably less for 3 steps.

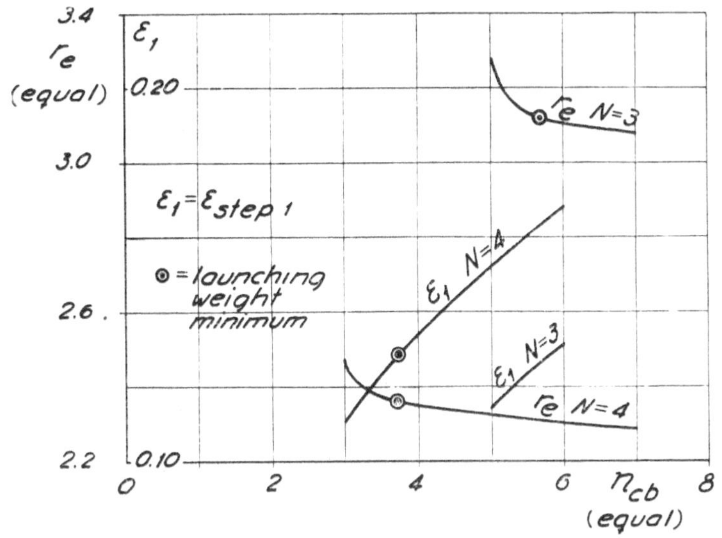

Fig. 3. Mass ratio and structural factor versus burnout load factor

Fig. 4 shows that the optimum characteristic velocity, far from reaching a minimum at minimum weight, increases more and more rapidly the more the burnout load factor decreases. If the latter one is slightly less than 3 for $N = 4$ and 5 for

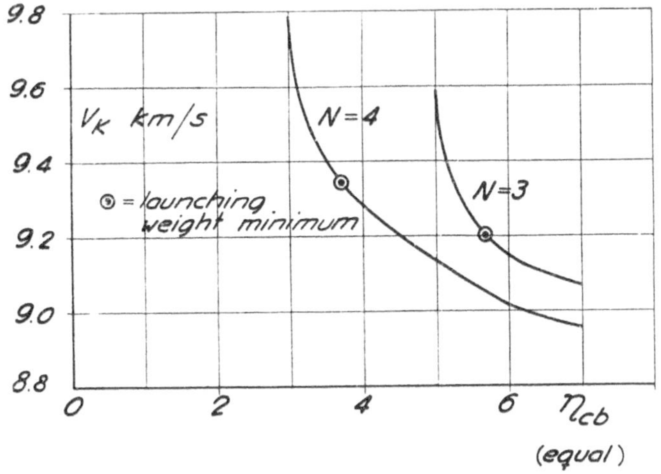

Fig. 4. Characteristic velocity versus burnout load factor

$N = 3$, very great velocity losses are obtained and a practical limit may be reached even before the critical value is reached below which the vehicle will not lift immediately after burning has begun.

The difference in n_{cb} for minimum energy, minimum launching weight and minimum propellant weight is easily explained from an analysis of the formulae giving the useful work and of the relationships between $r_{e\,opt}$, ε_{opt} and $n_{cb\,opt}$. These difference, so often neglected in rocket literature, should be remembered. The difference in n_{cb} for minimum cost and minimum weight is easily understood already from the simple fact that different vehicle components have very different prices per kilogramme. But more fundamental points of view in this connection are: maximum allowed accelerations for delicate instruments, estimating the increased lack in reliability when using 4 steps instead of 3 etc.; in short: what price higher reliability?

With launching weights of about 100 tons per ton payload, an enterprise of 1000 vehicles does not show a considerable difference in magnitude, as compared to single industrial undertakings nowadays. The total manufacturing cost, calculated with the aid of to-days experience with advanced military aircraft, would amount to something between 1,0 and 1,5 billion dollars.

6. Secondary Factors Varied One by One

1. Bi-propellant $LO_2 + C_2H_5OH$ assumed for $N = 3$, $n_{cb} = 6$ increases launching weight with 15,5% and structural factor for step 1 from 0,131 to 0,140.

2. Doubling the weight G_f of the functional system as related to the motor plant weight G_m in all steps for $N = 3$, $n_{cb} = 6$ increases launching weight with 14,5%.

3. Increasing the ratio motor nozzle area—throat area in step 2—4 from 20 to 30, which fully exploits the base area of step 2, decreases the launching weight by 1,5%, through the increase in exhaust velocity, for $N = 4$, $n_{cb} = 6$.

4. Increasing r_e for step 2—4 with 2% and decreasing r_e for step 1 with about 6%, for $N = 4$, $n_{cb} = 6$, decreases launching weight by about 2%, by virtue of shorter burning time at lower exhaust velocity of step 1.

5. Decreasing n_{cb} to 3 for *last step* for $N = 4$, $n_{cb} = 6$, decreases launching weight with 11,7%, by virtue of lighter motor plant and smaller accelerations for certain structural components in step 4.

The gain in economy may thus be considerable by some more detail calculations in the project stage. It is, nevertheless, certainly wise to remember that practice always shows that the weight tends to increase during the subsequent design phase and that it is as well to save modifications as those mentioned above to serve as reserve possibilities during the latter period.

7. Conclusions

Nowhere in literature we have found anything similar to our continuous weight minimizing procedure, which uses an organic synthesis between stress analysis and trajectory energy considerations. Without this new knowledge, the real insight into the problem of optimal variable combination and of best economic dimensioning can not be had. Probably, there are idealized stress models that can be used in the more complicated problems of elaborating weight minimum procedures for more economic thrust programmes and step configurations. We in S.I.S. recommend our procedure with this paper and hand it over to the astronautical fellowship.

Appendix Showing Calculations

1A. Variables Involved

Burning times, mass ratios and exhaust velocities

Main formulae:

Mass ratio $r_e = e^{\frac{k_1 v_k}{v_e}}$ (1)

Step burning time $t_s = \frac{v_e (r_e - 1)}{k_1 g n_{bs}}$ (2)

k_1 = correction factor for pump propellant (f.i. = 1,02 for $HNO_3 + H_4N_2$)
v_e = effective average exhaust velocity, without reduction for k_1

When applied to subrocket 1 (step 1 burning), (1) and (2) should be properly adjusted.

Load factors, accelerations and curvature, see Fig. 5

Load factors are given in terms of earth surface gravity

n_T = load factor available for velocity increase = momentaneous thrust load factor n_c diminished by losses deriving from air resistance T, gravity G and F [$1 - \cos(\alpha + \beta)$], see Fig. 5.

n_N (positive inwards) = load factor available for curvature inwards

$n_N = n_g \cos\delta - n_f \cos\delta - n_s - n_k - F \sin(\alpha + \beta)$

n_g = gravity reduced for height, force G

n_f = centrifugal acceleration, force C

n_s = CORIOLIS' acceleration (equal to zero when air density zero), force S

n_k = normal air force (K) load factor, proportional to $\sin 2\alpha$

α = angle of attack

β = thrust direction adjustment within vehicle, determines α — time programme

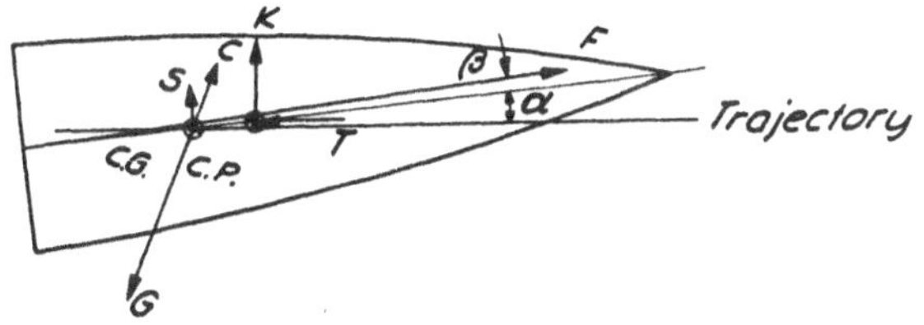

Fig. 5. Vehicle force system in ascending path

Curvature is calculated from

$$\frac{d\delta}{dt} = \frac{n_N g}{v}. \quad (3)$$

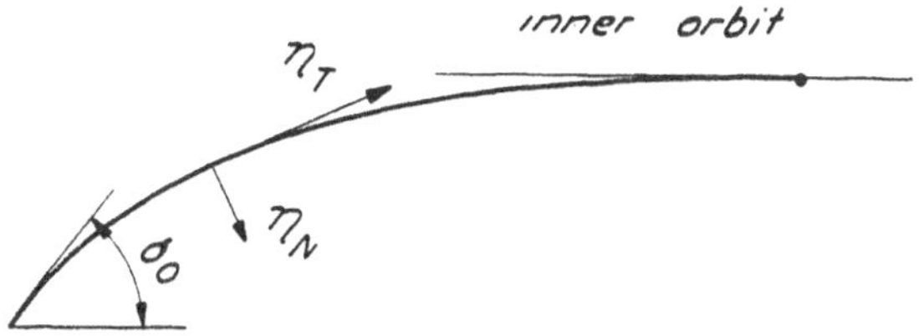

Fig. 6. Ascending path

Characteristic velocity

$v_k = v_{r70} - 232 + v_p + v_a + v_m + v_s + v_L \text{syn}$ (4)

v_{r70} = 7870 m/s = orbital velocity at 70 km height

232 m/s = earth rotation launching help; launching at 45° latitude and in a great circle cutting equator at 45° is assumed throughout

v_p = perigee increment in inner orbit for entering transferring HOHMANN ellipse to outer orbit = 261 m/s between 70 and 1000 km height, resp.

v_a = apogee increment in outer orbit when arriving from transferring ellipse to reach orbital velocity in outer orbit, = 253 m/s

v_m = increment for manoeuvering purposes, assumed to 150 m/s throughout

v_s = increment for compensation of overdimension in vehicle structural components (here 3 % margin has been assumed worthy to pay for) and inaccurate maintaining of propellant mixture ratio (here 1 % is considered possible). For each step:

$$\Delta v_s = \Delta G_{po} \frac{k_2 G_L v_e}{G_s G_o} \tag{5}$$

k_2 = empiric coefficient found from weight minimum calculations, = 1,18—1,21

ΔG_{po} = calculated increment in propellant weight G_{po} to ensure the same mass ratio as without malfunctions

G_L = subrocket payload

G_s = subrocket final weight

G_o = subrocket initial weight

v_{Lsyn} = total losses in synergic curve, see paragraph 4A

Vehicle weight equation for minimizing

$$G_x = G_L + G_{struct} = G_L + A + BD^2 + CD^3 + ED^2 \sqrt{G_x} + \frac{HG_x}{D} \tag{6}$$

D = diameter of idealized step cylinder

A, B, C, E, H constants that have different values for different steps

2A. General Assumptions

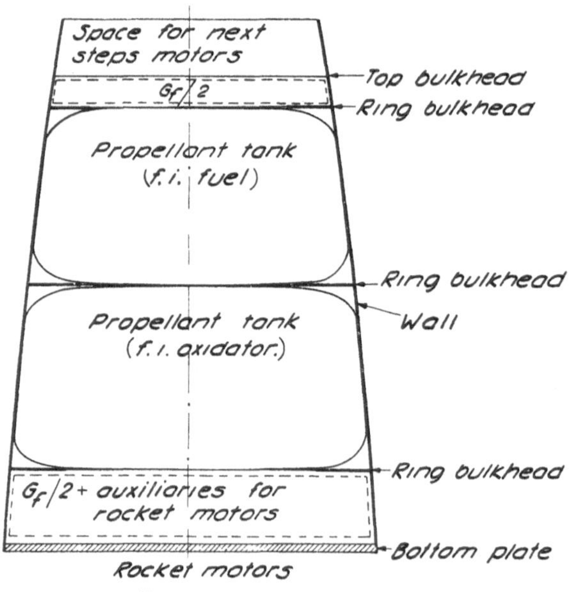

Fig. 7. Main step design scheme

1. 5 sec. pause in free flight, of practical reasons, between subsequent step burnings

2. Helium tanks for propellant tank pressurization attached inside top of latter one

3. Top bulkhead and bottom plate of double skin design when size makes this feasible, otherwise solid

4. Propellant tanks integral with hull

5. Vehicle structural material throughout: Cr-Mo-Steel, ultimate tensile strength 110 kg/mm² (156 ksi)

3A. Specified Initial Values

For each step:

v_{e1} = exhaust velocity for step 1, $HNO_3 + H_4N_2$, earth surface 2350 m/s

v_{e2} = do for step 2—4: 2850 m/s, about 0,5 % higher values for $LO_2 + C_2H_5OH$, whereby $k_1 = 1{,}025$

G_f = functional systems weight, divided equally between top and bottom, see Fig. 7.
G_m = motor plant weight
ξ = G_f/G_m
$\varkappa$ = $G_m/F = 0{,}025$ kg/kg
γ_m = motor plant specific weight = 0,42 kg/dm³
γ_f = functional systems specific weight = 0,6 kg/dm³
σ_a = allowed stress in bottom plate and top bulkhead = 160 kg/mm² (230 ksi) in double skin case and 300 kg/mm² (435 ksi) in solid case, according to new experiments on steel plate strength
t_w = wall skin thickness of the cellular core design, mm, see Fig. 8
h_w = wall total thickness of the cellular core design, mm, see Fig. 8

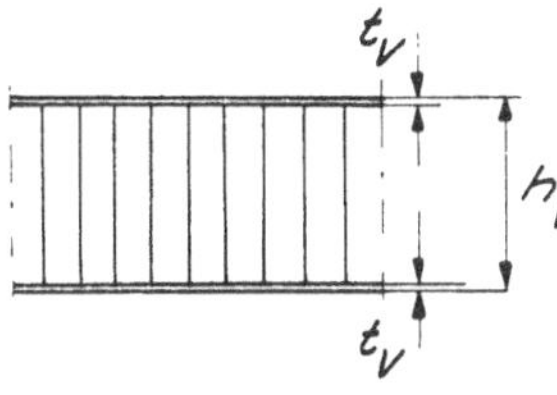

Fig. 8. Cellular core design parameters

N = 3

Step	ξ	t_w	h_w
1	0,10	1,3	31
2	0,20	1,0	24
3	0,25	0,7	17

N = 4

Step	ξ	t_w	h_w
1	0,10	1,2	29
2	0,15	1,0	24
3	0,20	0,8	19
4	0,25	0,6	14

4A. Procedure and Example

(only most important points). $N = 4n_{cb} = 4$

This was calculated as the third alternative in order and some experience of estimating values of different quantities at $n_{cb} = 6$ had already been had.

1. v_k assumed 9400 m/s by analyzing different kinds of v_k-losses in earlier calculations at $n_{cb} = 6$.
2. v_{el} *average* estimated on experience at 2495 m/s. $v_{e\,tot} = 2495 + 3 \cdot 2850 = 11045$
3. From (1): $r_e = 2{,}382$
4. From (6): $G_{t\,min} = 90{,}1$ (i.e. $G_o = 90{,}1$ tons)
5. Thrust increase from earth surface to height at burnout of step 1 estimated at 1,110. Later adjusted.
6. $n_{c\,initial} = \dfrac{4}{2{,}382 \cdot 1{,}110} = 1{,}513$
7. $\sin \alpha_o\,(\text{Fig. 1}) = \dfrac{\cos 50^\circ}{1{,}513}$; $\alpha_o = 25{,}1^\circ$
8. $n_{T\,init} = 0{,}602$
9. 10 equal time intervals for subrocket 1 and 6 for each of subrockets 2—4 give the following velocity losses (burning times not shown here):

Subrocket no	v_k available	Losses m/s					v_{total} at burnout
		air res.	grav.	$\alpha + \beta$ (Fig. 5)	free flight	total	
1	2142	54	532	33	19	638	1504
2	2426	11	134	10	0	155	3775
3	2426	0	11	35	0	46	6155
4	1493	0	0	10	0	10	7638
						$v_{L\,syn} = 849$	

$$7638 = 7870 - 232. \text{ Correct}$$

Rest of v_{k4} or $2426 - 1493 = 933$ must be available for v_p, v_a, v_m and v_s [Formula (4)].

10. From (5): $v_s = 137$
11. $v_{k\,needed} = 7870 - 232 + 261 + 253 + 150 + 137 + 849 = 9288$
12. $v_{k\,available} = 9420$ because $v_{el\,av}$ was underestimated with 20 m/s
13. $v_{e\,available} = 11045 + 20 = 11065$
14. From (1), 12. and 13.: $r_{e\,corr} = 2{,}354$
15. Repeated calculations and a second correction finally give:

$$r_e = 2{,}347 \quad v_k = 9280 \text{ m/s} \quad G_{t\,min} = 84{,}55$$

Reference

1. B. Bergqvist, An Engineering Approach to a Minimum Weight Design of Orbital Ferry Vehicles. Proceedings of the VIIth International Astronautical Congress, Rome, 17–22 September 1956, p. 859. Roma: Associazione Italiana Razzi, 1956.

"Biospheric Index", a Contribution to the Problem of Determination of the Existence of Extra-Solar Planetary Biospheres

By

Thomas P. Bún[1] and **Flávio A. Pereira**[1], SIB

The comparative examination of the different regions of the "Solar Ecosphere" [1] allows the classification of the planets of the Solar System in three groups, as concerns the presence of life:

1) Biosphere improbable (Mercury, Moon, Pluto);

2) Possibility of formation of a biosphere (Giant planets);

3) Biosphere probable (Venus, Earth, Mars).

The distinction of these three groups corresponds to the distinction among three types of planetary atmospheres:

1) No or insufficient atmosphere;

2) Gases as NH_3 and CH_4 where proteic molecules might be formed under the action of radiation as proved by experimental evidence [2];

3) Atmospheres chemically adequate for subsistance of organisms maintaining a metabolism.

Comparison of spectroscopic absorption lines permits the establishment of a "*Biospheric Index*" destined for the classification of a given planet within the indicated scale of groups.

One of the first fields of observational work in future satellite or lunar observatories will be the determination of this biospheric index for the nearest extra-solar planets. Absence of terrestrial atmospheric absorption, which so far prevents direct observation of these bodies, will allow the photographic and spectroscopic separation of these objects, so far only indirectly discovered by their perturbation effects on the primary star image [3].

A first extrapolation of connection between the value of the biospheric index and conditions found on planets of the Solar System will allow a preliminary classification of extra-solar planets as concerning the probability of presence of a biosphere.

References

1 H. Strughold, Ecological Aspects of Planetary Atmospheres. J. Aviat. Med. **23,** 130 (1952).

2. S. L. Miller, Production of Amino Acids under Possible Primitive Earth Conditions. Science **117,** 3046 (1953).

3. J. Dommanget, La Recherche de Systèmes Planétaires Inconnus. Ciel et Terre **9—10,** 229 (1953).

---

[1] Sociedade Interplanetária Brasileira, Caixa Postal 6450, São Paulo, Brasil.

Previsione tempestiva delle caratteristiche del moto di mobili aero-balistici nella cibernetica aeronautica

C. E. Cremona[1], AIR

Riassunto — Zusammenfassung — Abstract

Previsione tempestiva delle caratteristiche del moto di mobili aero-balistici nella cibernetica aeronautica. Se si limita il campo, della continuità, nel senso matematico, a brevi intervalli di tempo o lo si estende, rinunciando alla continuità stessa, fino a pervenire alla concezione di una *pseudocontinuità in un intervello* durante il quale si *ammetta,* anche se non dimostrabile, continua una funzione, e possibile applicare lo sviluppo in serie di Taylor nella determinazione preventiva del valore di una informazione o di un segnale (anche se frequenziato) al termine di un intervallo (*previsione*) se si conoscano o si calcolino per derivazione od integrazione (o con entrambe le operazioni) le n derivate dello sviluppo e se ne proceda alla sommatoria.

Si nota subito, nello studio del telecomandi di un mobile aereo o di un missile, come sia preferibile rilevare una sola informazione e calcolare le altre e come sia sufficiente fermarsi alla terza derivata (*attitudine*) in rapporto all'elevato grado di precisione raggiungibile con l'automazione.

Si accenna infine alla eventualità di considerare speciali cicli asincroni di frequenze (individuati, per esempio, dall' annullarsi delle derivate di un certo ordine e dal cambiamento di segno di quella dell'ordine successivo) i quali consentirebbero di realizzare meccanismi automatici di rara ed avvincente semplicità specie nel campo di applicazione alla tecnica della guida impulsiva.

Rechtzeitige Voraussicht der Charakteristika der Bewegung aero-ballistischer Flugkörper in der aeronautischen Kybernetik. Wenn das Kontinuitätsfeld im mathematischen Sinn auf kurze Zeitintervalle beschränkt wird oder wenn es unter Verzicht auf strenge Kontinuität bis zum Begriff einer *Pseudokontinuitat in einem Intervall* erweitert wird, in welchem eine, wenn auch unbeweisbare, kontinuierliche Funktion angenommen wird, ist es möglich, die Taylorsche Reihenentwicklung fur die Vorherbestimmung des Wertes einer Information oder eines Signals (auch mit Frequenz) am Ende eines Intervalls (Vorhersage) anzuwenden, wenn man die n Ableitungen der Reihenentwicklung kennt oder durch Ableitung oder Integration (oder durch beide Operationen) berechnet und dann die Summierung durchfuhrt.

Beim Studium der Fernlenkung eines Luftfahrzeuges oder eines Flugkörpers merkt man sofort, daß es vorteilhaft ist, eine einzige Information aufzunehmen und die anderen zu berechnen, und daß es genugt, im Hinblick auf den hohen, durch die Automation erreichbaren Präzisionsgrad sich mit der dritten Ableitung (*Lage*) zu begnugen.

Zum Schluß wird auf die Berücksichtigung spezieller synchroner Frequenzzyklen hingewiesen (die z. B. durch das Verschwinden der Ableitungen einer bestimmten Ordnung und durch den Vorzeichenwechsel der nächsthöheren Ableitung individualisiert werden). Diese wurden die Realisierung automatischer Mechanismen von seltener und faszinierender Einfachheit gestatten, insbesondere auf dem Anwendungsgebiet der Impulslenkungstechnik.

---

[1] Direttore dell'Istituto di Balistica dell'Università di Roma, Italia.

Well-Timed Foresight of the Characteristics of the Motion of Aeroballistic Missiles in Aeronautical Cybernetics. By limiting the field of continuity, in the mathematical sense, to short time intervals, or extending it, thus renouncing to the continuity, so as to reach a concept of a *pseudo-continuity in an interval* during which a continuous function *is assumed*, even if it cannot be demonstrated, it is possible to apply the series development of TAYLOR in the anticipated determination of the value of a certain information or signal (even if in frequency) at the end of an interval (*forecast*). This can be accomplished knowing or calculating by derivation or integration, or both, the n derivatives of the development, then proceeding to their summation.

Due to this high degree of accuracy obtainable with the automation, it is clear that, in studying remote controls of an aircraft or a missile, it is preferable to obtain one information only and to calculate the others, and that it is sufficient to arrive at the third derivative (*attitude*).

Mention is made to the eventuality of taking into consideration special asynchronous cycles of frequency (individuated, for instance, by the elimination of the derivatives of a certain order and by the change of sign of the derivative of the successive order), which would allow the realisation of automatic devices of an exceptional simplicity, particularly as far as impulse guidance technique is concerned.

1. — La necessità di sostituire le ormai inadeguate ed insufficienti possibilità di governo umano (*ανδϱος κυβεϱνησις*) — dovute alle sempre minori entità del tempo disponibile per la *esecuzione* di una manovra correttiva o per il *controllo* di una evoluzione di un mobile aerobalistico — con una indispensabile automazione tempestiva (*αυτοματου κυβεϱνησις*) esalta il problema della opportunità e della convenienza di sostituire la *intuizione* — generalmente legata alle caratteristiche intellettive degli esseri umani e variabile con esse — con le possibilità di *previsioni* degli automatismi specialmente se questi ultimi presentino un *grado di ritardo* inferiore a quello di un organismo umano e costante nel tempo.

2. — *L'automazione* di un qualsiasi apparato di *comando* o di *controllo* deve avere a base la possibilità di conoscere i valori dei parametri (che ne regolino la *manovra* o la *correzione*) relativi non all'istante nel quale pervengano le informazioni di essi ma, per lo meno, all'istante nel quale debba essere operata la loro variazione (o quella di alcuni di essi).

Questa difficoltà può essere agevolmente superata, in linea teorica, ricorrendo alla apposizione concettuale di considerare ammissibile applicare, ad una traiettoria *telecomandata* o *telecontrollata* lo sviluppo di TAYLOR scrivendo a titolo di esempio:

$$f(t+\tau) = f(t) + f'(t)\,\tau + f''(t)\,\frac{\tau^2}{2} + f'''(t)\,\frac{\tau^3}{6} \qquad (1)$$

nella quale il primo termine $f(t+\tau)$ può interpretarsi come un *segnale futuro* cioè come il segnale che giungerebbe dopo un tempo τ successivo a quello t nel quale sia invece pervenuta la informazione $f(t)$.

Per la applicabilità della (1) occorre che la funzione $f(t)$ sia continua in un intervallo comprendente almeno quello $t \to t+\tau$ e la approssimazione del valore della funzione nel tempo $t+\tau$ dipenderà dalla quantità dei termini trascurati nello sviluppo di TAYLOR.

In realtà, nel caso specifico delle traiettorie di missili, di velivolvi ecc. i parametri del moto non presentano normalmente variazioni eccessivamente rapide cosicché anche se le informazioni non abbiano un carattere di continuità matematico, esse, pseudocontinue o frequenziate a gruppi d'alta frequenza, possono assicurare una specie di continuità compatibile con il grado di rispondenza dei meccanismi asserviti e con la tempestività del loro intervento.

D'altra parte la condizione suddetta autorizza a limitare il numero delle informazioni potendosi ad alcune (od a tutte meno una) sostituire i valori —

calcolabili, con l'ipotesi della pseudo continuità — delle altre per integrazione o differenziazione sempre che, evidentemente, tali operazioni vengano conglobate e restituite dal sommatore in un intervallo di tempo inferiore a quello τ.

In conclusione appare sufficiente pervenire alla possibilità pratica di approssimare la funzione informativa $f(t)$ mediante un numero limitato di valori.

3. — A titolo esemplificativo si immagini di controllare o di dirigere una traiettoria contenuta in un piano verticale passante per la stazione di guida.

Ad un certo istante t sarà lo spazio percorso $s = f(t)$ e quello al tempo $t + \tau$ per la (1):

$$s_\tau = s + \dot{s}\tau + \frac{1}{2}\ddot{s}\tau^2 + \frac{1}{6}\dddot{s}\tau^3 + \ldots \tag{2}$$

cioè allo stesso tempo t, attraverso un calcolatore, è possibile conoscere la futura posizione al tempo $t + \tau$, del baricentro del mobile.

Se il moto fosse per esempio verticale le informazioni della quota raggiunta (*altimetro*), della velocità istantanea (*velocimetro*), della accelerazione (*accelerometro*) potrebbero essere sufficienti a fornire, con una approssimazione a priori valutabile, la posizione futura del mobile alla fine dell'intervallo di tempo τ preso in considerazione.

In realtà è possibile sia eliminare le correzioni da apportare a ciascuna informazione dovute ai difetti insiti negli apparati e nei dispositivi di rilevamento delle informazioni, sia estendere la somma ad un maggiore numero di termini anche se non si possa disporre o non convenga realizzare apparati di misure di quantità pluridimensionali.

Così, per esempio, è noto quanto poco conveniente sia ricorrere all'impiego degli altimetri usuali nella determinazione della *quota reale*; come sia sempre la *velocità* affetta dalla interferenza della densità dell'aria ecc. cosi ché rilevata e trasmessa per esempio la sola informazione della *accelerazione* questa integrata e derivata potrà fornire gli addendi dello sviluppo di Taylor fino al termine compatibile con la approssimazione che si vuole conseguire e con l'intervallo di previsione desiderata.

4. — Particolare importanza assume il quarto addendo dello sviluppo di Taylor nel caso in esame. Esso, infatti, fornisce una intuitiva caratteristica del moto che chiameremo *attitudine* in quanto con il segno di cui sarà affetto fornirà la immediata indicazione della variazione del parametro cui si riferisce.

Ad un primo esame appare come sia più concettualmente conveniente ottenere anche questa interessante informazione da un calcolatore (derivometro) anziché da un apparato di rilevamento diretto.

5. — In realtà una traiettoria spaziale necessita di un gruppo di almeno sei sviluppi di Taylor per una soddisfacente soluzione del problema della guida e del controllo. Sarà, cioè necessario considerare il sistema:

$$\begin{aligned}
x + x_\tau &= x + \dot{x}\tau + \frac{1}{2}\ddot{x}\tau^2 + \frac{1}{6}\dddot{x}\tau^3 + \ldots \\
y + y_\tau &= y + \dot{y}\tau + \frac{1}{2}\ddot{y}\tau^2 + \frac{1}{6}\dddot{y}\tau^3 + \ldots \\
z + z_\tau &= z + \dot{z}\tau + \frac{1}{2}\ddot{z}\tau^2 + \frac{1}{6}\dddot{\psi}\tau^3 + \ldots \\
\vartheta + \vartheta_\tau &= \vartheta + \dot{\vartheta}\tau + \frac{1}{2}\ddot{\vartheta}\tau^2 + \frac{1}{6}\dddot{\vartheta}\tau^3 + \ldots \\
\delta + \delta_\tau &= \delta + \dot{\delta}\tau + \frac{1}{2}\ddot{\delta}\tau^2 + \frac{1}{6}\dddot{\delta}\tau^3 + \ldots \\
\eta + \eta_\tau &= \eta + \dot{\eta}\tau + \frac{1}{2}\ddot{\eta}\tau^2 + \frac{1}{6}\dddot{\eta}\tau^3 + \ldots
\end{aligned} \tag{3}$$

delle tre coordinate baricentriche riferite ad una terna geografica di riferimento e delle tre coordinate angolari dell'orientamento di un asse di riferimento del mobile, per predeterminare posizione ed orientamento di questo alla fine di un intervallo di tempo prestabilito al fine di adeguare previsioni e manovre alla effettiva condizione dinamica istantanea quando le costanti di tempo del sistema rendano efficace il controllo o l'intervento.

6. — Conviene, sotto quest'ultimo aspetto, riferirsi alla insensibilità dei servosistemi di controllo (misura dell'errore ed esecuzione della manovra correttiva conseguente) alfine di non richiedere più di quanto sia effettivamente possible. Una informazione di entità numerica inferiore a tale insensibilità risulterebbe inutile mentre, il conoscere in quale istante un elemento del moto subisce una variazione "sensibile" nel significato suddetto, può offrire una tempestività d'intervento veramente razionale, anche prescindendo dal problema della "predizione". Si adombra in quanto sopra un tipo di prelievo di dati che potrebbe giustamente chiamarsi con intervallo "elastico" e che, almeno in linea teorica, non sembra offrire difficoltà particolari.

Bibliografia

C. E. Cremona, Corso di Balistica Aeronautica. Università di Roma, 1955.

S. Locke etc., Guidance. Princeton: D. Van Nostrand Co., 1955.

J. C. Gille, M. Pelerin e P. Decaulne, Asservissements. Paris: Dunod, 1956.

S. De Franceso, Quantizzazione dell'informazione nella cibernetica aeronautica. Napoli: Rassegna Tecnica dell'A.N.I.A.I., 1957.

Reliability Factors For Ground Electronic Equipment. New York: McGraw-Hill, 1956.

Research Goals in Astronautics

By

W. O. Davis[1]

Abstract — Zusammenfassung — Résumé

Research Goals in Astronautics. Progress takes place not only within scientific and engineering specialties but also in the interaction between these specialties. It is, in fact, precisely in the changing interrelationship between fields that the greatest opportunity for invention lies. Solid state electronics and magnetohydrodynamics are two recent examples.

This is particularly true in the field of space vehicles since a maximum of invention is desired. Here we must consider a broader pattern of interaction also, namely: the environment, vehicle technology, human factors, and the functions to be performed.

Research is required in all four areas. We are just beginning to have tools to examine the space environment and much remains to be done. In vehicle technology, propulsion represents the greatest challenge since we are reaching the limits of chemical fuels and new concepts must be sought in the boundaries between the sciences. In human factors the greatest remaining unknown is the effect of zero gravity. As to the functions to be performed, we have the areas of space weapons, eventual commercial application, and scientific exploration.

In summation, the field of astronautics will be best advanced by careful consideration to the synthesis of technology and function as well as the analysis of individual problems.

Forschungsziele in der Astronautik. Fortschritt wird nicht nur in den Spezialgebieten der Wissenschaft und Ingenieurkunst erreicht, sondern auch in den dazwischenliegenden Grenzgebieten. Es ist tatsächlich der Austausch wechselseitiger Beziehungen zwischen den beiden Gebieten, in dem die größte Erfindungsmöglichkeit liegt. Zwei neuerliche Beispiele sind die Elektronik des Festzustandes und die Magnetohydrodynamik.

Dies trifft besonders für das Gebiet der Raumfahrzeuge zu, da hier ein Höchstgrad von Erfindung erwünscht wird. Auch hier mussen wir eine Mannigfaltigkeit der Zusammenwirkung beachten, nämlich zwischen Umgebung, Fahrzeugtechnik, menschlichen Faktoren und taktischen Aufgaben.

Forschung ist auf allen vier Gebieten erforderlich. Forschungsmittel, um den Weltraum zu untersuchen, werden eben erst verfügbar und vieles bleibt zu tun übrig. In der Raumfahrzeugtechnik stellt die Antriebskraft das größte Problem dar, da wir die Grenzen der chemischen Treibstoffe bereits erreichen, so daß neue Ideen in den Grenzgebieten zwischen den Wissenschaften gesucht werden müssen. Unter den physiologischen Problemen bleibt der Effekt der Schwerelosigkeit (Andrucklosigkeit) die große Unbekannte. Was die zu lösenden Aufgaben betrifft, so haben wir die Gebiete der Fernwaffen, etwaige kommerzielle Verwertung und wissenschaftliche Forschungen.

Zusammenfassung: Auf dem Gebiete der Astronautik wird man am besten vorwärtskommen, indem man vorsichtig die Synthese der Technik und Aufgaben ebenso wie die Analyse der einzelnen Probleme in Erwägung zieht.

---

[1] Colonel, U.S.A.F.: Headquarters, Wright Air Development Center, Wright Patterson Air Force Base, Ohio, U.S.A.

Objectifs en recherche astronautique. Le progrès ne se fait pas seulement au sein des spécialisations propres aux sciences et aux arts de l'ingénieur mais aussi dans les interactions entre ces divers domaines. C'est précisément dans l'exploitation des relations entre ces divers champs que le génie intenvif trouve le plus d'occasions de se manifester. Les théories électroniques en physique de l'état solide et la magnéto-hydrodynamique en sont des exemples récents.

Cette remarque s'applique spécialement au domaine du vol spatial, où le génie inventif est fortement mis à contribution. Le champ d'interaction y est plus vaste, s'étendant notamment aux conditions environnantes, à la technologie de l'engin, aux facteurs humains et aux performances à réaliser.

Un effort de recherche est nécessaire dans ces quatre domaines. Les instruments pour l'exploration de l'espace environnant sont à peine forgés et beaucoup reste à faire. Dans la technologie de l'engin, c'est la propulsion qui représente le plus grand défi, car nous atteignons les limites des possibilités offertes par les réactions chimiques et de nouvelles conceptions doivent être cherchées aux frontières de la science. La plus grande inconnue des facteurs humains est l'effet d'une suppression de la pesanteur. Quant aux performances à réaliser nous avons les domaines des armes spatiales, les éventuelles applications commerciales et les recherches scientifiques.

En résumé, les plus grands progrès en Astronautique seront le fruit d'une intégration soigneuse de la technologie et des considérations fonctionnelles aussi bien que des avances dans l'analyse des problèmes particuliers.

In the year 1856 a French citizen named JEAN LASSIE patented a new approach to propulsion for manned flight. His balloon-like device was to have been 300 meters long and 30 meters in diameter, shaped like an Archimedian screw and powered by 300 men walking on a treadmill. Fortunately for the men, the device was never built and tested.

Here is an excellent example of an approach to flight which was probably correct in principle but deficient in both its power source and method of application. Many years later, manned flight became possible when a suitable source of power, the internal combustion engine, was used to turn a suitably designed airscrew. The important factor here was that a source of power and a method of application had been designed to function after a fashion in their movement.

Today, the state of the art in Astronautics is comparable to conditions existing in the early days of aeronautics in the late 19th century... say 1896 when the WRIGHT brothers first began to study it. Flight, in a heavier than air machine, was just around the corner and the great weight of theoretical data accumulated through decades, was awaiting practical application. The single elements were all available; PENAUD and GAUCHOT's patents, WENHAM's biplane surfaces, MAXIM's patents, HARGRAVE's box kite, LANGLEY's powered models, CHANUTE's adaptation of the PRATT truss—all these and more, offered a wealth of material from which to choose.

Using lighter-than-air craft, men had flown for over one hundred years. The reciprocating gasoline engine, which was to be married to the airplane for the next half century had just made its early beginning and man was ready to attempt flight in the earth's atmosphere, just as today he is making ready to fly outside the earth's atmosphere.

As in the case of aeronautics the propulsion system will probably be the critical point in the development of astronautics. No one questions that we must use a reaction motor of some kind to fly outside of the earth's atmosphere. The important point to consider at this time is that to develop the optimum concept for this new medium we must be sure that we have chosen the best source of power as well as the best application.

At the present time two systems of propulsion for extra-atmospheric flight have been extensively considered. The first of these is the chemical, liquid fueled rocket which is characterized by high thrust and high mass ratios. This type of propulsion is considered the only type feasible at the present time for departure from the earth and entry into an orbit.

The second type of propulsion being considered is of the socalled ion propulsion or electromagnetic acceleration type. This engine operates by the electrical acceleration of charged particles and is characterized by lower acceleration rates and lower mass ratios than the solid or liquid fueled chemical type rocket. Ion propulsion is being considered for possible departure from an orbit around the earth and eventual travel outside the earth's atmosphere to another location in space.

Unfortunately the chemical fuel rocket is severely limited by exhaust velocities. Since exhaust velocity is a function of the square root of the absolute temperature in the combustion chamber, we are limited by the temperature at which materials can stay solid. If present combustion chamber temperatures could be doubled this would only increase exhaust velocities by the square root of two or roughly give an increase of about 40 % over current values. Mass ratio will therefore always remain relatively high with a chemical fueled rocket and there is some question as to whether this type of propulsion system will ever be economically feasible for large scale flight outside the atmosphere.

The Air Force Office of Scientific Research has therefore considered in its research program planning the great importance of seeking an improved propulsion concept for this kind of flight.

The concept of electromagnetic acceleration has many advantages. In this system the particles are ejected in the desired direction with relatively little energy going into the structure in the form of heat. There are, of course, many practical problems concerned in this type of propulsion and the problem of producing ions rapidly and in great volume may lead to the result that this system of propulsion will be characterized by relatively low thrust. At the same time, however, and in view of the limitations of heat engines in general, it is clear that in the direction of linear acceleration lies the most promising future for space propulsion systems. If the concept of electromagnetic acceleration can be applied to a higher thrust engine and one which can be made to operate in the earth's atmosphere it will obviously have great impact on the future of Astronautics.

For this reason, the Air Force is supporting the basic study of magneto-hydrodynamics, plasma jets, and research activities such as those of the Gianinni Corporation at Santa Ana, California. Most of you are probably aware of GIANNINI's reports on his recent experiments with highly ionized gas jets having energy contents much greater than those of chemically accelerated jets. This work is an extension of some methods which were previously developed in Germany for the direct transfer of electrical energy into kinetic energy by means of gas discharge. Results to date have produced continuous operation at temperatures in excess of 10,000° Kelvin. Powers in excess of 50 kilowatts per square centimeter of nozzle area have been obtained. This plasma jet is now a flexible and powerful means of the study of gas dynamics.

The question, however, as to whether this type of jet would be useful for propulsion purposes is still open because of the problem of generating sufficient electrical power on board a vehicle to create these high-velocity jets. What we are currently trying to do is not to build a rocket but to study the basic properties of electrically charged gases at very high densities—how to handle them, how to accelerate them, and how to determine the efficiency of the process. This is work which can be done in the laboratory for a reasonably small amount of money. The plasma jet, therefore, is a very useful tool in this connection. The work of this nature is presently very basic, and we are presently attempting to prove theoretical calculation.

We are certainly not at the point where we can say our problems are solved. However, this particular study, though started quite recently, may hold great promise in six or eight months. It will certainly increase our understanding of just what problems are involved. The possibility of using an electrically accelerated ram jet in the ionosphere is very intriguing, particularly if this might be combined with a thermal ram jet to achieve flight from the lower atmosphere to the ionosphere. In addition to the linear acceleration of exhaust gases this propulsion concept offers maximum utilization of atmospheric gases as a working fluid. In general, this propulsion techniques would not require vertical take-off and direct opposition to the gravitational vector of the earth. It should therefore be safer than the conventional rocket.

We are not at the point where we can say our problems are solved or that this form of propulsion is possible, but it must be explored scientifically until necessary knowledge is available to answer the important questions of speed, altitude, and thrust. These answers may indicate that we will be interested in this as a successor to the chemical rocket sooner than we realize. The Air Force will continue to explore these and other concepts of propulsion to help set the stage for what is to follow.

Current trend curves show that we may want to know what will come after the chemical rocket much sooner than most people realize. Today we hear of many spectacular, forced, ad hoc, solutions of rocket engineering problems. Without an adequate theoretical understanding each cure becomes an individual problem. In keeping with our mission in the Air Force Office of Scientific Research we must talk and deal with the fundamental areas which will eventually produce for us new theories and techniques which will show dramatic progress and greater future capabilities.

A matter which confuses many is the length of time required to engineer something as against the length of time it takes to perfect a new idea. Mankind, for most of its history, traveled about using approximately one horsepower and dreaming of the time when his steed might sustain a velocity of thirty miles an hour for a mile. In the late 1800's the internal combustion engine came into being and in less than forty years man had not only attained one hundred miles an hour on the ground but had used the device to fly through the air—a very short time span when measured against the ages of dependence on the wind and on animals to achieve motive power.

A new concept may not in itself imply a longer development time—it may imply a shorter development time—it may mean a much simpler approach to arrive at the same result. Which is more complicated, a multicylinder turbo-supercharged engine or a solid rocket? Is it more complicated to try to build a four-stage liquid rocket and then combine twelve stages to build a space station, or to evolve a principle which permits you to do this with a single-stage device? If the latter concept is better, it might be assumed that we could get away with a simpler engineering cycle and therefore take less time to solve the same problem.

We must never fail to recognize the importance of understanding and supporting the very fundamental and basic areas of research. If we become overly involved at too early a phase with the engineering aspects of a problem, it is quite possible to get ourselves so oriented in a particular direction that we may miss some important possibilities.

I might describe the mission of the AFOSR as that of concentrating on a variety of possibilities rather than the directed inquiry into the expansion and engineering of present concepts.

Astronautics must be seen as a total environment involving aerodynamics, propulsion, guidance, communications, and human factors. We must recognize the foreseeable limits of current knowledge in all of these areas. Therefore, we support basic investigations into all their aspects. This is why we seek the answers in ion propulsion, magnetohydrodynamics, and the use of unexplored frequency ranges for communication purposes.

We have discussed engineering problems; we must also consider human factors. The work in the high-vacuum research laboratories of Litton Industries, California, may do much to produce answers to factors of human environment as a by-product of the basic area of electronics research. In the area of human factors much remains to be determined, but it can only be determined by basic research on all aspects of the problem.

Already in very high altitude flight heavy primary cosmic-ray particles may be of medical interest. It has been pointed out that such cosmic-ray primaries can produce severe damage on cellular structure. The total effects will inevitably depend on the frequency of hits. Very accurate data on the incidence of heavy primaries are now being collected for balloon altitudes.

Only informed guesses are possible for the upper atmospheric region. If man is to fly aboard an eventual space vehicle, these guesses are insufficient to insure his physical well being and safeguarding against the possible effects of cosmic rays. Recent observations during a giant solar flare have established that the sun at that time was responsible for over a 35-fold increase in cosmic-ray particle intensity and contributed protons of at least 30 billion electron volts energy.

This unexpected finding shows how limited is our present knowledge of the cosmic-ray phenomena. It is currently thought that the damage by primary cosmic radiation to the human organism, whatever its quantity, will not be dramatic or conspicuous or immediately apparent, but might be a slow deterioration of the individual and his germ plasm. It is indicated that we have a responsibility, both to our crews and to future generations to determine and control possible long range damaging effects to personnel.

We have spoken now both of propulsion and of human factors. These problems and their answers must be closely integrated. In order to launch a satellite, approximately 1,000 pounds of fuel and structure are required for each pound of payload held aloft. Although one certainly cannot and should not attempt to extrapolate direct data concerning the satellite into that of a manned space vehicle, certain logistics are brought into light by these figures. Each pound of man, food, oxygen, water, protective equipment, and environmental conditioning will cost us hundreds or thousands of pounds of launching weight. All these factors might best be summed up in the statement that so far as propulsion requirements and human factors are concerned, if space flight is to be achieved, obviously there must be a general rule that energy, personnel protection, and other factors must be optimized at a minimum weight and in minimum volume, and if consumed, utilized at the lowest consumption rate consistent with the job to be done.

To bring together all the possible areas of discussion, the Air Force Office of Scientific Research recognizes only too well the need for better communication between the military, the academic, and the industrial environments involved in astronautics research. Recognizing as we do this vital need for communication, AFOSR looks upon information handling and analysis and communication as a science in itself and is supporting research in several areas of communication and information handling.

However, we have found no substitute in information networks for the direct communication offered by meetings and symposia of international scope. At meetings such as this much exchange of ideas takes place to the mutual benefit of all concerned. The Air Force Office of Scientific Research sponsors and co-sponsors many symposia every year and regards these meetings of the International Astronautical Federation as of great value in the advancement of the vital field of Astronautics.

It is only by the free and open discussion of new concepts that we can move into the future rapidly and with full understanding of man's next great environment.

Instrumented Comets — Astronautics of Solar and Planetary Probes

By

Krafft A. Ehricke[1], ARS

(With 32 Figures)

Abstract — Zusammenfassung — Résumé

Instrumented Comets—Astronautics of Solar and Planetary Probes. A study of instrumented comets is presented for the purpose of analysing the aspects of interplanetary research by unmanned vehicles. The technical, scientific, political and cultural significance of interplanetary research is stressed and consideration of an International Astronautical Decade proposed. The basic performance data and the possibilities of gravitational navigation by close encounter with the Moon, Venus or Mars are analyzed. A discussion of accuracy requirements, both from the flight mechanical and the mission point of view, follows. The flight mechanics of the lunar probe is discussed, the need for 4-body treatment shown, and examples of orbits under 4-body conditions given, including lunar encounter for hyperbolic departure. Orbits involving close encounter between the artificial comet and Venus, Mars and Jupiter are presented and evaluated. A classification of unmanned space vehicle systems is proposed. A number of instrumented comet systems is investigated and specific mission proposals made, such as the exploration of interplanetary plasma characteristics by solar-triggered thermonuclear probes and the study of natural comets by thermonuclear probes or inflated tracer bodies. A radiation-propelled space probe is suggested which can be used as tracer body for the exploration of interplanetary dust or as a payload-carrying space probe.

Instrumentierte Kometen — Astronautik von Sonden zur Erforschung der Sonne und der Planeten. Um die Aussichten interplanetarischer Forschung mit Hilfe unbemannter Raumfahrzeuge zu studieren, werden „instrumentierte Kometen" vorgeschlagen. Der Verfasser betont die technische, wissenschaftliche, politische und kulturelle Bedeutung der interplanetarischen Forschung und schlagt eine „Internationale Astronautische Dekade" vor. In der Abhandlung werden die grundlegenden Daten fur die Ausfuhrung und die Möglichkeiten der Gravitationssteuerung durch enge Annäherung an Mond, Venus und Mars analysiert. Es folgt eine Diskussion der Genauigkeitserfordernisse, sowohl vom flugmechanischen als auch vom Zweckstandpunkt. Die Flugmechanik der „Mondsonde" wird erortert, die Notwendigkeit fur eine Vierkörperproblem-Behandlung gezeigt und Beispiele fur Bahnen bei Vierkörperproblembedingungen gegeben, die eine Begegnung mit dem Mond mit hyperbolischem Abflug einschließen. Die Arbeit zeigt Bahnen auf, bei denen es zu einer nahen Begegnung zwischen dem künstlichen Kometen und den Planeten Venus, Mars und Jupiter kommt. Diese Bahnen werden auch ausgewertet. Es wird ferner eine Klassifizierung unbemannter Raumfahrzeugsysteme vorgeschlagen. Eine Anzahl von Systemen instrumentierter Kometen wird untersucht und Vorschlage für besondere Aufgaben gemacht, wie die Erforschung der Charakteristika des interplanetarischen Plasmas

[1] Assistant to the Technical Director, Convair-Astronautics Division, San Diego, Calif., U.S.A.

durch von der Sonne ausgelöste thermonukleare Sondenfahrzeuge sowie das Studium natürlicher Kometen durch die thermonuklearen Sonden oder aufgeblasene Tracer-Körper. Auf eine strahlungsgetriebene Raumsonde wird hingewiesen, die als Tracer-Körper für die Erforschung des interplanetarischen Staubes oder als eine nutzlastbefördernde Raumsonde benützt werden kann.

Comètes instrumentées — Eléments astronautiques des sondes solaires et planétaires. Une étude des comètes instrumentées, présentée dans le but d'analyser les aspects de la recherche interplanétaire par des véhicules non habités. Les significations technique, scientifique, politique et culturelle de la recherche interplanétaire sont soulignées et l'auteur propose de prendre en considération la création d'une Décade Astronautique Internationale. Les performances de base et les possibilités de navigation gravitationnelle par pénétration proche des champs de la lune, de Vénus ou de Mars sont analysées. Suit une discussion de la précision requise, aussi bien du point de vue de la mécanique du vol, que du point de vue de la mission assignée. La mécanique du vol d'un véhicule-sonde lunaire est discutée et la nécessité du traitement du problème des quatre corps établie. Dans le cadre du problème des quatre corps, des exemples d'orbites sont donnés et notamment celui d'une rencontre lunaire avec départ hyperbolique. Des orbites pésentant des pénétrations proches de la comète artificielle dans les champs de Vénus, de Mars et de Jupiter sont évaluées. Une classification des véhicules spatiaux non-habités est proposée. Certains types de comètes instrument sont étudiées et des propositions faites pour des types nouveaux de missions spécifiques. Telle l'exploration de la matière interplanétaire par sondes thermonucléaires déclanchées par rayonnement solaire et l'étude des comètes naturelles par sondes thermo-nucléaires ou ballons-sonde. Une sonde spatiale propulsée par rayonnement est proposée comme élément traceur pour l'exploration de la poussière interplanétaire ou comme sonde spatiale instrumentée.

Nomenclature

A cross-sectional area
a semi-major axis
b semi-minor axis
C_D gasdynamic drag coefficient
c velocity of light
D distance Earth-Moon
D gasdynamic drag
d diameter
e eccentricity
$e_{\odot}$ eccentricity of heliocentric orbit of artificial comet
F force
G gravitational constant
g gravitational acceleration
h orbital energy
i orbital inclination
$K = G \cdot M$ gravitational parameter
K_r radiation acceleration parameter
M mass of Sun; generally, center mass
M momentum
m mass of planet; generally, mass of companion
m_0/m_1 mass ratio of rocket vehicle
n_r radial acceleration of sphere due to radiation pressure
p_r radiation pressure
R distance from the Sun; generally, distance from center body
R^* distance measured in astronomical units
R reflectivity
r distance from planet; generally, distance from companion
S solar constant
T period of revolution
t time
u orbital velocity of companion with respect to M
V velocity of space vehicle with respect to the Sun
v velocity of space vehicle in general, or with respect to planet
v_h hyperbolic departure velocity
v_{id} ideal velocity
v_∞ hyperbolic velocity excess over parabolic
v_c circular velocity
$(\Delta v)_{st} = v_h - v_c$, actual velocity increment required for hyperbolic departure if ship already has circular velocity
W weight

w_∞ relative velocity between space vehicle and m before encounter
w_{eff} effective exhaust velocity
β intersection angle between two orbits
ζ angle between vectors u and w_∞
η true anomaly
θ flight path angle to local horizon
μ mean daily motion
Φ half angle between symptotes of hyperbola
ϱ distance from the Moon; distance of libration point
ϱ radius of inflated sphere; radius of interplanetary dust particle
Ω position of ascending node

Subscripts

o beginning of burning
oo surface of the body
∞ value at infimite distance from the reference body
1 end of burning; conditions preceding gravitational encounter
2 conditions after gravitational encounter
A apo-distance
a azimuthal
c circular
h hyperbolic
M, m referring to center mass or companion
P peri-distance
p parabolic
r radial

1. Introduction

An elementary calculation of the velocity required to fire intercontinental ballistic missiles over a distance of 5000 to 6000 miles shows that the required ratio of cut-off velocity to local circular velocity is of the order of 0.9 to 0.93. It is therefore quite obvious that with comparatively small velocity increments not only the entire Earth-Moon field can be covered, but that it is also possible with such powerful boosters to penetrate into interplanetary space. The idea of using automatic vehicles for probing into interplanetary regions certainly is not new. It is a logical follow-up to the instrumental satellite and the lunar probe. A. C. CLARKE [1] specifically depicts an "automatic rocket surveying Mars".

The purpose of this paper is:

(1) to present a synoptic treatment of the astronautical significance (scientifically as well as technically) of these vehicles for which the designation "instrumented comets" (IC) is used to emphasize their interplanetary and solar research mission;

(2) to propose a new type of interplanetary research vehicle, namely, the sunlight-propelled balloon-type tracer body which does not require pressure stabilization to stay inflated and which can carry even a payload;

(3) to propose that serious consideration be given to the organization of an International Astrophysical Decade for the study of the interplanetary environment of our Earth and for the preparation of manned space flight. This operation should follow the International Geophysical Year after an appropriate time interval, and when the necessary technical prerequisites are available, perhaps from 1965 to 1975. During this period the asteroid Icarus (1566) approaches the Earth as close as 0.038 A.U. in 1968 and Geographos (1969) 0.032 A.U. in 1969[1]. Both asteroids could be intercepted with photographic equipment as part of the program.

---

[1] For this information I am greatly indebted to Prof. S. HERRICK of the University of California.

2. The Significance of Astronautic Research by Instrumented Comets

Here is a field of research and development which carries a tremendous potential, technically, scientifically, politically and culturally in the imminent future, mainly for the following reasons which will become more apparent further down in this paper:

(a) *Small payloads* (compared to manned space vehicles) are involved in most cases; thus, the missions can be carried out with propulsion systems presently under development in several countries.

(b) *Simple payloads and vehicle systems* are possible. The resulting reliability level is comparatively high, increasing the probability of a high percentage of successful missions.

(c) *Accuracy requirements* are in many cases quite relaxed; therewith guidance and control systems presently needed for ICBM's and IRBM's are adequate for a large number of missions.

(d) *Recovery* is not a prerequisite for the success of most flight purposes. In the case of IC's, this means three distinct advantages:

(d—1) no propellant required for return to Earth; this cuts in half the total energy requirement for the respective mission. Capture by the target planet is not needed for most research mission. Thereby, the overall propulsion energy requirement is reduced even further.

(d—2) operational lifetime is frequently restricted to a brief period (hours) in the target region of space so that the duration of power drainage is brief, although the degree of drainage may be high. Flight time to the target space region is directly proportional to the energy of departure. In missions to the neighboring planets, much of the travel time consists of waiting near the target planet and high energy departure therefore does not bring a corresponding reduction in time.

(d—3) atmospheric entry and descent to planetary surfaces either is not involved or—if attempted—would not require intact landing. The probe would transmit information until destroyed, or cause a nuclear explosion which can be evaluated spectroscopically on Earth.

(e) *Instrument and data transmission techniques* have been greatly perfected in connection with high-altitude sounding experiments and flight testing of guided missiles. Presently, the advent of minimum satellites provides another powerful stimulus to the miniaturization of equipment (measuring as well as transmission) and to "rugged" transistorization of space-borne circuitry. For many interplanetary missions these developments are entirely or largely adequate.

(f) *A world-wide net of tracking stations and a centralized system of immediate data evaluation and electronic computer orbit analysis* is being established at the present time in connection with Project Vanguard. In its wake improvements are being made in optical instrumentation for the specific purpose of observing extremely faint objects. In addition, the astronomical telescopes will be more readily available (at least in principle) for interplanetary and lunar experiments than for instrumental satellites, because the motion of the former occurs essentially in the ecliptic plane and is not retrograde (seen from the Earth) as are satellite motions in orbits below the 24-hour orbit. The number, size and sensitivity of radio telescopes increases every year. They represent a powerful tool for maintaining electronic contact with an IC. The global net of Vanguard tracking stations may be intended for the duration of the IGY only. However, we believe it will not be abandoned entirely; in any case, a precedence has been established which will, in several respects, simplify similar operations a few years later.

(g) *The scientific importance* of planetary and astrophysical research by means of instrumented comets is self-evident and specific cases will be outlined below. There are three principal areas of research to which the IC can be applied:

(g—1) planetology (gravitational and magnetic fields, atmospheres, surfaces),

(g—2) interplanetary matter (comets, meteors, dust, gas, plasma streams of solar origin, interstellar wind),

(g—3) solar and astrophysical research.

In each of these areas there are tests which can be done only with the IC because of:

the possibility of very close approach to Sun and planets, and because the IC offers a possibility of probing cosmic phenomena outside the specific gravitational, magnetic and electric conditions of the Earth-Moon field (representing a research area in its own right). This may be of similar importance as it was to escape the atmosphere of the Earth with high altitude rockets.

(h) *Politically*, one of the most effective and constructive methods of reducing tension and the danger of war is to increase the number of joint activities, preferably activities which aim high, idealistically and culturally. In this respect, solar system research with instrumental comets qualifies supremely. Firstly, such a program is elastic in that it permits of initially dealing only with the comet stages, leaving out the boosters and launching facilities, i.e., the potential weapon components which demand protection, not because of their performance capability which is known anyway, as pointed out in the first sentence of this paper, but because of the significance of technical details involved. Much of the research information is of a general scientific value, but the methods by which it is obtained (e.g. advanced techniques in auxiliary power supply, orbit accuracy, etc.), which need not readily be made available either, certainly can be as impressive (hence as deterring) a manifestation of missile superiority as a race in terrestrial range and payload capacity.

On Earth, the large rocket vehicles presently under development in the U.S.A. and the U.S.S.R. cannot find a much broader application than as weapon systems. Their use as mail or cargo carrier is expensive and therefore will meet severe competition from other means of transportation. Between the World Wars, potential bombers were used as passenger and cargo airliners. A similar trend can now be observed in regard to large jet airplanes. Yet, this potential is only to a very limited extend available to rocket-powered craft, namely, in the form of hypersonic passenger gliders and even this is considerably further in the future. Indeed, such craft belongs in the era of manned space craft rather than automatic vehicles. The only additional application conceivable is therefore space research, sending probes into terrestrial, cislunar, lunar and interplanetary space. We feel that here is a worthwhile field of investigation for national and international scientific and cultural organizations to explore in more detail realistic approaches to instrumented comet operations in interplanetary space. There are only a few years left until such launchings become a distinct technical possibility.

The successful growth from the International Polar Year to the International Geophysical Year despite major political and military conflict has demonstrated man's underlying devotion to expanding his knowledge of the universe, a devotion which extends far beyond the bounds of the scientific and engineering fraternity. In this spirit we propose that consideration be given by the I.A.F., the International Council of Scientific Unions and associated national organizations to the preparation of an International Astrophysical Decade (IAD), devoted to *systematic cislunar, lunar, solar and interplanetary research*, using the instrumented space probe as a tool for scientific research, for permanent cooperation among nations in an important field of human endeavor and for the advancement of astronautics by man himself.

Perhaps the most far-reaching *cultural significance* of large-scale cosmic experiments lies in their effect on the younger generation all over the world. The danger of nuclear wars and of mutual annihilation may not materialize after all. However, the mere existence of this threat, made possible only by science and technology, means — in the ultimate analysis — dangerous psychological warfare against the youth of this planet, because it denies them a safe future. Nuclear energy and rocket power are the two most formidable symbols of our time. Great efforts are being made to turn nuclear energy into a blessing for mankind; similar possibilities exist in the field of rocket power. While nuclear energy at present undoubtedly promises greater and more immediate material returns, the deeper spiritual significance and potential appear to lie in the cosmic use of rocket power. We believe that probing into cislunar and interplanetary space means more than an interesting scientific exercise. It is one of the most impressive ways of demonstrating a sane alternative to the young generation who will run the world of tomorrow.

3. Basic Performance Data

At the beginning of the discussion we place a survey of the velocities required for departure from the Earth to any point in the solar system. Fig. 1 shows these velocities. The four upper curves which should be read on the outer scale of the ordinates indicate the velocity of departure with respect to the center of the Earth when starting from the surface (r_{oo}, uppermost curve) or from one of three satellite orbits (500, 1,322 and 8,200 km altitude). This velocity is simply

$$(\Delta v_{st})_{r_y} = \sqrt{\frac{2 K_{\oplus☾}}{r} + \left| V_i^2 - \frac{K_\odot}{R_\oplus} \right|} \tag{1}$$

where the first term under the square root represents the escape velocity from the Earth-Moon system, the second the initial velocity with respect to the Sun in the circumsolar orbit and the third term the circular velocity of the Earth at the mean distance of one astronomical unit (A.U.) from the Sun. The absolute value of the last two terms is the hyperbolic excess with respect to the Earth.

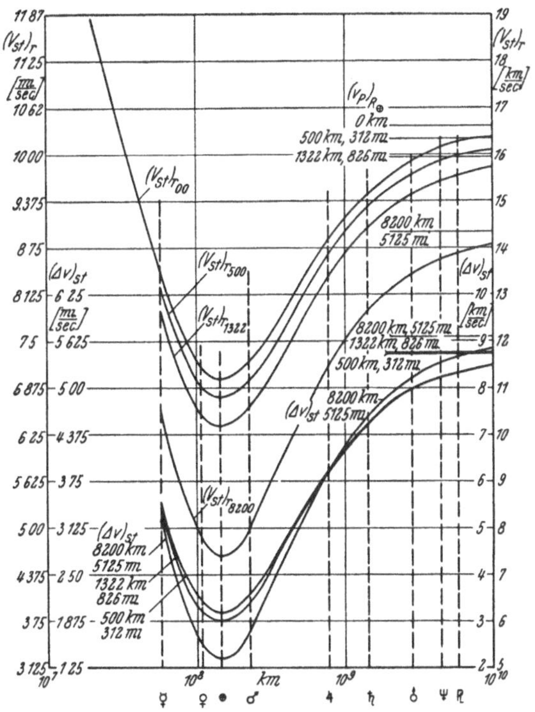

Fig. 1. Energy spectrum of one-way interplanetary missions

When starting from the Earth's surface no relief from circular velocity is available. However, for the three satellite orbits the actual departure impulse required is

$$(\Delta v)_{st} = V_{st} - \sqrt{\frac{K_\oplus}{r}}. \tag{2}$$

This velocity change is plotted in the three bottom curves of Fig. 1 (inner scale on the ordinates). The basic assumption for all curves in Fig. 1 is that one apsis of the circumsolar transfer ellipse lies at the departure point (i.e., in the Earth's orbit) while the other apsis lies at the distance from the Sun indicated on the abscissa. The horizontal lines on the right-hand side of the graph are the asymptotes of the respective v_{st} and $(\Delta v)_{st}$ curves, designating the value required for parabolic escape from the solar system. The minima of the curves at Earth distance designate the parabolic escape velocity from the Earth. The cross-over of the $(\Delta v)_{st}$-curves at about Jupiter distance is due to the effect to the departure orbit distance on the interplanetary transfer energy [6] [7].

The energy made available by the propulsion system (after consideration of all internal conversion efficiency factors) is distributed as follows: velocity at cut-off, v_1, gravitational loss

$$(\Delta v)_g = \int_o^{t_1} g \sin\theta \, dt \tag{3}$$

where θ is the instantaneous trajectory angle to the local horizon, and drag loss

$$(\Delta v)_d = \int_o^{t_1} \frac{D}{m} dt \tag{4}$$

where D is the vehicle drag and $m = m_o - \dot{m}t$ the instantaneous vehicle mass ($\dot{m}$ = constant). The mass ratio of the vehicle, m_o/m_1, must therefore be laid out so as to accommodate the following conditions:

$$v_{id} = \frac{m_o}{m_1} exp = \left[\frac{v_1 + \int_o^{t_1} g \sin\theta \, dt + \int_o^{t} \frac{D}{m} dt}{w_{eff}}\right]. \tag{5}$$

Here v_{id} is the ideal velocity, v_1 corresponds to $(V_{st})_{roo}$ or $(\Delta v)_{st}$ for orbital departure as given in Fig. 1 and w_{eff} is the effective exhaust velocity taking into account the increase in pressure thrust for ascent through the atmosphere. The losses $(\Delta v)_g$ are only partly recovered as potential energy, since powered motion in a potential field is not a conservative process.

For departure from a satellite orbit $(\Delta v)_d = 0$ and the penalty in mass ratio due to $(\Delta v)_g$ at low thrust acceleration has been plotted for chemical propulsion systems in [8]. For departure from the surface — a case which is of particular interest in connection with automatic space probes — the sum of $(\Delta v)_g$ and $(\Delta v)_d$ amounts to roughly 1.3 km/sec (4,000 ft/sec or 0.76 mi/sec) if circular velocity and horizontal flight direction are reached, at about 120 km (390,000 ft). Most of this (85 to 90 percent), is gravitational loss if large vehicles are considered.

The ideal velocity of an ICBM must be of the order of 8.6 km/sec (28,000 ft/sec) if its cut-off velocity is to be 7.2 km/sec or 0.91 $(v_c)_{oo}$ where v_c is the circular velocity at the surface. Since the escape velocity from the Earth is 11 to 11.1 km/sec (36,200 to 36,500 ft/sec or about 6.9 mi/sec), it is possible by means of one additional stage to increase the cut-off velocity of such vehicles (surface launched) not only to parabolic velocity but even higher. Fig. 1 shows that Venus and Mars lie very close to the minimum of the curves. A cut-off velocity of 11.48 km/sec

(37,600 ft/sec) carries the vehicle to Venus and 11.58 km/sec (about 37,000 ft/sec) to Mars. To all these numbers read from Fig. 1, about 1.3 km/sec (4000 ft/sec) must be added to obtain the required mass ratio for the given w_{eff}. Thus, a lunar probe whose payload capacity is not too marginal is also a potential Venusian or Martian probe.

4. Gravitational Navigation

Considerably greater energy is required to penetrate still deeper into the inner solar system, or into the outer solar system. However, for such missions we can take advantage of greatly relaxed accuracy requirements.

Considering first the inner solar system, it appears that operations in intra-Venusian space are of interest mainly in connection with solar research, not with the exploration of Mercury or the asteroid Icarus. The accuracy of a solar probe can obvious be very low, since it does not matter whether the perihelion lies a few hundred thousand miles closer or farther away from the Sun.

As far as the outer solar system is concerned, reaching out beyond Jupiter is of little practical value, since Jupiter has most characteristics of the outer planets incorporated in its system, (they are called Jovian planets), and also because technical difficulties — already very high for a Jovian prove — grow rapidly as more distant planets are considered. Confining, then, the present aim for outer solar system explorations to Jupiter, we can take advantage of its enormous gravitational pull whose focussing effect washes out almost all practical inaccuracies in the departure orbit.

If, then, the main navigational characteristic of research flights beyond the limits of Venus and Mars is very low flight accuracy for the missions under consideration, it appears realistic to consider the use of perturbative fields for relieving the energy requirements indicated in Fig. 1.

In particular, the utilization of lunar perturbation for increasing the orbital energy of the departing probe appears very attractive under these circumstances. It is long known in theoretical astronomy that perturbative gravitational fields can have the effect of increasing or decreasing the orbital energy and also change the eccentricity of the perturbed body. Jupiter, for example, is known to have comets in near-parabolic orbits thrown out of the solar system (e.g. Comet 1886 III after encounter with Jupiter departed from the solar system in a hyperbolic orbit of eccentricity 1.013). Such cases are rare, however. By the same token the eccentricity can be reduced and the orbital energy either increased or decreased. This results in a change toward increased circularity of the comet orbit. Astronomers call this process capture. The captured comet whose new aphelion lies not far from the orbit of the capturing planet is said to belong to the comet family of this planet. Each of the Jovian planets has its comet family (the famous HALLEY comet belongs to Neptune's family). The terrestrial planets are too small to affect natural comets significantly. However, artificial comets, sent along prescribed orbits, can be greatly affected by them as well as by our Moon or satellites of other planets.

Utilization of perturbative forces for astronautical purposes has been contemplated before. D. F. LAWDEN [9] devoted a brief analytical discussion to this subject and emphasized the savings in propulsion energy which could be achieved with a cotangential orbit encounter and passage at minimum (i.e. surface) distance from Moon or Mars.

The analytical treatment shall subsequently be extended to the determination of the optimum distance (maximum velocity pick-up) of passage in the case of intersecting orbits and shall be applied systematically to the case of instrumented comets.

The principal difficulty in using gravitational navigation lies in the fact that this method is very sensitive. The approach to the perturbing body must be quite accurate. Small errors at the entrance into the perturbative field result in much greater errors at the exit. With manned vehicles, especially those equipped with low-thrust propulsion systems, it will be possible by means of running path corrections to achieve a high accuracy of approach. With instrumented comets, whose mission requires close encounter or even collision with a terrestrial planet, this method may not be very promising. However, the energy requirements for Venus and Mars transfers do not make such maneuver necessary, and for the other missions inaccuracy is not a serious disadvantage.

This method of gravitational navigation does not, of course, violate the law of conservation of energy, since the orbit of the perturbing body is changed likewise by the vehicle, the changes in orbital energy and orbital elements being, of course, in proportion to the perturbing masses. Thus, obviously the Moon or a planet is not perceptibly perturbed by the vehicle. If after an encounter, orbital energy and eccentricity are increased and if the eccentricity of the original vehicle orbit was high, then the new orbit can be changed into a hyperbola with respect to the given center of attraction.

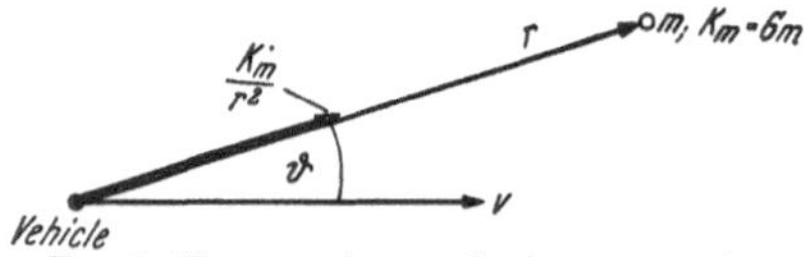

Fig. 2. Energy change during encounter

The instantaneous change in orbital energy during the encounter can be determined very easily. The orbital energy of the vehicle with respect to the center body M is

$$h = v^2 - \frac{2\,K_M}{R} = -\frac{K_M}{a} \tag{6}$$

where v, R and K_M are the vehicle velocity and distance with respect to the center body, and the gravitational parameter of the center body, respectively, while a is the semi-major axis of the orbit. Differentiation with respect to time yields

$$\frac{dh}{dt} = \frac{K_M}{a^2}\frac{da}{dt}. \tag{7}$$

In order to relate a to the velocity, Eq. (6) is written in the form

$$v^2 = K_M\left(\frac{2}{R} - \frac{1}{a}\right) \tag{8}$$

and differentiated with respect to time

$$\frac{da}{dt} = \frac{2\,a^2}{K_M}\,v\,\frac{dv}{dt}. \tag{9}$$

Since, from Fig. 2

$$\frac{dv}{dt} = \frac{K_m}{r^2}\cos\vartheta \tag{10}$$

one obtains finally from Eqs. (9), (8) and (6),

$$\frac{dh}{dt} = 2\,\frac{K_m}{r^2}\,v\,\cos\vartheta. \tag{11}$$

The component $K_m/r^2 \sin\vartheta$ is directed normally to the direction of motion and causes no change in orbital energy. The change in orbital energy can be positive or negative, depending on the direction of the perturbing force with respect to the direction of motion of the vehicle. The change in orbital energy is independent of K or a and, aside from the path velocity with respect to the center body, dependent only upon the gravitational strength K_m of the perturbing body as well as the distance from it. If $\vartheta = 90°$ the change in orbital energy is zero. The angle ϑ is counted from the radius vector r to the velocity vector v. The value of $\cos\vartheta$ should be considered negative when the perturbative force K_m/r^2 tends to cause a retrograde rotation of the major axis (i.e., a rotation opposite to the direction of revolution of the body).

In order to determine the direction of change of the eccentricity e, the tangential component of the perturbative force $(K_m/r^2)\cos\vartheta$ as well as the normal component $(K_m/r^2)\sin\vartheta$ must be taken into account. Indeed, it can be shown that

$$\left(\frac{de}{dt}\right)_v = \frac{2(e+\cos\eta)}{v}\,\frac{K_m}{r^2}\cos\vartheta \tag{12}$$

$$\left(\frac{de}{dt}\right)_{v_n} = -\frac{r}{a}\,\frac{\sin\eta}{v}\,\frac{K_m}{r^2}\sin\vartheta \tag{13}$$

where $(de/dt)_v$ and $(de/dt)v_n$ are the change in e due to the tangential and the normal perturbative force component, respectively. The normal component is taken as positive in the direction toward the major axis of the orbit. Thus, according to Eq. (13) the eccentricity is reduced if $(K_m/r^2)\sin\vartheta$ points towards the interior of the orbit. From this it can readily be seen that the orbital energy can be increased while the eccentricity is decreased. An example for this condition will be seen below in the case of a computed hyperbolic encounter with Mars.

The term "hyperbolic encounter" implies that solar (or terrestrial) gravitation may be neglected during the brief period of close passage and that therefore the path of the vehicle near the perturbing body is indeed an exact hyperbola. Let M be the center body (i.e., Earth or Sun), m the perturbing body, v the velocity of the space vehicle, u the velocity of m and w the relative velocity of the vehicle with respect to m. The encounter process involving a velocity increase of the space vehicle is illustrated in Fig. 3, showing the conditions before and after the encounter. Body m moves at velocity u in an orbit which is intersected at the angle β_1 by the orbit of the space vehicle moving at velocity v_1. The relative velocity w_∞

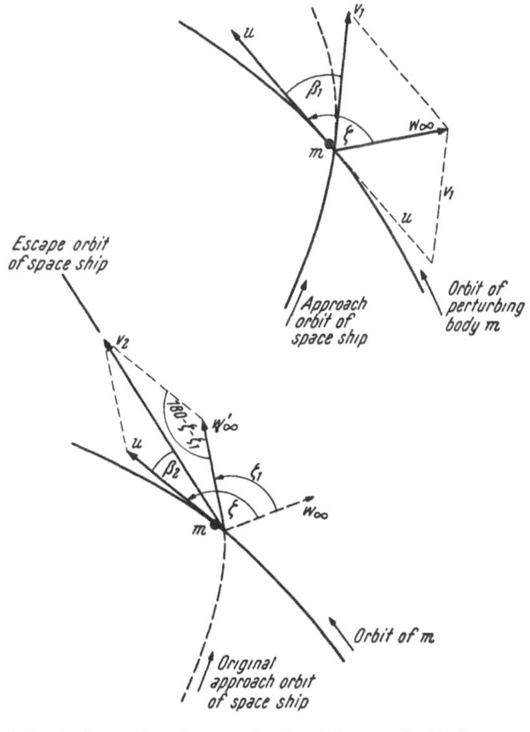

Fig. 3. Geometry of encounter involving a velocity increase of the space ship

can easily be constructed or computed from u, v_1 and β. The angle between w_∞ and u is ζ. After the encounter, w_∞, at constant scalar value, is turned by angle ζ_1 in direction w_∞' while u is practically unchanged. The new velocity v_2 can now be constructed or computed from u, $w_\infty' = w_\infty$ (scalar) and angle $(180 - \zeta + \zeta_1)$. Therewith the new intersection angle β_2 can be found. In transferring the information in Fig. 3 to a hyperbolic orbit, m must be regarded as stationary. Then, however, w_∞ and w_∞' become the asymptotes of the hyperbola, hence, the change in direction of flight due to the hyperbolic encounter. The stationary hyperbola and the associated nomenclature are shown in Fig. 4. The characteristic data can be computed as follows:

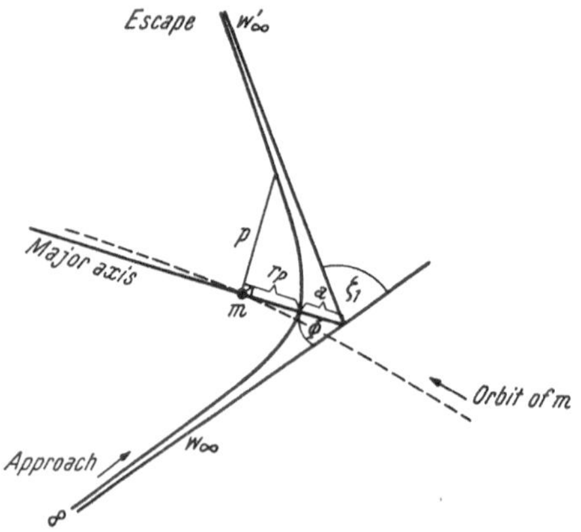

Fig. 4. Hyperbolic path followed during encounter according to Fig. 3

The relative velocity before the encounter is

$$w_\infty^2 = u^2 + v_1^2 - 2\,u\,v_1\,\cos\beta_1\,. \qquad (14)$$

The major axis of the hyperbola is

$$a = \frac{G_m}{h} = \frac{K_m}{h} = \frac{K_m}{w_\infty^2} = \frac{K_m}{(w_\infty')^2} \qquad (15)$$

since the orbital energy for the hyperbola $h = (2\;K_m/r + w_\infty^2) - 2\;K_m/r$ where the term in parenthesis is the hyperbolic escape velocity from m along an orbit which intersects that of m with the angle β_1. Thus w_∞ represents the hyperbolic excess and, since $h = 0$ for the parabola, w_∞^2 is the energy of a body in a hyperbolic orbit. The eccentricity of the encounter hyperbola is

$$e = 1 + \frac{r_P}{a} = \sec\varnothing = \sec\frac{180 - \zeta_1}{2} = \operatorname{cosec}\frac{1}{2}\zeta_1 \qquad (16\,\text{a})$$

since the eccentricity of a hyperbolic orbit is inversely proportional to the turning angle involved. With Eq. (15) the first Eq. (16a) becomes

$$e = 1 + w_\infty^2\frac{r_P}{K_m}\,. \qquad (17)$$

Thus, if the perturbing body is known (K_m), r_P selected and w_∞ computed from the known data in Eq. (14), a and ζ_1 can be calculated from (15) and (16a), respectively. The value of ζ follows immediately from the relation

$$\sin\zeta = \frac{v_1}{w_\infty}\sin\beta_1. \qquad (18)$$

It should be noted, in calculating ζ_1, that $\zeta > 90°$ if $v_1\cos\beta > u$, since under these conditions obviously w_∞ must have a component opposite to u. Thus, if under these conditions (18) results in, say, $\zeta = 40°$, the value $\zeta = 180° - 40° = 140°$ is to be taken. Knowing ζ, the angle $180° - \zeta + \zeta_1$ is known also, and the new path velocity and intersection angle follow from

$$v_2^2 = u^2 + w_\infty^2 + 2\,u\,w_\infty\,\cos(\zeta - \zeta_1) \qquad (19\,\text{a})$$

$$\sin\beta_2 = \frac{w_\infty}{v_2}\sin(\zeta - \zeta_1). \qquad (20)$$

Finally, the parameter of the hyperbola, i.e., the distance from m at $\eta = 90°$ is

$$p = a\,(e^2 - 1) = \frac{K_m}{w_\infty^{\,2}} \cot^2 \frac{1}{2}\,\zeta_1 = \frac{K_m}{\left(w_\infty \tan \frac{1}{2}\,\zeta_1\right)^2} \tag{21}$$

or, since $\tan \frac{1}{2}\,\zeta = \sin \zeta / (1 + \cos \zeta)$,

$$p = \frac{K_m}{w_\infty^{\,2} \sin^2 \zeta}\,(1 + \cos \zeta)^2 . \tag{22}$$

Thus with u, v_1, β_1 and r_P given, w_∞, a, e, ζ_1, ζ, v_2, β_2 and p can be computed from Eqs. (14), (15), (17), (16a), (18), (19a), (20) and (21). The parabolic velocity with respect to the center mass M at the distance of m is given by

$$v_P^2 = \frac{2\,K_m}{R} . \tag{23}$$

The interesting resulting data are therefore ζ_1, $v_2 - v_1$ and the new hyperbolic excess (or deficiency) *with respect to the center body* M,

$$v^2_{\ \infty} = v_2^{\,2} - v_P^{\,2} = v_2^{\,2} - \frac{2\,K_M}{R_m} . \tag{24}$$

Instead of using r_P as an independent variable, it is of interest to determine the value of r_P for which v_2 (hence v_∞) becomes a maximum. It follows immediately from Eq. (19a) that v_2 becomes a maximum if $\zeta_1 = \zeta$, in which case

$$(v_2)_{max} = u + v_\infty , \tag{19 b}$$

i.e., the relative velocity vector w'_∞ after the encounter must point in the same direction as the orbital velocity of m. Then, with u, v_1 and β_1 given, $(v_2)_{max}$ follows

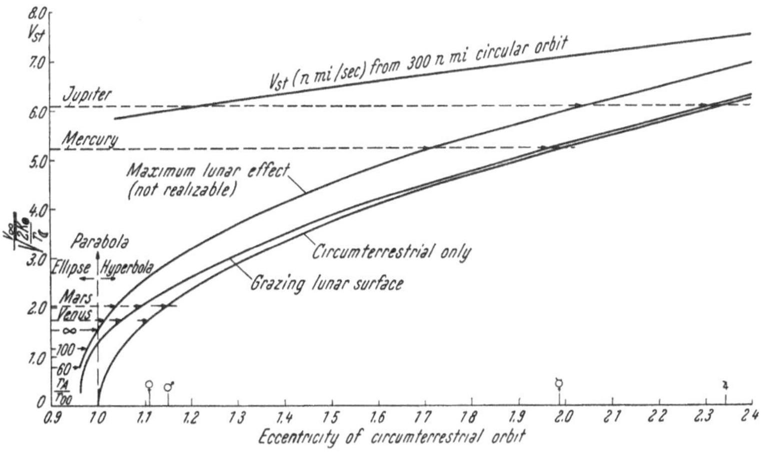

Fig. 5. Velocity increment due to lunar perturbation for elliptic, parabolic and hyperbolic departure orbits

immediately from Eq. (19b), ζ from (18) and the corresponding value of r_P from (16a)

$$r_P = \frac{K_m}{w_\infty^{\,2}}\,(e - 1) . \tag{16 b}$$

Fig. 5 shows the effect of a close encounter with the Moon on the circumterrestrial orbit of the vehicle. The three lower curves give the hyperbolic excess

Table I. *Gravitational Navigation Using the Moon*

Orbital velocity of Moon: $u = 0.550236$ n.mi./sec; Earth parabolic speed at Moon distance: $v_{p☾} = \frac{2K_{\oplus}}{r_{☾}} = 0.60552$ n.mi.²/sec²;
$v_{p☾} = 0.778$ n.mi./sec

A. Basic Data

r_A/r_{00}	60	63	70	80	90	100	∞	Hyperbolic escape paths to enter min. energy circumsolar transfer ellipse without the Moon: Hyp. to ♀	Hyp. to ♂	Hyp. to ♀	Hyp. to ♃
e (circumterrestrial orbit)	0.9643	0.966	0.9694	0.9732	0.9761	0.9785	1.0	1.1108	1.1508	1.9868	2.3444
η (intersection)	180	176°53′	174°58	172°12′	171°0′	170°10′	164°32′	149°59′	146°38′	118°26′	113°37′
$v_{st} = v_P$ (n.mi./sec)	5.7427	5.7442	5.7501	5.7556	5.7599	5.7634	5.7936	5.95	6.008	7.080	7.4918
β_1 (deg./min.)	0	61°02′	70°19′	74°50′	76°46′	77°53′	82°16′	86°04′	86°24	88°13′	88°23′
v_1 (n.mi./sec)	0.1043	0.2139	0.3100	0.3993	0.457	0.4987	0.78	1.571	1.772	4.144	4.8136
w_∞ (n.mi./sec)	0.466	0.4843	0.5329	0.589	0.63	0.6605	0.892	1.628	1.822	4.163	4.829

B. Data Pertaining to Encounter with Maximum Velocity Increase

$(v_2)_{max}$ (n.mi./sec)	0.996	1.0345	1.0831	1.14	1.18	1.21	1.442	2.179	2.372	4.713	5.38
ζ (deg/min.)	113°04′	157°15′	146°47′	146°04′	135°03′	132°25′	119°57′	105°46′	103°58′	95°49′	94°56′
ϱ_P (n.mi.)	$(\varrho_{00} = 938.445)$	65.67	118	99.8	159.6	163.8	150	73.8	62.47	15.44	11.8
e_h	1.199	1.02	1.0436	1.045	1.0827	1.0928	1.155	1.2541	1.269	1.347	1.3571

$(v_2)_{max} - v_1$ (n.mi./sec)	0.892	0.8205	0.7730	0.74	0.723	0.712	0.662	0.6078	0.600	0.569	0.566
$v_\infty = (v_2^2 - v_P^2)^{\frac{1}{2}}$ (n.mi./sec)	0.621	0.68117	0.753	0.833	0.887	0.926	1.214	2.035	2.24	4.467	5.32
v_∞ (without Moon) (n.mi./sec)	—	—	—	—	—	—	0	1.3636	1.5907	4.0696	4.750
$v_\infty/v_{p☾}$ (without Moon) (n.mi.sec)	—	—	—	—	—	—	0	1.752	2.0446	5.2308	6.105
$v_\infty/v_{p☾}$ (with Moon) (n.mi./sec)	0.789	0.8755	0.9678	1.0707	1.140	1.19	1.56	2.6154	2.879	5.973	6.838

C. Data Pertaining to Surface-Grazing Encounter

v_2 (n.mi./sec)	.8042	.9175	.9784	1.0017	1.0687	1.0942	1.2669	1.8205	1.9830	4.2054	4.8707
ζ_1 (deg./min.)	107°12′	102°6′	95°58′	89°16′	84°46′	81°31′	61°2′	27°20′	22°51′	5°11′	3°53.4′
e_h	1.2423	1.2858	1.3461	1.4232	1.4833	1.5317	1.9697	4.2322	5.0457	22.1245	29.4281
$v_2 - v_1$ (n.mi./sec)	.6998	.7035	.6683	.6023	.6168	.5956	.4870	24.96	.2114	.0617	.0571
$v_\infty = \sqrt{v_2^2 - v_p^2}$ (n.mi./sec)	.1961	.4832	.59078	.6287	.7307	.7675	.9984	1.6467	1.8232	4.1324	4.8078
$v_\infty/v_{p☾}$	.2515	.6196	.7575	.8062	.9369	.8842	1.2802	2.1115	2.3378	5.2988	6.1648

r = distance from Earth; r_P = perigee; r_A = apogee; r_{00} = radius of Earth; ϱ = distance from Moon;
$r_☾$ = distance of Moon from the Earth; η (intersection) = true anomaly of undisturbed circumterrestrial orbit at intersection with lunar orbit;
$v_{st} = v_P$ = starting (perigee) velocity; β_1 = angle of intersection of undisturbed orbit with lunar orbit;
v_1 = velocity of ship at intersection point without perturbation; w_∞ = apparent velocity of ship seen from Moon;
$(v_2)_{max}$ = maximum possible vehicle velocity after encounter; ζ = angle between w and u; u = velocity of Moon;
ϱ_P = closest distance from Moon; ϱ_{00} = radius of Moon; e_h = eccentricity of circumlunar hyperbola;
v_∞ = hyperbolic velocity excess after encounter; $v_{p☾} = \sqrt{\frac{2K_\oplus}{r_☾}}$ = parabolic speed with respect to earth at the Moon's distance;
v_2 = vehicle velocity after encounter; ζ_1 = angle by which w is turned; $v_2 - v_1$ = velocity change due to encounter;
(v_{2max}) = maximum post-encounter velocity possible

v_∞ in terms of the parabolic velocity at lunar distance $\sqrt{2\,K_\oplus/r_{☾}} = \sqrt{2\,K_M/R_m} =$ $= 0.778$ n.mi./sec $= 0.895$ mi/sec $= 1.41$ km/sec $= 4{,}625$ ft/sec. The lowest of these curves refers to the excess required for departure without lunar encounter at all. The second pertains to the closest possible lunar encounter, namely, grazing its surface. The upper of these three curves gives the greatest possible increase in velocity, i.e. $(v_2)_{max}$, which can be derived at all from the lunar encounter. This case is not realizable, however, since, as the supporting data in Table I indicate, the vehicle's orbit would have to pass the lunar center inside the Moon's radius. The curve on top of Fig. 5 shows the departure velocity, with respect to the Earth's center, from an altitude of 300 n.mi. This curve enables one to estimate immediately the reduction in departure velocity which can be derived from a given lunar encounter. All these velocity curves are plotted against the eccentricity of the circumterrestrial conic. The abscissa is shown to cover elliptic, parabolic and hyperbolic orbits with respect to the Earth.

As an example consider a departure for the orbit of Venus. The abscissa shows that without the Moon a hyperbolic orbit, $e = 1.11$, would be required. The corresponding departure velocity is $v_{st} = 5.95$ n.mi./sec $= 36{,}176$ ft/sec without encountering the Moon. The closest possible encounter with the Moon, passing just above its surface reduces the eccentricity of the circumterrestrial hyperbolic departure orbit to $e = 1.055$ and v_{st} to 5.86 n.mi. $= 35{,}629$ ft/sec. If the maximum possible benefit could be derived from the encounter one would obtain $e = 1.015$, $v_{st} = 5.8$ n.mi./sec $= 35{,}264$ ft/sec; but then, as shown in Table I, part B, line 3, the vehicle would have to pass somewhere between 150 and 74 n.mi. from the Moon's center. The differences in v_{st} do not seem impressive. However, their true significance becomes apparent if they are compared on the basis of the hyperbolic excess. This is more appropriate, since obviously a velocity close to parabolic velocity is required to make a lunar encounter possible in the first place. From Fig. 5 we find that the parabolic velocity ($e = 1.0$) is $v_{st} = v_p = 5.78$ n.mi./sec $= 35{,}142$ ft/sec. Thus, without the Moon, the hyperbolic excess to the brought up by propellant energy is $36{,}176 - 35{,}142 = 1{,}034$ ft/sec, while a surface-grazing encounter reduces this amount to 487 ft/sec and, in the extreme case, to 122 ft/sec. Thus, the optimum practical case already yields most of the obtainable benefit, saving $1{,}034 - 487 = 547$ ft/sec, i.e. more than half the velocity increment otherwise required. Depending on the marginality of the overall design, this velocity reduction means a gain of between some 500 to over thousand pounds of end weight, hence, mainly payload weight; or, conversely, for the same end weight, it reduces the take-off weight by some 70,000 to over a hundred thousand pounds [3]. The absolute hyperbolic excess required to reach the orbit of Venus is, of course, the same in all cases, namely, 1.75 times the parabolic velocity at lunar distance.

It is of particular interest to note that the gain which can be derived from the lunar encounter decreases rapidly for target distances from the Sun above Mars or beneath Venus.

The directional change ζ_1 associated with surface-grazing passage represents the maximum feasible turning angle under the given circumstances. It is not difficult to compute the turning angle for encounter at greater distance and to compute the starting time to obtain any desired direction of w_∞' after the given encounter.

For the conditions just discussed, the value of r_P lies always beneath the Moon's surface. Therefore, minimum distance of encounter yields maximum velocity increment (at correct approach). However, this is not necessarily always so. If the intersection angle is small and the velocities are not too different, so that $v_1 \cos\beta \sim u$, the best distance may be very large. Suppose the Moon's orbit ($u = 0.550236$ n.mi./sec) is intersected at $\beta = 5°$ by a body moving at $v_1 = 0.6$ n.mi./sec. One finds then that $(v_2)_{max} = 0.6209$ n.mi./sec, $\zeta = 47° 45'$ and $r_P = 226{,}700$ n.mi. (this, of course, neglects the Earth). Thus, the hypothetical body would best pass the isolated Moon at about Earth's distance, picking up 0.0209 n.mi./sec = 127 ft/sec in the process.

5. Accuracy Requirements

First, the accuracy requirements for an undisturbed orbit in a central force field shall be considered. The vehicle's position in space is described, in a rectangular coordinate system, by the coordinates x, y and z, where x, y may lie in the orbital plane, z is the orthogonal coordinate.

Any deviation from the correct coordinates at cut-off, i.e. any inaccuracy, can be considered as produced by an equivalent impulse which, at a given time during burning, caused the given change in cut-off coordinates. An orthogonal impulse does not change the major axis, eccentricity, mean motion or the perigee position of the orbit, but affects the orbital inclination and the longitude of the nodes. For solar probes or, generally, for "roving probes" whose mission does not require close encounter with a small terrestrial planet, an orthogonal error impulse has little or no practical significance. For a close encounter with Venus or Mars, such an error must be taken into account, but its effect may still not be very large. If v_w is the orthogonal velocity component ($v_w = 0$ in the correct orbit at the cut-off point) then it is easily shown that

$$\frac{di}{dv_w} = \frac{1}{v_a} \cos(\eta - 90 - \eta') \tag{25}$$

$$\frac{d\Omega}{dv_w} = \frac{1}{v_a} \frac{\sin(\eta - 90 - \eta')}{\sin i} \tag{26}$$

where v_a is the instantaneous azimuthal velocity component, η the instantaneous true anomaly and η' the angle between the major axis and the normal to the nodal line (measured from perigee in the direction of motion, like η) and i the orbital inclination with respect to the reference plane which specifies the nodal points. It can be seen that an orthogonal impulse does not change the orbital inclination if it occurs at the "lowest" point ($\eta - \eta' = 0$) or the "highest" point ($\eta - \eta' = 180°$) of the orbit with respect to the reference plane, while an orthogonal impulse given at the nodes does not change the orientation of the nodal line. Thus, in order to assess the effect of an orthogonal error impulse, the orbital elements must be known. This requires special consideration of a v_w error for any given case.

Errors in the orbital plane can be treated more generally. They affect the major axis, eccentricity, mean motion and the perigee position. Let the error impulses be expressed in terms of radial (v_r) and azimuthal (v_a) components, $v_r^2 + v_a^2 = v^2$. Then,

$$\frac{\partial a}{\partial v_r} = -\frac{2a^2 v_r}{K} \sin\eta \tag{27}$$

$$\frac{\partial a}{\partial v_a} = \frac{2a^2 v_a}{K} \tag{28}$$

$$\frac{\partial e}{\partial v_r} = -\frac{C}{K} \sin\eta \tag{29}$$

$$\frac{\partial e}{\partial v_a} = \frac{C}{K a e r}(b^2 - r^2) \tag{30}$$

$$\frac{\partial \mu}{\partial v_r} = \frac{3}{\sqrt{a p}} \sin \eta \tag{31}$$

$$\frac{\partial \mu}{\partial v_a} = -\frac{3}{r} \sqrt{\frac{p}{a}} \tag{32}$$

$$\frac{\partial \eta_o}{\partial v_r} = \frac{1}{e} \sqrt{\frac{p}{K}} \cos (\eta - \eta_o) \tag{33}$$

$$\frac{\partial \eta_o}{\partial v_a} = (p + r) \frac{\sin (\eta - \eta_o)}{K e}. \tag{34}$$

Here C is the KEPLER constant, p the parameter of the orbit, b the semiminor axis of the orbit, μ the mean angular motion and η_o the true anomaly of the perigee with respect to a fixed direction in space (vernal equinox).

If given at an apsis, a radial impulse will have no effect, except on the perigee position η_o. Again, the effect of a change in η_o depends upon the particular orbit and mission under consideration and shall not be treated further here. For the same reason $\delta\eta_o/\delta v_a$ will presently be disregarded. This then leaves the effect of an azimuthal impulse on a, e and μ, or, in other words, on r_A and r_P, the apogee (aphelion) or perigee (perihelion) distance of the respective orbit.

Equations describing the effect of radial and azimuthal velocity errors on the apsides of an orbit *in a central force field* were presented in [2]. Writing the polar equation of the conic in the form.

$$r = \frac{p}{1 + e \cos \eta} = \frac{r_1^2 v_1^2 \cos^2 \theta_1}{K (1 + e \cos \eta)} \tag{35}$$

and remembering that $K = r_1 v_c^2$ and

$$e^2 = \left(1 - \frac{v_1^2 \cos^2 \theta_1}{v_c^2}\right)^2 + \tan^2 \theta_1 \left(\frac{v_1^2 \cos^2 \theta_1}{v_c^2}\right)^2 \tag{36}$$

and that $\tan \theta = v_r/v_a$ and $v_1 \cos \theta_1 = v_{a1}$ one can write Eq. (35) in the form

$$\frac{r}{r_1} = \frac{(v_a/v_c)^2}{1 + X(v) \cos \eta} \tag{37}$$

where

$$X(v) = \sqrt{\left[1 - \left(\frac{v_a}{v_c}\right)^2\right]^2 + \left(\frac{v_a}{v_c}\right)^2 \left(\frac{v_r}{v_c}\right)^2}\,. \tag{38}$$

In the above equations the subscript 1 refers to the conditions at cut-off and beginning of celestical mechanical coasting. Eq. (37) gives the distance r in terms of the cut-off distance r_1 which is attained for a given cut-off velocity vector $v_1^2 = v_a^2 + v_\gamma^2$. The true anomaly η describes the point in the elliptic orbit at which the distance r is reached. If one is interested only in the apogee or perigee distance, $r = r_A$ or $r = r_P$, then $\eta = 0$, $\cos \eta = 1$. If any other point is to be investigated, η must be pregiven. The cut-off point r_1 obviously does not coincide with one of the apsides (i.e. $\eta' \neq 0°$), unless $v_r = 0$. If, for any particular cut-off condition the desired values are $v_a = v_a'$ and $X(v) = X(v')$ — which may involve $v_r = 0$ or $v_r \neq 0$ — and if the actual values, attained at the correct distance r_1, are v_a'' and $X(v'')$ then, for a given true anomaly η', the ratio of actual distance r'' to desired distance r' is given by

$$\frac{r''}{r'} = \frac{(v_a'')^2}{(v_a')} \cdot \frac{1 + X(v'') \cos \eta'}{1 + X(v') \cos \eta'}\,. \tag{39}$$

If the apsidal distances are required, Eq. (39) simplifies to

$$\frac{\gamma_{A'',P}}{\gamma_{A',P}} = \frac{(v_a'')^2}{(v_a')} \frac{1 + X(v'')}{1 + X(v')}\,. \tag{40}$$

Fig. 6 [2] is a generalized plot of Eq. (37) showing the variation of apogee distance and perigee distance in terms of cut-off distance (r_A/r_1; r_P/r_1) as functions of the ratio of azimuthal velocity component v_a at cut-off to the local circular velocity. The radial velocity component at cut-off, v_r/v_c, was taken as parameter. Thus, if $v_a/v_c = 1$ and $v_r/v_c = 0$, a circular orbit is obtained. For the case $v_a/v_c = 0$ the effect of a radial error v_r/v_c is seen to be greatest. If $v_a/v_c \lesssim 1$, the osculating orbit at cut-off is an ellipse, and if $v_r/v_c = 0$, the cut-off print coincides with one of the apsides of this ellipse. It can be seen that already at relatively small deviations from $v_a/v_c = 1$, the influence of a radial component is greatly reduced. This becomes even more apparent if the plot is extended not only over a range of ± 5 percent variation of v_a/v_c as in Fig. 6, but over the whole practical range of operation in a central force field. This case is presented in Fig. 7. The dash-lined square around $v_a/v_c = 1$ represents the range covered in Fig. 6. It follows from this graph that for central force field transfer orbits in geocentric or heliocentric space, the effect of v_r/v_c is small if $v_a/v_c < 0.9$ or $v_a/v_c > 1.1$, approximately. Fig. 7 is generally valid for any central force field transfer. Therefore, the geocentric field (r) as well as the heliocentric field (R) may be considered. Solar distances (R_A/R_1, R_P/R_1, $R_1 = R_\oplus$) as well as terrestrial distances are marked on the ordinate. On the curve itself, certain cases pertaining to circumterrestrial as well as circumsolar flights are indicated. The

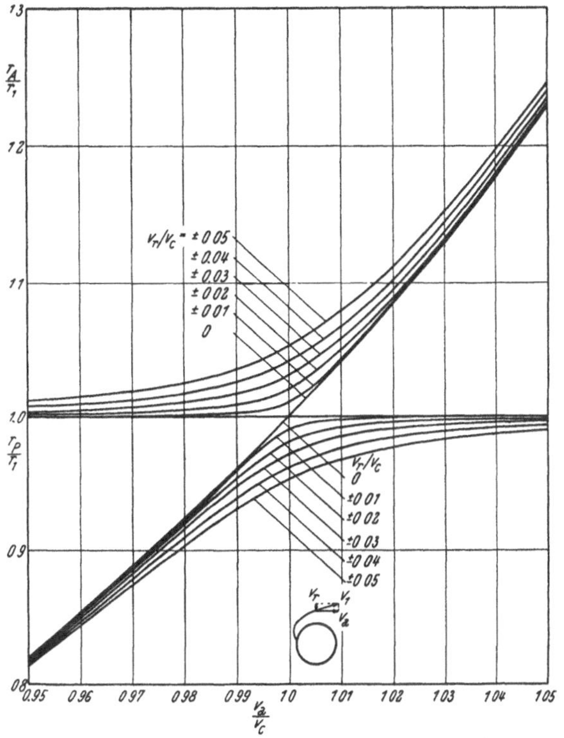

Fig. 6. Cut-off error spectrum with lines of constant radial velocity (distance error $\triangle r = 0$)

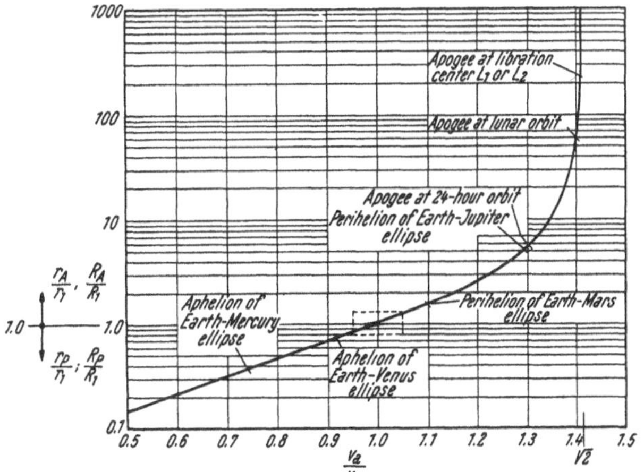

Fig. 7. Distance error as function of the cut-off velocity error for the Earth-Moon system and the solar system

libration centers L_1 and L_2 at the extreme right designate the two collinear Lagrangian libration centers near the Earth of the Sun-Earth system (not the Earth-Moon system).

The two points lie on the line Sun-Earth on both sides of the Earth at a distance of $r = 813{,}000$ n.mi.

The gradient of the curves in Figs. 6 and 7 represents the error sensitivity. Obviously, the error sensitivity is zero if the tangent is horizontal ($\partial (r/r_1)/$ $/\partial (v_a/v_c) = 0$), and infinite if the tangent is vertical ($\partial (r/r_1)/\, \partial (r_a/v_c) = \infty$). The general equation for the gradient is [2],

$$\frac{\partial\left(\dfrac{r}{r_1}\right)}{\partial\left(\dfrac{v_a}{v_c}\right)} = 2\,\frac{v_a}{v_c}\,\frac{1 + X(v)\cos\eta - [X(v)]^{-1}\left(\dfrac{v_a}{v_c}\right)^2 \cos\eta \left[\left(\dfrac{v_a}{v_c}\right)^2 + \dfrac{1}{2}\left(\dfrac{v_r}{v_c}\right)^2 - 1\right]}{[1 + X(v)\cos\eta]^2}\,. \tag{41}$$

Again, if $v_r/v_c = 0$, cut-off is given at one of the apsides. Then Eq. (41) simplifies to the equation for the error sensitivity in reaching the apsis opposite to the cut-off apsis,

$$\frac{d\left(\dfrac{r_{apsis}}{r_1}\right)}{d\left(\dfrac{v_a}{v_c}\right)} = \frac{4\,\dfrac{v_a}{v_c}}{\left[2 - \left(\dfrac{v_a}{v_c}\right)^2\right]^2}\,. \tag{42}$$

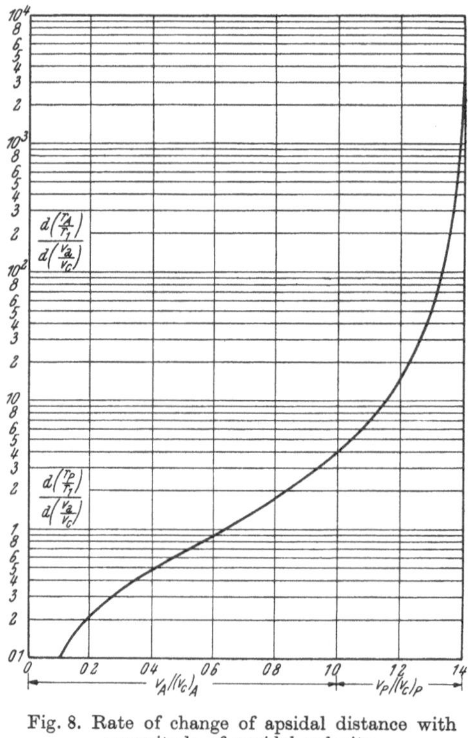

Fig. 8. Rate of change of apsidal distance with magnitude of apsidal velocity

This relation is shown in Fig. 8 for almost the whole range of $v_a/v_c = v_P/(v_c)_P$ and $v_a/v_c = v_A/(v_c)_A$.

If the error, Δv_a is very small, compared to v_c, the differential forms Eq. (41) or (42) can be used. The error follows then from

$$(\Delta r)_{\eta,\,\frac{v_r}{v_c}} = \frac{\partial\left(\dfrac{r}{r_1}\right)}{\partial\left(\dfrac{v_a}{v_c}\right)}\,\gamma_1\,\frac{\Delta v_a}{v_c} \tag{43a}$$

or

$$\Delta r_{apsis} = \frac{d\left(\dfrac{r}{r_1}\right)}{d\left(\dfrac{v_a}{v_c}\right)}\,r_1\,\frac{\Delta v_a}{v_c}\,. \tag{43 b}$$

Fig. 7 shows that the error sensitivity of the circumsolar part of the transfer ellipse to Venus is smaller than for the Earth-Mars orbit. Assuming the case of minimum energy transfer which is represented in Fig. 8, one finds that the aphelion displacement for the Earth-Mars orbit is 6,000 n.mi. per feet per second change in heliocentric departure velocity. For Venus the displacement is 2,000 n.mi per (feet per second). These numbers do not include the effect of departure velocity errors within the Earth's field nor the focussing effect of the target planet's gravitational attraction. They are, therefore, a measure of the effect of the accuracy with which the velocity of the Earth with respect to the Sun is known at the time of launching of the artificial comet. Specifically, if we launch the comet when the Earth stands at one of the intersections of the semi-minor axis with its orbit (at a distance of 1 A.U. and circular velocity with respect to the Sun, V_c, at 1 A.U. distance), it is necessary to know the exact value of 1 A.U. and the exact circular velocity, (hence, the value of $K_\odot = k^2 M_\odot$), in order to compute the correct hyperbolic departure velocity of the instrumented comet from its cut-off point near the Earth. The above values of 2,000 and 6,000 n.mi. show, in essence, the effect of such inaccuracy in V_c on the shift in perihelion or aphelion position for the Earth-Venus or Earth-Mars transfer orbits, respectively. At the same time, they show the effect of *technical* velocity errors *with respect to the Sun* as the comet leaves the Earth's field.

In analysing the error sensitivity of interplanetary orbits if the vehicle departs near the Earth, *the effect of the terrestrial as well as the solar field* must be taken into account. The hyperbolic velocity of departure with respect to the Earth from a point near the Earth is given by

$$v_h = \sqrt{v_p^2 + (V_1 - V_0)^2} = \sqrt{v_p^2 - (\Delta V)^2} \tag{44}$$

where v designates velocities measured with respect to the Earth, and V those measured with respect to the Sun. The velocity V_1 designates the velocity of the vehicle as it enters the circumsolar transfer orbit and V_0 is the velocity of the planet with which the vehicle moves around the Sun (in this case, the Earth). The problem is now, to find the effect of errors in v_h on the heliocentric distance R_A or R_P (or, more generally, on any other distance R at the true anomaly $\eta_\odot$ of the circumsolar orbit). The parabolic velocity v_p is given, within the accuracy with which $K_{\oplus☾}$ is known, by the distance of the cut-off point from the Earth. We assume for a moment that this value is known exactly (actually this is not true). Then all that is to be varied is v_h. Assume now that for a given transfer case a value v_h' yields the correct circumsolar transfer impulse $\Delta V'$; then, in terms of v_0,

$$\left(\frac{\Delta V'}{V_0}\right)^2 = \left(\frac{v_h'}{V_0}\right)^2 - \left(\frac{v_p}{V_0}\right)^2. \tag{45}$$

An actual, slightly erroneous cut-off velocity v_h'' yields the erroneous impulse

$$\left(\frac{\Delta V''}{V_0}\right)^2 = \left(\frac{v_h''}{V_0}\right)^2 - \left(\frac{v_p}{V_0}\right)^2. \tag{46}$$

Dividing these two equations, eliminating V_0 on the righthand side and dividing numerator and denominator of this part by v_c, the circumterrestrial circular velocity at the cut-off point, and replacing ΔV by $V_1 - V_0$, one obtains

$$\frac{\left(\frac{V_1''}{V_0} - 1\right)^2}{\left(\frac{V_1'}{V_0} - 1\right)^2} = \frac{\left(\frac{v_h''}{v_c}\right)^2 - 2}{\left(\frac{v_h'}{v_c}\right)^2 - 2}. \tag{47}$$

Assuming now the simplified case that the planetary orbits are circular, V_0 becomes V_c. Considering only minimum energy transfer ellipses (apsis to apsis, cotangential), V_1 is either equal to the perihelion velocity V_P or to the aphelion velocity V_A of the transfer orbit,

$$\left.\begin{aligned} \frac{V_P}{V_c} &= \sqrt{\frac{2 R_A}{R_P + R_A}} \\ \frac{V_A}{V_c} &= \sqrt{\frac{2 R_P}{R_P + R_A}} \end{aligned}\right\}. \tag{48}$$

Therewith Eq. (45) or (46) becomes

$$\frac{V_{A,P}}{V_c} = 1 + \sqrt{\left(\frac{v_h}{V_c}\right)^2 - \left(\frac{v_p}{V_c}\right)^2}\,. \tag{49}$$

Eq. (47) provides a relation for computing the effect of an error in v_h on the initial transfer velocity V_1 with respect to the Sun. In order to correlate our error in v_h with the resultant error in R at the destination we can write Eq. (37) for the heliocentric case

$$\frac{R}{R_1} = \frac{(V_a/V_c)^2}{1 + X\,(V)\cos\eta_\odot} \tag{50}$$

where $X(V)$ has the same definition as given by Eq. (38), replacing v by V. The azimuthal component is given by $V_a^2 = V_1^2 - V_r^2$. For a correct apsidal transfer it is $V_r^2 = 0$ and $V_a = V_1 = V_A$ or $_P$. From Figs. 6 and 7, it is seen that the effect of a radial velocity component as large as 5 percent of circular velocity (in this case approximately $0.05 \cdot 100{,}000 = 5{,}000$ feet per second) is negligible compared to errors in the azimuthal velocity component. Putting therefore, in the first approximation, $V_r = 0$ one obtains $X(V) = 1 - V_1/V_c$ and

$$\frac{R}{R_1} = \frac{(V_a/V_c)^2}{1 + \cos\eta_\odot\,[1 - (V_1/V_c)^2]} \tag{51}$$

or, since according to Eq. (49)

$$\frac{v_1}{v_c} = 1 + \sqrt{\left(\frac{v_h}{V_c}\right)^2 - \left(\frac{v_p}{V_c}\right)^2} \tag{52}$$

one can write, putting $v_h/V_c = v_h{}^*$ and $v_p/V_c = v_p{}^*$,

$$\frac{R}{R_1} = \frac{(1 + \sqrt{v_h{}^{*2} - v_p{}^{*2}})^2}{1 + \cos\eta_\odot\,[1 - (1 + \sqrt{v_h{}^{*2} - v_p{}^{*2}})]}\,. \tag{53}$$

The value of R for two different hyperbolic volocities of departure is therefore,

$$\frac{R''}{R'} = \left(\frac{1 + \sqrt{(v_h{}^{*\prime\prime})^2 - v_p{}^{*2}}}{1 + \sqrt{(v_h{}^{*\prime})^2 - v_p{}^{*2}}}\right)^2 \frac{1 + \cos\eta_\odot\,\{1 - [1 + \sqrt{(v_h{}^{*\prime})^2 - v_p{}^{*2}}]^2\}}{1 + \cos\eta_\odot\,\{1 - [1 + \sqrt{(v_h{}^{*\prime\prime})^2 - v_p{}^{*2}}]^2\}} \tag{54}$$

or since $\sqrt{v_h{}^{*2} - v_p{}^{*2}} = \Delta V/V_c = \Delta V^*$

$$\frac{R''}{R'} = \left(\frac{1 + \Delta V^{*\prime\prime}}{1 + \Delta V^{*\prime}}\right)^2 \frac{1 + \cos\eta_\odot\,[1 - (1 + \Delta V^{*\prime})^2]}{1 + \cos\eta_\odot\,[1 - (1 + \Delta V^{*\prime\prime})^2]} \tag{55}$$

where

$$\Delta V = \sqrt{\frac{K_\odot}{R_1}}\left(\sqrt{\frac{2R_{P,A}}{R_P + R_A}} - 1\right) = V_c\left(\sqrt{\frac{2R_{P,A}}{R_P + R_A}} - 1\right). \tag{56}$$

$R_{P,A}$ meaning that either R_P or R_A should be used, depending on whether the target orbit lies inside or outside the Earth's orbit. With V_c is meant here the circular velocity of the Earth at 1 A.U. ($V_c = 97{,}770$ feet per second).

Eq. (54) and (55) are valid for any difference $v_h{}^{*\prime\prime} - v_h{}^{*\prime}$. The error sensitivity is again given by the gradient $d(R/R_1)/d\,v_h{}^*$. We restrict ourselves presently to the case $V_r/V_c = 0$. Then, by differentiating Eq. (53) one obtains

$$\frac{d\left(\frac{R}{R_1}\right)}{d\,v_h{}^*} = 2\,v_h{}^*\,\frac{1 + \Delta V_{st}{}^*}{\Delta V_{st}{}^*}\,\frac{\{1 + \cos\eta_\odot\,[1 - (1 + \Delta V_{st}{}^*)^2]\} + \cos\eta_\odot\,(1 + \Delta V_{st}{}^*)^2}{\{1 + \cos\eta_\odot\,[1 - (1 + \Delta V_{st}{}^*)^2]\}^2} \tag{57}$$

whence, with $R_1 = R_\oplus$

$$(\Delta R)_{\eta_\odot} = \frac{d\left(\frac{R}{R_1}\right)}{d\, v_h^*}\, R_\odot\, \frac{\Delta\, v_h}{V_c} \tag{58}$$

or, if $\cos\eta_\odot = 1$, the apsidal shift is given by, for departure from the Earth's orbit,

$$\frac{d\left(\frac{R_{apsis}}{R_\oplus}\right)}{d\, v_h^*} = \frac{1+\Delta\, V_{st}^*}{\Delta\, V_{st}^*}\, \frac{4\, v_h^*}{[2-(1+\Delta\, V_{st}^*)^2]^2} \tag{59}$$

or, since $\Delta V_{st} = V_1 - V_c$, and $V_1^* = V_1/V_c$,

$$\frac{d\left(\frac{R_{apsis}}{R_\oplus}\right)}{d\, v_h^*} = \frac{V_1^*}{V_1^*-1}\, \frac{4\, v_h^*}{(2-V_1^{*2})^2}\,. \tag{60}$$

Since $V_1^* = \sqrt{2}$ for parabolic departure from the solar system, Eq. (60) shows that the error sensitivity in this case becomes infinitely large, a result which matches the single force field equations. The value V_1'' is given by

$$V_1^* = \frac{V_1}{V_c} = \sqrt{\frac{2\, R_{P,A}}{R_P + R_A}}\,. \tag{61}$$

The displacement of the apsis for a given error Δv_h in the hyperbolic departure velocity from the Earth is thus

$$\Delta R_{apsis} = \frac{d\left(\frac{R_{apsis}}{R_\oplus}\right)}{d\, v_h^*}\, R_\oplus\, \frac{\Delta\, v_h}{V_c}\,. \tag{62}$$

Specifically, the value of ΔR_{apsis} for $\Delta v_h = 1$ ft/sec is, with $R_\oplus = 80.8184 \cdot 10^6$ n.mi. and $\Delta v_h / V_c = 1/97{,}770$,

$$\Delta R_{apsis} = 826.6179\, \frac{d\left(\frac{R_{apsis}}{R_\oplus}\right)}{d\, v_h^*} \left[\frac{\text{n.mi.}}{\text{ft/sec}}\right]. \tag{63}$$

Eq. (59) or (60) is plotted in Fig. 9, showing the differential as function of the heliocentric distance of the target apsis. As can be seen immadiately from Eq. (60), the differential becomes infinite if $V_1^* = 1$. This means, for small changes of the artificial comet's orbit from the orbit of the Earth, the error sensitivity is extremely high. It falls off quickly, passes through a minimum and then climbs again. The right hand branch reaches infinity for $V_1^* = \sqrt{2}$, while the left branch eventually decreases to zero for vertical fall toward the center of the Sun ($V_1^* = 0$).

For interplanetary flights, accuracy requirements seem really important only of very close encounters with other planets are intended, but not for establishing a heliocentric interplanetary orbit. Therefore, only the values of the differential pertaining to the planetary orbits are of primary significance. These values are listed for the planets Mercury to Uranus in Table II, together with other pertinent data. It is seen that an error of 1 ft/sec at cut-off at 300 n.mi. altitude above the Earth results in a miss of the (theoretical) center of Venus by about 23,000 n.mi. Since the diameter of Venus is about 6,700 n.mi., the cut-off accuracy theoretically required for a collision with Venus is 0.29 ft/sec at a cut-off velocity of $v_h = 0.3695 \cdot 97{,}770 = 36{,}126$ ft/sec. For Mars, ΔR_{apsis} is slightly smaller; but due to the planet's smaller diameter, the theoretical cut-off velocity must be as high as 0.16 ft/sec out of 36,507 ft/sec.

It should be emphasized, however, that these values mean little, since a realistic appraisal must take the focussing gravitational pull of the target planet into account. Furthermore, it is not likely in practice, to send a probe to another planet without designing into it the capability of small corrective impulses which can be applied after an intermediate orbit has been established in cislunar space

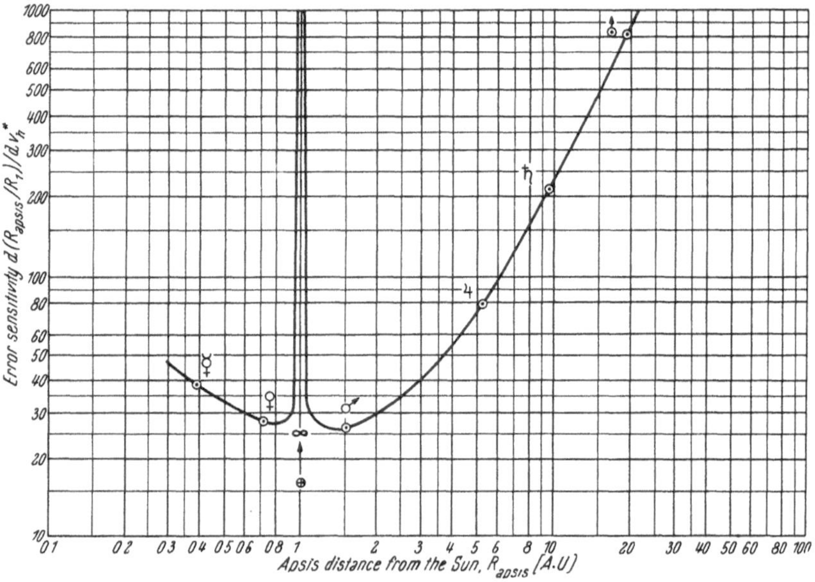

Fig. 9. Error sensitivity of a two-field transfer (Earth to Planet)

Table II. *Error Sensitivity of Interplanetary Transfer from the Earth (without attraction from Target Planet)*

1	2	3	4	5	6
Target Planet	a A.U.	$\Delta V_\infty{}^* = \dfrac{\sqrt{apsis}}{\sqrt{c,\oplus}} - 1$	$v_h{}^* = \dfrac{v_h}{\sqrt{c,\oplus}}$	$\dfrac{d\left(\dfrac{R_{apsis}}{R_1}\right)}{dv_h{}^*}$	$\Delta R_{apsis}/\Delta v_h$ n.mi./(ft/sec)
Mercury . . .	0.3871	0.25291	0.4401	47.12	38,950
Venus	0.7233	0.08377	0.3695	27.98	23,129
Mars	1.5237	0.09886	0.3734	26.43	21,847
Jupiter	5.2028	0.29520	0.4656	78.60	64,968
Saturn	9.5388	0.34514	0.4988	214.04	176,925
Uranus	19.1820	0.3794	0.5226	802.68	663,512

Col. 2 — mean distance of planet in A.U.

Reference Values:

Circular velocity at mean distance of Earth, $\sqrt{c,\oplus} =$ 97,770 ft/sec
Mean distance of the Earth, 1 A.U. $= 80.8184 \cdot 10^6$ n.mi.
Gravitational parameter of the Sun, $K_\odot = 2.0925 \cdot 10^{10}$ n.mi.^{3}/sec^2
Gravitational parameter of the Earth, $K_\odot = 62{,}767 \cdot 7$ n.mi.^{3}/sec^2
Altitude of departure of instrumented comet, $y = 300$ n.mi.
Corresponding parabolic velocity, $v_p = 35{,}208$ ft/sec

and determined accurately. A final correction may be applied after the vehicle has entered translunar space and is about to begin its heliocentric transfer orbit. At this point, it will be remembered, the error sensitivity is reduced to 2,000 n.mi. and 6,000 n.mi., respectively, for Venus and Mars. This permits for Venus an error of $\pm$ 1.5 ft/sec for collision, or even more if Venus' attraction is taken into account. A theoretical analysis of the focussing effect of the target body will be presented in a separate study. Special cases, for Venus, Mars and Jupiter, will be discussed below.

It is realized, of course, that corrections after departure from the immediate vicinity of the Earth, in fact, pass the problem on to the field of tracking, i.e. accurate measurement of position and velocity vector. This is pointed out here as a significant problem, involving even such fundamental factors as exact knowledge of the speed of light. A more detailed discussion of the tracking problems exceed the frame of this paper.

In conclusion, it should also be pointed out that the preceding analysis necessarily had to assume a set of fixed astronomical constants for numerical evaluation. It is generally recognized that the accurate values of these constants are not yet known. An authoritative discussion of this subject is left to the astronomer. However, it may briefly be noted here that the uncertainty in knowledge of the natural constants of the solar system appears to exceed, or at least equal the uncertainty in flight path to be expected on technical grounds.

For this latter reason, rather than because of lack of technical capability it is felt that early attempts for obtaining a collision course with other planets are not worthwhile. First, probes are needed which coast through interplanetary space, exploring environmental conditions and improving our knowledge of the Sun-Earth distance, of the Earth-Moon mass and of planetary masses and distances, notably those of Venus and Mars.

6. Classification of Artificial Unmanned Space Vehicles

Before entering into a discussion of individual probes we shall here name and appropriately define the pertinent systems. A comprehensive survey of automatic vehicles together with their orbital and performance specifications has been presented earlier [3]. Here, this classification will be condensed[1] and amended from the functional viewpoint, thus specifying the information quality of the probe.

As shown in Table III the classification distinguishes between 4 basic operational systems: circumterrestrial satellites, lunar probes, instrumented comets and circumplanetary satellites.

The terrestrial satellite operates in the immediate vicinity of the Earth where characteristic terrestrial effects, such as oblateness, atmosphere and electric and magnetic effects are the principal factors influencing the satellite's orbit. The cislunar satellite operates at greater distance from the Earth in the Earth-Moon field where solar and lunar perturbative forces are no longer negli-

[1] Introducing a few minor changes: In [3] instrumented comets and circumplanetary satellites were taken as two sub-groups of a main group called hyperbolic probes (with respect to the Earth). These two sub-groups are now listed as separate main groups. The term hyperbolic probe is applied to probes having a hyperbolic encounter with the target body, in contradistinction to collision or landing. The second modification is that the lunar hyperbolic probe which in [3] was included in the cislunar satellite class, is now listed under lunar probes, leaving the designation cislunar satellites exclusively for very high altitude satellites.

Table III. *Data on Silver-Coated Mylar Spheres*

$[x\ (-n) = x \cdot 10^{-n}]$

Sphere diameter (ft)	10	25	50	75	100	180	250	500	1000
Volume ($10^3 ft^3$)	0.5236	8.18	65.45	220.9	523.6	3,054	8,181	65,450	523,600
Surface area ($10^3 ft^2$)	0.314	1.961	7.855	17.67	31.42	101.79	196.1	785.5	3,142
Cross-sectional area = A (ft^2)	78.54	490.9	1.963	4.418	7.854	25.450	49.090	196.300	785.400
Weight of Mylar hull (lb)[1] (10^{-4} in thickness; 0.001 lb/ft^2)	0.314	1.961	7.855	17.67	31.42	101.79	196.1	785.5	3.142
Hydrogen gas pressure[2] (psia) (for constant skin stress of 4,000 psi)	1.3(-2)	5.3(-3)	2.65(-3)	1.76(-3)	1.3(-3)	7.4(-4)	5.3(-4)	2.65(-4)	1.3(-4)
Specific weight of H_2 (300°F = 150°C) (lb ft^{-3})	2.6(-6)	1.06(-6)	5.3(-7)	3.52(-7)	2.6(-7)	1.5(-7)	1.06(-7)	5.3(-8)	2.6(-8)
Weight of gas filling (lb)				negligible				3.5	14
Total weight + 10 % = W (lb)	0.35	2.2	8.6	19.5	35	112	216	868	3475
W/A (lb/ft^2)				4.4(—3)					
$C_D A/W$ (C_D = 2.2 (ft^2/lb)	494	490	502	498	494	500	500	498	497
Stellar magnitude m (Moon distance cf. Fig. 32)	14.3	12.3	10.8	9.9	9.2	8	7.5	5.8	4.3

[1] Density 1.39; therefore theoretical weight of 10^{-4} in thick Mylar 0.000719 lb/ft^2; rounded off to 0.001 including coating.

[2] Tensile strength of Mylar 10,000 psi at 300° F; 25,000 — 30,000 psi at 0° F.

gible. The translunar satellite moves in orbits beyond the lunar orbit. It is never exposed to the opposing influence of terrestrial and lunar gravitational forces and its orbital mechanics is therefore comparatively simpler than that of the cislunar satellite, especially at greater translunar distances.

Among the *lunar probes*, the *collision probe* which hits the Moon, and the *hyperbolic probe* which passes the Moon at close distance in a circumlunar hyperbolic orbit, probably will be the first ones to be tried out. The *lunar satellite* not only requires capability of powered flight, but also a very accurately working combination of guidance and attitude control. Still more sophisticated is the *lunar landing craft.*

Four types of instrumented comets have been listed, of which the first two are self-explanatory, while the third refers to probes which investigate the nature of space regions between the planets. More about this will be said below. Of particular interest is what might be called the *comet probe* which, at the appropriate time can be sent out to explore natural comets passing by in orbits whose nodal point is near the Earth.

The *circumplanetary satellite* represents a very advanced state of the art, in respect to many facets of the system; in fact, manned space flight may have become a reality, before one can fire a circumplanetary satellite from the Earth. They may be "planted" by manned expeditions, visiting these planets.

Functionally, one can distinguish between inert probes, telemetering probes and television probes. Powered probes have also been mentioned as a separate class, because the propulsion system would be their distinguishing mark. Since they would require guidance and attitude control in any case, their payload would almost certainly contain TV and would in general be of a quite advanced type, especially in regard to auxiliary power supply. Several applications and characteristics of TM and TV-probes for Venus and Mars have been discussed in [3]. Subsequently we will therefore mainly deal with thermonuclear probe and tracer body.

7. Flight Mechanics of the Lunar Probe

Various aspects of the lunar probe pertaining to performance, flight mechanics and its potential as a research instrument have been discussed in [2], [3], [4], [5], [10], [11], [12], and [13]. The lunar probes can be regarded as the forerunner of the instrumented comet.

The Moon rocket is often considered as an example for the restricted 3-body problem. This is roughly correct as far as the Earth-Moon transfer path is concerned. If one considers also the return path, the Sun's perturbation cannot be reglected a priori; because of the before mentioned sensitivity of the departure path to even small changes in the approach path during a hyperbolic encounter. One therefore deals with a restricted 4-body problem. Removal of the vehicle from the Earth-Moon barycenter, freely falling in its circumsolar orbit, produces a torque caused by the gradient of the solar field. This results in an acceleration of the vehicle which is proportional to the angular velocity of the barycenter about the Sun, ω, and to the distance of the vehicle from the barycenter, r, viz. $a = 2\, r\omega^2$. The torque is of course present, independent of the Moon's existence (in which case the barycenter would be the Earth's center) and would change the vehicle's motion about the Moon-less Earth as compared to motion about an isolated Earth. In addition, the Moon changes the effective local gradient of the terrestrial field. An analysis of the effect of the Sun's gravitational field on the lunar probe has been presented in [5]. The perturbing effect of the Sun is greatest if it is located in the direction of the Earth-Moon axis. In most cases, the solar effect therefore, cannot be neglected if the return path to the Earth is of interest. Maximum velocity changes due to the Sun are about 10 ft/sec. This suffices to alter the orbit completely compared to a 3-body orbit during and after the hyperbolic encounter with the Moon.

An example is shown in Figs. 10 and 11 [5]. The Sun stands in the extension of the positive x-axis, in order to have the reverse side of the Moon illuminated.

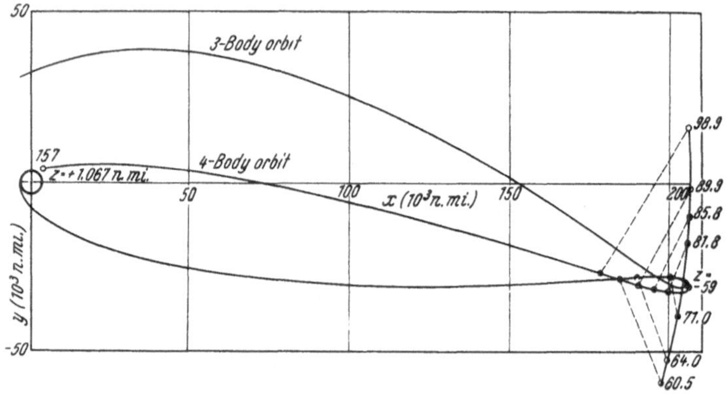

Fig. 10. Lunar orbit with and without consideration of the Sun

The motion is based on a coordinate system, rotating with the Earth-Moon axis. Without the Sun the vehicle would approach the Moon to a distance of $\varrho_{min} = 992$ n. mi., i.e. it would hit the surface. Due to the Sun's field, $\varrho_{min} = 2{,}231$ n. mi. From the discussion of gravitational navigation before, it will be apparent that such difference must change greatly the direction of the asymptotes of the encounter hyperbola so that, instead of returning to the Earth at a perigee distance of about 30,000 n.mi., it actually re-enters the atmosphere. A discussion of the work on Earth-Moon orbits has been presented in [5] and only the most important results will be given here.

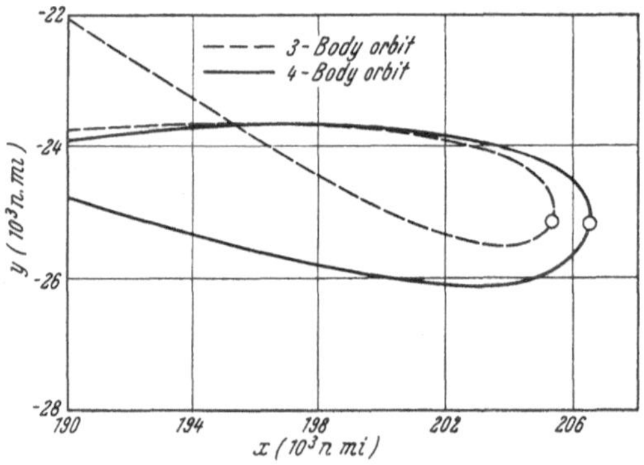

Fig. 11. Lunar orbit with and without consideration of the Sun. With ○ $\varrho_{min} = 2{,}231$ n.mi. (15.6 hours); without ○ $\varrho_{min} = 922$ n.mi. (15.6 hours)

Systematic 4-body computations in three dimensions for a given "Standard" condition, defined below reveal an error pattern as shown in Fig. 12. The path lies in the Earth-Moon plane. The Moon's orbit is assumed circular. The curves shown relate the variation in minimum distance during lunar encounter with deviations in x, y, and z from a standard case $\dot{y} = 34{,}970.53$ ft/sec, $\dot{x} = 0$, $\dot{z} = 0$ from a 300 n. mi. high circular satellite orbit. All basic trends specified in the preceding section are found here again: The reduced sensitivity against errors in v_a $(\delta\varrho/\delta y)$ compared to the circumterrestrial elliptical case $(\delta r_A/\delta y$; dashed line) due to the focussing effect of lunar attraction; the comparatively much smaller effect of a radial velocity error on ϱ $(\delta\varrho/\delta x)$; and the relatively minute effect of an orthogonal impulse even if its magnitude is several hundred feet per second. The most significant modification of these trends is the greater magnitude of a radial velocity error than might have been expected on the basis of Figs. 6 and 7. The difference is caused solely by the Moon. Since a radial component rotates the major axis [Eq. (33)], the vehicle is brought under lunar influence at different angles ϑ (Fig. 2) and therefore its orbit is changed differently during the encounter. It is interesting to note that nevertheless the vehicle returns into the atmosphere over a wide range of $\delta\varrho/\delta x$—values. Under standard conditions the vehicle would approach the Moon to $\varrho_{min} = 2{,}231$ n. mi. An excess in v_a at cut-off ($\Delta\, y$) quickly increases the minimum distance. The curve for v_a (i.e. for $\dot{y}_0$) consists of two parts. The solid line applies to retrograde orbits around the Moon, i.e., the type shown in Fig. 10. The dashed part pertains to direct orbits, the type shown in Fig. 13. In the latter case hyperbolic escape

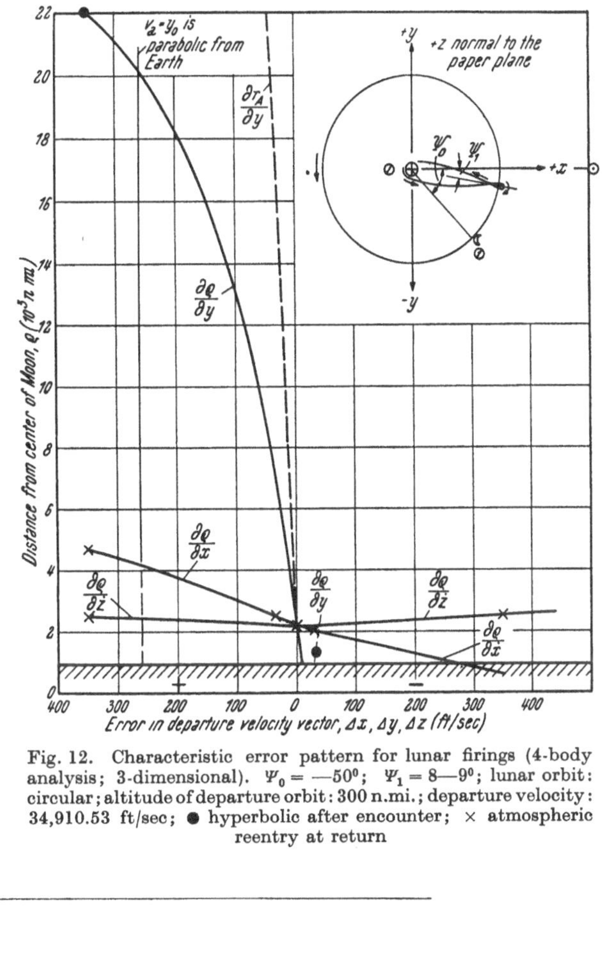

Fig. 12. Characteristic error pattern for lunar firings (4-body analysis; 3-dimensional). $\Psi_0 = -50^0$; $\Psi_1 = 8$—9^0; lunar orbit: circular; altitude of departure orbit: 300 n.mi.; departure velocity: 34,910.53 ft/sec; ● hyperbolic after encounter; × atmospheric reentry at return

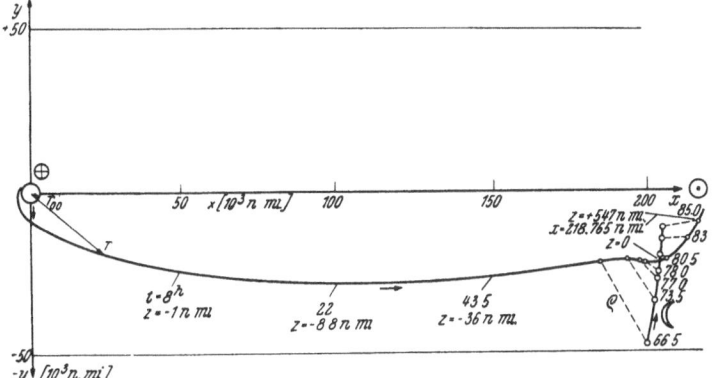

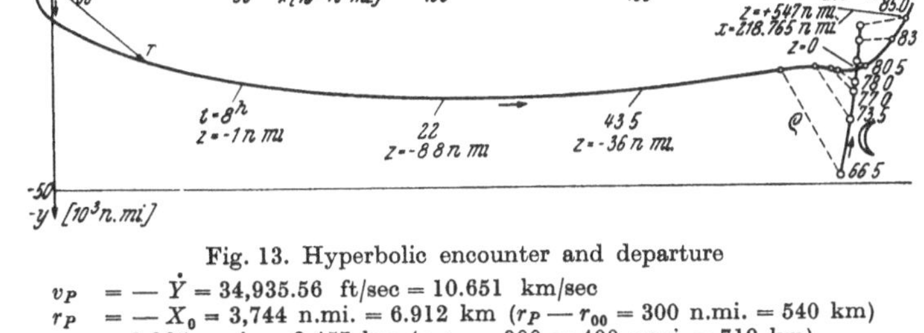

Fig. 13. Hyperbolic encounter and departure

$v_P \;= -\dot{Y} = 34{,}935.56$ ft/sec $= 10.651$ km/sec

$r_P \;= -X_0 = 3{,}744$ n.mi. $= 6.912$ km ($r_P - r_{00} = 300$ n.mi. $= 540$ km)

$\varrho_{min} = 1{,}365$ n.mi. $= 2{,}457$ km ($\varrho_{min} - 900 \simeq 400$ n.mi. $= 719$ km)

hyperbolic excess at $t = 85.0$h: $\dfrac{v_\infty}{\sqrt{\dfrac{2\,K_\oplus}{r_{85.0}}}} = 0.82389$ ($v_\infty = 1{,}054$ ft/sec $= 0.321$ km/sec) from the Earth-Moon system will result, if the encounter is sufficiently close. Large radial velocity errors are permissible until the vehicle hits the Moon. For the present case the results indicate considerable tolerance in radial and orthogonal velocity errors, but little tolerance in v_a[1]. Between $\Delta v_a = -10$ ft/sec and -30 ft/sec, lunar collision is indicated. At $\Delta v_a = -35$ ft/sec the orbit becomes direct, with hyperbolic escape. Thus, for hitting the Moon a tolerance in Δv_a of only 20 ft/sec is indicated in this case. For hitting the Earth a considerable smaller tolerance in Δv_a was indicated also, since already deviations of $\pm \Delta v_a = 35$ ft/sec resulted in no re-entry upon return. In fact, at $\Delta v_a = -35$ ft/sec the circumterrestrial orbit became hyperbolic after the encounter, as indicated in Fig. 12. The other hyperbolic encounter was found under conditions of slightly hyperbolic departure ($\Delta v_a = +350$ ft/sec; $e = 1.001$).

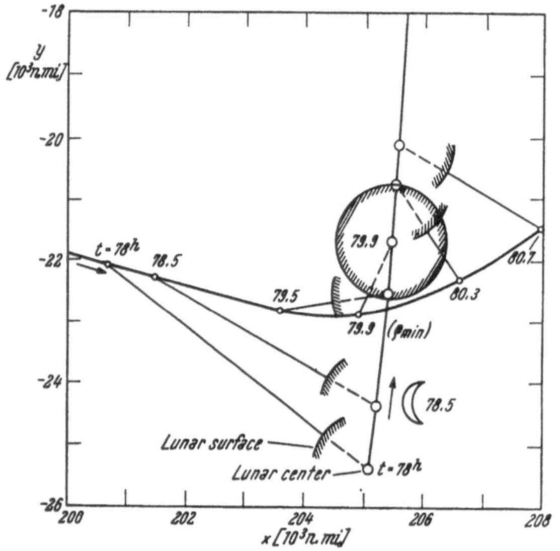

Fig. 14. Close passage of Moon

The two hyperbolic orbits are shown here, because of their interest for interplanetary research. Figs. 13 and 14 present the low velocity case. Here the vehicle's intersection with the lunar orbit leads it to a great proximity (about 400 n.mi. altitude) at about the 80th flight hour. The result is a hyperbola of the type computed before in Sect. 4. It is an almost grazing passage and perhaps as close as one would like to get in practice with an instrumented comet. The high speed case (Fig. 15) is interesting, because the Moon's effect almost cancels itself before and after encounter, thus leading to an almost straight line departure at near-parabolic velocity. Since the Sun is at $x = +\infty$, the vehicle departs in the direction of the inner solar system. The eccentricity of the circumsolar orbit ($e = 0.080168$) is about five times that of the Earth (0.016). Therefore the vehicle approaches the orbit of Venus slightly more than half way from the Earth, while its apogee stays in the immediate vicinity of the Earth's orbit. The orbit is shown in Fig. 16. The period is 322.7^d. It therefore requires theoretically 8.39 years for the probe and the Earth to meet again in relatively close encounter (about 500,000 n.mi.). Actually the time will be different, because of perturbation by Venus and Earth. During the encounter the vehicle could be re-captured by the Earth or thrown into a quite different orbit.

In the low-speed case (Figs. 13 and 14) the position of the vehicle at the end of computation is still too close to Earth to use the osculating elements

[1] During the Astronautics Symposium in San Diego, February 1957, H. LIESKE from RAND quoted a tolerance of ± 75 ft/sec for hitting the Moon and ± 4 ft/sec for hitting an area 1000 mi. in diameter. For hitting the Earth ± 150 ft/sec tolerance was given. However, these data apply to different "Standard" conditions. In general, it is useless to give tolerances without specifying the "Standard" conditions with respect to which these tolerances are measured.

at $t = 85.0^h$ for a determination of the unperturbed circumsolar orbit. In a rigorous sense, even in the high-speed case the osculating elements at $t = 73.5^h$ are not exactly those of the circumsolar orbit, mainly because of the effect of the combined Earth-Moon mass. In fact, the vehicle has not even left yet the activity

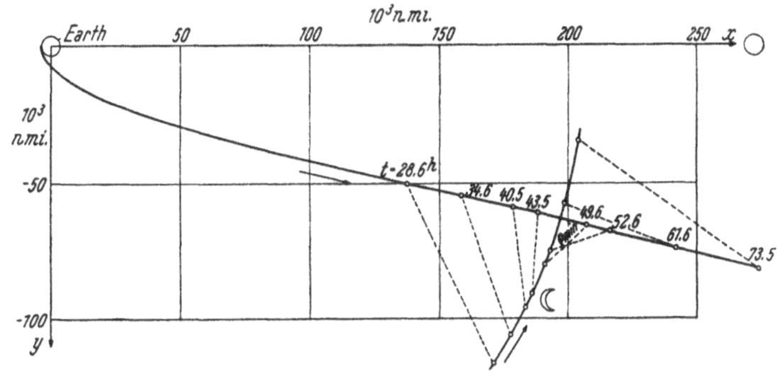

Fig. 15. Hyperbolic departure with Moon cancelling its gravitational effect. $v_P = 35{,}320.228$ ft/sec; $\varrho_{min} = 22{,}115$ n.mi.; $R_{P\odot} = 0.86$ A.U.; $e_\odot = 0.08$; $e_\oplus = 1.001$

sphere of the Earth-Moon system [4] which extends to a distance of 500,000 n.mi. (925,000 km) and within which, as LAPLACE has shown, the perturbation is comparatively so strong that the Earth (or planet) should be regarded as center of attraction rather than the Sun which, within this sphere, is reduced to the role of a perturbative force. TISSERAND [14] has shown that the radius of the sphere d'activité is with good approximation given by the relation

$$r_{act} = R\left(\frac{m}{M}\right)^{\frac{2}{5}} \tag{42}$$

where R is the distance of the planet m from the Sun M. The activity sphere is not to be understood as defining the limits within which the Earth can hold a satellite. It merely states that, for a body passing hyperbolically through this sphere, the ratio of center force to perturbative force in an Earth-fixed coordinate system is greater than the ratio of center force to perturbative force in a Sun-fixed coordinate system. The orbital computations are described in [5].

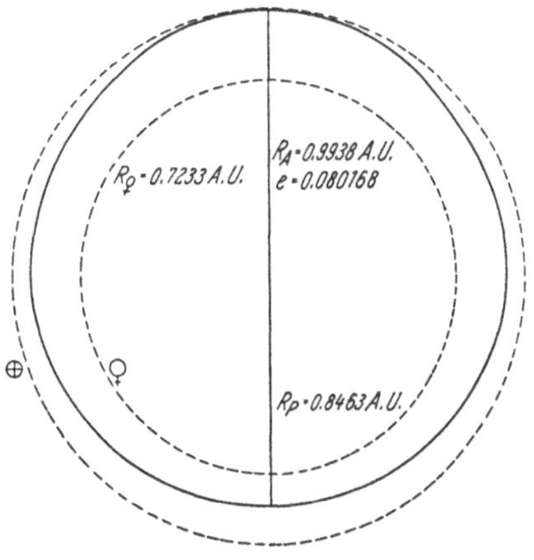

Fig. 16. Circumsolar orbit resulting from departure along orbit in Fig. 15

8. Flight Paths of Instrumented Comets to Venus, Mars and Jupiter

A few preliminary orbits were computed, in order to assess the effect of perturbations and the stay times and altitudes near the planet. In these calculations the planet orbits were assumed circular and in one plane. The value used for the Sun's gravitational parameter is

$$K_\odot = 1.3291976 \cdot 10^{11}\,\frac{\text{km}^3}{\text{sec}^2} = 2.0925055 \cdot 10^{10}\,\frac{\text{n.mi.}^3}{\text{sec}^2}. \tag{64}$$

The Earth's orbit was taken as semi-major axis $R_\oplus = 80.8184285 \cdot 10^6$ n.mi. $= 1.$O.A.U. In all cases the circumsolar transfer orbits were cotangential ellipses with apsides in the Earth's orbit and the orbit of the target planet (Fig. 17). The information required to read the subsequent graphs is presented in Fig. 17. The motion of the comet in cartesian coordinates was obtained by integration of the equations

$$x = -K_\odot \frac{x}{R^3} - K_1 \frac{x - x_1}{R_1^{\,3}}$$

$$y = -K_\odot \frac{y}{R^3} - K_1 \frac{y - y_1}{R_1^{\,3}} \qquad (65)$$

where x, y, z are the vehicle coordinates, x_1, y_1 the coordinates of the target planet. The distance of the vehicle from the Sun is $R^2 = x^2 + y^2$, the position of the target planet is $x_1 = R_1 \cos \varphi$, $y_1 = R_1 \sin \varphi$ where $\varphi = \varphi_0 + \mu\, t$, μ being the mean angular motion of the target planet and R_1 the distance of the target planet from the Sun. The orbit was computed as invariant conic until the position (R_1, η_0) was reached, with (R_1, φ_0) defining the simultaneous position of the target planet. At that time $t = t_0 = 0^h$ the perturbation was introduced. In each case the initial distance from the planet was $2 \cdot 10^6$ n.mi. or greater; in any case far beyond the activity sphere of the planet. Initial perturbation was negligible in all cases except in the case of Jupiter where the initial distance was almost 10^8 n.mi. (activity sphere radius $\sim 0.27 \cdot 10^8$ n.mi.) and where perturbation was noticeable from the start. After beginning of perturbation the following osculating elements were obtained by means of standard relations: orbital energy h, eccentricity e, semi-major axis a, parameter p, period T, time since perigee t_P, true anomaly η, perihelion and aphelion distance R_P, R_A and the area constant C.

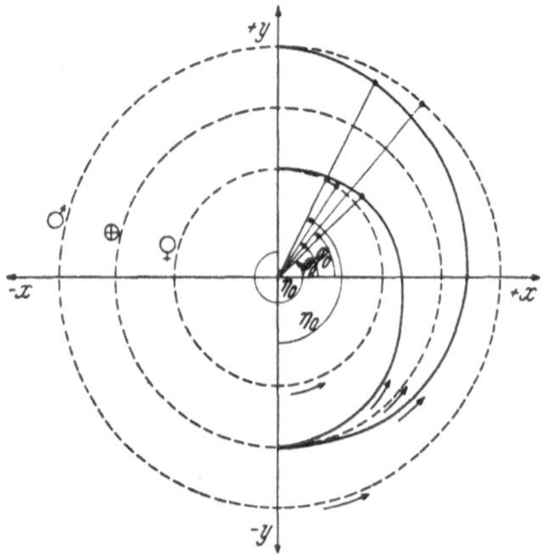

Fig. 17. Nomenclature for orbit calculations of instrumented comets. ● Beginning of perturbation calculation ($t = 0^h$)

Venus

The pertinent data for the circumsolar orbit to Venus are

$$R_♀ = 0.723332 \text{ A.U.} = 58.45855 \cdot 10^6 \text{ n.mi.}$$

$$R_P = 0.723332 \cdot R_\oplus + 2(r_{00})_♀ = 58.46525 \cdot 10^6 \text{ n.mi.}$$

where $2(r_{00})_♀$ is the diameter of Venus. The vehicle would therefore, if undisturbed, pass at an altitude of one radius above the surface.

$$K_♀ = 51{,}204 \text{ n.mi.}^3 \sec^2 = 325{,}256 \text{ km}^3/\text{sec}^2$$

mean daily motion $\mu_♀ = 5{,}767.''670$; apogee velocity $V_A = 14.7422$ n.mi./sec eccentricity $e_\odot = 0.1604866$; perturbation was introduced at $\eta = \eta_0 = 330°$ when $R = 59.56889 \cdot 10^6$ n.mi., or the distance from Venus about $2 \cdot 10^6$ n.mi. Of the total flight time of $T/2 = 146.0858$ days to the perihelion, at $\eta_0 = 330°$ only 17.6046 days or 422.51 hours were left to reach the unperturbed perihelion. At that time the position angle of Venus was $\varphi_0 = 61.^\circ 7951$.

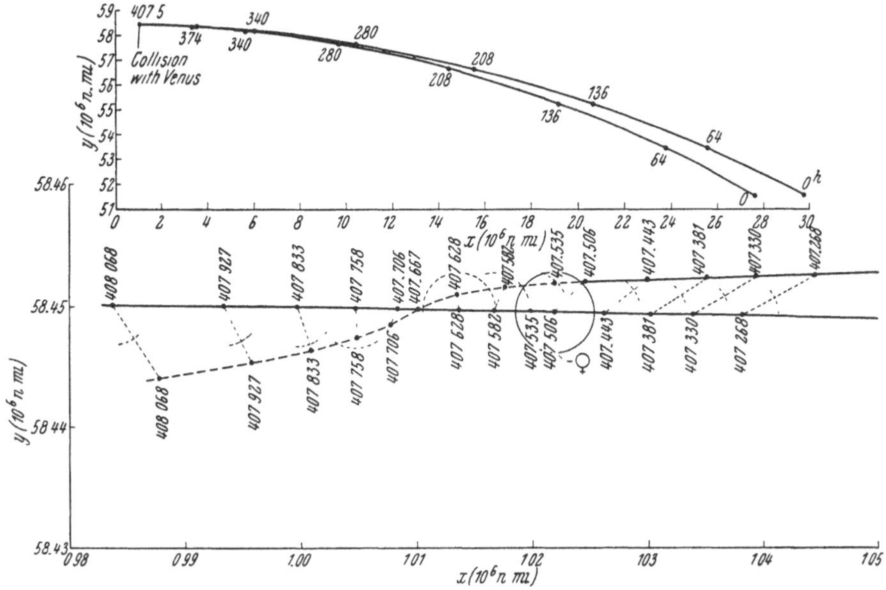

Fig. 18. Collision of instrumented comet with Venus
Venus: $R_P = 58{,}465{,}495$ n.mi.; $R_A = 80{,}818{,}246$ n.mi.; $\varphi_0 = 61.^\circ 795099$; $\eta_0 = 330^\circ$; t = hours

The subsequent encounter is shown in Fig. 18. It turned out to be a collision. The upper part of Fig. 18 shows the path from beginning of the perturbation

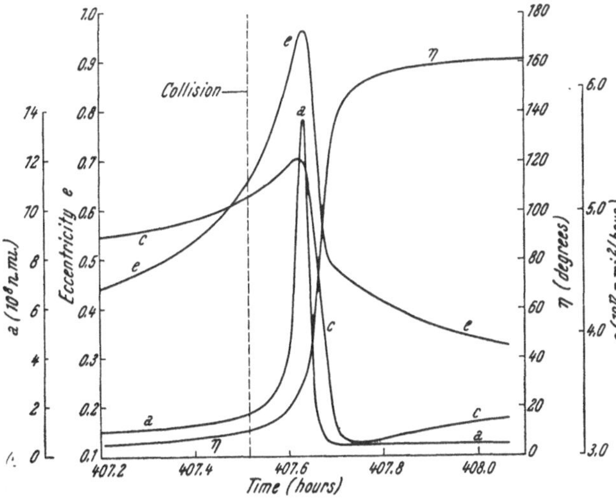

Fig. 19. Osculating orbital elements of vehicle's collision orbit with Venus
Venus: $R_P = 58{,}465{,}495$ n.mi.; $R_A = 80{,}818{,}246$ n.mi.; $\varphi_0 = 61.^\circ 795099$; $\eta_0 = 330^\circ$; t = hours

to collision. The lower portion shows the collision process in more detail. The collision is eccentric and it is obvious that a very small change in the initial course would have led to a close passage. The comet theoretically would have intersected the orbit of Venus at 407.667 hours, i.e. about 15 hours before the theoretical perihelion. The x-position at that time is $1.01 \cdot 10^6$ n.mi. ahead of the theoretical perihelion. In Fig. 19 which shows the osculating elements of the collision path, it is of interest to note that $e_{\odot} < 1$ at all times, although the vehicle "passes" almost through the center of Venus. Under no practical conditions, therefore, can Venus throw an artificial comet out of the solar system. Since $e_{\odot}$ after the encounter tends to be less than before the encounter, the subsequent circumsolar orbit would actually have been more circular than the original transfer path.

In Fig. 20 the same path was flown, except that $\varphi_0 = 61.°5$ instead of $61.°7951$. The difference of 1,080″ corresponds to a change in take-off time of 1,080/5,767.670 =

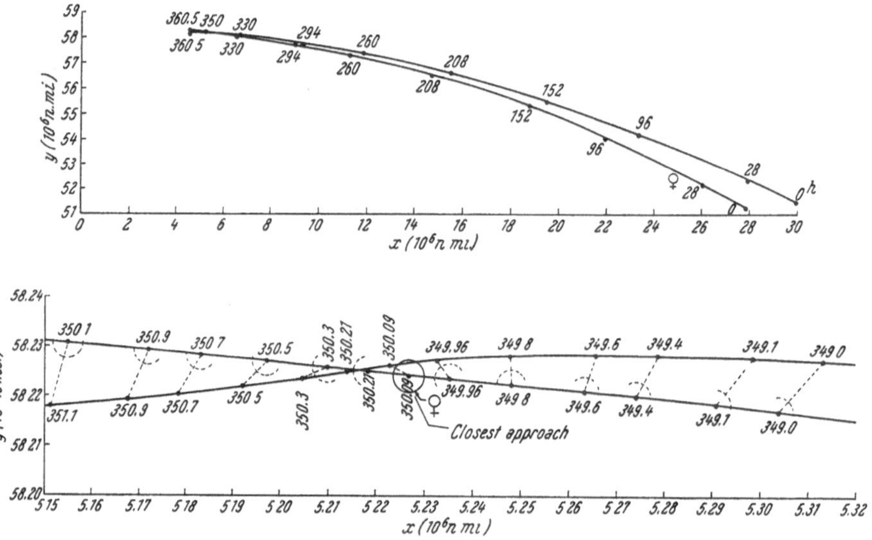

Fig. 20. Close encounter of instrumented comet with Venus
Venus: $R_P = 58{,}452{,}374$ n.mi.; $R_A = 80{,}819{,}785$ n.mi.; $\varphi_0 = 61.°5$; $\eta_0 = 330°$; t = hours

$= 0.18725\, d = 4.5$ hours. In this case, the vehicle passes at 350.09 hours over Venus at a minimum altitude of 1,118 n.mi. The comet passes the 10,000 n.mi. altitude line at 349.05, being over the night side of the planet. At 350.21, or 1.16 hour $\simeq$ 70 min. later, it crosses the terminator and swings over to the sunlit side. The altitude at that time is 1,250 n.mi. At 350.5 or 1.45 hours after passing the 10,000 n.mi. altitude line, the comet has completed about one-half revolution around Venus. Its altitude has now increased to 4,000 n.mi. and grows quickly, passing again the 10,000 n.mi. mark at 351.1 or 2.05 hours after passing it on the way toward Venus. During this period a revolution of 202 degrees about Venus was made, of which 67.5° was over the sunlit side. The selection of a 10,000 n.mi. altitude mark is representative for that portion of the path which might be called "close" to the target planet. This flight was particularly beautiful among those computed. The osculating elements in Fig. 21 indicate again a change toward greater circularity the of post-encounter orbit. Of course, Fig. 20 shows that the calculation was not extended sufficiently to obtain the new heliocentric orbital elements.

Fig. 19 shows, in fact, that a "miss" of Venus by $2(r_{00})_{♀} = 6{,}700$ n.mi. still produces a collision course. Since the collision is eccentric, the allowable miss cannot be much greater. Tentatively, an allowable perihelion displacement of about $\pm$ 10,000 n.mi. with respect to the center of Venus in a perturbation-free orbit is therefore postulated. Since $\Delta R_{apsis} \simeq$ 23,000 n.mi./ft (sec) for Venus (Table II), the above displacement would therefore indicate a hyperbolic departure velocity tolerance of almost $\pm$ 0.5 ft/sec if the standard case is dead center

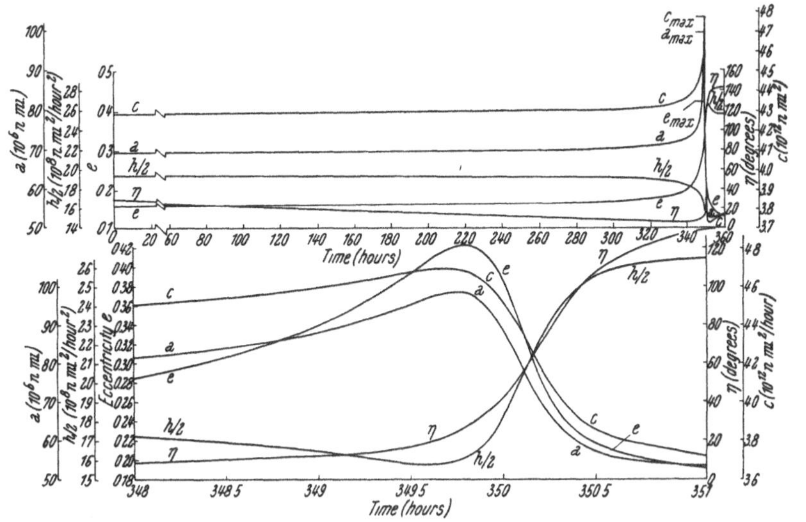

Fig. 21. Osculating elements of vehicle orbit during close encounter with Venus
Venus: R_P = 58,452,374 n.mi.; R_A = 80,819,785 n.mi.; $\varphi_0 = 61.^05$; $\eta_0 = 330^0$; t = hours

collision with Venus. This value includes the effect of Venusian attraction. Without this effect the tolerance would be $\pm$ 0.145 ft/sec based on the value given at the end of Section 5. Although, on a percentage basis, the increase is large, it still indicates a very small absolute tolerance.

Mars

The data for the circumsolar orbit to Mars:

$$R_{♂} = 1.523691 \text{ A.U.} = 123{,}142{,}312 \text{ n.mi.}$$

$$R_A = R_{♂} + 7{,}000 \text{ n.mi.} = 123{,}149{,}312 \text{ n.mi.}$$

The diameter of Mars is 3,574 n.mi. The undisturbed orbit would therefore lead the vehicle to a minimum altitude of about 5,210 n.mi. above Mars.

$$K_{♂} = 6{,}776 \text{ n.mi.}^3/\text{sec}^2 = 4.40439 \cdot 10^4 \frac{\text{km}^3}{\text{sec}^2}$$

$$\mu_{♂} = 1{,}886.''519 \text{ per day.}$$

$$\text{Perihelion velocity } V_P = 17.6819 \text{ n.mi./sec}$$

$$\text{eccentricity} = e_{\bigcirc} = 0.207537\ .$$

Perturbation was introduced at $\eta = \eta_0 = 170°$ when $R = 122.66218 \cdot 10^6$ n.mi., or the distance from Mars $2 \cdot 9 \cdot 10^6$ n.mi.[1] Of the total flight time of $T/^2. = 261.458$

[1] A greater distance was selected here than for Venus because of the slower motion at the aphelion, involving longer flight time in the vicinity of Mars.

days to the aphelion, only 21.574 days were left at that point. The corresponding position angle of Mars was $\varphi° = 78.°695$.

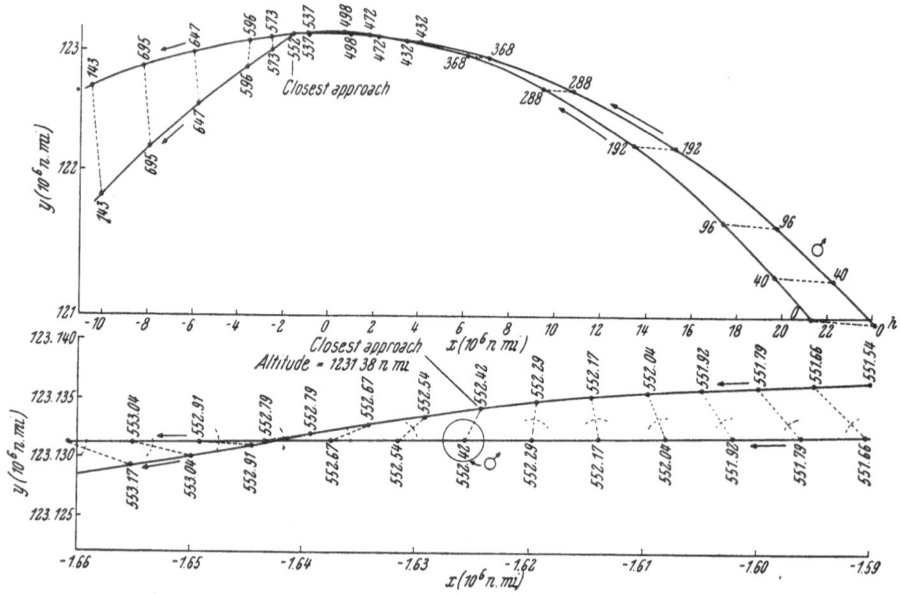

Fig. 22. Close encounter of instrumented comet with Mars
Mars: $R_P = 80{,}844{,}074$ n.mi.; $R_A = 123{,}150{,}950$ n.mi.; $\varphi_0 = 78.°694510$; $\eta_0 = 163°$; t = hours

This value of $\eta_0 \equiv 170°$ resulted in a particularly successful flight, as far as close passage is concerned, but not from the viewpoint of obervation. Fig. 22

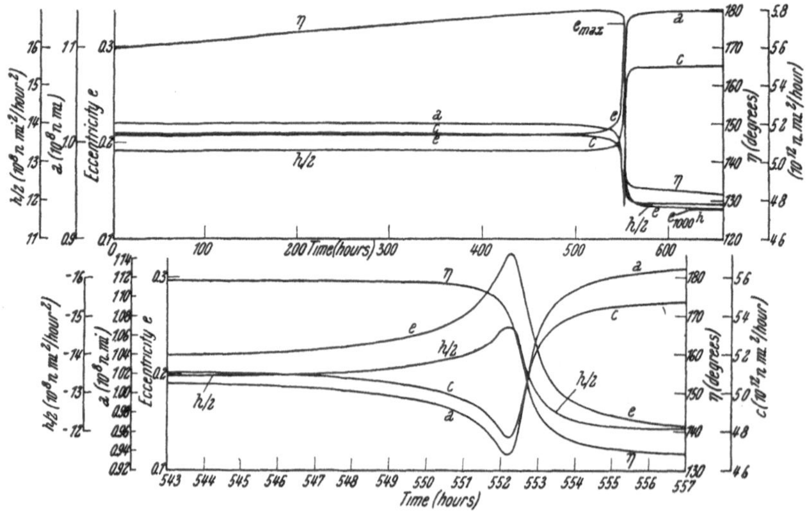

Fig. 23. Osculating elements of vehicle orbit during close encounter with Mars
Mars: $R_P = 80{,}844{,}074$ n.mi.; $R_A = 123{,}150{,}950$ n.mi.; $\varphi_0 = 78.°694510$; $\eta_0 = 163°$; t = hours

shows that after a gradual approach, which extends over more than 3 weeks, the vehicle passes the 4,000 n.mi. altitude mark at 551.79 on the night side.

At 552.41 or 0.63 hours = 38 min. later, the closest point is reached with an altitude of 1,231 n.mi., still over the night side of Mars. In fact, the vehicle moves retrograde, seen from Mars, because the planet has a higher orbital velocity. Finally, when the comet swings over to the sunlit side at 552.91, its altitude is 2,900 n.mi. The vehicle is over the subsolar point of Mars at 573, but then it is 550,000 n.mi. away. Clearly, the comet's path should have led inside the Martian orbit. Such an orbit was also computed, but resulted in a greater distance from the planet. The vehicle is inside the 10,000 n. mi. altitude line for 3 hours, about one hour longer than during the Venus passage which led to almost the same minimum altitude. Fig. 23 shows the variation of the osculating elements during the encounter. This time the path was followed long enough to compute the post-encounter circumsolar orbit which is shown in Fig. 24. It does not come close to the Earth again, except as a result of future encounters with Mars.

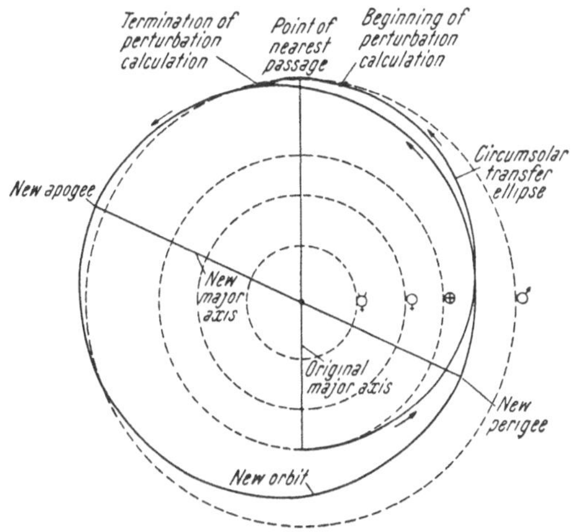

Fig. 24. Circumsolar orbits of instrumented comet before and after close encounter with Mars. ♂ Planet Mars; ● instrumented comet

The results of this flight show that the probe's perimartian point was changed from about 7,000 n.mi. to about 3,800 n.mi. distance, that is, by about 3,000 n. mi. It therefore appears that a "miss" of ± 6,000 n.mi. from the center of Mars might still result in a collision course. Since $\Delta R_{apsis} \simeq$ 22,000 n.mi./(ft/sec), this would indicate a hyperbolic departure velocity tolerance of about ± 0.27 ft/sec for collision course, if the standard case is dead-center collision with Mars. The value without Martian attraction was ± 0.08 ft/sec.

We will designate the allowable "miss" diameter as *effective collision* diameter δ. It is interesting to note that $\delta_{♂}/(r_{00})_{♀} = 12{,}000/3{,}574 = 3.4$ is larger than $\delta_{♀}/(r_{00})_{♀} = 20{,}000/6{,}700 = 3.0$, in spite of the stronger gravitational pull of Venus. The reason lies in the slower motion of the probe at the aphelion which gives Mars more time to deflect the comet's path. This result already gives an indication of what to expect in the case of Jupiter where tremendous gravitational pull and long apsidal "loitering" time are both effective.

Jupiter

As a final, somewhat theoretical example, the flight of an artificial comet to Jupiter is presented.

$$R_{♃} = 5.202803 \text{ A.U.} = 4.2048236 \cdot 10^8 \text{ n.mi.}$$

$$R_A = R_{♃} - 10^6 \text{ n.mi.} = 4.1948236 \cdot 10^8 \text{ n.mi.}$$

The undisturbed comet would therefore have passed Jupiter at a distance of 10^6 n.mi., i.e. at about the distance of J IV (Callisto), the outermost of the four Galilean moons. The effect of Jupiter's moons, incidentally, was not included in the analysis. It would have made very little difference, since the slowly moving artificial comet was literally sucked into Jupiter like a fly into the vortex of a tornado.

$$K_{♃} = 20{,}035{,}284 \text{ n.mi.}^3/\text{sec}^2 = 1.27268 \cdot 10^8 \text{ km}^3/\text{sec}^2$$

$$\mu_{♃} = 299''.128 \text{ per day}$$

$$V_P = 20{,}837 \text{ n.mi./sec}$$

$$e_{\bigcirc} = 0.67692.$$

Perturbation was introduced at $\eta = \eta_0 = 160°$ when $R = 3.72422 \cdot 10^8$ n.mi., or the distance from Jupiter $99 \cdot 10^6$ n.mi. The flight of the undisturbed vehicle

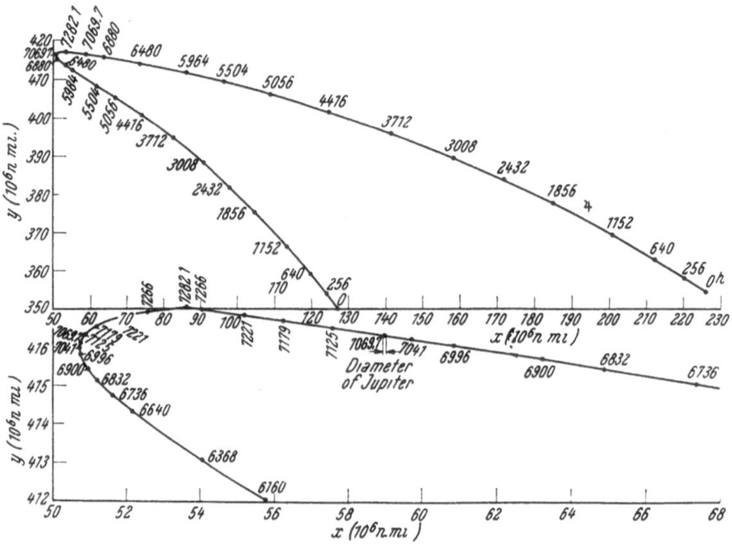

Fig. 25. Orbit of instrumented comet drawn into Jupiter
Jupiter: $R_P = 81{,}151{,}071$ n.mi.; $R_A = 419{,}766{,}470$ n.mi.; $\varphi_0 = 57.°46699$; $\eta_0 = 159.°89543$; t = hours
8760 hrs = 1 year

to R_A would be $T/2 = 994.6$ days. At $\eta = 160°$ a flight time of 391.54 days or 1.0727 years was left. At $t = 0$ Jupiter's acceleration was about $4 \cdot 10^{-7}$ g. Figs. 25 and 26 show the conditions for this flight. For 6,368 hours or 0.727 years Jupiter slowly, but irrevocably, "bent" the comet's path, until at that time an inversion point was reached. From then on reversal of the flight direction began and at 7,282.1 or 914 hours = 38 days later the comet collided almost head-on with Jupiter.

The enormous path deflection is of interest in that it emphasizes the aforementioned high tolerance in flight accuracy required. From Fig. 7, V_a/V_c for a Jupiter flight is about 1.29. According to Fig. 8 the corresponding apogee displacement factor is about 45. The Earth's velocity is close to 97,770 ft/sec and the distance $R_{\oplus} \simeq 80.8 \cdot 10^6$ n.mi. It follows, therefore, from Eq. (38) that for $V_p - V_p' = 1$ ft/sec the shift in R_A is $45 \cdot 80.8 \cdot 10^6/97{,}110$ or $3.1 \cdot 10^4$ n.mi., i.e. 6.1 times as much as for Mars. Fig. 25 indicates that even a few more million miles unperturbed aphelion distance would not have "saved" the comet. One million miles difference in unperturbed aphelion distance corresponds to a heliocentric departure velocity error of 27 ft/sec and a geocentric error of 13 ft/sec.

Thus, the comet's departure velocity (and correspondingly, its departure direction) can be inaccurate probably by over 100 ft/sec with respect to the Sun or some 50 ft/sec with respect to the Earth. This greatly facilitates gravitational navigation. Unfortunately, however, to reach Jupiter even with the help of the Moon,

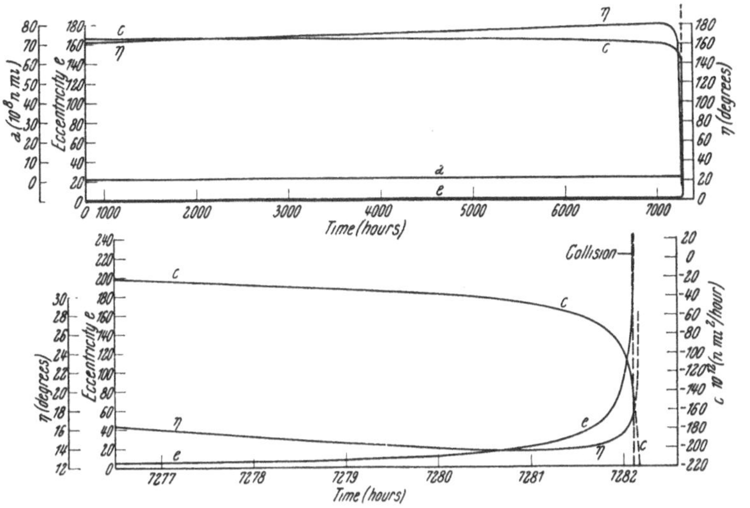

Fig. 26. Osculating elements of vehicle's collision orbit with Jupiter
Jupiter: $R_P = 81{,}151{,}071$ n.mi.; $R_A = 419{,}766{,}470$ n.mi.; $\varphi_0 = 57.^0 46699$; $\eta_0 = 159.^0 89543$; t = hours

the vehicle must be in a highly hyperbolic circumterrestrial orbit. As Fig. 5 shows, the gain which can be derived from the Moon is then quite small. Thus, unless Mars can be used properly, Jupiter flights remain largely an energy problem.

9. On the Application of Instrumented Comet Systems

Many of the principal fields of interest in research with lunar probes and instrumented comets will not be different from those motivating the Vanguard program today. They pertain to solar research, extraterrestrial matter, cosmic radiation and other astrophysical areas. Their scope can be considerably expanded. In other fields a shift will occur from geophysical research to selenophysical and planetophysical research. There are a great many individual research problems which could be considered for artificial comets capable of reaching Mars and Venus and possibly penetrating into intra-Venusian space by proper gravitational navigation near Venus. Their discussion and scientific value would by far exceed the scope of this paper and also should be left to more expert authors in the respective fields.

There are, however, a few applications of basic as well as astronautical interest which shall briefly be mentioned here. The discussion is general and the reader should keep in mind what was said before regarding the accuracy requirements.

1. Thermonuclear Probe

Ever since Goddard proposed to explode a powder charge on the Moon's surface [15], the concept of this simplest type of space probe plays a role in astronautical literature. It is obvious that the impact flash not only signals the flight mechanical success of the mission, but also can be evaluated spectroscopically to gain information about the nature of lunar soil. Today where the thermonuclear charge can replace GODDARD's powder charge, the possibilities for using this method of research are considerably increased. A powder charge would have sufficed for the Moon, but not for the planets.

Sending instrumented comets, which consist essentially of a thermonuclear charge, on collision courses with Venus and Mars and setting off the charge by thermal (entry heating) or pressure control, *a spark spectrum of the inert atmosphere of these planets* could be obtained which, by comparison with the spark centrum of air (showing O, N, H and A), would clarify the question as to their true composition. Taking, for example, the collision course with Venus (Fig. 18), the observational conditions are shown in Fig. 27. The charge explodes on the night side of Venus. This facilitates the observation, although the flash produced by a charge of several kt might also be visible on the sunlit side. Seen from the Earth, Venus is slightly beyond the last quarter. Observability is favorable. It can be expected that stratospheric balloon-suspended observation stations will be used, to become independent from ground weather. This will improve further the observability of the planet[1].

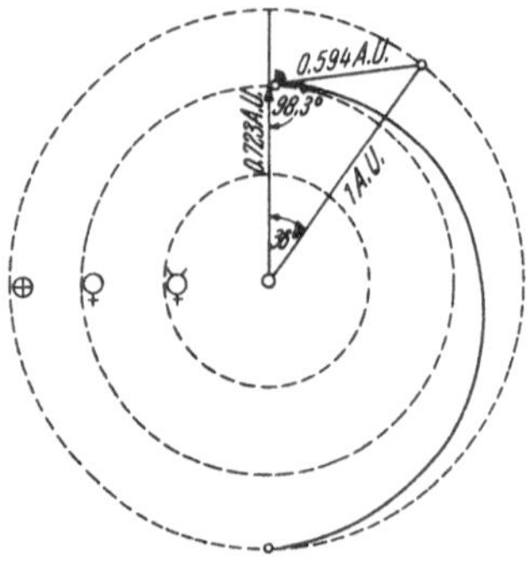

Fig. 27. Observation of exploding thermonuclear probe in Venus atmosphere following collision course according to Fig. 18

It can also be hoped that a few successful experiments with thermonuclear probes will solve the *enigma of the Venusian clouds* [17], [18]. It is unlikely that such a charge would succeed in tearing open the cloud cover and permitting a glimpse at the surface of Venus. If this were possible, the charge would preferably be set off on the sunlit side. To assure such accuracy of impact without the benefit of controlled approach maneuvering as in a manned expedition, would indeed be a formidable task. One of the basic missions of the Venusian probe would therefore be to be observable (optically or through radar contact) *after* the encounter to that the *planet's mass can be determined* more accurately from the elements of the new circumsolar orbit.

A particularly interesting application of the thermonuclear probe lies in the field of *solar research*, as a comparatively simple means of determining the particle flux density and hard radiation (x, γ) flux density distribution from the Sun at various distances and as function of solar activity. This can be done, for example, by sensitizing[2] the firing mechanism of the charge with respect to the nature and intensity of the flux, e.g., a γ-radiation counter, proton-counter, neutron-counter of given sensitivity. The *plasma characteristics in extra-terrestrial space* can thus be measured quantitatively, simply by observing the distance at which the thermonuclear probe of known threshold sensitivity is set off. At the same time, this information would be of great importance in clarifying the question of *plasma dynamics in interplanetary space*[3]. Magnetic storms and aurorae are proven—by identification of neutral hydrogen atoms in

[1] It will be noted in Fig. 27 that the distance from the Earth is favorably small, just about twice the minimum distance between the two planets.

[2] The writer is indebted to Dr. L. L. LOWRY, formerly of Los Alamos, now Convair Astronautics, for suggesting this mechanism.

[3] Another, though somewhat more complex approach would, of course, be a telemeterious probe, equipped with counters, transmitter and corresponding auxiliary power supply.

the auroral spectrum by STORMER and VEGARD—to be caused by a stream of corpuscles travelling at high speed (order of 60 times the Earth's orbital velocity) through interplanetary space. The physical nature of motion of these corpuscles through space is not fully understood. According to a theory by S. CHAPMAN [23], [24] and V. C. A. FERRARO [25], they are streams of protons and electrons ejected from the Sun. Since the number of positive and negative particles is assumed roughly equal, the ionized stream as a whole is neutral and remains focussed on its way through space. A different viewpoint is adopted by the magneto-hydrodynamicists (H. ALFEN) [26], [27], arguing that the corpuscles are accelerated by magnetic and electric fields in the vicinity of the Sun. In an attempt to explain the sudden commencement of magnetic storms, GOLD [28] has suggested the formation of an interplanetary shock wave. SINGER [29] estimates the speed of this wave to be 50 times the speed of an ALFEN wave (propagation speed of a magnetohydrodynamic disturbance in a motionless interplanetary gas). Experimental evidence for all these estimates throughout extra-terrestrial space is needed.

The performance feasibility of such a project would depend upon the distance from the Sun required for threshold neutron flux density. Since no particular point accuracy requirements exist[1], use could be made of Venus to hurl the probe into intra-Venusian space and inside the Mercury orbit.

In applying the thermonuclear probe to *comet research*, similar advantages can be gained as in the case of firing them into planetary atmospheres, except that the accuracy requirements are again lower, since the comet's coma, which would be the most interesting target, measures several hundred to more than a thousand miles. Most of the mass of the comet is concentrated in the nucleus which is generally much smaller. Their mass is of the order of 10^{-5} to 10^{-7} the Earth's mass, or $6 \cdot 10^{16}$ to $6 \cdot 10^{14}$ tons. Assuming, for the sake of argument, a mass of $5 \cdot 10^{15}$ tons and a spherical nucleus of 1000 miles diameter, the mean density of the nucleus would amount to 0.0025 ($H_2O = 1$) or twice the surface density of air (0.00112). The nucleus, of course, does not consist of gas, but, as WHIPPLE, UREY and other authors suspect, of ice and rock. The density is indicative, however, of the fact that the explosion of a thermonuclear probe would affect cometary material and make it available to spectroscopic analysis under different conditions of excitation than sunlight. This is particularly true because such probe would be fired into the comet along an intercepter path, to save energy, which would lead to an intersection point at relatively close distance from the Earth[2]. At this distance from the Sun the development of gas, forming the coma, is still fairly intense.

2. Inflated Bodies

The use of inflated spheres in space is a fairly obvious thought and has been considered for lunar probes in [2] and in [1]. It permits high accuracy optical tracking all the way to the Moon and beyond. In [2] the use of such spheres

[1] It is desirable, however, to place the orbit so that the comet is at not too small elongation from the Sun in the region where its explosion is expected. Where visibility difficulties exist, a TM-probe can be used.

[2] Since most comet orbits are highly inclined, interception would be feasible only if the encounter with the comet takes place at or near one of the nodes, to avoid sending the probe out of the orbital plane of the Earth to any great extent, since this would require much more energy.

Table IV. *Classification of Artificial Unmanned Space Vehicles*

Operational

Circumterrestrial Satellites	*Lunar Probes*	*Instrumented Comets*	*Circumplanetary Satellites*
Terrestrial Satellite	Collision Probe	Solar Probe	Venusian Satellite
Cislunar Satellite	Hyperbolic Probe	Planetary Probes	Martian Satellite
Translunar Satellite	Lunar Satellite	Collision Probe	
		Hyperbolic Probe	
	Lunar Landing Craft	Interplanetary Probes	
		Comet Probe	

Functional

Inert Probes	*TM—Probes*	*TV—Probes*	*Powered Probes*
Inflated bodies (no instrumentation)	Must carry:	Must carry:	Same equipment like TV-probes plus guidance system propulsion system
	End instrumentation	End instrumentation	
Thermonuclear probes (simple instrumentation and receiver required in some cases)	Small bandwidth transmitter	Optical equipment	
	Moderate power supply	Television camera	
	Attitude control not always required	Large bandwidth transmitter	
	Receiver not always required	Large power supply	
		Attitude control	
		Receiver	

was also contemplated to measure the possible resistance from a highly attenuated lunar atmosphere.

Using Mylar or polyethylene spheres [16] which are thinly sprayed with smooth silver coating, extremely light bodies can be comparatively easily produced in space. Such techniques, using aluminum spheres, have also been suggested in connection with Project Vanguard. However, aluminum foil would be too heavy for present purposes. Mylar 10^{-4} inch thick weighs only about 10^{-3} lb/ft². The sphere can be pressurized with hydrogen at very low pressure. A number of pertinent data for Mylar spheres has been computed and the results summarized in Table IV. It will be noted that tremendous cross-sectional areas can be obtained at very little weight. The gas weight is negligible in almost all cases, since the pressure must be lowered with increasing size, in order to keep the skin stress at tolerable

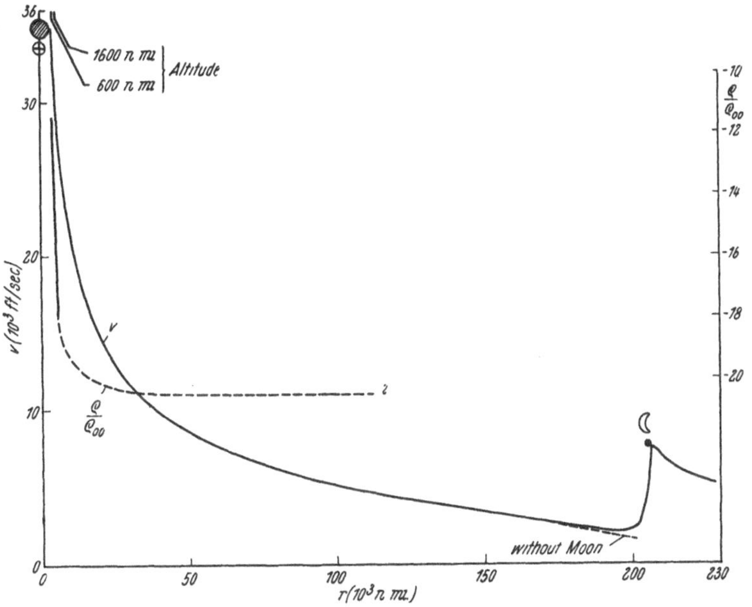

Fig. 28. Velocity profile of lunar probe and gas density in cislunar space

levels. A value of 4000 psi has been chosen, which appears safe, since even at 300° F (150° C) the tensile strength of Mylar is still 10,000 psi. Actually, the Mylar will be quite cold. Silver has a very low absorption coefficient in the visible light, but a high absorption (hence, also emission) coefficient in the far infrared. The silver coating therefore absorbs little radiation offered by the Sun in the region of main radiation energy, but it emits strongly at low temperatures. For this reason, the radiation equilibrium temperature of the Mylar must be expected to reach low temperatures, of the order of 170°—200° K (—150° to —100° F) at 1 A.U. distance, some 60 centigrades lower near Mars and some 40 centigrades higher near Venus.

The Mylar will therefore be quite strong. In order to keep the gas temperature up (if this is necessary), heat windows can be provided. They consist of circular areas of transparent Mylar which lets most of the incident radiation pass into the interior which acts like a black body. These heat windows do not have to be large, since 1 kw or about 1 Btu/sec solar energy is offered per square meter (33 ft²).

It may, however, not be necessary to do this, since most or all of the hydrogen may diffuse through the thin Mylar shell during the extended time periods under consideration here. If material is selected which becomes sufficiently brittle at these temperatures, continued gas pressure would not be required once the sphere is inflated. The reflectance of silver exceeds that of aluminum, being about 96 percent in the 6000 A region and 98 percent in the infrared (11,000—20,000 A), i.e. one has almost 100 percent reflectance, compared to aluminum with 72 to 82 percent. Assuming 100 percent reflectance, an atmospheric attenuation coefficient of 0.117, 90 degrees elevation angle of the object (zenith

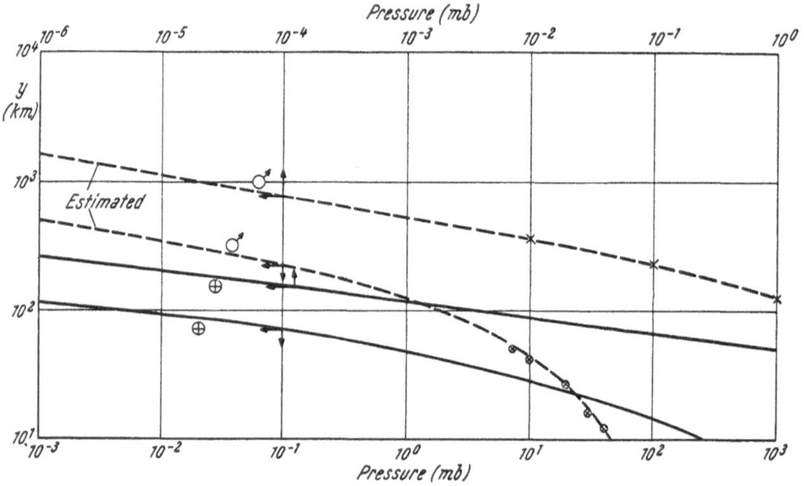

Fig. 29. Pressure in terrestrial and Martian atmosphere. ⊕ G. DE VAUCOULEURS [31]

position) full phase and a distance of 200,000 n.mi. (360,000 km) the stellar magnitudes given in Table III were computed in [3]. They represent about the most optimistic result and may be higher by as much as one magnitude. Tracking of a 25 to 50 ft. diameter sphere would be quite simple, even with smaller telescopes of 4 to 10 inches, and even if the body is close to the Moon, provided the lunar disk is eclipsed artificially to reduce the disturbing light effect.

It will be noted in Table IV that the drag parameter C_DA/W is consistently high, almost 500 ft²/lb. The reason for this constancy rather than decrease at the largest sizes is that the gas weight does not increase in proportion to the volume on account of the decreasing pressure.

In order to check the applicability of such bodies to drag measurements in cislunar space, the velocity profile of the lunar probe according to the orbit Fig. 13 has been plotted in Fig. 28. Many conditions, favoring high measuring sensitivity, are present: very large C_DA/W of the tracer bodies, high sensitivity of the orbit to small changes in velocity and considerable speed for producing dynamic pressure. It turns out, however, that the gas density is so low (at least this is the general assumption) that not far from the Earth only negligible deceleration is obtained. At the start where $v = 34{,}900$ ft/sec and $y = 300$ n.mi., a density of $\varrho = 2 \cdot 10^{-15}$ causes a deceleration of the 50 ft sphere of about $6 \cdot 10^{-4}$ g. At $y = 1{,}600$ n.mi., $v = 27{,}000$ ft/sec, $\varrho \sim 1.9 \cdot 10^{-19}$, the deceleration is $3.4 \cdot 10^{-8}$ g. At 50,000 n.mi., assuming interplanetary gas at a relative velocity of 100,000 ft/sec, the deceleration is down to about $5 \cdot 10^{-12}$ g. Even in view of the much longer flight times of instrumental probes to other planets, such small decelerations in interplanetary space would have no effect.

At *close distance from the Moon* or at appropriate distances from Venus and Mars, an effect could be shown which is just strong enough to change the sphere's orbit significantly, without destroying it. If the Moon has an atmosphere dense enough to be detected by deceleration of the sphere, this density must be of the order of $2 \cdot 10^{-10}$ g/cm^3 (10^{-10} slug/ft^3) in order to produce a deceleration of about 10^{-3} g ($C_D A/W = 500$; velocity near Moon about 7,500 ft/sec). This is a density which, in the Earth's atmosphere, exists at about 110 km (350,000 ft). However,

Fig. 30. Inflated tracer bodies leaving the Earth-Moon system

terrestrial aurorae have been observed at 100—400 km altitude and sunlit aurorae as high as 1,000 km. A lunar aurora has never been observed. This could mean either that the Moon has no magnetic field or that its atmospheric density is much lower than 10^{-10} g/cm^3 (or both). If a path change of the sphere near the Moon were observed corresponding to 10^{-10} g/cm^3 density, then this would be a strong indication of the absence of a lunar magnetic field. The Moon's atmosphere, if any, would be very "fluffy" because of the low gravitational pull. This means that the density gradient would be moderate at altitude and the flight accuracy would not have to be extreme.

In the case of *Venus and Mars*, great distance and denser atmospheres would combine to make such a test much more difficult from the viewpoint of accuracy. Of these two planets, Mars has the more "fluffy" atmosphere as shown in Fig. 29, in comparison with the Earth's atmosphere. In the previously discussed close encounter with Mars an altitude as low as about 1,200 n.mi. was attained. According to Fig. 29 a pressure of the order of 10^{-7} mb might be expected at this altitude, compared to 10^{-11} mb on Earth. The associated density would depend on composition and temperature, both being unknown. Assuming, presently, that the conditions are comparable to those on Earth at this altitude, a 10^4 times higher density, or $\varrho = 10^{-14}$ g/cm^3 ($5 \cdot 10^{-15}$ slug/ft^3) would exist. This would not cause enough deceleration of the Mylar sphere during the brief period of close encounter to measurably alter its orbit.

Thus, density measurements in the outer atmosphere of Mars or Venus would require very ambitious accuracy tolerances and certainly would constitute an advanced state of the art of comet launching. Measurements of planetary atmospheres with inflated tracer bodies may have to await the arrival of a manned expedition.

Fig. 30 depicts a set of three inflated tracer bodies deep in translunar space. In the background are Earth and Moon. The sphere in the center has been punctured by meteoritic material. Its surface is crumpled due to contractions by a spring mechanism (Ref. Section 9.4 below), shown in the cut-away view of the sphere in the right foreground, while the left sphere is still intact. Both spheres in the foreground show heat windows (dark spots).

3. Radiation Propelled Tracer Bodies

It is of interest and of great practical significance for the development of interplanetary research vehicles to consider the forces which, in addition to the resistance of interplanetary gas, act on a sphere of the described type.

There is, of course, first the gravitational force of the Sun and the planets. The solar force is given by

$$g_\odot = \frac{k^2 (M_\odot + m_{sphere})}{R^2} = \frac{k^2 M_\odot}{R^2} = \frac{K_\odot}{R^2} \tag{66}$$

where R is the distance of the sphere from the Sun. In terms of the absolute value of the Earth's gravity $g_\oplus = 980.66$ cm/sec^2 $= 32{,}174$ ft/sec^2 at 45° latitude the Sun's acceleration is at mean Venus distance: $1.2 \cdot 10^{-3} g_\oplus$; at Earth distance: $6.2 \cdot 10^{-4} g_\oplus$, at Mars distance: $2.7 \cdot 10^{-4} g_\oplus$ and at Jupiter distance: $1.5 \cdot 10^{-5} g_\oplus$.

Secondly, the sphere is acted upon by solar radiation pressure. If R^* designates the distance from the Sun in astronomical units, the solar constant in the sphere's orbit has the value

$$S' = \frac{S}{R^{*2}} \tag{67}$$

where the solar constant at 1 A.U. is:
$S = 1.36$ erg/cm^2 sec $= 1{,}386.6$ g — cm/cm^2 sec $= 92$ ft-lb/ft^2 sec.
According to EINSTEIN's mass-energy equivalence $E = m\, c^2$, the mass of a photon of solar radiation is

$$m = \frac{E}{c^2} = \frac{S' A t}{c^2} \left[\frac{force \cdot (time)^2}{length}\right]. \tag{68}$$

The momentum per unit area, $A/\cos\alpha$, is given by

$$m\, c = M = \frac{S' t}{c} \cos\alpha \left[\frac{force \cdot time}{(length)^2}\right], \tag{69}$$

where α is the angle between incident radiation and the normal to the irradiated plane. The force per unit area is equal to the temporal variation of the momentum, according to NEWTON's second law, for zero reflectivity,

$$\frac{d}{dt}(M) = \frac{S'}{c} \cos\alpha \left[\frac{force}{(length)^2}\right]. \tag{70}$$

If the body has the reflectivity $\bar{R}$, the force per unit area (radiation pressure) is given by

$$\frac{d}{dt}(M)\cos a = p_r = (1+\overline{R})\,\frac{S'}{c}\cos^2 a\left[\frac{force}{(length)^2}\right]. \tag{71}$$

For a perfect reflector, $\overline{R}=1$, and vertical incidence $a=0$, this pressure becomes $p_r = 2\,S/c$. For different planets this value of p_r is:

Distance	Mercury	Venus	Earth	Mars
p_r(lb/ft^2)	10^{-6}	$3.55\cdot10^{-7}$	$1.95.\cdot10^{-7}$	$9\cdot10^{-8}$

If the exposed body is spherical, the gravitational force is proportional to the cube of the body's radius, while the radiation pressure is proportional to its cross-sectional area. The ratio of outwardly directed radiation pressure and inwardly directed gravitational pull is therefore proportional to ϱ^{-1} if ϱ is the radius of the sphere. For a sphere of specific gravity equal to unity, both forces balance each other if $\varrho = 1.5$ microns (μ). At smaller size, the acceleration due to radiation n_r (in $g_\oplus$—units) is greater than $g_\odot$. The density of a 100 ft diameter sphere (Table III) is about $1.07\cdot10^{-6}$g/cm^3 and therefore its radius would have to be $\varrho < 1.6\cdot10^6\,\mu$ for the sphere to be blown out of the solar system by radiation pressure ($n_r > g_\odot$). Actually, the sphere's radius is $\varrho = 1.5\cdot10^7\mu$. Likewise, it can be shown that the other tracer bodies are above critical size. This means, they would fall toward the Sun if they were motionless, with respect to the Sun, in interplanetary space. However, having circular (or elliptic) velocity, they are rather in the position of a rocket in a circular orbit on which a radial thrust force acts. In the present case, using the above values for p_r since a silver sphere is almost a perfect reflector, the acceleration by radiation pressure is (Table IV)

$$n_r = \frac{p_r}{W/A} \approx \frac{p_r}{4.4\cdot10^{-3}}. \tag{72}$$

The resulting accelerations are

Distance	Mercury	Venus	Earth	Mars
$n_r\ (g_\odot)$	$2.73\cdot10^{-4}$	$8.07\cdot10^{-5}$	$4.4\cdot10^{-5}$	$2\cdot10^{-5}$

This is 10^7 times the deceleration due to drag from interplanetary gas.

This acceleration will lift the body out of the orbit computed for a non-radiating sun, into a new orbit which involves greater distances. The motion of the sphere is given by the equation

$$\ddot{R} = R\,\eta^2 - g_\odot + g_\oplus\,n_r \tag{73}$$

where $g_\odot = K_\odot/R^2$, $g_\oplus = K_\oplus/r_{00}{}^2$, and η the angular velocity (η = true anomaly)

$$g_\odot\,n_r = \frac{g_\oplus\,p_r}{W/A} = \frac{g_\oplus\,(1+\overline{R})\,S\,R^2_\oplus\cos^2 a}{c\,(W/A)\,R^2} = \frac{K_r}{R^2} \tag{74}$$

with K_r a constant, equal to

$$K_r = \frac{S\,(1+\bar{R})\,g_\oplus\,R_\oplus{}^2\cos^2 a}{c\,W/A} = \frac{S\,(1+R)\,K_\odot\cos^2 a}{c\,W/A}\left[\frac{(length)^3}{(time)^2}\right] \tag{75}$$

having the same dimensions as K, so that one can write

$$\ddot{R} = R\,\eta^2 - \frac{K_\odot - K_r}{R^2}. \tag{76}$$

The third phenomenon affecting the sphere's motion is the POYNTING-ROBERTSON effect, a relativistic force, directed azimuthally, caused by the DOPPLER effect of radiation being emitted (or reflected) in the direction of azimuthal motion and in the opposite direction [19], [20], [21]. WHIPPLE [21] states that this effect is quite independent of the sphere's albedo, hence, the equations can be applied to the bodies under consideration here. The radiation emitted in the direction of azimuthal motion has a slightly higher frequency than the radiation emitted in the opposite direction. Since the energy of radiation is a function of the frequency and since the energy determines the radiations pressure, the radiation pressure difference resulting from the azimuthal DOPPLER effect tends to reduce the angular momentum of the sphere, causing it to spiral towards the Sun. There exists also a radial DOPPLER effect which is proportional to $\dot{R}$ and which tends to counteract the outwardly directed radiation pressure. ROBERTSON [19] defines a coefficient

$$a = \frac{A\,S\,R_{\odot}^2}{m\,c^2} \left(\frac{\text{g-cm/cm}}{\text{g/sec}}\right) \tag{77}$$

where m is the mass of the sphere, A its cross-sectional area, S the solar constant, such that the retarding accelerations in azimuthal and radial directions can be expressed in the form $2a\dot{R}/R^2$ and $a\eta/R$, respectively. The equations of motion in polar coordinates are then written

$$\ddot{R} = R\,\eta^2 - \frac{K_{\odot} - K_r}{R^2} - \frac{2\,a\,\dot{R}}{R^2} \tag{78}$$

$$\frac{1}{R}\frac{d}{dt}(R^2\,\dot{\theta}) = -\frac{a\,\eta}{R}\,. \tag{79}$$

The third term in Eq. (78) can be neglected, since $\dot{R}$ is comparatively very small in the circumsolar orbits considered here. Eq. (79) which is zero for the motion of a large, heavy body in vacuo, represents more specifically the POYNTING-ROBERTSON effect. It will be shown here that this effect is negligibly small for the bodies under consideration here. The azimuthal retarding force per unit area is given by

$$\frac{F_a}{A} = \frac{a\,m}{A\,R^2}\,v_a = \frac{A\,S'}{c^2\,R^{2*}}\,v_a \tag{80}$$

where v_a is the azimuthal velocity and S' the solar constant at the distance R^* (A.U.). For the 100 ft diameter sphere (and similarly for the others) it follows $F_a/A = 4.6\cdot10^{-15}$ g/cm^2 $= 9\cdot10^{-15}$ lb/ft^2. Consequently, the azimuthal deceleration is about $2.3\cdot10^{-12}\,g_{\oplus}$. At Earth distance, $n_r \sim 0.07\,g_{\odot}$ [$g_{\odot}$ as defined in Eq. (66)]. Thus, as soon as the POYNTING-ROBERTSON effect has reduced the azimuthal velocity to such an extent that the apparent weight of the sphere due to loss in centrifugal relief becomes $1 - (v/v_c)^2 = 0.07$, the sphere will begin to spiral toward the Sun. To reach this point (assuming a near-circular orbit) $(v/v_c)^2 = 0.93$ or $v/v_c = 0.964$, corresponding to a deceleration of about 2,500 feet per second (at 1 A.U.). At a deceleration of $7.4\cdot10^{-11}$ ft/sec^2 it takes $1.4\cdot10^{10}$ years to accumulate a velocity decrement of one ft/sec. The POYNTING-ROBERTSON effect will therefore not noticeably affect the motion of the sphere. Eq. (79) becomes zero and (78) identical with (76). The solar gravitational parameter is $K_{\odot} = 1.3291976\cdot10^{26}$ cm^3/sec^2 while the radiation factor is $K_r = 0.093766\cdot10^{26}$ cm^3/sec^2 (assuming 100 per cent reflectivity, $W/A = 4.46\cdot10^{-3}$ lb/ft^2 $= 2.17648$ g/cm^2 and $\sphericalangle\,\alpha = 0$) so that $K_{\odot} - K_r = 1.2354316\cdot10^{11}$ km^3/sec^2. Assuming that $a = -K_{\odot}/h$ refers to the semi-major axis of the circumsolar orbit without considering radiation, and $a' = -(K_{\odot} - K_r)/h'$ refers to the final ellipse with radiation, then, if for example $a = R = 1.495\cdot10^8$ km, so that $h = -889$ km^2/sec^2, since the velocity in the Earth's orbit is 29.67 km/sec, it follows that, for the same velocity, but $(K_{\odot} - K_r)$, the value of $h' = -767$ km^2/sec^2 (i.e. greater) and a' becomes approximately $1.61 \cdot 10^8$ km. Since for Mars' orbit $a = 2.28 \cdot 10^8$ km, the sphere is still too heavy to reach this orbit on light-pressure alone.

This preliminary discussion does not give an account of the effect of interplanetary electron accumulation by the sphere as a result of its motion in the interplanetary plasma, nor of the electromagnetic forces if the sphere should move into a strong magnetic field. An exact analysis of these effects is not possible, because neither the particle density nor the distribution of the charged gas, nor the average relative velocity between sphere and gas, nor the behavior of the silver-coated Mylar itself under these conditions is known. Mylar, even if hardened to a certain extent by the low temperature, is comparatively soft material. The coating of silver—a soft metal—is quite thin. Ions, impacting at high speed may therefore stick. The interplanetary gas as a whole must contain about as many positive charges as there are electrons. A large number of these high-speed positive ions could be captured in the material, thereby preventing the build-up of an excessive negative charge.

More will be said on this subject in a subsequent investigation and reference must be made to SINGER's excellent analysis of the electric and electromagnetic forces on interplanetary dust particles [22].

For the present we conclude tentatively that such spheres constitute an interesting vehicle for interplanetary research, optically easily visible and capable of reducing the propellant requirement for missions outside the Earth's orbit due to the utilization of radiation pressure. In conclusion we note briefly that if the thickness of the Mylar could be reduced to 10^{-5} inch, critical conditions would be reached where gravitational and radiation acceleration balance each other. Thus, if wall thickness between 10^{-4} and 10^{-5} inch could be achieved, very much greater range benefits could be derived from radiation pressure than with the spheres considered above.

4. Measurement of Interplanetary Matter

It is of interest to use the inflated and well visible tracer bodies described above, to gain more information on interplanetary matter at greater distance from the Earth by measuring their lifetime, i.e., the time they exist in an inflated condition. Presently, we expect two causes of destruction: penetration by larger meteors, or gradual erosion by very fine dust particles. From statements made by Professor WHIPPLE during the USAF/OSR—Convair sponsored Astronautics Symposium in February 1957 in San Diego, it can be concluded that the very fine dust which seems to represent the bulk of interplanetary material would not be able to penetrate even 10^{-4} inch Mylar, but would gradually erode the extremely thin silver coating so that the balloon becomes increasingly transparent and finally is destroyed. Of course, proof of these assumption can be furnished only by experiments with such bodies in space.

If an inflated body in space is pierced by a meteorite, it would not immediately collapse, since there is no outside pressure, and radiation pressure and possible electric forces would require considerable time to change the shape. A contracting mechanism such as a few spring tension wires mounted across the interior of the sphere would be required to change the shape after depressurization. However, as pointed out before, the sphere may already essentially be evacuated by diffusion, and the hull is cold and rigid. A meteor penetration would thus probably not have an observable effect on the sphere if it is at great distance. If the sphere is in the closer vicinity of the Earth, smaller dimensions, greater skin thickness and an internal tension mechanism could be used to detect decompression of the sphere[1]. At Venus or Mars distance from the Earth, it would require a large meteor, smashing a good portion of the sphere, to produce a sudden disappearance.

The other tracer body, used for testing the "soft dust" theory would not have to be equipped with a contracting mechanism. It would not disappear if pierced. Continuous erosive action, affecting the whole surface, since the body doubtlessly would rotate, would in time reduce the reflectivity by increasing the transparency. Thus a gradual optical fading of these bodies would be indicative of the density and distribution of this material in different areas of the solar system as far as such a sphere can be observed.

5. Visibility of Interplanetary Tracer Bodies

Although their size and reflectance makes the inflated bodies bright objects, the large distances involved combined with lesser radiation density outside the Earth's orbit and phases inside the orbit tend to reduce their visibility. Using

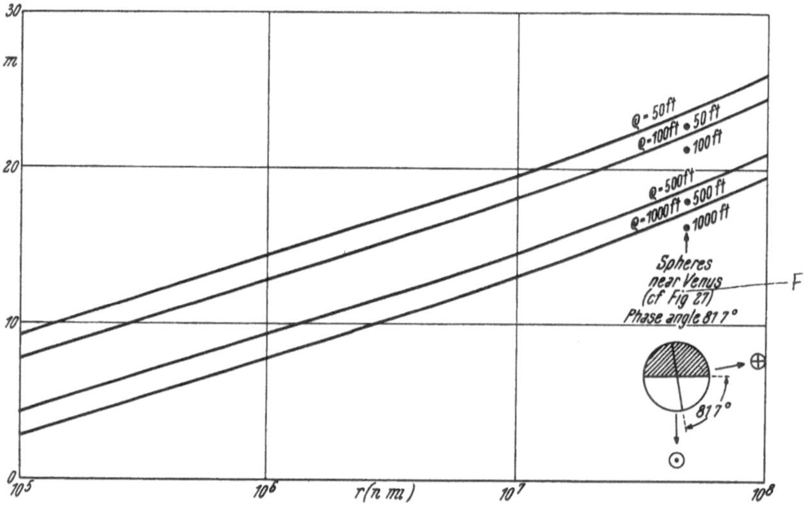

Fig. 31. Stellar magnitude of silver-coated mylar sphere. ● Sphere of given radius as seen near Venus for conditions Fig. 27. Assumptions: atmospheric attenuation coefficient = 0.117; elevation angle = = 90 degrees; phase of spherical body: full; reflectance: 100 per cent

TOUSEY's method [30] assuming 100 percent reflectance, very clear air, full phase of the tracer body, Fig. 31 was computed to obtain an idea what stellar magnitudes to expect. In this chart the diminuation of incident illumination was computed for the distance R from the Sun, assuming that the Earth is located on the line Sun—sphere, so that the distance of the sphere varies at the same rate from the Sun as well as the Earth. This permits a general diagram, but corresponds to the true conditions only when Earth and sphere are in opposition. The theoretical magnitudes of spheres near Venus under the conditions of Fig. 27 are shown also, taking both, the increase in brightness due to greater radiation intensity and the decrease due to phase into account. It is interesting to note that although more than half of the sphere is dark, as seen from Earth, the increase in brightness of the sunlit portion more than makes up for it, so that the spheres appear brighter than

[1] This decompression would certainly not be sudden in view of the large volumes and low pressures, but it would produce a comparatively rapid change in appearance.

if they *were at the same distance outside* the Earth orbit and in full phase[1]. The brightness of Martian spheres was computed for the conditions as shown in Fig. 32. The center angle of the dark portion of the disk was found to be 37° 20′ (phase angle 142° 40′ compared to 81.7° for Venus). Thus the Martian spheres under the given conditions would also not appear in full phase. The stellar magnitudes of the Martian spheres exceed the limits of Fig. 31. They are 25.86 (50 ft diameter), 24.36 (100 ft), 20.86 (500 ft), 19.36 (1000 ft). With the exception of the two largest spheres, these bodies could no longer be observed, optically or photographically. By comparison, one of the faintest moons in the solar system, Jupiter X, is of 17.8 magnitude.

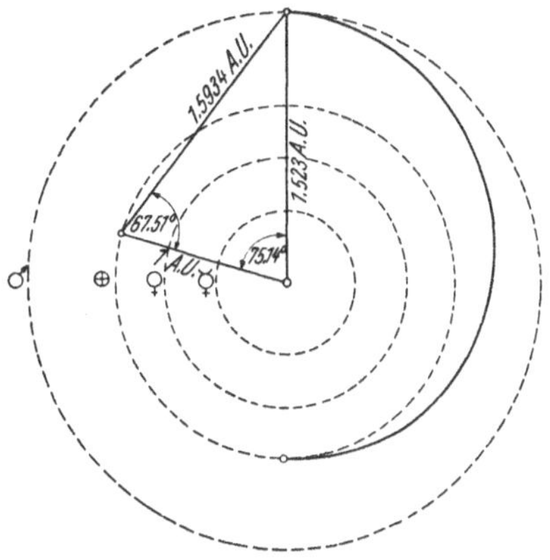

Fig. 32. Constellation at arrival of comet near Mars

6. Radiation-Propelled Space Vehicles with Payload

In proposing the very large sphere we should point out that this vehicle does not necessarily have to be inert. The larger spheres can carry a useful payload. For example, the 500 ft sphere weighs 868 lb (Table IV) and, near Earth, has an acceleration of $8 \cdot 10^{-5}$g by radiation. If this sphere would carry another 868 lb payload, a' would still be about $1.55 \cdot 10^8$ km, i.e., the increase over a for the Earth's orbit (cf. Section 9.3) would be cut in half, roughly, as the density is doubled. The payload can consist of attitude control, power supply, telescope, television camera, transmitter and receiver. Less ambitious versions could carry end instruments and telemetering equipment as well as a radar beacon for tracking, if fired into the inner solar system. The 1,000 ft sphere could carry correspondingly more weight, or the same weight as the 500 ft sphere at less penalty in range.

Since such spheres cannot reach Mars or the asteroid belt by means of radiation pressure alone, it will still be necessary to fire them into elliptic circumsolar orbits. However, the farther out the aphelion without the benefit of radiation pressure, the greater will be the increase in range caused by the radiation effect. This follows clearly from Fig. 7 as well as from general considerations regarding the motion of the apogee as the energy of the elliptic orbit increases.

It is important to remember that due to the rigidity of the cold sphere, pressure stabilization should not be required, once the sphere is inflated. This is a basic advantage over the author's solar powered space ship proposal [15] which, due to its higher acceleration and due to the bending moments involved in applying point source thrust devices, must have pressure stabilization. If the inside contains no pressure, meteor penetration can do no harm. Because of the very thin shell, the meteors pass right through the sphere without significant momentum exchange which could cause a random perturbation of the sphere's orbit. A greater potential danger would be rapid erosion of the reflecting coating, since this causes the sphere to become transparent, thereby losing the radiation propulsive force.

---

[1] The illumination due to the Sun is 12,500 f.c. near the Earth, 24,273 f.c. near Venus, and 3.242 near Mars.

10. Summary and Conclusions

Interplanetary flight is not restricted to manned operations only, nor are only flights from one celestial body to another of interest or worthwhileness. It has been emphasized in this paper that, through scientific space research, also operations between planets are drawn into the sphere of astronautical interest. In fact, such operations are of greatest significance, not only for science, but also for the development of manned space flight, which will have essentially a planet-to-planet mission character. A wide field of research activities offers itself to instrumented comets — especially the simpler systems — in the far-flung regions of interplanetary space, at greatly relaxed accuracy requirements.

An analysis of gravitational navigation, using a perturbative force field to increase the orbital energy of the instrumented comet, shows three fundamental results:

(a) It is not possible to use the lunar field to the maximum extent, because this requires passage near the Moon's center. However, a very close ("grazing") passage would yield, for flights to Venus and Mars, slightly more than one-half of the maximum theoretical gain. This means a significant saving in propulsive energy, resulting in an increase in vehicle end weight of 500 to over 1000 pounds.

(b) While the maximum theoretical gain obtainable from the lunar field increases with the heliocentric distance of the target apsis, the maximum actual gain obtainable from a surface-grazing encounter decreases with distance. In other words, lunar interaction is of little help for getting to Mercury or Jupiter.

(c) Perturbative interactions are characterized by an extreme error sensitivity.

Especially for general interplanetary missions it appears practical to apply the principle of gravitational navigation through properly directed, close encounter with the Moon for reaching Venus or Mars distance or intermediate regions; and through encounter with Venus for penetrating deeper into the inner solar system. For advancing into the asteroid belt and into Jovian space, a close encounter with Mars could be utilized. This is more difficult to do than with Venus, because of greater sensitivity of the transfer ellipse to errors and because of the weaker gravitational field of Mars.

An error analysis for single-force field and two-force field flights has been presented. In addition, flight paths of lunar probes, encounters with the Moon, Venus, Mars and Jupiter have been computed, to obtain preliminary data on the effect of the target body's attraction on the error sensitivity. It is shown that for lunar flights with return to Earth, restricted 4-body analysis must be applied in almost all cases. Lunar firing accuracy requirements for a specific standard case are summarized in Fig. 12. For transfer to Venus, Mars and Jupiter the following results were obtained:

1. Neglecting target planet's attraction:

	Venus	*Mars*	*Jupiter*
Allowable heliocentric error for collision course (±with respect to dead center collision) (ft/sec)	±1.68	±0.3	±1.0
Corresponding allowable geocentric error (ft/sec)	±0.145	±0.08	±0.48

2. Considering target planet's attraction:

	Venus	*Mars*	*Jupiter*
Allowable heliocentric error (ft/sec)	$\sim \pm 5$	$\sim \pm 1$	$> \pm 100$
Allowable geocentric error (ft/sec)	$\sim \pm 0.5$	$\sim \pm 0.27$	$\gtrsim \pm 50$

The "*error*" is defined as a change in azimuthal velocity from the velocity required for a theoretical dead-center collision with the target planet. Specifically, the *geocentric error* is defined as error in hyperbolic departure velocity from the Earth. The *heliocentric error* is defined as error in velocity at the beginning of the unperturbed heliocentric transfer ellipse.

The *effective collision diameter*, defined as allowable "miss" diameter without considering the planet's attraction, is for Venus approximately 20,000 n.mi., or 3 planet diameters, and for Mars 12,000 n.mi., or 3.4 planet diameters. It is emphasized that all results involving the effect of the planet's attraction are tentative only and must be refined. It is recognized that the "standard" values for dead-center collision are themselves uncertain to a degree exceeding the error limits given above, because of uncertainties in our present knowledge of masses and distances in the solar system.

Among the instrumented comets, classified and discussed, the thermonuclear probe and the radiation-propelled probe are of interest. The thermonuclear probe can be used for planetologic research, comet research and for solar research or interplanetary plasma research, using particle or photon streams from the Sun of selected type and intensity to set off the probe of pregiven sensitivity. The radiation-propelled probe offers a new and energetically attractive possibility for exploring the Earth-Mars region at reduced energy requirement. The probe is essentially a large, silver coated polyethylene sphere of extremely low weight. This vehicle is light, simple, inexpensive, much more readily observable than other artificial comet designs and not affected by meteor penetration. For successful operation in space it is necessary, however, that erosion of the reflective coating does not occur at too fast a rate. Sizes from 500 to 1,000 ft diameter can carry at least several hundred pounds of payload. These bodies are useful tools for exploring the distribution and density of interplanetary matter and for carrying light weight end instruments and transmitters for various measurements.

Man is presently engaged in building satellites. We know he can, and we hope he will, build artificial comets a few from now.

Acknowledgements

The perturbation calculations were run on the IBM 704 Digital Computer of *Convair-Astronautics* when programming the computer for 4-body satellite orbit studies.

Special acknowledgement is due Messrs. W. C. Ridell and J. Mueller for preparing and coordinating the computer operation.

Mrs. Helen Pence participated ably in preparing and evaluating the extensive numerical and graphical material of this paper.

To Professor George Gamow I am particularly indebted for reviewing and discussing critically the content of this study.

References

1. A. C. Clarke, The Exploration of Space. New York: Harper & Brothers, 1951.
2. K. A. Ehricke, Basic Aspects of Operations in Cislunar and Lunar Space. Amer. Rocket Soc., Paper No. 235 A—55, Nov., 1955.
3. K. A. Ehricke, Astronautical and Space Medical Research with Automatic Satellites, in: Earth Satellites as Research Vehicles. The Franklin Institute Monograph No. 2, June, 1956.
4. K. A. Ehricke, Cislunar Orbits, Paper presented before the American Mathematical Society, New York, April, 1957; to be published by McGraw-Hill.

5. K. A. EHRICKE, Cislunar Operations. Amer. Rocket Soc., Paper No. 467—57, June, 1957.
6. D. F. LAWDEN, Entry into Circular Orbits 2. J. Brit. Interplan. Soc. **13,** 27 (1954).
7. K. A. EHRICKE, A New Supply System for Satellite Orbits, Pt. 1. Jet Propulsion **24,** Sept.-Oct. (1954).
8. K. A. EHRICKE, On the Application of Solar Power in Space Flight. Proceedings of the VIIth International Astronautical Congress, Rome, 1956, p. 451. Roma: Associazione Italiana Razzi, 1957.
9. D. F. LAWDEN, Perturbation Maneuvers. J. Brit. Interplan. Soc. **13,** 329 (1954).
10. C. H. CLEMENT, The Moon Rocket, in: Earth Satellites as Research Vehicles, loc. cit. [3].
11. R. W. BUCHHEIM, Artificial Satellites of the Moon. The RAND Corp., Rep. P-873, June, 1956.
12. K. R. STEHLING and R. FOSTER, We Can Build a Moon Rocket Now. Missiles and Rockets, Oct., 1956.
13. G. GAMOW and K. A. EHRICKE, A Rocket Around the Moon. Scientific American No. 6, June, 1957.
14. F. TISSERAND, Traité de Mécanique Céleste, Tome IV, p. 198, 1889.
15. R. H. GODDARD, A Method of Reaching Extreme Altitudes. Smithsonian Miscellaneous Collection, **71,** No. 2, Washington, 1919.
16. K. A. EHRICKE, The Solar Powered Space Ship. Amer. Rocket Soc., Paper No. 310—56, June, 1956.
17. G. P. KUIPER, The Atmosphere of the Earth and the Planets, 2nd. ed. Chicago: University of Chicago Press, 1952.
18. D. H. MENZEL and F. L. WHIPPLE, The Oase for H_2O Clouds on Venus. J. Astron. Soc. Pacif. **67,** No. 396, June (1955).
19. H. P. ROBERTSON, Dynamical Effects of Radiation in the Solar System. Monthly Notices **97,** 423 (1937).
20. S. P. WYATT, JR. and F. L. WHIPPLE, The POYNTING-ROBERTSON Effect on Meteor Orbits. Astrophysic. J. **111,** 134 (1950).
21. F. L. WHIPPLE, A Comet Model, III, The Zodiacal Light. Astrophysic. J. **121,** 750 (1955).
22. S. F. SINGER, Measurements of Interplanetary Dust, in: Scientific Uses of Earth Satellites, J. A. VAN ALLEN, Edit. Ann Arbor: The University of Michigan Press, 1956.
23. S. CHAPMAN, Monthly Notices **106,** 218 (1946).
24. S. CHAPMAN and J. BARTELS, Geomagnetics. Oxford: University Press, 1940.
25. V. C. A. FERARRO, Amer. Geophys. **11,** 284.
26. H. ALFÉN, Cosmical Electrodynamics. Oxford: University Press, 1950.
27. H. ALFÉN, Tellus **7,** 50 (1955).
28. T. GOLD, in: Rocket Exploration of the Upper Atmosphere, p. 366. London: Pergamon Press, 1954.
29. S. F. SINGER, Measurement of the Earth's Magnetic Field, loc. cit. [22].
30. R. TOUSEY, The Visibility of an Earth Satellite. Astronaut. Acta **2,** 101 (1957).
31. G. DE VAUCOULEURS, Physics of the Planet Mars. New York: The MacMillan Co., 1954.

Die Sternenökosphären im Radius von 17 Lichtjahren um die Sonne

Von

Jan Gadomski[1]

(Mit 1 Abbildung)

Zusammenfassung — Abstract — Résumé

Die Sternenökosphären im Radius von 17 Lichtjahren um die Sonne. Es wird der Bereich der Ökosphären der benachbarten Sterne im Radius von 17 Lichtjahren um die Sonne berechnet. Mit Hilfe der vom Verfasser angeführten Formeln und auf Grund des Kataloges von van de Kamp wird eine Liste von 59 Sternenökosphären zusammengestellt. Die Zahlenwerte der berechneten Radien sowie der Massen der Sterne sind — als etwas unsicher — abgerundet.

Es hat sich gezeigt, daß nur 16 sonnenahnliche Sterne breite (etwa 200×10^6 km) Ökosphären besitzen, während die Ökosphären um die ubrigen 43 Unterzwerge sehr schmal (etwa 10×10^6 km) sind und nur Planeten mit „gehemmter" Rotationsdauer, also mit sehr begrenzten ökologischen Möglichkeiten, haben können.

The Star Ecospheres within a Radius of 17 Light-Years around the Sun. The range of the ecospheres of those stars is calculated that are situated in the neighbourhood of the Sun within a radius of 17 light-years. Using certain formulae and van de Kamp's catalogue the author has made up a list of 59 star ecospheres. Being somewhat uncertain the values of the calculated radii and of the masses of the stars are rounded off.

It follows that only 16 stars which are similar to the Sun show broad ecospheres (about 200×10^6 km), whereas the ecospheres around the other 43 sub-dwarfs are very narrow (about 10×10^6 km) and can possess only planets with stopped rotation periods, i.e. with very limited ecological possibilities.

Les écosphères stellaires dans un rayon héliocentrique de 17 années-lumière. Les dimensions des écosphères des étoiles voisines, dans un rayon héliocentrique de 17 années de lumière, sont évaluées. A l'aide des formules établies par l'auteur et du catalogue de van de Kamp une liste de 59 écosphères a été rassemblée. Les rayons et les masses calculées ont été arrondies, pour tenir compte du degré d'incertitude.

Il apparait que 16 soleils seulement possèdent une écosphère étendue (environ 200×10^6 km.), tandis que les 43 étoiles naines restantes n'ont que de très petites écosphères (10×10^6 km.) et ne peuvent avoir que des planètes d'une durée arrêtée de rotation, dont les possibilités écologiques sont donc fort limitées.

Der auf dem VII. Internationalen Astronautischen Kongreß in Rom im Jahre 1956 von H. Strughold (USA) eingeführte Begriff „Ökosphäre" des solaren Planetensystems bezeichnet den die Sonne umgebenden Raum, worin die thermischen und ökologischen Bedingungen für das Dasein belebter Organismen, die aus Eiweißverbindungen aufgebaut sind, vorhanden sind.

Die vorliegende Arbeit stellt den Versuch dar, die Lage und die Ausdehnung der Sonnen- sowie Sternenökosphären mittels mathematischer Formeln zu berechnen. Außerdem wurde ein Schätzungsversuch der Ausdehnung der „Sphäre der gehemmten Planeten" durchgeführt.

[1] Al. Ujazdowska 4, Obserwatorium, Warszawa, Polska.

Für das Dasein belebter Organismen fordern die Biologen:

1. den Zwischenraum der äußeren Temperaturen in den Grenzen von:

$$\begin{cases} T_A = 353° \text{ K} = +80° \text{ C bis} \\ T_E = 203° \text{ K} = -70° \text{ C};\end{cases} \tag{1}$$

2. das Vorhandensein einer Atmosphäre mit freiem Sauerstoff;

3. das Vorhandensein von Wasser in flüssiger Gestalt.

Ein vollständig schwarzer Körper, senkrecht von einem Stern bestrahlt, absorbiert alle Arten der elektromagnetischen Strahlung und wandelt sie in Wärme um. Die Temperatur (T) der bestrahlten Oberfläche eines solchen Körpers kann — nach Einstellung des thermischen Gleichgewichtes — mit Hilfe der nachstehenden Formel berechnet werden:

$$T = T_* \sqrt{\frac{R_*}{d}}, \tag{2}$$

wo T_* die Oberflächentemperatur des Sternes, R_* den Sternenradius, ausgedrückt in Sonnenradien ($R_\bigcirc = 1$) und d die Distanz des Körpers vom Mittelpunkt des Sternes bezeichnen.

Wenn der Körper kugelförmig ist und genügend rasch rotiert, dann nimmt Gl. (2) die Form (3) an:

$$T' = \frac{T}{\sqrt{2}} = T_* \sqrt{\frac{R_*}{2d}}. \tag{3}$$

Bei Einsetzen von Gl. (1) in Gl. (3) erhält man für den Anfang (d_A) und das Ende (d_E) der Ökosphäre die Formeln:

$$\begin{cases} d_A = \left(\frac{T^*}{T_A}\right)^2 \cdot \frac{R_*}{2} \\ d_E = \left(\frac{T^*}{T_E}\right)^2 \cdot \frac{R^*}{2}. \end{cases} \tag{4}$$

Es hat sich gezeigt, daß man durch Verwendung der Werte (1) folgendes findet:

$$d_E = 3{,}024 \, d_A,$$

und daß man mit großer Annäherung annehmen kann:

$$d_E = 3 \, d_A. \tag{5}$$

Auf Grund der Gl. (5) sehen wir auch, daß die Dicke der Ökosphäre ($d_E - d_A$) gleich ist der Entfernung der Mitte der Ökosphäre vom Mittelpunkt des Sternes $\frac{d_E + d_A}{2}$:

$$d_E - d_A = \frac{d_E + d_A}{2}. \tag{6}$$

Die Sonnenökosphäre

Wenn man in die Gl. (4) die Zahlenwerte:

$$T_\bigcirc = 5\,700° \text{ K und } R_\bigcirc = 695\,300 \text{ km}$$

einsetzt, erhält man:

$$\begin{cases} d_{A,\,\bigcirc} = 0{,}61 \text{ Astr. Einh.} = 91{,}5 \times 10^6 \text{ km} \\ d_{E,\,\bigcirc} = 1{,}83 \text{ Astr. Einh.} = 274{,}5 \times 10^6 \text{ km}. \end{cases}$$

Die Sonnenökosphäre umfaßt also die Planeten Venus, Erde und Mars (Abb. 1). Von den mit Planeten besetzten Räumen der Nachbarschaft der Sonne umfaßt die Sonnenökosphäre kaum 1/5000. Die „ökosphärischen" Planeten erhalten ungefähr den 1/800 000 000 Teil der gesamten Sonnenstrahlung.

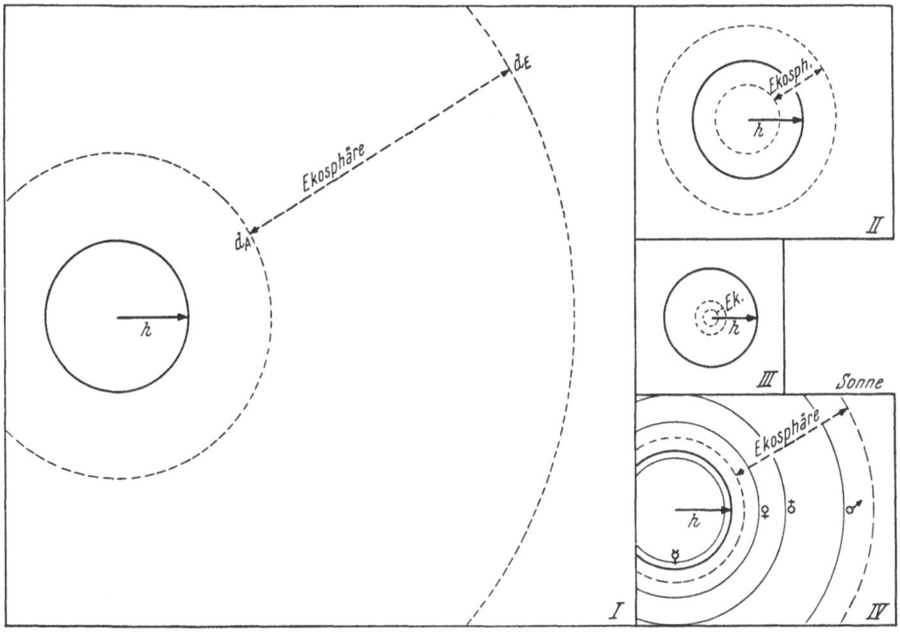

Abb. 1

Die Sternenökosphären im Radius von 17 Lichtjahren um die Sonne

Wir stützen uns auf den Sternkatalog der Nachbarschaft der Sonne von P. VAN DE KAMP (Publications of the Astronomical Society of Pacific, April 1953). Dieser Katalog enthält 11 verschiedene Parameter, die 42 bis jetzt in diesem Raume entdeckte Sternsysteme betreffen. Dieser Raum wird vielleicht während der Lebensdauer eines Menschen mit Hilfe der „Photonenrakete" erreichbar sein. Es fehlen aber im genannten Katalog die Werte der Parameter T_* und R_*, die zum Berechnen der Sternenökosphären nötig sind. Diese Werte wurden vom Verfasser mit Hilfe der in der Astrophysik benutzten Gln. (7) und (8) berechnet:

$$T_* = \frac{11\,700^\circ\,\mathrm{K}}{2{,}03 - \log B_*} \tag{7}$$

$$R_* = \sqrt{\frac{L_*}{B_*}}, \tag{8}$$

wo B_* die visuelle Flächenhelligkeit des Sternes, interpoliert mit Hilfe der Tab. I, und L_* die visuelle totale Helligkeit des Sternes unter der Voraussetzung $L_\odot = 1$ bezeichnen. Die Sternmassen wurden auch aus Tab. I interpoliert.

Tabelle I

Spektrum *	B_* ($B_\odot = 1$)	M_* ($M_\odot = 1$)	Spektrum *	B_* ($B_\odot = 1$)	M_* ($M_\odot = 1$)	Spektrum *	B_* ($B_\odot = 1$)	M_* ($M_\odot = 1$)
B0	32,4	16	F5	1,91	1,5	dM0	0,043	0,52
B5	17,8	7	dG0	1,02	1,25	dM2	0,039	0,44
A0	9,8	4	dG5	0,74	1,07	dM4	0,0354	0,40
A5	5,5	2,2	dK0	0,44	0,85	dM6	0,0170	0,36
F0	3,23	1,8	dK5	1,117	0,65	dM9	0,0135	0,32

Tabelle II

Nr.	*	Lichtjahre	Visuelle Helligkeit und Spektrum			Visuelle absolute Helligkeit			L_* ($L_\odot = 1$)		
			A	B	C	A	B	C	A	B	C
1	Sonne	—	—26^m,9 G0	—	—	4^M,7	—	—	1,0	—	—
2	α Cen	4,3	0,3 G0	1^m,7 K5	11^mM5e	4,7	6^M,1	15^M,4	1,0	0,28	0,000052
3	Barnard's Stern	6,0	9,5 M5	*	—	19,7	*	—	0,00040	*	—
4	Wolf 359	7,7	13,5 M6e	—	—	16,6	—	—	0,000017	—	—
5	Luyten 726—8	7,9	12,5 M6e	13,0 M6e	—	15,6	16,1	—	0,00004	0,00003	—
6	Lalande 21185	8,2	7,5 M2	*	—	10,5	*	—	0,0048	*	—
7	Sirius	8,7	—1,6 A0	7,1 wd	—	1,3	10,0	—	23	0,008	—
8	Ross 154	9,3	10,6 M5e	—	—	13,3	—	—	0,00036	—	—
9	Ross 248	10,3	12,2 M6e	—	—	14,7	—	—	0,00010	—	—
10	ε Eri	10,8	3,8 K2	—	—	6,2	—	—	0,25	—	—
11	Ross 128	10,9	11,1 M5	—	—	13,5	—	—	0,00030	—	—
12	61 Cyg	11,1	5,6 K6	6,3 M0	*	7,9	8,6	*	0,052	0,028	*
13	Luyten 789—6	11,2	12,2 M6	—	—	14,5	—	—	0,00012	—	—
14	Prokyon	11,3	0,5 F5	10,8 wd	—	2,8	13,1	—	5,8	0,00044	—
15	ε Ind	11,4	4,7 K5	—	—	7,0	—	—	0,12	—	—
16	Σ 2398	11,6	8,9 M4	9,7 M4	—	11,1	11,9	—	0,0028	0,0013	—
17	Groombridge 34	11,7	8,1 M2e	10,9 M4e	—	10,3	13,1	—	0,0058	0,00044	—
18	τ Cet	11,8	3,6 G4	—	—	5,8	—	—	0,36	—	—
19	Lacaille 9352	11,9	7,2 M2	—	—	9,4	—	—	0,013	—	—
20	BD + 5° 1668	12,4	10,1 M4	—	—	12,2	—	—	0,0010	—	—

21	Lacaille 8760	12,8	6,6 M1	—	—	8,6	—	—	0,028	—	—
22	Kapteyn's Stern	13,0	9,2 M0	—	—	11,2	—	—	0,0025	—	—
23	Krüger 60	13,1	9,9 M4	11,4 M5e	—	11,9	13,4	—	0,0013	0,00033	—
24	Ross 614	13,1	10,9 M5e	**	—	12,9	**	—	0,00052	**	—
25	BD—12° 4523	13,4	10,0 M5	—	—	11,9	—	—	0,0013	—	—
26	van Maanen's Stern	13,8	12,3 wdF	—	—	14,2	—	—	0,00016	—	—
27	Wolf 424	14,6	12,6 M6e	12,6 M6e	—	14,3	14,3	—	0,00014	0,00014	—
28	Groombridge 1618	14,7	6,8 K5	—	—	8,5	—	—	0,030	—	—
29	CD—37° 15492	14,9	8,6 M3	—	—	10,3	—	—	0,0058	—	—
30	CD—46° 11540	15,3	9,7 M4	—	—	11,3	—	—	0,0023	—	—
31	BD + 20° 2465	15,4	7,5 M4e	*	—	11,1	*	—	0,0028	*	—
32	CD—44° 11909	15,6	11,2 M5	—	—	12,8	—	—	0,00058	—	—
33	CD—49° 13515	15,6	9 M3	—	—	10,6	—	—	0,0044	—	—
34	A0e 17415—6	15,8	9,1 M3	—	—	10,7	—	—	0,0040	—	—
35	Ross 780	15,8	10,2 M5	—	—	11,8	—	—	0,0014	—	—
36	Lalande 25372	15,9	8,6 M2	—	—	10,2	—	—	0,0063	—	—
37	CC 658	16,0	11 wd	—	—	12,5	—	—	0,0008	—	—
38	o^2 Eri	16,3	4,5 K0	9,2 wdA	11,0M5e	6,0	10,7	12,5	0,30	0,0040	0,0008
39	70 Oph	16,4	4,2 K1	5,9 K5	—	5,7	7,4	—	0,40	0,083	—
40	Altair	16,5	0,9 A5	—	—	2,4	—	—	8,3	—	—
41	BD + 43° 4305	16,5	10,2 M5e	—	—	11,7	—	—	0,0016	—	—
42	AC 79° 3888	16,6	11,0 M4	—	—	12,5	—	—	0,0008	—	—

* — unsichtbarer Begleiter ** — beobachtet mit Hilfe des 5-Meter-Teleskops.

Fortsetzung der Tabelle II

Nr.	*	R_* ($R_\odot = 1$)			M_* ($M_\odot = 1$)			T_* °K			d_A 10^6 km			d_E 10^6 km			h_E 10^6 km		
		A	B	C	A	B	C	A	B	C	A	B	C	A	B	C	A	B	C
1	Sonne	1	—	—	1	—	—	5700	—	—	91,5	—	—	274,5	—	—	77	—	—
2	α Cen	0,99	1,58	0,045	1,25	0,65	0,38	5150	4950	3250	75,5	107,7	14,0	226,5	323,1	41,9	83	67	56
3	Barnard's-Stern	0,12	*	—	0,38	*	—	4500	*	—	6,9	*	—	20,7	*	—	56	*	—
4	Wolf 359	0,032	—	—	0,25	—	—	3100	—	—	0,8	—	—	2,5	—	—	49	—	—
5	Luyten 726-8	0,039	0,042	—	0,25	0,36	—	3100	3100	—	1,0	1,1	—	3,0	3,3	—	49	55	—
6	Lalande 21185	0,351	*	—	0,44	*	—	3200	*	—	9,4	*	—	28,2	*	—	59	*	—
7	Sirius	1,53	0,030	—	2,35	0,98	—	9000	8600	—	336,3	6,3	—	1008,9	18,8	—	102	76	—
8	Ross 154	0,12	—	—	0,38	—	—	3000	—	—	3,0	—	—	9,0	—	—	56	—	—
9	Ross 248	0,077	—	—	0,36	—	—	2900	—	—	1,8	—	—	5,3	—	—	55	—	—
10	ε Eri	0,89	—	—	0,77	—	—	4200	—	—	43,3	—	—	130,0	—	—	71	—	—
11	Ross 128	0,11	—	—	0,38	—	—	3000	—	—	2,7	—	—	8,1	—	—	56	—	—
12	61 Cyg	0,71	0,81	*	0,63	0,52	*	3600	3200	*	25,2	22,9	*	75,6	68,7	*	66	62	*
13	Luyten 789-6	0,084	—	—	0,38	—	—	2900	—	—	2,0	—	—	6,0	—	—	56	—	—
14	Prokyon	1,74	0,007	—	1,8	0,46	—	6700	7500	—	217,6	1,9	—	652,0	5,7	—	94	59	—
15	ε Ind	1,01	—	—	0,65	—	—	3950	—	—	44,1	—	—	132,3	—	—	67	—	—
16	Σ 2398	0,28	0,19	—	0,41	0,41	—	3400	2300	—	8,9	2,7	—	26,7	8,1	—	57	57	—
17	Groombridge 34	0,49	0,12	—	0,44	0,40	—	3200	3100	—	10,7	3,0	—	32,1	9,0	—	59	57	—
18	τ Cet	0,57	—	—	1,1	—	—	5400	—	—	42,8	—	—	129,4	—	—	79	—	—
19	Lacaille 9352	0,28	—	—	0,44	—	—	3200	—	—	7,7	—	—	23,1	—	—	59	—	—
20	BD + 5° 1668	0,17	—	—	0,40	—	—	3350	—	—	5,3	—	—	15,9	—	—	57	—	—

21	Lacaille 8760	0,83	—	—	0,48	—	—	3200	—	—	23,2	—	—	69,6	—	—	60	—	—
22	Kapteyn's Stern	0,24	—	—	0,52	—	—	3200	—	—	6,9	—	—	20,6	—	—	62	—	—
23	Krüger 60	0,19	0,11	—	0,40	0,38	—	3100	3000	—	5,2	2,8	—	15,6	8,5	—	57	56	—
24	Ross 614	0,14	**	—	0,38	**	—	3000	**	—	3,6	**	—	10,7	**	—	56	**	—
25	BD—12° 4523	0,22	—	—	0,38	—	—	3000	—	—	5,6	—	—	16,9	—	—	56	—	—
26	van Maanen's Stern	0,009	—	—	>0,11	—	—	7000	—	—	1,2	—	—	3,7	—	—	>37	—	—
27	Wolf 424	0,091	0,091	—	0,36	0,36	—	3100	3100	—	2,4	2,4	—	7,2	7,2	—	55	55	—
28	Groombridge 1618	0,50	—	—	0,65	—	—	3600	—	—	18,5	—	—	55,6	—	—	67	—	—
29	CD—37° 15492	0,40	—	—	0,42	—	—	4300	—	—	20,2	—	—	60,7	—	—	57	—	—
30	CD—46° 11540	0,26	—	—	0,40	—	—	4800	—	—	15,8	—	—	47,4	—	—	57	—	—
31	BD + 20° 2465	0,28	*	—	0,40	*	—	3300	*	—	8,8	*	—	26,6	*	—	57	*	—
32	CD—44° 11909	0,15	—	—	0,38	—	—	3200	—	—	4,4	—	—	13,1	—	—	56	—	—
33	CD—49° 13515	0,34	—	—	0,42	—	—	3400	—	—	11,0	—	—	32,9	—	—	57	—	—
34	AOe 17415—6	0,33	—	—	0,42	—	—	3400	—	—	10,5	—	—	31,4	—	—	57	—	—
35	Ross 780	0,15	—	—	0,38	—	—	3200	—	—	6,8	—	—	20,3	—	—	56		—
36	Lalande 25372	0,40	—	—	0,44	—	—	3400	—	—	13,0	—	—	38,9	—	—	59	—	—
37	CC 658	0,058	—	—	0,7	—	—	8000	—	—	10	—	—	30	—	—	68	—	—
38	o^2 Eri	0,83	0,059	0,17	0,85	1,3	0,38	4400	5900	3200	44,9	8,3	5,1	134,7	24,9	15,3	73	84	56
39	70 Oph	1,04	0,84	—	0,81	0,65	—	4800	3700	—	63,3	37,9	—	190,0	113,7	—	72	67	—
40	Altair	1,23	—	—	2,2	—	—	9100	—	—	282,1	—	—	846,3	—	—	100	—	—
41	BD + 43° 4305	0,25	—	—	0,38	—	—	3100	—	—	6,6	—	—	19,8	—	—	56	—	—
42	AC + 79° 3888	0,15	—	—	0,40	—	—	3400	—	—	4,7	—	—	14,2	—	—	57	—	—

Statistik

Aus den Zahlen der Tab. II geht hervor, daß im betrachteten Weltraum bisher 42 Sternsysteme, die insgesamt 59 Sterne enthalten, festgestellt wurden. Man hat durchschnittlich ein Sternsystem auf je 500 Kubiklichtjahre entdeckt. Es hat sich erwiesen, daß 27 (d. h. 64%) Einzelsterne, 13 (31%) Doppelsterne und 2 (5%) Dreifachsterne sind.

Im betrachteten Weltraum fehlen die Riesen gänzlich. Nur 12 (22%) Sterne sind mit bloßem Auge sichtbar. Kaum 16 (27%) Sterne zeigen Dimensionen ähnlich der Sonne. Der Rest, d. h. 43 (73%) Sterne, sind „Unterzwerge". Die Unterzwerge sind also die Hauptbeherrscher dieses Raumes. Der mittlere Radius der Sterne, die der Sonne ähnlich sind (von 1,53 $R_\odot$ bis 0,50 $R_\odot$), beträgt 1,00 $R_\odot$ und ihre mittlere Masse 1,02 $M_\odot$. Für die Unterzwergsterne (von 0,49 $R_\odot$ bis 0,007 $R_\odot$) erhalten wir Mittelwerte: 0,155 $R_\odot$ und 0,41 $M_\odot$.

Sphäre der Planeten mit gehemmter Rotation

Die thermischen und ökologischen Bedingungen für das Dasein des Lebens auf der Oberfläche der Planeten mit durch die Flutkräfte des Sternes ausgeglichener Rotations- und Revolutionsdauer sind auf einen schmalen Gürtel begrenzt, der auf der stetig bestrahlten Halbkugel liegt. Dieses Problem erfordert ein spezielles Studium. Wir wollen uns hier auf eine angenäherte Berechnung des Endes (h_E) der Sphäre der Planeten mit gehemmter Rotation und seiner Lage betreffs der Ökosphäre der Sterne beschränken.

Die Intensität der Flutkraft (S), die der Stern auf die Planeten ausübt, berechnet man aus der Formel:

$$S = 2\,k^2 \frac{M_* r}{D^3}, \tag{9}$$

wo k die Gravitationskonstante, M_* die Masse des Sternes, r den Radius des Planeten und D seine Distanz vom Zentrum des Sternes bezeichnen.

Wenn wir die Gl. (9) auf die Planeten und die „gehemmten" Monde des Sonnensystems anwenden, erhalten wir:

$$\begin{aligned} S_{☿} &= 2{,}19 \times 10^{-11}\ \text{cm/sec}^2\ \text{gr} \qquad & S_{♂} &= 0{,}04 \times 10^{-11}\ \text{cm/sec}^2\ \text{gr} \\ S_{♀} &= 0{,}84 \times 10^{-11}\ \text{cm/sec}^2\ \text{gr} & S_{☾} &= 16{,}28 \times 10^{-11}\ \text{cm/sec}^2\ \text{gr} \\ S_{♁} &= 0{,}34 \times 10^{-11}\ \text{cm/sec}^2\ \text{gr} & S_{Iapetus} &= 2{,}38 \times 10^{-11}\ \text{cm/sec}^2\ \text{gr}. \end{aligned} \tag{10}$$

Der kleinste Planet unseres Sonnensystems, der auf seiner Oberfläche noch genügende Gravitationskraft zur konstanten Festhaltung der Atmosphäre besitzt, ist Mars. Wenn wir in die Gl. (9) auf Grund der Zahlenwerte (10) für:

$$S = \frac{S_{☿} + S_{♁}}{2} = 1{,}27 \times 10^{-11}\ \text{cm/sec}^2\ \text{gr} \tag{10a}$$

(d. h. den Mittelwert zwischen den Werten für den sicher „gehemmten" Planeten ☿ und den sicher nicht gehemmten ♁) einsetzen, erkennen wir, daß in der Entfernung von der Sonne 77×10^6 km der Planet Mars noch nicht gehemmt wäre. Man kann also die Distanz 77×10^6 km von der Sonne als das Ende des Bereiches der Sphäre der gehemmten Planeten ($h_{E,\odot}$) annehmen[1]. Unter der Annahme,

[1] Die neuen radioteleskopischen Beobachtungen von J. KRAUS geben für die Rotationsdauer der Venus: $22^h\ 17^m - 10^m$. Wenn wir, diese Ergebnisse berücksichtigend, in Gl. (10a) die Venus anstatt des Planeten Erde annehmen, erhalten wir für diese Distanz 74×10^6 km, was unsere Ergebnisse nur wenig ändern würde.

daß die Sterne der betrachteten Umgebung der Sonne ungefähr in demselben Zeitraume entstanden sind, daß also die hemmenden Flutkräfte der Sterne auf die dortigen eventuellen Planeten ungefähr solange wie in unserem Sonnensystem ihren Einfluß ausüben, erhalten wir durch Analogie:

$$h_{E,*} = 77 \sqrt[3]{\frac{M_*}{M_\odot}} \times 10^6 \text{ km}. \qquad (11)$$

Tab. II enthält in der letzten Kolonne die Zahlenwerte des maximalen Bereiches der Sphäre der gehemmten Planeten. (Die Planeten mit Radien, die größer als der Radius des Mars sind, werden schon in einer Distanz gehemmt, die kleiner als $h_{E,*}$ ist.)

Folgerungen

Die hellsten Sterne der Tab. II, nach sich verminderndem Bereich der Ökosphären gereiht, sind in Tab. III zusammengestellt. Die Mittelwerte der Tab. III sind in Tab. IV angegeben (Abb. 1).

Tabelle III

Nr.	*	d_E-d_A 10^6 km	d_A 10^6 km	d_E 10^6 km	R_* $(R_\odot=1)$	M_* $(M_\odot=1)$	h_E 10^6 km	$d_A=h_E$ 10^6 km
7	Sirius A	678	336	1009	1,53	2,35	102	234
40	Altair	564	282	846	1,23	2,2	100	182
14	Prokyon	435	218	653	1,74	1,8	94	224
2	α Cen B	216	107	323	1,58	0,65	67	40
1	Sonne	183	92	275	1,00	1,00	77	15
2	α Cen A	151	76	227	0,99	1,25	83	— 7
39	70 Oph A	127	63	190	1,04	0,81	72	— 9
38	o^2 Eri A	90	45	135	0,83	0,85	73	—28
15	ε Ind	88	44	132	1,01	0,65	67	—23
10	ε Eri	87	43	130	0,89	0,77	71	—28
18	τ Cet	86	43	129	0,57	1,1	79	—36
39	70 Oph B	76	38	114	0,84	0,65	67	—29
12	61 Cyg A	51	25	76	0,71	0,63	66	—41
21	Lacaille 8760	47	23	70	0,83	0,48	60	—37
12	61 Cyg B	46	23	69	0,81	0,52	62	—39
29	CD — 39° 15492	41	20	61	0,40	0,42	57	—37

Tabelle IV (Mittelwerte)

	n	d_A 10^6 km	d_E 10^6 km	R_* $(R_\odot=1)$	M_* $(M_\odot=1)$	h_E 10^6 km	d_A-h_E 10^6 km
I	5	207	621	1,42	1,6	88	139
II	11	40	121	0,81	0,65	69	—29
III	39	7,4	18,7	0,169	0,43	58	—51

Es sind also nur fünf der hellsten Sterne der Tab. II mit breiten Ökosphären umgeben, worin sich keine „gehemmten" Planeten befinden sollen. (Die Sonne nimmt unter ihnen den letzten Platz ein.) Dagegen besitzen 11 Sterne verhältnismäßig geringe, schmale Ökosphären, die teilweise sich mit einer „gehemmten" Sphäre bedecken. Der Rest, die 39 Unterzwergsterne, besitzen ganz winzige und sehr der Oberfläche der Sterne nahe Ökosphären, die vollständig im Innern der „gehemmten" Sphäre gelegen sind. In solchen Ökosphären sind natürlich die ökologischen Möglichkeiten sehr beschränkt. Vom astronautischen Standpunkt am interessantesten sind die Sterne des Typus I und II.

So stellt sich das Ergebnis unserer Überlegungen dar.

Zum Schluß noch eine Bemerkung: Unsere Formeln betreffen die Planeten, die als vollständig schwarze Körper betrachtet wurden. Das ist natürlich in der Natur nicht verwirklicht. Es hat sich aber erwiesen, daß die von Atmosphären umhüllten Planeten sich sehr dieser idealen Vorstellung nähern, wie das aus dem Beispiel der Erde ersichtlich ist. Gl. (3) gibt für die Erde: $T' = +4°C$, während die Klimatologie den Wert $+14°C$ zeigt.

Recovery of a Circum-Lunar Instrument Carrier

By

Carl Gazley, Jr. and **David J. Masson**[1]

(With 5 Figures)

Abstract — Zusammenfassung — Résumé

Recovery of a Circum-Lunar Instrument Carrier. The possibility of the physical recovery of a circum-lunar vehicle widens the scope of scientific investigations possible for a vehicle with lunar capabilities. While a very high guidance capability is necessary to impact such a recoverable vehicle within a given area on the earth's surface, only moderate accuracy is required to effect just a return to earth. Radio tracking during return would enable prediction of the approximate impact point. Ultimate recovery could be accomplished through a radio beacon and overflight search.

While the penetration of the earth's atmosphere involves more severe decelerations and heating than in the case of the recovery of a scientific satellite, the magnitude of the deceleration poses few structural difficulties and the use of a vaporizing surface for heat absorption does not require an excessive weight penalty.

Wiedererlangung eines mit Instrumenten versehenen, den Mond umfahrenden Flugkörpers. Die Möglichkeit der physischen Wiedererlangung eines den Mond umfahrenden Flugkörpers erweitert den Bereich der ausführbaren wissenschaftlichen Untersuchungen fur ein Raumfahrzeug, das den Mond erreichen kann. Während fur das Auftreffen eines solchen wiedergewinnbaren Fahrzeuges innerhalb eines gegebenen Gebietes der Erdoberfläche ein Steuerungssystem von sehr großer Präzision erforderlich ist, wäre eine nur mäßige Genauigkeit notwendig, um eine bloße Rückkehr zur Erde zu bewirken. Bahnverfolgung mit Radio während des Rückfluges wurde die Vorhersage eines angenäherten Auftreffpunktes gestatten. Die schließliche Wiedererlangung wäre mit Hilfe eines Radiosenders und Verfolgung des Überfliegens auszufuhren.

Während die Durchdringung der Erdatmosphäre beträchtlich größere Verzögerungen und Erhitzung mit sich bringt als im Fall der Wiedererlangung eines Forschungssatelliten, verursacht die Größe der Verzögerung nur wenig konstruktive Schwierigkeiten. Auch die Verwendung einer verdampfbaren Oberfläche verursacht keinen Verlust durch überschüssiges Gewicht.

Récupérabilité d'un satellite instrumental circum-lunaire. Les possibilités scientifiques d'un satellite circum-lunaire seraient élargies si ses instruments pouvaient être récupérés. S'il faut un guidage très fin pour limiter le point d'impact à une aire suffisamment restreinte, une précision modérée suffirait à assurer simplement le retour sur terre. Une prédiction du point d'impact pourrait être obtenue en suivant radio-électriquement la phase du retour. La récupération serait finalement accomplie par l'emploi d'une radio-balise aidant les recherches aériennes.

Quoique le retour dans l'atmosphère terrestre implique une décélération et un échauffement cinétique plus sévères que pour un satellite instrumental ordinaire,

[1] The RAND Corporation, 1700 Main St., Santa Monica, Calif., U.S.A.

la grandeur de la décélération ne pose que peu de difficultés structurales et l'emploi de surfaces d'évaporation pour la dissipation des calories ne pénaliserait pas exagérément le poids du véhicule.

Symbols

A_c frontal (cross-sectional) area of body
A_s surface area
C_D drag coefficient
d characteristic dimension of body
e base of natural logarithms
g gravitational acceleration
h altitude
M MACH number
q heating rate
Re REYNOLDS number, $du\varrho/\mu$
u velocity
u_i velocity at initial entry condition
W mass of body
x distance
a exponential coefficient in density relation, $\sigma = e^{-ah}$
ϱ atmospheric density
ϱ_{SL} atmospheric density at sea level
σ density ratio, ϱ/ϱ_{SL}
θ path angle with local horizontal
μ viscosity

Subscripts

FM free molecule
LC laminar convection
R radiation

Introduction

Several recent studies [1] [2] indicate the feasibility of sending an instrumented vehicle to the moon or to the vicinity of the moon. Vehicle weights of several hundred pounds were considered practicable in these studies. While much useful astronomical and geophysical information could be obtained by radio observation and communication with such a vehicle, actual physical recovery would enable a wider scope of investigations. For example, the oft-mentioned photographic observation of the other side of the moon would be possible. Furthermore, the physical recovery of an instrumented lunar vehicle would be a reasonable prerequisite for recovery of interplanetary vehicles, and eventually manned vehicles.

In a previous paper [3] the recovery of a scientific satellite was considered and found feasible. The natural decay of a satellite orbit involves a rather gradual entry into the earth's atmosphere with a maximum deceleration of about nine g's and aerodynamic heating rates which are low enough to be dissipated by thermal radiation from a thin metallic skin at a tolerable temperature. As discussed in a general study of entry into planetary atmospheres [4], a direct entry from a parabolic orbit (an elliptical circum-lunar orbit is close to parabolic) involves much more severe entry deceleration and heating than does entry by orbital decay. The heating is too great to be dissipated by thermal radiation from a surface whose temperature is in the range of current materials. Some other means of heat absorption and surface protection must be used. An expendable surface coating utilizing the material's latent heat of vaporization or sublimation offers promising possibilities. Several plastic materials have a sufficiently high heat of ablation so that the surface-protection system does not involve a severe weight penalty.

Circum-Lunar Trajectories

A circum-lunar trajectory with a return to earth can essentially be considered as an elliptical orbit relative to the earth whose perigee lies near or within the earth and has a sufficiently high eccentricity so that its apogee lies outside the moon's orbit. A launching velocity of about 35,000 ft/sec is required for a circum-lunar path [5]. Such a trajectory is illustrated in Fig. 1.

The required initial accuracy for measurement and control of velocity and direction for various lunar paths have been investigated by LIESKE [5], and are summarized in the following tabulation:

Mission	Δu, ft/sec	$\Delta\theta$, degrees
Just hit the moon	75	0.5
Hit moon, $\Delta x = 100$ miles	4	0.01
Just return to earth	150	10
Return to earth, $\Delta x = 1{,}000$ miles	0.25	0.03
Return to earth tangentially, $\Delta h = 50{,}000$ ft	1.0	0.001
Vanguard: perigee 200 mi alt, apogee 1,400 mi alt	1,000	4

These figures correspond to conditions at 350 miles altitude and a launching in an easterly direction. For purposes of comparison, the figures which have been inferred by LIESKE [5] for the Vanguard launching of the IGY Scientific

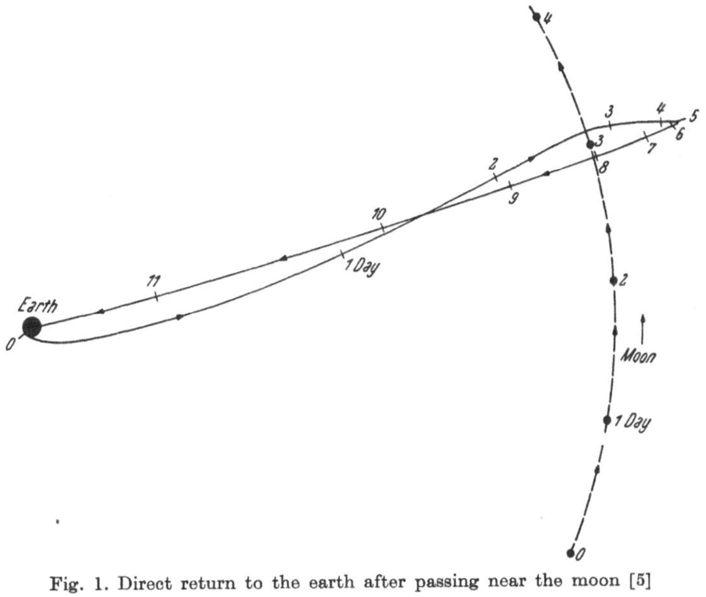

Fig. 1. Direct return to the earth after passing near the moon [5]

Satellite are also given. It is apparent that these apparently obtainable tolerances are similar to those required for a flight to the vicinity of the moon and a return to earth. Thus if one is satisfied with a return to *anywhere* on earth, present guidance capability is sufficient.

The path of a lunar vehicle does, of course, deviate from an ellipse due to the presence of the moon and represents an example of the classic three-body problem. Since no analytic solution is available for the general case of the three-body problem, the prediction of such trajectories is a matter of numerical computation. Such calculations have been presented by LIESKE [5], EHRICKE and GAMOV [2] [6], and EGOROV [7]. Close to the earth, however, the disturbing effects of the moon are negligible and the path may be considered elliptical. The effects of the moon can be accounted for rather simply by saying that the return portion of the ellipse may have a different orientation and energy than the ascent portion of the ellipse due to the disturbing effect of the moon. Thus, without precise computation, it is possible to say that the return path of the vehicle is an elliptical path with respect to the earth with a velocity close to the earth about the same as the launching velocity—35,000 ft/sec. However, it is not possible, analytically, to define the orientation, eccentricity, or total energy of this ellipse. As described in the next section, these can be ascertained by radio observation and communication with the vehicle during the return path with sufficient accuracy so as to yield an approximate impact location[1].

Tracking and Impact Location

Of the means available for tracking the circum-lunar instrument carrier during its flight a radio-tracking system appears to be the most suitable [8] [9]. Optical tracking would require a very large body, a light source, a gas cloud, or some other means of making the vehicle visible.

In a radio-tracking system for predicting the location of the impact area, more than one type of measurement is possible. For example one could measure:

(1) Two angular positions and the corresponding range with the times of observation.

(2) Three angular positions.

(3) One angular position and corresponding range together with the angular and range rates to give velocity.

For the purpose of this report, it is assumed that the first set of measurements are made, i.e., angle and range are measured at two points of the vehicle trajectory. It is anticipated that more than one tracking station will be used, and that actually several measurements will be made. Multiple measurements will make it possible to reduce the error by using data smoothing and also will yield a satisfactory impact-point prediction a reasonable time before the impact actually occurs.

The equipment in the vehicle necessary to accomplish these measurements consists of a transponder emitting on the order of one watt. With a one watt signal the vehicle can be tracked during its entire flight except for the relatively short time that the moon is directly between the vehicle and the earth. In order to keep the vehicle's antenna correctly oriented, as well as to correctly orient other instrumentation and photographic equipment, a moderate spin rate could be used.

Angle accuracies of 0.2 mil or 40 seconds of arc and range to one nautical mile appear feasible for ranges half the distance to the moon or less and would yield a prediction error of about $\pm$ 150 n mi in the plane of the trajectory, and considerably less or about 50 n mi perpendicular to the plane of the trajectory [8]. This assumes that the last measurement is made at a range of about 3,000 n mi, or about 20 min before impact in a typical case. It is interesting to note that the predicted impact position error is somewhat smaller than in the case of a recoverable scientific satellite [3]. By flying over the impact area with a direction finder, the package can be more precisely located to within a circle of about 50 ft radius [10].

[1] Of course, with sufficiently accurate knowledge of the launching (burn-out) conditions, and with sufficiently accurate knowledge of the astronomical constants, one could predict (by numerical integration) the return trajectory—at least accurately enough for acquisition and perhaps ultimately for impact prediction. However, as indicated by the tabulation of LIESKE [5] presented above, the accuracy requirements would be rather severe.

Atmospheric Entry

The effects of the earth's atmosphere on the vehicle's path and velocity is dependent both on the approach path and on the vehicle's mass-drag characteristics. Because of the guidance uncertainties outlined above, one must be content with a random approach to earth, and the vehicle must be designed for the most severe entry—a direct approach at 90° to the local horizontal.

Except for the unlikely possibility of an essentially tangential approach to the atmosphere with a resulting shallow path through the atmosphere, the portion of the elliptical path within the atmosphere is essentially a linear path

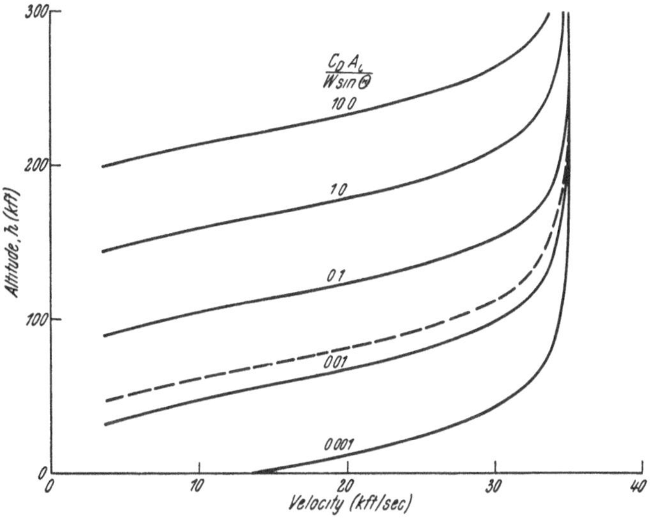

Fig. 2. Velocity variation with altitude for direct entry into the earth's atmosphere from a circum-lunar orbit

during most of the deceleration. For such a direct entry into the atmosphere, an approximate analysis [4] yields a velocity variation for a simple drag (i.e., non-lifting) body

$$\frac{u}{u_i} = e^{-\frac{C_D A_c}{W \sin \theta} \frac{\varrho_{SL}}{2a} \sigma} . \tag{1}$$

The derivation of Eq. (1) involves the following assumptions:

(a) gravitational force small compared to aerodynamic drag force

(b) path angle (θ), drag coefficient (C_D), frontal area (A_c), and mass (W) are constant

(c) air density variation with altitude is given by $\sigma = e^{-ah}$.

Comparison of the results of Eq. (1) with exact numerical computation indicates good agreement until the velocity has been reduced to perhaps 5,000 ft/sec where the path angle begins to deviate from its initial value. Since the major heating and deceleration associated with atmospheric entry occur at higher velocities, this equation is adequate for heating and dynamic calculations.

The velocity variation with altitude for entry into the earth's atmosphere with an initial velocity of 35,000 ft/sec is shown in Fig. 2 for several values of the drag-mass parameter, $C_D A_c/W \sin\theta$. It is seen that similar curves result, with relative displacements for various values of the drag-mass parameter.

One interesting facet of these results is that the major deceleration of a body during entry takes place in a stratum of the atmosphere about 100,000 ft thick. The level of this stratum in the atmosphere is, however, dependent on the value of $C_D A_c/W \sin\theta$. Large values of this parameter cause deceleration high in the atmosphere, while smaller values delay deceleration until the lower atmosphere is reached.

The maximum deceleration occurs when the velocity has been reduced to about 61 per cent of its initial value

$$\frac{u}{u_i} = \frac{1}{\sqrt{e}} = 0.607 .$$

This maximum deceleration

$$\left(-\frac{du}{dt}\right)_{max} = \frac{a\, u_i^2 \sin\theta}{2\, e} \tag{2}$$

is independent of vehicle's size, shape, and mass and dependent only on the initial velocity and on the entry angle. For the initial velocity of about 35,000 ft/sec of the present case, this yields a maximum deceleration of about 300 g's for a direct 90° entry into the earth's atmosphere. For more shallow entries this would be reduced by the sine of the entry angle, as Eq. (2) indicates. The altitude of maximum deceleration is dependent on the entry angle as well as on the drag-mass characteristics of the body.

As in any preliminary design, the tentative choice of parameters involves a cyclic process in which considerations of structure, heating, recovery, etc. all contribute to the final choice. Although the reasons for choosing various phases of the vehicle design will be developed in later sections, it is necessary for discussion purposes to describe that design (shown schematically in Fig. 5). It consists essentially of a spherical body with a heat absorbing coating over the front face, having a diameter of 3.0 ft and a weight of 400 lb. The weight was chosen as representative of those discussed as possible for lunar vehicles [1] [2]. The size and shape evolved from heating and recovery considerations. Instrumentation, radio equipment, batteries, and impact structure are arranged so as to enable entry stability and recovery. The drag-mass parameter for this vehicle has a value of $C_D A_c/W = 0.0177$ sq ft/lb and its velocity variation is indicated by the dashed curve in Fig. 2. This results in a maximum deceleration for direct entry at an altitude of about 80,000 ft.

As previously noted the orientation of the vehicle is controlled by a moderate spin rate so that the vehicle is properly oriented in the vicinity of the moon for photography, etc. This means that the original orientation will be approximately preserved during the return trajectory, so that initial entry may be rear end first. As described in a later section, the center of gravity is displaced toward the front of the body. Thus aerodynamic forces will tend to orient the body during entry. The design should, of course, be such that the orientation is completed before the heating is appreciable. A dynamic analysis of the postulated vehicle indicates that, for an essentially backward initial entry, the vehicle becomes righted at an altitude of about 250,000 ft. It then oscillates about the correct orientation; as the altitude decreases, the oscillations decrease in magnitude and increase in frequency. In the region of maximum heating and deceleration, the amplitude is about 10° and the frequency about 15 cycles/sec.

Ultimately, the body falls vertically at terminal velocity

$$u = \frac{1}{\sqrt{\frac{C_D A_c}{W} \frac{\varrho_{SL}}{2g} \sigma}} \tag{3}$$

which results in an impact velocity of about 300 ft/sec for the body described above.

Heating During Entry

During penetration of the atmosphere, a vehicle's kinetic energy is converted into thermal energy of the surrounding air. Some of this thermal energy is transferred to the body as heat. The rate of this transfer varies during descent both with air density and vehicle velocity. Heat is transferred by both convection and radiation from the hot gas 'cap' over the front of the body to the body's surface. The rates of both convective and radiative heat transfer increase with air density and vehicle velocity, and are thus most severe when high velocities are allowed to persist into the lower atmosphere.

Below about 300,000 ft altitude the atmosphere is dense enough to give an effective continuum type of flow. Here a shock wave occurs ahead of the body and the thermal energy appears in the hot 'shocked' air between the shock wave and the body. Passage through the shock wave increases the air density by a factor of ten or so, increases the temperature 10 to 50 fold, and causes appreciable dissociation and some ionization. Heat is transferred from this heated region to the vehicle surface by convection (and conduction) through the viscous boundary layer and by radiation from the hot gas. When the boundary layer is of the laminar type (say above about 100,000 ft altitude) the convective heating rate per unit *frontal* area may be approximated, for relatively blunt bodies, as [4]

$$\left(\frac{q}{A_c}\right)_{LC} = \frac{1}{\sqrt{\left(\frac{Re}{Md}\right)_{SL} d}} \frac{\varrho_{SL}}{2} \sqrt{\sigma}\, u^3 . \tag{4}$$

It will be noted that this indicates a variation in heating rate per unit *surface* area with the sine of the angle of surface inclination. A turbulent boundary-layer condition, occuring in the lower atmosphere, results in heating rates which are higher by about an order of magnitude.

Surface heating by radiation from the hot-gas region is still somewhat a matter of conjecture. Preliminary work [11] [12], however, indicates that it is appreciable only in the region of the stagnation point; here it can be expressed approximately as

$$\left(\frac{q}{A_s}\right)_R \sim \varrho^{3/2}\, d\, u^{10} . \tag{5}$$

Aft of the stagnation point it falls off much more rapidly than the convective heating—approximately as the fourth power of the surface inclination. Thus even though radiation may be as large as convection in the region of the stagnation point, its contribution to the overall heat load may be small.

The heating rate increases during the initial stages of entry because of the increasing atmospheric density. However, it reaches a maximum value when the velocity has been reduced to about 85 per cent of its initial value where the rate of decrease of velocity overcomes the rate of increase of atmospheric density. Thereafter, the heating decreases as additional deceleration occurs. As described more extensively elsewhere [4] the aerodynamic heat input can be balanced by thermal radiation from a thin metallic skin providing the heating rate is low enough so that the maximum equilibrium surface temperature is allowable for the surface material. For direct 90° entry into the earth's atmosphere from space, drag-mass ratios ($C_D A_C/W$) greater than about 10 sq ft/lb are required [4] to reduce the heating rates sufficiently. This would require a very light drag brake such as a parachute. In the present case, such complexity was not desired. Therefore the radiation technique of heat rejection was discarded and some means of heat absorption sought.

One possibility is the use of a thick metallic skin to absorb the heat by temperature. Such a system is not very efficient from a weight standpoint, however, absorbing only about 100 Btu/lb [13]. More efficient is the use of a surface material which absorbs heat by a phase change. Heat absorptions in the order of 1,000 Btu/lb or more appear to be obtainable through the use of a material which vaporizes [13]. For example, it has been estimated that the depolymerization of Teflon will absorb somewhat more than 1,000 Btu/lb [14]. The required weight of such a vaporizing surface material depends on the total heat input during atmospheric penetration. If we integrate, for example, the laminar convective heating rate [Eq. (4)] over the time of entry, the total heat load is found to be

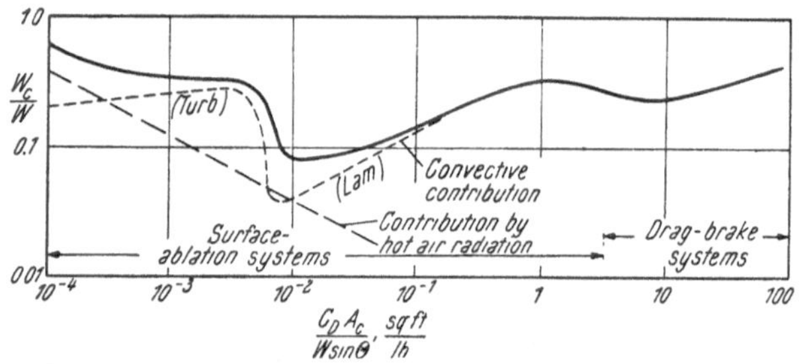

Fig. 3. Cooling system weight requirements for direct entry from a lunar orbit

$$\left(\frac{Q}{A}\right)_{LC} = \frac{\sqrt{\pi}\,\varrho_{SL}\,u_i^2\,\operatorname{erf}\left(\sqrt{\frac{C_D A_c}{W\sin\theta}\,\frac{\varrho_{SL}}{a}}\right)}{2\,a\sin\theta\sqrt{\frac{C_D A_c}{W\sin\theta}\,\frac{\varrho_{SL}}{a}}\sqrt{\left(\frac{Re}{Md}\right)_{SL} d}} \tag{6}$$

and for a surface material having a heat absorption per unit weight, λ, the weight of cooling system is found to be, in ratio to the total weight,

$$\frac{W_c}{W} = \frac{\sqrt{\pi}\,u_i^2\sqrt{\frac{C_D A_c}{W\sin\theta}\,\frac{\varrho_{SL}}{a}}\,\operatorname{erf}\sqrt{\frac{C_D A_c}{W\sin\theta}\,\frac{\varrho_{SL}}{a}}}{2\,C_D\lambda\sqrt{\left(\frac{Re}{Md}\right)_{SL} d}}\,. \tag{7}$$ [1]

This equation yields the rather surprising conclusion that the cooling system weight requirement increases with the drag-mass parameter. A somewhat similar expression can be derived for turbulent convective heating. The cooling system weight required to counter the hot-air radiative heating, however, exhibits the opposite trend with drag-mass parameter. A qualitative picture of the weight requirements for a cooling system or drag brakes is indicated in Fig. 3 as a function of the drag-mass parameter. Two minima appear: one where the weight required

[1] This, of course, presupposes that the ratio W_c/W is relatively small compared to unity, since the velocity variation, Eq. (1), assumes a constant mass. For larger values of the weight ratio, a meteor-type analysis which takes into account the changing mass would have to be used.

to absorb the sum of the convective and radiative heating is a minimum, the other where the drag-mass parameter is just large enough so that radiative cooling is possible and yet not large enough to require a large weight in drag brakes. To avoid the complexity of drag brakes the first of these minima was chosen. The choice, as described previously, was a spherical body 3 ft in diameter and weighing 400 lb. The average heating rate variation during descent for this body is shown in Fig. 4 as the total of laminar convective and radiative heating. It is seen that, for this case, radiative heating is appreciable compared to the convective heating.

Using a vaporizing material with a heat-absorbing ability of 1,000 Btu/lb, about 40 or 50 lb is predicted to vaporize during descent. Using a relatively generous factor of safety, 125 lb of heat-absorbing surface has been postulated for this circum-lunar vehicle.

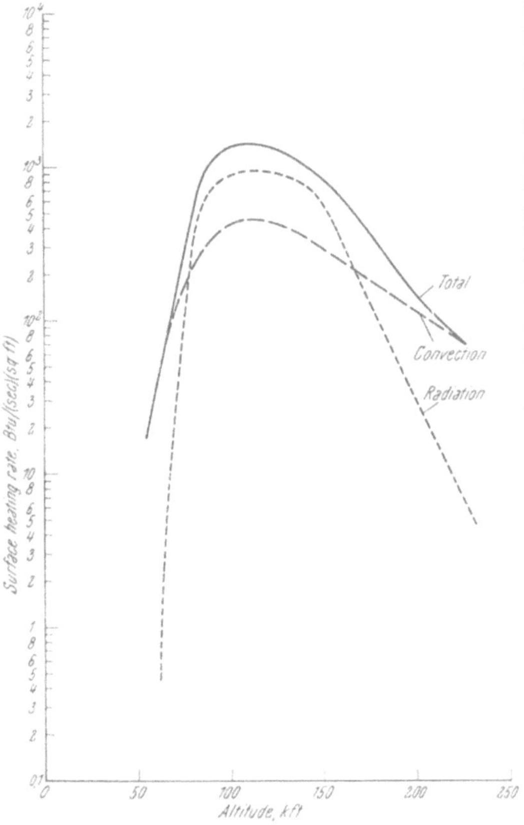

Fig. 4. Heating rate during direct entry of circum-lunar vehicle

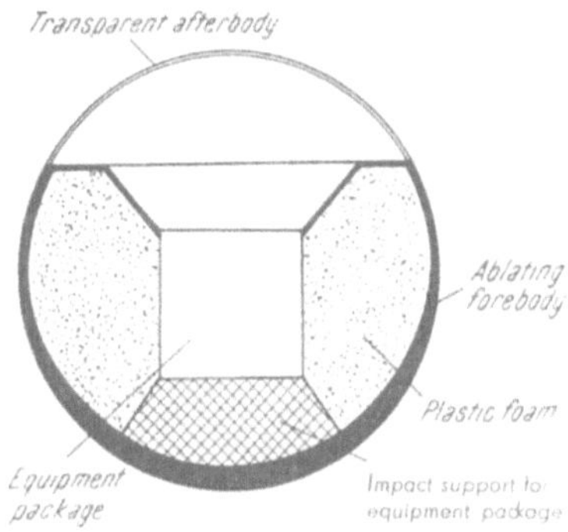

Fig. 5. Schematic of circum-lunar instrument carrier

Vehicle Design Considerations

In an initial program for the recovery of a circum-lunar instrument carrier, it would be desirable to maintain the greatest possible degree of simplicity of the return package. This would increase our ability to predict reliably the behavior of the vehicle during the various phases of its operation. Therefore, there should be a minimum of reliance on such devices as drag brakes at re-entry, reverse rockets just before impact, and the like. The package must survive its impact on its return to earth and the radio beacon must continue to operate after impact.

The proposed vehicle is shown schematically in Fig. 5. The outside configuration consists of a sphere of 3 ft diameter. Since the total weight is to be on the order of 400 lb, a 3 ft diameter is necessary to give the proper drag-mass ratio and average density low enough to insure proper flotation after impact:

Here, the spherical forebody is made up of some 125 lb of plastic (e.g., Teflon) to protect the vehicle against the high heat inputs it will be subjected to during re-entry into the earth's atmosphere. This forebody portion actually extends over the forward 240° of the sphere in order to insure heat protection to the package while it is still oscillating during the high heat-transfer portion of its trajectory.

The antennas are located on the flat plastic disc used to close off the segmented spherical forebody. The afterbody consists of a spherical segment of the same diameter as the forebody. Transparent windows are provided in order to allow photographic observation.

The instruments, radio, batteries, etc. are connected to the forebody by means of a properly chosen plastic with properties compatible with the deceleration loads that can be tolerated by the internal components (several plastic materials with suitable combination of properties are available). The void space in the instrument carrier would be filled with a foaming plastic with sufficient rigidity and strength that the package would float with the proper orientation after impact. Electronic equipment, required to operate after impact would be potted in order to insure its proper operation after having been subjected to large g-loads during a water impact.

On re-entry a maximum deceleration load of about 300 g's takes place at about 80,000 ft altitude for a vertical (90°) entry. Since 70 to 80 per cent of the earth's surface is covered with water, the vehicle is designed for a water impact. In this case an analysis shows that the internal equipment can be protected and the deceleration loads kept to below about 1,500 g's. Electronic equipment can easily be made to withstand the forces involved.

References

1. G. H. CLEMENT, The Moon Rocket. Franklin Institute Monograph No. 2, June 1956.
2. K. A. EHRICKE and G. GAMOV, A Rocket Around the Moon. Scientific American **196**, No. 6, 47 (1957).
3. C. GAZLEY, JR., and D. J. MASSON, A Recoverable Scientific Satellite. The RAND Corporation, Paper P-958, October 1956.
4. C. GAZLEY, JR., Deceleration and Heating of a Body Entering a Planetary Atmosphere from Space. Convair-OSR Astronautics Symposium, San Diego, February 1957.
5. H. A. LIESKE, Accuracy Requirements for Trajectories in the Earth-Moon System. Convair-OSR Astronautics Symposium, San Diego, February 1957.
6. K. A. EHRICKE, Cislunar Operations. Paper presented at American Rocket Society Meeting, San Francisco, June 12, 1957.
7. V. A. EGOROV, Some Questions on the Dynamics of Flight to the Moon. Doklady Academii Nauk SSSR **113**, No. 1, 46 (1957).
8. R. T. GABLER and H. R. O'MARA, The RAND Corporation, Personal communication, August 1957.
9. R. T. GABLER and H. R. O'MARA, Tracking and Communication for a Moon Rocket. Convair-OSR Astronautics Symposium, San Diego, June 1957.
10. A. W. HARBAUGH, The RAND Corporation, Personal communication, August 1957.
11. J. KECK, B. KIVEL, and T. WENTINK, JR., Emissivity of High Temperature Air. Heat Transfer and Fluid Mechanics Institute Preprints, June 1957.
12. R. C. MEYEROTT, Absorption Coefficients of Air from 6000° K to 18000° K. The RAND Corporation, Research Memorandum RM-1954, September 1955.
13. D. J. MASSON and C. GAZLEY, JR., Surface-Protection and Cooling Systems for High-Speed Flight. Aeron. Engng. Rev. **15**, No. 11, 46 (1956).
14. R. W. PORTER, Recovery of Data in Physical Form. Franklin Institute Monograph No. 2, June 1956, pp. 103—112.

On the Generation of Temperatures to 30,000° K

By

Peter E. Glaser[1]

(With 9 Figures)

Abstract — Zusammenfassung — Résumé

On the Generation of Temperatures to 30,000° K. The attainment of high temperatures under stable, steady-state conditions was once limited not only by the materials themselves but also by the means of generating such temperatures. Thus research was handicapped, since materials could not be exposed to these high temperatures and their properties studied.

The drawbacks of conventional methods of heating materials are briefly touched upon, and various techniques for heating materials without crucibles and in controlled atmosphere are pointed out.

The convenience of using electromagnetic radiation from a suitable source instead of the source environment itself, and the research applications of furnaces using solar energy and high-intensity electric arcs are discussed in some detail.

In addition, the techniques of the stabilized-gas vortex arc are described and the temperature limits mentioned. Future developments in even higher temperature ranges are touched upon.

Über die Erzeugung von Temperaturen bis 30000° K. Die Erreichung höherer Temperaturen unter stabilen, gleichmäßigen Bedingungen war einst nicht nur durch die Materialien selbst beschränkt, sondern auch durch die Methoden der Erzeugung solcher Temperaturen. Diese Forschungsrichtung war beeinträchtigt, da die Materialien nicht diesen hohen Temperaturen ausgesetzt und ihre Eigenschaften dabei studiert werden konnten.

Auf die Nachteile konventioneller Methoden der Erhitzung von Materialien wird kurz eingegangen und verschiedene Arbeitsweisen zur Erhitzung von Materialien ohne Tiegel und in kontrollierter Atmosphäre werden hervorgehoben.

Etwas eingehender wird der Vorteil der Verwendung elektromagnetischer Strahlung aus einer geeigneten Strahlungsquelle statt der Umgebung der Strahlungsquelle selbst besprochen und ebenso die Forschungsverwendung von Öfen, die Sonnenenergie und elektrische Bögen hoher Intensität benützen.

Des weiteren wird die Technik des stabilisierten Gaswirbel-Bogens beschrieben und auf die Temperaturgrenzen eingegangen. Es werden auch künftige Entwicklungen im Bereich noch höherer Temperaturen berührt.

Sur l'obtention de températures atteignant 30000° K. L'obtention de hautes températures dans des conditions permanentes et stables était autrefois limitée non seulement par les matériaux eux-mêmes mais aussi par les méthodes utilisées. Ceci constituait une limitation aux recherches sur les propriétés des matériaux aux hautes températures.

Les inconvénients des méthodes conventionnelles sont brièvement esquissées et plusieurs techniques proposées pour chauffer les matériaux sans creuset en atmosphère contrôlée.

---

[1] Arthur D. Little, Inc., Cambridge 40, Massachusetts, U.S.A.

Les facilités présentées par le rayonnement électromagnétique de sources appropriées et les applications de fours à rayonnement solaire ou utilisant des arcs électriques de haute intensité sont discutées en détail.

En outre les techniques utilisées dans l'arc à tourbillon gazeux stabilisé sont décrites ainsi que les limitations de température. Les extensions futures aux températures encore plus élevées sont esquissées.

High-temperature research has as its goal the increased understanding of materials for technological processes at temperatures far above those which nature provides on earth. A great deal of physical data on the behavior of materials at high temperatures remains to be gathered to improve current designs for missile skins, gas turbines, rocket motors, and nuclear applications, and to establish criteria for materials that will ultimately spearhead thermonuclear-power developments.

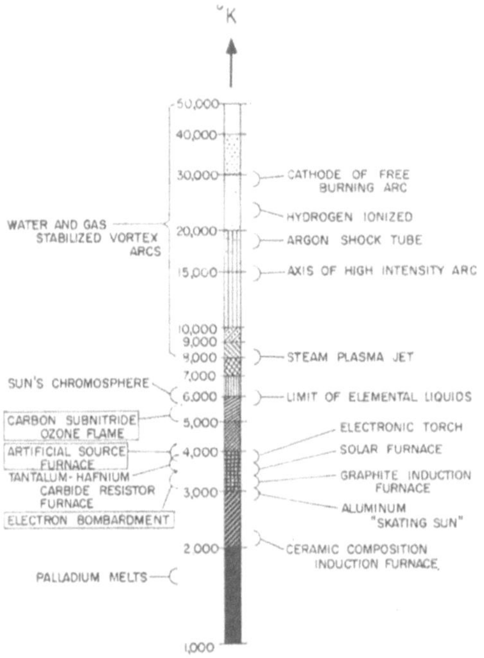

Table I

The attainment of high temperatures from 2000° K to 30,000° K, under stable steady-state conditions, was limited not only by the available materials but also by the means of generating such temperatures. Thus construction materials could not be exposed to these temperatures, and their properties could not be studied. This vicious circle was broken only within the last few years, when new research tools began to push the temperature barrier upward, as Table I shows. Fundamental research now extends to within 1,000,000° K.

The accelerated use of missiles as a practical mode of bridging vast distances highlighted the need for the study of phenomena that occur during a missile's re-entry into the atmosphere. The rate of heat transfer and the functioning of materials at re-entry temperatures of the order of 30,000° K can now be studied in the laboratory. The availability of such temperatures for research has extended the interest in gaseous electronics, and allowed the experimental study of the magnetic-hydrodynamics of plasmas.

The information available on materials at more conventional temperatures, including the physical properties of the more common materials [1] is rather limited. Emphasis has therefore been placed on research efforts covering a temperature range of 2000° K to 5000° K.

The methods that have been used in high-temperature research utilize electrical and chemical energy, and electromagnetic radiation from a high-temperature source. The method to be used for a particular high-temperature-research problem depends upon the nature of the material to be tested, the type of measurements to be made, and the temperatures to be reached.

Electrical Methods

Electrical-resistance furnaces [1] are one of the oldest methods for heating various materials; uniform heating can be obtained with accurate temperature control. Metal, refractory-oxide, carbide, graphite, and granular resistors have been developed. With resistors made of a mixture of tantalum and hafnium carbide, temperatures of 3900° C [2] have been reached in a furnace rich in carbon vapor.

Induction furnaces [1] have found wide use in high-temperature research. Closely controlled graphite induction furnaces have reached 3600° C [3]. These temperatures cannot yet be obtained in an oxidizing or neutral atmosphere. Ceramic-composition susceptors can operate in an oxidizing atmosphere at 2200° C [4].

One special application of induction heating is the drop-by-drop fusion of compressed powdered metal [5] which is gradually introduced into an inductive field created in the interior of acoil. Another is an elegant method of levitation [6], in which a metal is heated while suspended in space by an electromagnetic field generated by a high-frequency current flowing through two separated coils.

Fig. 1. ADL solar furnace in operation

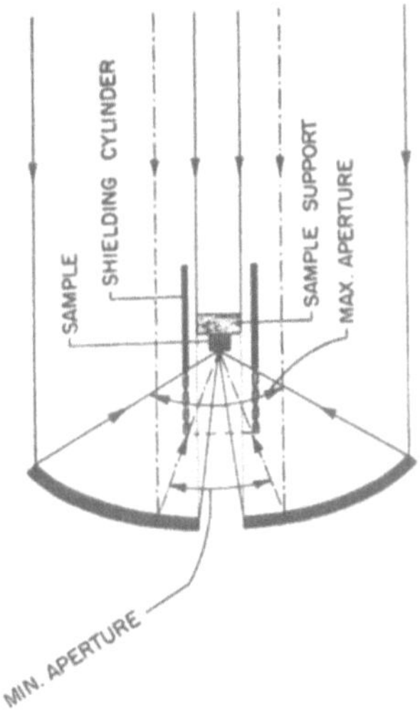

Fig. 2. Principle of optical system

Electric arcs in intimate contact with the material to be heated have found applications in the metallurgical and chemical process fields [7]. In the laboratory they have been used to melt highly refractory metals [8] and study the behavior and various states of atomic or molecular particles.

Electron bombardment [9] has been used to melt and recrystallize small samples in vacuum; an electron gun directs and focuses the electrons into the sample. So that the required current may be reduced, the samples should not be too highly conducting. Electrical insulators cannot be treated in this manner, however, since the charges built up tend to repel the electrons.

Electromagnetic-Radiation Methods

The conditions essential to high-temperature research can be met conveniently when the electromagnetic radiation of a suitable high-temperature source is substituted for the source environment itself. Radiation emitted by a source, and the utilization of its image form in an optical system, have been found a promising approach for the study of high-temperature phenomena.

Fig. 3. View of sample-moving mechanism

One source that can be used very easily is the radiation from the sun. The surface of the sun has an equivalent-black-body temperature of around 5800° C; even with an inefficient optical system, temperatures up to 3500° C can be reached.

The concentration of the sun's rays to generate high temperatures over a limited area overcomes container-contamination problems, since the material to be heated is not exposed to an electric or magnetic field, which in certain investigations may be undesirable. That this method of reaching high temperatures is a most useful one is demonstrated by the increased use of solar furnaces for high-temperature research. Several years ago, the large solar furnace of Professor TROMBE [10] was about the only research installation. Since then, many more have been built in several parts of the world and over thirty are now nearing completion in the United States. Fig. 1 shows the solar furnace designed by Arthur D. Little, Inc., as a tool for high-temperature investigations.

The principle of this solar furnace is shown in Fig. 2. The shielding cylinder is used for achieving fine temperature control of the sample, while the focusing provides for a more sudden temperature change. The sample can be held in any desired position with an accuracy of $^1/_{10}$ millimeter; movement is provided in three directions parallel to the coordinates, as Fig. 3 shows. Electric motors are used to achieve the traversing motion; their speed is varied by controls at the base of the mirror support. One of the advantages of the imaging type of furnace is that instruments can be placed quite close to the sample and the sample observed during the heating cycle (see Fig. 4).

The successful development of high-intensity electric arcs [11, 12] provides a promising alternative to the use of the sun as a source of radiation for the generation of high temperatures. When combined with a suitable optical system, such arcs can heat a sample to a temperature approximately 1000 degrees C higher than the temperature obtained with the sun as a source of radiation. The advantage of this approach is that experiments can be conducted indoors continuously, independent of the vagaries of the weather. The arrangement by which such arcs can be utilized in an artificial-source furnace is shown in Fig. 5. Two parabolic mirrors on a common optical axis are used; the arc is formed in the focus of one mirror, and the sample is placed at the focus of the other mirror. An arc suitable for this purpose has a radiation-flux density of 400 calories per square centimeter per second, as measured in the focal plane of the collecting mirror.

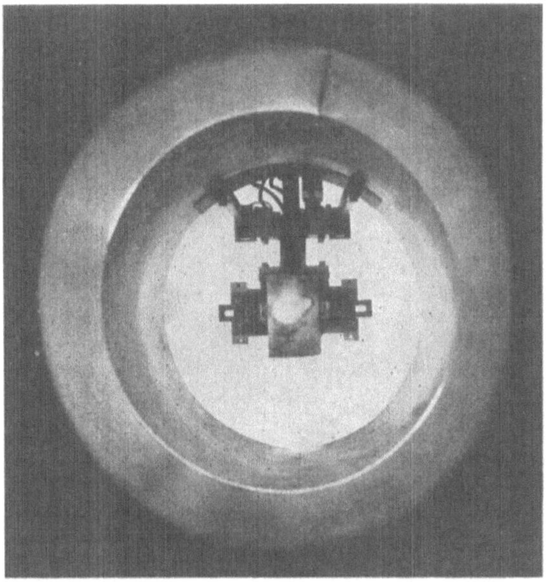

Fig. 4. View of sample through observation post

Another method of focusing an arc's radiation on a sample is shown in Fig. 6. An elliptical mirror [13] is used; the arc is formed in one of the foci, and the sample is placed in the other. The advantage of this arrangement is that the radiation arrives at the sample at a smaller angle of incidence. The modulation of energy arriving at the sample can be accomplished with a unit termed the Mitchell source [14] (see Fig. 7). This unit employs an optical system con-

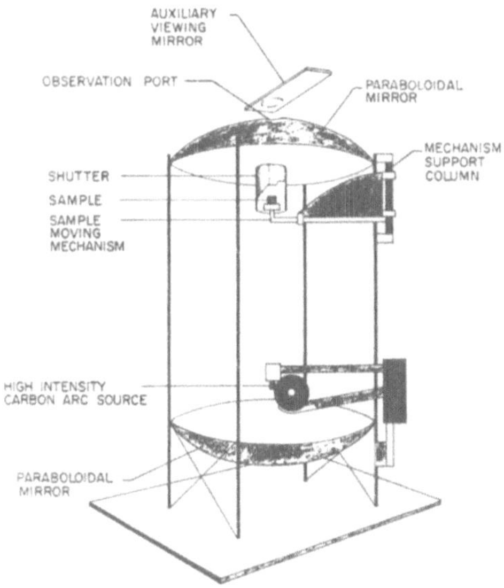

Fig. 5. High intensity arc imaging furnace—two parabolic mirrors

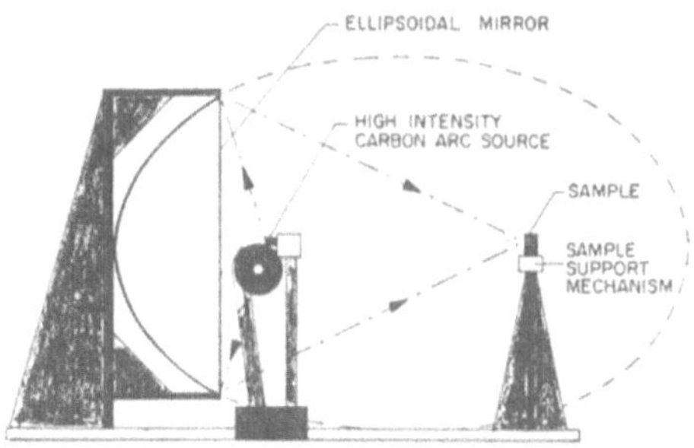

Fig. 6. High intensity arc imaging furnace—one elliptical mirror

sisting of a pair of condenser lenses, which form a prime image of the positive crater; the image is then used as a source for a second pair of lenses. This type of source provides a peak irradiance of 30 calories per square centimeter per second, the maximum image size being $3^1/_2$ centimeter in diameter.

Chemical Methods

Chemical energy has long been a source of heat. The temperatures that can be reached are limited by the dissociation of the molecular species taking part in the chemical reaction, since the energy source itself then ceases to exist. Flames have reportedly reached a temperature of 5500° C [15]. The direct use

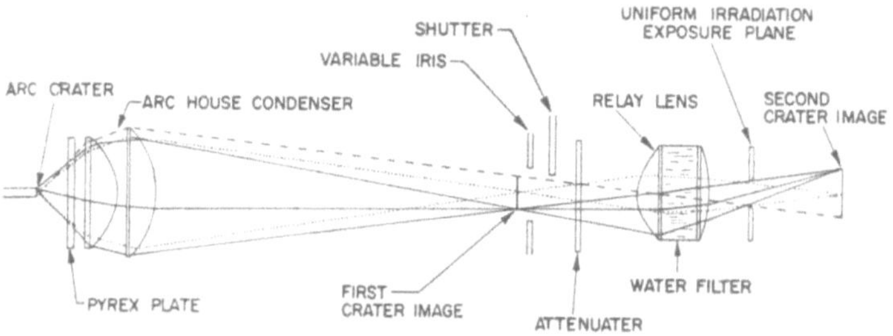

Fig. 7. Mitchell optical system

of flames for heating substances is not convenient, because of the chemical reactions between the sample and the flame. Flames as sources of infra red radiation have the drawback of radiating only in narrow spectral bands; they do not appear to be useful for imaging purposes.

The combustion of metals [16] to provide an intense source of radiation has been an interesting development. The most promising method seems to be the aluminium "skating sun," where the molten metal floats on the liquid oxide, burning in an atmosphere of oxygen with a brightness temperature of 3200° K.

Methods of Heating Gases

The generation of temperatures above those we have mentioned can only be accomplished by the use of materials in their gaseous state, since elemental solids do not exist above 3500° C, and elemental liquids above 5900° C.

To bring a gas stream to a high temperature requires special techniques; normal heat-transfer methods are not satisfactory. The compression of gases by a large-amplitude pressure wave in a shock tube is one of the methods being used. In argon, temperatures of 18,000° K have been measured behind the shock front [17]. Similarly produced high temperatures have permitted the study of the physical behavior of gases and the effect on models installed in shock tubes.

The phenomena last only a few microseconds and the instrumentation that can measure the various parameters is complex. A promising development is the production of continuous shocks, whereby test samples can be subjected to high-temperature gas flows over longer periods.

The steady-state-equilibrium conditions desirable for high-temperature research can be obtained more readily from the energy of electric-arc discharges. An interesting technique is the high-frequency electronic torch [18], where an ultra-high-frequency arc discharge is used to heat polyatomic gases to temperatures of 4000° K.

The temperatures near the cathode of a free burning arc have been measured to be near 30,000° K [19]. Along the axis of a high-intensity arc temperatures reach a value of 15,000° K [20]. The high temperatures obtained in these arcs are due to the constrictions formed by the geometry of the arc itself.

A further increase in temperature can be obtained by the mechanical constriction of the arc. The arc is forced through a narrow orifice cooled by a water film between the orifice surface and the arc [21]. In another improvement, the arc is stabilized by being forced to burn through a narrow column whose walls are water [22]. The cooler shell of this column is relatively nonconducting; the interior parts of the column reach a higher temperature

Fig. 8. Water stabilized arc in operation

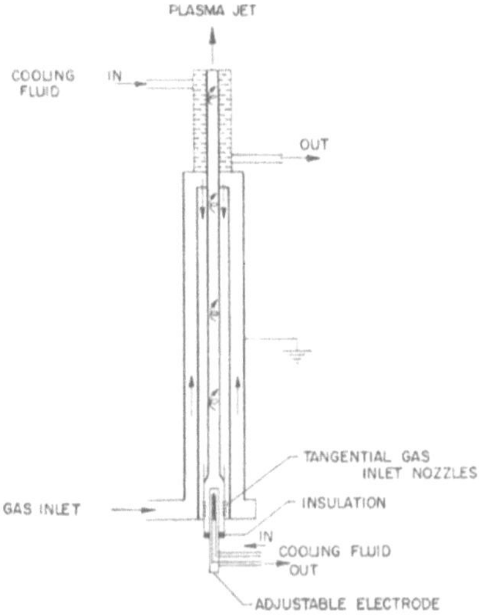

Fig. 9. Diagram of gas stabilized arc column

because of the increase in current density. In addition, the columnar arc can be made quite small in diameter; thus an added constriction is formed and the arc temperature raised.

Water vapor generated by the intense heat is ejected through the orifice at high velocity. Temperatures over 50,000° K have been measured when such a water-stabilized arc [23] has been used with a 2.5-mm-diameter orifice. According to one report, a LAVAL-nozzle and a pressure of 50 atmospheres inside the water-filled arc chamber were used; a jet of plasma at 8000° K issued from the nozzle at a velocity of 6 km per second [24].

In this country, similar water-stabilized arcs have been developed. Fig. 8 shows the plasma jet issuing from an arc at 14,000° K at a velocity of 4000 ft per second [25]. The input power required for the operation of this arc with a 1 $^1/_4$-inch orifice was 3000 kW. The heat-transfer rate was measured to be 2000 Btu per square foot per second, at a distance of 4 inches from the orifice.

The upper temperature limit of the water-stabilized arc is given by the physical strength of the orifice in withstanding the erosion of the high-velocity gas, and is a function of the current density and the pressure developed in the arc.

When gas is used as the stabilizing medium, even higher plasma-jet velocities and temperatures are expected to be reached. The principle of the gas-stabilized arc is similar to that of the water-stabilized arc; it was used 50 years ago as part of a research program on nitrogen fixation [26]. Fig. 9 illustrates the arrangement used in the gas-stabilized arc. The gas is introduced tangentially; the rotation of the gas around the axis stabilizes the arc.

Various types of gas-stabilized arcs are now under development; the aim is to supplant the hypersonic wind tunnel for model tests under the extreme conditions encountered by a missile in flight. With this research tool, basic studies can be made of the chemistry and physics of plasmas, e.g., transport properties, temperature measurements, and ionization states. In addition, it promises to be useful for propulsion studies on the feasibility of using ionized gases as the fuel in rocket engines.

Future Developments

The development of the energy sources inherent in a thermonuclear reaction depends on the methods of confining such a reaction and making it continuously controllable. The generation of temperatures even higher than those we have discussed has been possible for a few fleeting moments. Thus short-time coaxial, electrical discharges have produced gas temperatures of 250,000° K [27] in helium at 30 atm. By a similar coaxial discharge method in deuterium at low pressures, gas temperatures close to 1,000,000° K have reportedly been reached [28]. Exploding wires [29] create metal vapors at very high temperatures, and thus permit investigations not possible by other means. Shock waves travelling at Mach 200 generated by accelerating an ionized gas with an electromagnetic field, produce temperatures of several hundred thousand degrees [30].

As the techniques of generating and stabilizing high temperatures for some length of time are refined, our knowledge of the behavior of materials and our understanding of their use will increase. Then we shall be in a position to assure that thermonuclear power will make us independent of our dwindling supply of fossil fuels, and usher in a truly atomic age.

References

1. I. E. CAMPELL, High Temperature Technology. New York: Wiley & Sons, Inc., 1956.
2. C. AGTE and H. ALTHERTUM, Z. techn. Physik **11**, 182 (1930).
3. Ajax Electrothermic Corp.
4. W. H. DAVENPORT, S. S. KISTLER, et al., Amer. Ceram. Soc. **33**, 333 (1950).
5. T. T. MAGEL, P. A. KULIN. and A. R. KAUFMAN, J. Metals **4**, 295 (1952).
6. D. M. WROUGHTON, E. K. OKRES et al., J. Electrochem. Soc. **99**, 205 (1952).
7. W. E. KUHN, Arcs in Inert Atmospheres and Vacuum. New York: Wiley & Sons, Inc., 1956.
8. J. W. PUGH, R. L. HADLEY and R. W. HESSING, Metal Progr. **63**, 70 (1953).
9. M. DAVIS, A. CALVERLEY and R. F. LEVER, J. Appl. Physics **27**, 195 (1956).
10. F. TROMBE and M. FOEX, Métaux **31**, 126 (1956).
11. O. BECK, Elektrotechn. Z. **42**, 993 (1921).
12. W. FINKELNBURG and J. P. LATIL, J. Opt. Soc. America **44**, 1 (1954).
13. T. P. DAVIS, L. J. KROLAK, et al., J. Opt. Soc. America **44**, 766 (1954).
14. T. R. BROIDA, U. S. Naval Radiological Defense Laboratory Report USNRDL-417 (1952).
15. A. D. KIRSHENBAUM and A. V. GROSSE, J. Amer. Chem. Soc. **78**, 2020 (1956).
16. J. B. CONWAY and A. V. GROSSE, Second Technical Report, Office of Naval Research, High Temperature Project, Contract N 9-onr-87301 (1952).

17. H. E. PETSCHEK, P. H. ROSE, et al., J. Appl. Physics **26**, 83 (1955).
18. J. D. COBINE and D. A. WILBUR, J. Appl. Physics **22**, 835 (1951).
19. G. BUSZ and W. FINKELNBURG, Naturwiss. **40**, 550 (1953).
20. W. FINKELNBURG, Physik und Technik des Hochstromkohlebogens. Leipzig: Akad. Verlagsges., 1944.
21. H. GERDIEN and A. LOTZ, Wiss. Veröff. Siemens-Konzern **2**, 489 (1922).
22. H. MAECKER, Z. Physik **129**, 108 (1951).
23. F. BURHORN, H. MAECKER and TH. PETERS, Z. Physik **131**, 28 (1951).
24. TH. PETERS, Naturwiss. **24**, 571 (1954).
25. TH. R. HOGENESS, University of Chicago, Midway Laboratories.
26. O. SCHONHERR, Elektrotechn. Z. **30**, 365 (1909).
27. H. FISCHER, Proc. Symposium "High Temperature", Stanford University, Cal., June, 24–26 (1956).
28. I. V. KURCHATOV, Discovery **1956**, 227.
29. B. EISELT, Z. Physik **132**, 54 (1952).
30. A. C. KOLB, Naval Research Laboratory, Office of Naval Research (1956).

A Method of Integrating the Equations of Motion of a Body Entering an Arbitrary Atmosphere with an Automatic Error Analysis

By

F. G. Gravalos[1]

(With 7 Figures)

Abstract — Zusammenfassung — Résumé

A Method of Integrating the Equations of Motion of a Body Entering an Arbitrary Atmosphere with an Automatic Error Analysis. A very simple method for the integration of ordinary differential equations is developed. The range of applicability is not restricted beyond the CAUCHY-LIPSCHITZ condition. Since the utilization of the method is dependent upon the use of digital, electronic computing machines, the manner of programming—and related questions—for an I.B.M. 704 is given.

The method is applied to the equations of motion of a body entering an arbitrary atmosphere and numerical examples are worked out. The salient aspect of this method is the continuous tracing of the error committed in each step. The author's view on the problem of error propagation as a disturbance is presented.

Eine Methode zur Integration der Bewegungsgleichungen eines in eine Modellatmosphäre eindringenden Körpers mit automatischer Fehleranalyse. Eine sehr einfache Methode zur Integration gewöhnlicher Differentialgleichungen wird entwickelt. Der Anwendungsbereich ist nicht durch die CAUCHY-LIPSCHITZ-Bedingung eingeschränkt. Da die Anwendung der Methode von der Benützung elektronischer Rechenmaschinen (digital computers) abhängt, werden die Art der Programmierung und dazugehörige Fragen für eine I.B.M. 704 besprochen.

Das Verfahren wird auf Bewegungsgleichungen eines Körpers angewendet, der in eine Modellatmosphäre eindringt; numerische Beispiele werden ausgearbeitet. Der Hauptgesichtspunkt dieser Methode ist die kontinuierliche Kontrolle des bei jedem Schritt begangenen Fehlers. Die Anschauung des Verfassers über das Problem der Fehlerfortpflanzung als Störung wird dargelegt.

Une méthode d'intégration des équations du mouvement d'un corps pénétrant dans une atmosphère arbitraire avec contrôle automatique des erreurs. Une méthode simple pour l'intégration des équations différentielles ordinaires est développée. Son champ d'application n'est pas restreint par les conditions de CAUCHY-LIPSCHITZ. Comme la méthode implique l'utilisation de machines électroniques digitales, les méthodes de programmation et questions connexes sont données pour un ordinateur I.B.M. 704.

La méthode est appliquée aux équations du mouvement d'un corps pénétrant dans une atmosphère arbitraire et de nombreux exemples sont traités. L'aspect saillant de cette méthode est le contrôle constant de l'erreur commise à chaque pas. Les vues de l'auteur sur la propagation de l'erreur en tant que perturbation sont présentées.

---

[1] General Electric Company, Missile and Ordnance Systems Department, Philadelphia 4, Pennsylvania, U.S.A.

1. Statement of the Problem and Acknowledgements

Using the nomenclature and sign conventions of Fig. 1, the scalar equations of the plane motion of a rigid body along the tangent and the normal directions to the flight path are

$$\dot{\theta} = \frac{V}{R} \sin \beta \tag{1}$$

$$\dot{R} = -V \cos \beta \tag{2}$$

$$m V \dot{\beta} = \frac{m}{R} \left[V^2 \sin \beta - \frac{K}{R} \sin \beta \right] + [X \sin \alpha + N \cos \alpha] \tag{3}$$

(A)

$$m V = \frac{K m}{R^2} \cos \beta + [X \cos \alpha - N \sin \alpha] \tag{4}$$

to which the angular momentum equation,

$$m I^2 (\ddot{\alpha} + \ddot{\beta}) = M \tag{5}$$

must be added and, where

K is the product of the universal gravitational constant times the mass of the Earth (or planet),
m the mass of the moving body,
I the radius of gyration of the body,
X the component of the external resultant force along the longitudinal axis of the body,
N the component of the external resultant force along the normal to the longitudinal axis,
M the resultant moment.

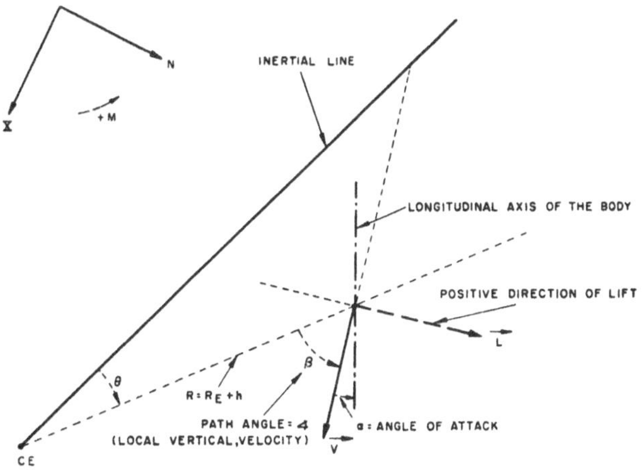

Fig. 1

The problem at hand is the numerical integration of the system of equations (A) with X, N and M the force components and moment of the aerodynamic reaction as they obtain in the so called "NEWTONian" theory for a blunt body of revolution, namely,

$$X = -2\pi\varrho V\left[2A(\dot{\alpha}+\dot{\beta})\sin\alpha + \frac{B}{V}(\dot{\alpha}+\dot{\beta})^2 + V(D\cos^2\alpha + C\sin^2\alpha)\right] \quad (6)$$

$$N = 2\pi\varrho V[2A(\dot{\alpha}+\dot{\beta})\cos\alpha + CV\sin 2\alpha] \quad (7)$$

$$M = -2\pi\varrho V[AV\sin 2\alpha + 2B(\dot{\alpha}+\dot{\beta})\cos\alpha] \quad (8)$$

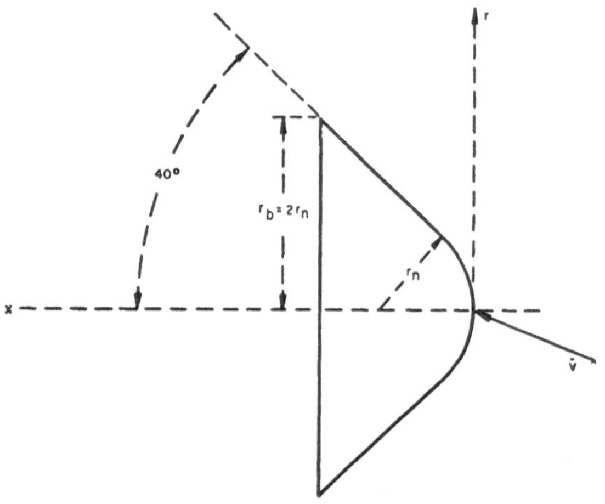

Fig. 2

ϱ, the density of the atmosphere.

The constants A, B, C, D are given by

$$A = \int_0^L \frac{ff'(f' + x - x_{c.g.})}{f'^2 + 2f}\, dx$$

$$B = \int_0^L \frac{ff'(f' + x - x_{c.g.})^2}{f'^2 + 2f}\, dx$$

$$C = \int_0^L \frac{ff'}{f'^2 + 2f}\, dx$$

$$D = \int_0^L \frac{f'^3}{f'^2 + 2f}\, dx$$

with the equation defining the shape of the body in the form $\frac{1}{2}r^2 = f(x)$, the r and x axes as in Fig. 2, $x_{c.g.}$ the x — coordinate of the c.g.[2] and L the axial length of the body, or that part of it "exposed" to the air flow.

It is the writer's pleasure to acknowledge the work of those in M.O.S.D. who collaborated in this work. Mr. A. J. DENNISON checked all the algebra, drew the flow diagram of Section 3, and was most valuable in designing the logical model to test the program for an I.B.M. "704", where the computations of Section 4 were carried out. Mr. E. V. DAMON took complete charge of obtaining the numerical results given in Section 4.

2. The Method of Integration

Let y be a function of t defined by the differential equation,

$$y' = f(y, t) \quad (9)$$

and assume the necessary and sufficient conditions for the existence of an analytic solution $y = y(y_o, t)$ such that $y = y_o$ at $t = t_o$.

The step by step method suggested here is repetitive. Simultaneously with the description of the method an upper bound will be found for the error committed in the step from the zero point — no error — to the first computed value. Then the question of error propagation will be examined.

The values of y_1, y_1', y_1'' of y and its first two derivatives given in the expansions,

$$y_1 = y_o + \frac{\Delta t}{1!}y_o' + \frac{\Delta t^2}{2!}y_o'' + \frac{\Delta t^3}{3!}y_o''' + \frac{\Delta t^4}{4!}y_o^{IV} \quad (10)$$

$$y_1' = y_o' + \frac{\Delta t}{1!}y_o'' + \frac{\Delta t^2}{2!}y_o''' + \frac{\Delta t^3}{3!}y_o^{IV} \quad (11)$$

$$y_1'' = y_o'' + \frac{\Delta t}{1!}y_o''' + \frac{\Delta t^2}{2!}y_o^{IV} \quad (12)$$

are obtained from a function y of t as defined by (10) in the closed interval $t_o \leq t \leq t_1$ with $\Delta t = t - t_o$. Elimination of y_o''' and y_o^{IV} among (10), (11) and (12) yields the following relation,

$$y_1 - y_o = \frac{\Delta t}{2}(y_1' + y_o') - \frac{\Delta t^2}{12}(y_1'' - y_o''). \tag{13}$$

This formula is used to compute y_1 by an iterative procedure. The first trial value of y_1 is obtained in

$$y_{10} = y_o + \frac{\Delta t}{1!} y_o' + \frac{\Delta t^2}{2!} y_o'' \tag{14}$$

from which the first trial values of y_{10}' and y_{10}'' are obtained in (9) and in

$$y'' = f_Y(y, t)\, y' + f_t(y, t). \tag{15}$$

This set of trial values,

$$y_{10}, y_{10}', y_{10}''$$

in (13) yields a new y_{11} value from which, by use of (9) and (15), a new y_{12} is obtained. A tolerance, or relative error ε_y, is fixed a priori so that this iterative process is continued until

$$\left| \frac{y_{1n} - y_{1n+1}}{y_{1n+1}} \right| < \varepsilon_y.$$

The process described for the interval (t_o, t_1) applies unaltered to any other interval and will be analyzed now. The maximum accuracy obtainable is naturally determined by the polynomial approximation (10). An upper bound for the truncation error is

$$\varepsilon_y^* < M^* \frac{1}{720} \Delta t^5$$

where $M^* \leq |f^V(\bar{t})|$ $\bar{t}$ being an adequate value of t in the interval under consideration. Assuming $\frac{M^*}{y} \leq 1$ — something quite conservative if the system is stable — the selection of Δt defines the condition for ε_y

$$\varepsilon_y \leq \frac{\Delta t^5}{720}. \tag{16}$$

The iteration process itself corrects the errors $\varepsilon_y^{*\prime}$ and $\varepsilon_y^{*\prime\prime}$ with which y_1' and y_1'' are computed in (9) and (15) because of the error ε_{y1}^* in y_1, namely,

$$\varepsilon_{y_1}^{*\prime} = f_Y \varepsilon_{y_1}^*$$

$$\varepsilon_{y_1}^{*\prime\prime} = [f f_{YY} + f_Y^2 + f_{Yt}]\, \varepsilon_{y_1}^*$$

if the round-off error is assumed zero.

The question left is how the error ε_{yi}^* — at the ith step — is propagated to the nth (final) step.

If instead of y_i a value of y in the interval $(y_i, y_i + \varepsilon_{yi}^*)$ is taken, then at $t = t_n$:

$$\varepsilon_{y_{ni}}^* = \varepsilon_{y_i}^* e^{\int_{t_i}^{n} f_y \, dt} \tag{17}$$

where f_y is the function of t given by $f_Y\,[y(t), t]$, with $y(t)$ the solution under study. The total error at $t = t_n$ is given by

$$\varepsilon_{y_n}{}^* = \sum_{i=1}^{n}{}' \varepsilon_{y_{ni}}{}^*. \tag{18}$$

The important thing to bear in mind is that no complete error analysis is possible without studying the variational equation — of which (17) is the solution. Again, if the ε_y is fixed and the equation — or system — is stable, a preliminary examination will show how the error dies out, i.e., in how many steps the error becomes smaller than 10^{-s}, if s is the number of digits carried. This in turn will provide the total upper bound for the error at any step.

3. Application of the Method to Higher Order Systems

The method of the previous section will be applied to system (A); however, its applicability is independent of the specific form of the system, as long as the conditions imposed on (9) are satisfied by all the equations. In describing how the method is applied, the logical questions that arise when the problem is programmed on an electronic, digital machine, such as the I.B.M. "704", will be dealt with. For, the analysis and the programming are fundamentally inseparable. The flow diagram of Fig. 3 indicates how this was done. An inspection of the "logic" involved shows the number of choices that must be made for a complete definition of the problem. The initial $\Delta_0 t$ interval and the tolerance or error limit, ε, must be fixed. Immediately the question arises, suppose that formula (13) cannot yield this accuracy for the selected $\Delta_0 t$. Therefore, a limit k must be set for the number of iterations allowed. If after "k" iterations, still the error limit is not attained, the interval $\Delta_0 t$ is divided by a fixed number, say "l", and the process is repeated. However, both the point of view of economy and of analysis, too small an interval of time may ultimately — as it was pointed out at the end of Section 2 — cancel, throughout the integration, the high local accuracy obtained. Hence, a limit must be put on the number of subdivisions permitted, say "p", so that if after k iterations for $\Delta t = \frac{\Delta_0 t'}{l^p}$ the condition $\left|\frac{y_{1n+1} - y_{1n}}{y_{1n+1}}\right| < \varepsilon$ is not satisfied, ε is changed to ε', $\varepsilon' = \frac{\varepsilon}{10}$. Since for $\varepsilon(r) \geq N^*$, the goal of the work cannot essentially be reached, when $\varepsilon(r) \geq N^*$ the machine is instructed to iterate k times, print the errors, then proceed to the next step. In total the following magnitudes must be selected:

I. An initial $\Delta_0 t$
II. The ε for each of the dependent variables
III. The number k of iterations allowed
IV. The divisor l of $\Delta_0 t$
V. The number p of permitted subdivisions of $\Delta_0 t$
VI. The number of times r the error limit may be down-graded.

Finally, a brief sketch of how a "program" can be tested will be given:

The algebra corresponding to blocks II, IV and VI of Fig. 3 is identical and may be tested in conjunction with (14). Eq. (13), block V, also may be tested separately.

The cycle of blocks VI, VII, VIII and IX, or the flow of blocks VI, VII, VIII and X, may be tested by selection of arbitrary inputs that satisfy and do not satisfy the error restrictions. By assuming $\varepsilon < 10^{-s}$, s larger than the number of digits carried in the computations, the part of the program corresponding to the proper use of the six magnitudes $\Delta_0 t$, ε, k, l, p and r can be tested.

For an accurate determination of the errors, the variational equations corresponding to (A) should be studied.

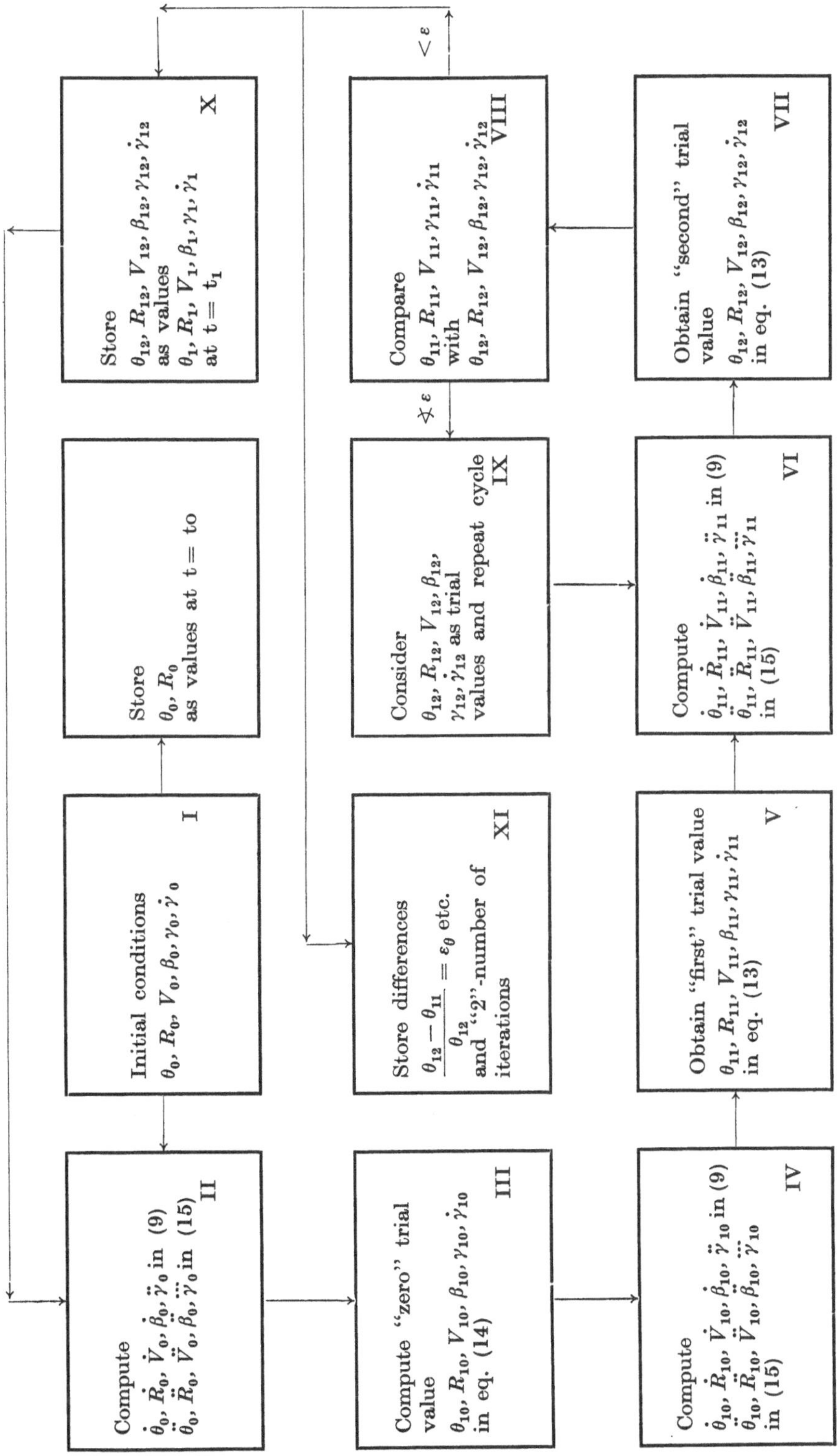

Fig. 3. Flow diagram

4. Numerical Examples

To study the motion of the body geometrically described in Fig. 2, for which four different motions were computed, certain magnitudes must be specifically given.

I. Physical Constants (Earth)

R_E, radius of the earth $= 2.09029 \times 10^7$ ft.
K, gravitational constant $= 1.40816 \times 10^{16}$ ft.^{3}/sec.^{2}
ϱ, atmospheric density taken as the function of h given in "Atmospheric Models", USAF Cambridge Research Center, Cambridge, Massachusetts (1956).

II. Configuration Constants

$$A = .216\,\frac{r_b^2 L}{4} \qquad B = 1.966\,\frac{r_b^2 L^2}{8}$$
$$C = .543\,\frac{r_b^2}{4} \qquad D = .91\,\frac{r_b^2}{4}$$
$$r_b = 16.4135 \text{ ft} \qquad L = 15 \text{ ft}$$
$$x_g = .8 r_b, \quad m = 310.2853 \text{ slugs}, \quad I = 5 \text{ ft}.$$

III. Initial Conditions

$\theta_o = 0$, $h_o = 299{,}000$ ft., $V_o = 15{,}000$ ft/sec., $\beta_o = 60°$, $\gamma_o = 80°$, $\dot{\gamma}_o = -.0011958$ rad/sec; where $\gamma = \alpha + \beta$ so that $\alpha_o = 20°$ and $\dot{\alpha}_o = 0$. (The computations were carried out in γ and $\dot{\gamma}$.)

IV. Integration Constants

$\Delta_o t = .027$ seconds, $\varepsilon < 10^{-7}$ for all variables, $k = 25$, $l = 3$, $p = 3$, $r = 2$.

In the four examples worked out only the physical constants were changed. *Case 1*, with R_E, K and ϱ as above, for a body entering the earth's atmos-

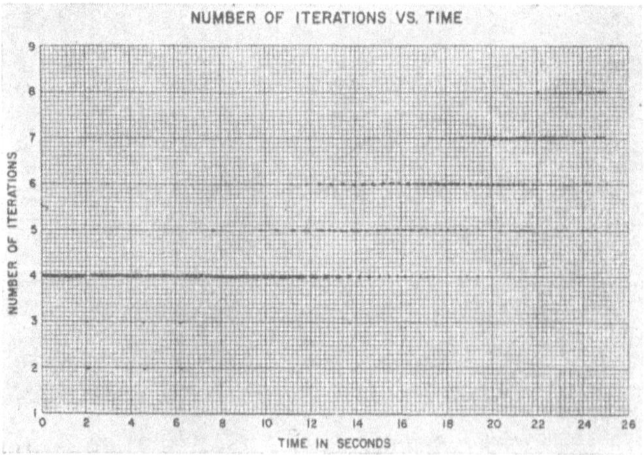

Fig. 4

phere, designated in the figures with (K, ϱ). *Case 2*, same R_E, the gravitational constant equals $.6K$, same ϱ, $(.6K, \varrho)$. *Case 3*, same R_E, K as for the earth, and half the atmospheric density of the earth, (K, $\varrho/2$). *Case 4*, same R_E and the other two constants, $.6K$ and $\varrho/2$, ($.6K$, $\varrho/2$). The total time interval of integration was 25 seconds in all cases. All the remaining constants were also the same for all cases.

Examination of the results showed no appreciable differences with the variation in K. Maximum dynamic pressure ($q = \frac{1}{2}\varrho V^2$) is reached at about 24 seconds from the start. The corresponding Mach number is in the neighbor-

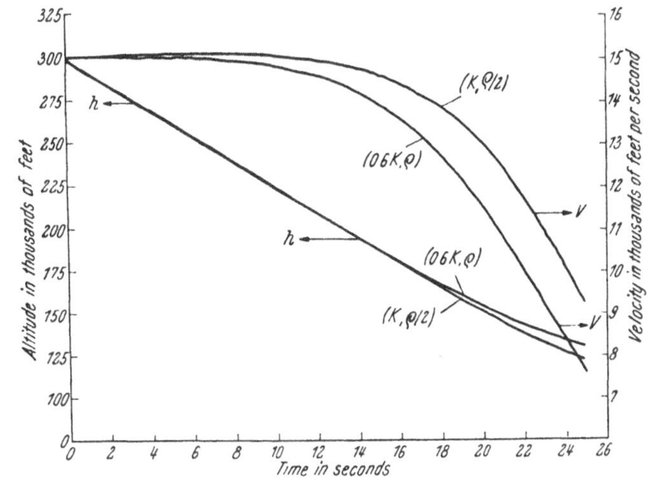

Fig. 5

hood of 8, so that the expressions for the aerodynamic reaction given in Section 1 are quite valid. Only the extreme physical cases (K, $\varrho/2$) and ($.6K$, ϱ)

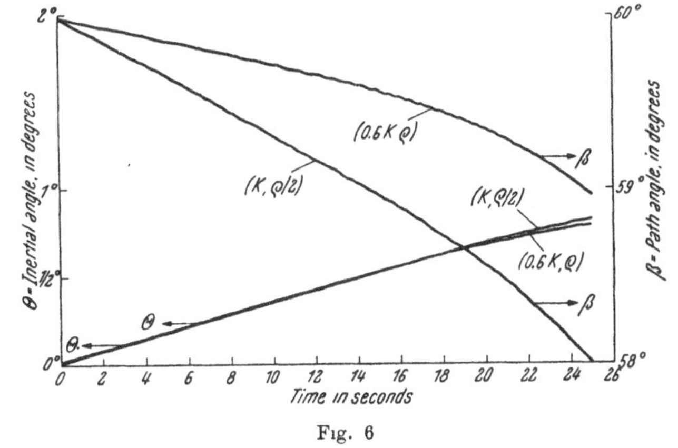

Fig. 6

were plotted. In Fig. 4 the number of iterations at each step is shown; no change of time interval or error limit took place. Fig. 5 shows the variation of velocity and altitude with time. In Fig. 6 the inertial angle θ and the path angle β are given. In Fig. 7 the envelopes of the variation of α with time are plotted with the zero and the minima points marked.

From preliminary studies showing the system to be very stable, it was concluded that the errors will die out in no more than $3\ \Delta_0 t$. Consequently, at each point the dependent variables are computed here with a relative error less than 3×10^{-7}.

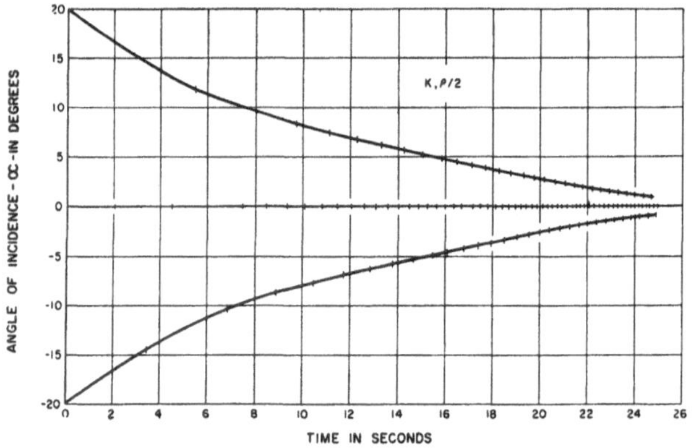

Fig. 7a

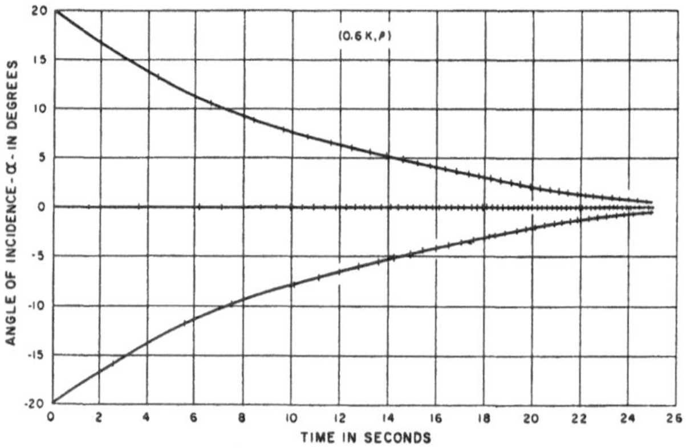

Fig. 7b

References

1. G. E. SOLOMON, Aerodynamic Coefficients of an Axially Symmetric Body. Ramo-Wooldridge Corporation Report CMCC 271.528 (November 1954).
2. W. E. MILNE, A Note on the Numerical Integration of Differential Equations. J. Res. Nat. Bur. Standards **43** (1949).

Pilotage d'un astronef par des moyens radioélectriques

Par

H. Gutton[1]

Résumé — Zusammenfassung — Abstract

Pilotage d'un astronef par des moyens radioélectriques. Un astronef libéré de l'attraction terrestre s'éloignera de plusieurs millions de kilomètres. Son pilotage à distance est possible en établissant une liaison radio-électrique bilatérale entre l'engin et la terre. La mesure de la position de l'engin dans l'espace est nécessaire. Les moyens à mettre en œuvre pour résoudre cette question sont étudiés dans ce rapport.

Les études de la radioastronomie, des echos sur la lune et les sondages de l'ionosphère fixent le choix de la fréquence porteuse utilisable entre 100 et 300 Mc. Le bruit radioélectrique galactique augmente avec la fréquence.

La liaison est possible à 300 Mc avec à bord de l'astronef un émetteur et un récepteur de rayonnement isotropique, à terre avec un système directif. Le bruit galactique parasite limite le gain de l'aérien terrestre, de telle sorte que le diamètre d'un paraboloïde émetteur récepteur doit être inférieur à 20 mètres.

Ce paraboloïde sera orientable vers l'émetteur de l'engin par un procédé analogue à celui utilisé généralement dans les radars. Le calcul de la liaison faite dans ces conditions, montre la possibilité de communiquer à un million de kilomètres avec un émetteur à impulsions de 2 microsecondes de 100 kilowatts de puissance de crête. La fréquence de répétition est définie par la quantité d'informations à transmettre par seconde. Pour un engin très éloigné la densité de ces informations est assez faible pour que la puissance moyenne de l'émetteur situé à bord de l'engin soit de quelques dizaines de watts, puissance compatible avec les sources d'énergie de l'engin.

L'ensemble de ces raisonnements, montre que dès maintenant, un engin robot peut être lancé dans les espaces interplanétaires, suivi en permanence par une ou plusieurs stations terrestres, et par suite, piloté par un observateur terrestre.

Lenkung eines Raumfahrzeuges mit Hilfe radioelektrischer Methoden. Ein dem irdischen Schwerebereich entronnenes Raumfahrzeug wird sich mehrere Millionen Kilometer entfernen. Seine Steuerung auf Distanz ist aber möglich, wenn man eine zweiseitige radioelektrische Verbindung zwischen dem Fahrzeug und der Erde herstellt. Hierauf ist die Bestimmung der Position des Fahrzeuges im Raum erforderlich. Die zur Lösung dieser Frage anzuwendenden Mittel werden in der vorliegenden Arbeit untersucht.

Das Studium der Radioastronomie, vom Mond reflektierte Echos und die Ionosphärenlotungen legen die Wahl der verwendbaren Trägerfrequenz in den Bereich zwischen 100 und 300 Megahertz. Das radioelektrische Störgeräusch aus der Milchstraße verstärkt sich mit der Frequenz.

Die Verbindung ist bei 300 Megahertz möglich, wenn sich an Bord des Raumfahrzeuges ein Sender und ein Empfänger fur isotrope Strahlung befinden, auf der Erdoberfläche ein Richtsystem. Wegen des parasitären galaktischen Störgeräusches

[1] Compagnie Générale de Télégraphie Sans Fil, Paris, France.

muß die Antenne so begrenzt sein, daß ein Sender- und Empfänger-Paraboloid einen Durchmesser unter 20 Meter hat.

Das Paraboloid wird gegen den Sender des Raumfahrzeuges nach einem analogen Verfahren orientierbar sein, wie es allgemein beim Radarverfahren verwendet wird. Die Berechnung der unter diesen Bedingungen möglichen Verbindungen zeigt die Möglichkeit von Mitteilungen auf eine Entfernung von einer Million Kilometer mit einem Sender, der Impulse von 2 Mikrosekunden mit 100 kW Spitzenleistung ausstrahlt. Die Wiederholungsfrequenz wird durch die Menge an Mitteilungen begrenzt, die in der Sekunde zu übertragen sind. Für ein sehr weit entferntes Raumfahrzeug ist die Dichte dieser Informationen ziemlich gering, damit die mittlere Stärke des Senders an Bord des Flugkörpers einige Zehner von Watt betragen kann, eine Leistung, die mit den Energiequellen des Raumfahrzeuges vereinbar ist.

Die Gesamtheit dieser Gründe zeigt, daß schon jetzt eine Robotermaschine in den interplanetarischen Raum entsandt, dauernd von einer oder mehreren irdischen Stationen verfolgt und daher durch einen irdischen Beobachter gesteuert werden kann.

The Navigation of a Space Vehicle by Means of Radio-Electric Methods. The navigation of a space vehicle travelling many millions of kilometers away from the earth is possible by means of radio communication between the vehicle and the earth. The method of resolving the problem of establishing the position of the vehicle in space is studied in the report.

Investigations in radioastronomy, of lunar echoes and ionospheric sounding have fixed the choice of usable carrier frequencies in the range between 100 and 300 Mc. Galactic radio noise increases with frequency.

Communication is possible at 300 Mc using non-directional transmitting and receiving antenna in the spaceship and a directional system on the earth. Galactic noise sets a limit to the maximum gain of the antenna so that the diameter of a paraboloid for transmission and reception should not exceed 20 metres.

Standard radar techniques would be used to align the paraboloid with the vehicle. Calculations relating to communication established under such conditions demonstrate the possibility of communicating over a distance of a million kilometers with a transmitter of 2 microsecond pulses of 100 kilowatts peak power. The repetition frequency is determined by the amount of information to be transmitted per second. In the case of a very distant vehicle the density of information would be fairly small because the mean power of the shipboard transmitter must be limited to a few tens of watts compatible with the energy sources on the vehicle.

These arguments show that, even at the present time, a vehicle may be launched into space and controlled by a terrestrial observer through the medium of radio stations on the earth which constantly follow it.

La réalisation d'un voyage interplanétaire nécessite encore de gros progrès techniques; le travail minimum à fournir pour échapper à l'attraction terrestre est en effet considérable. Nous supposerons ces progrès possibles et nous nous attacherons au problème des liaisons nécessaires au pilotage d'un engin à travers l'espace.

Une liaison bilatérale entre la terre et l'engin est indispensable pour permettre à un opérateur terrestre de piloter un engin robot. Cette première phase d'expérimentation utilisant un engin non habité est obligatoire pour de nombreuses raisons dont la plus essentielle est que le poids d'un être humain est très largement supérieur à celui d'un pilote robot. La liaison engin terre permet de connaître la situation dans l'espace de l'engin, ainsi que son état c'est-à-dire le comportement des systèmes de propulsion et de pilotage, le cap de l'engin par rapport au vecteur engin terre, la distance de la terre, etc. ... Le sens inverse permet d'envoyer en fonction des renseignements recueillis par la première liaison, les ordres de pilotage. La permanence de la liaison est de toute évidence, une condition impérative. Le choix d'une onde porteuse se porte sur une onde hertzien-ne de courte longueur d'onde. En effet, la lumière visible ne peut être retenue malgré les facilités qu'elle donnerait pour la localisation précise de l'engin en utilisant les moyens des observatoires terrestres; les interruptions des communications résultant d'un temps couvert ou du brouillage de la liaison de jour par la diffusion de la lumière solaire par l'atmosphère, interdisent son emploi.

Les études poursuivies depuis de nombreuses années sur l'ionosphère montrent que seules des ondes de fréquence supérieures à 20 mégacycles sont utilisables. Les ondes hertziennes sont en effet réfléchies ou réfractées par la couche ionisée de la haute atmosphère. Cette réfraction est due au mouvement des électrons dans le champ hertzien. Ce mouvement contribue à modifier la constante diélectrique apparente du milieu dont la valeur est:

$$K = 1 - \frac{\frac{Ne^2}{m}}{\pi\,(f^2 - f_0^2)}\,.$$

N nombre d'électrons par cm^3.
e et m charge et masse de l'électron.
f fréquence de l'onde porteuse.
f_0 fréquence critique fonction croissante de l'ionisation.

De nombreuses observations sur l'ionosphère ont permis de conclure que le nombre N dépasse très rarement la grandeur de 10^6 par cm^3. La fréquence critique correspondante est de 9 mégacycles. Pour des fréquences beaucoup plus élevées supérieures à 100 mégacycles et pour cette densité d'ionisation, la constante diélectrique devient indépendante de f_0; elle peut s'écrire:

$$K = 1 - \frac{80}{f^2}\,.$$

f est exprimé en mégacycles.
Pour $f = 100$ Mc, $K = 0{,}992$ très voisine de l'unité. Il en résulte qu'un rayonnement de fréquence égal ou supérieur à 100 Mc traverse l'ionosphère sans se réfléchir et pratiquemment sans se réfracter même sous des angles très inclinés sur l'horizon.

Ces calculs indiquent l'intérêt, pour assurer des liaisons avec un engin interplanétaire, de l'utilisation de fréquences supérieures à 100 Mc. Cette conclusion est confirmée par les nombreuses observations du rayonnement radioélectrique des astres qui constituent depuis quelques années les bases d'une nouvelle science, la radioastronomie. La comparaison entre les pointés radioélectriques des nébuleuses sources de bruit hertzien et les visés optiques de ces mêmes nébuleuses prouvent que pour ces fréquences élevées la réfraction de l'ionosphère peut être négligée. La fenêtre radioélectrique ouverte sur le monde extraterrestre est cependant limitée aux fréquences supérieures à 10.000 mégacycles, par l'absorption atmosphérique due à l'effet des résonances moléculaires de l'oxygène et de la vapeur d'eau.

La radioastronomie et notamment l'étude radioélectrique du soleil montrent la croissance de l'intensité du rayonnement radioélectrique des astres avec la fréquence. Ainsi le rayonnement radioélectrique du soleil décelé par un récepteur terrestre est en moyenne de 10^{-22} watts par cycle et par mètre carré de surface éclairée. Cette intensité croît régulièrement avec la fréquence, elle atteint dans les mêmes conditions 10^{-22} watts à 3000 Mégacycles. Il faut tenir compte dans l'établissement d'une liaison avec un engin extraterrestre du brouillage provoqué par ce rayonnement. Nous verrons dans la suite que ce brouillage limite les possibilités de liaison et favorise l'emploi de fréquences porteuses de l'ordre de quelques centaines de mégacycles.

Ceci posé, il reste à déterminer par le calcul, les conditions requises pour établir une liaison permanente utilisable pour piloter un engin extraterrestre situé à une distance au moins égale à un million de kilomètres.

Nous porterons notre choix sur une fréquence porteuse de 300 Mégacycles. L'émetteur et le récepteur situés à bord de l'engin doivent émettre et capter des ondes dans toutes les directions. Nous admettrons que le rayonnement des aériens correspondants est isotrope, c'est-à-dire que le gain de ces aériens est égal à l'unité. Par contre, le système émetteur récepteur de terre est orientable; nous lui donnerons un gain G élevé.

La puissance reçue à terre ou sur l'engin P_r est fonction de la puissance d'émission P_E:

$$P_r = P_E \frac{G \lambda^2}{4 \pi^2 d^2}$$

λ est la longueur d'onde de la fréquence porteuse.
d est la distance qui sépare la terre de l'engin.

Le gain G de l'aérien terrestre est dans le cas d'un projecteur parabolique excité à son foyer égal à $\frac{2 \pi S}{\lambda^2}$, S est la surface de l'ouverture de la parabole; cette formule tient compte d'un rendement radioélectrique du projecteur égal à 50%.

En fonction de cette surface S la puissance reçue devient indépendante de la longueur d'onde:

$$P_r = P_E \frac{S}{8\pi d^2} .$$

La puissance de bruit du récepteur P_B est définie par:

$$P_B = F \times K T \text{ par cycle.}$$

F, facteur de bruit du récepteur est de l'ordre de 2 à 3 dans la bande de 300 mégacycles, $K = 1{,}4 \times 10^{-23}$ joule par degré centigrade, est la constante de BOLTZMANN, T température absolue du récepteur est choisie égale à 300°, d'où:

$$P_B = 1{,}2 \times 10^{-20} \text{ watts par cycle.}$$

Le brouillage dû au bruit radioélectrique des astres augmente avec la surface S du projecteur. Ce bruit P_A peut atteindre à 300 Mc un niveau égal à:

$$P_A = 0{,}5 \times 10^{-22} \times S \text{ watt par cycle.}$$

Le choix de la surface S résulte de la comparaison des bruits P_B et P_A, le meilleur compromis est obtenu en écrivant que P_B est égal à P_A d'où la valeur optimum de $S = 240$ mètres carrés. Un aérien de plus grande surface ne permettrait pas d'obtenir une réception plus intense de l'émission de l'engin. Cette réception serait limitée par le bruit radioélectrique stellaire, qui nous le savons croît avec la fréquence.

Ces raisonnements justifient l'emploi de la fréquence relativement basse de 300 mégacycles. Ils conduisent à l'utilisation d'un projecteur parabolique orientable de 30 mètres de diamètre. A 300 mégacycles l'ouverture utile du faisceau diffracté par ce projecteur est de 6 degrés.

La station terrestre munie de ce projecteur parabolique comporte un émetteur et un récepteur pour l'orienter. Le récepteur permet la mesure de la direction de l'engin au moyen d'un dispositif classique dans la technique des radars. Ce système commande les moteurs d'entraînement du projecteur. Ce dispositif de poursuite fournit à chaque instant à l'observateur terrestre la direction de l'engin avec une précision supérieure au dizième de degré.

Il est bien évident que plusieurs stations terrestres liées par des lignes de transmission, observeront l'engin céleste, de façon à maintenir en permanence le contact malgré la rotation de la terre.

Le choix du système de modulation de l'onde porteuse est fonction du nombre d'informations à transmettre par seconde. Il est en général plus simple d'utiliser un système à impulsion lorsque ce nombre est réduit. Ce choix est également conditionné par les possibilités de stabilité de fréquence à bord de l'engin, à l'instabilité s'ajoute l'effet DOPPLER dû à une vitesse de l'engin supposée atteindre 100 kilomètres par seconde. Cet effet modifie la fréquence porteuse de $\pm$ 100 kilocycles.

Nous ferons le calcul de la liaison pour une bande passante égale à 500 kilocycles permettant le passage d'impulsions de durée égale à 2 microsecondes.

La puissance du signal reçu à une distance d exprimée en millions de kilomètres est

$$P_r \# P_E \times \frac{10^{-17}}{d^2}.$$

Le seuil de réception est égal à

$$P_B + P_A = 2{,}4 \times 10^{-20} \times \Delta F = 1{,}2 \times 10^{-14} \text{ watt.}$$

En adoptant une puissance de crête de 100 kilowatts relativement aisée à produire à bord de l'engin

$$P_r = \frac{10^{-12}}{d^2} \text{ watt}.$$

Pour $d = 1$ million de kilomètres le rapport signal à bruit en admettant une atténuation de propagation au passage à travers l'atmosphère terrestre de 10 décibels est égal à:

$$\frac{P_r}{P_B + P_A} = \frac{10^{-13}}{1{,}2 \times 10^{-14}} \# 8.$$

Ce rapport de l'ordre de grandeur de 10 décibels permet une très bonne utilisation du signal reçu.

Ces calculs démontrent la possibilité de liaison à plusieurs millions de kilomètres, utilisant un système électronique concevable dès à présent.

Ce genre de transmission ne permet pas d'assurer une liaison téléphonique, mais les informations transmises ont une rapidité suffisante pour assurer le pilotage de l'engin à très grande distance.

On peut conclure que l'électronique est techniquement capable d'assurer dès maintenant le pilotage d'un engin robot, à des distances de quelques millions de kilomètres, les frais occasionnés par les installations terrestres nécessaires ne dépasseraient pas sensiblement les dépenses d'une station radar de surveillance à grande distance.

Un engin robot pourrait être lancé, suivi et piloté jusqu'à des distances considérables, l'électronique est ainsi dans le domaine de l'astronautique certainement en avance sur le système de propulsion de l'engin lui même.

Space Law—The Development of Jurisdictional Concepts

By

Andrew G. Haley[1], ARS

(With 3 Figures)

Abstract — Zusammenfassung — Résumé

Space Law—The Development of Jurisdictional Concepts. The nations of the world through participation in the IGY program have assented to the peaceful flight of earth satellites above their sovereign territories. A valid world agreement as to this limited aspect of the law of space has thus evolved—not through the formalities of a treaty, but simply by unanimous consent of the nations. The author urges the International Astronautical Federation to formulate a committee of scientists and jurists to consider the broader aspects of space law. This Committee would define "airspace" and recommend to the appropriate constituent of the United Nations a rule delimiting air space jurisdiction. The Committee should determine where national sovereignty over air space ends, and where freedom of space—analogous to freedom of the seas—begins.

The author advances the theory that the line of demarcation is the so-called "Kármán Primary Jurisdiction Line" where the conditions for accomplishing aerial flight end. He deplores the use of the London Disarmament Conference (1957) as the forum for the first international consideration of the question of space law jurisdiction. This important question should be explored in the peaceful atmosphere of the United Nations, rather than in a conference concerned only with the regulation of instruments of war. As to the substantive law which should govern the conduct of men of earth in dealing with other beings in space, he discusses the concept of metalaw—the law governing the rights of intelligent beings of different natures and existing in an indefinite number of different frameworks of natural law. He delineates the underlying concept of metalaw—"Do unto others as they would have done unto them". He describes some of the problems relative to application of that concept to extraterrestrial beings who may exist in two-dimensional or multi-dimensional frames of reference.

Weltraumrecht — Die Entwicklung von Begriffen der Rechtsprechung. Durch die Beteiligung am Programm des Internationalen Geophysikalischen Jahres haben die Nationen der Erde dem friedlichen Flug von Erdsatelliten uber ihre souveränen Territorien zugestimmt. Eine gültige Weltzustimmung zu diesem begrenzten Gesichtspunkt eines Weltraumgesetzes hat sich auf diese Weise ergeben — nicht durch die Formalitäten eines Vertrages, sondern einfach durch die einstimmige Billigung der Nationen. Der Autor fordert die Internationale Astronautische Föderation auf, ein Komitee von Wissenschaftern und Juristen einzusetzen, das die breiteren Gesichtspunkte eines Weltraumgesetzes betrachten solle. Dieses Komitee würde den „Luftraum" definieren und der zuständigen Stelle innerhalb der Vereinten Nationen eine Verfügung empfehlen, welche die Luftraumgesetzgebung abgrenzen soll. Das Komitee

[1] General Counsel of the American Rocket Society; President of the International Astronautical Federation. 1735 De Sales Street N.W., Washington 6, D.C., U.S.A.

sollte bestimmen, wo die nationale Herrschaft uber den Luftraum endet und wo die Freiheit des Weltraums — analog zur Freiheit der Meere — beginnt.

Der Autor bringt die Theorie vor, daß die Demarkationslinie die sogenannte „KÁRMÁN-Linie primärer Jurisdiktion" sei, wo die Bedingungen für die Ausführbarkeit von Luftflügen endet. Er bedauert die Betrauung der Londoner Abrüstungskonferenz (1957) als Forum für die erste internationale Diskussion der Frage einer Raumrechts-Gesetzgebung. Dieses wichtige Problem sollte eher in der friedlichen Atmosphäre der Vereinten Nationen als in einer Konferenz geklärt werden, die bloß mit der Beschränkung von Kriegswerkzeugen befaßt ist. Was das wesentliche Gesetz betrifft, welches das Verhalten von Menschen der Erde beim Verkehr mit anderen Lebewesen im Weltraum regeln sollte, diskutiert der Autor den Begriff des „Metagesetzes" („Metarechtes") — jenes Gesetzes, durch das die Rechte intelligenter Wesen geregelt werden, die von verschiedener Natur sein und in unbegrenzter Zahl unter verschiedenen Systemen des Naturrechtes leben können. Er entwirft die zugrundeliegende Bestimmung des Metagesetzes: „Verhalte Dich gegen andere so, wie diese wollen, daß man sich gegen sie verhalte!". Der Autor beschreibt einige der Probleme, die mit der Anwendung dieser Auffassung gegenüber außerirdischen Wesen zusammenhängen, die in zwei- oder vieldimensionalen Bezugssystemen existieren mögen.

L'élaboration de concepts juridiques. En participant au programme de l'année géophysique internationale les nations ont consenti au survol pacifique de leur territoire souverain par des satellites artificiels. Un accord valable sur ce point particulier de droit spatial a donc résulté, non des formalités de traités, mais par un consentement unanime. L'auteur prie instamment la Fédération Internationale d'Astronautique de constituer un comité de savants et de juristes pour envisager les aspects plus étendus du droit spatial. Ce comité serait chargé de fournir une définition du terme „espace aérien" et de soumettre pour recommandation aux Nations Unies une règle délimitant la juridiction de l'espace aérien. Ce comité devrait déterminer les limites de souveraineté nationale de l'espace aérien, où commence la liberté de l'espace, analogue à la liberté des mers.

L'auteur avance l'opinion suivant laquelle la ligne de démarcation serait la soi-disant "limite de juridiction primaire de KÁRMÁN" où cesseraient les conditions de réalisation du vol aérien. Il déplore que les premières considérations de juridiction spatiale aient eu pour cadre la conférence du désarmement de Londres (1957). Cette question importante devrait être examinée dans l'atmosphère pacifique des Nations Unies plutôt qu'au cours d'une conférence uniquement préoccupée de la réglementation d'engins de guerre. Quant aux règles de conduite des terriens, dans leurs rapports avec d'autres êtres, il met en discussion le concept de métaloi, implicite dans un grand nombre de codes de loi naturelle, et qui règle les droits des êtres intelligents de nature différente. Il souligne le concept sous-jacent "traite les autres comme ils voudraient être traités par toi". Il décrit quelques-uns des problèmes relatifs à l'application de ce concept aux êtres extra-terrestres qui pourraient exister dans un espace de référence bi- ou multi-dimensionnel.

In Rome, on this same occasion last year, it seemed appropriate to introduce the concept of metalaw and its relation to exploration in space. We have gathered this year in Barcelona, the famed center of Catalonia, whose people are noted for being practical and efficient, and so, on this occasion, we attempt to clarify practical jurisdictional questions.

Never before in the history of mankind has the ncecessity arisen so quickly to state legal parameters in connection with a vast new area of social change. The legal problems presented by the advent, within the past decade, of space flight instrumentalities have been climacteric, and technology and other manifestations of the natural sciences have far outstripped the formulation of the legal rules, and the gap has widened to the point that the peace of the world is threatened. Before I discuss this point further, I desire to state that it is quite impossible in this paper to review the history and background of the inception of air law, to review the opinions of legal commentators, or even to review in detail the current legal situation. To those who so request, I will furnish reprints of a series of writings in which I have reviewed background history; in which I have discussed the works of such authorities as Professor JOHN C. COOPER, Dr. ALEX MEYER, Dr. WELF HEINRICH PRINCE OF HANOVER, C. WILFRED JENKS, OSCAR SCHACHTER, P. K. ROY, and many others; and in which I have made the point, *inter alia*, that the consent of all nations to the satellite program of the International Geophysical Year and the active cooperation therein of most nations have resulted under familiar and accepted principles of international law in having the legal effect of a valid world agreement which now cannot be challenged by any nation[1].

The most unfortunate development in connection with the orderly statement of space law has been the involvement of the problem in the proceedings of the United Nations Subcommittee on Disarmament in London. I suppose this involvement was inevitable because the question of control of objects entering outer space was bound to arise. I deeply regret, however, that the problems of space law found their first and most critical examination in the intense political atmosphere of the London Disarmament Conference. The statement of space law problems and the formulation of jurisdictional concepts and regulatory rules should long since have been undertaken by appropriate juridical bodies of the United Nations and of the International Civil Aviation Organization (ICAO). The urgency and extent of the problem is illustrated by colloquies which took place during press briefings between the chiefs of delegations and reporters in London. The proceedings of the Disarmament Conference are still confidential so one must look to paraphrases of press statements for information. I submit the following paraphrases (without correcting grammar or technical inaccuracies) and from which the names of the reporters and the heads of delegations are omitted.

The Delegate commenced the briefing by stating:

We had special matter for consideration on those missiles that had not yet been perfected, and this meant missiles that would move through outer space at very high speeds and great distances. He felt if we reached first step agreement we should also address ourselves to handling this problem. He said that uncontrolled race into outer space in years ahead might lead to great tragedy for mankind, that proposal we made we thought was very practical and very modest, it was one on which it would seem

---

[1] *International Cooperation in Rocketry and Astronautics,* Jet Propulsion **25**, No. 11 (1955); *Basic Concepts of Space Law*, Jet Propulsion **26**, No. 11 (1956); *Space Law—Basic Concepts*, Tennessee Univ. Law Rev. **24**, No. 4 (1956); *Space Law and Metalaw—A Synoptic View*, Harvard Univ. Law Rec. **23**, No. 6 (1956); *The Present Day Developments in Space Law and the Beginnings of Metalaw*, Harvard Univ. Law Rec. **24**, No. 2 (1957); *The Present Day Developments in Space Law*, Canad. Oil J. **8**, No. 7, 8, 9 (1957); *Seventh IAF Congress Stresses Cooperation*, Jet Propulsion **27**, No. 1 (1957); *Space Law and Metalaw—Jurisdiction Defined*, Amer. Rocket. Soc. April 1957; *The International Situation and Legal Involvements with Respect to Long-Range and Earth-Circling Objects*, U.S. Air Force Office of Scientific Research, Washington, D.C., February 1957; *Weltraumrecht und Recht außerhalb der Erde*, Z. Luftrecht, Issue 2 (1957); *Space Law and Metalaw—Jurisdiction Defined* (revised), Deutsche Gesellschaft für Raketentechnik und Raumfahrt, at the Technical University of Stuttgart, April 13, 1957; *Droit de L'Espace et "Metadroit" (Limites de Juridiction)*, Rev. Gén. air (1957); *Space Law and Metalaw—Jurisdiction Defined*, J. Air Law and Commerce, September 1957; *Weltraumrecht und Recht außerhalb der Erde*, Weltraumfahrt, April 1957.

that all of states that were interested in control of armaments for sake of peace should be able to agree. In first step agreement there should be provision that within three months after effective date of agreement parties would cooperate in establishment of technical committee to study design of inspection system which would make it possible to assure that sending of objects through outer space should be exclusively for peaceful and scientific purposes.

* * *

A Delegate stated:

The very nature of outer space objects means it takes a great deal of testing and experimentation to perfect them and this could only be done through actual use of outer space. To get to outer space you have to get from earth up to it. It is not matter that would be very easily concealed, and it should be possible with good scientific advice to work out way of getting control of this, and the working out of way of getting control of it should be one of tasks in first step agreement.

* * *

Question: Would control of missiles imply control of launching bases?

Answer: That would be one element of it. It would be matter of whether you get control of them before they are perfected. No one has perfected them as yet. There has been lot of work on them, experiments and testing, but no country has thus far perfected them.

* * *

We decided concrete thing that ought to be done in first step would be to bring scientists together to design system, rather than trying to say in advance what system of control would be.

* * *

Question: Did you spell out some of specific matters Committee would want to take up?

Answer: I said this involved things that are called space weapons, platforms, military satellites, business of outer space.

Question: Could you tell us whether today it was specified what were missiles in outer space?

Answer: We said one of first things for scientific committee to agree on was what was distance from earth you would define as outer space. We thought of it generally as distance beyond earth at which you no longer have friction of air to delay and retard the speed. Consequently it is at any distance where objects can fly at tremendous speeds, thousands of miles an hour.

Question: In order for scientists to participate in committee, would law have to be changed?

Answer: No, there would have to be a treaty. Treaty itself establishes power to participate in committee.

* * *

You are setting course of events for objective of having all movements through outer space exclusively for peaceful and scientific purposes. This is relatively modest practical step which we believe best way to set in motion forces for peaceful side of outer space development.

* * *

Question: How did this rather revolutionary idea of bringing scientists together on committee come about?

Answer: Suggestions came from scientists themselves who have knowledge of this outer space business, who believe if scientists get together they can work this problem out.

Question: What happens in fact if the scientists are not able to agree on any kind of control system? Does that then cause disarmament agreements to collapse or stop going to next stage, or what?

Answer: Not necessarily. Another factor here will be how rapidly missile research itself goes ahead. You read all kinds of estimates but nobody in fact knows whether these intercontinental missiles are three years off or thirteen years off. When you start trying to solve some of these unresolved problems you may solve them rapidly, you may solve them slowly. Conceivably you would not solve them at all.

* * *

The foregoing paraphrases rather vividly illustrate the fact that the officials of the chancellories of the nations of the world are groping for correct answers. They must have a reasonable definition of the limitation of "airspace," a term which now appears in many treaties and in the municipal statutes of most of the nations of the world. They desire help from qualified social and natural scientists. I urge that at a plenary session of this Congress, the President of the International

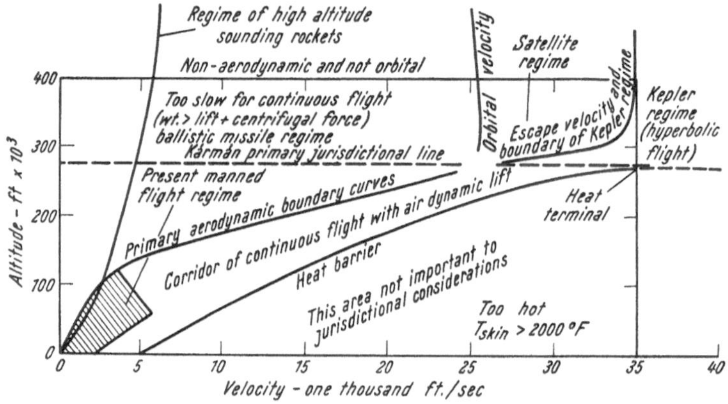

Fig. 1. Diagram showing regimes of atmospheric and extra-atmospheric flight and depicting the jurisdictional boundary lines

Astronautical Federation be authorized to appoint a Committee of seven persons, consisting of four physicists and three lawyers, who will draft a definition of "airspace" and recommend a rule delimiting airspace jurisdiction, such definition and rule to be supported by a statement of Findings of Fact and Conclusions of Law. The resolution should be transmitted to the attention of the Secretary-General of the United Nations and to the Secretary-General of the International Civil Aviation Organization (ICAO), with the statement that the Committee will cooperate with the appropriate officials of said organizations.

As I have pointed out in earlier papers, Dr. VON KÁRMÁN has suggested methods of formulating the jurisdiction of space law. Last spring he told me that he had delivered a paper at a luncheon at the University of California, Berkeley, entitled "Aerodynamic Heating—the Temperature Barrier in Aeronautics," and in that paper he had occasion to use a diagram made by MASSON and GAZLEY of the Rand Corporation showing the possible ranges for continuous flight in the velocity-altitude coordinate system. Later on, he sent me a copy of his paper which contains the MASSON and GAZLEY diagram. He said that this diagram, although designed to show the variation of velocity versus altitude for various values of dynamic pressure and equilibrium pressure, in the hands of a skilled person could readily be used to show the regimes of atmospheric and extra atmospheric flight and to depict the jurisdictional boundary lines thereof.

I have unskilfully redone the MASSON and GAZLEY diagram (see Fig. 1) to indicate curves showing the high altitude sounding rocket regime, the earth orbital satellite regime and the KEPLER regime (earth escape velocity), and some supernumerary information, but most importantly, I have shown what I now call the KÁRMÁN primary jurisdiction line.

To establish sound bases for demarcation of air and space jurisdiction it is necessary to consider that the conditions for accomplishing aerial flight, that is to circle at constant altitude, are weight equals aerodynamic lift plus centrifugal force. The aerodynamic lift decreases with altitude because of the decreasing density of the air and in order to maintain continued flight beyond zero air lift, centrifugal force must take over. Consider the flight of Captain IVAN C. KINCHELOE, in which he took the X2 rocket plane to 126,000 feet altitude. His flight was strictly an aeronautical adventure and did not partake of space flight. At the altitude indicated aerodynamic lift carries 98% of the weight and only 2% is centrifugal force, or "KEPLER force." It will be noted that in the corridor of continuous flight when an object reaches approximately 275,000 feet and is traveling at 25,000 feet per second, the KEPLER force takes over and aerodynamic lift is gone. This is a critical jurisdictional boundary.

I have reproduced the MASSON and GAZLEY right side curve—the so-called temperature barrier, or heat barrier—simply to show the present state of the art, and thus arbitrarily to delimit the corridor of continuous flight. This line has nothing to do with the jurisdictional question as improved techniques in cooling and discovery of heat resisting materials will undoubtedly change this curve.

Fig. 1 is intended to be illustrative, and it is not presented as an apodeictic solution of jurisdictional boundary lines. The KÁRMÁN primary jurisdictional line may eventually actually remain as shown on Fig. 1 or, after due consideration, the line may be significantly changed. In any event, this is the line at which "airspace" terminates.

I quite agree with Professor COOPER that any such definition should be determined under the aegis of the United Nations and that the pertinent regulations should be promulgated by the International Civil Aviation Organization (ICAO). In determining the KÁRMÁN line, the United Nations will require the advice of a Committee, such as I have suggested *supra*, and ICAO in drafting detailed regulations will further require the advice and findings of those possessed of a large number of scientific skills, including medical doctors, biologists, aerodynamicists stress analysts, aeronautical engineers, upper atmosphere physicists, economists and lawyers. Incidentally, the basic statutes of ICAO will have to be broadened by international agreement and the name undoubtedly must be changed.

It would be senseless to build a surface trans-atlantic steamship to perform the undersea functions of a submarine. The functions of the aircraft and the rocket ship are essentially even more disparate. In arriving at a reasonable KÁRMÁN line, physicists and lawyers inevitably will reach agreement as to the point where the aeronautical vehicle no longer may perform efficiently and within reasonable physical and engineering parameters. It may be useful to examine, momentarily, some of these parameters:

A. M. MAYO has pointed out that control of the pilot's immediate environment from the standpoint of pressure and composition would become increasingly difficult as a function both of the length of time of flight and of the pressure reduction and change of atmospheric composition[1].

---

[1] The statements from MAYO, STRUGHOLD, HULBURT, BUETTNER and SALTER are taken from "Physics and Medicine of the Upper Atmosphere. A Study of the Aeropause", Report of a symposium sponsored by the USAF School of Aviation Medicine and the Lovelace Foundation for Medical Education and Research, composed and printed at the University of New Mexico Press, Albuquerque.

He goes on to state that at altitudes below approximately 70,000 feet the problem of pressurization and composition is and can be taken care of relatively easily by pressurizing outside air. At higher altitudes, pressurization of outside air becomes increasingly difficult both from the standpoint of the power required and from that of handling the very high temperatures resulting from extreme ratios.

He also states that as outside pressures become negligible with respect to cockpit pressure, the problem of explosive decompression or even gradual loss of pressure becomes so acute that until pressurized cockpits are as highly reliable as the wings and basic structure of present aircraft, we will need to provide some sort of pressurization safety equipment.

MAYO states that information as to the intensity and scope of cosmic radiation, together with data on their effect on human beings, is needed by the engineer. Questions as to the possible existence of dangerous levels of radiation, such as X-rays from the sun, should also be surveyed. No completely practical approach to methods of protecting occupants of aircraft against high-energy radiations has yet been outlined. The problem might be reasonably simple were weight not such a primary consideration in all aircraft design problems.

Many other considerations will enter into the final determination of the KÁRMÁN line, such as the danger of material collisions with the airframe, escape problems, the problems posed by combined stresses and, indeed, multifold fundamental questions of the construction of aircraft, as such, will all enter into the final decision.

H. STRUGHOLD points out that with increasing altitudes, some of the biological effects creep in gradually, while others rise at sharply defined levels. On the whole, the road from the surface of the earth to free space displays characteristic ecological stages. These stages are determined by the functions which the atmosphere has for man and craft.

Going more into detail, he points out that we must first consider the atmosphere as milieu for respiration, or, in other words, the oxygen component in the chemical constitution of the air. In this respect, only the lower half of the troposphere can be designated as the physiological atmosphere or the ecosphere of the air. It is in this narrow zone that the stage for the drama of life on our planet is normally set. Only this layer deserves the name "atmosphere" which, from the Greek word "atmos", means "breath."

STRUGHOLD summarizes the biological jurisdictional line by pointing out that we face complete biological anoxia at about 16 km (52,000 ft.), despite the occurrence of free molecular oxygen in the atmosphere up to 90 km (295,000 ft.). It is only above this level that molecular oxygen vanishes completely from the atmosphere and dissociates into atomic oxygen by radiation. The alveolar air in the lungs, however, and not the ambient air, is decisive from the biological point of view. The first determines the physiological oxygen dividing line between atmosphere and space. STRUGHOLD's conclusions agree generally with the KÁRMÁN line on Fig. 1.

He also states that on the basis of the explosive decompression experiments, performed in the altitude range between 8 and 17 km (26,000 and 56,000 ft.), we can predict the course of events for the upper atmosphere. If a meteorite hits a rocket ship cruising in these altitudes, and the crew members are suddenly exposed to the ambient air, they have only 15 seconds at their disposal for action, from the standpoint of oxygen. Aeroembolism, body fluid boiling effects, and expansion of gas-filled cavities within the body, will come into play at the same time in a most fulminating manner and aggravate the situation. This is a problem of mutual interest to the engineer and medical man and demands their common attention and cooperation.

E. O. HULBURT points out that there is evidence that above 90 or 100 km (56 or 62.1 mi.) a rapid transition from molecular to atomic oxygen occurs, such that at about 120 km (74.5 mi.) all oxygen is dissociated to the atomic condition. The evidence was obtained from a rocket flight of September 29, 1949, by means of measurements of ultraviolet sunlight with photon counters. The intensity of solar light in a spectral band from about 1450 to 1530 A, which is near the center of the great absorption band of the oxygen molecule, was observed to increase very rapidly as the rocket rose from 90 to 120 km, indicating that in this region the rocket was passing entirely above molecular oxygen.

He also states that the region of the atmosphere from 0 to 10 km (6.2 mi.), where the temperature falls rapidly with increase of altitude, has long been called the troposphere, and the region from about 10 to 20 km (33,000 to 66,000 ft.), where the temperature is approximately constant, is called the stratosphere. The ionosphere has its own well-accepted nomenclature, the terms D, E, F_1 and F_2 designating the four ionized regions with maxima of ionization at about 70, 100, 200 and 300 km (43.5, 62.1, 124.3, 186.4 mi.), respectively. Aside from these, there is no generally accepted terminology of upper atmospheric regions. HULBURT goes on to discuss definitions. The terms upper and outer atmosphere are used with different meanings depending on the context, and it is best to keep their meanings fairly elastic. The region from about 20 to 35 km (12.4 to 21.7 mi.) which embraces most of the ozone has been called the ozone layer or ozonosphere. It has been proposed that the region from the top of the stratosphere, at about 20 km (12.4 mi.), to the minimum of temperature, at about 70 km (43.5 mi.), be called the mesosphere and the region of increasing temperature, somewhere above 100 km (62.1 mi.), the thermosphere. The exosphere has been used to refer to the outer fringe of the atmosphere, where the air particles execute long elliptical orbits bouncing outward from impacts with other particles and falling back under gravity. In general, HULBURT concludes, the physical properties of the various regions are not yet well enough known to permit their fixation by an accepted terminology.

K. BUETTNER summarizes a few of the problems of an airframe in the aeropause. The necessity of repainting the Skyrocket after each flight merits attention in this connection and poses a challenge to the paint industry. For example, BUETTNER says, assuming the cabin wall to be part of the outer hull of a non-atomic rocket ship, the cabin-hull combination would have to withstand: high temperatures and strong draft during ascent; and during flight in the aeropause structures would have to stand nearly an absolute vacuum, weightlessness, meteoritic dust, an abundance of ionizing radiation such as cosmic ray primaries, solar X-rays, ultraviolet, and solar protons. Also, the absence of a force retaining surface particles which happen to be driven off by electro-static repulsion might be a factor of significance.

On the element of reasonableness, R. M. SALTER points out that some of the first questions to be answered are, "how high," "how fast," and "how long" can flight be sustained in the atmosphere? Emphasis must be given the last item if a pilot is carried in the vehicle. Obviously, other than for the purposes of physiological experimentation or establishing a record, one would not conceive of a manned sounding rocket. Here, there is not time or need to supplant electronic equipment for making observations. However, in cases where flight duration is of sufficient length that electronic reliability is a problem, where computer operations (such as having adequate "memory" included) are too complex, and, in particular, where judgment in unforeseen circumstances is needed, then the participation of man will be required. It may be seen that the first two requisites are limited only by prevailing engineering development, while the last is clearly a basic constraint.

SALTER continues to state that the employment of pilots in supersonic rocket planes and in balloons is an example of present approaches to the problem. In the case of air-borne vehicles (those using forward motion to derive lift from the atmosphere) we must consider duration of flight as well as altitude. It is convenient to subdivide this class of vehicles into those using rocket engines and those using air-breathing power plants. This latter type is represented by the various jet propelled aircraft and missiles. In order to fly at very high altitudes it is necessary for such a vehicle to operate at supersonic speeds, not only to provide sufficient lift but also for adequate thrust. At an altitude of 20 mi. (32.2 km), for example, the required MACH number for a ramjet is over 5 and the resultant incoming air has a stagnation temperature of the order of 2000° F. Since energy must be imparted to this air at higher temperatures, it may be seen that a present engineering limitation on suitable fuels and materials is approached. This is particularly true with the use of nuclear heating. Thus the air-breathing vehicle is limited in altitude.

In their most recent writings, COOPER and MEYER have logically sought actually to locate airspace jurisdiction at a point in the ocean of air surrounding the earth where aerodynamic lift is gone. These authorities have carefully avoided the mistakes of some who are attempting to locate airspace jurisdiction in such heterodox and altogether eclectic regions as the thermosphere, exosphere, mesosphere and ozonosphere, even the nomenclature of which is doubtful, and none of which has any reference to the problem at hand.

I have at last resolved my problem with Professor COOPER's theory of "contiguous space." At the Fiftieth Annual Meeting of the American Society of International Law, Professor COOPER proposed a new convention which might include these solutions:

(a) Reaffirm Article I of the Chicago Convention, giving the subjacent state full sovereignty in the areas of atmospheric space above it, up to the height where "aircraft" as now defined, may be operated, such areas to be designated "territorial space."

(b) Extend the sovereignty of the subjacent state upward to 300 miles above the earth's surface, designating this second area as "contiguous space," and provide for a right of transit through this zone for all nonmilitary flight instrumentalities when ascending or descending.

(c) Accept the principle that all space above "contiguous space" is free for the passage of all instrumentalities.

Basing a jurisdictional area on the concept of "contiguous space," an area as noted *supra* from the subjacent state upward to 300 miles, is quite unnecessary, although wisely precautionary, and is not susceptible of implementation. On the other hand, under certain conditions national jurisdiction will be *quite indirectly* but effectively maintained over what might be called "contiguous space." In the near future the nations of the earth will be offered point-to-point rocket communications involving many services. Professor COOPER, in London, authorized me to state that he believed the first commercial use of rocket vehicles will be for rocket mail transportation between New York and London. This service and other services will gradually be extended to the four corners of the world—Melbourne to London; Moscow to Los Angeles; Buenos Aires to Chicago; Los Angeles to

New York. The trajectories of each of these routes will be different, and will involve different altitudes. Some of these rockets will describe a trajectory requiring heights of 300 miles or less, and others will probably require heights in excess of 1000 miles. National jurisdiction will be effectively maintained by the granting of launching and landing rights, and thus there will be indirect national control with respect to point-to-point earth rockets over contiguous space. With the advent of manned rocket ships, this control undoubtedly will become more severe because of vastly increased considerations of safety and other nationalistic and even ethnological problems.

The international regulation of point-to-point earth rocket vehicles will be under the jurisdiction of a successor International Governmental Organization to ICAO. The very problem of locating instrumentation along the manifold aerodynamic and nonaerodynamic routes, orbits and trajectories, will require the highest degree of international cooperation and regulation for the operational efficiency and safety of all concerned. The aircraft, the point-to-point earth rocket ship, and the space ship capable of freely maneuvering in outer space when navigating around earth or landing on earth, will each and all need navigation aids, anti-collision devices, secondary radar, communications systems, meteorological services, and many other international aids and services, and in every instance of movement to or from any point on earth, the national licenses and permits which are indeed the essential prerequisites of national society as we know it today. Mankind must mature appreciably to create the International Authority which I discussed in the Rome paper, and I expect such achievement must await the wise action of future generations.

No single nation has a paramount claim to outer space nor a monopoly on the scientific genius which will soon make its exploration and exploitation a reality. Therefore it is axiomatic that the field of astronautics will progress only as rapidly as international cooperation in the field is accomplished. It has been suggested that because of the hates, fears and prejudices of people, such startling advances of science will never be brought under a system of world cooperation based on international law. Certainly, the treatment of the hydrogen and atomic bombs lends credence to such a pessimistic prediction. Yet we may find, in the words of one of the greatest scientists of our time, the basis for a contrary position. Nearly a dozen years ago Dr. EINSTEIN said, "The ability to think is also a part of human nature. It is intelligence, which is the ability to learn from experience, to plan ahead. It includes the capacity to give up immediate, temporary benefits for permanent ones. This part of human nature recognizes that a man's security and happiness depend on working society; and that working society depends on laws; and that men must submit to these laws in order to have peace."

I conclude my discussion on the subject of airspace jurisdiction by stating that national sovereignty ends for all purposes with the KÁRMÁN line, and by adding by way of analogy the STEPHEN DECATUR Doctrine demonstrated 150 years ago that

> The seas beyond reasonable coastal areas (space beyond the KÁRMÁN line) are free and subject to control by no single despot or nation, and
>
> The sponsors of ships at sea (spaceships in space) must be responsible for the conduct of their vessels[1].

And now a few more words about the satellite program. More than fifty nations are participating in the IGY. The participation of these nations in the IGY

[1] The DECATUR Doctrine is quoted from PH. B. YEAGER and J. R. STARK, U.S. Naval Inst. Proc. 83, 931 (1957).

Program involves very extensive utilization of governmental and non-governmental facilities and personnel. Active participation in the program is required of each nation, of its army, navy, air force and coast guard personnel and facilities; of such governmental agencies as those concerned with standards, radar, radio, meteorology, weather, coast and geodetic surveys, geological surveys, and all types of official scientific and research organizations. In addition, parallel, non-official institutions are involved, including universities and observatories in those few countries where such institutions are not controlled by the State. By agreeing to support actively the satellite program the foregoing nations also agreed to the legal validity of the project.

On the basis of sound principles of international law, the nations of the world may not protest the flight of a non-military artificial satellite over their territories—when the purpose of such flight is the accumulation and dissemination of scientific data which shall be made available without restriction to all the nations of the world.

No single formal treaty, signed at a single council table, emerged from the myriad agreements involved in the IGY. Nevertheless, a valid and binding world pact emerged from these acts of agreement and cooperation. The international pact, in written form, may be abstracted from the thousands of documents and exchanges from which the living IGY has evolved. There is nothing about a single formal treaty which makes it sacrosanct—which makes it even an essential source of international law. A treaty is merely a formal expression of the will of the contracting states—a formal method by which the nations involved show their consent to some act or agreement or series of acts and agreements. In many instances the principle set forth in the treaty itself may have been established in international law long prior to the signing of the formal document itself. A rule of international law does not receive its validity from its enactment into a legal instrument such as an international treaty. There are rules of international law which are valid although not enacted in such legal instruments, and there are rules of international law which are not valid, although enacted in such instruments. Enactment, therefore, is no objective criterion for the alleged validity of a rule of international law.

The social scientist, just as clearly as the natural scientist, has the constant duty to acquire base-line data and then to implement such data in the dynamic evolution of society. The advent of the rocket motor, as a great prime mover with its potential of unlimited access on the earth and in space, presents a great new mutation requiring the immediate and careful attention of the social and natural scientists. The base-line material of the lawyer has always been invested in the mores of mankind from which it must extrapolate principles of justice.

I am deeply disturbed that the concept of metalaw has not generally been understood, even by some great scientists, as witness the complete lack of understanding voiced by the distinguished Professor Antonio Ambrosini in his comments at the Rome Congress.

There can be no doubt that in the not distant future, social and natural scientists will undertake studies of a substantive statement of metalaw and as a by-product of these studies our own anthropocentric law will be improved.

I pointed out at the Rome Congress that even at this time we may postulate a profoundly necessary rule of space exploration, namely, in any instance where there is reason to believe that intelligent life exists on a planet, no earth spaceship may land without having satisfactorily ascertained that (1) the landing and contact will injure neither the explorer nor the explored; and, (2) until the earth ship has been invited to land by the explored.

Many of my friends have protested that this is like a requirement that man must make intelligible communication with an amoeba! The answer is that nevertheless the regulation must be adhered to without exception, or we will project into space and perpetuate the bleak and devastating geocentric crimes of mankind. Further, we must conquer certain problems of semantics before we are worthy of space travel beyond our Solar System. The regulation is so necessary that it would be better for mankind to perish than allow its violation.

The principle I have stated is just as old and simple as the basic idea of justice itself. EPICURUS said that justice is never anything in itself, but the dealings of men with one another in any place whatever and at any time. It is a kind of compact not to harm or be harmed (Principle Doctrine XXXIII). The maxims we live by are: *Justitia debet esse libera, quia nihil iniquius venali justitia; plena, quia justitia non debet claudicare; et celeris, quia dilatio est quaedam negatio.* Justice ought to be free, because nothing is more iniquitous than venal justice; full, because justice ought not to halt; and speedy, because delay is a kind of denial. *Justitia nemini neganda est.* Justice is to be denied to none.

In my Rome paper I pointed out that all of the statements of the Golden Rule, even that of the heterodox Hindu religion of Jainism, are, in each case, starkly anthropocentric. I stated that during all the long centuries of human civilization no lawgiver has framed the Golden Rule in the language of the Great Rule of metalaw. This is probably due to some inherent necessity to relate all human law to oneself. It is a law for one frame of existence. But with equal positiveness the Golden Rule has no application whatsoever in the field of metalaw. In metalaw we deal with all frames of existence—with intelligent beings different in kind. *We must do unto others as they would have done unto them.* To treat others as we would desire to be treated might well mean their destruction. We must treat them as they desire to be treated. This is the simply expressed but vastly significant premise of metalaw.

Even at this time it may be possible to arrive at some ideas with respect to the physical characteristics of sapient extraterrestrial beings, and having done so, conjecture on a negative principle of metalaw.

Merely by way of example, let us very briefly consider in our own frame of reference the kind of phenomena associated with man's flight in space. What might be the effects on other three dimensional creatures of (1) Communications systems; (a) electromagnetic waves; (b) light signalling (photons); (c) in an atmosphere—of pressure waves (2) Propulsion; (a) infrared rays, from heat; (b) radiation from a nuclear process; (c) in an atmosphere — of pressure waves (3) Man's physical and mental properties: (a) parapsychological or telepathic impacts (b) body offenses and germ dissemination; (c) impact of ideas and customs of man.

In order to proceed further, it is necessary to make some assumptions as to the nature of the extraterrestrial beings with whom space exploration may bring us in contact. In one frame of reference the minimum assumption is that they are composed of the same elementary substances that are now known to us. The next assumption in the same frame of reference is that they are large aggregates of atoms capable of sensation, locomotion, and thought. From considerations of biophysics and the theory of logical machines, it may be possible to arrive at a statement of the minimum size and weight of such extraterrestrial beings. From considerations of the dynamics of bodies and structural analysis, it may be possible to arrive at a statement of the maximum size and weight of such extraterrestrial beings. In this way, it may be possible to bracket between lower and upper limits the probable size of extraterrestrial beings.

Flights into interplanetary and interstellar space in the future will have an appreciable effect on possible extraterrestrial beings only if they intercept a minimum amount of the energy radiated from an earth space vehicle. In considering each form of transference of energy, there is a minimum condition which just barely allows the radiated energy to be noticed above the background noise. Thus, the particles radiated from a space vehicle have to be

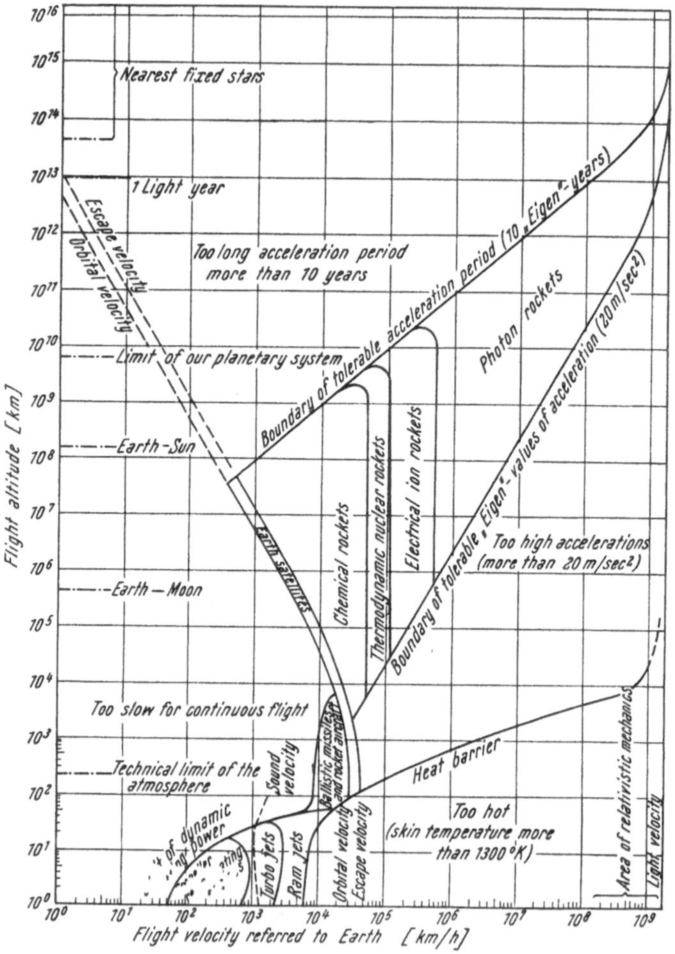

Fig. 2. The E. SANGER schemating diagram illustrating all possible types of flight regimes (altitude versus velocity)

noticeable against a background of cosmic rays in order to represent an "interference" on our part. Similarly, thermal radiation must be noticeable against the thermal background of the prevailing temperature, and radio waves must be noticeable against the background radio noise.

Considering the probable limits of size and weight of such beings regarded as receivers of physical radiation, and the signal to noise problem outlined above, it should be possible to calculate for each mode of energy transfer the limits of free space, that is, the closest distance of approach outside of which no possible effect can be exerted upon the hypothetical being. In other words, it should be possible to estimate the size of the sphere surrounding each individual that may be called his zone of sensitivity.

The size of this zone will depend on the size of the transmitter, i.e., the spaceship and what energy it radiates and in what form.

Hence, for this question in metalaw as applied in this simple frame reference, the rule should be that space outside an individual's zone of sensitivity is free space to which the traditional freedom of the seas may apply, as demonstrated by STEPHEN DECATUR. The shores of earth become the undreamed of shores of space.

I have defined metalaw at Rome, San Diego, Washington, Stuttgart and Paris, somewhat ineptly, as the law governing the rights of intelligent beings

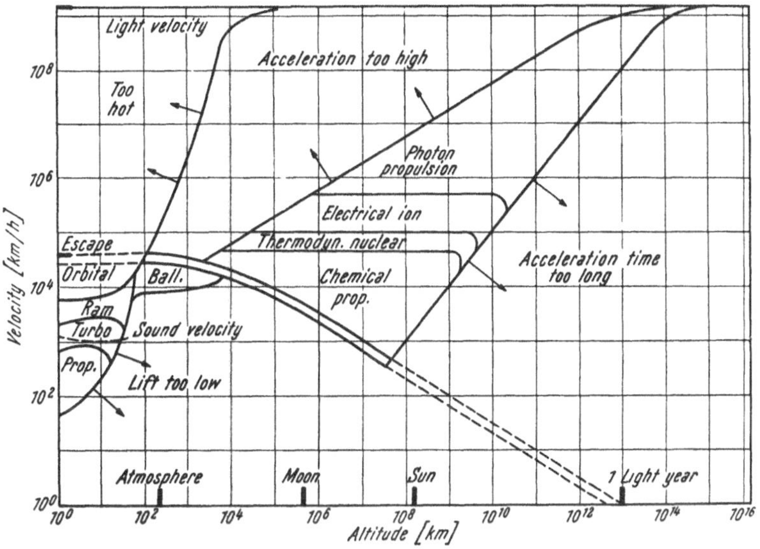

FIG. 3. The S. F. SINGER adaptation of the SANGER schematic diagram illustrating all possible types of flight regimes (velocity versus altitude)

of different natures and existing in an indefinite number of different frameworks of natural law. It will be noted that the problems I have discussed *supra* have dealt in the areas of three-dimensional frames of reference. There remain a multitude of other frames of reference requiring our apprehension. I do not intend to discuss these concepts beyond pointing out that an intelligent multidimensional creature existing in a creation involving only three dimensions would probably be so inferior as to require most considerate and sympathetic understanding and treatment by three-dimensional creatures existing in a three-dimensional universe, such as we conceive of mankind. The intelligent mankind concept runs into deep trouble vis à vis an intelligent two-dimensional creature existing in a three-dimensional continuum and problems of metalaw become truly complicated when one considers man's relationship with intelligent two- and one-dimensional creatures existing in a two-or one-dimensional universe, or any combination thereof.

I had the privilege of presenting a paper before the Deutsche Gesellschaft für Raketentechnik und Raumfahrt, at the Technical University of Stuttgart on April 13, 1957, and upon the conclusion of the program I discussed at length with Dr. EUGEN SÄNGER and Dr. IRENE SÄNGER-BREDT the possible types of flight regimes, including those regimes which might philosophically spell the elimination of a dimension. The most comfort I could obtain from Dr. SÄNGER was a hand-drawn diagram illustrating all possible types of flight regimes which he brought to Paris on April 15th. Later, Dr. SÄNGER was kind enough to refine his diagram and I now present it as Fig. 2. Dr. S. F. SINGER of the University of Maryland has adapted the SÄNGER schematic diagram for slide purposes. I present his adaptation as Fig. 3. I borrow the words of Dr. SÄNGER in describing Fig. 2:

The abscissa plots are in logarithmic scale the flight velocities (in kilometers per hour) relative to the surface of the Earth, up to their largest possible value, the velocity of light of 1.08×10^9 km/h. Shortly over 103 km/h, the altitude dependence of sonic velocity is indicated.

The ordinate, also in logarithmic scale, shows the altitudes of flight (in km) from the surface of the Earth up to the nearest fixed stars.

On the ordinate scale, particularly characteristic altitudes are specially marked, such as the technically dependent and argued limit of national sovereignty at an altitude of 80 km, the boundary of the technically sensible atmosphere at 200 km, the distance of the Moon over the surface of the Earth at 400,000 km, that of the Sun at 2×10^8 km, the limit of our planetary system at approximately 10^{10} km, etc.

Within this coordinate system, the domains of aeronautics, the transition domain from aeronautics to astronautics and the domain of astronautics are plotted.

Proper aeronautics, characterized by air-breathing propulsion systems, reaches up to about 60 km altitude. It lies between the two well-known limiting curves, namely the limit of aerodynamics lifting power and the heat barrier. This domain plot shows an onion-peel shape (attaching one onion-peel to the next by increasing flight velocities and altitudes) when progressing from propeller reciprocating engines to turbojets and then to ramjets.

Though aerodynamic lifting power is gradually replaced by the centrifugal force from the trajectory curvature beyond a velocity of 10^4 km/h, the intersection of the curves of limit of aerodynamic lifting power and heat barrier is nevertheless physically real and constitutes as the utmost limit of the ramjet also the definite limit of aeronautics.

Contrary to this, ballistic rockets and rocket aircraft are not limited by the limit of aerodynamic lifting power due to their "non-air-breathing" propulsion systems and they can fully exploit the lifting support of the inertial forces due to the trajectory curvature from about 6,000 km/h on, so that their possible altitudes of flight increase to several thousand kilometers and their flight velocities approach orbital velocity, whence the aerodynamic lifting power then is completely replaced by the inertial forces of the circular orbit about the earth.

These ballistic missiles and rocket aircraft form the transition domain from aeronautics to astronautics which flows into the domain of pure astronautics with the reaching of the orbital velocity. Hence, only a very small corridor connects aeronautics with astronautics.

Within the narrow band between orbital velocity and the constant $\sqrt{2}$-times larger escape velocity lie the artificial satellites of the Earth. This band curves to decreasing flight velocities with increasing altitude due to the decrease of the gravitational acceleration by the square of the altitude over the Earth.

Beyond this band there open the immense vistas of interplanetary, interstellar and intergalactic space flight, in a structural shape surprisingly similar to that of aeronautics according to present concepts.

This domain of pure astronautics is again bounded by two limiting curves, this time due to biological considerations.

The lower "boundary of tolerable values of acceleration" results from the physiological possibly still tolerable uniform constant acceleration of $b = 20 \mathrm{m/sec}^2$, presented as the parabola $h = v^2/2\,b$ (h = altitude, v = velocity, b = acceleration) which appears as a straight line on the logarithmic scales in the range of non-relativistic mechanics.

Certain velocities at low flight altitudes are not obtainable any more due to biologically too high accelerations (more than $20\ \mathrm{m/sec}^2$).

The upper "boundary of tolerable acceleration period (10 years)" results from the consideration that for low values of the uniform acceleration b the duration t of acceleration for achieving certain altitudes of flight h becomes too large with respect to the natural lifetime of human beings and thus cannot be considered longer than 10 years. Thus follows the relation $h = vt/2$ as a limiting curve in the (h, v)-diagram against the inadmissible range of too long acceleration periods, more than 10 years.

Between these two biological boundary curves of maximum applicable constant acceleration and maximum possible duration of the acceleration periods, there are grouped the "onion peels" of the increasingly fast astronautical vehicles, chemical rockets (up to approximately 50,000 km/h), thermodynamic nuclear rockets (up to approximately 100,000 km/h) and electrical ion rockets (up to approximately 500,000 km/h).

The increasingly narrow corridor between the two boundary curves for still higher flight velocities is reigned by the photon rockets at the end of astronautics in analogy to the rocket aircraft at its beginning.

The domain of the photon rockets on the one hand reaches up to the altidudes of fixed stars, on the other into the wonderland of relativistic mechanics.

* * *

On this trip to Europe I hope to have the honor and privilege of further conferences with Dr. SÄNGER and Dr. SÄNGER-BREDT, with the thought that in the great complex of space and metalaw jurisdiction I will be able to evolve a satisfactory statement of the concept of the SANGER regime.

Über Stabilitätsuntersuchungen an flüssigkeitsgetriebenen Raketenmotoren mit Hilfe des Verfahrens der „Harmonischen Balance"

Von

G. Heinrich [1] und **W. Peschka** [2]

(Mit 5 Abbildungen)

Zusammenfassung — Abstract — Résumé

Über Stabilitätsuntersuchungen an flüssigkeitsgetriebenen Raketenmotoren mit Hilfe des Verfahrens der „Harmonischen Balance". In vorliegender Arbeit wird das Verhalten flüssigkeitsgetriebener Raketenmotoren (monopropellant systems) mit Hilfe des Verfahrens der „Harmonischen Balance" untersucht. Es zeigt sich, daß der numerische Aufwand nur bei zwei Spezialfällen in erträglichen Grenzen gehalten werden kann. In diesen beiden Fällen sind jedoch allgemeine Aussagen über das Stabilitätsverhalten möglich, die weitergehender sind als die mit Hilfe des Verfahrens der kleinen Störungen erhaltenen.

Investigation of the Stability of Liquid Propellant Rocket Motors, Using the Method of the "Harmonical Balance". The authors have investigated the behaviour of liquid propellant rocket motors (monopropellant systems) using the method of the "harmonical balance". It is shown that the numerical expenditure can be tolerably limited only in two special cases. Then general statements concerning the stability behaviour are possible which are more far reaching than those statements that are gained by the method of perturbations.

Recherches sur la stabilité des moteurs-fusée à monergol par la méthode de "l'équilibre harmonique". Le fonctionnement des moteurs-fusée utilisant un monergol est analysé par la méthode de "l'équilibre harmonique". L'application numérique de la méthode n'est raisonnablement limitée que dans deux cas particuliers. Des conclusions générales sur la stabilité du fonctionnement sont cependant possibles. Les plus avancées rejoignent celles obtenues par la méthode des petites perturbations.

Im Gegensatz zu den bisher erschienenen Arbeiten, worin das Verhalten im kleinen behandelt wurde, soll in vorliegender Arbeit das Verhalten von Raketenmotoren im großen untersucht werden. Um den großen numerischen Aufwand, den das Verfahren der „harmonischen Balance" in diesem Falle mit sich bringt, klein zu halten, wurde als Sonderfall ein Raketenmotor untersucht, dessen Treibstoffversorgung konstant und von den Brennkammerzuständen unabhängig ist.

Eine Verallgemeinerung ist jedoch möglich, wenn zur numerischen Rechnung programmgesteuerte elektronische Rechenmaschinen zur Verfügung stehen.

---

[1] Vorstand des I. Instituts für Allgemeine Mechanik an der Technischen Hochschule, Wien IV, Karlsplatz 13, Österreich.

[2] ÖGfW; Fellow BIS; Assistent am II. Institut für Allgemeine Mechanik an der Technischen Hochschule, Wien IV, Karlsplatz 13, Österreich.

Über das Problem der Verbrennungsinstabilität bei Raketenmotoren von Flüssigkeitsraketen liegen bereits sehr viele Veröffentlichungen vor. Die Experimente zeigen bei den meisten Raketenmotoren ein Oszillieren des Brennkammerdrucks und der damit gekoppelten Größen [8]. Im Frequenzspektrum tritt vor allem ein niederfrequenter Anteil (50—100 Hz) und ein „hochfrequenter" Anteil im Gebiet von 500—1000 Hz hervor. Vorliegende Arbeit befaßt sich mit dem niederfrequenten Anteil dieses Phänomens.

Der Gedanke, diese Erscheinungen als Instabilität des Raketenmotors, also der Gruppe: Treibstoff-Fördersystem—Leitungen—Brennkammer—Düse, zu erklären, ist naheliegend. Allerdings ergeben sich beträchtliche mathematische Schwierigkeiten, wollte man alle Einflüsse streng erfassen. Um nun auch einigermaßen allgemeine Aussagen machen zu können, muß man sich vereinfachte Modelle schaffen, welche die wesentlichen Eigenschaften eines Raketenmotors beschreiben. Durch Diskussion der Eigenschaften dieser Modelle erhält man dann Resultate, die mit den experimentellen Ergebnissen verglichen werden müssen. Aus dem Grad der Übereinstimmung kann dann ein Modell gefunden werden, das den Tatsachen am besten entspricht.

Ein solches Modell, das sich auch bewährt hat, ist das von L. CROCCO [1]. Die wichtigste Annahme der Theorie von CROCCO ist, daß zwischen dem Einbringen des Treibstoffes in die Brennkammer und seiner Entzündung eine gewisse Zeit verstreicht, die im wesentlichen vom Brennkammerdruck und der Tröpfchengröße der Treibstoffe nach Einbringen in die Brennkammer abhängt. Diese Zeitverzögerung (time-lag) bewirkt, daß beispielsweise bei Erhöhung des Förderdruckes die damit verbundene Änderung der Brennkammerzustände erst um eine Zeit $\overline{\tau}$ später erfolgen wird[1]. (In der Regelungstheorie nennt man eine solche Erscheinung: Regelstrecke mit Totzeit.)

Es kann nun Werte für $\overline{\tau}$ geben, für welche die Änderung der Brennkammerzustände gerade unterstützend auf die Ursachen dieser Änderung zurückwirkt, womit bereits eine Instabilität gegeben ist (innere Instabilität — intrinsic instability).

Die Aufgabe besteht nun darin, die Gleichungen, die das Verhalten eines solchen Systems beschreiben, aufzustellen und dann mit Hilfe dieser Gleichungen die Stabilität zu untersuchen. Dies wurde in [1] für den Fall kleiner Störungen mit der Methode der kleinen Schwingungen gemacht. Die Ausgangsgleichungen wurden also linearisiert. Auch in den übrigen Veröffentlichungen wird von den linearisierten Gleichungen ausgegangen. Auf diese Weise kann aber nur eine Entscheidung über die Stabilität im kleinen getroffen werden. Ferner kann keine Aussage über die Amplitude der etwa entstehenden Schwingungen gemacht werden. Will man darüber Auskunft erhalten, dann muß von den strengen, nichtlinearisierten Gleichungen ausgegangen werden. Dies soll in vorliegender Arbeit geschehen.

---

[1] Die Zeitverzögerung $\overline{\tau}$ ist, wie schon erwähnt, von den Brennkammerzuständen abhängig. Im allgemeinen wird nur die Druckabhängigkeit berücksichtigt. Die Abhängigkeit von der Temperatur kann vernachlässigt werden. Für die Abhängigkeit vom Druck wird in [1] folgendes Gesetz verwendet:

$$\int_{t-\tau}^{t} p^n \, dt' = \overline{\tau} \cdot p^n = \text{const.}$$

wobei $\overline{\tau}$ die Zeitverzögerung im stationären Fall und n eine empirisch zu ermittelnde Konstante ist.

Den Untersuchungen wird vorerst ein „Monopropellant-System“ zugrunde gelegt, also ein Raketenmotor, dessen Treibstoff aus einem homogenen Gemisch eines Sauerstoff- und eines „Energieträgers“ besteht. Die Ausgangsgleichungen lauten dann nach [1] in dimensionsloser Form folgendermaßen:

$$\frac{d\varphi}{dz} + \varphi + 1 = \left[\frac{1+\varphi(z)}{1+\varphi(z-\delta)}\right]^n \cdot [1 + \mu(z-\delta)]$$

$$a\frac{d\mu}{dz} = f^*(1+\mu) - (1+\varphi)$$

$$\text{mit:}\ \varphi = \frac{p_K - \bar{p}_K}{\bar{p}_K};\ \mu = \frac{\dot{m} - \bar{\dot{m}}}{\bar{\dot{m}}};\ z = \frac{t}{\vartheta_g},\ \delta = \frac{\bar{\tau}}{\vartheta_g} \quad (1)$$

$$a = \frac{\bar{\dot{m}} \cdot l}{\bar{p}_K \cdot F \cdot \vartheta_g},\ f^*(1+\mu) = \frac{1}{\bar{p}_K} \cdot f(\dot{m}),$$

wobei p_K den Druck in der Brennkammer, $\dot{m}$ die Fördermenge der Treibstoffpumpe in der Zeiteinheit, ϑ_g eine Bezugszeit für die Zeit t und die Zeitverzögerung τ bedeutet. (Es wurde mit [1] für ϑ_g die mittlere Laufzeit eines Gaselementes in der Brennkammer gewählt.)

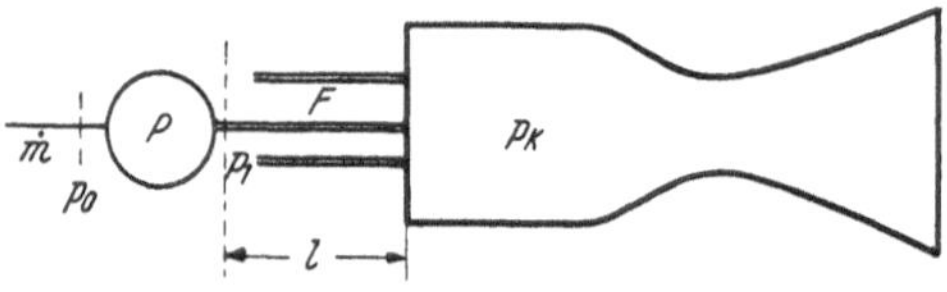

Abb. 1

l ist die Länge der Rohrleitung von der Pumpe bis zur Brennkammer, F der gesamte Querschnitt der Rohrleitung (wenn beispielsweise mehrere Rohre parallel verlegt sind) (Abb. 1). f^* ist die mit p_K dimensionslos gemachte Pumpencharakteristik bei konstanter Drehzahl.

Sind die Größen p_K, $\dot{m}$, τ mit einem Querstrich versehen, so bezeichnet dies ihren stationären Wert.

Die Annahme konstanter Drehzahl ist ohne weiteres zulässig. Es ergeben sich aber Schwierigkeiten bei der Pumpencharakteristik. Bekannt ist nur die stationäre Kennlinie. Über die instationäre Kennlinie ist derzeit noch nichts bekannt. Man könnte die stationäre Kennlinie verallgemeinern, indem man setzt $p_1 = f(\dot{m}, \ddot{m})$, aber hat man keine Klarheit, wie man diese Abhängigkeit ansetzen soll, da keine experimentellen Resultate darüber vorliegen. Daher wurde vorerst angenommen, daß der Einfluß bei Änderungen von niedriger Frequenz gering sei, und es wurde deshalb in vorliegender Arbeit die statische Pumpencharakteristik verwendet (Beschränkung auf niederfrequente Instabilitäten).

Eine weitere Schwierigkeit bringt das Verhalten der an die Brennkammer angeschlossenen Düse. In [1] wird die Annahme gemacht, daß die Düsenströmung den von der Brennkammer diktierten Zuständen praktisch trägheitslos folgen kann. Nach [2] ist diese Annahme nicht richtig, trifft aber im Fall niederfrequenter Schwingungen, wie sie hier vorliegen, zu.

Die Stabilitätsuntersuchung geht nun folgendermaßen vor sich: Wenn das System eine oszillatorische Instabilität besitzt (nur eine solche soll hier untersucht werden), dann müssen an der Stabilitätsgrenze ungedämpfte Schwingungen auftreten. Innerhalb des Stabilitätsbereiches muß die Dämpfung negativ, außerhalb hingegen positiv sein. Es muß also an der Stabilitätsgrenze das Gleichungssystem (1) eine periodische Lösung besitzen. Es wird daher eine periodische Lösung von der Form $A \cdot \sin \Omega z$ angesetzt[1]. Nun wird zu (1) ein lineares Ersatzsystem gesucht, das dieselbe Lösung besitzt. Dieses Ersatzsystem wird nun

[1] $\Omega = \omega \cdot \delta_g$, ω = Kreisfrequenz.

weiter untersucht, und mit Hilfe der ROUTH-HURWITZschen Kriterien kann dann die Stabilität beurteilt werden.

Obige Methode, einem nichtlinearen Gleichungssystem ein lineares zuzuordnen, stammt von N. KRYLOW und N. BOGOLJUBOW und ist unter dem Namen „Verfahren der harmonischen Balance" bekannt [3], [4]. Die Kombination mit der Methode der Stabilitätsuntersuchung nach A. A. HURWITZ stammt von K. MAGNUS [5], [6].

Um nun die Verallgemeinerungen gegenüber der Methode der kleinen Schwingungen zu zeigen, und der Übersichtlichkeit halber, werden in vorliegender Arbeit zwei Spezialfälle, denen aber eminente Bedeutung zukommt, untersucht. Zuerst wird ein System betrachtet, wo die Zeitverzögerung $\bar{\tau}$ und damit $\delta = O$ ist, dabei jedoch die Pumpencharakteristik berücksichtigt. Im zweiten Fall wird konstante Treibstoff-Förderung angenommen, die Zeitverzögerung soll aber dabei von Null verschieden sein.

I. Systeme ohne Zeitverzögerung

In diesem Fall reduzieren sich die Ausgangsgleichungen auf:

$$\begin{aligned} \frac{d\varphi}{dz} &= -\varphi + \mu \\ \frac{d\mu}{dz} - &= \frac{1}{a} \cdot \varphi + \frac{1}{a}\left[f^*(1+\mu) - 1\right]. \end{aligned} \tag{2}$$

Diesem Gleichungssystem wird nun ein lineares Ersatzsystem der Form (3) zugeordnet (vgl. [3], [4], [5], [6]):

$$\begin{aligned} \frac{d\varphi^*}{dz} &= -\varphi^* + \mu^* \\ \frac{d\mu^*}{dz} &= -\frac{1}{a}\varphi^* + \bar{a}_{22} \cdot \mu^*. \end{aligned} \tag{3}$$

Es soll nun sowohl das System (2) als auch das System (3) für φ und μ die Lösungen (4) besitzen:

$$\varphi = A\varkappa_1 \sin\Phi_1, \ \mu = A\varkappa_2 \sin\Phi_2; \ \Phi_\nu = \Omega z - \psi_\nu. \tag{4}$$

Die Phasenwinkel ψ_ν sind in diesem Fall ohne Belang und können gleich Null gesetzt werden.

Weiters wird nun $f^*(1+\mu)$ an der Stelle $\mu = O$ in eine Potenzreihe entwickelt:

$$\begin{gathered} f^*(1+\mu) - 1 = \alpha_1\mu + \alpha_2\mu^2 + \alpha_3\mu^3 + \ldots \\ f^*(1) = 1, \alpha_1 = f^{*\prime}(1), \alpha_2 = \frac{1}{2} f^{*\prime\prime}(1), \alpha_3 = \frac{1}{6} f^{*\prime\prime\prime}(1). \end{gathered} \tag{5}$$

Aus (4) und (5) erhält man dann mit Hilfe von (2) und (3) unter der Forderung, daß (2) und (3) die Lösung (4) besitzen sollen, den Koeffizienten a_{22}, der eine Funktion von A ist, folgendermaßen:

$$a_{22} = \frac{1}{a}\alpha_1 + \frac{1}{a} \cdot \frac{1}{\pi A \varkappa_2} \cdot$$

$$\cdot \int_O^{2\pi} \left[\alpha_2 A^2 \varkappa_2^2 \sin^2\Phi_2 + \alpha_3 A^3 \varkappa_2^3 \sin^3\Phi_2 + \alpha_4 A^4 \varkappa_2^4 \sin^4\Phi_2\right] \cdot \sin\Phi_1 \cdot d\Phi_1$$

$$a_{22} = \frac{1}{a}\alpha_1 + \frac{3\alpha_3}{4a} \cdot A^2. \tag{6}$$

Vgl. hierzu [3] bis [6].

Bezieht man die Amplitude A auf μ, dann kann $\varkappa_2 = 1$ gesetzt werden, was hier auch getan wurde.

Das Ersatzsystem lautet also

$$\begin{aligned} \frac{d\varphi}{dz} &= -\varphi + \mu \\ \frac{d\mu}{dz} &= -\frac{1}{a}\varphi + a_{22}\mu; \\ a_{22} &= \frac{1}{a}a_1 + \frac{3a^3}{4a}A^2, \end{aligned} \tag{7}$$

wobei das Zeichen (*) [vgl. (3)] wieder weggelassen wurde, da keine Verwechslung möglich sein dürfte.

Die Lösung von (7) kann nun in der Form (8) geschrieben werden:

$$\begin{aligned} \varphi &= \varkappa_1 A \cdot e^{\lambda z}, \\ \mu &= A e^{\lambda z}. \end{aligned} \tag{8}$$

Aus (7) erhält man mit (8):

$$\begin{aligned} &\varkappa_1 A(\lambda + 1) - A = 0 \\ &\frac{1}{a}\varkappa_1 A + (\lambda - a_{22}) A = 0 \end{aligned} \tag{9}$$

Eine Lösung des homogenen Gleichungssystems existiert nur dann, wenn seine Determinante verschwindet. Daraus erhält man die Säkulargleichung:

$$\lambda^2 + (1 - a_{22})\lambda - \left(a_{22} - \frac{1}{a}\right) = 0 \tag{10}$$

und weiters $\lambda_{1,2} = -\frac{1 - a_{22}}{2} \pm \sqrt{\frac{(1 - a_{22})^2}{4} + \left(a_{22} - \frac{1}{a}\right)}$. (11)

Stabilität ist nun vorhanden, wenn die Lösung (8) beschränkt ist und mit wachsender Zeit gegen Null geht, wenn also der Realteil von λ oder λ selbst [falls die Wurzel in (11) reell] negativ ist. Daraus ergibt sich die

$$\text{Stabilitätsbedingung: } -(1 - a_{22}) < 0 \text{ oder } a_{22} < 1. \tag{12}$$

Schwingungen sind allerdings nur dann möglich, wenn gilt:

$$\frac{(1 - a_{22})^2}{4} + \left(a_{22} - \frac{1}{a}\right) < 0. \tag{13}$$

In Abb. 2 sind die Ergebnisse (12) und (13) in einem a_{22}, a-Diagramm aufgetragen. Das Gebiet unterhalb I ist also stabil. Schwingungen können in dem schraffierten, durch II begrenzten Gebiet auftreten. In Abb. 2 sind ferner die Linien konstanter Frequenz Ω aufgetragen. Ω ist einfach der Imaginärteil von λ.

Aus (11) erhält man: $\Omega^2 = \left(\frac{1 - a_{22}}{2}\right)^2 + \left(a_{22} - \frac{1}{a}\right)$;

daraus ergibt sich $a_{22} = -1 \pm 2\sqrt{\frac{1}{a} + \Omega^2}$.

Es kann daher zu jedem Wertepaar a_{22}, a die dazugehörige Frequenz aus Abb. 2 entnommen werden.

Zuletzt ist noch A durch a_{22} und a auszudrücken. Dies geschieht mit Hilfe von (6):

$$a_{22} = \frac{1}{a}\,a_1 + \frac{3\,a_3}{4\,a}\,A^2.$$

Es gehen also noch die Parameter a_1 und a_3 ein; daher ist es nicht möglich, allgemeine A-Kurven zu zeichnen. Man muß für jeden gegebenen Betriebsfall, also für gegebene Werte a, a_1 und a_3 die A-Kurve berechnen und in das Diagramm eintragen. Dies bereitet hier keine großen Schwierigkeiten, weil $a_{22}(A^2)$ eine Gerade wird.

Damit ist nun eine einfache Diskussion aller vorkommenden Fälle möglich:

1. $a_1 < 0$, $a_3 < 0$:

a) Der stationäre Betriebsfall entspricht P_1^{I}.[1]

Bei Zunahme von A geht $a_{22} \to -\infty$, bleibt also im Stabilitätsbereich. Das System ist im kleinen wie im großen stabil. Ein ähnliches Verhalten tritt für

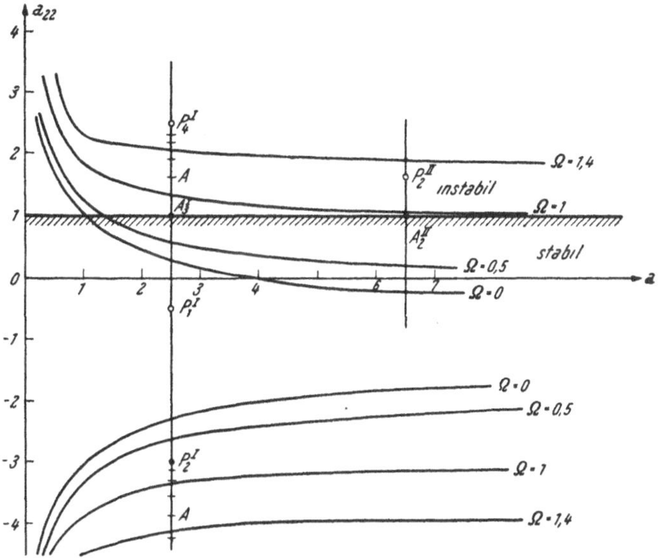

Abb. 2. Stabilitätsdiagramm eines „Monopropellant“ ohne Zeitverzögerung, aber mit Berücksichtigung der Pumpencharakteristik

Betriebspunkte P_2^{I} usw. auf. Nur klingt hier die Störung als gedämpfte Schwingung ab. Die Frequenz kann unmittelbar entnommen werden.

b) Der stationäre Betriebsfall entspricht P_4^{I} oder P_2^{II}.

Das System ist im kleinen instabil. Im großen führt es Schwingungen an den Punkten A_3^{I} oder A_2^{II} aus. Frequenz und Amplitude können aus dem Diagramm entnommen werden.

[1] Die Anwendung des Verfahrens der harmonischen Balance ist nicht mehr gerechtfertigt, wenn das System keine Schwingungen ausführen kann. In unserem Fall ist eine Aussage über das Verhalten des Systems in dem Bereich, in dem keine reellen Werte für Ω existieren, nicht möglich. Da aber nur oszillatorische Instabilitäten in diesem Fall von Interesse sind, stellt diese Tatsache keine Einschränkung unserer Untersuchungen dar.

2. $\alpha_1 > 0,\ \alpha_3 > 0$:

Ist $\frac{1}{a}\alpha_1 < 1$, dann ist das System im kleinen stabil, im großen aber instabil.

Ist hingegen $\frac{1}{a}\alpha_1 > 1$, dann ist das System im kleinen wie im großen instabil.

3. $\alpha_1 > 0,\ \alpha_3 < 0$:

Ist $\frac{1}{a}\alpha_1 < 1$, dann ist das System im kleinen wie im großen stabil.

Ist $\frac{1}{a}\alpha_1 > 1$, dann tritt der Fall 2 auf.

4. $\alpha_1 < 0,\ \alpha_3 > 0$:
Das System ist im kleinen stabil, im großen jedoch instabil.

Das Stabilitätsdiagramm ist also auch für Untersuchungen im kleinen, also $A \to 0$, geeignet. Die Stabilitätsbedingung lautet hier einfach nach (12)

$$\frac{1}{a}\alpha_1 < 1 \quad \text{oder} \quad \alpha_1 < a = \frac{\bar{\dot{m}} \cdot l}{\dot{p}_K \cdot F \cdot \vartheta_g}. \tag{14}$$

Werden die Parameter der Rohrleitung klein $\left(\frac{l}{F} \to 0\right)$, dann reduziert sich dies auf die Bedingung:

$$\alpha_1 < 0. \tag{14a}$$

Diese Bedingung ist von den Kreiselpumpen und Axialverdichtern her bekannt, wonach ein stabiler Betrieb im schraffierten Bereich in Abb. 3 nicht möglich ist.

Aus (14) sieht man jedoch, daß eine Rohrleitung im kleinen stabilisierend wirkt.

Um also stabilen Betrieb zu erhalten, muß der Arbeitspunkt erstens einmal in einem Bereich liegen, wo $\alpha_1 < a$. Dann ist das System im kleinen stabil (Fall 1 und 4). Nun hängt es ganz von α_3, also im wesentlichen von $f'''^*_{(1)}$, der dritten Ableitung der Pumpencharakteristik, ab, ob das System auch im großen stabil wird.

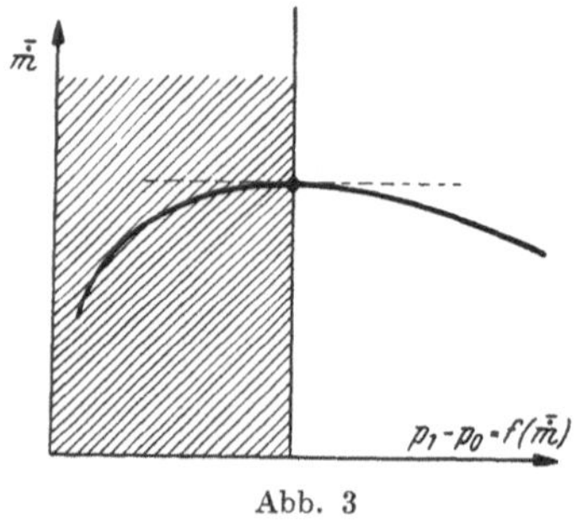

Abb. 3

Stabilität ist immer vorhanden, wenn $\alpha_3 < 0$. Ist jedoch $\alpha_3 > 0$, dann ist das System im großen instabil.

Dieser Fall verdient besondere Beachtung: Wenn die Kennlinien der Treibstoffpumpen Parabeln sind — wie es bei Radialpumpen meist angenommen wird — dann ist $f'''^* \equiv 0$ und das System verhält sich dann im großen ebenso wie im kleinen. Im allgemeinen treten aber stets Abweichungen von diesem Verhalten auf, und besonders bei Grenzleistungsverdichtern, wie sie gerade im Raketenbau vorkommen, ist es möglich, daß die Kennlinien Wendetangenten besitzen und also zumindest die dritte Ableitung von Null verschieden ist (wenn auch noch die höheren Ableitungen berücksichtigt werden). Interessant ist übrigens, daß die geradzahligen Ableitungen der Kennlinie keinen Einfluß auf die Stabilität besitzen.

Es können aber auch Abweichungen vom Parabelverlauf der Kennlinie auftreten, wenn die dynamische Kennlinie mit der statischen nicht mehr übereinstimmt. Deswegen wäre es nötig, dem dynamischen Kennlinienverlauf mehr Aufmerksamkeit zu widmen.

II. Systeme mit Zeitverzögerung und konstanter Treibstoff-Förderung

Die Ausgangsgleichungen (1) reduzieren sich hier auf eine einzige:

$$\frac{d\varphi}{dz} + \varphi + 1 = \left[\frac{1 + \varphi(z)}{1 + \varphi(z - \delta)}\right]^n. \tag{15}$$

Entwicklung der rechten Seite in eine Potenzreihe bis einschließlich Glieder vierter Ordnung ergibt:

$$\begin{aligned}\frac{d\varphi}{dz} + \varphi + 1 = {} & 1 + n\,[\varphi(z) - \varphi(z-\delta)] + \\ & + \frac{n}{2}\,[(n-1)\cdot\varphi^2(z) + (n+1)\,\varphi^2(z-\delta)] - n^2\,\varphi(z)\cdot\varphi(z-\delta) + \\ & + \left[\frac{n}{6}(n-1)(n-2)\,\varphi^3(z) - \frac{n}{6}(n+1)(n+2)\,\varphi^3(z-\delta) + \right.\\ & \left. + \frac{n^2}{2}(n+1)\,\varphi(z)\,\varphi^2(z-\delta) - \frac{n^2}{2}(n-1)\,\varphi^2(z)\,\varphi(z-\delta)\right] + \\ & + \left[\frac{n}{24}(n-1)(n-2)(n-3)\,\varphi^4(z) + \frac{n}{24}(n+1)(n+2)(n+3)\,\varphi^4(z-\delta) - \right.\\ & - \frac{n^2}{6}(n-1)(n-2)\,\varphi^3(z)\,\varphi(z-\delta) - \frac{n^2}{6}(n+1)(n+2)\,\varphi(z)\,\varphi^3(z-\delta) + \\ & \left. + \frac{n^2}{4}(n^2-1)\,\varphi^2(z)\,\varphi^2(z-\delta)\right].\end{aligned} \tag{16}$$

Ferner muß $\varphi(z) = A \cdot \sin \Omega z$ an der Stabilitätsgrenze wieder eine Lösung sein. Man kann nun (16) wieder ein lineares Ersatzsystem zuordnen, welches dieselbe Lösung hat.

$$\frac{d\varphi(z)}{dz} + (1-n)\,\varphi + n\cdot\varphi(z-\delta) = a^*_{11}\frac{d\varphi(z)}{dz} + \bar{a}_{11}\cdot\varphi(z). \tag{17}$$

Die Koeffizienten ergeben sich dann wiederum nach [5], [6]:

$$\bar{a}_{11} = \frac{n A^2}{8}\left\{(n-1)(n-2) - 2(2n^2+1)\cos\Omega\delta + n(n+1)(1 + 2\cos^2\Omega\delta)\right\} \tag{18a}$$

$$a^*_{11} = \frac{n A^2}{8\Omega}\left\{2(n^2+n+1)\sin\Omega\delta - (n+1)\cdot n\cdot\sin 2\Omega\delta\right\}. \tag{18b}$$

(17) lautet dann:

$$(1 - a^*_{11})\frac{d\varphi(z)}{dz} + (1 - n - \bar{a}_{11})\cdot\varphi(z) + n\cdot\varphi(z-\delta) = 0. \tag{19}$$

Da in (19) nicht nur die Funktionswerte an der Stelle z, sondern auch an der Stelle $z - \delta$ vorkommen, ist dies keine gewöhnliche Differentialgleichung, sondern eine „Differenzendifferentialgleichung", allgemein „Hysterodifferentialgleichung" genannt. Die Lösung ist dann eindeutig gegeben, wenn $\varphi(z)$ im Bereich z_0 bis $z_0 + \delta$ gegeben ist (z_0 beliebig) [7].

Gl. (19) geht in die Gleichung der linearisierten Theorie [1] über, wenn man A und damit a_{11}^* und $\bar{a}_{11}$ gegen Null gehen läßt.

Es wird nun der Lösungsansatz gemacht:

$$\varphi(z) = e^{\lambda z}. \tag{20}$$

Die charakteristische Gleichung lautet dann:

$$(1 - a^*_{11})\,\lambda + (1 - n - \bar{a}_{11}) + n\cdot e^{-\lambda\delta} = 0. \tag{21}$$

Mit $\lambda = \lambda_0 + i\Omega$ erhält man aus (21) die beiden Gleichungen:

$$(1 - a^*_{11})\,\lambda_0 + 1 - n - \bar{a}_{11} + n \cdot e^{-\lambda_0\delta} \cdot \cos \Omega\delta = 0, \qquad (22a)$$

$$(1 - a^*_{11})\,\Omega - n \cdot \sin \Omega\delta \cdot e^{-\lambda_0\delta} = 0. \qquad (22b)$$

Aus (22a) und (22b) kann λ_0 eliminiert werden, und man erhält:

$$\frac{1 - a^*_{11}}{\sin \Omega\delta}\,\Omega\delta = n \cdot \delta \cdot e^{\delta \frac{1 - n - \bar{a}_{11}}{1 - a_{11}^*} + \frac{\Omega\delta}{\operatorname{tg}\Omega\delta}}. \qquad (23)$$

Setzt man (18a) und (18b) ein, so erhält man eine Beziehung zwischen den Größen n, δ, Ω, A.

(23) gestattet also, für gegebenes n und δ die Werte von Ω als Funktion von A anzugeben.

Eine graphische Darstellung für alle Werte n, δ ist nur im R 4 möglich. Es soll jedoch später ein Diagramm für ein festes n mit δ als Parameter angegeben werden.

Die Stabilitätsgrenze erhält man aus (22a) und (22b), indem $\lambda_0 = 0$[1] gesetzt wird. [Die Stabilitätsgrenze sei durch (′) angedeutet.]

$$\frac{\Omega'\,\delta'}{\operatorname{tg}\Omega'\,\delta'} = -\frac{1 - n' - \bar{a}_{11}'}{1 - a^*_{11}{}'}\,\delta'. \qquad (24)$$

$1 - a^*_{11}{}'$ kann mit Hilfe einer der beiden Gln. (22) eliminiert werden, und man erhält dann:

$$\operatorname{tg}^2 \Omega'\,\delta' = \frac{2n' - 1 + \bar{a}_{11}' \cdot [2(1 - n') - \bar{a}_{11}']}{(1 - n' - \bar{a}_{11}')}. \qquad (25)$$

Setzt man $\bar{a}_{11}'$, also den Wert von $\bar{a}_{11}$ an der Stabilitätsgrenze aus (18a) ein, so erhält man eine Beziehung zwischen $\Omega'\delta'$, n' und A'. Demnach könnte die Stabilitätsgrenze in einem $\Omega\delta$, n, A-Raum graphisch wiedergegeben werden.

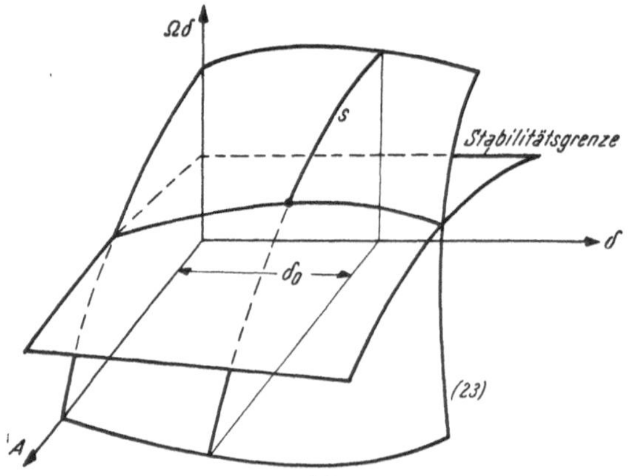

Abb. 4

Zur Beurteilung über das Verhalten eines Systems muß jedoch ein Parameter fix gewählt werden (beispielsweise n). Als Koordinatenachsen seien A und $\Omega\delta$ gewählt, δ sei Parameter. Die Stabilitätsfläche (25) ist in diesem Raum ein gerader Zylinder, da δ explizit in (25) nicht vorkommt. Mit Hilfe dieser Fläche und mit Gl. (23) kann nun wieder sehr einfach das Verhalten des Systems beurteilt werden (Abb. 4). Gl. (25) stellt also im $\Omega\delta$, δ, A-Raum eine Fläche dar. Gl. (23) ist bei gegebenem n ebenfalls eine Fläche in diesem Raum. Ist nun δ für einen Betriebsfall gegeben, und sei gleich δ_0, so wird das Verhalten des Systems beschrieben durch die Schnittkurve der Ebene $\delta = \delta_0 =$ konst. mit der Fläche (23). Je nachdem, ob bei von 0 anwachsendem A die Schnittkurve s

[1] $\lambda_0 \ldots$ Dämpfung, bzw. Anfachung.

innerhalb des stabilen oder instabilen Bereiches bleibt oder die Stabilitätsfläche durchsetzt, ergibt sich dann daraus das Vorhandensein von Stabilität oder nicht.

(23) kann in der $\Omega\delta$, A-Ebene durch seine Schichtenlinien (δ als Parameter) dargestellt werden. Dies ist in Abb. 5 für den Fall $n = 0{,}75$ durchgeführt worden.

Die numerische Berechnung von (25) und (23) erfordert eine enorme Arbeitszeit, selbst wenn man eine der modernsten Tischrechenmaschinen zur Verfügung hat. Die Berechnung würde sich aber bei Verwendung einer programmgesteuerten elektronischen Rechenmaschine sehr einfach gestalten.

Eine Diskussion über das Verhalten eines gegebenen Systems ist nun mit Hilfe des Stabilitätsdiagramms leicht durchführbar. Man sieht aus Abb. 5, daß beispielsweise für $\delta = 0{,}25$ das System instabil ist, und da die Kurve $\delta = 0{,}25$ die Stabilitätsgrenze schneidet, wird dort das System mit der zugehörigen Amplitude A Schwingungen ausführen, weil die Kurve mit wachsendem A aus dem instabilen in den stabilen Bereich eintritt. Die Frequenz kann an der $\Omega\delta$-Achse abgelesen werden. Interessant ist, daß die Kurven $\delta = 0{,}25$ usw. die $\Omega\delta$-Achse nicht erreichen, das heißt, daß der Wert $A = 0$ im großen nicht möglich ist. Dies deutet auf einen harten Schwingungseinsatz hin. Es kann nämlich das System im kleinen stabil sein, aber bei einer endlichen Störung setzt plötzlich eine Schwingung von endlicher Amplitude ein.

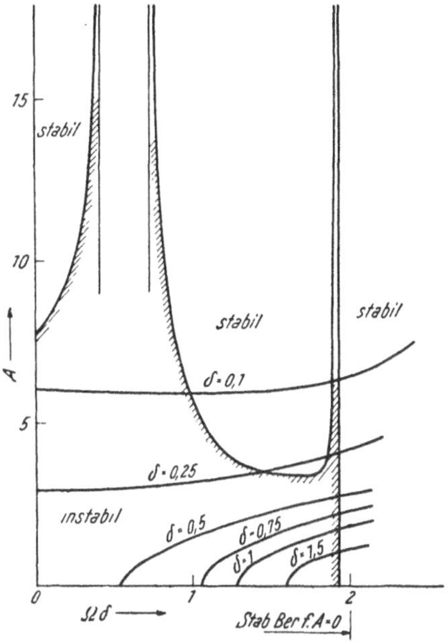

Abb. 5. Stabilitatsdiagramm eines „Monopropellant“ mit konstanter Treibstoff-Forderung, aber mit Berucksichtigung der Zeitverzogerung, fur den Fall $n = 0{,}75$

Eine Untersuchung der Stabilitätsgrenze zeigt ferner, daß ihre Minima für A im Fall $n < {}^1/_2$ folgende Werte nicht unterschreiten können:

$$A^{2\prime}_{min} = \frac{4\,(1 - 2\,n)}{(2\,n^2 + 1)\cdot 2\,n}\,. \qquad (26)$$

Die Stabilitätsgrenze muß in diesen Fällen also oberhalb der $\Omega\delta$-Achse verlaufen. Sollte nun in den Fällen $n < {}^1/_2$ das Gebiet zwischen $\Omega\delta$-Achse und Stabilitätsgrenze instabil sein, dann würde dies ebenfalls einen harten Schwingungseinsatz zulassen, da der Fall $n < {}^1/_2$ im kleinen immer stabil ist [1].

In Abb. 5 ist ferner der Stabilitätsbereich, der sich nach [1] ergibt, eingetragen. Er entartet in diesem Diagramm in eine Strecke entlang der $\Omega\delta$-Achse ($A = 0$). Als wichtigstes Ergebnis dieser Untersuchung wäre zu nennen, daß ein System mit Zeitverzögerung und konstanter Treibstoff-Förderung für $\Omega\delta <$ im kleinen für den Fall $n = 0{,}75$ stabil ist, während es im großen stets oszillatorisch instabil ist. Das System wird also stets Schwingungen ausführen. Die Schwingungsamplituden werden um so kleiner, je größer δ wird, wie man aus Abb. 5 entnehmen kann. Um also kleine Schwingungsamplituden zu erhalten, muß man große Zeitverzögerungen τ und kleine Brennkammerlaufzeiten δ_g anstreben $\left(\delta = \frac{\tau}{\delta_g}\right)$.

Die Annahme konstanter Treibstoff-Förderung ist allerdings eine Idealisierung. Ferner werden auch im Falle eines „Bipropellant" Abweichungen von obigem Verhalten auftreten. Um hier Aussagen machen zu können, müssen die Gleichungen für ein „Bipropellant" aufgestellt und ihre Lösungen auf Stabilität untersucht werden. Auch die Elastizität der Rohrleitungen, die hier vernachlässigt wurde, kann noch Korrekturen zu obigen Ergebnissen liefern.

Vorliegende Ergebnisse zeigen jedoch, daß das Verfahren der harmonischen Balance auf Stabilitätsuntersuchungen bei Raketenmotoren im Prinzip anwendbar ist. Der numerische Aufwand ist allerdings schon in den stark vereinfachten Fällen, die hier behandelt wurden, erheblich. Wenn sich in den erwähnten Fällen noch allgemeine Aussagen machen lassen, so ist dies im Fall: Triebwerk mit Zeitverzögerung und Berücksichtigung der Pumpencharakteristik nicht mehr möglich.

Im Falle eines „Bipropellant" erhält man bei Berücksichtigung der Variabilität des Mischungsverhältnisses ein Simultansystem von acht nichtlinearen Differentialgleichungen erster Ordnung. Macht man nun Ansätze der Form (4), so ergibt sich ein verwickeltes transzendentes Gleichungssystem für x_u und ψ_u, was elementar nicht mehr auflösbar ist. Ferner ist wegen der Zahl der vorkommenden Parameter eine graphische Darstellung nicht mehr möglich. Es bleibt hier nur die Möglichkeit, einen gegebenen Raketenmotor auf Stabilität zu untersuchen, was mit Hilfe von programmgesteuerten Rechenmaschinen oder mit Analogrechnern möglich ist [10].

Interessant wäre eine Untersuchung darüber, inwieweit eine Instabilität des Fördersystems die innere Stabilität beeinflussen kann, und umgekehrt. Es könnte beispielsweise eine Schwingung zufolge innerer Instabilität als Störung für das Fördersystem auftreten und umgekehrt. Sollte nun der Schwingungseinsatz zufolge innerer Instabilität hart sein, so ist wohl anzunehmen, daß ein Raketenmotor, bei dem dies der Fall ist, solchen Beanspruchungen nicht lange gewachsen sein wird.

Literaturverzeichnis

1. L. Crocco, Aspects of Combustion Instability in Liquid Propellant Rocket Motors. Jet Propulsion **21**, 163 (1951); **22**, 7 (1952).
2. H. S. Tsien, The Transfer Function of Rocket Nozzles. Jet Propulsion **22**, 139 (1952).
3. N. Krylow und N. Bogoljubow, Einführung in die nichtlineare Mechanik. Kiew, 1937. Auszugsweise Übersetzung: Princeton, 1949.
4. N. Minorsky, Introduction to Non Linear Mechanics. Ann Arbor: J. W. Edwards, 1947.
5. K. Magnus, Über ein Verfahren zur Untersuchung nichtlinearer Schwingungen und Regelsysteme. VDI-Forschungsheft 451.
6. W. Hahn, Nichtlineare Regelvorgänge. Vorträge der Tagung in Darmstadt am 8. 9. 1955. Daraus: K. Magnus, Das A-Kurvenverfahren zur Berechnung nichtlinearer Regelvorgänge. München: R. Oldenbourg, 1956.
7. N. Minorsky, Self Excited Oscillations in Dynamical Systems Possessing Retarded Actions. J. Appl. Mechanics **9**, 67 (1942).
8. M. Barrère und A. Moutet, Low-Frequency Combustion Instability in Bipropellant Rocket Motors—Experimental Study. J. Amer. Rocket Soc. **26**, 9 (1956).
9. L. Crocco, J. Grey and G. Matthews, Measurements of the Combustion Time Lag in a Liquid Bipropellant Rocket Motor. J. Amer. Rocket Soc. **26**, 20 (1956).
10. B. N. Smith, Perturbation Analysis of Low Frequency Rocket Engine Dynamics on an Analog Computer. J. Amer. Rocket Soc. **26**, 40 (1956).

Gravitational and Related Constants for Accurate Space Navigation[1]

By

S. Herrick[2,3], ARS, **R. M. L. Baker, Jr.**[2,3], ARS, and **C. G. Hilton**[2], ARS

(With 1 Figure)

Abstract — Zusammenfassung — Résumé

Gravitational and Related Constants for Accurate Space Navigation. The paper first compares the GAUSSIAN gravitational constant, k_s, and the astronomical unit with the cgs value of G and the meter, showing that the latter two are valueless for heliocentric orbits. For geocentric orbits a quasi-GAUSSIAN geocentric gravitational constant, k_e, is determined from the equatorial acceleration of gravity, g_e, the equatorial radius, a_e, the flattening, f, and the related coefficients, J and K, of the second and fourth harmonics in the potential of the terrestrial ellipsoid, after improved values of these data are ascertained and adopted:

$$g_e = 9.780{,}368\left(1 + g' + \frac{1}{3} f'\right) \text{meters/sec}^2 \qquad g' = 0 \pm 3 \times 10^{-6}$$

$$a_e = 6{,}378{,}270\left(1 + a' + \frac{4}{3} f'\right) \text{meters} \qquad a' = 0 \pm 10 \times 10^{-6}$$

$$f = +0.003{,}367{,}00 + f' \qquad f' = 0 \pm 4 \times 10^{-6}$$

$$J = +0.001{,}638{,}08 + f' \qquad K = +9.04 \times 10^{-6}$$

The value of k_e is ascertained for three units of distance, the megameter $= 10^6$ meters, the "q-radius" = the equatorial radius, and the "g-radius" = the "gravitational radius", so defined by $a_e = 1 + \frac{3}{1} a' - \frac{1}{3} g' + \frac{3}{2} f'$ that the uncertainty of k_e will be zero:

$$k_e = 1.197{,}918{,}5\left(1 + a' + \frac{1}{2} g' + f'\right) \text{megameters}^{3/2}/\text{min}, \qquad \frac{\Delta k_e}{k_e} = \pm 11 \times 10^{-6}$$

$$k_e = 0.074{,}365{,}74\left(1 - \frac{1}{2} a' + \frac{1}{2} g' - f'\right) q\text{-radii}^{3/2}/\text{min}, \qquad \frac{\Delta k_e}{k_e} = \pm 6\frac{1}{2} \times 10^{-6}$$

$$k_e = 0.074{,}365{,}74\ g\text{-radii}^{3/2}/\text{min}, \qquad \frac{\Delta k_e}{k_e} = 0\,.$$

Basic expressions due to DE SITTER, LAMBERT, and others, that are alternative to those used in the foregoing determinations are compared with them in the appendix.

The moon's parallax, as an alternate source of k_e, is shown to be inferior to g_e. A new value of the dynamical parallax is ascertained to be $\pi_☾ = 3422.''650\left(1 \pm 4\frac{1}{2} \times 10^{-6}\right)$.

---

[1] Revision of a January 1957 paper circulated under the title "Units and Constants for Geocentric Orbits" and supported in part by the International Geophysical Year Satellite program, through the Smithsonian Astrophysical Observatory.

[2] Department of Astronomy, University of California, Los Angeles 24, California, U.S.A.

[3] Presently also associated with Aeronutronic Systems, Inc., Glendale, California.

Other possible sources are considered; the effects of the constants on ICBM and satellite trajectories are developed; extensive tables of conversion factors are supplied for mass, length, time, velocity, angular velocity, acceleration, and period.

Gravitationskonstante und verwandte Konstanten für exakte Navigation im Raum. Die vorliegende Arbeit vergleicht die GAUSSsche Gravitationskonstante k_s und die Astronomische Einheit mit den CGS-Werten von G und dem Meter. Dabei wird gezeigt, daß die beiden letztgenannten für heliozentrische Bahnen ohne Wert sind. Für geozentrische Bahnen wird eine quasi-GAUSSsche geozentrische Gravitationskonstante k_e aus der äquatorialen Schwerebeschleunigung g_e, dem äquatorialen Radius a_e, der Abplattung f und den entsprechenden Koeffizienten J und K der zweiten und vierten Harmonie im Potential des Erdellipsoids bestimmt. Hierauf werden verbesserte Werte dieser Daten ermittelt und angenommen:

$$g_e = 9{,}780\,368\left(1 + g' + \frac{1}{3}f'\right)\text{m/sec}^2 \qquad g' = 0 \pm 3\times 10^{-6}$$

$$a_e = 6\,378\,270\left(1 + a' + \frac{4}{3}f'\right)\text{m} \qquad a' = 0 \pm 10\times 10^{-6}$$

$$f = +0{,}003{,}367{,}00 + f' \qquad f' = 0 \pm 4\times 10^{-6}$$

$$J = +0{,}001{,}638{,}08 + f' \qquad K = +9{,}04\times 10^{-6}$$

Der Wert von k_e wird für drei Entfernungseinheiten ermittelt: für das Megameter $= 10^6$ Meter, den „q-Radius“ = Äquatorradius und den „g-Radius“ = „Gravitationsradius“, durch $a_e = 1 + \frac{1}{3}a' - \frac{1}{3}g' + \frac{2}{3}f'$ so definiert, daß die Unsicherheit von k_e Null wird:

$$k_e = 1{,}197\,918\,5\left(1 + a' + \frac{1}{2}g' + f'\right)\text{ Megameter}^{3/2}/\text{min} \qquad \frac{\Delta k_e}{k_e} = \pm 11\times 10^{-6}$$

$$k_e = 0{,}074\,365\,74\left(1 - \frac{1}{2}a' + \frac{1}{2}g' - f'\right)q\text{-Radien}^{3/2}/\text{min} \qquad \frac{\Delta k_e}{k_e} = \pm 6\frac{1}{2}\times 10^{-6}$$

$$k_e = 0{,}074\,365\,74\; g\text{-Radien}^{3/2}/\text{min} \qquad \frac{\Delta k_e}{k_e} = 0.$$

Grundgleichungen nach DE SITTER, LAMBERT und anderen Autoren, die neben den in den vorstehenden Bestimmungen verwendeten möglich sind, werden mit diesen im Anhang verglichen.

Eine andere Berechnungsquelle für k_e, die Mondparallaxe, erwies sich als dem g_e unterlegen. Als ein neuer Wert der dynamischen Parallaxe wird: $\pi_{☾} = 3422.''650 \left(1 \pm 4\frac{1}{2}\times 10^{-6}\right)$ ermittelt. Noch andere Berechnungsquellen werden betrachtet; die Auswirkungen der Konstanten auf interkontinentale ballistische Geschosse und Satellitenbahnen werden entwickelt; umfangreiche Tabellen der Umrechnungsfaktoren für Masse, Länge, Zeit, Geschwindigkeit, Winkelgeschwindigkeit, Beschleunigung und Periode werden aufgestellt.

Constantes de gravitation et constantes associées pour une navigation spatiale de précision. L'article compare d'abord la constante de gravitation k_s de GAUSS et l'unité astronomique avec l'unité cgs de G et le mètre, montrant que ces deux dernières sont sans signification pour les orbites héliocentriques. Une constante de gravitation quasi-gaussienne k_e est déterminée pour les orbites géocentriques à partir de l'accélération équatoriale de la pesanteur g_e, du rayon équatorial a_e, de l'applatissement f et des coefficients J et K relatifs au second et au quatrième harmonique du potentiel du géoide. Les valeurs améliorées et finalement adoptées pour ces grandeurs sont les suivantes:

$$g_e = 9.780{,}368\left(1 + g' + \frac{1}{3}f'\right) \text{ mètres/sec}^2 \qquad g' = 0 \pm 3 \times 10^{-6}$$

$$a_e = 6\,378\,270\left(1 + a' + \frac{4}{3}f'\right) \text{ mètres} \qquad a' = 0 \pm 10 \times 10^{-6}$$

$$f = +0.003\,367\,00 + f' \qquad f' = 0 \pm 4 \times 10^{-6}$$

$$J = +0.001\,638\,08 + f' \qquad K = +9.04 \times 10^{-6}$$

La valeur de k_e est obtenue pour trois unités de distance: le mégamètre (10^6 mètres) le "rayon q" ou rayon équatorial et le "rayon g" ou rayon gravitationnel défini par la formule $a_e = 1 + \frac{1}{3}a' - \frac{1}{3}g' + \frac{2}{3}f'$ telle que l'inexactitude sur k_e soit nulle:

$$k_e = 1.197\,918\,5\left(1 + a' + \frac{1}{2}g' + f'\right) \text{ mégamètres}^{3/2}/\text{min} \qquad \frac{\Delta k_e}{k_e} = \pm 11 \times 10^{-6}$$

$$k_e = 0.074\,365\,74\left(1 - \frac{1}{2}a' + \frac{1}{2}g' - f'\right) \text{ rayons-}q^{3/2}/\text{min} \qquad \frac{\Delta k_e}{k_e} = \pm 6\frac{1}{2} \times 10^{-6}$$

$$k_e = 0.074\,365\,74 \text{ rayons-}g^{3/2}/\text{min} \qquad \frac{\Delta k_e}{k_e} = 0.$$

Les expressions fondamentales de DE SITTER, LAMBERT et autres, qui peuvent remplacer celles utilisees dans la détermination précédente, leurs sont comparées en appendice.

La parallaxe de la lune, autre source possible de détermination de k_e, est montrée être inférieure à g_e. Une nouvelle valeur de la parallaxe dynamique est établie: $\pi_{☾} = 3422.650\left(1 \pm 4\frac{1}{2} \times 10^{-6}\right)$. D'autres sources possibles sont envisagées; l'influence de ces constantes sur les trajectoires des engins balistiques intercontinentaux et des satellites artificiels est analysée. Enfin des tables de conversion étendues sont fournies pour la masse, la longueur, le temps, la vitesse, la vitesse angulaire, l'accélération et la période.

A. The Evaluation of Gravitational Constants

The evaluation of gravitational constants is a first requirement for serious work with heliocentric and geocentric orbits. The necessary constants start with the constant of gravitation, k^2, and the masses, m_1 and m_2, that appear in the two-body potential function, which may be written

$$\Phi = k^2\,(m_1 + m_2)/r \tag{A 1}$$

wherein, of course, r is the distance between the two bodies. It will become evident, however, that satisfactory constants for the earth's gravitational attraction cannot even be evaluated without reference to additional constants, such as those that appear in the potential for a spheroid, which, when m_2 is negligibly small, may be written

$$\Phi = \frac{k^2 m_1}{r}\left\{1 + \frac{J}{3}\left(\frac{a_e}{r}\right)^2(1 - 3\sin^2\delta) + \frac{K}{30}\left(\frac{a_e}{r}\right)^4(3 - 30\sin^2\delta + 35\sin^4\delta) + \ldots\right\} \tag{A 2}$$

where δ is the geocentric declination of m_2, a_e is the earth's equatorial radius. and J and K are the coefficients of the second and fourth order terms (or "second and fourth harmonics"). The coefficients that enter into other perturbation terms, such as the masses of planets and satellites, some of the thrust and drag coefficients, etc., are likewise a part of the evaluation and compilation of constants that bear upon the orbits of planets, comets, satellites, and rockets.

We shall be concerned herein primarily with the selection of appropriate units of time, mass, and distance and with the determination of k and associated constants. The units of time, mass, and distance will be found to determine not only the numerical value of k, but also its accuracy and suitability for heliocentric and geocentric orbits (cf. Secs. C and H). We find, in fact, that we are concerned with two principal values of k: the GAUSSian gravitational constant, which is usually designated simply k, but which we shall call k_s because it applies to orbits centered in the sun; and a "geocentric gravitational constant," k_e, for satellite orbits and other orbits for which the principal focus is the center of the earth. Neither value can be determined with sufficient accuracy from the centimeter-gram-second value of k^2, which is usually designated G (cf. Sec. C).

It is often possible to include some perturbations in two-body formulae by specifying that the mass function shall be

$$\mu = m_1 + m_2 + \Delta m \tag{A 3}$$

where Δm is an appropriate augmentation of the two masses ordinarily considered. If in these simple circumstances it is possible to measure the acceleration of gravity, g, for a given r, k may be determined from

$$g = k^2 \mu / r^2. \tag{A 4}$$

It may also be determined from such an integral as KEPLER's third law,

$$k^2 \mu P^2 = (2\pi)^2 a^3 \tag{A 5}$$

where P is the period of revolution and a is the mean distance of m_2 referred to m_1. Conceivably, k might be determined from another integral of the two-body problem such as the *vis-viva* integral,

$$V^2 = \left(\frac{ds}{dt}\right)^2 = k^2 \mu \left(\frac{2}{r} - \frac{1}{a}\right) \tag{A 6}$$

but it is unlikely that the velocity, V, or other such quantities, would be determined accurately enough for this purpose from measurement alone.

Each of the foregoing expressions will have modifications or additional terms, of course, if there are involved perturbations that cannot be included in Δm.

In the determination and use of k_s, two devices have been employed that serve as useful guides in our consideration of k_e:

(1) The numerical value of GAUSS's k is preserved without change, when new determinations of the basic constants would otherwise require one, by an alteration instead of the relationship between the unit of distance and the actual planetary distances, especially the mean distance of the earth from the sun (cf. Sec. B).

(2) The day, as unit of time, is supplemented by an auxiliary unit, of which it may be said either that it makes k equal to unity or that time, τ, expressed in it is related to time, t, expressed in days by the definition

$$\tau = k(t - t_0) \tag{A 7}$$

where t_0 is an arbitrary epoch.

For the second device we shall adopt the latter mode of expression, thus somewhat formalizing and extending a usage that has been employed in the determination of orbits to a limited degree but effectively for many years. We shall accordingly speak of the "unit of τ," which for heliocentric orbits and definition (A 7) is k_s^{-1} or 58.132,44087 days. For geocentric orbits the "unit of τ" tends to de-emphasize the selection of a reference unit of time and the resulting numerical value of k_e. For example, we shall herein evaluate k_e first for the second of time, which would be a first choice to many. Then, multiplying by 60, we shall obtain k_e for the minute of time, which value we shall thereafter prefer chiefly because of convenience in the location of decimal points in k_e and other constants. The "unit of τ" will be the same for either choice.

Some question has been raised as to the need for an accurate determination of k_e or an equivalent constant, on the grounds that a definitive determination of the orbit of an artificial satellite will lead to improved values of basic constants even if we employ initially values that are less accurate than the best ones that we can determine in advance, so long as the squares of the discrepancies between initial and final values are negligible. Such an argument is valid only if it is possible to obtain from a definitive satellite orbit values of constants that are more accurate than the values we can obtain from presently available data. For the coefficient J of eq. (A 2) it seems very likely that we can do so, and our ultimate choice of J is accordingly governed by considerations of precedent as well as accuracy. Sections M and N of this paper, however, indicate that it will be extremely difficult or impossible to obtain as accurate a value of k_e as that we can obtain beforehand from the acceleration of gravity. The determinations from a definitive orbit calculation of all of the remaining constants will accordingly be improved if we eliminate k_e from among them by using the most accurate value we can now obtain[1].

The foregoing discussion, moreover, is limited to the matter of obtaining the best set of constants for a particular definitive and lengthy orbit determination. Many geocentric orbits are such that we cannot wait, but must use the best values of basic constants that can be determined in advance. Notable among such orbits are intercontinental trajectories, for which it would be absurd to reject the best available information for any reason whatsoever (cf. Sec. K).

The proposal that the best available values of the basic constants be rejected stems from the fact that certain other values have been adopted for the international ellipsoid, for the international gravity formula, or for the "system of astronomical constants." The proposal, however, modifies the international gravity formula, and is accordingly inconsistent in refusing to accept a more accurately known correction to the international ellipsoid. The accepted modification of the gravity formula, moreover, is based upon a consideration of only one of the two factors that cause it to be in error, and upon obsolescent data for that. Most of the "system of astronomical constants" has no bearing on geocentric orbits; the constants that bear upon observational data, such as the constants of precession and nutation, have no firm connection with the gravitational constants that concern us. In fact there are inconsistencies between most of the astronomical constants that indicate the presence of systematic errors or neglected terms in the theoretical relationships between them. Finally, there is no general agreement as to the values that make up the "system of astronomical constants."

B. The Gaussian or Heliocentric Gravitational Constant

The Gaussian or heliocentric gravitational constant, k_s, was evaluated by Gauss by means of Kepler's third law, eq. (A 5), for the following units:

Time: the mean solar day, or m.s.d., now (see below) replaced by the "ephemeris mean day," or e.m.d.

[1] The foregoing remarks were written before they were so strikingly confirmed by the lack of accurate orbital and observational data on Sputniks I and II.—*Note added in proof.*

Mass: the mass of the sun, $m_\odot$.

Distance: the astronomical unit, or "a.u."

Since the known data for any planet would serve for the evaluation, Gauss selected, as best established, those pertaining to the earth. These he assumed or determined to be

$$a = a_\oplus \text{ (in a. u.)} = 1$$

$$\mu = m_\odot + m_\oplus \text{ (in } m_\odot) = 1 + \frac{1}{354{,}710} = 1.000{,}002{,}819$$

$$P = \text{sidereal year (in m.s.d.)} = 365.256{,}3835.$$

With these values,

$$k_s = \frac{2\pi a^{3/2}}{P\sqrt{\mu}} = 0.017{,}202{,}098{,}95.$$

The values of P and $m_\oplus$ have been redetermined a number of times since the calculation of Gauss. In order to avoid intermittent changes in k_s (cf. Sec. A), it became customary and finally it was agreed internationally (I.A.U. 1938, pp. 20, 336) to redefine the astronomical unit, defined in Gauss's time as the mean distance of the earth, as the mean distance instead of a fictitious unperturbed planet having exactly the mass and period used by Gauss, or equivalently as the radius of a circular orbit in which a body of negligible mass and free of perturbations would revolve around the sun in a "Gaussian" year,

$$P_o = 2\pi/k_s = 365.256{,}898{,}3263 \text{ m.s.d.}$$

The practice of holding k to a fixed value is in accord with usual astronomical practice in the correction of an orbit, in which the elements and perhaps the masses peculiar to the orbit are modified to bring them into agreement with observation and with certain physical constants and equations that are left unchanged.

In determining the numerical value of the earth's mean distance with the present definition of the astronomical unit, it is interesting to find that the inaccuracies or errors of the values that Gauss used for P and μ cancel one another in part. If we may take his words literally, Gauss used the mass of the earth in μ; he should have used the mass of the earth-moon system, since it is the center of mass of the latter that most nearly describes an elliptic orbit around the sun in one sidereal year. With Newcomb's (1898) values,

$$P = 365.256{,}36042$$

$$\mu = 1 + \frac{1}{329{,}390} = 1.000{,}003{,}036$$

we find that the mean distance of the earth-moon system from the sun, neglecting perturbations, is

$$a = 1.000{,}000{,}030$$

Increasing this value for the action of certain planets, Newcomb adopted

$$a = 1.000{,}000{,}230$$

and so, effectively,

$$\mu = 1.000{,}003{,}636.$$

The "mean solar day" was precisely described until recently as the "mean solar day (1900.0)," but increasing accuracy in the determination of time has finally deprived the latter term of the precision that it seemed to have. The fundamental unit of time was shifted first (I.A.U. 1952, p. 66) to the sidereal year (1900.0), and then (SPENCER JONES 1956, p. 22) to the tropical year (1900.0). The "mean solar day" is replaced by the "ephemeris mean day" (or the "ephemeris mean second," SADLER 1954), which is the tropical year (1900.0) divided by 365.242,198,79 (NEWCOMB's 1898 value).

In estimating the accuracy of the foregoing determination of k_s, we return to the concept wherein it is an observational constant, but with a chosen arbitrarily so that k_s will have its GAUSSian value. Then from eq. (A 5),

$$\frac{2\Delta k_s}{k_s} = -\frac{2\Delta P}{P} - \Delta\mu.$$

The uncertainty in the sidereal period, P, is now due only to the uncertainty of the general precession in longitude, which we may estimate from de SITTER-BROUWER (1938). The uncertainty in μ is owing to the uncertainties in the masses of the planets, and deducible from NEWCOMB (1898). These uncertainties may be taken as the probable errors given by SPENCER JONES (1941), RABE (1950), HERTZ (1953), and others, as modified by CLEMENCE and BROUWER (1955). The deduced relative uncertainty $\Delta k_s/k_s$ is of the order of $\pm 2 \times 10^{-9}$, so that k_s may be regarded as known nearly to its 9th significant figure!

C. The Centimeter-Gram-Second Gravitational Constant

The centimeter-gram-second gravitational constant, G, has been used in orbit work by some of the writers in the field of rockets, and may at first glance appear to be a more fundamental constant then the one derived in the preceding section. With such an interpretation

$$k^2 = G m_1.$$

It is profitable to compare the determination of k_s from G with GAUSS's determination (Sec. B).

Using (HEYL 1930, BIRGE 1941),

$$G = 6.670\,(1 \pm 0.0007) \times 10^{-8}\ \text{cm}^3\,\text{gm}^{-1}\,\text{sec}^{-2}$$

and (RDS 1945)

$$m_{\oplus} = 5.975\,(1 \pm 0.0007) \times 10^{27}\ \text{gm}$$

and (RABE 1950)

$$\frac{m_{\odot}}{m_{\oplus} + m_{☾}} = 328{,}452\,(1 \pm 0.00013), \frac{m_{\oplus}}{m_{☾}} = 81.375\,(1 \pm 0.0003)$$

we obtain

$$\frac{m_{\odot}}{m_{\oplus}} = 332{,}488\,(1 \pm 0.00013)$$

$$m_{\odot} = 1.9866\,(1 \pm 0.0007) \times 10^{33}\ \text{gm}$$

so that, finally

$$k_s^2 = G m_{\odot} = 1.3251\,(1 \pm 0.0010) \times 10^{26}\ \text{cm}^3/\text{sec}^2$$

$$k_s = 1.1511\,(1 \pm 0.0005) \times 10^{13}\ \text{cm}^{3/2}/\text{sec}.$$

This value of k_s has 3 significant figures instead of the preceding section's nearly 9 significant figures, which make the GAUSSian k_s known more accurately than any other physical constant. Thus it becomes apparent that G is nearly valueless to us in heliocentric orbit problems.

As a corollary to this conclusion on G, we find that the centimeter or any related laboratory unit is likewise nearly valueless in heliocentric orbit problems. Even if we adopt Gauss's k and attempt to adapt it to the centimeter, we find that we are again reduced to 3- or 4-figure accuracy because of the limited accuracy of the solar parallax, which is the connecting link between miles or kilometers and the astronomical unit. Adopting

$$1 \text{ a. u.} = 1.4950\,(1 \pm 0.00010) \times 10^{13} \text{ cm}$$

we find

$$k_s = 1.1509\,(1 \pm 0.00015) \times 10^{13} \text{ cm}^{3/2}/\text{sec}.$$

By contrast we find that the units of mass and time may be chosen arbitarily. The unit of mass is conveniently taken as $m_{\odot}$, but if one prefers to think of the unit as the gram he absorbs $m_{\odot}$ into k_s^2. The other masses are then divided by $m_{\odot}$—an action that has the same effect as expressing them in units of the sun's mass in the first place. The unit of time presents no problem because we have as yet no way of defining fundamental time units that cannot be reconciled with the ephemeris mean day and the tropical year.

From the foregoing basic data we may obtain also

$$k_e^2 = Gm_{\oplus} = 3.9853\,(1 \pm 0.0010) \times 10^{20} \text{ cm}^3/\text{sec}^2$$
$$k_e = 1.9963\,(1 \pm 0.0005) \times 10^{10} \text{ cm}^{3/2}/\text{sec}\,.$$

The accuracy of this determination is evidently about the same as that of k_s, and we are led immediately to expect that there are other means that will yield more accurate values of k_e. We are led also to inquire into the relationship between the adoption of a unit of distance and the accuracy of k_e. As in the determination of k_s, the units of mass and time may be selected arbitrarily.

D. The Evaluation of the Geocentric Constant

The evaluation of the geocentric constant, k_e, from g_e, the equatorial value of the acceleration of gravity, requires that we take into account the rotation of the earth and the extent to which it cannot be represented as homogeneous in spherical concentric layers. If we assume that the earth is a rotating oblate spheroid with the inertial-field potential of eq. (A 2), eq. (A 4) may be written

$$g_e = k_e^2\, a_e^2\,(1 - A - \tilde{\omega} + J + \frac{1}{2} K). \tag{D 1}$$

Herein g_e is the equatorial value of the acceleration of gravity and a_e is the equatorial radius. As in eq. (A 2), m_2 is negligible. The mass of the earth is assumed to be unity or to be absorbed into k_e, so that m_1 is replaced by $1—A$, with A representing the mass of the atmosphere, which, insofar as it approximates a set of homogeneous and similar ellipsoidal shells, will have no attraction on a particle on the surface of the earth (Newton 1687, Book 1, Prop. 91, Cor. 3; Moulton 1914, pp. 101—102). The centrifugal acceleration resulting from the earth's angular velocity of rotation, ω, is included in

$$\tilde{\omega} = a_e^3\, \omega^2 / k_e^2. \tag{D 2}$$

Finally J and K are the coefficients in the second and fourth harmonics in the attraction of the spheroid [cf. eq. (A 2)].

The quantities appearing in eq. (D 1, 2) are sometimes replaced by other quantities, especially those associated with the mean latitude, $\varphi_1 = \arcsin(1/3)$, (cf. DE SITTER 1915, 1924, 1927, DE SITTER-BROUWER 1938). We shall discuss the use of such quantities and evaluate the formulae containing them in the appendix (Sec. Z).

Eq. (D 1) might be modified also to include the effect of a third axis in the reference ellipsoid, or, without modification, it can include the slight departure of the spheroid from a perfect ellipsoid of revolution, a depression reaching its maximum at latitude 45°, that is used by DE SITTER (1924, 1938), following DARWIN and WIECHERT, to satisfy the requirements of fluid equilibrium. Both of these effects tend to be lost in local undulations of the geoid (approximately the mean sea-level surface of the earth, carried through the continents), and in "local anomalies" or "station errors" in gravity, latitude, and longitude. The observational evidence for a third axis has never been strong (JUNG 1956), and is contradicted by recent measurements (CHOVITZ and FISCHER 1956). Accordingly, it is advisable for the present to seek parameters associated with the spheroid rather than the triaxial ellipsoid that best fits the geoid. The depression of the spheroid, embodied in the $\varkappa$-terms of the following equations, can be taken into account without complications, and is taken into account not only because of theoretical indications of its existence but also because it serves to indicate how the principal anomalies might affect basic relationships.

The parameters g_e, a_e, $\tilde{\omega}$, J, K, and $\varkappa$ are associated with one another, with f, the "flattening" of the spheroid, with β and γ, coefficients in the gravity formula, and with g and R, the gravity and geocentric distance at any point on the spheroid, through the following array of formulae:

$$R = a_e\left[1 - f\sin^2\varphi + \left(\frac{5}{8}f^2 - \varkappa\right)\sin^2 2\varphi + \ldots\right] \tag{D 3}$$

$$g = g_e\left[1 + \beta\sin^2\varphi + \gamma\sin^2 2\varphi + \ldots\right] \tag{D 4}$$

$$\beta = -f + \frac{5}{2}\tilde{\omega} - \frac{26}{7}f\tilde{\omega} + \frac{15}{4}\tilde{\omega}^2 + \frac{8}{7}\varkappa + 0(f^3) \tag{D 5}$$

$$\gamma = \frac{1}{8}f^2 - \frac{5}{8}f\tilde{\omega} - 3\varkappa + 0(f^3) \tag{D 6}$$

$$J = f - \frac{1}{2}\tilde{\omega} - \frac{1}{2}f^2 + \frac{9}{14}f\tilde{\omega} + \frac{4}{7}K + 0(f^3) \tag{D 7}$$

$$K = 3f^2 - \frac{15}{7}f\tilde{\omega} + \frac{24}{7}\varkappa + 0(f^3). \tag{D 8}$$

Eqs. (D 5, 6, 7, 8) are derived for the spheroid; expressions for the ellipsoid of revolution may be obtained from them by setting $\varkappa = 0$. For these two figures, then, it will be possible to obtain J from either f or β, the former being determined along with a_e from geodetic surveys, and the latter being determined along with g_e from gravimetric surveys. It is possible also to obtain J from astronomical sources of information: from the motion of the moon's node and perigee, from the monthly term in the moon's latitude, and from the precessional constant (Sec.G). The three sources of information do not agree, partly because of unavoidable errors in individual determinations of f, β, J, but also because the geoid is neither the spheroid nor the ellipsoid of revolution described above. The $\varkappa$-terms, representing the difference between spheroid and ellipsoid, indicate something of the effects of other departures of the geoid from the ellipsoid. Even in the relationship between β and J, when f is eliminated between eqs. (D 5) and (D 7), though both are gravitational in nature, a $\varkappa$-term remains.

If our aim were simply to determine the best available values of g_e, a_e, and J, accordingly, we might be inclined to discount eqs. (D 5) and (D 7) and to rely upon strictly gravitational sources to obtain g_e and β, upon strictly geodetic sources to obtain a_e and f, and upon strictly astronomical sources to obtain J. But in the observation of objects traveling in geocentric orbits, some of these quantities are again tied together, and in the anticipation that these observations will be used to improve geodetic information it becomes evident that they must be united firmly in theory. With a consistent basic model for the earth, whether it be the spheroid, the ellipsoid of revolution, or something else, it becomes possible to seek out the departures of the geoid therefrom in the perturbations that they produce in geocentric orbits. Our goal, accordingly, is the determination of values of a_e, g_e, f, β, J, etc., for the spheroid or other simple figure from which the geoid departs least and preferably in such a fashion that its astronomical effects are as nearly random as possible.

For reference in the sections that follow, we compute β, γ, J, K for

$$f = 0.003{,}367{,}00 = 1/297$$

$$\tilde{\omega} = 0.003{,}461{,}49\,.$$

The first of these is the value adopted for the International Ellipsoid (PERRIER 1925); the second will be derived later (Sec. J). Then, for three values of $\varkappa$ we obtain the values of β, γ, J, K shown in Table I.

Table I

$10^6\,\varkappa$	0.00	0.50	0.68
$10^6\,\beta$	+ 5288.4	+ 5288.9	+ 5289.1
$10^6\,\gamma$	—5.9	—7.4	—7.9
$10^6\,J$	+ 1638.08	+ 1638.36	+ 1638.47
$10^6\,K$	+ 9.04	+ 10.75	+ 11.37

E. The Equatorial Value of the Acceleration of Gravity

The equatorial value of the acceleration of gravity, g_e, and the coefficients β and γ of eq. (D 4) are given by the International Gravity Formula (CASSINIS 1930, LAMBERT 1931, 1945) as follows:

$$g_e = 978{,}049 \text{ milligals} \qquad 1 \text{ milligal} = 0.001 \text{ cm/sec}^2$$

$$\beta = +\,0.005{,}288{,}4 \qquad \gamma = 0.000{,}005{,}9.$$

The values of β and γ are clearly those associated with the International Ellipsoid (cf. the values in Table I under $\varkappa = 0$). The value of g_e was derived by HEISKANEN (1928) from a worldwide set of values of g based upon the classic 1906 determination of absolute gravity at Potsdam, g_P, which is revealed to be 10 to 17 milligals too high by recent determinations of g_W (Washington) at the National Bureau of Standards, and of g_T (Teddington, England) at the National Physical Laboratory. Our reexamination of the problem is accordingly made in two steps: (1) the estimation of a correction to g_P, and (2) the estimation of corrections to g_e, β γ, on the Potsdam standard. The value of g_e that we ultimately adopt will have both corrections applied to it.

For revaluation of the Potsdam standard the following list of absolute and differential determinations and conclusions therefrom is in milligals:

$g_P = 981{,}274 \pm 3$	KUHNEN and FURTWÄNGLER	(E 1)
$g_W = 980{,}080 \pm 3$	HEYL and COOK 1936	(E 2)
$g_T = 981{,}181.5 \pm 1.5$	CLARK 1939	(E 3)
$g_W - g_P = -1{,}177.1 \pm 1.0$	E. J. BROWN 1933 (revised by JEFFREYS 1948, rejecting bronze pendulums)	(E 4)
$g_W - g_T = -1{,}096.9 \pm 0.9$	BULLARD and BROWNE 1940	(E 5)
$g_T - g_P = -79.9 \pm 0.8$	JEFFREYS 1948	(E 6)

$$\Delta g_{PW} = g_W - (g_W - g_P) - g_P = -16.9$$

$$\Delta g_{PT} = g_T - (g_T - g_P) - g_P = -12.6\,.$$

Earlier versions of these discrepancies, —20.0 and —13.8 respectively, led DRYDEN to reexamine the Potsdam determination and to reject a correction for bending of the knife-edges, with the result:

$$g_P = 981{,}262.3 \qquad \text{DRYDEN 1942} \qquad \text{(E 7)}$$

Results (E 2, 3, 7) were revised by JEFFREYS (1948), with reduction of all indices of precision to standard errors, to

$$g_W = 980{,}080.2 \pm 1.5 \qquad \text{(E 8)}$$

$$g_T = 981{,}181.5 \pm 1.0 \qquad \text{(E 9)}$$

$$g_P = 981{,}263.3 \pm 2.2. \qquad \text{(E 10)}$$

These and eqs. (E 5, 6, 7) he solved by least squares to obtain (JEFFREYS 1948)

$$g_P = 981{,}260.6 \pm 1.0 \qquad \Delta g_P = -13.4\,.$$

In the following year, however, a new examination of the problem (JEFFREYS 1949) led him to revise results (E 8, 9) to

$$g_W = 980{,}081.6 \pm 1.2 \qquad \text{(E 11)}$$

$$g_T = 981{,}183.2 \pm 0.6\,. \qquad \text{(E 12)}$$

Solving these and eqs. (E 4, 5, 6, 10) in the same manner as JEFFREYS (1948) we obtain

$$g_P = 981{,}263.1 \pm 0.9 \qquad \Delta g_P = -10.9\,.$$

In the course of our investigation two more recent papers were uncovered and yielded the following:

$g_T - g_P = -78.0 \pm 0.5$	COOK 1952	(E 13)
$g_W - g_T = 1096.6 \pm 0.5$	GARLAND and COOK 1955	(E 14)

Introducing these equations into the least squares solution along with eqs. (E 11, 12) we obtain

$$g_P = 981{,}260.9 \pm 0.7 \qquad \Delta g_P = -13.1\,.$$

Accordingly we adopt:

$$\Delta g_P = -13.1 \pm 1.0 . \tag{E 15}$$

Returning for the moment to an acceptance of the Potsdam standard, we list in Table II the recent determinations of g_e and β, omitting those that include longitude terms. The values of γ, derived from theory rather than observation, indicate the presence or absence of $\varkappa$-terms in the reductions.

Table II

	g_e (mgal)	$10^6\,\beta$	$10^6\,\gamma$
HEISKANEN 1928[1]	978,049 ±1	+5289	—7
DE SITTER-BROUWER 1938[2]	978,053.0 ±2.0	+5286.12 ±0.99	—7.34 ±0.30
JEFFREYS 1936	978,051	+5282 ±6	—7
JEFFREYS 1941	978,050	+5286.6 ±8.1	—5.9
JEFFREYS 1948	978,051.3	+5285.9 ±5.0	—5.9
SCHUTTE (ZHURAVLEV) 1950 (JUNG 1956)	978,052.0 ±3.3	+5282.7 ±6.0	—5.9
HEISKANEN 1956[3]	978,049.6		

The given values of β indicate that this quantity corresponds to a value of f somewhat in excess of 1/297 (cf. Table I). The values of g_e, however, are not strictly comparable because of their dependence upon the associated value of β. From eq. (D 4) we may derive a mean value,

$$g_m = g_e\,(1 + \beta\,[\sin^2\varphi]_m + \gamma\,[\sin^2 2\,\varphi]_m + \ldots)$$

so that

$$\frac{\Delta g_m}{g_m} = \frac{\Delta g_e}{g_e} + \Delta\beta\,[\sin^2\varphi]_m + \Delta\gamma\,[\sin^2 2\,\varphi]_m + \ldots .$$

If the observations were distributed uniformly over the surface of the earth, we should have

$$[\sin^2\varphi]_m = \frac{\int_0^{\pi/2} \sin^2\varphi\cos\varphi\,d\varphi}{\int_0^{\pi/2}\cos\varphi\,d\varphi} = \frac{1}{3};\ [\sin^2 2\,\Phi]_m = \frac{8}{15}.$$

Two samples from JEFFREYS (1941, p. 17; 1943, p. 61) indicate that these figures are not far wrong in spite of the non-uniform distribution of the observations, since they yield $[\sin^2\varphi]_m = 0.357$ and 0.349, and $[\sin^2 2\,\varphi]_m = 0.609$ and 0.645, respectively. Adopting, then, $\Delta g_m = 0$, $\Delta g_e/g_e = \frac{1}{3}\,\Delta\beta - 0.6\,\Delta\gamma$ and reducing the first six values in the foregoing to the International Ellipsoid values, $\beta = +0.005{,}2884$ and $\gamma = -0.000{,}0059$, we obtain $g_e = 978048.6$, 978051.4, 978048.3,

[1] Assumed $f = 1/297$.

[2] Not purely gravimetric.

[3] Communicated verbally, without β and γ.

978049.4, 978050.5, 978050.1. In the light of these data and the seventh value, 978049.6, we estimate that 978049.9 ± 2.5 is the best value to which to apply the correction of eq. (E 15).

As a result of the two investigations, then, we adopt tentatively the value (in milligals)

$$g_e = 978{,}036.8\,(1 + g' + \frac{1}{3} f') \qquad \text{(E 16)}$$

where our best estimate of g' is

$$g' = 0 \pm 3 \times 10^{-6} \qquad \text{(E 17)}$$

and we define f', which enters the formula through β, from eq. (D 5), by

$$\Delta\beta = -\Delta f = -f'$$

$$f' = f - 1/297 = f - 0.003{,}367{,}00. \qquad \text{(E 18)}$$

From eq. (D 4) we may obtain

$$\frac{\Delta g}{g} = \frac{\Delta g_e}{g_e} + \Delta\beta \sin^2\varphi + \ldots = \frac{\Delta g_e}{g_e} - f' \sin^2\varphi + \ldots.$$

At the "mean latitude", $\varphi_1 = 35\frac{1}{4}^\circ$, $\sin^2\varphi = 1/3$, and we have

$$\frac{\Delta g_1}{g_1} = g'\,.$$

It is the freedom of g_1 from f' that led DE SITTER (1924, 1927, 1938) to prefer it to g_e (cf. Secs. F, H).

F. The Earth's Equatorial Radius

The earth's equatorial radius, a_e, in meters, and the flattening, $f = 1 - b_e/a_e$, where b_e is the polar radius, are given by the International Ellipsoid (HAYFORD 1909, PERRIER 1925) as follows:

$$a_e = 6{,}378{,}388 \text{ meters} \qquad f = 1/297 = 0.003{,}367{,}00$$

Determinations of a_e and f, or of a_e with an assumed value of f, are based upon least-squares reductions of triangulation and astronomical data from the widest available areas. If the reduction is preceded by an attempt to estimate the effects of topography and isostatic compensation, it is termed "isostastic"; if the only corrections are for height above sea level, the reduction is termed "free-air." Notable determinations are listed in Table III (cf. also LAMBERT 1924, JUNG 1956).

It is clear from a comparison of the tabular values of a_e that the accidental errors indicated by most of the given probable or standard errors contribute less to the uncertainty of a_e than systematic effects introduced by topography and variations in crustal density, whether or not there is an attempt to estimate them isostatically or otherwise. It will become evident in what follows that the choice or derivation of a value of f will be reflected in a_e. These effects are realistically estimated in the standard errors of CHOVITZ and FISCHER; the credibility of their values of a_e is increased by the close agreement of their free-air and isostatic reductions and by the fact that they were able to use, for the first time, meridional arcs running from Canada to Chile and from Finland to the Cape of Good Hope.

Discussion of f we shall reserve to Sec. G. The dependence of a_e upon it we may estimate, for the meridional arc determinations that are now of paramount interest, from the following approximate expressions for a measured arc, s:

$$s = a_e (\varphi - \varphi_o)(1 - fx). \tag{F 1}$$

Table III

	a_e (meters)		$1/f$	
CLARKE 1866	6,378,258[9]		294.98	
CLARKE 1880	6,378,301[9]		293.465	
HAYFORD 1907	6,378,283	±34[1]	297.8	±0.9
HAYFORD 1909	6,378,388	±18[1]	297.0	±0.5
	or	±53[4]		±1.5[4]
	6,378,062[2]	±33[5]	298.15	±1.02[5]
HEISKANEN 1929	6,378,400[3]		298.2	
KRASSOVSKI 1942	6,378,293	±16	298.4	±0.4
(ZVEREV 1950; CHOVITZ and FISCHER 1956)	6,378,245	±15[3]	298.40	±0.4
JEFFREYS 1948	6,378,117	±119[2,6]	298.40	±2.61
	6,378,099	±116[2,7]	297.10	±0.36
LEDERSTEGER 1951	6,378,298	±34	297.0	(assumed)
HAYFORD Revision (SCHMID 1953)	6,378,228[1]		299.5	
	6,378,240		297.0	(assumed)
"Hough Ellipsoid"	6 378,250	±95[2]	297.0	(assumed)
(CHOVITZ and FISCHER 1956)	6,378,240	±100[2,8]		
	6,378,285	±100[1]		

where

$$x = \frac{1}{2} + \frac{3}{2}\,\frac{\sin(\varphi - \varphi_o)}{\varphi - \varphi_o}\cos(\varphi + \varphi_o)\,.$$

The quantities φ and φ_o are ideally the geodetic latitudes of the northern and southern extremities of s, but practically they are the astronomical latitudes with an estimate of the effects of topography and isostacy, or without any such compensation in a free-air reduction. For the meridional arcs used by CHOVITZ and

[1] Isostatic.
[2] Free air.
[3] Triaxial.
[4] Revised by HELMERT (1911).
[5] Supplied by JEFFREYS (1948).
[6] Survey data only.
[7] Including gravitational data.
[8] With JEFFREYS' components.
[9] The equatorial radius is converted from British feet by the ratio: 1 international meter = 39.370,113 British inches. In the "CLARKE spheroid of 1866" (U.S.C.G.S. Spec. Rep. no. 5), the value a_e = 6,378,206 meters is that determined by CLARKE with his ratio: 1 legal meter = 39.370,432 British inches. CLARKE's ratio and the legal meter now are separately superseded, but the "CLARKE spheroid", with its a_e treated as if it were in international meters, is very much alive as a reference spheroid in North American geodetic work. Used in conjunction with it is the ratio that defines U.S. statute units: 1 international meter = 39.370,000 U.S. statute inches.

FISCHER, we find that x in each instance differs but little from 4/3. From eq. (F 1) we may derive, accordingly,

$$\frac{\Delta a_e}{a_e} = a' + \frac{4}{3} f'$$

where a' is the statistical uncertainty resulting from errors in s, φ, and φ_o and f' is again [cf. eq. (E 18)]

$$f' = f - 1/297 = f - 0.003{,}367{,}00. \qquad \text{(F 2)}$$

More detailed expressions given by CHOVITZ and FISCHER enable us to find a' and verify the coefficient 4/3:

$$a_e = 6{,}378{,}250 \pm 20 - 95\left(\frac{1}{f} - 297\right) \qquad \text{free-air}$$

$$a_e = 6{,}378{,}240 \pm 40 - 95\left(\frac{1}{f} - 297\right) \qquad \text{free-air with JEFFREYS' components}$$

$$a_e = 6{,}378{,}285 \pm 15 - 100\left(\frac{1}{f} - 297\right) \qquad \text{isostatic}$$

The corresponding values of a' are $\pm 3.1 \times 10^{-6}$, $\pm 6.3 \times 10^{-6}$, $\pm 2.4 \times 10^{-6}$ and of x are 1.31, 1.31, 1.38.

CHOVITZ and FISCHER recommend the compromise figure 6,378,260, but we are inclined to favor a slightly larger value in the light of the foregoing discussion and because we read into their own study a somewhat more favorable case for isostasy than they did. Accordingly, we adopt:

$$a_e = 6{,}378{,}270\left(1 + a' + \frac{4}{3} f'\right) \text{ meters} \qquad \text{(F 3)}$$

where we estimate, after consultation with J. A. O'KEEFE,

$$a' = 0 \pm 10 \times 10^{-6} \qquad \text{(F 4)}$$

and f' is given by eqs. (F 2), (E 18). The value of a' corresponds to an uncertainty of $\pm$ 64 meters in a_e, in good agreement whith JEFFREYS' table (1948, p. 229).

It is well to note, in view of the prominence of f' in eq. (F 3), that other radii of the earth are now better determined than the equatorial one. From eq. (D 3) we derive

$$\frac{\Delta R}{R} = \frac{\Delta a_e}{a_e} - f' \sin^2\varphi = a' + f'\left(\frac{4}{3} - \sin^2\varphi\right) \qquad \text{(F 5)}$$

and in particular

$$\varphi_0 = 0^\circ \qquad \sin^2\varphi_0 = 0 \qquad \Delta R_0/R_0 = a' + \frac{4}{3} f'$$

$$\varphi_1 = 35\frac{1}{4}^\circ \qquad \sin^2\varphi_1 = 1/3 \qquad \Delta R_1/R_1 = a' + f'$$

$$\varphi_2 = 54\frac{3}{4}^\circ \qquad \sin^2\varphi_2 = 2/3 \qquad \Delta R_2/R_2 = a' + \frac{2}{3} f'$$

$$\varphi_3 = 90^\circ \qquad \sin^2\varphi_3 = 1 \qquad \Delta R_3/R_3 = a' + \frac{1}{3} f'.$$

Thus it appears that the polar radius is more accurately determined than any other by measurements such as those employed by CHOVITZ and FISCHER. The mean radius preferred by DE SITTER does not appear to offer a substantial advantage over the equatorial radius (cf. Sec. H and JEFFREYS 1948, p. 238).

G. The Remaining Constants

The remaining constants that enter into the determinaticn of k_e by eq. (D 1) include J, $\tilde{\omega}$, A, $\varkappa$. We may dismiss A and $\varkappa$ by computing the former from the mean atmospheric pressure (in grams/cm² rather than millibars) and by reluctantly rejecting BULLARD's 1948 value of the latter (cf. also SPENCER JONES 1954) in favor of the international ellipsoid:

$$A = +0.88 \times 10^{-6} \tag{G 1}$$

$$\varkappa = 0. \tag{G 2}$$

By eq. (D 2) $\tilde{\omega}$ depends upon the earth's angular velocity of rotation, referred to an inertial framework,

$$\omega = 1.002{,}737{,}803 \text{ rotations/day}$$

$$= 72.921{,}1508 \times 10^{-6} \text{ radians/sec}$$

from which

$$\omega^2 = 5.317{,}4942 \times 10^{-9}. \tag{G 3}$$

The remaining steps in the calculation of $\tilde{\omega}$, requiring successive approximations, result in the value given in Sec. D, of which only the last step, or verification, will be shown hereafter.

The monthly term in the moon's latitude and the motions of the moon' snodes and perigee, in discussions involving also the inclination of the moon's axis, other lunar data, and perhaps some terrestrial data, yield the following values of $10^6\ J$:

$+1641.5 \pm 6.5$	JEFFREYS 1936
$+1637.7 \pm 6.2$ (*s. e.*)	JEFFREYS 1941a (including gravity data)
$+1641.46 \pm 3.60$ (*p. e.?*)	SPENCER JONES 1941
$+1637.0 \pm 4.1$	JEFFREYS 1948

The last of these determinations rejects the monthly term in the moon's latitude, used theretofore by both JEFFREYS and SPENCER JONES, because the latter had found that systematic errors might enter through the obliquity of the ecliptic; terrestrial data were substituted for it, but terrestrial data that alone gave the same value.

The precessional constant, somewhat admixed with other data, yields the following values of $10^6\ J$:

$+1639.2$	DE SITTER 1924
$+1638.8 \pm 1.0$	DE SITTER 1927
$+1641.12 \pm 0.97$	DE SITTER-BROUWER 1938
$+1647.32$	CLEMENCE 1948
$+1634.6 \pm 3.1$	JEFFREYS 1948 (deduced from other data)

JEFFREYS distrusts this source of information because of its dependence upon the assumption of hydrostatic equilibrium.

It has been noted that the values of β cited in Sec. E correspond to a value of f somewhat greater than the reference value, 1/297. The same increase would be found in the values of J given at the end of Sec. D. For example, the JEFFREYS' 1948 value,

$$10^6\beta = +5285.9 \pm 5.0$$

corresponds to

$$10^6 J = +1641.7 \pm 5.0\,.$$

The values of f displayed in Sec. F, however, suggest that f is somewhat smaller than 1/297. If we take

$$1/f = 298 \pm 1$$

or

$$10^6 f = +3{,}355.7 \pm 11.3$$

we obtain

$$10^6 J = +1{,}627.2 \pm 11.3\,.$$

We are inclined somewhat toward JEFFREYS' preferred 1948 value, $+1637.0 \pm 4.1$, but do not find sufficient evidence in its favor to overcome the many advantages to be gained by adhering to the value, only slightly different, that is consistent with the reference $f = 1/297$. Accordingly we adopt, using $\varkappa = 0$,

$$f = +3{,}367.00 \times 10^{-6} + f' \tag{G 4}$$

$$\beta = +5{,}288.4 \times 10^{-6} - f' \tag{G 5}$$

$$\gamma = -5.9 \times 10^{-6} \tag{G 6}$$

$$J = +1{,}638.08 \times 10^{-6} + f' \tag{G 7}$$

$$K = +9.04 \times 10^{-6} \tag{G 8}$$

$$f' = 0 \pm 4 \times 10^{-6}\,. \tag{G 9}$$

We may hope for a strong determination of J, and thence f, from the motion of the nodes of the orbit of an artificial satellite, which will be roughly 3600 times the corresponding effect in the regression of the nodes of the moon's orbit. The accuracy of such a determination will depend, of course, upon the number and accuracy of the observations and the length of time over which they are made, and upon the sufficiency of the perturbation theory employed and the accuracy to which the coefficients of the other perturbation terms, especially the drag terms, can be ascertained.

The adopted value of f' is chosen in consultation with J. A. O'KEEFE, and of course depends more upon present astronomical sources than geodetic or gravimetric ones.

H. The Calculation of k_e

The calculation of k_e from the acceleration of gravity, then, involves eq. (D 1), or

$$k_e^2 = g_e a_e^2 / \left(1 - A - \tilde{\omega} + J + \frac{1}{2} K\right), \tag{H 1}$$

and the determinations of a_e, g_e, A, J, K, and ω^2 given in eq. (E 16), (F 3), (G 1, 3, 7, 8). It involves also a choice of units, but it will be observed that A, $\tilde{\omega}$, J, K are dimensionless, and that the unit of mass does not enter through either g_e or a_e (cf. Sec. C). The unit of time we shall take first as the second, because of its common use with accelerations, and then shift to the minute for convenience in the decimal point (cf. Secs. A, B, C). For unit of length, however, we must choose between the meter, or rather the megameter for convenience in the decimal point, or other laboratory unit, and the various radii discussed in Sec. F.

At first glance it would seem that, reserving judgment in the matter of megameter vs. the various radii, we should prefer the polar radius among the latter, because it is better determined (Sec. F), but further investigation reveals that the radius for $\sin^2\varphi = 2/3$ yields a stronger determination of k_e. Of course neither of these radii, less familiar than and so psychologically at a disadvantage as compared with the equatorial radius, would have to be used numerically as the actual unit of distance; either could be multiplied by a factor that would make the unit of distance always approximately equal to the equatorial radius, and initially exactly so. When this concept is pursued to its logical conclusion, moreover, and our experience and practice with the GAUSSian k_s are taken into consideration (Sec. B), we are led to the definition of what we shall call the "gravitational radius," which enables us to fix upon a permanent value for k_e. For comparison, then, and for various needs the future may bring forth, we shall determine k_e for each of the three following units of distance:

$$\left.\begin{aligned} 1 \text{ megameter} &= 10^6 \text{ meters} \\ 1 \text{ ``}q\text{-radius''} &= 1 \text{ equatorial radius} \\ &= 6.378{,}270\left(1 + a' + \tfrac{4}{3} f'\right)\text{megameters} \\ 1 \text{ ``}g\text{-radius''} &= 1 \text{ gravitational radius} \\ &= \left(1 - \tfrac{1}{3} a' + \tfrac{1}{3} g' - \tfrac{2}{3} f'\right) q\text{-radii} \\ &= 6.378{,}270\left(1 + \tfrac{2}{3} a' + \tfrac{1}{3} g' + \tfrac{2}{3} f'\right)\text{megameters.} \end{aligned}\right\} \quad \text{(H 2)}$$

In the following quantities, whether they represent basic data from Secs. E, F, and G, or are derived therefrom with the aid of eqs. (H 1, 2), we shall carry a', g', f' for (1) estimation of present uncertainties or for (2) future corrections based upon improved data. The present relative uncertainties, calculated from $a' = \pm 10 \times 10^{-6}$, $g' = \pm 3 \times 10^{-6}$, $f' = \pm 4 \times 10^{-6}$ are given in parentheses at the right, with unit 10^{-6}:

$$\left.\begin{aligned} a_e &= 6.378{,}270\left(1 + a' + \tfrac{4}{3} f'\right) \text{ megameters} && (\pm 11) \\ &= 1\ q\text{-radius} && (0) \\ &= \left(1 + \tfrac{1}{3} a' - \tfrac{1}{3} g' + \tfrac{2}{3} f'\right) g\text{-radii} && (\pm 4) \end{aligned}\right\} \quad \text{(H 3)}$$

$$\left.\begin{aligned} g_e &= 9.780{,}368\left(1 + g' + \tfrac{1}{3} f'\right) \times 10^{-6} \text{ megameters/sec}^2 && (\pm 3) \\ &= 1.533{,}3888\,(1 - a' + g' - f') \times 10^{-6}\ q\text{-radii/sec}^2 && (\pm 11) \\ &= 1.533{,}3888\left(1 - \tfrac{2}{3} a' + \tfrac{2}{3} g' - \tfrac{1}{3} f'\right) \times 10^{-6}\ g\text{-radii/sec}^2 && (\pm 7) \end{aligned}\right\} \quad \text{(H 4)}$$

$$A = +0.88 \times 10^{-6} \qquad \text{(H 5)}$$

$$J = +1{,}638.08 \times 10^{-6} + f' \qquad (\pm 4) \qquad \text{(H 6)}$$

$$\tfrac{1}{2} K = +4.52 \times 10^{-6} \qquad \text{(H 7)}$$

$$\omega^2 = +5{,}317.49 \times 10^{-12} \text{ radians/sec}^2. \qquad \text{(H 8)}$$

To these we add a value of $\tilde{\omega}$ that is obtained by successive approximations; it may be verified after the determination of k_e^2:

$$\tilde{\omega} = a_e^3\, \omega^2 / k_e^2 = +\, 3{,}461.49 \times 10^{-6}\,. \tag{H 9}$$

Then

$$1 - A - \tilde{\omega} + J + \frac{1}{2} K = +\, 0.998{,}180{,}23 + f' \qquad (\pm 4) \tag{H 10}$$

and

$$\left.\begin{aligned} k_e^2 &= 3.986{,}1353\,(1 + 2a' + g' + 2f') \times 10^{-4} \text{ megameters}^3/\text{sec}^2 && (\pm 22)\\ &= 1.536{,}1843\,(1 - a' + g' - 2f') \times 10^{-6}\, q\text{-radii}^3/\text{sec}^2 && (\pm 13)\\ &= 1.536{,}1843 \times 10^{-6}\, g\text{-radii}^3/\text{sec}^2 && (0) \end{aligned}\right\} \tag{H 11}$$

$$\left.\begin{aligned} k_e &= 1.996{,}5308\left(1 + a' + \frac{1}{2} g' + f'\right) \times 10^{-2} \text{ megameters}^{3/2}/\text{sec} && (\pm 11)\\ &= 1.239{,}4290\left(1 - \frac{1}{2} a' + \frac{1}{2} g' - f'\right) \times 10^{-3}\, q\text{-radii}^{3/2}/\text{sec} && \left(\pm 6\frac{1}{2}\right)\\ &= 1.239{,}4290 \times 10^{-3}\, g\text{-radii}^{3/2}/\text{sec} && (0) \end{aligned}\right\} \tag{H 12}$$

or, for the aforementioned convenience in the decimal point,

$$\left.\begin{aligned} k_e &= 1.197{,}9185\left(1 + a' + \frac{1}{2} g' + f'\right) \text{megameters}^{3/2}/\text{min} && (\pm 11)\\ &= 0.074{,}365{,}74\left(1 - \frac{1}{2} a' + \frac{1}{2} g' - f'\right) q\text{-radii}^{3/2}/\text{min} && \left(\pm 6\frac{1}{2}\right)\\ &= 0.074{,}365{,}74\; g\text{-radii}^{3/2}/\text{min}\,. && (0) \end{aligned}\right\} \tag{H 13}$$

As for the relative uncertainties it will be observed that those for the megameter are larger for both a_e and k_e than those for either of the radii, although the discrepancy is far less than that found for heliocentric orbits. It will be noted also that the g-radius gives a zero-uncertainty in k_e, while the q-radius, of course, gives a zero-uncertainty in a_e. The uncertainties in g_e are not of concern in orbit problems, but those in a_e enter into comparisons of an assumed orbit with observational data, and those in k_e permeate the whole of the calculations and tables of an orbit. We are led to prefer the gravitational radius to the equatorial radius because it is easier to revise only the comparison with observation than to revise first all of the orbit calculations and then the comparisons with observation as well, if and when we are required to do so by revision of the basic constants. The situation is precisely analogous to that which led to the adoption of the present definition of the astronomical unit (Sec. B).

It is apparent that improved data on a_e, g_e, or f will supply us with values of a', g', or f' through eqs. (H 3) and (H 4) or

$$f = 0.003{,}367{,}00 + f'\,. \tag{H 14}$$

The International Geophysical Year satellites, however, will almost certainly give us a better value of J, from which f' may be obtained through eq. (H 6); f' may then be used to improve f through eq. (H 14). If the observations of the IGY satellite are sufficiently numerous and accurate, they will justify an attack upon the basic gravitational constant itself, as represented by k_e or the gravitational radius. In simplified version, the following equations, or similar ones, would be used:

$$k_e^2\, \mu\, P^2 = (2\pi)^2 a^3 \tag{H 15}$$

$$a_e = a \sin HP \tag{H 16}$$

where HP is the horizontal parallax, P is the period, a is the mean distance, etc.

With the gravitational radius, one would use P and eq. (H 15) to determine a, and then eq. (H 16) to determine a_e in gravitational radii. Eq. (H 3) would then be solved, probably in conjunction with other information, for appropriate values of a', g', f' to be used in the correction of other data.

With the megameter or equatorial radius, eq. (H 16) would be used to determine a, and then eq. (H 15) to determine k_e. Then eq. (H 12) or (H 13) would be solved for a', g', f' for corrective purposes.

As an example of the use of the gravitational radius, we may consider the changes that would be necessary to return to the international ellipsoid and gravity formula, for which $a_e = 6.378{,}388$ megameters, $g_e = 9.780{,}49$ meters/sec², $f = 1/297$. Then from eqs. (H 3, 4, 14), $f' = 0$, $a' = +18.5 \times 10^{-6}$, $g' = +12.5 \times 10^{-6}$, and $a_e = 1.000{,}002$ g-radii and, from eq. (H 2), 1 g-radius $= 6.378{,}375$ megameters.

J. The Moon's Parallax

The moon's parallax may be used to determine k_e, or may be determined from it. The former process does not yield results that can compete with those of Sec. H because of the uncertainty of the moon's parallax when it is determined observationally. The latter process results in the moon's *dynamical parallax*, which is accordingly stronger than the observational parallax.

To determine k_e from the moon's parallax, we employ eq. (A 5), which we rewrite, according to custom,

$$k_e^2 = n^2 a^3/\mu \tag{J 1}$$

where μ is a mass-factor, a is the moon's mean distance, and, for $P =$ the moon's period, its mean angular motion,

$$n = 2\pi/P = 47{,}434''.890{,}9701/\text{day} \atop \pm 50$$

according to the DE SITTER-BROUWER 1938 revision of BROWN's figure. The uncertainty in this figure is evidently quite negligible in a determination of k_e, and we may write

$$n = 2.661{,}699{,}55 \times 10^{-6} \text{ radians/sec}$$
$$n^2 = 7.084{,}644{,}49 \times 10^{-12} \text{ radians}^2/\text{sec}. \tag{J 2}$$

The constant effect of perturbations on the moon we shall treat as diminishing μ, in accordance with the remarks on eq. (A 3). The effect is often thought of, however, as diminishing a or n, or equivalently increasing P. These different concepts we may summarize by rewriting eq. (J 1) in the alternate forms

$$k_e^2 = \frac{n^2 a^3}{\mu} = \frac{n^2 a_c^3}{m_\oplus + m_☾} = \frac{n_c^2 a^3}{m_\oplus + m_☾}$$

where $m_\oplus$ and $m_☾$ are the actual masses of the earth and the moon, μ being diminished by perturbations as compared with their sum; where a_c would be the mean distance of the moon for its given n if there were no perturbations, the actual mean distance, a, being smaller because of the perturbations; and where n_c would be the mean motion of the moon at its actual mean distance, a, if there were no perturbations, the actual mean motion, n, being less, and P greater, because of the perturbations. Accordingly,

$$\frac{\mu}{m_\oplus + m_☾} = \left(\frac{a}{a_c}\right)^3 = \left(\frac{n}{n_c}\right)^2.$$

E. W. BROWN (1899, p. 89) treated the effect of the perturbations as in a; accordingly, adopting his coefficient, we may write

$$\mu = (0.999{,}093{,}14)^3\,(m_\oplus + m_☾)$$

wherein for the masses of the earth and the moon, $m_\oplus$ and $m_☾$, we adopt

$$m_\oplus = 1 \qquad m_☾ = (1/81.375 \pm 0.026)$$

of which the latter is the value given by RABE (1950), which differs from that of SPENCER JONES (1941) chiefly because of the difference between their values of the solar parallax. Thus we may write

$$m_\oplus + m_☾ = 1.012{,}288{,}8(1 + \mu')$$

$$\mu = 1.009{,}537{,}3\,(1 + \mu') \tag{J 3}$$

$$\mu' = 0 \pm 4 \times 10^{-6}\,. \tag{J 4}$$

Investigation shows that the dependence of μ upon the oblateness is such that the uncertainty of μ due to f', the uncertainty of f and J, is negligible in comparison with μ'.

We have two observational sources for the moon's mean distance, a. One source is by way of the observational determination of the moon's equatorial horizontal parallax, π, reported by CHRISTIE and GILL (1911) — cf. eq. (J 7) —other determinations of the moon's parallax, we should note, are principally dynamical. The other source is a direct determination of the moon's a in meters by O'KEEFE and ANDERSON (1952).

CROMMELIN's reductions of the observations of the crater Mösting A made at Greenwich and the Cape from 1906 to 1910 (CHRISTIE and GILL 1911) yielded a correction,

$$\delta\pi_☾ = +0''.49 - 0''.057\,(f^{-1} - 293.5) \pm 0''.06\,(p.e.) \tag{J 5}$$

that was to be applied to the value of π basic to the Berliner Jahrbuch ephemerides that were used. This value is referred to as "Hansen's mean parallax," and is assumed by LAMBERT (1928) to be NEWCOMB's 1912 revision, 3422".23, but DE SITTER (1927) concludes that it is 3422".27 on the basis, apparently, of a reference in the Berliner Jahrbuch that we find to be inconclusive. Applying the correction of eq. (J 12) to the DE SITTER choice, nevertheless, we obtain

$$\pi_☾ = 3422''.56 - 0''.057\,(f^{-1} - 297.0) \pm 0''.06$$

or

$$\sin\pi_☾ = 0.016{,}592{,}278\left(1 + \frac{3}{2}f' + p'\right)$$

where f' is defined by eq. (E 18) or (F 2) and is given a value in eq. (G 9), and, derived from the foregoing,

$$p' = \pm 18 \times 10^{-6}\,. \tag{J 6}$$

Then, if we take a_e from eq. (H 3),

$$a = \frac{a_e}{\sin\pi_☾} = 60.269{,}00\left(1 - \frac{3}{2}f' - p'\right)\ q\text{-radii} \tag{J 7}$$

or

$$a = 384.412{,}0\left(1 + a' - \frac{1}{6}f' - p'\right)\ \text{megameters}\,. \tag{J 8}$$

O'KEEFE and ANDERSON (1952) give three solutions for the distance of the moon, the third of which is apparently free from dynamical considerations. It states directly

$$a = 384{,}406.7 \pm 4.7 \text{ km}$$

from which we may write

$$a = 384.406{,}7\,(1 + d') \text{ megameters} \tag{J 9}$$

or

$$a = 60.268{,}18 \left(1 - a' - \frac{4}{3} f' + d'\right) \text{q-radii} \tag{J 10}$$

where

$$d' = \pm 12 \times 10^{-6}\,. \tag{J 11}$$

We may judge from eqs. (J 8, 9) that f' is only moderately present in d', if at all.

Adopting the O'KEEFE and ANDERSON values as the most favorable, we obtain the following values, the relative uncertainties being shown in parentheses, as in Sec. H, with unit 10^{-6}:

$$k_e^2 = 3.986{,}2868\,(1 + 3d' - \mu') \times 10^{-4} \text{ megameters}^3/\text{sec}^2 \quad (\pm 36)$$

$$k_e = 1.996{,}5688 \left(1 + \frac{3}{2} d' - \frac{1}{2} \mu'\right) \times 10^{-2} \text{ megameters}^{3/2}/\text{sec.} \quad (\pm 18).$$

The determination is inferior to those of eq. (H 12), and shares the uncertainty of a_e expressed in megameters, eq. (H 3), but it indicates clearly that a radar determination of the distance of the moon to ± 500 meters would be of great value (cf. Sec. M).

The determination of the moon's *dynamical parallax* from k_e, a_e, n and μ is a corollary to the foregoing. Identical results, so far as parallax is concerned, are obtained from the several pairs of k_e and a_e; we shall accordingly use the values for the gravitational radius, in terms of which the moon's mean distance, a, is best determined.

From eqs. (H 3, 11),

$$k_e^2 = 1.536{,}1843 \times 10^{-6} \text{ } g\text{-radii}^3/\text{sec}^2$$

$$a_e = \left(1 + \frac{1}{3} a' - \frac{1}{3} g' + \frac{2}{3} f'\right) g\text{-radii.}$$

With n^2 and μ from eqs. (J 2, 3), then,

$$a^3 = \frac{k_e^2 \mu}{n^2} = 218{,}900.94\,(1 + \mu')\, g\text{-radii}^3$$

$$a = 60.267{,}412 \left(1 + \frac{1}{3} \mu'\right) g\text{-radii} \qquad \left(\pm 1 \frac{1}{3}\right)$$

$$\sin \pi_{☾} = \frac{a_e}{a} = 0.016{,}592{,}715 \left(1 + \frac{1}{3} a' - \frac{1}{3} g' + \frac{2}{3} f' - \frac{1}{3} \mu'\right)$$

from which the moon's *dynamical parallax*, with its relative uncertainty $\times 10^{-6}$, is

$$\pi_{☾} = 3422''.650 \left(1 + \frac{1}{3} a' - \frac{1}{3} g' + \frac{2}{3} f' - \frac{1}{3} \mu'\right) \qquad \left(\pm 4 \frac{1}{2}\right).$$

K. The Accuracy of a Point-To-Point Trajectory

The accuracy of a point-to-point trajectory, e.g., the trajectory of an "intercontinental ballistic missile" (ICBM), is dependent upon the accuracy of the gravitational constants, of course. It is a matter of interest, accordingly, to investigate the differences that would result in an ICBM trajectory if the constants adopted herein were replaced by those of the international ellipsoid and international gravity formula:

$$f = 1/297$$

$$a_e = 6{,}378{,}388 \text{ meters}$$

$$g_e = 9.780{,}490 \text{ meters/sec}^2.$$

Substituting these values into eqs. (H 3, 4, 14), as at the end of Sec. H, we obtain

$$f' = 0, \qquad a' = +18.5 \times 10^{-6}, \qquad g' = +12.5 \times 10^{-6}.$$

Since $f' = 0$, we are happily able to simplify our developments by its omission in what follows, and especially to write, from eqs. (H 3, 11), using the equations involving megameters,

$$\frac{\Delta a_e}{a_e} = a', \qquad 2\frac{\Delta k_e}{k_e} = g' + 2a'.$$

Since the quantities ultimately sought are dimensionless, the following analysis would give the same result regardless of the unit of distance adopted, as may be shown by parallel developments.

Assuming that burnout is essentially at the surface of a spherical earth, and that the velocity at burnout, $\frac{ds}{dt}$, is correctly known in meters/sec, we have

$$r = a_e, \qquad \frac{\Delta r}{r} = \frac{\Delta a_e}{a_e} = a'$$

and

$$\dot{s} = \frac{1}{k_e}\frac{ds}{dt}, \qquad \frac{\Delta \dot{s}}{\dot{s}} = -\frac{\Delta k_e}{k_e} = -\frac{1}{2}(g' + 2a').$$

Likewise, the direction of takeoff being assumed known,

$$\frac{\Delta \dot{r}}{\dot{r}} = \frac{\Delta(r\dot{v})}{r\dot{v}} = \frac{\Delta \dot{s}}{\dot{s}} = -\frac{1}{2}(g' + 2a').$$

Then, the mass factor, μ, also being assumed to be correct,

$$p = r^2(r\dot{v})^2/\mu \qquad \frac{\Delta p}{p} = 2\frac{\Delta r}{r} + 2\frac{\Delta(r\dot{v})}{r\dot{v}} = -g'$$

and from

$$1 + e\cos v = \frac{p}{r} \qquad e\sin v = \dot{r}\sqrt{p/\mu}$$

we obtain

$$e\Delta v = +\sin v\,(g' + a').$$

Since, for a point-to-point trajectory on a spherical earth, the distance in radians,

$$d = 2\pi - 2v$$

we have

$$\Delta d = -2\Delta v = -2e^{-1}\sin v\,(g' + a').$$

The uncertainty in the end point will also be affected by the uncertainty in the time of flight, however, because of the rotation of the earth under the trajectory. Accordingly, we need

$$\frac{\dot{s}^2}{\mu} = \frac{2}{r} - \frac{1}{a} \qquad \frac{\Delta a}{a} = \frac{2a\,\dot{s}^2}{\mu}\,\frac{\Delta\dot{s}}{\dot{s}} + \frac{2a}{r}\,\frac{\Delta r}{r}$$

$$= -\left(\frac{2a}{r} - 1\right) g' - \left(\frac{2a}{r} - 2\right) a'$$

$$n = \frac{k_e\sqrt{\mu}}{a^{3/2}} \qquad \frac{\Delta n}{n} = \frac{\Delta k_e}{k_e} - \frac{3}{2}\,\frac{\Delta a}{a}$$

$$= \left(\frac{3a}{r} - 1\right) g' + \left(\frac{3a}{r} - 2\right) a' .$$

Then from

$$1 - e\cos E = r/a\,, \qquad e\sin E = r\dot{r}/\sqrt{\mu a}$$

we obtain

$$\Delta e\cos E - e\,\Delta E\sin E = \frac{ra}{a}\left(\frac{\Delta a}{a} - \frac{\Delta r}{r}\right) = -\left(2 - \frac{r}{a}\right)(g' + a')$$

$$\Delta e\sin E + e\,\Delta E\cos E = e\sin E\left(\frac{\Delta r}{r} + \frac{\Delta\dot{r}}{\dot{r}} - \frac{1}{2}\,\frac{\Delta a}{a}\right) = e\sin E\left(\frac{a}{r} - 1\right)(g' + a')$$

and

$$e\,\Delta E = +\frac{a}{r}\sin E\,(g' + a')\,, \qquad \Delta e = -\frac{a}{r}(1 - e^2)\cos E\,(g' + a') .$$

From KEPLER's equation,

$$M = E - e\sin E\,, \qquad \Delta M = \frac{r}{a}\Delta E - \Delta e\sin E$$

or

$$2\,\Delta M = 2\,\frac{\sin E}{e}\left(\frac{1 - e^3\cos E}{1 - e\cos E}\right)(g' + a') .$$

This expression is useful because the time of flight,

$$t_{BE} = \frac{2\pi - 2M}{n}$$

and

$$\Delta t_{BE} = -\frac{2\,\Delta M}{n} - t_{BE}\,\frac{\Delta n}{n} .$$

For numerical example, let us take a 90° minimum velocity ICBM trajectory, for which

$$a = +0.854 \qquad n = +0.0943 \text{ radians/min}$$

$$e = +0.414 \qquad \frac{3a}{r} = +2.562$$

$$r = 1.000$$

$$\sin v = +0.707 \qquad 2\,e^{-1}\sin v = +3.41$$

$$\sin E = +0.910 \qquad 2\,e^{-1}\sin E = +4.39$$

$$\cos E = -0.414$$

$$t_{BE} = 32.2 \text{ minutes} \qquad 2\,\frac{\sin E}{e}\left(\frac{1 - e^3\cos E}{1 - e\cos E}\right) = +3.86 .$$

Then, given

$$a' = +18.5 \times 10^{-6} \qquad g' = +12.5 \times 10^{-6}$$

$$\left(\frac{3a}{r} - 1\right) g' + \left(\frac{3a}{r} - 2\right) a' = \frac{\Delta n}{n} = +29.9 \times 10^{-6}$$

$$2 \frac{\sin E}{e} \left(\frac{1 - e^3 \cos E}{1 - e \cos E}\right) (g' + a') = +2\,\Delta M = +119.7 \times 10^{-6}$$

$$-\frac{2\,\Delta M}{n} - t_{BE} \frac{\Delta n}{n} = \Delta t_{BE} = -0.002232 \text{ min.}$$

The maximum effect of Δt_{BE}, through the rotation of the earth, is 204 feet, to be added to or subtracted from

$$-2\, e^{-1} \sin v\, (g' + a') = \Delta d = -105.7 \times 10^{-6} \text{ radians}$$
$$= -2212 \text{ feet}.$$

Since the effect of the negative Δt_{BE} will be to bring the rocket down early, tending to make it land east of its target, the correction from the adopted set of constants to those of the international formulae would be about 2000 feet for a rocket launched eastward and about 2400 feet for a rocket launched westward, on the equator. In these two circumstances, the uncertainties that we calculate from $a' = \pm 10 \times 10^{-6}$, $g' = \pm 3 \times 10^{-6}$ are $\pm$ 700 and $\pm$ 800, so that the refinement of the international values to our adopted values is justified.

L. The Accuracy of Satellite Positions

The accuracy of satellite positions is difficult to assess with any generality. Simplifying assumptions can be made: (a) that the orbit is nearly circular, (b) that the period, P, is determined by observations with relatively negligible error, (c) that the perturbations are accurately known with the exception of the coefficient J as it affects the mean motion, (d) that the corrections of the observer's position to the spheroid are accurately known, and (e) that the observed object is nearby, or overhead, so that its geocentric coordinates are approximately proportional to those of the observer. Such a model will provide at least a reference frame for more complicated ones. We cannot make the assumption that the earth is spherical, however, because $f' = f - 1/297$ is critical to all of our discussions of accuracy thus far, and it cannot be carried in the part if not in the whole.

We may think of geocentric positions first as referred to a mean orbit plane and specified by

$$x = a \cos b \cos n\,(t - t_o),\; y = a \cos b \sin n\,(t - t_o),\; z = a \sin b\,.$$

The x-axis and the epoch t_o are not yet specified, b is the latitude and is known accurately according to assumption (c), n is the mean angular motion and is known accurately in accordance with assumption (b) through

$$n = 2\pi/P$$

and a, the semi-major axis, is determined by

$$a^3 = k_e^2 \mu / n^2\,. \tag{L 1}$$

Then

$$\frac{\Delta x}{x} = \frac{\Delta y}{y} = \frac{\Delta z}{z} = \frac{\Delta a}{a} = \frac{2}{3}\frac{\Delta k_e}{k_e} + \frac{1}{3}\frac{\Delta \mu}{\mu}\,. \tag{L 2}$$

It is evident that we may generalize x, y, z to any axes.

With $m_1 = m_\oplus = 1$ and $m_2 = 0$ we may specify μ, inclusive of the principal perturbation term for a close satellite, by

$$\mu = 1 + c^2 J\left(1 - \frac{3}{2}\sin^2 i\right) \qquad c = \frac{a_e}{a} \tag{L 3}$$

where a_e is the earth's equatorial radius and i is the inclination of the orbit plane to the equator plane. Then, remembering that $\Delta J = f'$,

$$\frac{\Delta\mu}{\mu} = c^2 f'\left(1 - \frac{3}{2}\sin^2 i\right) \tag{L 4}$$

and, with (Sec. H)

megameters	*g-radii*
$2\frac{\Delta k_e}{k_e} = g' + 2a' + 2f'$	$2\frac{\Delta k_e}{k_e} = 0$

we have

$i = 0°$:	$\frac{\Delta a}{a} = \frac{1}{3}[g' + 2a' + (2 + c^2) f']$	$\frac{\Delta a}{a} = \frac{1}{3} c^2 f'$
$i = 35\frac{1}{4}°$:	$\frac{\Delta a}{a} = \frac{1}{3}\left[g' + 2a' + \left(2 + \frac{1}{2} c^2\right) f'\right]$	$\frac{\Delta a}{a} = \frac{1}{6} c^2 f'$
$i = 54\frac{3}{4}°$:	$\frac{\Delta a}{a} = \frac{1}{3}[g' + 2a' + 2f']$	$\frac{\Delta a}{a} = 0$
$i = 90°$	$\frac{\Delta a}{a} = \frac{1}{3}\left[g' + 2a' + \left(2 - \frac{1}{2} c^2\right) f'\right]$	$\frac{\Delta a}{a} = -\frac{1}{6} c^2 f'$.

For the geocentric position of the observer we may use, deferring to an astronomical convention on notation and sign to avoid confusion between observer and object [cf. eq. (L 5)].

$$-X = a_e \cos\varphi \cos\lambda\,(1 + f\sin^2\varphi)$$
$$-Y = a_e \cos\varphi \sin\lambda\,(1 + f\sin^2\varphi)$$
$$-Z = a_e \sin\varphi\,(1 + f\sin^2\varphi - 2f)$$

where φ is the latitude and λ is the longitude (here measured eastward in the geodetic convention, but with no necessary specification as to its origin). In accord with assumption (d) above, φ and λ are treated as accurately known. O'KEEFE points out that arc distances are more accurately known than φ and λ, but the lack of a round-the-globe tie-in complicates developments without adding accuracy or enlightenment beyond those of the simpler assumption (d). Then

$$\frac{\Delta X}{X} = \frac{\Delta Y}{Y} = \frac{\Delta a_e}{a_e} + \Delta f \sin^2\varphi$$
$$\frac{\Delta Z}{Z} = \frac{\Delta a_e}{a_e} + \Delta f(\sin^2\varphi - 2)$$

and with [cf. eq. (H 3)]

megameters	*g-radii*
$\frac{\Delta a_e}{a_e} = a' + \frac{4}{3} f'$	$\frac{\Delta a_e}{a_e} = \frac{1}{3} a' - \frac{1}{3} g' + \frac{2}{3} f'$

we have, noting that $\Delta f = f'$,

$$\frac{\Delta X}{X} = a' + f'\left(\sin^2\varphi + \frac{4}{3}\right) \qquad \frac{\Delta X}{X} = \frac{1}{3}a' - \frac{1}{3}g' + f'\left(\sin^2\varphi + \frac{2}{3}\right)$$

$$\frac{\Delta Z}{Z} = a' + f'\left(\sin^2\varphi - \frac{2}{3}\right) \qquad \frac{\Delta Z}{Z} = \frac{1}{3}a' - \frac{1}{3}g' + f'\left(\sin^2\varphi - \frac{4}{3}\right).$$

For the position of the observed object referred to the observer we may use

$$\xi = x + X \qquad \eta = y + Y \qquad \zeta = z + Z. \tag{L 5}$$

Remembering that in difference equations we may set, by assumption (e),

$$c = \frac{a_e}{a} = -\frac{X}{x} = -\frac{Y}{y} = -\frac{Z}{z}$$

we have

$$\Delta\xi = \Delta x + \Delta X = x\left[\frac{\Delta a}{a} - c\frac{\Delta X}{X}\right]$$

$$\Delta\eta = \Delta y + \Delta Y = y\left[\frac{\Delta a}{a} - c\frac{\Delta Y}{Y}\right]$$

$$\Delta\zeta = \Delta z + \Delta Z = z\left[\frac{\Delta a}{a} - c\frac{\Delta Z}{Z}\right].$$

The various combinations of c, φ, and i are too numerous to summarize; as a sample we set

$$c = \frac{a_e}{a} = 1, \qquad \varphi = i = 54\frac{3^\circ}{4}, \qquad \sin^2\varphi = \frac{2}{3}$$

to obtain, for both meters and g-radii,

$$\frac{\Delta\xi}{x} = \frac{\Delta\eta}{y} = \frac{1}{3}[g' - a' - 4f'] = \pm 6.4 \times 10^{-6}$$

$$\frac{\Delta\zeta}{z} = \frac{1}{3}[g' - a' + 2f'] = \pm 4.4 \times 10^{-6}$$

wherein the numerical evaluations are for $a' = \pm 10 \times 10^{-6}$, $g' = \pm 3 \times 10^{-6}$, $f' = \pm 4 \times 10^{-6}$; so also in what follows.

With a satellite at 2000 miles above the earth's surface, however,

$$c = \frac{a_e}{a} = \frac{2}{3}$$

and with the same values of φ and i,

megameters	*g-radii*
$\frac{\Delta\xi}{x} = \frac{1}{3}[g' - 2f']$	$\frac{\Delta\xi}{x} = \frac{2}{9}[g' - a' - 4f']$
$= \pm 2.9 \times 10^{-6}$	$= \pm 4.2 \times 10^{-6}$
$\frac{\Delta\zeta}{z} = \frac{1}{3}[g' + 2f']$	$\frac{\Delta\zeta}{z} = \frac{2}{9}[g' - a' + 2f']$
$= \pm 2.9 \times 10^{-6}$	$= \pm 2.9 \times 10^{-6}.$

With a satellite at 8000 miles above the earth's surface,

$$c = \frac{a_e}{a} = \frac{1}{3}$$

and with the same values of φ and i,

megameters	*g-radii*
$\frac{\Delta\xi}{x} = \frac{1}{3}[g' + a']$	$\frac{\Delta\xi}{x} = \frac{1}{9}[g' - a' - 4f']$
$= \pm 3.5 \times 10^{-6}$	$= \pm 2.1 \times 10^{-6}$
$\frac{\Delta\zeta}{z} = \frac{1}{3}[g' + a' + 2f']$	$\frac{\Delta\zeta}{z} = \frac{1}{9}[g' - a' + 2f']$
$= \pm 4.4 \times 10^{-6}$	$= \pm 1.5 \times 10^{-6}$.

M. Determinations of k_e from Radar Distances

Determinations of k_e from radar distances will doubtless be possible in the not-too-distant future. To investigate the potentialities of such determinations let us assume, as in Sec. L, that the orbit is circular, with the period, P, and mean motion, n, determined beyond needed accuracy. Then, from eqs. (L 1—4)

$$k_e^2 = n^2 a^3/\mu \qquad \text{(M 1)}$$

$$2\frac{\Delta k_e}{k_e} = 3\frac{\Delta a}{a} - \frac{\Delta\mu}{\mu} \qquad \text{(M 2)}$$

$$\frac{\Delta\mu}{\mu} = c^2 f'\left(1 - \frac{3}{2}\sin^2 i\right) \qquad c = a_e/a\,. \qquad \text{(M 3, 4)}$$

Let us further assume that the object is "overhead" (strictly on the line through the observer and the center of the earth) so that eqs. (L 5) reduce to

$$a = \varrho + R \qquad \frac{\Delta a}{a} = \frac{\varrho}{a}\frac{\Delta\varrho}{\varrho} + \frac{R}{a}\frac{\Delta R}{R}$$

where R is the observer's geocentric distance, given by eq. (D 3), and ϱ is the observed radar distance or "range". Both ϱ and R we shall suppose to be expressed in megameters, their uncertainties being expressed as

$$\frac{\Delta\varrho}{\varrho} = \varrho'$$

and, from eq. (F 5),

$$\frac{\Delta R}{R} = a' + f'\left(\frac{4}{3} - \sin^2\varphi\right). \qquad \text{(M 5)}$$

In difference equations let us further set

$$\frac{\varrho}{a} = b, \frac{R}{a} = \frac{a_e}{a} = c\,. \qquad \text{(M 6, 7)}$$

Then we may write,

$$\frac{\Delta a}{a} = b\varrho' + ca' + cf'\left(\frac{4}{3} - \sin^2\varphi\right)$$

$$2\frac{\Delta k_e}{k_e} = 3b\varrho' + 3ca' + cf'\left[4 - 3\sin^2\varphi - c\left(1 - \frac{3}{2}\sin^2 i\right)\right]$$

wherein we may take $a' = 0 \pm 10 \times 10^{-6}$, $f' = 0 \pm 4 \times 10^{-6}$, in accordance with eqs. (F 4), (G 9).

If additionally, for an artificial satellite, we set $\varphi = i = 54\frac{3}{4}^\circ$, $b = 0.1$, $c = 0.9$, and $\varrho' = \pm 12 \times 10^{-6}$, corresponding to ± 25 feet,

$$2\frac{\Delta k_e}{k_e} = 3b\,\varrho' + 3ca' + 2cf'$$
$$= (\pm 3.6 \pm 27 \pm 7.2) \times 10^{-6}$$
$$= \pm 28.1 \times 10^{-6}.$$

It is evident that the uncertainty of the earth's equatorial radius makes it desirable to utilize a satellite whose mean distance, a, is larger, so that c and the effect of a' will be smaller.

Turning to the moon, accordingly, as an approximation we set $\varphi = i = 0°$, $b = 1$, $c = 1/60$, and $\varrho' = \pm 1.3 \times 10^{-6}$, corresponding to ± 500 meters, and replace eq. (M 3) by eq. (J 4) or

$$\frac{\Delta\mu}{\mu} = \mu' = \pm 3.9 \times 10^{-6}$$

$$2\frac{\Delta k_e}{k_e} = 3b\,\varrho' + 3ca' + 4cf' - \mu'$$
$$= (\pm 3.9 \pm 0.5 \pm 0.3 \pm 3.9) \times 10^{-6}$$
$$= \pm 5.5 \times 10^{-6}.$$

If radar observations of the moon determine its distance to $\pm$ 500 meters, accordingly, it promises to yield the best determination of k_e for the meter as unit of distance.

N. The Parallax of a Near Satellite

The parallax of a near satellite yields a determination of its distance and thence k_e. Again assuming that the orbit is circular, we adopt eqs. (M 1 — 4). Let us also assume that we have simultaneous symmetrical observations from two stations at the same distance, R, from the center of the earth, and in one plane with the object observed. Then, for altitude h and parallax p, we have (Fig. 1)

$$a = R \cos h / \sin p\,.$$

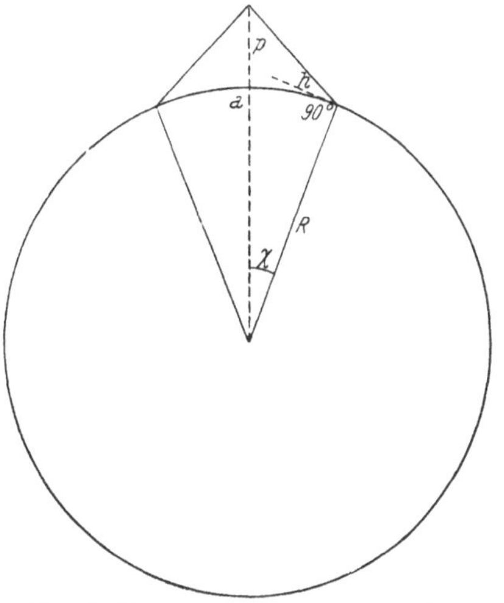

Fig. 1. The parallax of a near satellite

Let us further assume that the angle between the two stations, 2χ, is known accurately. O'Keefe points out that it is better to assume that the distance between the two stations is known accurately — but this assumption complicates the discussion without greatly altering the result. With the simpler assumption, then

$$\Delta p = -\Delta h \equiv -h'$$

and

$$\frac{\Delta a}{a} = \frac{\Delta R}{R} + h'(\operatorname{ctn} p - \tan h).$$

One may show that the coefficient of h' in this equation is a minimum when [cf. eq. (M 6)]

$$p = h \qquad \tan p = \tan h = \frac{R}{a} = \frac{a_e}{a} = c.$$

Then

$$\frac{\Delta a}{a} = \frac{\Delta R}{R} + h' \frac{1-c^2}{c}$$

and from eqs. (M 2, 3, 5), we have, for megameters

$$2\frac{\Delta k_e}{k_e} = 3h' \frac{1-c^2}{c} + 3a' + f'\left[4 - 3\sin^2\varphi - c^2\left(1 - \frac{3}{2}\sin^2 i\right)\right].$$

If $\varphi = i = 54\frac{3^\circ}{4}$, $c = 0.9$, $h' = \pm 10 \times 10^{-6}$, corresponding to $\pm 2''$, and, as before (Sec. M), $a' = 0 \pm 10 \times 10^{-6}$, $f' = 0 \pm 4 \times 10^{-6}$, we have

$$\begin{aligned} 2\frac{\Delta k_e}{k_e} &= 0.63\,h' + 3a + 2f' \\ &= \pm 32 \times 10^{-6}. \end{aligned}$$

P. Conversion Factors

Conversion factors may be computed usefully for any adopted set of units and constants. Some of them are the units themselves. We shall include conversion factors in Table IV for three systems:

(1) The heliocentric or Gaussian system based upon the astronomical unit, the day, and from Sec. B:

$$k_s = 0.017{,}202{,}09895 \text{ (a.u.)}^{3/2}/\text{day}$$

(2) The geocentric system based upon the radius, the minute, and from Sec. H:

$$k_e = 0.074{,}365{,}74 \text{ radii}^{3/2}/\text{min}.$$

(3) The geocentric system based upon the megameter, the minute, and the corresponding value of k_e from Sec. H that we shall now designate k_f:

$$k_f = 1.197{,}9185 \text{ megameters}^{3/2}/\text{min}.$$

Systems (2) and (3) may be converted from the minute to the second as basic unit by the following two changes only: divide k by 60; divide unit of t by 60.

In Table IV the foot and the mile are based upon the probable United States and Great Britain compromise standard[1]:

1 foot = 0.3048 international meters
1 mile = 1609.344 international meters

In addition we adopt the definitions

1 international nautical mile = 1852 international meters
1 megameter = 10^6 international meters

---

[1] Present U.S. standard: 1 foot = 0.304,800,61 international meters (approx.)
Present British standard: 1 foot = 0.304,799,73 international meters (approx.)

Q. The Adopted Units of Mass, Length, and Time

The adopted units of mass, length, and time are expressed in terms of other such units in the tabulations of Sec. P. The tabulated quantities will serve as conversion factors, of course, from one of the adopted systems to another, or to many other systems.

The adopted units of mass, $m_{\odot}$ and $m_{\oplus}$, are expressed in terms of one another, for use with perturbations or for transitions from geocentric to heliocentric orbits and vice versa. They are also given in grams, but more for descriptive than actual mechanical uses.

The adopted units of length, the "astronomical unit," the "radius," and the "megameter" are expressed in terms of one another and of the kilometer, the mile, the nautical mile, and the foot.

The adopted units of time, the day or the minute, must be used in any equation in which k appears, but other units of time may be used with the other constants (cf. Sec. S).

The "unit of τ" is k^{-1} units of t. For $t - t_o$, a time interval measured from an arbitrary epoch, t_o, in days or minutes,

$$\tau = k\,(t - t_o),$$

but for other units of time use the values in the table.

R. The Unit of Velocity

The unit of velocity, V_{co}, is used to convert velocities, $\dot{s}$, $\dot{x}$, etc., expressed in the adopted units, to velocities, ds/dt, dx/dt, etc., expressed in any desired units, and vice versa. To understand how it is used, and how the values in Sec. P are derived, we turn to the *vis-viva* integral (cf. Sec. A),

$$V = \frac{ds}{dt} = k\dot{s} = k\sqrt{\mu\left(\frac{2}{r} - \frac{1}{a}\right)}\,. \tag{R 1}$$

For V_c, the velocity in a circular orbit or circular velocity, $a = r$; for V_p, the parabolic velocity or "velocity of escape," $a = \infty$; for V_∞, the "velocity at infinity," real in parabolic and hyperbolic orbits but imaginary in elliptic ones, $r = \infty$; and

$$\begin{aligned} V_c &= k\sqrt{\mu/r} \\ V_p &= k\sqrt{2\mu/r} \\ V_\infty &= k\sqrt{-\mu/a}\,. \end{aligned} \tag{R 2}$$

When, additionally, we set $\mu = 1$, $r = 1$, we obtain V_{co} and V_{po}, the circular and parabolic velocities at unit distance (and for unit mass),

$$V_{co} = k \qquad V_{po} = k\sqrt{2}\,.$$

These quantities may now be substituted into eqs. (R 1, 2) and into similar equations to yield such forms as

$$\begin{aligned} &V = V_{co}\dot{s} = V_{co}\sqrt{\mu\left(\frac{2}{r} - \frac{1}{a}\right)} = V_{po}\sqrt{\mu\left(\frac{1}{r} - \frac{1}{2a}\right)} \\ &V_x = V_{co}\dot{x}, \qquad V_y = V_{co}\dot{y}, \qquad V_z = V_{co}\dot{z} \\ &V_c = V_{co}\sqrt{\mu/r}\,, \qquad V_p = V_{po}\sqrt{\mu/r}\,, \qquad V_\infty = V_{co}\sqrt{-\mu/a}\,. \end{aligned} \tag{R 3}$$

Table IV. *Conversion Factors*

System	Heliocentric "k_s"	Geocentric "k_e"	Geocentric "k_f"	
k	0.017,202,098,95	0.074,365,74	1.197,9185	
k^2	0.000,295,912,2083	0.005,530,2633	1.435,0087	
Unit of mass	1	$3.007,63 \times 10^{-6}$	$3.007,63 \times 10^{-6}$	$m_{\odot}$
	$0.332,488 \times 10^6$	1	1	$m_{\oplus}$
	1.987×10^{33}	5.975×10^{27}	5.975×10^{27}	grams
Unit of length	1	42.644×10^{-6}	6.6889×10^{-6}	a. u.
	23.439×10^3	1	0.156,7823	radii
	149.50×10^3	6.378,270	1	megameters
	149.50×10^6	$6.378,270 \times 10^3$	1×10^3	kilometers
	92.90×10^6	$3.963,273 \times 10^3$	621.371,19	miles
	80.72×10^6	$3.443,990 \times 10^3$	539.956,80	int. naut. miles
	490.49×10^9	$20.926,083 \times 10^6$	$3.280,840 \times 10^6$	feet
Unit of t	1	$0.694,444 \times 10^{-3}$	$0.694,444 \times 10^{-3}$	days
	1.440×10^3	1	1	minutes
	86.400×10^3	60	60	seconds
	0.017,202,098,95	$11.945,902 \times 10^{-6}$	$11.945,902 \times 10^{-6}$	k_s^{-1} day
	107.086,67	0.074,365,74	0.074,365,74	k_e^{-1} min
	$1.725,0026 \times 10^3$	1.197,9185	1.197,9185	k_f^{-1} min
Unit of $\tau = k^{-1}$ unit of t	1	$160.637,17 \times 10^{-6}$	$9.972,216 \times 10^{-6}$	k_s^{-1} day
	$6.225,209 \times 10^3$	1	0.062,079,13	k_e^{-1} min
	$100.278,61 \times 10^3$	16.108,473	1	k_f^{-1} min
	58.132,440,87	0.009,338,231	$0.579,7093 \times 10^{-3}$	days
	1395.178,581	0.224,117,54	0.013,913,022	hours
	$83.710,715 \times 10^3$	13.447,052	0.834,7813	minutes
	$5.022,6429 \times 10^6$	806.823,14	50.086,880	seconds

V_{co} = unit of velocity	1	0.265,590	0.670,754	a.u./k_s^{-1} day
= circular velocity at	3.765,21	1	2.525,5238	radii/k_e^{-1} min
unit distance	1.490,86	0.395,957,46	1	megameters/k_f^{-1} min
$= \dfrac{\text{unit of distance}}{\text{unit of } \tau}$	0.017,202,098,95	4.5687×10^{-3}	0.011,5384	a.u./day
	0.280,00	0.074,365,74	0.187,812,45	radii/min
$= \dfrac{k \text{ unit of distance}}{\text{unit of } t}$	4.6667×10^{-3}	$1.239,4290 \times 10^{-3}$	$3.130,2075 \times 10^{-3}$	radii/sec
	1.785,93	0.474,3248	1.197,9185	megameters/min
	29.766	7.905	19.965,308	kilometers/sec
	18.495	4.912	12.406	miles/sec
	66.584×10^3	17.684×10^3	44.661×10^3	miles/hour
	57.860×10^3	15.367×10^3	38.809×10^3	knots
	97.656×10^3	25.936×10^3	65.503×10^3	feet/sec
$V_{po} = V_{co}\sqrt{2}$	42.095	11.180	28.235	kilometers/sec
= parabolic	26.156	6.947	17.545	miles/sec
velocity	94.163×10^3	25.009×10^3	63.160×10^3	miles/hour
at unit	81.826×10^3	21.732×10^3	54.885×10^3	knots
distance	138.106×10^3	36.680×10^3	92.635×10^3	feet/sec
n_{co} = unit of angular velocity	1	$6.225,209 \times 10^3$	$100.278,61 \times 10^3$	radians/k_s^{-1} day
	$0.160,637,17 \times 10^{-3}$	1	16.108,473	radians/k_e^{-1} min
	$9.972,216 \times 10^{-6}$	0.062,079,13	1	radians/k_f^{-1} min
	0.017,202,098,95	107.086,67	$1.725,0026 \times 10^3$	radians/day
	0.000,716,754,123	4.461,9444	71.875,110	radians/hour
	$11.945,902 \times 10^{-6}$	0.074,365,74	1.197,9185	radians/min
	$199.098,37 \times 10^{-9}$	$1.239,4290 \times 10^{-3}$	$19.965,308 \times 10^{-3}$	radians/sec
"k^0" {	0.985,607,6685	$6.135,6140 \times 10^3$	$98.835,371 \times 10^3$	degrees/day
	0.041,066,986	255.650,58	$4.118,1405 \times 10^3$	degrees/hour
	$684.449,77 \times 10^{-6}$	4.260,8430	68.635,674	degrees/min
	$11.407,496 \times 10^{-6}$	$71.014,051 \times 10^{-3}$	1.143,9279	degrees/sec

Table IV. *Conversion Factors (Continued)*

System	Heliocentric "k_s"	Geocentric "k_e"	Geocentric "k_f"	
$A_o =$ unit of	1	$1.653,35 \times 10^3$	67.263×10^3	a.u./$(k_s^{-1}$ day$)^2$
acceleration	604.83×10^{-6}	1	40.682,332	radii/$(k_e^{-1}$ min$)^2$
= acceleration at	14.867×10^{-6}	0.024,580,70	1	megameters/$(k_f^{-1}$ min$)^2$
unit distance	$295.912,2083 \times 10^{-6}$	489.25×10^{-3}	19.904	a.u./(day)^2
$= \dfrac{\text{unit of distance}}{(\text{unit of } \tau)^2}$	3.3449×10^{-6}	$5.530,2633 \times 10^{-3}$	0.224,984,01	radii/(min)^2
$= \dfrac{k^2 \text{ unit of distance}}{(\text{unit of } t)^2}$	929.13×10^{-12}	$1.536,1843 \times 10^{-6}$	$62.495,556 \times 10^{-6}$	radii/(sec)^2
	21.335×10^{-6}	0.035,273,51	1.435,0087	megameters/(min)^2
	0.005,854	9.798,1979	398.613,53	meters/(sec)^2
	0.019,443	32.146,319	$1.307,78 \times 10^3$	feet/(sec)^2
$P_o =$ unit of period	6.283,1853	$1.009,3131 \times 10^{-3}$	$62.657,281 \times 10^{-6}$	k_s^{-1} day
$= 2\pi$ units of τ	39.114,143	6.283,1853	0.390,054,68	k_e^{-1} min
$= 2\dfrac{\pi}{k}$ units of t	$630.069,11 \times 10^3$	101.212,52	6.283,1853	k_f^{-1} min
	1	$160.637,17 \times 10^{-6}$	$9.972,216 \times 10^{-6}$	"GAUSSIAN" year
	1.000,0402	$160.643,64 \times 10^{-6}$	$9.972,618 \times 10^{-6}$	tropical year
	1.000,0015	$160.637,41 \times 10^{-6}$	$9.972,231 \times 10^{-6}$	sidereal year
	365.256,898,3263	0.058,673,835	$3.642,4207 \times 10^{-3}$	days
	$8.766,165,556 \times 10^3$	1.408,1720	0.087,418,097	hours
	$525.969,9335 \times 10^3$	84.490,322	5.245,0858	minutes
	$31.558,196,01 \times 10^6$	$5.069,4193 \times 10^3$	314.705,15	seconds

Thus the quantities V_{co} and V_{po}, expressed in many different units in Sec. P, now serve clearly as factors for converting velocities such as $\dot{s}, \dot{x}, \dot{y}, \dot{z}$ from adopted unit of distance per unit of τ to any other unit of distance per any unit of time. It will be noted that the quantities r and a must be expressed in the adopted unit of distance in eq. (R 3); but in the following derived equations any units whatsoever may be used:

$$V_\infty = V_c \sqrt{-r/a} = V_p \sqrt{-r/2a}$$
$$V^2 = V_p^2 + V_\infty^2.$$

For an object of negligible mass, orbiting unperturbed around the sun or the earth, $\mu = 1$, provided the appropriate system of units and constants is employed. When another planet or satellite is the central body, μ is its mass expressed in units of the sun's mass or of the earth's mass, depending upon the system used. For example, the velocities of escape from the surfaces of the planets and satellites may be calculated from the values of V_{po} in the geocentric k_e column of the table in Sec. P, provided μ is the mass of the object expressed in units of the earth's mass, and provided r is the radius of the planet expressed in units of the earth's radius. The values of V_{po} in the heliocentric k_s column may be used, on the other hand, if μ is in units of the sun's mass and r is in astronomical units.

S. The Unit of Angular Velocity

The unit of angular velocity, n_{co}, as given in Sec. P, is the mean angular velocity or "mean motion" of a hypothetical object whose mean distance, a, equals the adopted unit of distance. It provides for the expression of the mean anomaly, M, or mean longitude, L, in terms of non-radian angular measure:

$$n = n_{co} \sqrt{\mu} / a^{3/2} \tag{S 1}$$
$$M = M_o + n\,(t - t_o) = n\,(t - T) \tag{S 2}$$
$$L = L_o + n\,(t - t_o)\,. \tag{S 3}$$

It will be noted that $n_{co} = k_s$ when the units are radians and days, to be used in eq. (S 1) with mean distances expressed on astronomical units; $n_{co} = k_e$ for radians and minutes of time, with a in radii; $n_{co} = k_f$ for radians and minutes of time, with a in megameters. When M or L is to be expressed in degrees, $n_{co} = 57°.295,7795\, k$ is sometimes called $k°$; similarly k'' has been used in connection with seconds of arc.

T. The Unit of Acceleration

The unit of acceleration, A_o, converts accelerations $\ddot{s}$, $\ddot{x}$, etc., expressed in the adopted system of units, into accelerations d^2s/dt^2, d^2x/dt^2, etc., expressed in some conventional system of units, and vice versa:

$$\frac{d^2s}{dt^2} = A_o \ddot{s}, \qquad \frac{d^2x}{dt^2} = A_o \ddot{x}, \text{ etc.}$$

For gravitational acceleration

$$A_G = A_o \mu / r^2, \qquad \frac{d^2x}{dt^2} = -A_o \mu x / r^3, \text{ etc.}$$

A_o is thus the acceleration of gravity at unit distance, for the usual $\mu = 1$. An advantage of the "radius" as unit of distance, at least for descriptive purposes, is that A_o is one "g", if we adopt a precise definition of that term that is free of the effect of the earth's rotation, and is associated with the equatorial rather than the mean radius of the earth. The same A_o is equal to the acceleration of gravity on the rotating earth in latitude 36°, very near the mean latitude $35\frac{1}{4}^\circ$.

U. The Unit of Period

The unit of period, P_o, is the period of a hypothetical object of negligible mass whose mean distance equals the unit of distance. It is used to determine periods corresponding to other mean distances through

$$P = P_o a^{3/2}/\mu .$$

For the "radius" P_o is the well-known 84-minute period of a satellite just grazing the surface of the earth, or of the SCHULER pendulum.

Z. The De Sitter Formulae

The DE SITTER formulae, involving the mean radius and associated quantities (DE SITTER 1915, DE SITTER-BROUWER 1938), have been the basis of a number of important investigations. They are alternative to the formulae of Sec. D, although they have been used, not to obtain k_e as such, but to connect the acceleration of gravity with the parallax of the moon (cf. Sec. J), essentially through the elimination of k_e. We shall show here that neither the formulae nor the quantities used influence the accuracy of the determination of k_e, whether it is sought for its own sake or as a connecting link between geophysical and astronomical quantities. The comparison we shall extend to two other formulae, one used in substance by LAMBERT (1939), the other involving the acceleration of gravity at the pole.

It will be sufficient to consider terms involving only the first order in f, since the uncertainties in the remaining terms are negligible by comparison. From eqs. (D 3, 4, 5, 7) we may then write

$$\beta = -f + \frac{5}{2}\tilde{\omega} \qquad J = f - \frac{1}{2}\tilde{\omega}$$

$$R_1 = a_e\left(1 - \frac{1}{3}f\right) \qquad \sin^2\varphi_1 = 1/3$$

$$g_1 = g_e\left(1 + \frac{1}{3}\beta\right) \qquad \varphi_1 = 35\,\frac{1}{4}^\circ$$

$$R_3 = a_e(1 - f) \qquad \sin^2\varphi_3 = 1$$

$$g_3 = g_e(1 + \beta) \qquad \varphi_3 = 90^\circ$$

Eq. (D 1), which we shall term the "equatorial" equation, may then be abbreviated, and from it may be derived the "DE SITTER," the "LAMBERT," and the "polar" equations:

$$\text{Equatorial:}\quad k_e^2 = g_e\, a_e^2\,(1 + \tilde{\omega} - J) = g_e\, a_e^2\left(1 - f + \frac{3}{2}\tilde{\omega}\right)$$

$$\text{DE SITTER:}\quad k_e^2 = g_1 R_1^2\left(1 + \frac{2}{3}\tilde{\omega}\right)$$

$$\text{LAMBERT:}\quad k_e^2 = g_e R_1^3 a_e^{-1}\left(1 + \frac{3}{2}\tilde{\omega}\right)$$

$$\text{Polar:}\quad k_e^2 = g_3 R_3^2\,(1 + 2J) = g_3 R_3^2\,(1 + 2f - \tilde{\omega}) .$$

From these equations we derive by differentiation, omitting the negligible $\Delta\,\tilde{\omega}$ (see below),

$$\begin{aligned}
&\text{Equatorial:} && 2\frac{\Delta k_e}{k_e} = \frac{\Delta g_e}{g_e} + 2\frac{\Delta a_e}{a_e} - \Delta f \\
&\text{DE SITTER:} && 2\frac{\Delta k_e}{k_e} = \frac{\Delta g_1}{g_1} + 2\frac{\Delta R_1}{R_1} \\
&\text{LAMBERT:} && 2\frac{\Delta k_e}{k_e} = \frac{\Delta g_e}{g_e} + 3\frac{\Delta R_1}{R_1} - \frac{\Delta a_e}{a_e} \\
&\text{Polar:} && 2\frac{\Delta k_e}{k_e} = \frac{\Delta g_3}{g_3} + 2\frac{\Delta R_3}{R_3} + 2\,\Delta f\,.
\end{aligned} \tag{Z 1}$$

The seeming freedom of the DE SITTER and LAMBERT equations from

$$\Delta f = f'$$

is illusory. Restricting ourselves for the moment to the megameter as unit of distance, we take from Secs. E, F,

$$\begin{aligned}
&\frac{\Delta g_e}{g_e} = g' + \frac{1}{3}f' && \frac{\Delta g_1}{g_1} = g' && \frac{\Delta g_3}{g_3} = g' - \frac{2}{3}f' \\
&\frac{\Delta a_e}{a_e} = a' + \frac{4}{3}f' && \frac{\Delta R_1}{R_1} = a' + f' && \frac{\Delta R_3}{R_3} = a' + \frac{1}{3}f'\,.
\end{aligned} \tag{Z 2}$$

Substituting eqs. (Z 2) into eqs. (Z 1) *each* of the four forms yields [cf. eq. (H 13)]:

$$2\frac{\Delta k_e}{k_e} = g' + 2a' + 2f'.$$

When the process is followed through for the other units of distance considered in Sec. H, each of the four forms likewise gives the result that is appropriate to the adopted unit, as set down in eq. (H 13). It is evident, accordingly, that the adoption of the unit of distance affects the accuracy of the determination of k_e, but the choice of a formula, or of the quantities appearing in it, does not. This conclusion holds, by the way, whatever the coefficients of f' in eqs. (Z 2), provided those in the g-equations, and separately those in the a-equations, are consistent.

This discussion would not be complete without reference to the quantities that are alternative to the centrifugal acceleration coefficient [cf. eq. (D 2)],

$$\tilde{\omega} = a_e^3\,\omega^2/k_e^2\,.$$

These are $\tilde{\omega}_1$, preferred by DE SITTER, whose notation is ϱ_1, and defined by

$$\tilde{\omega}_1 = R_1^3\ \omega^2/k_e^2 = \tilde{\omega}\,(1 - f)$$

and a quantity often designated m, but which we shall call $\tilde{\omega}_g$,

$$\tilde{\omega}_g = a_e\ \omega^2/g_e = \tilde{\omega}\left(1 - f + \frac{3}{2}\,\omega\right).$$

This quantity is easy to calculate at the start, without successive approximations, but somewhat clutters eq. (D 1). To these we shall add, for comparison,

$$\tilde{\omega}_2 = R_2^3\,\omega^2/k_e^2 = \tilde{\omega}\,(1 - 2f).$$

These quantities are dimensionless and so it is no surprise that the various units of distance all give the same results:

$$\Delta\tilde{\omega} = \tilde{\omega}(a' - g' + 2f')$$

$$\Delta\tilde{\omega}_1 = \tilde{\omega}_1(a' - g' + f')$$

$$\Delta\tilde{\omega}_g = \tilde{\omega}_g(a' - g' + f')$$

$$\Delta\tilde{\omega}_2 = \tilde{\omega}_2(a' - g').$$

To the approximation necessary to these formulae,

$$\tilde{\omega} = \tilde{\omega}_1 = \tilde{\omega}_g = \tilde{\omega}_2 = 0.003$$

so that it is evident that any of these $\Delta\tilde{\omega}$'s is quite negligible by comparison with $\Delta f = f'$, and it is quite immaterial which $\tilde{\omega}$ is used in such formulae as eqs. (D 1, 5, 6, 7, 8). Thus, in the light of recent data, we fail to substantiate DE SITTER's preference for $\tilde{\omega}_1$ and the comments of JEFFREYS (1948, pp. 238—239). If preference should be given to any of the $\tilde{\omega}$'s in the score of accuracy, it should be given to $\tilde{\omega}_2$.

Acknowledgements

Acknowledgements have been made in the foregoing to the International Geophysical Year Satellite Program and Smithsonian Astrophysical Observatory for financial support. Other such support has been generously given by the University of California and by Mrs. DOT LEMON. Advice and criticism has been received from J. A. O'KEEFE of the Army Map Service, from Admiral ROBERT W. KNOX and his associates at the Coast and Geodetic Survey, and from Mrs. SILVIA R. MARCUS, who aided in the preparation of the manuscript. To Aeronutronic System, Inc., we are indebted for the manuscript assistance of Mrs. JOANNE CATE and Miss JEAN WILLIAMS.

References

R. T. BIRGE, Rep. Progress Physics **8**, 90 (1941).
D. BROUWER see W. DE SITTER and G. M. CLEMENCE items.
E. J. BROWN, U.S.Coast and Geodetic Survey Spec. Publ. 204 (1933).
E. W. BROWN, Mem. Roy. Astronom. Soc. **53**, 39 (1899).
E. C. BULLARD, Monthly Notices Roy. Astronom. Soc., Geophysic. Suppl. **5**, 186 (1948)
E. C. BULLARD and B. C. BROWNE, Proc. Roy. Soc. London, Ser. A, **175**, 110 (1940).
G. CASSINIS, Bull. Geod. **26**, 40 (1930).
B. CHOWITZ and IRENE FISCHER, Trans. Amer. Geophysic. Union **37**, 534 (1956); Abstract **37**, 339 (1956).
W. CHRISTIE and D. GILL, Monthly Notices Roy. Astronom. Soc. **71**, 526 (1911).
J. S. CLARK, Philos. Trans. Roy. Soc. London, Ser. A, **238**, 65 (1939).
A. R. CLARKE, Geodesy. Oxford, 1880.
G. M. CLEMENCE, Astronom. J. **53**, 169 (1948).
G. M. CLEMENCE and D. BROUWER, Astronom. J. **60**, 118 (1955).
A. H. COOK, Proc. Roy. Soc. London, Ser. A, **213**, 408 (1952).
W. DE SITTER, Proc. Kon. Akad. Wetensch. Amsterdam **27**, 1291 (1915).
W. DE SITTER, Bull. Astronom. Inst. Netherlands **2**, 97 (1924).
W. DE SITTER, Bull. Astronom. Inst. Netherlands **129**, 57 (1927).
W. DE SITTER and D. BROUWER, Bull. Astronom. Inst. Netherlands **8**, 213 (1938).
H. L. DRYDEN, J. Res. Nat. Bur. Standards **29**, 303 (1942).

G. D. GARLAND and A. H. COOK, Proc. Roy. Soc. London, Ser. A, **229**, 445 (1955).
J. F. HAYFORD, 1907, The Figure of the Earth and Isostasy. Washington, 1909.
J. F. HAYFORD, 1909, Supplementary Investigation of the Figure of the Earth and Isostasy. Washington, 1910.
W. HEISKANEN, Gerlands Beitr. Geophysik **19**, 356 (1928).
W. HEISKANEN, Publ. Finnish Geod. Inst. No. 12, (1929).
F. R. HELMERT, S.B. preuß. Akad. Wiss. **1911**, 10.
H. G. HERTZ, Astronom. Pap., Washington **15**, part 2 (1953); Astronom. J. **58**, 43 (1953) (Abstract).
P. R. HEYL, J. Res. Nat. Bur. Standards **5**, 1243 (1930).
P. R. HEYL and G. S. COOK, J. Res. Nat. Bur. Standards **17**, 805 (1936).
I. A. U.: Trans. Internat. Astronom. Union **6** (1938), **7** (1950), **8** (1952).
H. JEFFREYS, Monthly Notices Roy. Astronom. Soc. **97**, 3 (1936); Monthly Notices Roy. Astronom. Soc., Geophysic. Suppl. **4**, 1 (1936).
H. JEFFREYS, Monthly Notices Roy. Astronom. Soc., Geophysic. Suppl. **5**, 1 (1941).
H. JEFFREYS, Monthly Notices Roy. Astronom. Soc., Geophysic. Suppl. **101**, 34 (1941a).
H. JEFFREYS, Monthly Notices Roy. Astronom. Soc., Geophysic. Suppl. **5**, 55 (1943).
H. JEFFREYS, Monthly Notices Roy. Astronom. Soc., Geophysic. Suppl. **5**, 219 (1948).
H. JEFFREYS, Monthly Notices Roy. Astronom. Soc., Geophysic. Suppl. **5**, 398 (1949).
K. JUNG, Encycl. Physics **47**, 534 (1956).
T. N. KRASSOVSKY, Manuel de Géodésie Supérieure, part 2 (1942).
F. KÜHNEN and P. FURTWANGLER, Veröff. preuss. Geodat. Inst., N.F. No. 27 (1906).
W. D. LAMBERT, U.S.Coast and Geodetic Survey Spec. Publ. 100 (1924).
W. D. LAMBERT, Astronom. J. **38**, 181 (1928).
W. D. LAMBERT, Trans. Amer. Geophysic. Union **12**, 40 (1931).
W. D. LAMBERT, Physics of the Earth (New York) **7**, 329 (1939).
W. D. LAMBERT, Amer. J. Sci. **243**-A, 360 (1945).
K. LEDERSTEGER, Österr. Z. Vermessungswesen, Sonderheft 12 (1951).
F. R. MOULTON, Celestial Mechanics, 2nd ed. New York, 1914.
S. NEWCOMB, Astronom. Pap., Washington **6**, 7 (1898).
S. NEWCOMB, Astronom. Pap., Washington **9**, 1 (1912).
I. NEWTON, Principia. 1687.
J. A. O'KEEFE and J. P. ANDERSON, Astronom. J. **57**, 108 (1952).
G. PERRIER, Bull. Géod. **7**, 552 (1925).
E. RABE, Astronom. J. **55**, 112 (1950).
RDS 1926, 1927, 1938, 1945 = H. N. RUSSEL, R. S. DUGAN and J. Q. STEWART, Astronomy. New York.
D. H. SADLER, Occ. Notes Roy. Astronom. Soc. **3**, 103 (1954).
E. SCHMID, Bull. Géod. **30**, 412 (1953).
H. SPENCER JONES, Mem. Roy. Astronom. Soc. **66**, part 2 (1941).
H. SPENCER JONES, The Earth as a Planet. Ed. G. P. KUIPER, p. 1—41. Chicago, 1954.
H. SPENCER JONES, Encycl. Physics **47**, 1 (1956).
U.S. Coast and Geodetic Survey Spec. Publ. 5.
M. S. ZVEREV, Bull. Astronom. **15**, 243 (1950).

Alterations in Some Blood Reactions and in the White Cell Count during the Total Eclipse of the Sun in Poland 1954

By

J. Kaulbersz[1], **R. Bilski, I. Kocyan, A. Ogiński, D. Wiecha,** and **J. Zbiegień**

(With 5 Figures)

Abstract — Zusammenfassung — Résumé

Alterations in Some Blood Reactions and in the White Cell Count during the Total Eclipse of the Sun in Poland 1954. The total eclipse of the Sun, visible in Poland on 30 June 1954, gave some opportunity to investigate the influence of the disappearance of the Sun's rays on human body. An attempt was made to confirm or disprove the results obtained by Takata, concerning the changes of human serum flocculation related directly to an unidentified component of solar radiation. Also the blood clotting time was observed and white cells were counted. The average values of the Takata reaction showed a greatest decline when investigated 17 min. after the total eclipse. In comparison with those obtained before and after that time the result was close to the statistical significance. The number of lymphocytes decreased slightly, that of the segmented leukocytes rose insignificantly. The blood coagulation time was indeed decreased but in comparison with the daily variations no significant changes could be established. Low activity of the Sun in 1954 and a small number (3 to 6) of investigated persons might have been responsible for the failure to confirm the decisive influence of a total eclipse of the Sun on the Takata reaction and to find some indisputable changes in the results of other blood investigations.

Änderungen einiger Blutreaktionen und der Leukocytenzahl während der totalen Sonnenfinsternis in Polen am 30. Juni 1954. Die in Polen am 30. Juni 1954 sichtbare totale Sonnenfinsternis gab uns Gelegenheit, den Einfluß eines plötzlichen Verschwindens der Sonnenstrahlen auf den menschlichen Körper zu untersuchen. Ein Versuch wurde unternommen, die Resultate Takatas, die sich auf Änderungen der Flockung des menschlichen Serums unter dem Einfluß eines nicht näher identifizierten Bestandteiles der Sonnenstrahlung bezogen, zu bestätigen oder zu widerlegen. Auch die Blutgerinnungszeit wurde beobachtet und die weißen Blutkörperchen gezählt. Die durchschnittlichen Werte für die Takata-Reaktion zeigten die größte Verminderung, wenn die Blutentnahme 17 Minuten nach der totalen Sonnenfinsternis erfolgte. Im Vergleich zu den Werten, die vor und nach dieser Zeit erhalten wurden, waren die Resultate nahe der statistischen Wichtigkeit. Die Lymphocytenzahl verminderte sich etwas, die der segmentierten Leukocyten stieg unbedeutend. Die Blutgerinnungszeit war in der Tat verkürzt, doch im Vergleich zu den taglichen Schwankungen konnten keine zuverlässigen Änderungen festgestellt werden. Die geringe Sonnentätigkeit im Jahre 1954 und die Unmöglichkeit, eine größere Anzahl

---

[1] Zaklad Fizjologii U. J., Kraków, Grzegórzecka 16, Poland.

Personen zu untersuchen, waren vielleicht für das Ausbleiben der Bestätigung eines entscheidenden Einflusses der totalen Sonnenfinsternis auf die TAKATA-Reaktion und für das Fehlen unbestreitbarer Änderungen in den Ergebnissen anderer Blutuntersuchungen verantwortlich.

Altérations observées dans les globules blancs et dans cértaines réactions du sang pendant l'éclipse totale de soleil en Pologne en 1954. L'éclipse de soleil observable en Pologne le 30 juin 1954 a fourni l'occasion d'examiner l'influence sur le corps humain d'une suppression du rayonnement solaire. Une tentative fut faite de confirmer ou d'improuver les résultats de TAKATA sur les modifications de floculation du serum dues à une composante non identifiée du rayonnement solaire. Simultanément le temps de coagulation et le nombre de globules blancs furent comptés. Les moyennes statistiques de la réaction de TAKATA ont subi leur plus grande déviation 17 minutes après l'éclipse totale. Comparées à celles obtenues avant et après, le résultat est presque statistiquement significatif. Le nombre de lymphocytes fut légèrement décroissant, celui des leucocytes segmentés subit un accroissement imperceptible. Le temps de coagulation décrut effectivement sans écarts marquants par rapport aux modifications journalières. La faible activité du soleil en 1954 et le petit nombre de personnes examinées (3 à 6) sont peut être responsables de l'échec de cette tentative de confirmer de façon décisive l'influence du rayonnement solaire sur la réaction de TAKATA et sur d'autres modifications du sang.

Investigations on the influence of different kinds of solar radiation on the organism are of extreme importance for all physiological problems connected with space flights. If it is possible to prove that a certain reaction changes with altitude, but is independent of pressure, of the O_2 content, of the temperature and of the infrared, visible and ultraviolet radiation, then there is a probability that it may be due to an unidentified component of solar radiation. Such a reaction is, according to TAKATA, the reaction that bears his name and consists of the flocculation of the diluted and alkalized serum under the influence of a minimal amount of a fuchsin-sublimate mixture. It increases at high altitudes, is really independent of all known physical factors, like low pressure, low oxygen content, and according to TAKATA it shows a daily rhythm. Several minutes before astronomical sunrise the flocculation curve rises suddenly, increasing later during the day and slowly obtaining its highest point just after Sunset. As night approaches the curve drops, reaching its minimum a few minutes before Sunrise. The form of the curve does not depend either on the place where the person is situated or on the meteorological conditions. Its 24-hour rhythm reveals some similarity to the variation in the F_2 layer of the ionosphere. It does not show the same variations in earthed persons. During an augmented activity of the Sun the amplitude of the daily alterations is much greater than at the time of minimal Sun activity. An 11-year rhythm was found, possibly related to the cycles of increased Sun activity. Some 28-day variations were also observed, probably depending on the rotation of the Sun around its axis. There is no dependence on cosmic radiation.

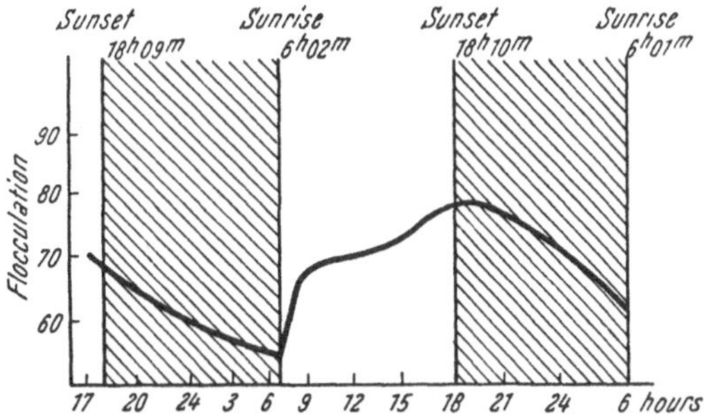

Fig. 1. Daily variations of the flocculation curve obtained by TAKATA

The daily changes are presented in Fig. 1.

The solar origin of these flocculation changes seemed to be confirmed by TAKATA in his observations during solar eclipses. A dropping of the reaction coincided with the totality of the eclipse (Fig. 2). On the basis of these findings TAKATA concluded that there is an emission of a new kind of radiation energy, which is biologically active and has not yet been established by physical methods. This radiation is marked by its great permeability and it evokes ionization phenomena in living organisms. According to TAKATA it has a corpuscular nature and a velocity similar to that of light.

The daily variations were not confirmed by SARRE from Freiburg.

The total eclipse of the Sun of 30 June 1954, $13^h 57^m 59^s$ — $13^h 59^m 29^s$ (that just touched the north-east of Poland) gave some opportunity to investigate the influence of a disappearance of all the Sun's rays on this reaction. Additionally, as the TAKATA reaction may depend on physico-chemical changes in the serum proteins and is usually accompanied by change in the lymphocyte count, we included in our investigations the white blood cell count and coagulation time. A team of Polish physicians (Mrs. KOCYAN, Mrs. ZBIEGIEŃ, Messrs. OGIŃSKI, BILSKI, WIECHA and myself) was cooperating in Suwałki at the time of the eclipse. The technique which we used was identical with that of TAKATA. Five cc. of blood, taken from the cubital vein, were kept for 12 hours in an ice box at a temperature of 10° C. After that time the sample was centrifuged for about 10 minutes at a velocity of 2,000 rpm. A 1:10 dilution of the serum was made with a physiological saline solution and divided into nine test tubes. In each tube 0.25 cc. of 10% Na_2CO_3 were added and two to three hours later the TAKATA reagent was applied consisting of equal volumes of 0.5% sublimate and 0.02% fuchsin solution. Adding different quantities of the reagent to each of the tubes the smallest amount of the mixture to initiate just perceptible flocculation was determined. The determinations were performed in a dark room using artificial light passing as a band through the test tube.

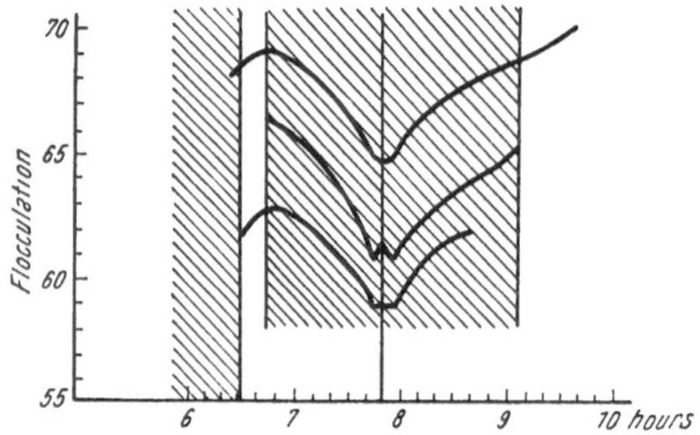

Fig. 2. Flocculation curve during the total eclipse of the Sun obtained by TAKATA in Japan 1943

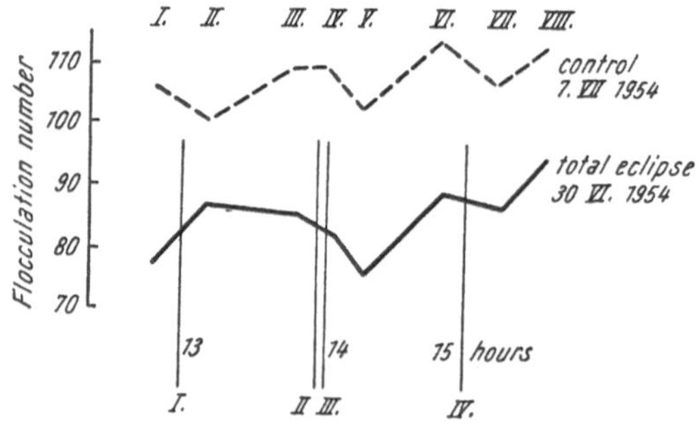

Fig. 3. Variations of the TAKATA reaction during the total eclipse of the Sun on 30 June 1954 in Suwałki

Results were presented as the flocculation number, i.e. the minimal quantities in cc. of the TAKATA reagent multiplied by 100. They were proved by statistical analysis. The blood coagulation time was determined by a modified BÜRKER method, using hollowed object glasses in a moist chamber. All persons with exception of one (G) were electrically isolated at the time of investigation.

Table I presents our results:

Table I. *Takata reaction on 6 electrically isolated and 1 earthed individuals during total solar eclipse and one day before*

Individual	29. VI. 1954 13^{58}	30. VI. 1954 12^{35}	12^{57}	13^{45}	13^{59}	14^{17}	14^{55}	15^{27}	15^{48} hour
		Sequency of the tests (see figures) I	II	III	IV	V	VI	VII	VIII
A	90	95	88	84	86	77	83	—	79
B	75	73	72	73	71	55	—	74	72
C	83	81	101	68	73	72	—	—	101
D	65	54	72	70	83	71	73	63	84
E	96	70	84	102	72	83	82	100	101
F	101	83	85	101	93	79	107	104	115
Average	85	76	84	83	80	73	86	85	92
G, earthed	101	71	88	85	84	85	100	74	101

One person, *E*, showed a definite decrease of the flocculation number during the eclipse, and three persons, *A*, *B*, and *F* — 17 minutes after the totality of the eclipse; one person, *C*, — 13 minutes before that time; one earthed person, *G*, did not exhibit any changes at all. On the average curve of six persons (Fig. 3) there is a drop a few minutes after the total solar eclipse. The samples were taken at the time marked on the curve, both on the day of the eclipse and at the corresponding time one week later. The flocculation reaction (*Fl*) proved to be 80 at the time of the totality of the eclipse, and 85 one day before, while 17 minutes after the eclipse it dropped to 73. In comparison with the results obtained one day before, this would be a difference of 12, corresponding to 14 %. The difference between *Fl* during the total eclipse and before the eclipse did not exceed 4 %. As compared with the average flocculation number before the partial eclipse, there was a diminution of 11, i.e. 13 %. After the eclipse there was a rise to 92. The statistical analysis

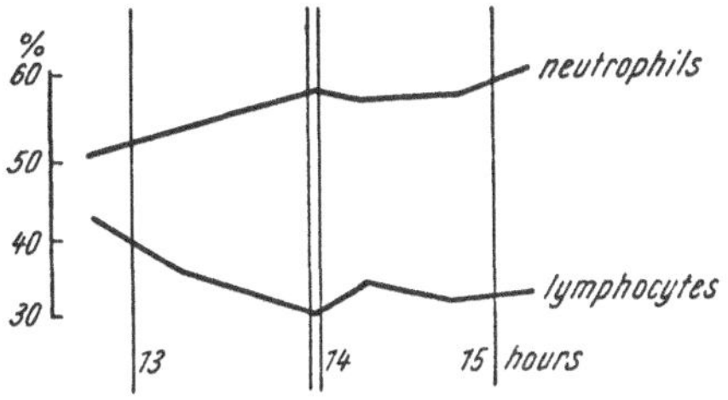

Fig. 4. Neutrophils and lymphocytes at the time of total eclipse of the Sun on 30 June 1954 in Suwałki. Averages of 6 individuals

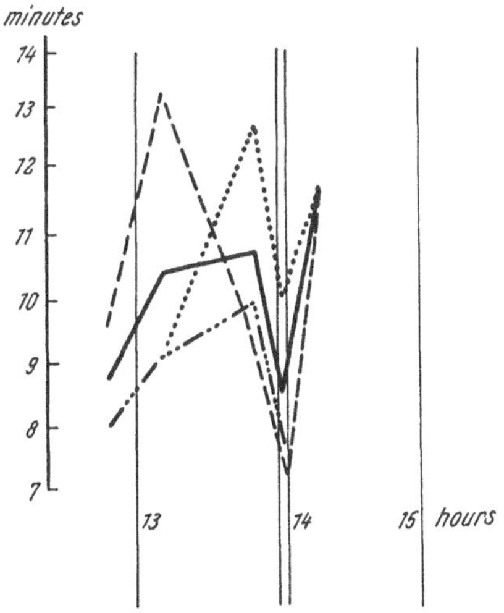

Fig. 5. Blood clotting time during the total eclipse of the Sun on 30 June 1954 in Suwałki. Results on 3 persons and their average (continued line)

did not reveal any significant differences, $t=0.46$; $P=0.66$. Only if we compare the results obtained 17 minutes after the total eclipse with the average of the first and last determinations, the probability by chance is smaller and amounts to $t=2.46$, $P=0.065$. This result is close to significance. The Sun's activity in 1954 was small, this may be a cause of the small variations. Nevertheless investigations performed one week after the eclipse also revealed some variations corresponding to the time of the total eclipse.

The lymphocytes showed a drop, the segmented leucocytes a slight rise (Fig. 4), the neutrophil band cells, the eosinophils and the monocytes varied only insignificantly.

The blood coagulation time investigated on three individuals (Fig. 5) was indeed decreased at the time of the eclipse, but in comparison with the daily variations the results were not significant either.

Our investigations can hardly be considered as a positive proof of the influence of a solar eclipse on the TAKATA reaction, and on the blood clotting time, but it is not impossible that the short time of the totality of the eclipse and the low activity of the Sun in the period of the tests might have been responsible for the failure to confirm the decisive influence of a total eclipse of the Sun on the investigated blood reactions.

Optimization Considerations for Orbital Payload Capabilities

By

H. H. Koelle[1], ARS, GfW, BIS

(With 3 Figures)

Abstract — Zusammenfassung — Résumé

Optimization Considerations for Orbital Payload Capabilities. Various phases of orbital techniques and optimization of orbital carrier vehicles are discussed in detail. Design criteria for "optimum" solutions are mentioned. Special emphasis is being placed on the selection of orbit and ascent trajectory parameters. The difference in requirements for elliptical and circular orbits is explained. Means and procedures how (nearly) circular orbits can be obtained are discussed in detail. Three diagrams are illustrating the influence of the main parameters on Payload-Lifetime capability. It is finally concluded that the Vanguard can be considered as a starting point only and that at least two decades will pass by until we can claim that orbital techniques are well under control, reliable, economically reasonable and approaching a certain perfection which is desirable and necessary for a large scale space flight development.

Betrachtungen über die Erreichung optimaler Nutzlast bei Umkreisungsbahnen. In Einzelheiten erörtert werden verschiedene Phasen der Umkreisungsbahntechnik und der besten Lösungen von Umkreisungsbahn-Trägerraketen. Kriterien der Planung fur „optimale" Lösungen werden erwähnt. Besonderes Gewicht wird auf die Auswahl der Parameter fur die Umkreisungsbahn und die Aufstiegsbahn gelegt und der Unterschied in den Erfordernissen fur elliptische und kreisförmige Bahnen erlautert. Eingehend werden Mittel und Verfahren besprochen, auf welche Weise nahezu kreisförmige Bahnen erzielbar sind. Drei Diagramme illustrieren den Einfluß der Hauptparameter auf die Möglichkeiten von Nutzlast und Lebensdauer. Es ergibt sich schließlich, daß die Vanguard nur als ein Anfang betrachtet werden kann und daß mindestens zwei Dekaden vorubergehen werden, bis wir werden sagen können, daß die Umkreisungsbahntechnik befriedigend unter Kontrolle, verläßlich, ökonomisch vernünftig und einem gewissen Maß von Vollendung nahe ist, wie es wunschenswert und notwendig fur eine Entwicklung des Raumfluges auf breiter Basis ist.

Recherche de la plus grande charge orbitale utile. Une discussion détaillée des diverses techniques de lancement sur une orbite et de l'optimisation des engins relativement à leur capacité en charge utile. Des critères possibles de solution optimale sont proposés. L'attention est spécialement portée sur le choix des paramètres de l'orbite et de la trajectoire ascensionnelle. Une orbite elliptique et une orbite circulaire ont des exigences différentes; les méthodes pour l'obtention d'une orbite quasi-circulaire sont expliquées en détail. Trois diagrammes illustrent l'influence

---

[1] Chief, Preliminary Design Section, (DOD), Army Ballistic Missile Agency, U. S. Army Ordnance Corps, Huntsville, Alabama, U.S.A.

des paramètres principaux sur la charge et la vie utiles. En conclusion le projet Vanguard doit être considéré comme un point de départ seulement; deux décades au moins passeront avant que les techniques de placement sur une orbite ne soient complètement maîtrisées, raisonnablement économiques et approchant la perfection désirable pour le développement du vol interplanétaire sur une grande échelle.

I. Introduction

Our present wishes and needs for artificial satellites with a large payload of scientific instruments are, and always will be, limited by the orbital payload capability of available carriers.

One way of doing the job is by using existing hardware which becomes available in the course of military development of long range missiles. This is a very convenient, logical and economical way, because military necessity has sponsored the development of rocket components which can be used also for orbital carriers and space vehicles.

One typical example of using components from military developments is the Russian design which is reported in newspapers occasionally. It would be logical to assume that the Russians are using the powerful 264,000 lb. thrust booster of their intermediate range ballistic missile as the basic booster of their three stage orbital carrier and the improved 77,000 lb. thrust V-2 missile as the second stage. The development of this combination goes back as far as 1948. This approach is very attractive because it has the advantage of using proven hardware with established reliability. It also promises a schedule which seems to be realistic. The job remaining to be done is to develop a third stage and to mate it properly with the other two stages. Even for this third stage, proven hardware seems to be in stock since the improved Wasserfall engine with about 20,000 lb. thrust, under development for almost a decade, might be a proper choice. Disadvantages of this solution are three different propellant combinations for the missile, the limited specific impulse in the upper stages, and perhaps a total structural factor (ratio of hardware-weight over take-off weight) which is not representative of the present state of the art. This will probably result in a growth factor (ratio of take-off weight over payload weight) which will be somewhat larger than that of an optimized missile of similar size designed by standards of the present state of the art. Nevertheless, this three-stage configuration will offer up to several hundred pounds orbital payload capability which is at least 10 times as much as the first American orbital vehicle can carry. But the growth factor of a missile is not the only, not even the most important, criterium of an orbital carrier. Schedule, cost and reability are at least of equal importance.

The other approach for an efficient orbital carrier is that of starting from scratch. If time allows a more or less optimistic view to be taken on the present state of the art and on components available now or in the years of development to come, it will possibly result in the so-called "optimum missile" which is actually nonexistent. It can be no more than an individual or "partial optimum" design. As a matter of fact, there are a large number of "optimums". For example: minimum hardware weight per weight unit payload; minimum take-off weight per weight unit payload (growth factor); maximum use of reliable and available components; shortest possible development time; maximum use of existing facilities and manpower (experienced teams); most desirable launching and handling characteristics; mass production aspects; smallest dimensions of vehicle; simplicity and reliability and many other partial optima.

Missile design and engineering involves nothing more than the ability to find a compromise which satisfies the most important requirements. During the development it is also the "know how" to eliminate all the "bugs". This is more complicated but little different from any other engineering field. There is no absolute yardstick to an "optimum solution". Opinions, beliefs, estimates and judgment of many individuals are involved, not only of design engineers, but also of scientists, business men, lawyers, administrative people or whoever, by merit or accident, might be in the committee which makes the final selection of the "optimum missile" for a certain mission.

I am rather sure that the Russians believe that their approach for a satellite carrier is the optimum solution, as most of the people working on the American Vanguard project probably believe that their approach with its compromise in optimum growth factor, reliability and schedule, is the "optimum" approach. Both groups might be right from their point of view, but only history will prove which approach was the better one! Certainly the party will be right who accomplishes the job of establishing a satellite with a reasonable lifetime and obtains a reasonable amount of scientific information at the earliest date. They will be honored by history as being the first to accomplish the breakthrough in space flight which, we must believe, will be for the benefit of the human race. (The answer to this question was not available at the time this paper was written but it might be available by the time the paper is presented. This paper was submitted in March 1957).

II. Selection of Orbit and Trajectory

Fundamentally we have known the answers on satellite orbits for many decades from studies in the field of celestial mechanics, however, there were many more contributions during the last ten years which helped considerably to give us a more complete picture.

There are fundamentally three kinds of orbiters (which have been discussed so far) with different major missions:

1. Scientific orbiters,
2. Military orbiters,
3. Space stations.

The mission and its detailed requirement of any orbiting vehicle is used in determining the selection of the orbit and the initial conditions to enter the selected orbit.

Scientific and military orbiting vehicles require orbits with various inclinations to the plane of the equator. Space stations, however, which in many cases will be relay and supply stations for individual flight missions into space, will preferably be located in the plane of the final orbit to the destination. Therefore such a space station might have an orbit which is in or close to the ecliptic (23.5° inclination to the earth equator). Most planets and the moon have orbits only slightly inclined to the ecliptic. Thus the orbit in the ecliptic seems to be a good selection for a standard orbit which is needed for comparison purposes.

The Vanguard orbit was mentioned to have an inclination of about 25 degrees against the equator. This will be a compromise of desired visibility at high latitudes, acceptable firing azimuth angle for the launching site selected (requirements of range instrumentation and range safety) and of performance considerations.

Normally, the following initial parameters will have to be selected before the payload capability of an individual carrier vehicle can be calculated:

1. Orbit inclination (against equator),
2. Launching site (Northern or Southern latitude),
3. Earth fixed firing azimuth angle,
4. Initial orbit altitude,
5. Initial orbit velocity,
6. Desired cut-off angle,
7. Number and kind of individual manœuvers.

As far as the orbit is concerned, the first decision to be made is whether a circular orbit or an elliptical orbit (controlled or uncontrolled) is desired. Studying the problem a little closer, one finds that it is extremely difficult or almost impossible to produce an accurate circular orbit. The following considerations are of interest in this connection:

1. If maximum lifetime-payload capability is the main objective, an elliptical orbit should be chosen. This approach is much simpler because it does not require accurate velocity and attitude control of the payload stage. Also, as a

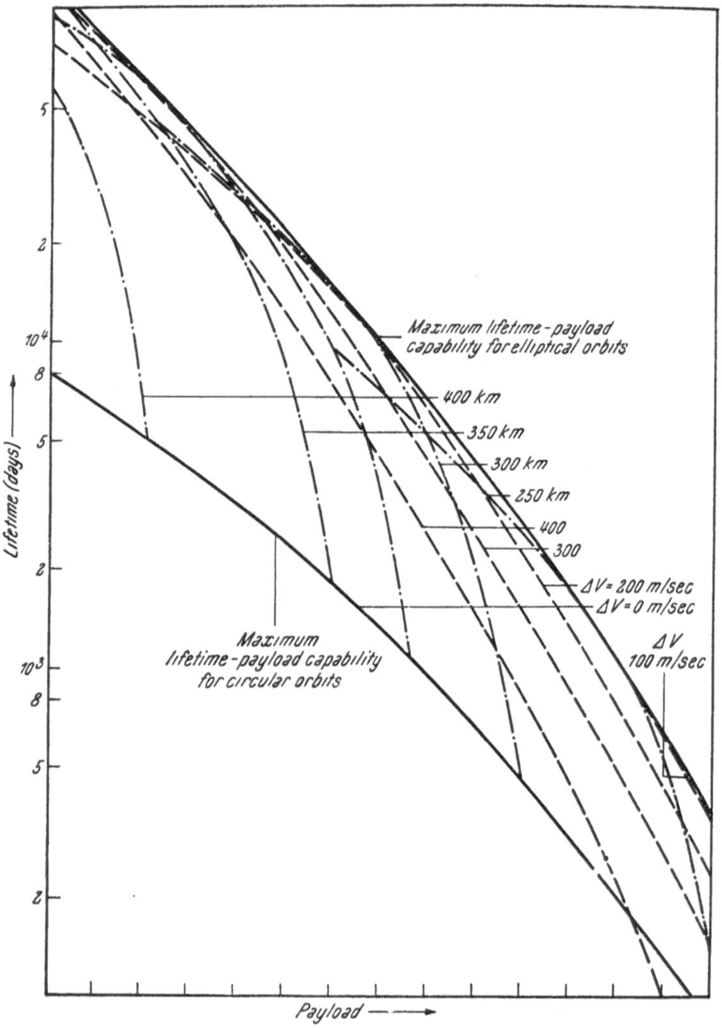

Fig. 1. Typical lifetime-payload capability of an orbital carrier vehicle with initial altitude and excess velocity as parameters

matter of fact, and as shown in the following diagram, the elliptical orbit promises more lifetime for an individual payload weight as the corresponding circular orbit.

Fig. 1 is a typical lifetime-payload diagram of an arbitrary selected orbital carrier. It is valid for the controlled and uncontrolled variation of the carrier vehicle under consideration. The lower curve indicates the lifetime-payload capability for circular orbit. The diagram contains lines of equal altitude and equal excess velocity at cut-off of the payload stage. The upper envelope of these curves is the maximum lifetime payload capability of this orbiter. As can be seen from the diagram, there is always for any selected payload one combination of altitude and excess velocity (over local circular velocity) which gives the longest lifetime and should definitely be selected as the design point. However, desiring or selecting this optimum combination as a design point is one thing, and obtaining this desired combination by an actual flight another. If the cut-off conditions (6 parameters: position x, y, z; velocity $\dot{x}$, $\dot{y}$, $\dot{z}$) of the payload stage cannot be controlled, as for example with the Vanguard vehicle, the expected maximum dispersions in the cut-off position are the determining factors for the payload capability of that carrier. The main emphasis must be placed on the velocity vector control which has the greatest influence on lifetime.

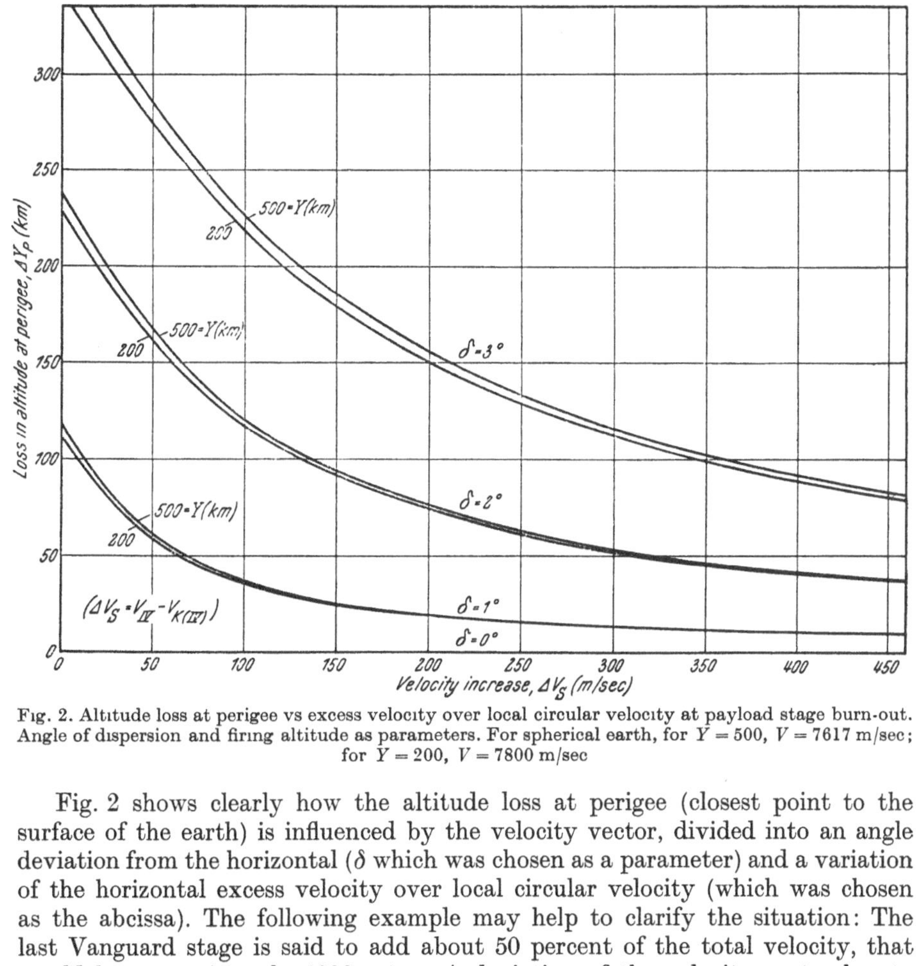

Fig. 2. Altitude loss at perigee vs excess velocity over local circular velocity at payload stage burn-out. Angle of dispersion and firing altitude as parameters. For spherical earth, for $Y = 500$, $V = 7617$ m/sec; for $Y = 200$, $V = 7800$ m/sec

Fig. 2 shows clearly how the altitude loss at perigee (closest point to the surface of the earth) is influenced by the velocity vector, divided into an angle deviation from the horizontal (δ which was chosen as a parameter) and a variation of the horizontal excess velocity over local circular velocity (which was chosen as the abcissa). The following example may help to clarify the situation: The last Vanguard stage is said to add about 50 percent of the total velocity, that would be approximately 4000 m/sec. A deviation of the velocity vector has to be expected as this stage has a solid propellant motor which has only limited control in respect to total impulse and attitude. If, for example, one percent is considered a representative figure for the deviation of the total impulse, then the resulting contribution to the velocity deviation of this stage alone is in the order of 40 m/sec. Other contributions to the dispersion in velocity and attitude will be added by the first two booster stages if their cut-off is not precisely controlled. If now a total dispersion of 100 m/sec is expected in velocity from the target excess velocity of 200 m/sec at 300 km altitude and a deviation of two degrees from the horizontal, a loss in pergee altitude up to 120 km will have to be expected. In other words, the expected deviation in initial conditions determines the most unfavorable orbit, therewith the lifetime and/or the playload capability. Summarizing Fig. 1 it can be stated that an elliptical orbit results always in more lifetime with the same payload but maximum use of this advantage can only be made if the cut-off of the payload stage can be controlled. Unfortunately, control components have some weight and they are adding complexity. But, in any case, there is always an option of control, limited control, or no control, depending on available hardware and individual mission.

2. If the mission requires a circular orbit within close tolerances, a fully controlled payload stage is the only answer. The sensitivity of the apogee altitude (farthest point from the earth) with deviations from the local circular velocity is shown in Fig. 3. A deviation of 50 m/sec (at about 7800 m/sec), which is less than one percent, produces an apogee which is 175 km higher than the perigee. Judging from these diagrams it seems to be desirable to control the velocity of a payload stage to better than 1/4 of a percent, the initial altitude to better than one percent (this is about the order of magnitude of dispersions to be expected from the oblateness of the earth), and the attitude or angle deviation to about 1/4 of a degree if a good circular orbit is expected at cut-off of the payload stage.

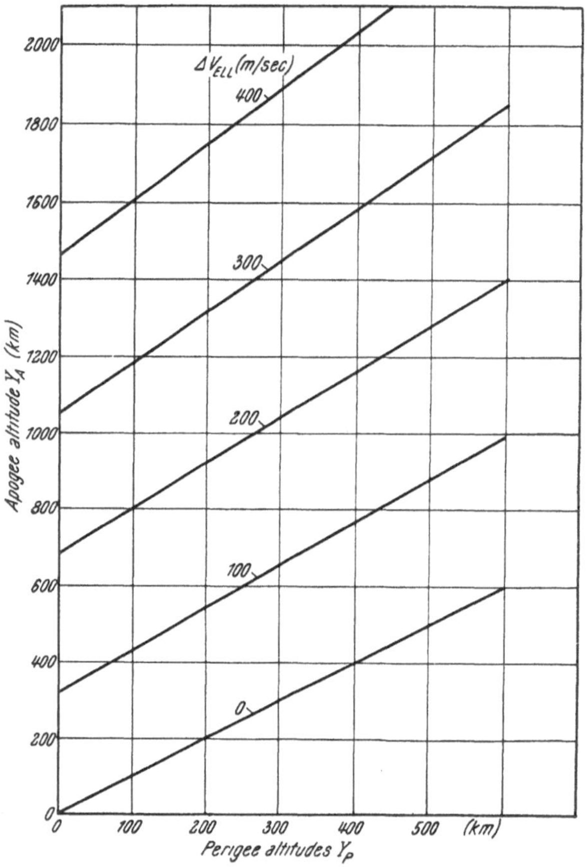

Fig. 3. Satellite apogee altitude vs perigee altitude with local excess velocity as a parameter. ΔV_{ELL} = Velocity in excess of circular velocity at perigee

To produce an orbit at all is a relatively easy task by providing a vehicle with a proper number of stages, with possibly proven hardware (engines, structure, guidance and control components), and burn each stage until burn-out. If all stages work properly and an ample amount of propellants was provided and used, an elliptical orbit will be the result. But the orbit parameters and the corresponding lifetime cannot be predicted with sufficient accuracy. In the worst case, if the payload was properly chosen, the lifetime might be one day or 100 years! Such an unpredictable orbit and lifetime will undoubtedly still be of interest and bring useful scientific results, however, it can only be considered as the beginning of the development of orbital techniques and a certain perfection of this technique is a necessity for further exploration of space. Thus, the controlled orbit has to be considered as the desired goal. In practice many possibilities exist to reach this goal. It is not possible to give a full description of all methods which seem to be feasible within this paper, but two examples will be shortly discussed.

One way to obtain a circular (or close circular) orbit is easily described but only with difficulties to materialize. The payload stage would have to have a full guidance and control system which controls the attitude of the payload stage into a horizontal position at cut-off of this stage resulting in only one velocity component ($\dot{x} = V_c$, $\dot{y} = O$, $\dot{z} = O$). The proper cut-off position and horizontal velocity is obtained by controlling the powered trajectories of the individual stages with respect to cut-off velocity and altitude, also with some relaxed accuracy requirements, cut-off distance, and azimuth angle. However, this approach requires a rather accurate, complex and heavy guidance and control system to be carried along in the payload stage. This requires orbital carriers with payload capabilities much larger than the Vanguard vehicle. In other words, a satisfactory control of the orbit can be expected to be accomplished without unsurmountable difficulties if large orbital carriers with corresponding large payload capabilities become available.

The trickiest problem is to obtain an almost circular orbit in desired altitudes with small carriers and corresponding small payload capabilities. The following approach which seems to be promising will be explained with the example of the Vanguard orbital carrier with the assumption that this is supposed to produce controlled circular orbits.

The first requirement is to select an orbital altitude where the air drag is negligible and lifetimes of several years with almost unchanged orbital parameters can be guaranteed. This, of course, depends on the ballistical factor of the satellite which is the hypersonic drag coefficient divided by the cross sectional load of the satellite. An altitude of 570 km or 355 miles resulting in a period of revolution of 90 minutes might be a proper choice for the final orbit as this even number of revolutions per day simplifies the tracking procedures and ground facilities. Furthermore, it is assumed that the second booster stage houses the guidance system, that the payload stage is spin stabilized and that an additional controllable motor will be provided whether within the payload or attached to the payload. The flight plan would then look as follows:

The first and second stage are cut-off controlled and will produce a proper combination of cut-off velocity, cut-off altitude and cut-off angle at the end of the powered ascent trajectory, which results in the desired initial conditions for firing of the third stage. This is assumed to be spin stabilized and powered by a solid rocket motor. The guidance compartment carrying the third stage and the altitude control system is separated after cut-off of the second stage and flies along in the free flight trajectory. The period of free flight without power between the controlled cut-off of the second stage and the ignition of the last stage will be used for proper alignment of the payload stage. An apex computer gives the proper ignition signal to the third stage shortly before reaching the apex and the third stage clears out of the launcher which is mounted on the instrument compartment of the second stage. Due to technical tolerances there will be some deviation in altitude, velocity and trajectory angle at cut-off of the third stage. They are held to a minimum as there was full control of the second stage cut-off. This procedure now allows a deviation of the cut-off altitude from the finally desired orbit altitude as the adjustment of orbital parameters will be done with an additional fourth stage. This allows the choice of a maximum energy trajectory and tilting program resulting in an altitude large enough to keep air drag small, but on the other hand small enough to minimize the earth gravity losses. For example, an altitude of about 150 km for cut-off of the third stage can be selected. This altitude, however, requires an excess velocity of about 120 m/sec over local circular velocity in order to enter a transfer ellipse, bringing the last stage with the payload to the desired altitude of about 570 km at the apogee of the transfer ellipse. The arrival at the apogee is now the proper time to ignite the last stage which gives the payload an impulse just big enough to increase the velocity to local circular velocity in order to stay in that altitude in the desired circular orbit.

In order to really obtain an orbit with this procedure several requirements must be met.

1. The dispersions at cut-off of the third stage at the apex of the booster trajectory must be kept small in order to reach the desired orbit altitude at apogee. This can be accomplished by having full control of the second stage cut-off, as described before.

2. There must be a possibility of accurate trajectory measurements after cut-off of the third stage in order to predetermine the apogee altitude, the magnitude of the impulse necessary to obtain circular velocity at this predetermined altitude, and finally, the time when this additional impulse has to be initiated. The calculation of orbit parameters can be carried out on the ground and the ignition signal of the fourth stage will be relayed to the payload stage by radio. An on board computer to accomplish this procedure in the payload is another possibility but can be used only if very large payloads are considered.

3. The fourth stage must have a controllable motor if a good circular orbit is desired. An uncontrolled charge will improve the eccentricity of the orbit but produce a close circular orbit only with a small probability. This so-called "*Kick*" procedure has the advantage not only to give the possibility of controlling initial orbital parameters, but also increases the payload capability by a more favorable ascent trajectory as a welcome by-product.

III. Fundamental Missile Parameters

The required orbit parameters, as discussed in the preceding chapter, will determine the missile parameters and, in turn, some missile characteristics will influence some orbit parameters. For example, if the launching site and azimuth angle are given parameters, the cut-off distance of the payload stage determines the X position where the orbit will be entered. The design of the carrier vehicle is more or less a process of successive approximation of optimum trajectory configuration and vehicle parameters. The total energy of the vehicle in the selected orbit (potential and kinetic energy) converted into a velocity plus 10 per cent of its value is already a good approximation value for the characteristic velocity the orbital carrier has to be designed for. The 10 per cent value is a good standard figure for covering velocity losses due to air drag and earth gravity during the powered phase of the trajectory. Characteristic velocities of orbital carriers are between 9,000 and 10,000 m/sec or even larger for higher altitudes. The theory of multistage rockets, discussed by several authors, requests three to four stages for orbital missions if chemical propellants are considered. Four stages are advantageous for manned carriers as they reduce maximum accelerations and also the weight of the returning stage which is quite a simplification for the wings or other recovery gear used. Low specific impulses and comparatively large structural weights require a larger number of stages. The optimum number of stages can also be different, whether the take-off weight or the hardware weight is to be kept to a minimum. Other items having a bearing on the selection of the number of stages are the rocket engines available or to be used, schedule and reliability.

To accertain extent the selection of the propulsion systems to be used in the individual stages also establishes the power plant weights, the specific impulses and the structural loads. After a preliminary design of the structure and guidance and control system, preliminary weight data for the overall vehicle becomes available for the first trajectory calculations.

The final amount of propellant weights and volumes, fixing also the burning times, will be found by a variation method. This variation process is relatively easy with electronic computers and a preliminary trajectory. A detailed load investigation and stress analysis will then result in an improved structure weight breakdown. Finally, an effort will be made to optimize the ascent trajectory by a variation of the tilting program and possibly the expansion ratio of the individual nozzles as well as the thrust level if deemed necessary.

The resulting standard trajectory must always be checked whether the requirements of the flight plan are fulfilled or a compromise could be made in some of them. It should always be kept in mind that the mission is the main objective, even if an adjustment of the carrier vehicle or the trajectory to the proper flight plan means a weight penalty or other drawbacks. Questions of control parameters and aerodynamic heating can be investigated and answered as soon as dimensions and weights as well as the design of the guidance and control system are frozen.

Other important items influencing the design concept of the orbital carrier system are:

1. Total number of flights and/or vehicles planned for development phase and actual mission.
2. Required schedule.
3. Payload size, shape and weight requirements.
4. Required reliability.
5. Required accuracies.
6. Facilities.
7. Launching and handling characteristics of vehicle and propellants.
8. Ground and flight testing programs.
9. Available experience and manpower.
10. Logistics.

All these questions and problems influence the design of the vehicle, specifically the growth factor of the carrier vehicle, the overall system, and last but not least, the overall cost of the project which very often is a limiting or decisive factor in selling a project on a competitive basis.

Finally, it should not be forgotten that after completion of the design the most difficult part of a missile development (often underestimated) begins; to cut the hardware and get thousands of parts working together in the way they are supposed to work. The failure of only one link in the chain, which might cost only a few cents, can wreck the effort of millions of working hours. The crucial point in this art is "to know how" to avoid these failures.

IV. Summary

The optimization of an orbital carrier system normally begins with the optimization of the flight plan on the basis of the flight mission. In lower orbit altitudes lifetime-payload capabilities of the carrier have a strong influence on the flight plan and missile. At high orbit altitudes the lifetime practically becomes indefinite and mainly the payload capability of the carrier vehicle is of primary importance.

In low orbit altitudes uncontrolled or controlled elliptical orbits have greater lifetimes than circular orbits for the same payload weights and configurations. Several orbital missions can be carried out only if circular orbits can be produced. This requires a rather complicated and accurate guidance and control system in the carrier vehicle and/or in the payload. Controlled circular orbits require a much larger technical and financial effort, but they will be mandatory for certain missions.

Optimizing an orbital carrier system requires the consideration of a very large number of parameters and can be accomplished only for each individual mission. However, certain procedures and methods are under development which will allow an economical calculation and comparison of similar orbital carriers. With the Vanguard project a starting point for the development of orbital techniques has been reached and at least two decades will pass by until we can claim that orbital techniques are well under control, reliable, economically reasonable and approaching a certain perfection which is desirable and necessary for a large scale space flight development.

References

A. Orbits and Trajectories

1. F. R. Moulton, Celestial Mechanics. New York: MacMillan, 1914.
2. D. F. Lawden, General Motion of a Rocket in a Gravitational Field. J. Brit. Interplan. Soc. **6**, 187 (1947).
3. F. J. Malina and M. Summerfield, The Problem of Escape from the Earth by Rocket. J. Inst. Aeronaut. Sci. **14**, No. 8 (1947).
4. D. F. Lawden, Initial Arc of the Trajectory of Departure. J. Brit. Interplan. Soc. **7**, 119 (1948).
5. J. M. J. Kooy and J. W. H. Uytenbogaart, Ballistics of the Future. Haarlem: The Technical Publishing Co. H. Stam, 1946.
6. E. Saenger, Laws of Motion of Space Travel. Interavia **4**, July 1949.
7. P. Blanc, Balistique Extérieure des Fusées dans le vide et avec champ de pesanteur constant. Mém. Artill. Franç. **24**, Fasc. 1 (1950).
8. H. S. Tsien and R. C. Evans, Optimum Thrust Programming for a Sounding Rocket. J. Amer. Rocket Soc. **21**, 99 (1951).
9. W. Schaub, Die Flutkräfte auf der Außenstation. Weltraumfahrt, **1951**, No. 1.
10. R. A. Smith, Establishing Contact Between Orbiting Vehicles. J. Brit. Interplan. Soc. **10**, 295 (1951).
11. E. Saenger, Atlas konkreter Bahnen von Raketenflugzeugen bis zur Außenstation. Forschungsreihe Nwd. GfW Bericht No. 3, April 1951.
12. D. F. Lawden, Entry Into Circular Orbits. J. Brit. Interplan. Soc. **10**, 5 (1951).
13. H. G. L. Krause, Die Kinematik einer Außenstation in einer zur Äquatorebene geneigten elliptischen Bahn. GfW Research Report No. 10, 1951 Stuttgart/Germany (and Weltraumfahrt **1952**, 17, 74).
14. R. H. Bacon, Motion Relative to the Surface of the Rotating Earth. Amer. J. Physics **19**, 52 (1952).
15. W. Schaub, Die himmelsmechanischen Grundlagen der Raumfahrt (in: Raumfahrt-Forschung, ed. by H. Gartmann, p. 27). Muenchen/Germany: R. Oldenbourg, 1952.

16. R. ENGEL, U. T. BOEDEWADT und K. HANISCH, Die Außenstation (in: Raumfahrt-Forschung, ed. by H. GARTMANN). Muenchen/Germany: R. Oldenbourg, 1952.
17. D. F. LAWDEN, Orbital Transfer via Tangential Ellipses. J. Brit. Interplan. Soc. **11**, 278 (1952).
18. D. F. LAWDEN, Inter-Orbital Transfer with Minimum Propellant Expenditure (in: Probleme aus der Astronautischen Grundlagenforschung, ed. by H. H. KOELLE). Stuttgart/Germany, 1952.
19. H. G. L. KRAUSE, Die säkularen Störungen einer Außenstation-Bahn (in: Probleme aus der Astronautischen Grundlagenforschung, ed. by H. H. KOELLE). Stuttgart/Germany, 1952.
20. A. R. HIBBS, Optimum Burning Program for Horizontal Flight. J. Amer. Rocket Soc. **22**, 204 (1952).
21. J. A. VAN ALLEN, The Angular Motion of High-Altitude Rockets (in: Physics and Medicine of the Upper Atmosphere, pp. 411-431). Albuquerque: University of New Mexico Press, 1952.
22. D. F. LAWDEN, The Determination of Minimal Orbits. J. Brit. Interplan. Soc. **11**, 216 (1952).
23. W. SCHAUB, Moglichkeiten des Überganges aus einer Ellipsenbahn in eine Kreisbahn und umgekehrt. Weltraumfahrt **1952**, 106.
24. N. V. PETERSEN, General Characteristics of Satellite Vehicles, I and II. J. Astronautics **2**, 41, 105 (1955).
25. D. F. LAWDEN, Optimal Programming of Rocket Thrust Direction. Astronaut. Acta **1**, 41 (1955).
26. K. A. EHRICKE, On the Descent of Winged Orbital Vehicles. Astronaut. Acta **1**, 137 (1955).
27. D. F. LAWDEN, Optimum Launching of a Rocket Into an Orbit About the Earth. Astronaut. Acta **1**, 185 (1955).
28. K. A. EHRICKE, Aero-Thermodynamics of Descending Orbital Vehicles. Astronaut. Acta **2**, 1 (1956).
29. F. M. PERKINS, Flight Mechanics of Ascending Satellite Vehicles. Jet Propulsion **26**, No. 5 (1956).
30. J. JENSEN, Satellite Ascent Mechanics. Jet Propulsion **26**, No. 5 (1956).
31. N. V. PETERSEN, Lifetimes of Satellites in Near-Circular and Elliptic Orbits. Jet Propulsion **26**, No. 9 (1956).
32. L. BITZER, et al., Perturbations of a Satellite's Orbit Due to the Earth's Oblateness. J. Appl. Physics **1956**, 1141.
33. P. F. WINTERNITZ, The Physical and Chemical Fundamentals of Satellite Flight I and II., J. Astronautics **3**, 43, 65 (1956).
34. K. A. EHRICKE, Ascent of Orbital Vehicles. Astronaut. Acta **2**, 175 (1956).
35. M. W. ROSEN, Placing the Satellite in its Orbit. J. Astronautics **3**, 61 (1956).

B. Carrier Vehicles and Satellites

36. K. W. GATLAND, Rockets in Circular Orbits. J. Brit. Interplan. Soc. 8, 52 (1949).
37. A. V. CLEAVER, Mass Ratios. J. Brit. Interplan. Soc. 8, 173 (1949).
38. A. V. CLEAVER, The Calculation of Take-off Mass. J. Brit. Interplan. Soc. **9**, 5 (1950)
39. R. ENGEL, Leistungsnomogramm fur Stufenraketen. GfW Research Report No. 6, 1950, Stuttgart/Germany.
40. H. H. KOELLE, Verfahren zur Bestimmung der minimalen Startgewichte und der günstigsten Konstruktionsgrundwerte von Raumfahrzeugen. GfW Research Report No. 5, 1950, Stuttgart/Germany.
41. H. H. KOELLE, Der Beweis der Möglichkeit der Weltraumfahrt. GfW Research Report No. 7, 1950, Stuttgart/Germany.
42. H. HOEPPNER und H. H. KOELLE, Die optimale Lastrakete zur Außenstation in 1669 km Höhe. GfW Research Report, 1951, Stuttgart/Germany.

43. H. H. KOELLE, Der Einfluß der konstruktiven Gestaltung der Außenstation auf die Gesamtkosten des Projektes. GfW Research Report No. 9, 1951, Stuttgart/Germany.
44. K. W. GATLAND, Orbital Rockets, Part I: Some Preliminary Considerations. J. Brit. Interplan. Soc. **10**, 97 (1951).
45. A. E. DIXON, Orbital Rockets. Part II: The Rocket Structure, with Special Reference to Expendable Construction. J. Brit. Interplan. Soc. **10**, 107 (1951).
46. A. M. KUNESCH, Orbital Rockets. Part III: Conception of an Instrument-Carrying Orbital Rocket. J. Brit. Interplan. Soc. **10**, 115 (1951).
47. W. SCHAUB, Die Außenstation als kräftefreier Kreisel. Weltraumfahrt **1951**, 103.
48. W. VON BRAUN, The Importance of Satellite Vehicles in Interplanetary Flight. J. Brit. Interplan. Soc. **10**, 237 (1951).
49. L. R. SHEPHERD, The Artificial Satellite. J. Brit. Interplan. Soc. **10**, 245 (1951).
50. W. SCHAUB, Die Raumstation als schwerer Kreisel. Weltraumfahrt **1951**, 121.
51. K. W. GATLAND, A. M. KUNESCH and A. E. DIXON, Minimum Satellite Vehicles. J. Brit. Interplan. Soc. **10**, 287 (1951).
52. H. H. KOELLE, Graphisches Verfahren zur Abschätzung der optimalen Konstruktionsgrundwerte von Raumfahrzeugen. Weltraumfahrt **1952**, No. 2.
53. C. T. AUBRY, Droppable Stages May Boost Rockets to Earth Circling Orbits. SAE Journal **60**, No. 9, p. 18 (1952).
54. W. VON BRAUN, Das Marsprojekt. Frankfurt/Germany: Umschau-Verlag, 1952.
55. C. RYAN, W. VON BRAUN, W. LEY, et al., Across the Space Frontier. New York: The Viking Press, 1952.
56. H. H. KOELLE, Zur Bestimmung des optimalen Brennkammerdruckes von Raketentriebwerken (in: Probleme aus der Astronautischen Grundlagenforschung, ed. by H. H. KOELLE). Stuttgart/Germany, 1952.
57. H. HOEPPNER, Die Satellitenrakete 1952 (in: Probleme aus der Astronautischen Grundlagenforschung, ed. by H. H. KOELLE). Stuttgart/Germany, 1952.
58. K. A. EHRICKE, Establishment of Large Satellites by Means of Small Orbital Carriers (in: Probleme aus der Astronautischen Grundlagenforschung, ed. by H. H. KOELLE). Stuttgart/Germany, 1952.
59. R. A. CORNOG and F. L. VAN DER WAL, On Optimizing the Component Proportions of High Performance Rockets (in: Probleme aus der Astronautischen Grundlagenforschung, ed. by H. H. KOELLE). Stuttgart/Germany, 1952.
60. P. BLANC, Le Calcul des Fusées à Etages, Mém. Artill. Franç. **26**, 705 (1952).
61. H. G. L. KRAUSE, Allgemeine Theorie der Stufenraketen. Weltraumfahrt **4**, 52 (1953).
62. W. VON BRAUN, The Early Steps in the Realization of the Space Station. J. Brit. Interplan. Soc. **12**, 23 (1953).
63. K. W. GATLAND, A. M. KUNESCH and A. E. DIXON, Fabrication of the Orbital Vehicle. J. Brit. Interplan. Soc. **12**, 274 (1953).
64. S. F. SINGER, A Minimum Orbital Instrumented Satellite. J. Brit. Interplan. Soc. **13**, 74 (1954).
65. S. F. SINGER, A Minimum Orbital Instrumented Satellite (in: Space-Flight Problems). Biel-Bienne/Switzerland: Laubscher & Cie, 1955.
66. S. F. SINGER, Studies of a Minimum Orbital Unmanned Satellite of the Earth (MOUSE). Astronaut. Acta **1**, 171 (1955); **2**, 125 (1956).
67. F. R. FURTH, Project Vanguard. Aeronaut. Engng. Rev. pp. 55–59, March 1956.
68. S. F. SINGER, Design Criteria for Minimum Satellites. Aero Digest, Apr. 1956, pp. 36-37.
69. W. LEY and W. VON BRAUN, The Exploration of Mars. New York: The Viking Press, 1956.
70. K. A. EHRICKE, The Satelloid. Astronaut. Acta **2**, 63 (1956).
71. Orbital and Satellite Vehicles, Vol. 1 and 2, Dept. of Aeron. Engng., MIT, 1956.

Sodium Emission at 140 km

By

E. R. Manring and **J. F. Bedinger**[1]

Abstract — Zusammenfassung — Résumé

Sodium Emission at 140 km. The efficiency of chemiluminescent processes occurring at night between atmospheric constituents and sodium atoms has been determined by releasing known amounts of atomic sodium in the form of a filament from 50 to 140 km. Photometric and photographic recordings indicate that a strong persistent glow occurs at about 90 km as expected; that above 105 km the process becomes inefficient; and that above 130 to 140 km, the rocket's zenith, a different process yields a strong emission of 5890 Å radiation.

It is proposed that this last process is due to transfer of energy from atmospheric atomic nitrogen in the highly metastable ^2D state to the released sodium.

Natriumemission bei 140 km. Die Wirksamkeit von Chemilumineszenzvorgängen, die sich nachts zwischen atmosphärischen Bestandteilen und Natrium-Atomen abspielen, wurde gemessen, indem bekannte Mengen atomaren Natriums in der Form eines Streifens von 50 bis 140 km Höhe ausgestoßen wurden. Photometrische und photographische Aufnahmen zeigen, daß, wie angenommen, starkes, anhaltendes Leuchten bei ungefähr 90 km vorkommt; daß über 105 km hinaus der Vorgang schwächer wird; und daß uber 130 bis 140 km, im höchsten Punkt der Rakete, ein anderer Vorgang eine starke Aussendung einer Strahlung von 5890 Å liefert.

Es wird vermutet, daß der letztgenannte Vorgang auf die Übertragung von Energie von atmosphärischem, atomarem Stickstoff im metastabilen ^2D-Zustand auf freies Natrium zurückzufuhren ist.

Emission de sodium à 140 kms. L'efficacité du phénomène de luminescence chimique, observable la unit, entre des éléments constitutifs de l'atmosphère et des atomes de sodium a été mesuré en libérant des quantités connues de sodium atomique sous forme de filements entre 50 et 140 km d'altitude. Les résultats photométriques aussi bien que photographiques indiquent l'occurrence d'une forte luminescence persistente vers 90 km comme prévue; au-dessus de 105 km le phénomène devient plus faible, au-dessus de 130 à 140 km, zénith de la fusée, un phénomène différent provoque une émission forte de radiation de 5890 Å.

On suppose que ce dernier phénomène se produit par le transfert au sodium libéré d'énergie provenant de l'azote atomique atmosphérique dans l'état fortement métastable ^2D.

Introduction

In a series of experiments the upper atmosphere has been studied by releasing sodium vapor over heights ranging from 50 to 140 km from Aerobee rockets. The experiments have been conducted during twilight when the sodium cloud

[1] Air Force Cambridge Research Center, Air Research and Development Command, L. G. Hanscom Field, Bedford, Massachusetts, U.S.A.

is illuminated [1] by sunlight, and during the night. The density of released sodium vapor is known approximately, and light produced by the ensuing reactions can be measured to determine the conversion of atmospheric chemical energy to light. Where specific reactions are known, this result may be used to determine the density of reacting components. In some cases it may be used to select as probable or to reject as impossible certain of the possible reactions.

Experimental Procedure

Metallic sodium in the form of small pellets is mixed with thermite which is composed of iron oxide and finely divided metallic aluminium. The mixture is then packed into a cylinder under considerable pressure [2]. When a magnesium ignitor placed at one end of the cylinder is fired electrically, the reaction

$$Fe_2O_3 + 2\,Al \rightarrow Al_2O_3 + 2\,Fe$$

liberates sufficient heat to vaporize and dissociate the sodium which escapes through vents in the cylinder.

The vaporizers are carried by Aerobee rockets. They are ignited at about 50 km by a timer mechanism. The coarseness and degree of packing of the thermite mixture determines the rate at which the above reaction proceeds along the length of the cylinder. In our case the sodium was vaporized at an approximately uniform rate for 150 seconds, or over heights from 50 km to the rocket's zenith at 140 km. Two kgm of sodium was released over this height range.

Observations

The ejected sodium underwent reactions which were observed visually. By relating the time of these observations to the corresponding rocket heights, it is determined that a trail of one or two seconds duration was visible from the time of ignition at 50 km, increasing in brightness and duration up to about 100 km where the duration of visible radiation was about 15 sec. becoming invisible above this height until about 130 km where it again appeared with relatively high intensity and again for about 15 sec. duration.

Photographs employing multiple exposures were made by the F/0.8 meteor cameras operated by Harvard University. The exposures were of 2 second duration and the camera indexed in declination between exposures to separate the various images. Two sets of photographs taken at Sacramento Peak, and Mayhill, New Mexico, were used to determine accurate heights utilizing the photographed stellar background. These photographs were also used to determine relative brightness. In Table I is listed this information.

A photometer of high spectral purity [3] was used to determine the intensity of 5890 Å and the absence of other wavelengths in the emission. The photometer scanned the region in azimuth angle at each 5° of zenith angle. A total survey required from one to two minutes, increasing to about 10 minutes as winds carried the cloud requiring that the entire azimuth be scanned. Emission was detected photometrically for about 30 minutes before the sodium cloud became too dispersed to distinguish from the normal night background. The measured surface brightness during the first part of the experiment in the brighter regions was about 10^{10} photons/cm^2/sec.

Interpretation

On the many reactions proposed to explain the 5890 Å emission of the night airglow, perhaps the most accepted are [1]

$$
\begin{aligned}
&(1)\ Na + O_2 + M \rightarrow NaO_2 + M, \text{ rate } 5 \times 10^{-30} \text{ cm}^6/\text{sec}\\
&(2)\quad NaO_2 + O \rightarrow NaO + O_2\\
&(3)\ \text{or } NaO_2 + H \rightarrow NaH + O_2\\
&(4)\quad NaO + O \rightarrow Na(^2P) + O_2\\
&(5)\ \text{or } NaH + O \rightarrow Na(^2P) + OH.
\end{aligned}
$$

Coefficients for the last two reactions are not known, but may be assumed at about 10^{-12} to 10^{-13} cm^3/sec. Table II [1] lists in the right hand column the surface brightness expected from the ejected cloud using one set of reactions, and presently available distributions of molecular and atomic oxygen [4]. It is noted that the calculated intensities agree very roughly with the observed values at 100 km, however, the intensity distribution with height does not agree at all well. Reactions involving hydrogen could not account for the observed brightness at 100 km unless the hydrogen density there is considerably greater than expected.

The observed intensity is a measure of reaction efficiency. Height of the normal 5890 Å airglow is also a function of the naturally occurring sodium distribution. However, the low efficiency at 85 km as noted in Table I would indicate that the heights for this portion of the night airglow may be above 85 km.

The strong emission appearing at 140 km is difficult to explain by processes proposed in the past. The intensity and density in these regions probably require that the photochemical reaction be a single one of the two body type. Assuming a coefficient of about 10^{-12} cm^3/sec for such a reaction, the reacting density need be about 10^7 per cm^3 to explain the observed luminosity. To account for the energy required to excite sodium to the (^2P) state it is necessary to find a mechanism such as energy transfer from a metastable level of the second body. Atomic nitrogen in the (^2D) state with radiative half life of 8 hours has been observed in aurora and it may exist in the atmosphere [5] in the required densities. The energy of this state is 2.37 volts, sufficient to excite the 2.11 volt sodium line. Very little is known about the atomic nitrogen distribution, the effective lifetime of the metastable state, the rate of formation by dissociative recombination [5], or the solar radiation during the daytime.

The high efficiency for excitation of atomic sodium at these heights calls for a reappraisal of the normal night airglow at 5890 Å. The relatively high maximum during November [6] occurs at about the same time that the maximum number of visible meteors is recorded. The amount of sodium which could be brought into the atmosphere by such meteors is, however, small. It is proposed that the number of micro-meteorites is also maximum in November, and that their penetration of the atmosphere as well as their sodium content is sufficient to account for the seasonal variation in the intensity of 5890 Å. The proposal that sodium enters the atmosphere by way of meteors is not a new one, but has not seemed feasible because the number of meteorites penetrating to the 85 km region could not account for the sodium content of the atmosphere, and calculation showed that the emission probably came from this region.

Table I

Height	*Relative Intensity*	*Height*	*Relative Intensity*
65 km	64	110	128
70	16	115	96
75	16	120	64
80	16	125	64
85	8	130	32
90	16	135	64
95	64	140	128
100	256	142	256
105	192		top of trajectory

Table II

Reaction	*Rate Coefficient*
(1) $Na + O_2 + M \rightarrow NaO_2 + M$	5×10^{-30} cm³/sec
(2) $NaO_2 + O \rightarrow NaO + O_2$	4×10^{-13} cm³/sec
(4) $NaO + O \rightarrow Na^2 (P) + O_2$	4×10^{-12} cm³/sec

Sodium Concentration = 10^9 Na/cm³
Layer Thickness = 0.5 km

	Formation per cm³ per sec			
Altitude	NaO_2 from reaction (1)	NaO from (2)	$Na(^2P)$ from (4)	Surface Brightness photons /cm²·sec
50 km	10^9	4.8×10^6	2.3×10^5	10^{10}
60	10^9	8×10^6	6.4×10^5	3.2×10^{10}
70	10^9	2×10^7	4×10^6	2×10^{11}
80	6×10^7	1.5×10^6	3.8×10^5	2×10^{10}
90	9×10^5	5.8×10^4	3.7×10^4	2×10^9
100	2.5×10^4	6.3×10^4	6.3×10^4	3×10^9
110	1.3×10^2	10^2	10^2	5×10^6

References

1. J. F. BEDINGER, E. R. MANRING, and S. N. GHOSH, Emission from Sodium Ejected from Rockets: Conference on Chemical Aeronomy, Harvard University, June 1956 (London and New York: Pergamon Press, to be published).
2. H. D. EDWARDS, J. F. BEDINGER, E. R. MANRING, and C. D. COOPER, The Airglow and Aurorae. London and New York: Pergamon Press, 1956.
3. R. B. DUNN and E. R. MANRING, J. Opt. Soc. America **46**, 572 (1956).
4. D. R. BATES, The Physics of the Upper Atmosphere: The Earth as a Planet. Chicago: The University Press, 1954.
5. S. DEB, J. Atmos. Phys. **2**, 309 (1952).
6. E. R. MANRING and H. B. PETTIT, A Study of the Airglow Emissions at 5577, 5890, and 6300 Å with a Photometer of High Spectral Purity: Conference on Chemical Aeronomy, Harvard University, June 1956 (London and New York: Pergamon Press, to be published).

Optimum Burning Program as Related to Aerodynamic Heating for a Missile Traversing the Earth's Atmosphere

By

Angelo Miele[1], ARS

(With 13 Figures)

Abstract — Zusammenfassung — Résumé

Optimum Burning Program as Related to Aerodynamic Heating for a Missile Traversing the Earth's Atmosphere. The burning program for a rocket-powered missile moving along a rectilinear path is investigated in the light of the thermal effects induced by such a program on the missile skin.

Part I considers the problem of determining the thrust-time relationship which minimizes the difference ΔG between the final and initial values of an arbitrarily specified function $G(t, m, h, V, T)$ of time, mass, altitude, velocity and skin temperature, for the case where the so-called induced drag is negligible with respect to the zero-lift drag. It is shown that the totality of extremal arcs is composed of zero-thrust sub-arcs, sub-arcs to be flown with maximum engine output and variable-thrust sub-arcs. For problems not involving time, an explicit solution is obtained for the optimizing mass flow as a function of the local coordinates of the missile.

In Part II closed form solutions are derived under the assumptions of isothermal atmosphere, negligible drag, negligible irradiation and constant STANTON number. Particular problems such as the one of minimum increase in skin temperature are treated within the general frame of the present theory. The boundary value problem is investigated and useful criteria are supplied for constructing extremal paths under various types of boundary conditions. Several numerical examples are included, illustrating the effect of important design parameters on the physical nature of the optimum burning program.

Optimales Brennprogramm in Beziehung auf die atmosphärische Erhitzung eines die Erdatmosphäre durchquerenden Geschosses. Das Brennprogramm für ein mit Raketenantrieb geradlinig fortbewegtes Geschoß wird im Licht der thermischen Effekte untersucht, die durch ein solches Programm auf die Geschoßhülle ausgeübt werden.

Im Teil I wird das Problem betrachtet, die Abhängigkeit Schub-Zeit zu bestimmen, welche ein Minimum der Differenz ΔG zwischen den End- und den Anfangswerten einer willkürlich spezifizierten Funktion $G(t, m, h, V, T)$ der Zeit, Masse, Höhe, Geschwindigkeit und Hullentemperatur fur den Fall bewirkt, daß der sogenannte induzierte Widerstand vernachlässigbar klein in Beziehung zum Widerstand bei Auftrieb Null ist. Es wird gezeigt, daß die Gesamtheit der extremen Kurvenstücke sich zusammensetzt aus Kurvenunterteilungen bei Schub Null, aus solchen, die mit maximalem Ausstoß des Triebwerkes geflogen werden, und schließlich aus Unterteilungen bei variablem Schub. Fur Probleme, bei denen die Zeit nicht mitspielt,

---

[1] Douglas Aircraft Co., Inc., Santa Monica, California, U.S.A.; Engineering Consultant, Guided Missiles Division; also, Associate Professor of Aeronautical Engineering, Purdue University, Lafayette, Indiana, U.S.A.

wird eine explizite Lösung für den optimalen Massenfluß als Funktion der örtlichen Koordinaten des Geschosses erhalten.

Im Teil II werden Lösungen in geschlossener Form unter der Annahme einer isothermen Atmosphäre, eines vernachlässigbaren Widerstandes, unerheblicher Bestrahlung und konstanter STANTON-Zahl abgeleitet. Besondere Probleme wie das des geringsten Anstiegs der Hüllentemperatur werden innerhalb des allgemeinen Rahmens der vorliegenden Theorie behandelt. Auch das Grenzwertproblem wird untersucht und Kriterien gewonnen, die für die Erreichung extremer Flugbahnen unter verschiedenartigen Grenzbedingungen nutzlich sind. Einige numerische Beispiele sind eingeschlossen, welche die Wirkung der wichtigen Konstruktionsparameter auf die physikalische Natur des optimalen Brennprogramms verdeutlichen.

Programmation optimale de la poussée d'un engin traversant l'atmosphère terrestre en relation avec son échauffement cinétique. Examen de la relation entre une programmation de la poussée d'un moteur-fusée et les effets thermiques induits sur la paroi de l'engin en vol rectiligne.

La première partie considère le problème de la détermination d'une loi, liant la poussée au temps, telle que la différence ΔG entre les valeurs terminales d'une fonction arbitraire G (t, m, h, V, T) du temps, de la masse, de l'altitude, de la vitesse et de la temperature de paroi soit rendue minimum en l'absence de traînée induite. Les extrémales comportent des arcs à poussée nulle, des arcs à poussée maximum et des arcs à poussée variable. Pour les problèmes où le temps n'intervient pas directement, une solution explicite est obtenue pour le débit optimum en fonction des coordonnées locales de l'engin.

Dans la seconde partie des solutions sous forme finie sont établies dans l'hypothèse où l'atmosphère est isotherme, la traînée négligeable ainsi que le rayonnement, et pour un nombre de STANTON constant. Le problème d'un accroissement minimum de la température de paroi rentre dans le cadre de cette théorie. Les conditions aux limites du problème sont analysées et des critères utiles fournis pour la construction des trajectoires extrémales. Plusieurs exemples numériques sont inclus; ils illustrent les effets de paramètres importants sur la nature physique de la programmation optimale.

1. Introduction

The analysis of the optimum burning program for a rocket-powered vehicle has received considerable attention in recent years. The problem can be stated as follows: "A rocket engine, capable of delivering a variable thrust, is applied to a vehicle of given configuration moving along a path of *prescribed* geometry. It is required to determine the thrust programming technique which minimizes the difference $\Delta G = G_f - G_i$ between the final and initial values of a specified function G of the coordinates of the vehicle."

Concerning vertical flight, previous investigations have been carried out by HAMEL [1] and by TSIEN and EVANS [2] for problems where $G = -m$. A broader treatment is due to MIELE [3] for problems where $G = G(h, V, m, t)$. With regard to horizontal flight, the papers by HIBBS [4] and by CICALA and MIELE [5] must be mentioned in connection with problems where $G = -X$, X denoting horizontal distance. More general questions, of the form $G = G(X, V, m, t)$, have been considered by the writer in [6].

In all of the above investigations, an optimum burning program was determined regardless of thermal effects induced by such a program on the missile skin. Mathematically speaking, the only auxiliary conditions accounted for in the variational problem were the equations of motion.

It is to be noted, however, that in the upward flight phase of sounding rockets or of long-range ballistic missiles and — in general — for all vehicles traversing the Earth's atmosphere at high speed, important heat transfer phenomena occur from and to the surrounding medium. More specifically, the missile skin receives heat energy from the boundary layer by convection (for instance) and from solar, terrestrial and interstellar sources by radiation; in turn, the skin emits radiant energy according to the well-known Stefan-Boltzmann law.

The increase in temperature of the skin is functionally related to the maneuver performed by the vehicle, i.e., functionally related to the thrust program for the case where the geometry of the flight path (or an equivalent condition) is prescribed. For missiles designed for very high Mach numbers, the temperature rise may be such as to dictate structural design, because of the loss of strength of metals. Thus, it becomes important to determine a thrust program which is optimum in the light of heat transfer phenomena. Mathematically speaking, the auxiliary conditions to be considered for the Mayer problem are the equations of motion *and* the heat balance equation [7].

2. Fundamental Hypotheses and Equations of Motion

The following hypotheses are used throughout the paper:

(a) the rocket-powered vehicle is ideally regarded as a particle of mass (m) variable with the time (t); (b) the thrust (T_*) is tangent to the flight path; (c) the equivalent exit velocity (V_e) of the rocket engine is regarded as a constant, independent of the operational condition of the engine; (d) the engine is capable of delivering all mass flows (β) bounded between a lower value ($\beta = 0$) and an upper value ($\beta = \beta_{max}$); (e) the aerodynamic lag is disregarded, i.e., the aerodynamic forces are calculated as in unaccelerated flight; (f) the acceleration of gravity is a constant; (g) the trajectory is straight and inclined at an angle θ with respect to a horizontal plane; (h) the "induced" drag is negligible with respect to the zero-lift drag[1].

In view of the hypotheses (a), (b), (g) and (h) the dynamic behaviour of the rocket-powered vehicle is represented with the following set of differential equations:

$$J_1 \equiv \dot{m} + \beta = 0 \tag{1}$$

$$J_2 \equiv \dot{h} - V \sin\theta = 0 \tag{2}$$

$$J_3 \equiv \dot{V} + g \sin\theta + \frac{D - T_*}{m} = 0 \tag{3}$$

where h is the altitude, V the velocity, g the acceleration of gravity and D the drag. The dot sign denotes derivative with respect to time.

2. 1. *The Thrust Function*

Because of the hypothesis (c) the thrust T_* is expressed as a linear function of the mass flow β:

$$T_* = \beta V_e. \tag{4}$$

In turn, hypothesis (d) implies that the following inequality

$$0 \leq \beta \leq \beta_{max} \tag{5}$$

be satisfied by the class of all arcs investigated in the present analysis.

---

[1] Notice that hypotheses (g) and (h) limit the applicability of the present theory to vertical or near-vertical trajectories, as it is the case with a sounding rocket and with some types of ballistic missiles.

The limitation (5) which the size of the engine imposes on the mass flow β has a critical importance in determining the behaviour of the solution, as has been recognized in [3], [5] and [6]. Several devices have been proposed in the literature in order to handle the above inequality. The one which the writer considers most appropriate is based on the idea of a parametric representation of the engine characteristics [3], [6]. Such a technique is developed in the present article.

The mass flow β is represented as a function of a parameter α having the following properties: (a) for $-\infty \le \alpha \le \alpha_1$, the mass flow is $\beta = 0$; (b) for $\alpha_2 \le \alpha \le +\infty$, the mass flow is $\beta = \beta_{max}$; (c) for $\alpha_1 \le \alpha \le \alpha_2$, a one-to-one correspondence is assumed between β and α.

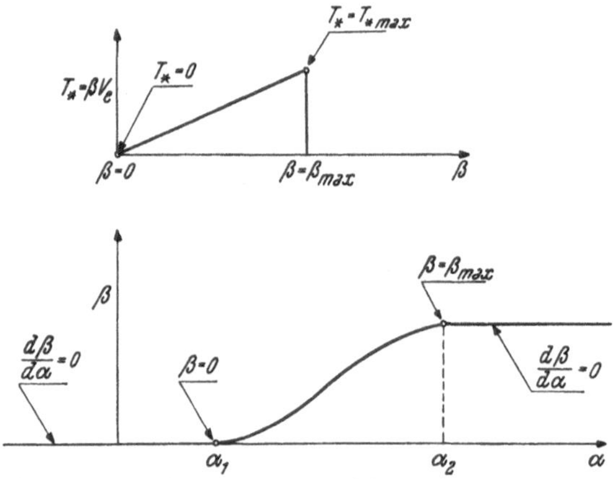

Fig. 1. Parametric representation of the engine characteristics

With the above scheme, α is considered as the *independent parameter of the rocket engine* and is allowed to vary between $-\infty$ and $+\infty$. The mass flow β becomes a *dependent* quantity, varying between 0 and β_{max}, according to the scheme of Fig. 1. In turn, the thrust T_* is limited between 0 and T_{*max} according to Eq. (4) and Fig. 1. Notice that the condition $\frac{d\beta}{d\alpha} = 0$ represents either a coasting flight or a flight with maximum engine output. On the other hand, $\frac{d\beta}{d\alpha} \neq 0$ represents any other operating conditions of the engine intermediate between the two limiting ones. Notice also that α is *only* a parameter and that there is no necessity of attributing to it any special physical meaning.

2. 2. *Atmospheric Properties*

In Part I of the present article it is assumed that the absolute temperature (τ) of the atmosphere in which the missile is flying is an arbitrarily specified function of the altitude (h). The distribution of static pressure (p) and speed of sound (a) versus altitude are consequently determined. The particular sub-case of an isothermal medium is considered in Part II.

2. 3. *The Drag Function*

Because of the hypothesis (h) the total drag of the missile (D) is identical with the zero-lift drag. The latter can be decomposed into pressure drag and friction drag. The pressure drag depends on velocity and altitude, even accounting for compressibility and viscosity effects. The friction drag, in turn, depends on velocity, altitude, and on the distribution of skin temperature of the missile. In the present analysis the effects of variations of frictional drag caused by variations in the distribution of skin temperature are neglected. As a consequence, the drag function is represented as:

$$D = D\ (V, h)\ . \tag{6}$$

3. Heat Transfer Equation

In accordance with the established practice, the calculation of the transient temperature at a point of the missile skin is carried out by assuming that heat transfer is polarized, in the sense that the speed of propagation of thermal effects is infinite in the direction normal to the skin and zero in any direction tangential to the skin.

In the light of the above hypotheses, a one-dimensional analysis becomes possible for *each point* of the missile skin, according to the following differential equation:

$$C_s \varrho_s \delta_s \dot{T} = Q + \varepsilon G_* - \varepsilon \sigma_* T^4 \tag{7}$$

where the first member denotes heat gained by the unit area of missile skin per unit time. Terms appearing on the right have the following significance: Q is the heat transferred from boundary layer to skin per unit time and unit area; εG_* is the heat received by the skin per unit time and unit area because of irradiation associated with solar, terrestrial and interstellar sources; $\varepsilon \sigma_* T^4$ is the radiant heat emitted by the missile skin per unit time and unit area[1].

In Eq. (7) T is the skin temperature; $C_s = C_s(T)$, ϱ_s and δ_s are, respectively, the specific heat (referred to the unit mass), density and thickness of the skin; $\varepsilon = \varepsilon(T)$ is the emissivity, assumed equal to the absorptivity; σ_* is the Stefan-Boltzmann radiation constant; G_* is the incident irradiation associated with solar, terrestrial, and interstellar sources and depends, in general, on the altitude (h) and on the angle of incidence between surface and solar rays. For a given missile design and for a given point of the missile skin, the parameter Q is a complicated function of free-stream Reynolds number, free-stream Mach number and ratio of wall temperature to free-stream temperature. Since the Reynolds number and the Mach number depend on velocity (V) and altitude (h) only, one concludes that—for the present problem—Eq. (7) has the form:

$$J_4 \equiv \dot{T} + \Phi(h, V, T) = 0 \tag{8}$$

$$\Phi = \frac{\varepsilon(\sigma_* T^4 - G_*) - Q}{C_s \varrho_s \delta_s}. \tag{9}$$

In the present article, a given point of the missile skin is considered and Eq. (8) applied to determine the transient temperature at such a point. In the case where the location in question is the one where the most unfavorable thermal effects are expected, it may be anticipated that the predicted skin temperature is higher than the actual one. In fact, the one-dimensional hypothesis embodied in Eq. (8) rules out any alleviation effect which the most critical parts of the missile skin may receive from the surrounding elements of matter.

I. Generalized Solutions Valid for Arbitrarily Specified Atmospheric Properties, Drag Function and Heat Transfer Equation

In this section of the paper, the optimum burning program is studied in connection with an arbitrary distribution of atmospheric temperature (τ) versus altitude (h), a drag function of the form $D = D(V, h)$ and a thermal function of the form $\Phi = \Phi(h, V, T)$.

---

[1] Eq. (7) neglects heat transfer phenomena between the skin and other components of the missile.

4. Variational Approach

The set of differential equations (1), (2), (3) and (8) is now considered, where $T_* = \beta V_e$, $\beta = \beta(\alpha)$ and $D = D(V, h)$. The time (t) is assumed as the independent variable. The dependent variables are the mass (m), the altitude (h), the velocity (V), the skin temperature (T), and the engine parameter (α). The number of differential equations is four while the number of unknown functions is five. One degree of freedom is left and an optimum requirement can be, therefore, imposed.

After prescribing the initial coordinates $t_i = 0$, $m_i = m\ (0)$, $h_i = h\ (0)$, $V_i = V\ (0)$, $T_i = T\ (0)$, and some, but not all, of the final coordinates t_f, m_f, h_f, V_f, T_f, the variational problem (of MAYER type) is stated as follows: "Amongst all sets of functions $m\ (t)$, $h\ (t)$, $V\ (t)$, $T\ (t)$, $\alpha\ (t)$ satisfying Eqs. (1), (2), (3), (8) and the prescribed end-conditions, to determine the special set which minimizes the difference ΔG between the final and the initial value of an arbitrarily specified function $G = G\ (t, m, h, V, T)$ of the coordinates of the missile". Particular cases of the above general problem are, for instance, the following: (a) the case $G = T$, minimizing the increase in skin temperature; (b) the case $G = -m$, minimizing the propellant consumption.

4. 1. *Euler Equations*

A set of variable LAGRANGE multipliers $\lambda_1\ (t)$, $\lambda_2\ (t)$, $\lambda_3\ (t)$, $\lambda_4\ (t)$ is introduced and the following expression formed:

$$F = \sum_{k=1}^{4} \lambda_k J_k \tag{10}$$

where J_1, J_2, J_3, J_4 denote, respectively, the first members of Eqs. (1), (2), (3) and (8). Since the unknown functions are five in number, one must write five EULER equations. These are written as follows:

$$\frac{d}{dt}\left[\frac{\partial F}{\partial \dot{z}_J}\right] = \frac{\partial F}{\partial z_J}\ (J = 1, \ldots, 5) \tag{11}$$

where $z_1 = m$, $z_2 = h$, $z_3 = V$, $z_4 = T$, $z_5 = \alpha$. The explicit form of Eqs. (11) is the following:

$$\dot{\lambda}_1 = \frac{V_e \beta - D}{m^2} \lambda_3 \tag{12}$$

$$\dot{\lambda}_2 = \lambda_3 \frac{D_h}{m} + \lambda_4 \Phi_h \tag{13}$$

$$\dot{\lambda}_3 = -\lambda_2 \sin\theta + \lambda_3 \frac{D_v}{m} + \lambda_4 \Phi_v \tag{14}$$

$$\dot{\lambda}_4 = \lambda_4 \Phi_T \tag{15}$$

$$0 = \frac{d\beta}{d\alpha}\left[\lambda_1 - \lambda_3 \frac{V_e}{m}\right] \tag{16}$$

$$D_h = \frac{\partial D}{\partial h};\ D_v = \frac{\partial D}{\partial V} \tag{17}$$

$$\Phi_h = \frac{\partial \Phi}{\partial h};\ \Phi_v = \frac{\partial \Phi}{\partial V};\ \Phi_T = \frac{\partial \Phi}{\partial T}. \tag{18}$$

4. 2. First Integral

Since the fundamental function F, given by Eq. (10), is formally independent of the time, the following first integral holds:

$$-\lambda_1 \beta + \lambda_2 \, V \sin\theta + \lambda_3 \left[\frac{V_e \beta - D}{m} - g \sin\theta\right] - \lambda_4 \, \Phi = C_1 \tag{19}$$

where C_1 is an integration constant.

4. 3. Discontinuity of the Eulerian Solution

The EULER equation (16) is particularly interesting because it shows that the extremal arc is discontinuous, being generally composed of:

(a) sub-arcs of equation

$$\frac{d\beta}{d\alpha} = 0 \tag{20}$$

(b) sub-arcs of equation

$$\lambda_1 - \lambda_3 \frac{V_e}{m} = 0. \tag{21}$$

According to the parametric representation indicated in Fig. 1, Eq. (20) represents either a coasting flight or a flight with maximum engine output. Eq. (21), on the other hand, represents a flight condition with a continuous variable-thrust, as shown in the following sections.

4. 4. Erdmann-Weierstrass Corner Conditions

In view of the discontinuous character of the solution, the ERDMANN-WEIERSTRASS corner conditions [8] must be applied. These are continuity conditions to be satisfied *at every corner* of the discontinuous extremal solution by the following six quantities:

$$\frac{\partial F}{\partial m} \, ; \; \frac{\partial F}{\partial \dot h} \, ; \; \frac{\partial F}{\partial \dot V} \, ; \; \frac{\partial F}{\partial \dot T} \, ; \; \frac{\partial F}{\partial \dot\alpha} \, ; \; \sum_{J=1}^{5} \frac{\partial F}{\partial \dot z_J} \dot z_J - F. \tag{22}$$

Since F is independent of $\dot\alpha$, the continuity of $\frac{\partial F}{\partial \dot\alpha}$ is inherently verified at all corner points. The analogous requirement for the remaining five quantities (22) leads to:

$$(\lambda_K)_- = (\lambda_K)_+ \quad (K = 1, 2, 3, 4) \tag{23}$$

$$(C_1)_- = (C_1)_+ \tag{24}$$

where the subscript (—) denotes a condition *immediately before* the junction and the subscript (+) a condition *immediately after* the junction. The ERDMANN-WEIERSTRASS conditions require, therefore: (a) that the multipliers λ_k ($k = 1, 2, 3, 4$) be *continuous* at all junction points; (b) that the integration constant C_1 be the *same* for all sub-arcs of the discontinuous extremal solution; this circumstance implies that the quantity $\lambda_1 - (\lambda_3 \, V_e/m)$ be *zero* immediately before and immediately after *each* junction point.

4. 5. *Boundary Conditions*[1]

For the problem under consideration, the boundary conditions include a number of fixed end-point conditions plus a number of natural conditions. The latter must be obtained from the following general *transversality conditions*, which is to be *identically satisfied* [8] for all systems of variations (δt, δz_j) consistent with the prescribed end-conditions:

$$[\delta G + \lambda_1 \delta m + \lambda_2 \delta h + \lambda_3 \delta V + \lambda_4 \delta T - C_1 \delta t]_i^f = 0. \tag{25}$$

5. Closed Form Expressions for the Distribution of Lagrange Multipliers along the Variable-Thrust Sub-Arc

As the analysis of [9] shows, closed form expressions can be derived, yielding the LAGRANGE multipliers λ_k ($k=1, \ldots 4$) at all points of the variable-thrust sub-arc, defined by Eq. (21). In this connection, extensive manipulations yield the following results:

$$\frac{\lambda_1}{C_2} = V_e \, exp\left(\frac{V + g t \sin\theta}{V_e}\right) \tag{26}$$

$$\frac{\lambda_2}{C_2} = \frac{\frac{C_1}{C_2}\Phi_v + \left[\Phi_v \left(D + m g \sin\theta\right) - \Phi\left(\frac{D}{V_e} + D_v\right)\right] exp\left(\frac{V + g t \sin\theta}{V_e}\right)}{\left(V \Phi_v - \Phi\right) \sin\theta} \tag{27}$$

$$\frac{\lambda_3}{C_2} = m \, exp\left(\frac{V + g t \sin\theta}{V_e}\right) \tag{28}$$

$$\frac{\lambda_4}{C_2} = \frac{\frac{C_1}{C_2} + \left[D + m g \sin\theta - V\left(\frac{D}{V_e} + D_v\right)\right] exp\left(\frac{V + g t \sin\theta}{V_e}\right)}{V \Phi_v - \Phi} \tag{29}$$

where C_2 is an integration constant. Notice that the above equations hold regardless of the boundary conditions and whatever be the shape of the function G whose difference ΔG is to be minimized.

5. 1. *Problems where no Time Condition is Imposed*

For the particular case where the G-function has the form $G = G\,(m, h, V, T)$ and the final time instant is free of choice ($\delta t_f \equiv$ arbitrary), the transversality condition (25) yields $C_1 = 0$.

5. 2. *Problems not Involving Heat Transfer*

If the G-function has the form $G = G\,(t, m, h, V)$ and the final temperature is free of choice ($\delta T_f \equiv$ arbitrary), the transversality condition (25) yields $\lambda_{4f} = 0$. In account of the above condition at the final point, the EULER equation (15) leads to $\lambda_4 = 0$, at all time instants. As a consequence, Eq. (29) reduces to:

$$D + mg \sin\theta - V\left(\frac{D}{V_e} + D_v\right) + \frac{C_1}{C_2} exp\left[-\frac{V + g t \sin\theta}{V_e}\right] = 0 \tag{30}$$

and therefore yields, for $\theta = \frac{\pi}{2}$, a result already obtained by the writer in [3].

[1] The present article is a condensed form of the investigation described in [9], where more extensive analytical details are available.

5. 3. Problems not Involving Heat Transfer where no Time Condition is Imposed

For the sub-case where the G-function has the form $G = G(m, h, V)$ and both the final skin temperature T_f and the final time instant t_f are free of choice, the simultaneous conditions $\lambda_4 = 0$ and $C_1 = 0$ must be applied. As a consequence, Eq. (30) leads to:

$$D + mg \sin\theta - V\left(\frac{D}{V_e} + D_v\right) = 0. \tag{31}$$

For $\theta = \frac{\pi}{2}$, Eq. (31) supplies a result which is implied in the equations already derived by Tsien and Evans, for one particular set of boundary conditions, in [2].

6. Optimum Mass Flow at Points of the Variable-Thrust Sub-Arc for Problems where no Time Condition is Imposed

An explicit expression for the optimum acceleration (or for the optimum thrust) can be obtained if the Lagrange multipliers are eliminated. In this connection, Eqs. (15) and (29) yield (for $C_1 = 0$) the following result:

$$\frac{\dot{V}}{g} = \frac{\Psi_1 + \Psi_2}{\Psi_3 + \Psi_4} \equiv \Psi \tag{32}$$

$$\Psi_1 = -\sin\theta \frac{V(D + V_e D_v) + \frac{V_e V}{g}[V D_h + V V_e D_{vh} - V_e D_h]}{V(D + V_e D_v) - V_e[D + mg \sin\theta]} \tag{33}$$

$$\Psi_2 = \frac{V V_e}{g} \frac{(\Phi_h - V\Phi_{vh})\sin\theta + \Phi\Phi_{vT} - \Phi_v \Phi_T}{\Phi - V\Phi_v} \tag{34}$$

$$\Psi_3 = V \frac{D + 2 V_e D_v + V_e^2 D_{vv}}{V(D + V_e D_v) - V_e(D + mg \sin\theta)} \tag{35}$$

$$\Psi_4 = V V_e \frac{\Phi_{vv}}{\Phi - V\Phi_v} \tag{36}$$

$$D_{vv} = \frac{\partial^2 D}{\partial V^2}\,;\; D_{vh} = \frac{\partial^2 D}{\partial V\, \partial h} \tag{37}$$

$$\Phi_{vv} = \frac{\partial^2 \Phi}{\partial V^2}\,;\; \Phi_{vh} \frac{\partial^2 \Phi}{\partial V\, \partial h}\,;\; \Phi_{vT} = \frac{\partial^2 \Phi}{\partial V\, \partial T}\,. \tag{38}$$

The drag D and its derivatives depend on velocity and altitude only. In turn, the thermal function Φ and its derivatives depend on velocity, altitude, and skin temperature. As a consequence, it is concluded that the Ψ-function has the form $\Psi = \Psi(m, h, V, T)$. The instantaneous mass flow at points of the variable-thrust sub-arc of the extremal solution is consequently given by:

$$\frac{V_e \beta}{mg} = \sin\theta + \frac{D}{mg} + \Psi(m, h, V, T). \tag{39}$$

In conclusion, the optimizing mass flow β can either be $\beta = 0$ or $\beta = \beta_{max}$ or can have the value which satisfies Eq. (39). The latter holds whatever be the distribution of atmospheric properties versus altitude, the drag function D and the thermal function Φ; it also holds for all G-functions of the form $G = G(m, h, V, T)$, provided the final time instant is free of choice.

II. Closed Form Solutions Valid for an Isothermal Atmosphere, Negligible Drag, Negligible Irradiation, in Connection with Problems where no Time Condition is Imposed

In Part I of the present paper an explicit solution has been obtained for the optimum mass flow. For the variable-thrust sub-arc, this optimum mass flow is defined by Eqs. (32) to (39). Its form, however, is such that analytical solutions of the system composed of heat transfer equation and equations of motion are out of reach. Numerical analyses are in order and digital computers must be used. Prior to any attack of the problem with digital computing equipment, however, it appears desirable to achieve some qualitative understanding of the physical nature of the optimum thrust program, even if drastic hypotheses are necessary to attain such an objective.

7. Additional Hypotheses

In the following section, closed form solutions are derived for problems where $G = G\,(m, h, V, T)$. The below indicated hypotheses are used in combination with those listed in paragraph 2: (a) an ideally isothermal atmosphere is assumed, i.e., an atmosphere where the speed of sound is constant; (b) the drag of the missile is considered negligible with respect to thrust and weight ($D = 0$); (c) irradiation terms are neglected in the heat balance equation; (d) use is made of a *modified Stanton number* (St), based on free stream density (ϱ), free stream velocity (V) and on a conventional adiabatic wall temperature (T_a) corresponding to free-stream conditions; such a STANTON number[1] is regarded as a constant; moreover, constant values are also assumed for the thermal properties of air and skin and for the recovery factor (r).

In view of the hypothesis (a) the distribution of relative density (σ) versus altitude (h) is given by:

$$\sigma = \frac{\varrho}{\varrho_0} = exp\,(-\gamma \bar{h}) \tag{40}$$

$$\bar{h} = \frac{g\,h}{a^2} \tag{41}$$

where $\bar{h}$ is a non-dimensional altitude, ϱ is the absolute density of the air at altitude h and ϱ_0 the absolute density at sea-level. Because of the neglect of irradiation terms, the thermal function Φ takes the form:

$$\Phi = -\frac{Q}{C_s\,\varrho_s\,\delta_s} \tag{42}$$

where Q is the heat input from boundary layer to skin per unit time and unit wetted area. The latter can be rewritten as:

---

[1] For a given missile design and for a given point of the missile skin, the STANTON number (St) is a complicated function of free-stream REYNOLDS number (Re), free-stream MACH number (M) and ratio of wall temperature (T) to free-stream temperature (τ). In spite of this, the following sections of this paper refer to the ideal case where the STANTON number is a constant, independent of M, Re and T/τ. It cannot be denied that the above hypothesis is quite drastic in its nature. Nevertheless, the writer feels that the essential facts of the mechanism of heat transfer are retained in the present idealized scheme. The latter—on the other hand—has the merit of leading to closed form solutions and, therefore, to clear-cut information on the qualitative nature of the optimum burning program. Moreover, as the subsequent analysis shows, the optimum burning program is affected only in a minor way by the hypothesis concerning the STANTON number.

$$\frac{Q}{C_p \varrho V (T_a - T)} = St \tag{43}$$

$$T_a = \tau \left(1 + \frac{\gamma - 1}{2} r M^2\right) \tag{44}$$

where M is the free-stream MACH number, τ the free-stream temperature and γ the ratio of specific heat at constant pressure (C_p) to specific heat at constant volume (C_v). Simple manipulations yield the following useful form for the thermal function:

$$\Phi = K_1 \sigma M \left[\frac{T}{\tau} - 1 - K_2 M^2\right] \tag{45}$$

$$K_1 = \frac{C_p}{C_s} \frac{\varrho_0}{\varrho_s} \frac{a}{\delta_s} \tau\, St \tag{46}$$

$$K_2 = \frac{\gamma - 1}{2} r. \tag{47}$$

8. Integration of the Equations of Motion

8. 1. Sub-Arc Flown with Maximum Engine Output

Because of the hypotheses of paragraph 7 the sub-arc $\beta = \beta_{max}$ is identified by:

$$M + \bar{t} + u \log \bar{m} = C_3 \tag{48}$$

$$\frac{\bar{m} u}{K_3} + \bar{t} = C_4 \tag{49}$$

$$\bar{h} + \frac{\bar{t}^2}{2} + \frac{u^2}{K_3} \bar{m} (1 - \log \bar{m}) = C_5 \tag{50}$$

where $\bar{m} = \frac{m}{m_i}$ is a non-dimensional mass, $M = \frac{V}{a}$ the MACH number, $u = \frac{V_e}{a}$ the ratio of exit velocity to atmospheric speed of sound, and $\bar{t} = (tg \sin \theta)/a$ a non-dimensional time. The terms appearing on the right of Eqs. (48) to (50) are integration constants and:

$$K_3 = \frac{V_e \beta_{max}}{g\, m_i \sin \theta}\,. \tag{51}$$

For the ideal case of a pulse-burning sub-arc ($K_3 = \infty$), Eqs. (48) to (50) imply the constancy of the following three quantities: $\bar{h}$, $\bar{t}$, and $M + u \log \bar{m}$.

8. 2. Coasting Sub-Arc

For $\beta = 0$ the equations of motion yield the following obvious results:

$$\bar{m} = C_6 \tag{52}$$

$$M^2 + 2 \bar{h} = C_7 \tag{53}$$

$$\bar{t} + M = C_8 \tag{54}$$

where C_6, C_7, C_8 are integration constants.

8. 3. Variable Thrust Sub-Arc

For the variable thrust sub-arc, the optimum acceleration reduces to:

$$\frac{\dot{V}}{g} = \frac{\gamma \sin \theta}{3} M^2 [1 - 3 K_4 \sigma] \tag{55}$$

$$K_4 = \frac{C_p}{C_s} \frac{\varrho_0}{\varrho_s} \frac{a^2}{g\,\delta_s} \frac{St}{3\,\gamma \sin\theta}\,. \tag{56}$$

At this point a change of independent variable is operated in the sense that the relative density σ is now assumed as the new independent variable. Eq. (55) is rewritten as:

$$\frac{d\,M}{M} = K_4 - \frac{1}{3\,\sigma} \quad d\,\sigma \tag{57}$$

admitting, therefore, the following general integral:

$$M = C_9 \frac{exp\,(K_4\,\sigma)}{\sqrt[3]{\sigma}} \tag{58}$$

where C_9 is a constant. The differential relationship between time and relative density is stated as:

$$d\bar{t} = -\,C_{10} \frac{d\,\sigma}{\sigma^{2/3}\,exp\,(K_4\,\sigma)} \tag{59}$$

$$C_{10} = \frac{1}{\gamma\,C_9}\,. \tag{60}$$

The integration process for Eq. (59) leads to[1]:

$$\bar{t} = C_{11} - C_{10}\,A\,(\sigma) \tag{61}$$

$$A\,(\sigma) = 3\,\sqrt[3]{\sigma} \sum_{n=o}^{\infty} (-\,1)^n \frac{(K_4\,\sigma)^n}{n\,!\,(3\,n+1)} \tag{62}$$

where C_{11} is a constant. The distribution of mass along the variable-thrust sub-arc is calculated from:

$$\bar{m} = C_{12}\,exp\left(-\,\frac{M+\bar{t}}{u}\right) \tag{63}$$

where C_{12} is a constant.

8. 3. 1. Discussion of Results. As the previous analysis shows, the thermal parameter K_4, defined by Eq. (56), is systematically present in the solutions. Such a parameter depends on the thermal properties of air and skin and on the STANTON number, which characterizes the convective process from boundary layer to skin. The lowest value for the above thermal parameter is $K_4 = 0$: it corresponds to the ideal case of a skin of infinite thermal capacity. Increasing values of K_4 are associated with decreasing thickness of skin.

Mach Number-Altitude Relationship. The ratio $\frac{M}{M_o}$ of MACH number at altitude to MACH number at sea-level is plotted in Figs. 2 and 3 as a function on the non-dimensional altitude $\bar{h}$, and of the thermal parameter K_4. For $K_4 < {}^1/_3$ (skin of relatively high thermal capacity) the MACH number increases monotonically as the altitude increases. On the other hand, for $K_4 > {}^1/_3$ (skin of relatively low thermal capacity) the MACH number initially decreases, has a minimum at the altitude where $\sigma = 1/(3\,K_4)$ and afterwards increases as the altitude increases; in other words, after entering the variable-thrust sub-arc, the velocity of the missile must be decreased to limit the amount of heat transferred from boundary layer to skin in the low altitude region.

[1] A discussion of the manipulations leading to Eqs. (61) and (62) can be found in [9]. In many practical cases, the term $K_4\,\sigma$ is at most of order one. As a consequence, no appreciable error is introduced into the results by neglecting the terms of exponent larger than 3 in the expression for $A\,(\sigma)$.

Acceleration-Altitude Relationship. The ratio $\frac{\dot{V}}{\dot{V}_0}$ of optimum acceleration at altitude to optimum acceleration at sea-level is indicated in Fig. 4 as a function of the altitude $\bar{h}$, for two values of the thermal parameter K_4. The diagram shows that the acceleration increases with such rapidity, as $\bar{h}$ increases, that there is a very definite altitude above which the optimum burning program may become impractical from a structural standpoint. Consider, for instance, a missile whose thermal constant is $K_4 = 0.2$ and assume that the initial acceleration is $\frac{\dot{V}_0}{g} = 0.5$. Assume also that the maximum permissible acceleration is $(\dot{V}/g)_{max} = 15$. As Fig. 4 shows, the optimum burning program can only be used at altitudes below[1] 100,000 ft.

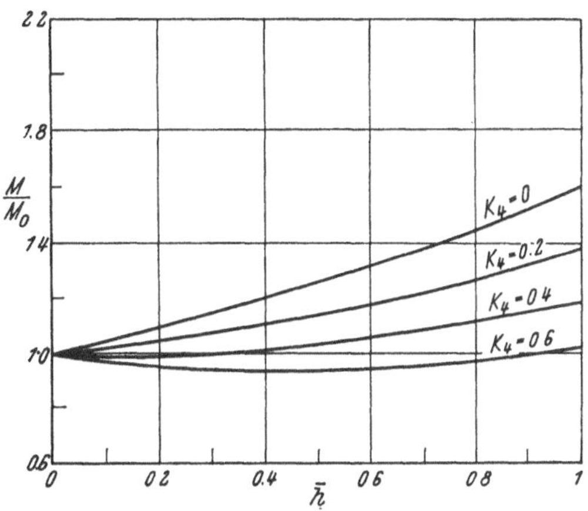

Fig. 2. Mach number-altitude relationship along the variable-thrust sub-arc

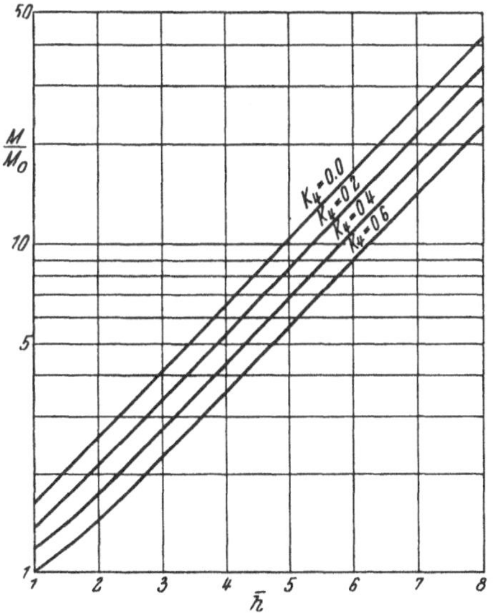

Fig. 3. Mach number-altitude relationship along the variable-thrust sub-arc

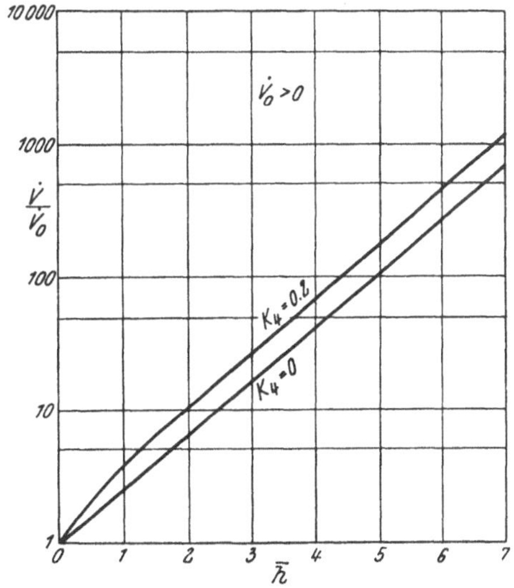

Fig. 4. Acceleration-altitude relationship along the variable-thrust sub-arc

9. Integration of the Heat Transfer Equation

After assuming the relative density σ as the independent variable the heat transfer equation (8) can be rewritten as:

$$\frac{d\bar{T}}{d\sigma} - 3\,K_4\,\bar{T} + 3\,K_4\,(1 + K_2\,M^2) = 0 \tag{64}$$

[1] The figure in question is based on an ideal isothermal atmosphere where $a \simeq 1000$ fps.

where $\bar{T} = \frac{T}{\tau}$ is the ratio of skin temperature to free-stream temperature and K_4 is defined by Eq. (56). For a prescribed MACH number-density relationship $M(\sigma)$, the above equation can be regarded as linear in the unknown function $\bar{T}(\sigma)$, yielding the following general integral:

$$\bar{T} = 1 + [\text{Const.} - 3 K_2 K_4 \int M^2 \exp(-3 K_4 \sigma) \, d\sigma] \exp(3 K_4 \sigma) \cdot \tag{65}$$

9. 1. Sub-Arc Flown with Maximum Engine Output

For the sub-arc discussed in section 8.1, the temperature equation must be dealt with, in general, by approximate methods. An important exception is represented by the pulse-burning case in which Eq. (65) reduces to:

$$\bar{T} = C_{13} \tag{66}$$

where C_{13} is a constant.

9. 2. Coasting Sub-Arc

For the coasting sub-arc, Eq. (65) yields the following result:

$$\bar{T} = 1 + K_2 C_7 + C_{14} \exp(3 K_4 \sigma) + \frac{2}{\gamma} K_2 C(\sigma) \tag{67}$$

$$C(\sigma) = \log \sigma + [\log(3 K_4) - E i(-3 K_4 \sigma)] \exp(3 K_4 \sigma) \tag{68}$$

where C_{14} is a constant and $E i(-3 K_4 \sigma)$ is the so-called exponential-integral function, defined as follows:

$$E i(-3 K_4 \sigma) = \int_{\infty}^{3 K_4 \sigma} \frac{\exp(-x)}{x} dx = 0.5772 + \log(3 K_4 \sigma) + \sum_{n=1}^{\infty} (-l)^n \frac{(3 K_4 \sigma)^n}{n! \, n} . \tag{69}$$

9. 3. Variable-Thrust Sub-Arc

For the variable-thrust sub-arc the skin temperature-density relationship is supplied by:

$$\bar{T} = 1 + [C_{15} - 3 K_2 K_4 C_9^2 A(\sigma)] \exp(3 K_4 \sigma) \tag{70}$$

where C_{15} is an integration constant.

9. 3. 1. An Upper Limit for the Increase in Skin Temperature. As the subsequent analysis shows, an upper limit can be detected for the increase in skin temperature of a missile which flies according to the optimum burning program. Assume, for instance, that at the moment of entering the variable-thrust sub-arc, the flight condition is represented by $\bar{h} = 0$, $\sigma = 1$, $M = M_o$, $\bar{T} = 1$. The integration constants C_9 and C_{15} are subsequently calculated and Eq. (70) rewritten as follows:

$$\frac{\bar{T} - 1}{K_2 M_0^2} = 3 K_4 [A(1) - A(\sigma)] \exp[K_4 (3\sigma - 2)]. \tag{71}$$

To investigate whether Eq. (71) has any stationary point, the condition $\frac{d\bar{T}}{d\sigma} = 0$ is imposed, yielding:

$$3 K_4 \sigma^{2/3} \exp(K_4 \sigma) [A(1) - A(\sigma)] = 1. \tag{72}$$

The above equation is plotted in Fig. 5 from which the following conclusions are drawn:

(a) For a thermal constant $K_4 < 0.99$ the skin temperature is a monotonically increasing function of the altitude. As a consequence, an upper limit to the skin

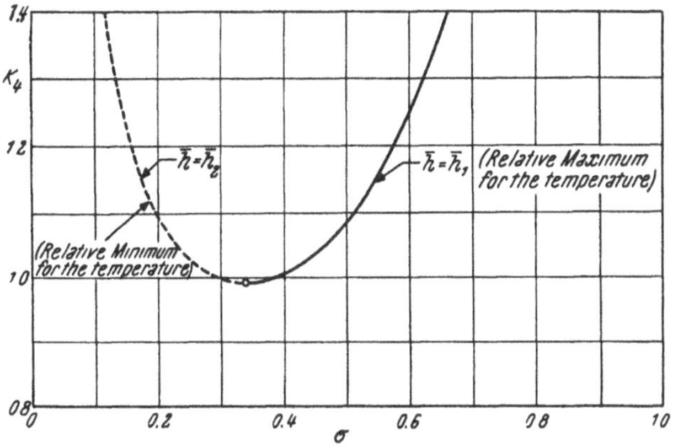

Fig. 5. Altitude where the skin temperature is stationary (variable-thrust sub-arc)

temperature is supplied by the ideal condition which the missile would attain, should the optimum burning program be continued up to $\bar{h} = \infty$, i.e., $\sigma = 0$:

$$\frac{\bar{T}-1}{K_2\, M_0^{\,2}} = 3\, K_4\, A\, (1)\, exp\, (-\, 2\, K_4). \qquad (73)$$

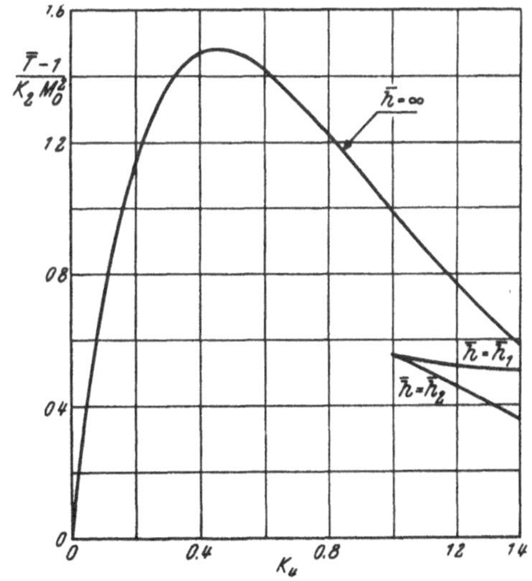

Fig. 6. Skin temperatures at $\bar{h} = \infty$ and at stationary points (variable-thrust sub-arc)

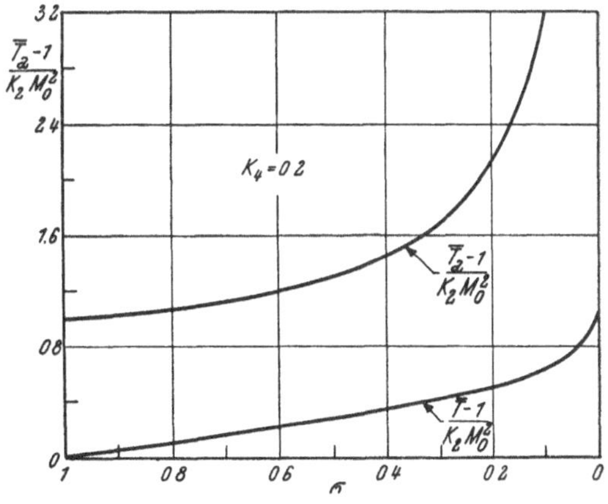

Fig. 7. Adiabatic wall temperature and skin temperature at points of the variable-thrust sub-arc

(b) For a thermal constant $K_4 > 0.99$ there are two altitudes at which the skin temperature is stationary, the lower altitude ($\bar{h}_1$) corresponding to a relative maximum for the temperature ($\bar{T}_1$), the higher altitude ($\bar{h}_2$) to a relative minimum ($\bar{T}_2$). The temperature rise between sea-level and the altitude defined by Eq. (72) is supplied by:

$$\frac{\overline{T}-1}{K_2 M_0^2} = \sigma^{-2/3} \, exp\,[2\, K_4 (\sigma - 1)] \tag{74}$$

where the density σ is to be consistent with Eq. (72).

In Fig. 6 the temperatures defined by Eqs. (73) and (74) are plotted as a function of the thermal constant K_4. As the graph indicates, for $0.99 < K_4 < 1.4$, the stationary condition reached at $\overline{h} = \overline{h}_1$ is not critical and the absolute maximum for the temperature is reached at $\overline{h} = \infty$. To emphasize the significance of the above results, Figs. 7 and 8 have been prepared. They yield the non-dimensional skin temperature ($\overline{T} = T/\tau$) and boundary layer temperature ($\overline{T}_a = T_a/\tau$) as a function of the relative density σ at points of the variable-thrust sub-arc. For a thermal constant $K_4 < 1/3$ [for instance $K_4 = 0.2$] the optimum MACH number is a monotonically increasing function of the altitude [Figs. 2 and 3]. As a consequence, the boundary layer temperature $\overline{T}_a$ increases with the altitude: heat energy is continuously transferred from boundary layer to missile skin, as shown in Fig. 7. On the other hand, for a thermal constant $K_4 > 0.99$, i.e., for a skin of relatively low thermal capacity, the MACH number-altitude diagram presents a minimum point. The diagram of the boundary layer temperature, therefore, has also a minimum point. The skin temperature rise becomes such that an *inversion* occurs in the heat transfer process. This interesting phenomenon is shown in Fig. 8 which refers to a thermal constant $K_4 = 1.4$: heat energy is released from *skin to boundary layer* in the altitude interval where $0.112 < \sigma < 0.657$.

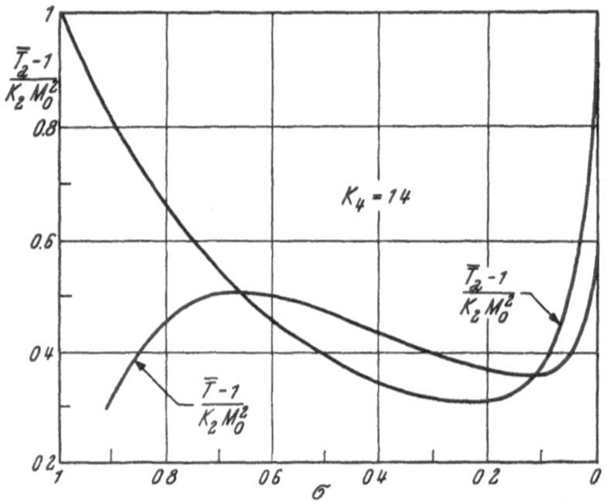

Fig. 8. Adiabatic wall temperature and skin temperature at points of the variable-thrust sub-arc

10. Solution of the Boundary Value Problem

In the previous sections, general solutions have been derived for the problem of minimizing the difference ΔG between the final and the initial value of an arbitrarily specified function $G = G\,(m, h, V, T)$. It has been shown that the extremal arc is composed of sub-arcs $\beta = 0$, sub-arcs $\beta = \beta_{max}$ and sub-arcs to be flown with regulated thrust.

In the present paragraph, a particular form of the G-function is considered, namely, $G \equiv T$ (minimum increase in skin temperature). In this connection, the boundary value problem is investigated: it consists of determining the special combination of sub-arcs which satisfy a set of prescribed end-conditions, more specifically: $\overline{t}_i = 0$, $\overline{m}_i = 1$, $\overline{h}_i = 0$, $M_i = 0$, $\overline{T}_i = 1$; $\overline{m}_f \equiv$ given, $M_f \equiv$ given, $\overline{h} \equiv$ $\equiv$ given.

Two main types of extremal trajectories may exist [9]. They are indicated in Fig. 9 with the Roman numerals I and II. The trajectories of type I include: an initial sub-arc IA flown with maximum thrust; a central sub-arc AB flown with variable thrust; and a final sub-arc BF flown with maximum thrust. The trajectories of type II differ from the previous ones insofar as the final sub-arc BF involves a coasting flight.

In the following sections the ideal case of an engine capable of delivering all mass flows between 0 and ∞ is considered, for simplicity. The reader, however, may find an extensive treatment in [9], for the case where the maximum thrust is finite.

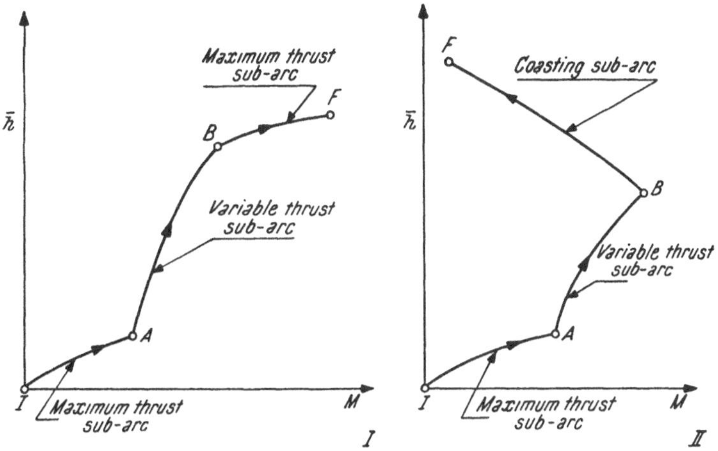

Fig. 9. Combination of sub-arcs for minimum increase in skin temperature

10. 1. *The ω-Function*

The following function of the local coordinates of the missile is now defined:

$$\omega = \frac{exp\,(K_4\,\sigma)}{\sqrt[3]{\sigma}}\,[A\,(\sigma) - A\,(1)] - \gamma\, M\,[M + u \log \overline{m}]. \tag{75}$$

Let the symbol ω_f denote value of ω calculated at final point. As the analysis of [9] points out, the final point F must be reached: (a) with a coasting flight if $\omega_f < 0$; (b) with engine operating at maximum output if $\omega_f > 0$; (c) with regulated thrust if $\omega = 0$.

10. 2. *Determination of Corner Points*

A fundamental aspect of the boundary value problem is to determine the corner points A and B of the extremal solution. In general, an indirect procedure must be used [9]. For the particular case $\beta_{max} = \infty$, however, a direct approach is possible.

10. 2. 1. Case (I) where the Final Sub-Arc BF is Flown with Maximum Engine Output. As the analysis of [9] points out the Mach numbers at points A and B are supplied by:

$$M_A = \frac{exp\,(K_4)}{\gamma}\,\frac{A(\sigma_f) - A\,(1)}{M_f + u \log \overline{m}_f} \tag{76}$$

$$M_B = \frac{1}{\gamma}\,\frac{A\,(\sigma_f) - A\,(1)}{M_f + u \log \overline{m}_f}\,\frac{exp\,(K_4\,\sigma_f)}{\sqrt[3]{\sigma_f}}\,. \tag{77}$$

10. 2. 2. Case (II) where the Final Sub-Arc BF is Flown with Engine Shut-Off. As it is shown in [9], the following transcendental equation determines the relative density at point B:

$$-\gamma\sqrt{M_f^2+\frac{2}{\gamma}\log\frac{\sigma_B}{\sigma_f}}\left[u\log\overline{m}_f+\sqrt{M_f^2+\frac{2}{\gamma}\log\frac{\sigma_B}{\sigma_f}}\right]+ \\ +[A(\sigma_B)-A(1)]\frac{exp(K_4\sigma_B)}{\sqrt[3]{\sigma_B}}=0\,. \tag{78}$$

Once σ_B is known, the MACH numbers at points A and B are determined from:

$$M_A=\sqrt{M_f^2+\frac{2}{\gamma}\log\frac{\sigma_B}{\sigma_f}}\sqrt[3]{\sigma_B\; exp[K_4(1-\sigma_B)]} \tag{79}$$

$$M_B=\sqrt{M_f^2+\frac{2}{\gamma}\log\frac{\sigma_B}{\sigma_f}}\,. \tag{80}$$

11. Numerical Examples

With the object of illustrating the previous approximate theory, several numerical examples have been carried out. These examples are now described.

11. 1. Burning Program for Minimum Temperature Increase in a Sounding Rocket, for Several Values of the Thermal Parameter K_4

The problem of minimizing the temperature increase in a sounding rocket is now considered in connection with the following set of conditions:

$$\begin{array}{llll} \overline{m}_i=1, & \overline{h}_i=0, & \overline{T}_i=1, & M_i=0 \\ \overline{m}_f=0.4, & \overline{h}_f=12.88, & & M_f=0. \\ K_2=0.19, & u=7, & & \end{array} \tag{81}$$

Several values are considered for the thermal parameter K_4, namely $K_4=0.2$, $K_4=0.4$ and $K_4=0.6$. The first step is to determine the sign of the function ω, defined by Eq. (75), at the final point. Due to the fact that $\omega_f<0$, one concludes that the final point is to be reached with a coasting flight. The second step is to determine the altitude $\overline{h}_B$ where the transition from regulated thrust to coasting occurs. In this connection, the transcendental Eq. (78) must be solved. The MACH numbers M_A and M_B are calculated with Eqs. (79) and (80). The subsequent step is to determine the distribution of MACH number, mass, time, and temperature with the equations of paragraphs 8 and 9. The EULERian paths include a pulse-burning sub-arc IA, a central sub-arc AB flown with regulated thrust and a final coasting sub-arc BF, and are described in Fig. 10 from which the following conclusion is drawn: *the optimum burning program is quite insensitive to changes in the thermal properties of air and skin.* The temperature rise $\overline{T}-1$, on the contrary, strongly depends on the value of the thermal parameter K_4 (Fig. 11).

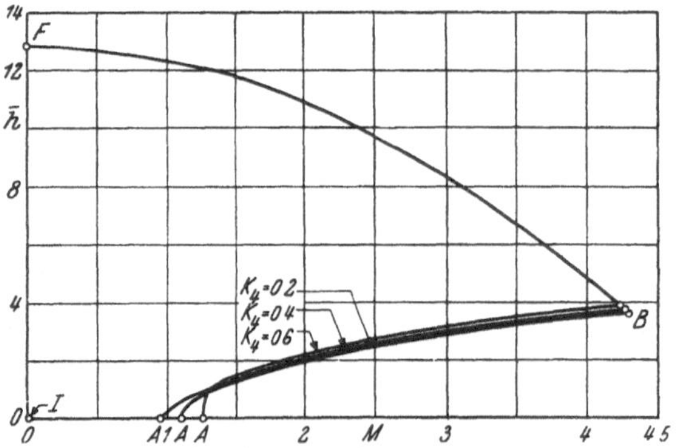

Fig. 10. Extremal trajectories of type II, for different values of the thermal parameter K_4

In the light of the above statement, the writer feels that the hypothesis of constancy of the Stanton number is well-justified from the point of view of predicting the optimum mass flow β as a function of the time t. The errors introduced by the Stanton number hypothesis in the calculation of skin temperatures, however, should be corrected. In this connection, the following iterative procedure appears to be logical: (a) calculate the optimum burning program in the $(\bar{h}, M, \bar{m}, \bar{t})$ space by assuming a constant value for the parameter K_4; (b) determine the distribution of temperatures $\bar{T}(\bar{h})$ for constant value of K_4; (c) starting from

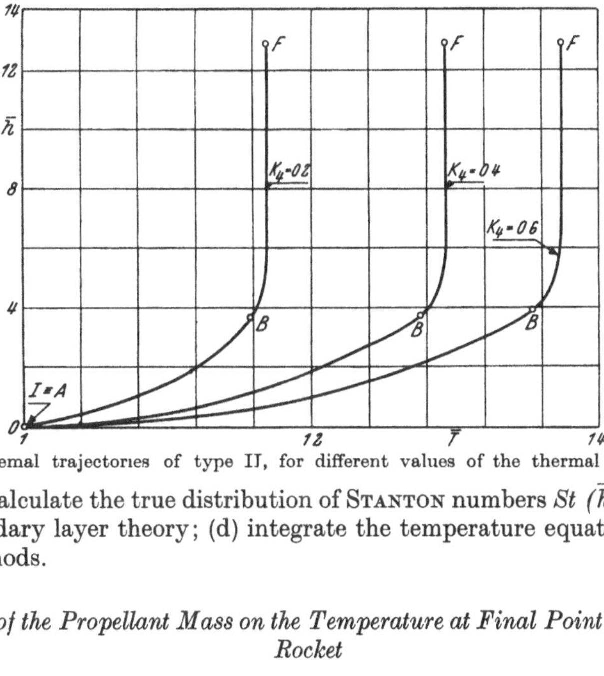

Fig. 11. Extremal trajectories of type II, for different values of the thermal parameter K_4

(a) and (b), calculate the true distribution of Stanton numbers $St(\bar{h})$, as supplied by the boundary layer theory; (d) integrate the temperature equation by approximate methods.

11. 2. Effect of the Propellant Mass on the Temperature at Final Point in a Sounding Rocket

The problem of a sounding rocket is now considered in connection with the following set of conditions:

$$\begin{array}{llll} \bar{m}_i = 1, & \bar{h}_i = 0, & \bar{T}_i = 1, & M_i = 0 \\ \bar{h}_f = 12.88, & M_f = 0, & & \\ K_4 = 0.20, & K_2 = 0.19, & u = 7 & \end{array} \qquad (82)$$

The extremal path includes a pulse-burning sub-arc IA, a central sub-arc AB flown with regulated thrust and a final coasting sub-arc BF, and is shown in Fig. 12 for several values of the ratio of propellant mass to take-off mass. Inspection of Fig. 12 shows that an increase in propellant mass causes a shifting of the optimum distribution of speeds towards the region of lower velocities. As a consequence, lower final temperatures must be expected when $\bar{m}_p = 1 - \bar{m}_f$ increases. This effect is clearly shown in Fig. 13.

The particular trajectory associated with $\bar{m}_f = 0.484$ (Fig. 12) has the property of minimizing the propellant consumption for the case where the drag is ideally zero and no condition of a thermal nature is imposed. The temperature rise associated with such a trajectory, however, is considerable (Fig. 13). In this connection, Fig. 13 points out one important concept: a slight increase in propellant mass over the minimum value necessary to reach the prescribed altitude $\bar{h}_f$ determines a sharp decrease in the skin temperature at final point. The use of extra-amounts of propellant, therefore, may be an effective means for preventing the occurrence of undesirable effects due to aerodynamic heating.

III. Conclusions and Recommendations for Further Research Work

The burning program minimizing an arbitrarily specified function of the final values of time, mass, altitude, velocity, and skin temperature is investigated. For problems where no time condition is imposed, explicit solutions are obtained for the optimizing mass flow as a function of the local coordinates of the missile. Closed form solutions are derived under the assumption of isothermal atmosphere, negligible drag, negligible irradiation and constant STANTON number. Numerical examples show that the STANTON number hypothesis originates only minor errors in the prediction of the optimum burning program.

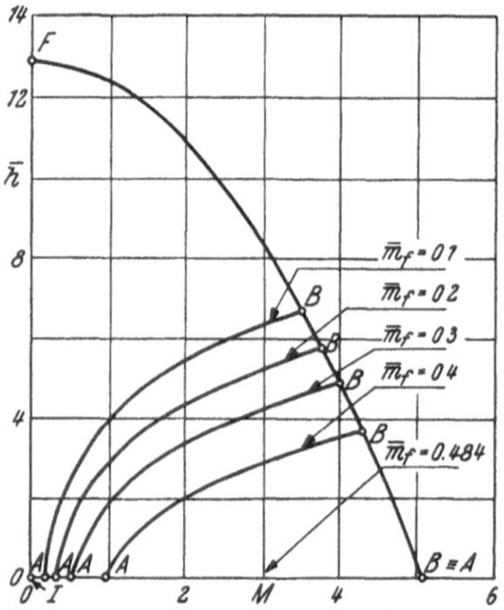

Fig. 12. Extremal trajectories of type II, for different values of the propellant mass

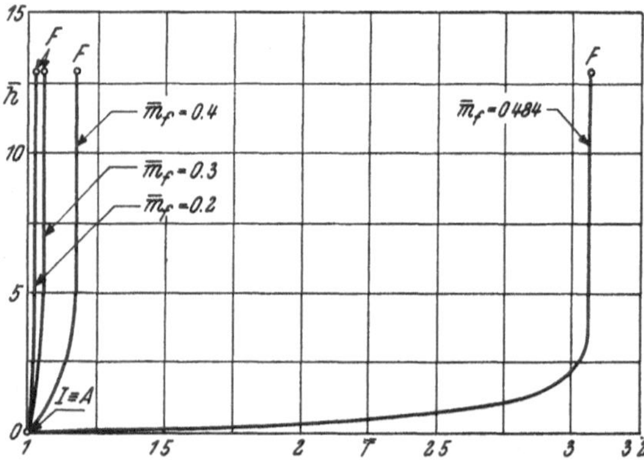

Fig. 13. Extremal trajectories of type II, for different values of the propellant mass

Due to the approximate character of the theory developed in Part II, further research is necessary to reach a full understanding of the problem of the optimum burning program. Because of the practical impossibility of obtaining closed form solutions for the general case, two lines of approach should be explored. In the first place, digital computing equipment should be used to integrate the equations of the optimum path by accounting for the effects of the aerodynamic drag (neglected in Part II), for irradiation phenomena and by lifting the hypotheses of constancy of the STANTON number and the assumption of isothermal atmosphere. In the second place, an approach should be undertaken with the so-called direct methods of the Calculus of Variations, such as—for instance—the RAYLEIGH-RITZ method. In this connection, the writer feels that much can be done by exploring the class of arcs composed of: sub-arcs $\beta = 0$, sub-arcs $\beta = \beta_{max}$, and sub-arcs where the acceleration is assumed to be proportional to some power of the MACH number.

References

1. G. HAMEL, Über eine mit dem Problem der Rakete zusammenhängende Aufgabe der Variationsrechnung. Z. angew. Math. Mechan. **7**, 451 (1927).
2. H. S. TSIEN and R. C. EVANS, Optimum Thrust Programming for a Sounding Rocket. J. Amer. Rocket Soc. **21**, 99 (1951).

3. A. Miele, Generalized Variational Approach to the Optimum Thrust Programming for the Vertical Flight of a Rocket. Part I. Necessary Conditions for the Extremum. Purdue University, School of Aeronautical Engineering, Report A-57-1, March, 1957 (AFOSR-TN-57-173).
4. A. R. Hibbs, Optimum Burning Program for Horizontal Flight. J. Amer. Rocket Soc. **22,** 206 (1952).
5. P. Cicala and A. Miele, Generalized Theory of the Optimum Thrust Programming for the Level Flight of a Rocket-Powered Aircraft. Jet Propulsion **26,** 443 (1956).
6. A. Miele, An Extension of the Theory of the Optimum Burning Program for the Level Flight of a Rocket-Powered Aircraft. Purdue University, School of Aeronautical Engineering, Report A-56-1, June, 1956 (AFOSR-TN-56-302).
7. L. H. Abraham, Structural Analysis of High Speed Vehicles. Douglas Aircraft Company, Inc., Report No. SM-18375, June, 1954.
8. G. A. Bliss, Lectures on the Calculus of Variations. Chicago: University Press, 1946.
9. A. Miele, Optimum Burning Program as Related to Aerodynamic Heating for a Missile Traversing the Earth's Atmosphere. Douglas Aircraft Company, Inc., Report No. SM-27236, April, 1957.

The Problem of Variable Thrust

By

W. N. Neat[1]

(With 3 Figures)

Abstract — Zusammenfassung — Résumé

The Problem of Variable Thrust. A review of the requirements for manned flight into space reveals the need for rocket engines capable of variable thrust. This arises from the need to keep any acceleration to an acceptable value and also to permit delicate manoeuvres such as the landing on a lunar surface to be carried out.

Thrust variation can be obtained by the use of a large number of small engines, but if this is done it may not be possible to obtain the necessary degree of control so even then some thrust variation of individual engines would still be required.

Thrust variation introduces special problems with regard to performance and combustion chamber cooling. It also complicates the control and propellent systems which then have to be designed to function satisfactorily over a wide range of operating conditions. This problem has been largely solved for aircraft type rocket engines and the techniques employed can possibly be adopted on larger engines for interplanetary purposes.

Whereas the Paper only discusses chemical rockets it is made clear that the variable thrust problem will also apply to nuclear powered rocket engines as well.

Das Problem des variablen Schubes. Ein Überblick uber die Erfordernisse des Raumfluges mit Bemannung zeigt das Bedurfnis, Raketentriebwerke mit variablem Schub zu verwenden. Dies geht auf die Notwendigkeit zurück, jede Beschleunigung auf ein annehmbares Maß zu beschränken und auch schwierige Manöver, wie eine Landung auf der Mondoberfläche, zu ermöglichen.

Die Regulierbarkeit des Schubes kann durch Verwendung einer großen Zahl kleiner Triebwerke bewirkt werden, doch ist in diesem Fall möglicherweise nicht das erforderliche Maß von Kontrolle erzielbar, so daß auch dann noch einige Variationen des Schubes der Einzeltriebwerke notwendig wäre.

Regulierbarkeit des Schubes bringt Spezialprobleme mit sich hinsichtlich der Konstruktion und Kuhlung der Brennkammern. Sie erschwert auch die Kontrolle und erfordert die Planung von Treibstoffsystemen, die uber einen weiten Bereich der Arbeitsbedingungen zufriedenstellend funktionieren sollen. Dieses Problem wurde zum großen Teil für Raketentriebwerke vom Flugzeugtyp gelöst und die dort benützte Technik kann möglicherweise den fur interplanetarische Fahrten nötigen größeren Triebwerken angepaßt werden.

Während die vorliegende Arbeit nur chemische Raketen in Betracht zieht, ist doch ersichtlich, daß das Problem des veränderlichen Schubes sich ebensosehr auf Kernraketentriebwerke bezieht.

Le problème de la poussée variable. Un recensement des exigences du vol interplanétaire avec passagers met en évidence la nécessité de disposer de moteurs-fusée à poussée variable. Les accélérations doivent en effet être maintenues à un niveau acceptable et des manoeuvres délicates exécutées, telles qu'un atterrissage sur la lune.

---

[1] Chief Engineer, Rocket Division, The De Havilland Engine Company Limited, Stag Lane, Edgware, Middlesex, England.

Une variation de la poussée peut s'obtenir par l'usage d'un grand nombre de petits moteurs; cependant, même dans ce cas, la précision du contrôle exige pour être suffisante la variabilité de la poussée de quelques éléments.

Cette variabilité pose des problèmes particuliers pour les performances et le refroidissement de la chambre de combustion. Elle complique aussi les dispositifs de contrôle et d'alimentation, dont la conception doit permettre un fonctionnement satisfaisant dans une large gamme de conditions opératoires. Ces problèmes sont résolus dans une large mesure pour les moteurs-fusée utilisés en aviation et les mêmes techniques peuvent être étendues aux gros moteurs pour les besoins du vol interplanétaire.

Quoique l'article ne mette en discussion que la propulsion chimique il est entendu que le problème d'une poussée variable se présente aussi pour les techniques de propulsion nucléaires.

Introduction

A critical review of the requirements for manned flight into space reveals the need for rocket engines capable of giving thrusts variable over a considerable range.

The basic reason for such a requirement is in order to limit the acceleration of the vehicle to a value acceptable to its crew, to its structure and to the delicate apparatus it will contain.

The more ambitious and far reaching missions into space become, the greater will be the need for thrust variations. The ratio of take-off weight to final weight will then be greater and the final acceleration will become even more unacceptable if means are not provided to limit the thrust producing it.

Furthermore when critical operations such as the manoeuvring of a space vehicle alongside its parent earth satellite or when retarding it on to the surface of the moon or planet are being performed, variable thrust will again be required. Whereas the former operation may just be carried out by means of relatively small steering rockets, there can be no doubt that the landing operation will require the use of the vehicle's main engines. In such cases it is evident that continuous and accurate control of thrust will be needed and that a high response rate will also be necessary.

Multi-Engines or Variable Thrust

Considerable thought has been given to whether sufficient control can be obtained from the use of a relatively large number of small rocket engines which can be turned on or off as required to give an approximation to variable thrust without actually varying the thrust for each individual rocket engine.

Such a methcd would give a discontinuous variation with definite steps in thrust as each rocket engine is shut down or started up. In view of this, it is doubtful whether such control would be fine enough for use, say when using the engines to retard a vehicle on to a landing surface. For other purposes, however, such an arrangement has certain advantages and for that reason the characteristics of the use of multi-engines are discussed here.

Apart from the fact that such an arrangement does not give infinitely variable thrust, there is the added disadvantage that the need to maintain the thrust line through the vehicle's centre of gravity, however many engines are in operation, rather limits the use of such an arrangement. The need to obviate any turning moment to the vehicle, implies that rocket engines in the battery must be operated in pairs or otherwise in some special sequence and unless a very large number of independent rocket engines are employed this must reduce the effectiveness of control.

Furthermore, the piping and mounting system for a multi-engine arrangement would be inevitably complex and heavy. The method of control for operating such a system would be complicated, especially if it is required to sequence each engine's performance to avoid any out of balance thrust components.

Although theoretically it can be shown that a small rocket engine might be made for a better specific weight than a large one, there are definite limits to this trend and it is doubtful if an arrangement employing a large number of small engines could be made as light as one employing a smaller number of larger engines.

On the other hand, by using a number of small engines, some measure of thrust variation could be achieved without any performance loss, since to meet different overall thrust requirements each engine will either not be working at all, or will be working at its optimum design condition under which its best specific consumption performance will be obtained. On the other hand throttling of larger engines will inevitably incur some performance loss as indicated in Figs. 1 and 2 although it can be seen that the loss so incurred is small. The main disadvantages of throttling large engines are the practical difficulties of control, cooling and combustion, all of which will be discussed later.

A further practical advantage of the use of a greater number of smaller engines is the fact that being smaller, each engine would be easier to make and develop and once developed could be used in different numbers and configurations for different purposes.

It is the author's view that although for some purposes, the use of a number of separate engines might be advantageous, even then some degree of individual throttling will be desirable. The use of a number of engines might mean that individual engines might not have to operate over such a wide thrust range as would be required if a single very large engine or a small number of quite large engines were used. In this way the advantages of both systems might be obtained and the most serious drawbacks avoided. Whatever the ultimate solution, the problem of some degree of throttling must be faced.

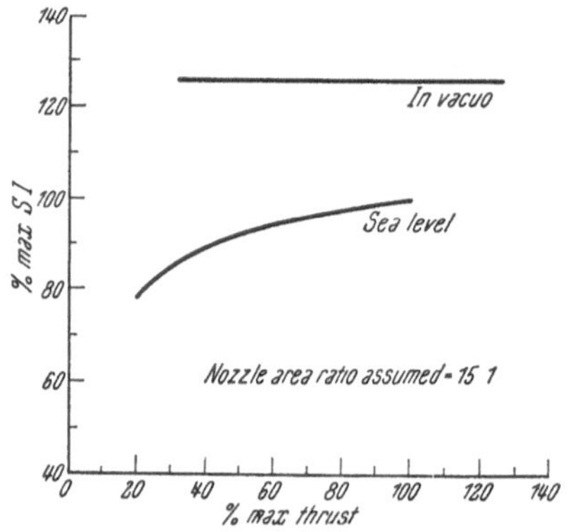

Fig. 1. Theoretical variation of specific impulse due to reduction in thrust

Effect on Performance

Excluding the possibility of varying the throat size of the combustion chamber, thrust reduction must be effected by reducing the combustion pressure and consequently the pressures through the expansion nozzle. The effect on specific impulse, due to this process, is twofold, being firstly a cycle efficiency effect and secondly the effect of reduced pressure on combustion efficiency, nozzle losses, etc.

The magnitude of the first effect depends on the ambient conditions under which the engine is throttled. For a rocket engine employing a reasonable practical maximum combustion chamber pressure and nozzle expansion ratio, the effect of throttling on specific impulse at sea level is quite considerable as illustrated in the lower curve of Fig. 1. This results from the fact that there occurs in the nozzle either considerable over-expansion, or gas breakaway, each resulting in a less efficient conversion of thermal to kinetic energy and some sacrifice in performance.

Under conditions in space, however, where the ambient pressure is to all intents and purposes zero, specific impulse is theoretically independent of throttling, as shown in the upper curve of Fig. 1. This results from the fact that it is never practicable to make an expansion nozzle big enough to expand down to zero ambient conditions. However much the chamber pressure is reduced, the expanding gases will always fill the nozzle, the pressure ratio through the

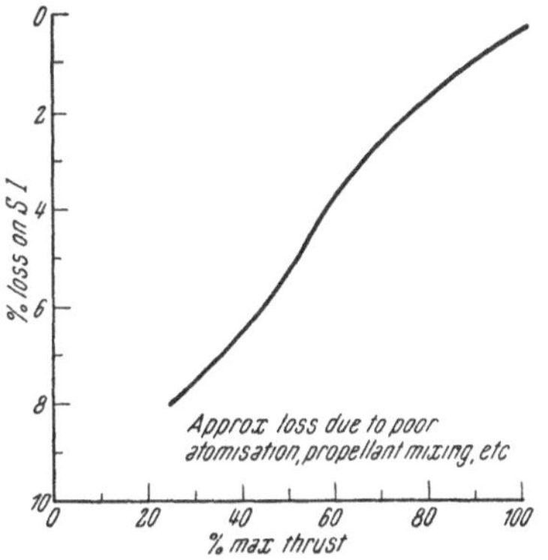

Fig. 2. Estimated additional losses due to reduction in thrust

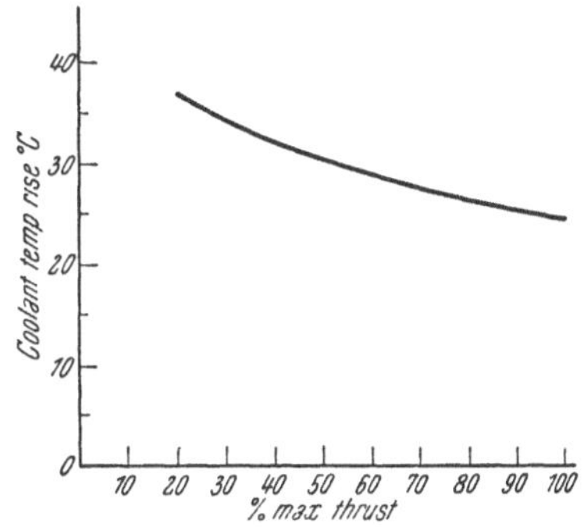

Fig. 3. Typical variation of coolant temperature rise due to reduction in thrust

nozzle and hence the cycle efficiency therefore remaining the same. This simple analysis ignores the effects of dissociation, boundary layer growth, etc., but these effects are hardly likely to have much influence on the result.

The second effect is a strictly practical one and will apply to the case of throttling, regardless of ambient conditions. As the chamber pressure is reduced, the time taken for combustion processes to be accomplished increases and the chance of complete combustion taking place within the combustion chamber is reduced. Furthermore, as pressures throughout the propellent system are reduced the pressure available for injection is also reduced and unless specific steps are taken to overcome it, poorer atomisation of the propellent droplets will be obtained. This will probably lead to poorer mixing of the propellents, which again will lead to a reduced combustion efficiency and an inferior performance. Additionally, as pressure are reduced, a greater degree of dissociation occurs resulting in a smaller energy release and a corresponding reduction in efficiency. An approximate indication of the percentage loss in specific impulse which might be expected from these considerations is given in Fig. 2. Further discussion later in this paper will indicate means by which this loss in performance can be reduced.

Effect on Chamber Cooling

The effect of throttling on the regenerative cooling of the combustion chamber must be considered. The chamber wall is kept sufficiently cool by the transport of heat from the combustion gases, through the gaseous boundary layer and the chamber wall itself into the propellent which is acting as a coolant, the temperature of which rises in the process.

Unless a considerable performance loss can be tolerated the temperature in the combustion chamber must be maintained as the engine is throttled, but as this is being done the coolant flow is reduced. The temperature rise of the coolant as it passes around the chamber walls therefore increases, a typical variation of temperature rise against thrust being indicated in Fig. 3. The safe upper temperature of the coolant may decide the amount of throttling which can be safely achieved. This limit is very dependent on the nature of the propellent being used, such as its capacity for absorbing heat, together with its tendency to boil, decompose or crack as it gets hot.

It is fortunate that as the chamber pressure is reduced, the stagnant boundary layer of gas lining the chamber wall increases in depth. The heat flow through the chamber wall into the coolant is therefore reduced. It is equally fortunate that due to longer reaction times at reduced pressures the flame front moves away from the head of the chamber thus leaving less of the chamber walls subjected to the maximum combustion temperature. These two effects, taken together, explain why it is possible to achieve a thrust reduction down to about 20% with only a 50% increase in coolant temperature rise. Were it not for the mitigating factors described above, an increase in temperature rise nearer 300% could be expected.

If the coolant in use will tolerate only a limited rise in temperature (as may well be the case) or if a very considerable degree of throttling is required, it may be necessary to resort to certain alleviating measures. For example, it might be necessary to adjust the propellent mixture ratio in order to reduce combustion temperatures ar very low thrusts. This will inevitably result in some performance loss, but if this occurs only when the consumption of propellents is in any case small, the overall effect on total propellent consumption may be acceptable. Otherwise it might be possible by a biased injector system to obtain cooler gaseous conditions near the chamber wall, or even wash the walls with one of the propellents at very low thrusts. The same effect could be obtained with a sweat cooling system which might be arranged to have greater effect at low thrusts, where the attendant performance loss would be acceptable.

Work which has already been carried out on small rocket engines has indicated that even without the use of such refinements a quite considerable thrust range can be achieved without running into any severe troubles due to cooling. To what extent new techniques will need to be developed for large engines using less suitable coolants remains to be established.

Chamber Pressure

It has been noted that a reduction in chamber pressure will result in longer reaction times which in turn might lead to a reduction in combustion efficiency. To offset this it may be necessary, when a considerable thrust range is required, to use a very high maximum chamber pressure at full thrust, so that even at the minimum thrust condition, the operating pressure has still not fallen to a value where combustion losses become serious. A further argument for maintaining as high pressures as possible is that if the pressure is reduced too much the combustion process as well as becoming less efficient, also tends to become unstable, the flame front continually changing its position and giving rise to an unsteady thrust.

A high chamber pressure, on the other hand, will mean larger pumps, a bigger turbine to drive them, together with generally heavier pipes and fittings. In an assessment of the design, this factor must be taken into account, the greater weight of the engine being considered in relation to the better performance which higher working pressures will undoubtedly give.

With certain engines employing particular propellent combinations the problem of instability under low thrust conditions might be a very serious one, perhaps the most important of all those introduced by the throttling requirement. Its magnitude will be greatly influenced by such factors as injector design and combustion chamber size and shape. It is perhaps less likely to be troublesome in engines (such as the thermal ignition hydrogen peroxide type or the vapourising liquid oxygen type) in which one of the propellents is injected as a gas, than those in which both propellents are injected as liquids, as is more usual.

Propellent Injection

As the margin of pressure available for the injection of propellents into the combustion chamber falls, so will coarser atomisation be obtained with a resultant performance loss.

Such an occurrence can be minimised by the use of an injection system specially designed for use under varying conditions of flow.

For example, the injectors can be arranged in stages so that as the rocket engine is throttled, the number of injectors in use is reduced, and those that are still working are doing so under near-optimum conditions. Such a system has the objection that under part thrust conditions, unless a very large number of injectors are employed, some combustion asymmetry may be obtained which not only fails to use the combustion chamber space to its best advantage but might lead to uneven chamber heating and eventual failure. Furthermore, those injectors which are not operating still have to be kept cool although they are passing no flow. Finally any staged system must result in a complex pipe and valve system which must add weight and possibly unreliability to the engine.

As an alternative to staging it is possible to employ variable orifice injectors which give the required pressure drop characteristic over a wide flow range. Although more complicated in themselves, such injectors overcome many of the objections of staging insofar as they are all always working and maintaining uniform combustion conditions.

A type of variable flow injector which might profitably be employed, is that of the spill-flow type commonly used in gas turbines. Such injectors always use full flow for atomisation but under part load conditions return some of this flow back to the tank. For the rocket engine application, they might have the especial advantage that the constant fuel flow up to the injector head can be used to keep it adequately cool under all conditions.

Pumps and Valves

The variable thrust requirement will inevitably result in the control system of the rocket engine becoming more complicated.

As the thrust is varied it will first of all be necessary to vary the flow of the different propellents in their correct relation one with another. Whether a constant mixture ratio or an intentionally variable one is required will depend on the particular application. It may be found that the mixture ratio must be varied as the thrust is reduced to maintain acceptable combustion chamber cooling or to get the optimum performance at each thrust.

Whatever the mixture ratio requirement, phased valves will be needed which must be capable of rapid and precise operation. Such valves are most likely to be servo-operated probably using one of the propellents themselves for this purpose.

As well as the propellent valves themselves, the turbo-pump system will need some measure of control to meet variable thrust conditions. Whether or not the turbo-pump system is allowed to run at constant speed regardless of the degree of throttling will again depend on the application. Little or no speed reduction will result in an extravagant turbine consumption under throttled conditions which may not be acceptable. On the other hand, it may be necessary to maintain a fairly high pump output under all conditions to provide a sufficient reserve of pressure for good atomisation under part load conditions. It may also be necessary to do so in order to maintain hydraulic stability on the propellent system and if a volatile coolant is used, to suppress boiling in the coolant jacket.

In practice it is likely that some reduction in the speed of the turbo-pump system will be permitted as the engine is throttled and to allow this a control valve will be required which must be linked up so that it will operate in phase with the propellent control valves.

Conclusions

As a summary to this short paper the following general conclusions can be enumerated:

(1) Rocket engines giving variable thrust will be required for any of the more ambitious journeys into space.

(2) The use of a large number of relatively small rocket engines might reduce the amount of thrust control necessary, but they will not obviate the requirement for variable thrust completely.

(3) For a number of reasons, a smaller number of rocket engines capable of variation over a greater thrust range are more likely to be used.

(4) In vacuo there is no theoretical performance loss due to throttling, although a practical loss due to incomplete combustion, etc., is likely to occur.

(5) This practical loss can be reduced by the use of a high chamber pressure, and due attention being given to injector design, etc.

(6) The coolant temperature rise consequent upon throttling may necessitate special techniques and may decide the degree of throttling possible.

(7) The throttling of large rocket engines may result in new problems, the experience to date having been obtained with relatively small units mainly for aircraft applications.

(8) Although consideration has been given here to chemical rocket engines, the variable thrust requirement will also apply to nuclear powered rocket engines. A number of the problems discussed such as the cooling of the reaction chamber will apply to both types of rocket engine.

Vertical Recovery

Feasibility of the Physical Recovery of Scientific-Research Payloads from Very-High-Altitude Near-Vertical Trajectories[1]

Prepared by

R. T. Patterson[2]

(With 14 Figures)

Abstract — Zusammenfassung — Résumé

Vertical Recovery. Feasibility of the Physical Recovery of Scientific-Research Payloads from Very-High-Altitude Near-Vertical Trajectories. This report presents the results of a study to determine the feasibility of recovering scientific-research payloads from very-high-altitude, near-vertical trajectories.

It cites emulsions as an example of a research payload which must be recovered for developing and analyzing.

The report then considers the parameters which define a recoverable payload; the features which govern the payload's structure; the design of the various payload components; and finally, it considers other problems associated with recovery.

It concludes that the recovery of such a scientific-research payload is feasible today.

Wiedergewinnbarkeit wissenschaftlicher Nutzlasten aus nahezu vertikalen Bahnen sehr großer Höhe. Die vorliegende Arbeit bringt die Ergebnisse einer Studie zur Bestimmung der Wiedergewinnbarkeit wissenschaftlicher Nutzlasten aus nahezu vertikalen Bahnen in sehr goßer Höhe.

Als Beispiel fur eine Forschungsnutzlast werden Photoemulsionen angefuhrt, die zum Zweck der Entwicklung und Analyse wieder erlangt werden müssen.

Der Bericht erörtert hierauf die Parameter, die eine wiedergewinnbare Nutzlast definieren; die Merkmale, welche die Struktur der Nutzlast beherrschen; die Planung der verschiedenen Nutzlastkomponenten und schließlich noch andere Probleme, die mit der Wiedererlangung zusammenhängen.

Die Arbeit zieht die Schlußfolgerung, daß die Wiedergewinnung einer solchen wissenschaftlichen Forschungsnutzlast gegenwärtig bereits möglich ist.

Récupérabilité des charges utiles scientifiques provenant de trajectoires quasi-verticales terminées à très haute altitude. Ce rapport présente les résultats d'une étude entreprise sur la possibilité de récupérer l'instrumentation constituant la charge utile des fusées-sonde de très haute altitude.

Les émulsions qui doivent être développées, puis analysées après récupération en sont un exemple.

Le rapport examine alors les paramètres définissant une charge utile récupérable, les caractéristiques qui gouvernent la structure de la charge, la conception des divers éléments qui la constituent et finalement d'autres problèmes annexes.

On peut en conclure que la récupération d'une telle charge utile scientifique est réalisable à l'heure actuelle.

---

[1] ER 9225.

[2] The Glenn L. Martin Company, Baltimore 3, Maryland, U.S.A.

Foreword

This report presents the results of a study by Martin Company personnel to determine the feasibility of the physical recovery of scientific-research payloads from very-high-altitude, near-vertical trajectories. It is concluded, without designing any specific payload, that such recovery is feasible today.

I. Introduction

It can be said about man, without justifying his ambitions, that he wishes to increase his knowledge of space and that, eventually, he plans to travel there. These ambitions create a need, today, for very-high-altitude scientific research. Such research, to date, has been limited to peak altitudes on the order of 200 miles, while requirements extend to 2000 miles and higher, as will be indicated.

Requirements for scientific-research payloads are, first, to get them up there, and, second, in many cases, to get them back again. The rocket vehicles for very-high-altitude research payloads can be provided with current engineering knowhow. It is the purpose of this report to further stimulate interest in very-high-altitude research by showing that the physical recovery of scientific-research payloads from these very high altitudes is feasible.

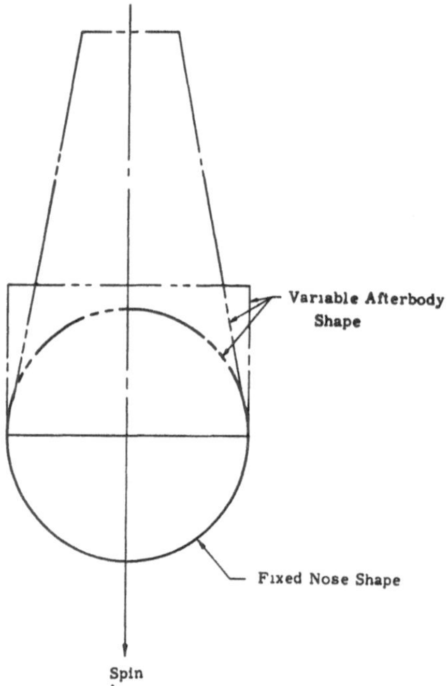

Fig. 1. The design shape concept

Recovery implies first that the payload can withstand the deceleration and heat input of a high-speed re-entry into the earth's atmosphere, and secondly that the landed payload can be located.

The subject of recovery is considered in two parts:

1. Recovery from near-vertical trajectories, such as sounding rockets and missiles.

2. Recovery from near-horizontal or orbital trajectories, such as the IGY satellites.

In this report the recovery from near-vertical trajectories, only, is discussed.

II. Types of Research

The list of desired research tasks for high-altitude research vehicles is long. From such a list, a few examples are chosen of the type of research tasks which can be conducted with near-vertical trajectories, and which require recovery:

1. Primary cosmic ray research with nuclear emulsions.

2. Geodetic, weather, astronomic and spectroscopic research with photographic emulsions.

3. Environmental research with animal tissues, structural materials, and equipment.

Emulsions, which must be recovered for developing and analyzing, provide an excellent example.

III. Payload Parameters

The parameters which define a recoverable payload are its shape, weight, and size, its velocity upon entering the atmosphere, and the angle at which it enters.

It is necessary to be somewhat arbitrary in the initial stages of the design of such a device because these parameters are inter-related. For this report a design concept is selected and followed. No attempt has been made to optimize this design. Thus, final designs might be even more feasible than this preliminary design.

A. Shape

Shape, as a parameter, applies chiefly to the nose of the body. A fixed nose shape and a variable afterbody shape are selected as illustrated in Fig. 1. Less

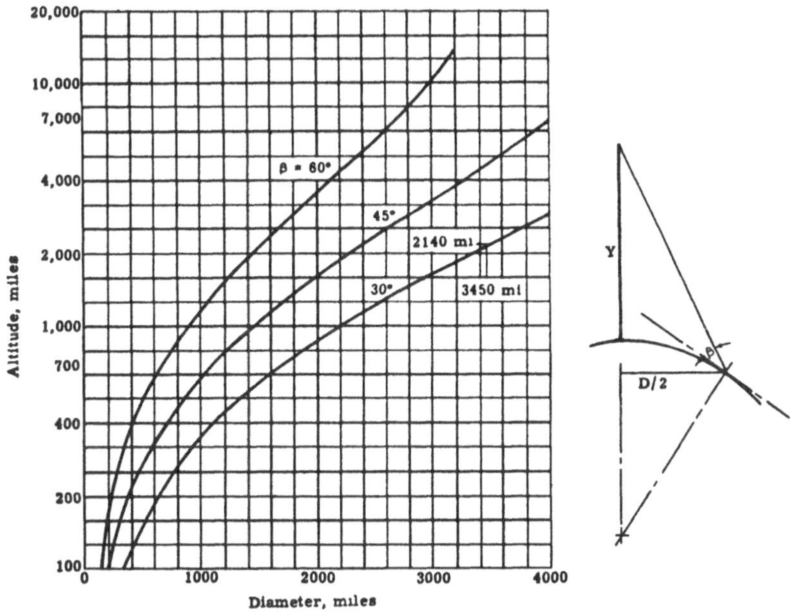

Fig. 2. Geometry of high-altitude photography

structural weight is required if the payload is stabilized in such a manner that the nose can be defined. Spin stabilization is desirable because it requires no increase in payload weight for stability equipment. A spherical nose shape is desirable for recoverable payloads, both for maximum strength and for minimum concentration of heat input during re-entry. The shape of the afterbody is of lesser importance, and may be tailored to suit the useful payload.

B. Weight

The total payload consists of a useful payload (i. e. the experimental apparatus), and the structure and the equipment required for its recovery.

It is generally agreed that the weights of useful payloads for scientific research vary from about 25 pounds for simple experiments to about 200 pounds for elaborate experiments.

The ratio of useful-payload weight to total-payload weight determines the feasibility of very-high-altitude research and recovery. Thus, a logical determination of this ratio is the real product of this report.

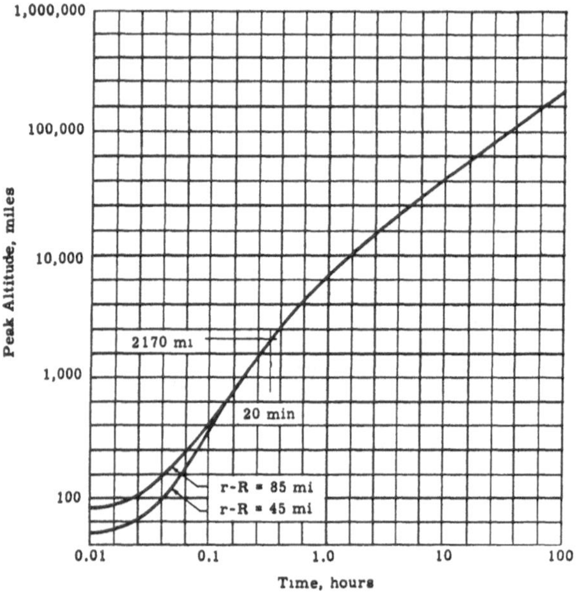

Fig. 3. Time to fall from peak altitude

$$= \frac{h}{R}\left(\frac{h}{2\,g_0}\right)^{1/2}\left[\frac{r}{h}\left(\frac{h}{r}-1\right)^{1/2}+\tan^{-1}\left(\frac{h}{r}-1\right)^{1/2}\right]$$

h = Radius to peak altitude
r = Radius to entrance altitude
R = Earth radius

C. Size

Size, as a parameter, applies to the diameter of the body. For the range of useful-payload weights and functions being considered, bodies between one and two feet in diameter appear to be adequate. As will become evident, the smallest reasonable diameter should be used.

D. Peak Altitudes

Normally, altitude is considered as a distance above the earth's surface. To illustrate distance requirements, Fig. 2 shows the variation of altitude required with diameter of observation for high-altitude geodetic and weather photo-

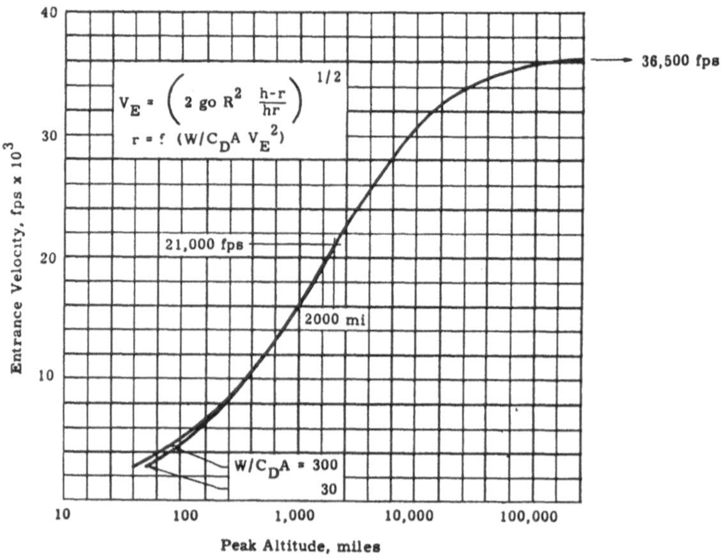

Fig. 4. Entrance velocity

graphs. Altitudes of 2000 miles and greater would be highly desirable. For example, a photograph of the earth, taken from 2140 miles altitude, centered over 60° N latitude and 15° W longitude, and with a peripheral viewing angle of at least 30 degrees, would include Iceland, Greenland, Norway, Sweden, Leningrad, Poland, Rome, Gibraltar, the Azores, and Eastern Newfoundland.

Altitude may be considered, also, as a time above the earth's surface or its atmosphere. Fig. 3 shows the variation of peak altitude with the time to fall from peak altitude to an entrance altitude. Remember that the total time above the atmosphere is twice the time to fall. The analytical expression for the curve of Fig. 3 expresses the time to fall in a NEWTONian field in a vacuum. Desired exposure times for nuclear emulsions range from 30 or 40 minutes to several hours. Thus, altitudes of 2000 miles and greater are desirable for primary cosmic-ray research.

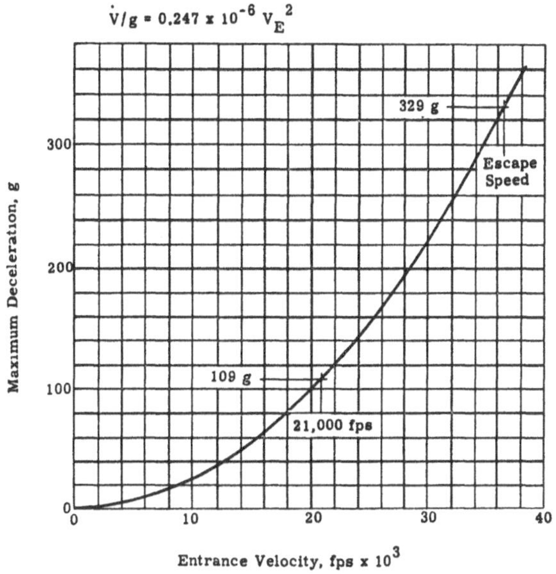

Fig. 5. Maximum deceleration

E. Entrance Velocity

Fig. 4 shows the variation of entrance velocity with peak altitude for a falling body. Entrance velocity is defined as the maximum velocity attained in the fall, which is the velocity attained during re-entry when the deceleration due to drag becomes equal to the acceleration due to gravity. The dependence of entrance velocity on the body weight, size, and shape is insignificant at the peak altitudes being considered, as shown. For peak altitudes of 2000 miles and greater the entrance velocity will range from 21,000 feet per second up toward escape velocity.

IV. Payload Structural Requirements

The factors which govern the payload structure are the maximum value of deceleration and the total heat input to the body during re-entry.

A. Maximum Deceleration

Fig. 5 shows the variation of maximum deceleration with entrance velocity. The curve presents a very good approximation, obtained by using the analytical expression shown. The maximum deceleration is essentially independent of body weight, size, and shape. For entrance velocities corresponding to peak altitudes of 2000 miles and higher, the maximum deceleration ranges from about 109 to 329 g.

B. Total Heat Input

Fig. 6 shows the variation of total heat input to the body with entrance velocity for two values which bracket the probable total payload weight and size. The empirical expression used was derived from the results of computed trajectories and is considered to be sufficiently accurate for the ranges of variables being considered.

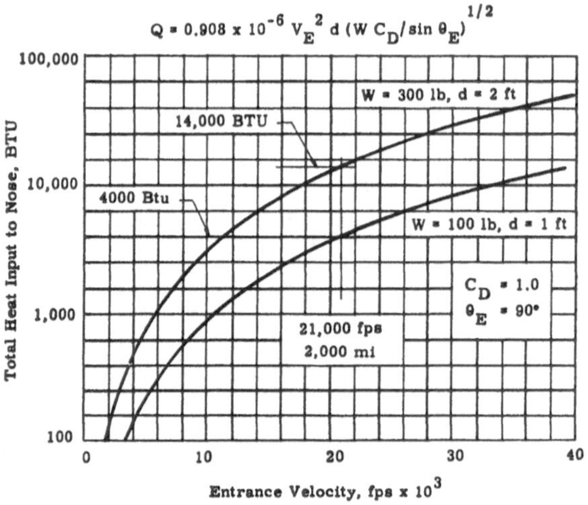

Fig. 6. Total heat input to nose

It is emphasized, here, that the evaluation of heat is the weakest part of any current recovery study, due to the lack of flight data. However, as will become evident, a false evaluation of the heat problem would change the ratio of useful-payload weight to total-payload weight, but not enough to make recovery unfeasible. It may be noted that the total heat input is a function of entrance velocity, entrance angle, and body diameter, weight and shape. The curves are presented for a spherical-nosed body re-entering vertically. For entrance velocities corresponding to peak altitudes of 2000 miles and higher, the total heat input ranges upward from 4000 Btu.

Heat input, rather than temperature, is the significant parameter. Surface equilibrium temperatures are of little concern since they exceed the melting temperatures of practical nose materials.

It may be of interest to note that the total heat input to the body is a small part of the total kinetic energy of the body, almost all of which must be dissipated aerodynamically. For example, the kinetic energy of a 200-pound body falling from 2000 miles is about 1 3/4 million Btu.

V. Payload Design

At this point enough information is available to design the payload. The total weight W_{tot}, is determined as the sum of the weights of the following components:

1. The useful payload (W_{use}).
2. The afterbody, (W_{aft}) which weight is arbitrarily assumed to be half the weight of the useful payload; (this assumption is not too critical since it is the sum of W_{use} and W_{aft} which governs other weights).
3. The recovery equipment, (W_{equip}) such as parachute and beacon.
4. The nose, ($W_{struc} + W_{insul} + W_{melt}$ — see Fig. 7) which weight depends on the above weights and on re-entry conditions.

A. Recovery Equipment

Various recovery equipment items and their weights are listed in Table I. These weights are based on a current state-of-the-art, and are considered to be independent of the total payload weight for the weight range being considered.

Table I. *Recovery Equipment Items*

Item	Function	Weight (lb)
Beacon, including batteries and antenna	Homing signal during and after landing	3
Parachute system	Low-speed final descent	11
g-switch	Initiate landing operations	2
Timer	Sequence landing operations	2
Mechanisms	Actuate landing operations	4
Power supply	Energize landing devices	3
	Total	25

Flotation equipment, for water landings, is not included in Table I. This will be discussed separately.

Any power, mechanisms, etc. required by the useful payload are considered a part of its weight.

B. Nose Design

The nose design illustrated in Fig. 7 is considered. Various parts of this structure are assigned independent functions. The structural shell is metal,

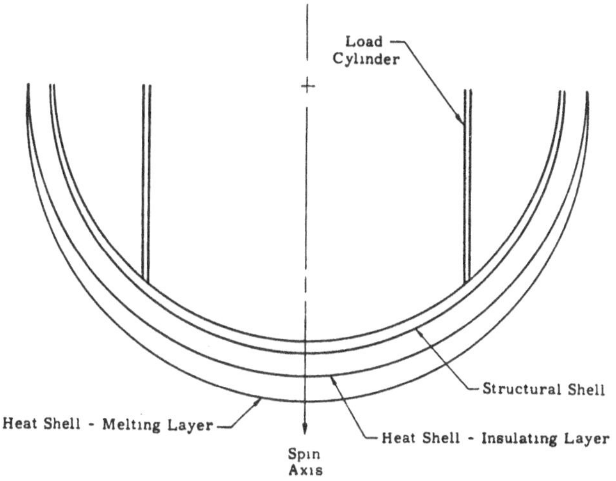

Fig. 7. Nose design

designed to take all of the load resulting from weight and maximum deceleration. Its thickness varies for optimum weight. The load cylinder is a part of the structural shell and transmits loads to it. Recovery equipment, with the exception, probably, of the parachute, is mounted on or in the load cylinder. The useful payload is located in or above the load cylinder. The heat shell removes all of the heat input to the body by melting and it insulates the body interior from the high surface temperatures. The heat shell consists of a constant-thickness insulating layer and a varying-thickness melting layer.

It is apparent that this design has not been optimized. For example, the heat shell takes some of the structural shell loads.

C. Nose Structural Shell

Fig. 8 shows the variation of nose structural shell weight with the weight of the payload less the nose for several peak altitudes and for several diameters. The structural analysis used was applied to 70,000 psi aluminium shells. The diameters shown allow for the thickness of the insulating layer.

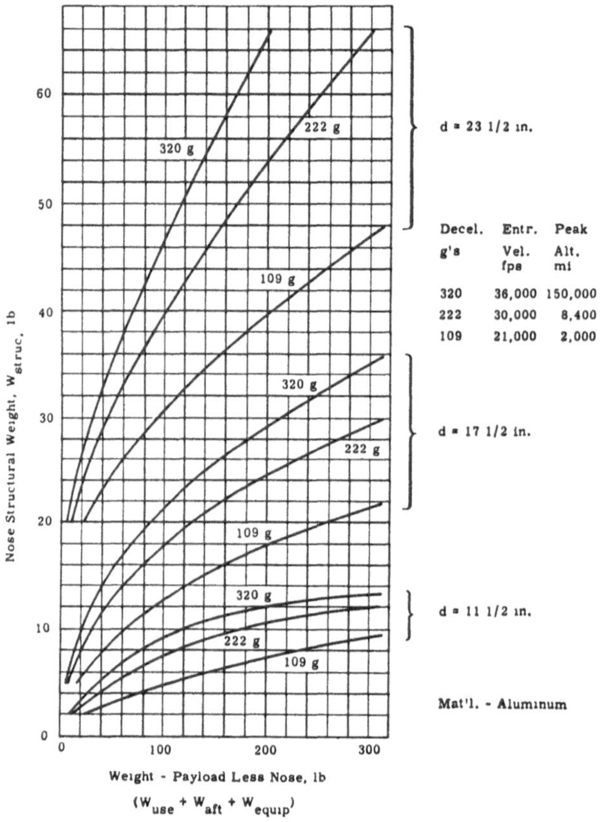

Fig. 8. Nose structural weight

It may be noted that nose structure weight increases rapidly with diameter. It may be noted also, that the required nose-structure weight becomes a lesser portion of total payload weight as the total payload increases.

D. Nose Heat-Shell Insulating Layer

It is calculated that a 1/4-inch thick insulating layer is more than adequate under the following conditions:

1. The insulating material is fiberglas-reinforced phenolic resin.
The temperature of its outer surface is about 1600° F.

2. The structural shell is aluminum.
It, alone, absorbs the conducted heat.
Its temperature does not exceed about 170° F.

3. The duration of the heat problem is about 10 seconds.

Most of the above conditions are conservative. Once again, an optimum design has not been sought.

The thickness of the insulating layer is independent of re-entry conditions. Its weight varies with diameter as follows:

Diameter (inches)	Weight (pounds)
12	4
18	9
24	16

E. Nose Heat-Shell Melting Layer

Fig. 9 shows the variation of the weight of the melting layer with the weight of the total payload less the melting layer for the several peak altitudes and for a two-foot diameter. Note that melting layer weight is directly proportional

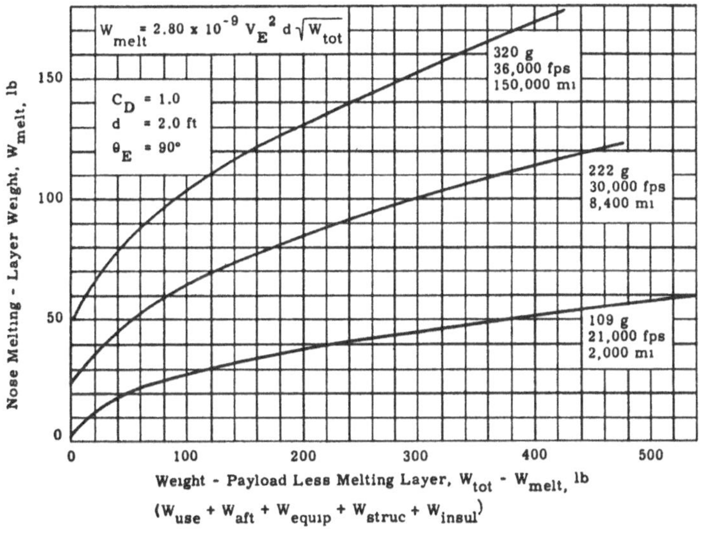

Fig. 9. Nose melting-layer weight

to diameter. The analytical expression shown in Fig. 9 combines the heat-flow expression of Fig. 6 with the specific-heat characteristics of fiberglas-reinforced phenolic resin. The expression is based on the assumption that all of the heat input goes into raising the temperature of the material to the melting point for the following reasons:

1. The material is amorphous, with indeterminate heat of fusion. (This assumption is conservative because the heat of vaporization of volatiles is neglected.)

2. The thermal conductivity is very low, such that a very small portion of the total heat input flows to the interior.

3. The melting temperature of the material is relatively low, such that radiated heat is negligibly small.

Other materials for the noses of re-entering bodies are being investigated. These materials must have suitable combinations of the following properties:

1. Conductivity;
2. Specific heat;
3. Density;
4. Melting temperature;
5. Heat of fusion, if applicable;
6. Strength;
7. Resistance to thermal shock.

It is emphasized, however, that the fairly common material used as an example in this report indicates that recovery is feasible. Better materials will increase the ratio of useful payload to total payload.

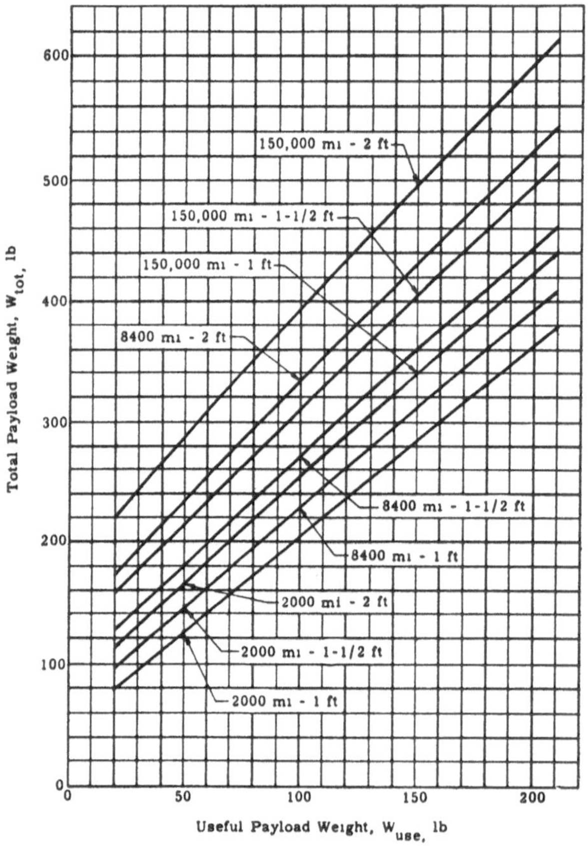

Fig. 10. Total payload weight

It may be noted in Fig. 9, just as for the nose structure weight, that the required nose melting-layer weight becomes a lesser portion of total payload weight as the payload increases.

It may also be seen that, for a given entrance velocity and material, there is a minimum payload weight below which the entire payload will melt.

F. Total Weight

The weights of all components which have been obtained, along with their sums, for the several peak altitudes, are given in Table II for diameters of one, one and a half, and two feet. Fig. 10 shows the total payload weights as a function of useful payload weights. The relative effects of diameter and peak altitude on total payload weight are immediately apparent. For a given useful payload weight, for example, a decrease in diameter from two feet to one foot is more than equivalent to a peak altitude increase from 2000 to 150,000 miles.

Table II also gives the sums of the total payloads less the nose melting layer weights, which are the total parachute-suspended loads. Ideally, the parachute weight should vary with this weight, five percent being a practical figure. A constant parachute weight was used for simplicity.

It is emphasized, again, that the weights shown are for a specific design concept, subject to the conditions, expressions, and assumptions which have been used and discussed, and that this design has not been optimized.

Table II. *Weights of the Various Components and their Sums*

Peak Altitude, mi	W_{use}	W_{aft}	W_{equip}	W_{struc}	W_{insul}	Total Parachute-Suspended Load	W_{melt}	W_{tot}
				$d = 1$ ft				
2,000	50	25	25	5	4	109	15	124
	100	50	25	7	4	186	18	204
	150	75	25	9	4	263	21	284
	200	100	25	10	4	339	24	363
8,400	50	25	25	8	4	112	34	146
	100	50	25	10	4	189	41	230
	150	75	25	12	4	266	48	314
	200	100	25	13	4	342	53	395
150,000	50	25	25	9	4	113	54	167
	100	50	25	12	4	191	64	255
	150	75	25	13	4	267	73	340
	200	100	25	14	4	343	81	424
				$d = 1$—$1/2$ ft				
2,000	50	25	25	13	9	122	23	145
	100	50	25	17	9	201	28	229
	150	75	25	20	9	279	33	312
	200	100	25	22	9	356	37	393
8,400	50	25	25	18	9	127	54	181
	100	50	25	23	9	207	65	272
	150	75	25	27	9	286	74	360
	200	100	25	30	9	363	82	445
150,000	50	25	25	21	9	130	85	215
	100	50	25	27	9	211	100	311
	150	75	25	32	9	291	113	404
	200	100	25	36	9	370	126	496
				$d = 2$ ft				
2,000	50	25	25	31	16	147	33	180
	100	50	25	37	16	228	40	268
	150	75	25	43	16	309	46	355
	200	100	25	49	16	390	51	441
8,400	50	25	25	40	16	156	77	233
	100	50	25	51	16	242	92	334
	150	75	25	60	16	326	105	431
	200	100	25	68	16	409	117	526
150,000	50	25	25	47	16	163	122	285
	100	50	25	61	16	252	143	395
	150	75	25	71	16	337	160	497
	200	100	25	79	16	420	177	597

VI. Descent Characteristics

The descent history of a payload depends on its shape, weight, and size, and on its entrance velocity and flight-path-angle. Fig. 5 and 6 have shown two of the characteristics of descent; namely, maximum deceleration and total heat input. Several additional characteristics may be of interest, and are discussed below.

A. Altitudes of Peak Heat Input Rate and Maximum Deceleration

Fig. 11 shows the altitude of peak heat input rate as a function of body weight. It may be noted that this altitude also depends on body size, shape, and entrance angle, but is independent of the entrance velocity. The approximating expression used is derived from computed trajectories.

The altitude of maximum deceleration is just slightly lower than the altitude of peak heat input rate.

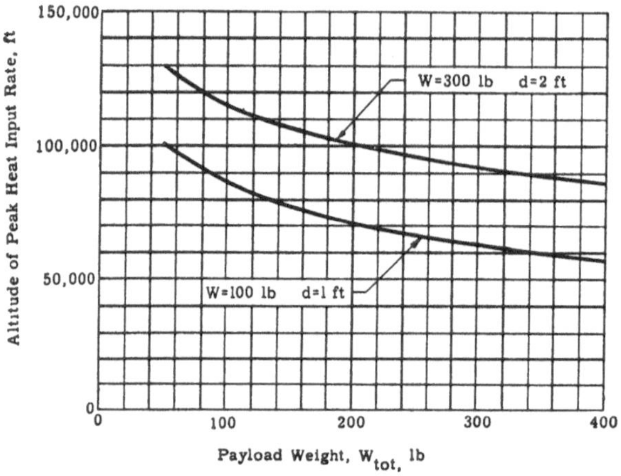

Fig. 11. Altitude of peak heat input rate
$y = 20{,}750 \ln 8210\, C_D A\, W \sin\theta$
$C_D = 1.0$
$\theta_E = 90°$

B. Significant Times

The time of significant heat input is of concern when calculating the melting-layer weight. Fig. 12 shows the time of significant heat input as a function of entrance velocity for a vertical re-entry. The empirical expression used is derived from computed trajectories.

The total heat input to the body occurs in about twice the time shown, as does the major portion of the deceleration.

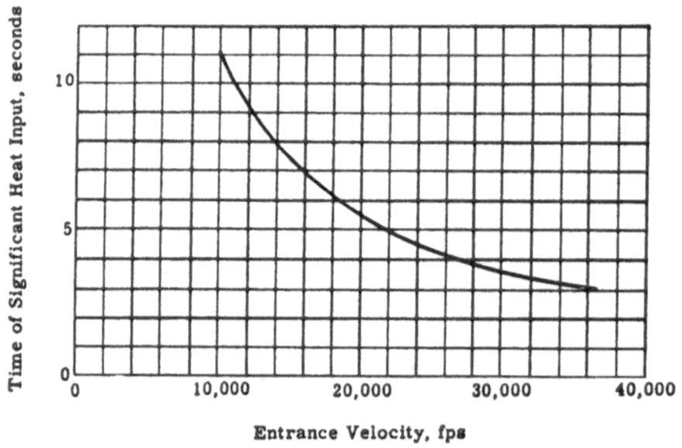

Fig. 12. Time of significant heat input
$t = 111{,}500/V_E \sin\theta_E$
$\theta_E = 90°$

C. Velocities at Parachute Deployment and Impact

At altitudes as low as 10,000 feet the deceleration has reduced to small magnitude and velocities may be estimated by equating weight to drag. Fig. 13 shows first, the velocity of the 2-foot diameter body at an altitude of 10,000 feet, which is indicative of parachute deployment velocity, and, second, the velocity at sea-level impact for the body suspended by an eight-foot diameter parachute. A drag coefficient of 1.0 is used in both cases.

VII. Water Recovery

About 80 per cent of the earth's surface is covered with water. Over-water ascents and descents of research payloads are desirable for the following reasons:

1. A greater choice of latitudes and longitudes may be chosen for the research.

2. Large safe-landing areas are available.

3. Ships are well-equipped and mobile devices for launching, tracking, and locating.

4. Water impacts are uniform and, in general, of less magnitude than land impacts.

5. Location on water is subject to less natural irregularities than location on land.

6. The attitude of the loaded payload is more easily predicted in water than on land because of buoyancy.

Water landings imply flotation for successful recovery. Either the landed payload is buoyant or flotation equipment is required. The former is preferred to save weight. The weight of a flotation balloon and compressed-gas bottles for the payloads discussed is about 15 pounds.

A buoyant payload, on the other hand, must have suitable weight distribution, such that its center of gravity is below its center of buoyancy, for the desired flotation attitude. In this attitude it is required that the beacon antennas be above the waterline.

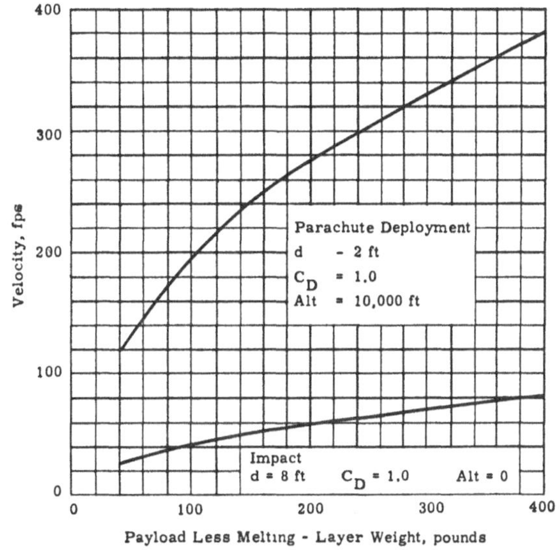

Fig. 13. Velocities of parachute deployment and impact

VIII. Ranges Required

Guidance and control system will introduce some error into the desired flight-path angle during ascent. Thus, the trajectory will include an error in horizontal velocity component, which will cause an error in the selected landing position. Fig. 14 shows range error as a function of peak altitude. The analytical expression used combines the time expression of Fig. 3 with the horizontal velocity component. The flight-path angle error shown represents the current state-of-the-art. Large landing areas are not required for recovery from altitudes of about 2000 miles. Of chief concern is the maximum allowable peak altitude for any research. For example, the 1170-mile range error, for a 46,000 mile peak altitude, represents the radius of one of the largest uninhabited sea areas on the earth, an area bounded by Newfoundland, and the Bermuda, Bahama, Cape Verde, and Azore Islands.

The range problem may be alleviated in several ways:

1. Reduce guidance and control system error.

2. Provide descent flight-path control.

3. Accept landings in unknown and/or inhabited areas.

IX. Location

Finding the landed payload is apt to depend much on probabilities. Considering recovery from peak altitudes on the order of 2000 miles, the following factors indicate reasonably high probability:

1. The range error due to flight-path angle error will be small (about 20 miles).
2. The impact position can be predicted closer than the range error with adequate tracking of the payload, both ascending and descending.
3. The payload will be aloft over 30 minutes, which is time for mobile tracking devices to converge onto the predicted impact position.

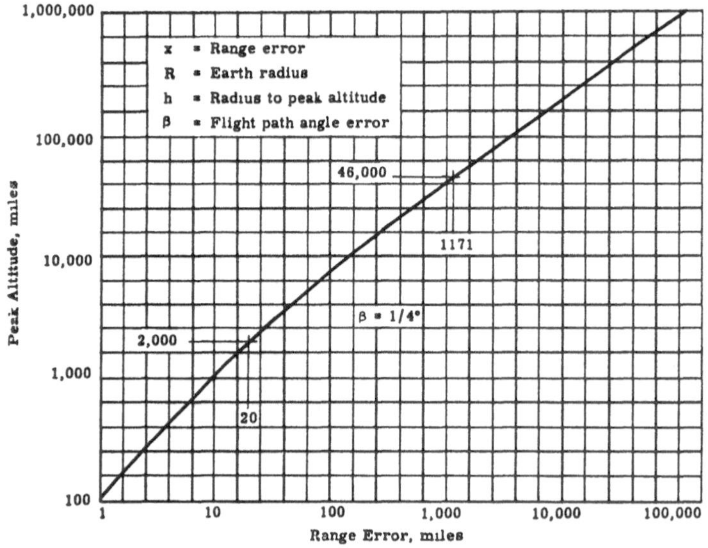

Fig. 14. Range error

$$x = \sqrt{\frac{h}{r} - 1}\,(h \sin\beta)\left(\frac{R}{h}\sqrt{\frac{h}{R} - 1} + \tan^{-1}\sqrt{\frac{h}{R} - 1}\right)$$

4. The parachute-suspended payload should present a fair visual and radar target from a reasonable area about the actual impact position.
5. The beacon has a range well in excess of the surface distances involved.
6. Suitable afterbody designs, not developed in this report, should permit good floating visual and radar targets, like buoys.
7. Auxiliary locating devices, like flares and dye-markers, can be employed.

X. Conclusions

The results of this study, applied to useful payloads of 25 to 200 pounds, in spherically-nosed bodies between one and two feet in diameter, falling vertically from altitudes of 2000 miles and higher, are summarized as follows:

1. Entrance velocities range upward from 21,000 feet per second.
2. The useful payload must withstand decelerations of at least 109 g.
3. The total heat input to the body ranges upward from 4000 Btu. This is a small part of the total kinetic energy of the system. This heat input is accommodated by permitting a part of the payload to melt.

4. The useful-payload weights range up to 55 percent of the total-payload weights.

5. Peak heat input and deceleration occur at altitudes of 60,000 feet or higher.

6. The durations of heat input and deceleration are brief.

7. The velocities of parachute deployment and impact are moderate.

8. Water recovery is advantageous.

9. Large landing areas are not required for the lower peak altitudes, but may be for higher peak altitudes.

10. The probability of finding the landed payload, at least from the lower altitudes, appears to be good.

From the above it is concluded that the physical recovery of such scientific-research payloads is feasible today.

Recovery Techniques for Manned Earth Satellites[1]

By

Norman V. Petersen[2], AAS

(With 14 Figures)

Abstract — Zusammenfassung — Résumé

Recovery Techniques for Manned Earth Satellites. The primary objective of this brief study is to propose and illustrate a possible recovery technique of retrieving manned earth satellites by highspeed aircraft. No specific conclusions are made as to feasibility with the exception that ample time to achieve payload transfer (approximately 3 minutes in this case) may be realizable. However, the programming of the two flight paths to coincidence would necessarily require a complex homing guidance system.

Verfahren zur Wiedererlangung bemannter Erdsatelliten. Das Hauptziel dieser kurzen Studie ist der Vorschlag und die Beschreibung eines möglichen Verfahrens der Wiedererlangung bemannter Satelliten mit Hilfe sehr schneller Flugzeuge. Keine speziellen Folgerungen, was die Verwirklichbarkeit anbelangt, werden gezogen mit Ausnahme dieser, daß genügend Zeit verfügbar sein *kann,* um die Überführung der Nutzlast auszufuhren (im vorliegenden Fall ungefähr drei Minuten). Trotzdem würde die Programmierung der beiden Flugbahnen, damit eine Koinzidenz erreicht werden kann, ein kompliziertes Steuerungssystem für die Heimkehr erfordern.

Techniques de récupération des satellites habités. L'objet principal de cette courte étude est de proposer et d'illustrer une technique possible de récupération de satellites habités au moyen d'avions rapides. Aucune conclusion spécifique n'est tirée quant aux possibilités de réalisation à l'exception du fait que le temps requis pour le transfert de la charge utile (environ trois minutes) est acceptable. Cependant la programmation nécessaire pour la coincidence des deux trajectoires nécessiterait un guidage final complexe.

Introduction

I should like to present a brief discussion on the several recovery techniques for manned earth satellites that have been proposed to date, and to suggest, for your consideration, a somewhat different approach; the use of high-speed satellite-retrieving aircraft.

In considering the operational aspects of space flight, one generally considers the phases to involve a two-way system. That is, ascent to sub-satellite, satellite, or escape speeds with the attendant free-fall condition existing (Phase 1), a stay

---

[1] Based on Luncheon Meeting Presentation to the Space Medicine Section of the Aero-Medical Association Annual Convention, May 6—10, 1957, Denver, Colorado.

[2] Senior Development Engineer, Sunnyvale Development Center of Sperry Gyroscope Company, Division of Sperry Rand Corporation, Sunnyvale, California, U.S.A.

period on orbit (Phase 2), and the descent, recovery or re-entry period (Phase 3). However, an alternative approach may be considered; that is, a one-way system of space flight. Thus, we can overlook the need for recovery or earth return. Many people, I know, frown on this point of view, though at the present time, with the re-entry problem being the major stumbling block to full-scale space flight missions, perhaps it should not be construed as totally undesirable. There are probably a few very serious-minded people — pilots in this case — who might not consider this one-way space flight system unworkable. I am not urging that we, or anyone, consider this approach, though I do believe that it is worth bearing in mind.

At the present time, nearly all of the technical and material ingredients are currently available in the field of high-speed flight to assure successful completion of many astronautical missions.

At the present level of achievement, we can proceed with earth satellite operations, lunar satellite and controlled landings, and satellite probes to any of the planets that one may care to contemplate.

Components having the desired performance capability are available for the guidance, and propulsion systems. Miniaturized computer and electronic systems are available, as required for navigation, instrumentation, communications and telemetry. Proof testing and operational use of the long range ballistic vehicles in the next few years will make available the prime movers for the first space vehicles.

Manned capsules with adequate heat resistance qualities for on-orbit operation, could be substituted for existing warheads of the long-range missiles to provide the first manned satellites.

Major General SCHRIEVER commented, at the recent Astronautical Symposium sponsored by OSR-Convair, that the same propulsive unit that boosts a heavy nose-cone warhead to 25,000 f/sec, could boost a somewhat lighter body to an orbital path around the earth or even to the escape velocity of 36,000 f/sec.

It is apparent that no severe problem exists to prohibit the launching and orbiting of satellites — though a major problem does exist for the orbit-to-surface transit, i.e., recovery. Although direct re-entry of manned spherical satellites, or other shapes with aerodynamic lifting surfaces, may eventually be the best approach, there is a lack of fundamental knowledge of the atmosphere and mechanics of fluid flow at extreme altitudes and speeds to permit this to be accomplished until much further research is made. Considerable debate exists as to whether laminar or turbulent flow conditions will exist — the most optimistic design studies to date seem to indicate that recovery of manned spherical satellites are possible, assuming laminar flow, thus giving maximum temperature levels of the order of only 2500°F. Although, on the other hand, if turbulent flow exists, the heat input will be perhaps many times greater.

Special techniques of minimizing the heat input through the use of special coatings of metallic oxides as vaporizing ablative mass, the use of thick walled heat sink configurations, or the use of lifting surfaces to prolong the flight duration at high altitudes and consequently, slow the vehicle to a safe re-entry speed may be favorable approaches.

All of these methods however, in operational use of orbit to surface vehicles, additionally require internal insultation or coolant systems. One must consider the initial effort necessary to put all of this resistance capability on orbit in the first place. The on-orbit phase by itself, and for that matter, the ascent phase on the launch trajectory, will require only a nominal heat resistivity. The heat inputs on orbit are insignificant in comparison with those of re-entry. For extended stay-times on orbit the satellite structural design and wall thickness may well be related more to meteoric penetration protection than to temperature control. Thus, let us consider the possibility of devising a recovery system for manned earth satellites, wherein no weight burden is placed on the satellite to cope with the enormous heat inputs of aerodynamic re-entry.

The purpose of my talk is to consider the feasibility of alleviating the problem

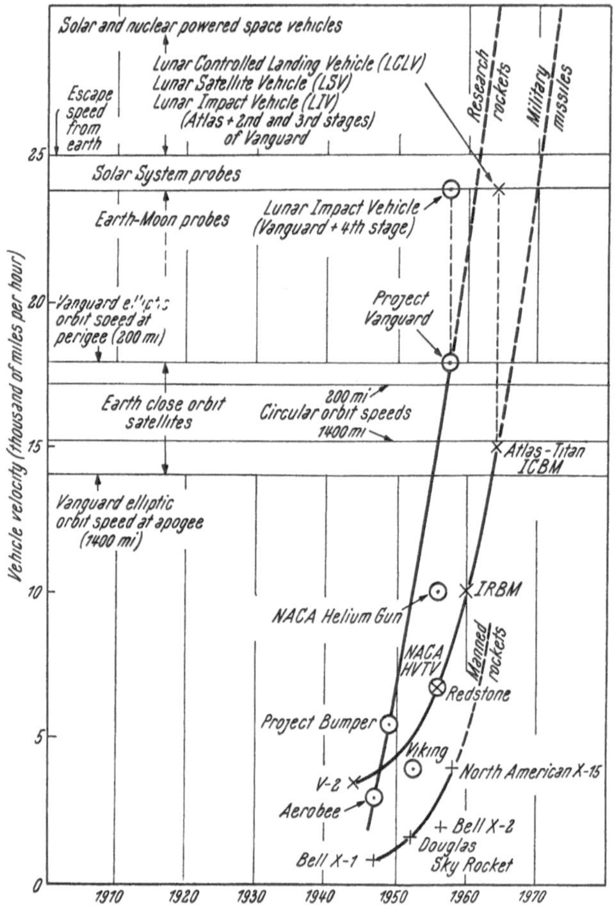

Fig. 1. Progress in high speed flight

of high-speed re-entry of manned satellites into the earth's atmosphere by the use of satellite retrieving aircraft.

Should one be concerned with the hazards in event of system failure these will be no more catastrophic than the case of system failure for controlled landing vehicles on airless bodies, specifically the moon.

I feel that this recovery technique falls somewhere between the two extremes of the possible spectrum of re-entry systems one can dream up. On one hand, the work-horse way is simply to use full rocket braking and the other is to permit the satellite to plummet into the atmosphere with aerodynamic braking. Somewhere in between, is this retrieving technique by high speed aircraft. To the best of my knowledge, this system has never been proposed in the literature—I believe it is worth discussing at this time.

Progress Toward Space Flight

There are perhaps three major unknown quantities in the progress toward space flight; these are, (1) the aerodynamic heating and (2) acceleration-time characteristics occuring during re-entry and (3) the little studied effects of zero or low-level gravity during extended freefall conditions.

All three of these effects are being pursued in a gradual step-by-step basis. The use of manned research rockets has progressed from NACA X-1 and X-2 to the forthcoming vehicle, the X-15, designed for speeds to 4000 mph. The use of the first two vehicles, the X-1 and X-2, have provided the first full-scale attack on these three major problems. The X-15 will increase our knowledge considerably. Fig. 1 illustrates the progress of high-speed flight vehicles.

However, there still remains a great separation between the nearcurrent 4000 mph capability and the lower limit of re-entry at satellite speeds. A number of manned research rockets may be needed to bridge the gap of information prior

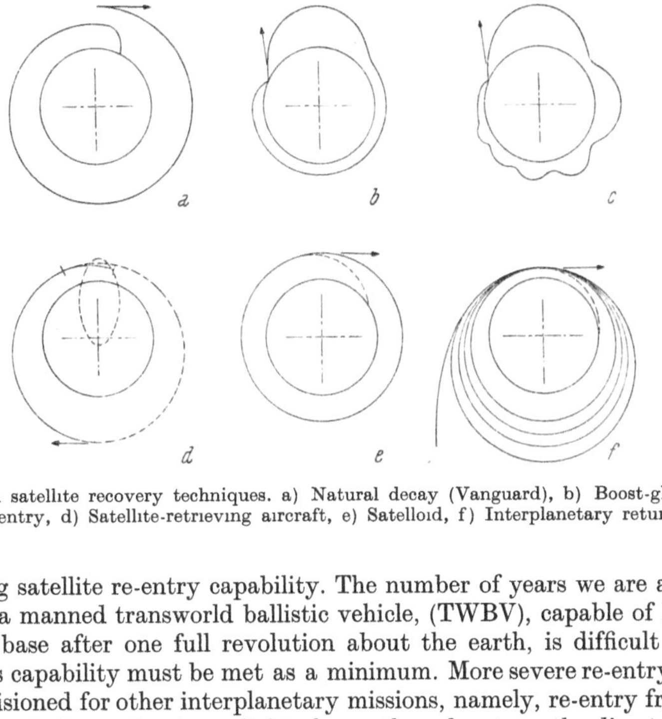

Fig. 2. Earth satellite recovery techniques. a) Natural decay (Vanguard), b) Boost-glide, c) Skip re-entry, d) Satellite-retrieving aircraft, e) Satelloid, f) Interplanetary return

to attaining satellite re-entry capability. The number of years we are away from test flying a manned transworld ballistic vehicle, (TWBV), capable of landing at its launch base after one full revolution about the earth, is difficult to forsee, though this capability must be met as a minimum. More severe re-entry problems can be envisioned for other interplanetary missions, namely, re-entry from escape or even hyperbolic earth return flights from other planets or the direct entry into more dense atmospheres of larger massed planets.

It seems reasonable to assume that the research and development toward effecting the recovery and/or re-entry of manned satellite or space vehicles, may well proceed along a rather straightforward predictable course. Considerable foundation has already been made in the field of re-entry aerodynamics for ballistic vehicles. Wind tunnel facilities now exist, having operating ranges up to Mach 20—30 for continuum flow studies in the pertinent atmospheric layer from sea level to 50 miles altitude. Representative laboratory facilities include the light gas gun, hypersonic wind tunnels, and shock tubes and ionic plasma jets. Direct re-entry vehicles now exist to give full scale data from Jupiter, Thor, and several re-entry test vehicles with the higher speed vehicles, Atlas and Titan to be available in the near future.

In regard to manned re-entry vehicles the field of astronautics is only beginning to scratch the surface on the problems of aerodynamic heating limits and acceleration levels acceptable for manned vehicles. At the present time, the best and only information has stemmed from the high-speed flight research program of the NACA. Starting with the capabilities of the X series rocket-powered vehicles, we have data from the X-1 with the first records over 1,000 mph in 1948 to the most recent X-2 vehicles in 1956, attaining speeds up to 2,178 mph and apogee altitudes of 25 miles.

Recent announcements in the Press state that the current development of the NACA X-15 research aircraft will provide a vehicle having a maximum speed of 4,000 mph. The performance of such a vehicle might be compared with that of the performance of the Viking high-altitude rocket. The Viking attains about the same cutoff velocity, 4,500 mph, and attains peak apogee altitudes from 100 to 160 miles. Thus, the X-15, or any other such rocket-powered vehicle having this cutoff velocity, may be programmed to ballistic paths having a 100 mile apogee.

An extensive program has been maintained on determining the effects of extended zero-gravity flights through the use of the T-33 and X-3 research aircraft. Characteristics for the T-33, X-3, Aerobee, V-2, Viking, and a typical 4000 mph vehicle, are shown on Table I.

Table I. *Zero Gravity Duration for High Speed Vehicles*

Vehicle	*G*-Level	Cutoff Velocity	Zero Gravity Duration (Approx.)	Peak Altitude
NACA T-33	0	,600 mph	12 sec	8 mi
NACA X-3	0	1,000 mph	80 sec	8 mi
Aerobee	0	3,000 mph	4 min	80 mi
V-2	0	3,400 mph	6 min	100 mi
X-15	0	4,000 mph	7 min	100 mi
Viking	0	4,500 mph	8 min	158 mi
X-?	0	7,000 mph	10 min	200 mi
IRBM	0	10,000 mph	15 min	400 mi
ICBM	0	15,000 mph	30 min	600 mi
TWBV	0	17,000 mph	90 min	1000 mi
SSBV	milligees	17,500 mph	90 + min	100 mi
Satellite	0	18,000 mph	90 + min	200 + mi

Typical flight paths for recovery of satellite and space vehicles are illustrated on Fig. 2, including those for the following missions:

a) Re-entry sphere (natural decay from close orbits) (Vanguard and others).
b) Boost-glide (Trans-World Ballistic Vehicle, TWBV).
c) Skip re-entry (SAENGER Skip Bomber).
d) Sub-satellite ballistic Vehicle (SSBV — Satelloid).
e) Satellite-Retrieving Aircraft System.
f) Interplanetary return braking ellipses.

Estimated maximum temperatures are illustrated on Fig. 3, for several proposed low-density re-entry configurations. Maximum temperatures range from 2500°F for low-density configurations, as shown on Fig. 3, to several times that for high-density bodies. Accelerations along the flight path vary from a few g's to 30—50 g's depending on re-entry angle. Though it has been estimated, a

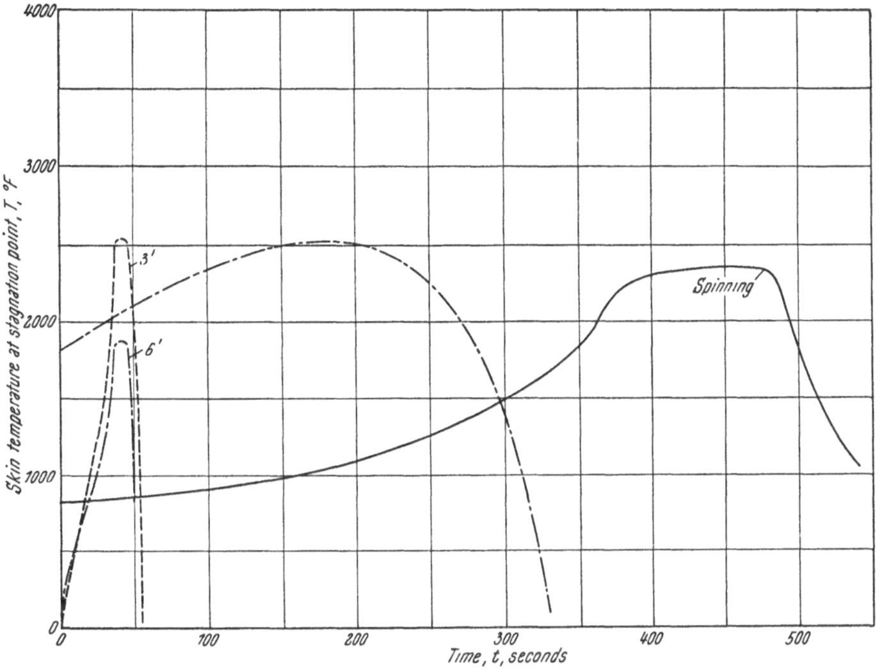

Fig. 3. Temperature vs time for low-density re-entry spheres. GAZLEY and MASSON for Vanguard; $d = 20''$, $W = 21.5$ lbs, $W/A = 9.87$ lbs/ft² (natural decay from 67.8 mi to 21.8 mi), $\varepsilon = 0.8$

PORTER; Skin temperature at stagnation point; $d = 6'$, $W = 7$ lbs, $W/A = .248$ lbs/ft² (decay from 5000 ft/s retardation from circular orbit at 76 miles)

PORTER; Skin temperature at stagnation point; $d = 3'$, $W = 7$ lbs, $W/A = .99$ lbs/ft²

EHRICKE; Sphere re-entry from circular velocity at 69.1 miles (plot shown from 56.8 mi to 18.9 mi); $W/A \approx 11.4$, temperature at stagnation point

Non-spinning	$T Max = 2880°$ F	$T Max = 3060°$ F
	$\varepsilon = 1.0$	$\varepsilon = 0.8$
	$T Max = 3520°$ F	$T Max = 4460°$ F
	$\varepsilon = 0.5$	$\varepsilon = 0.1$

600 lb., sphere can be designed to permit manned re-entry, with temperatures of the order of 2500°F resulting, actual proof-testing of such manned vehicles may be many years in the future.

Characteristics of a Satellite-Retrieving Aircraft System

It is the intent of this paper to show that manned satellite operations are currently feasible with existing equipment and to suggest that this system of recovery, by retrieving-aircraft, may be the most economical method for orbit-to-surface transport of personnel.

This design study considers the minimum manned configuration practicable wherein no weight burden, in regard to the atmospheric re-entry heat resistance, is placed on the satellite. The assumption is made that the first manned earth satellites need be provided with only sufficient heat transfer control equipment to account for the following heat sources or conditions on-orbit:

(1) direct solar radiation

(2) reflected solar radiation from the earth's surface and upper atmosphere

(3) direct earth radiation
(4) internal power supplies
(5) impingement of micro-meteorites, and
(6) day-night operation (periodic earth eclipse).

Fig. 4 illustrates these heat sources.

This study illustrates the recovery system for a minimum manned earth

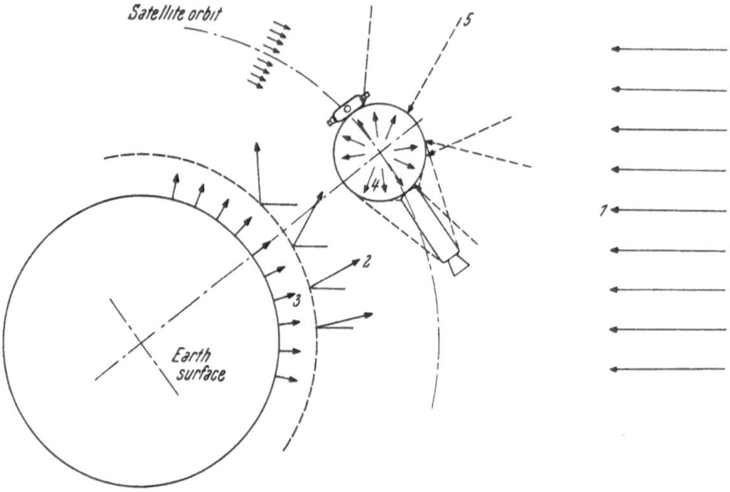

Fig. 4. On-orbit heat inputs. *1* Direct solar radiation, *2* Reflected solar radiation, *3* Direct earth radiation, *4* Internal power supplies, *5* Micro-meteorite impacts

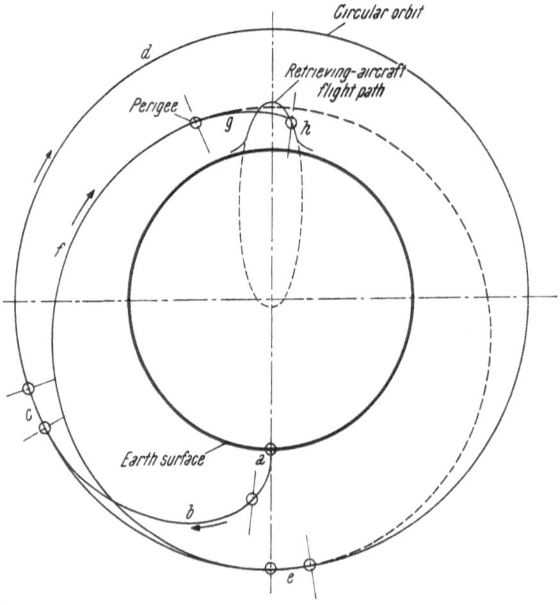

Fig. 5. Launch and recovery system for manned earth satellite

satellite by programming the path of a high-speed aircraft, capable of atmosphere re-entry, to a point of coincidence with the satellite (perhaps at 200 miles or greater altitude). The coincidence point, or point of capture, is established as the terminal point on the satellite retardation path. Sufficient propellant is assumed on board the satellite for retarding its perigee velocity to a value permitting coincidence and payload transfer on the downward path. A payload transfer time of several minutes, from coincidence to descent to a safe altitude above 40 miles, appears realizable.

The probable error limits in velocity and position for capture, based on existing techniques, need not be excessive. The time available from the coincidence point to affect payload transfer from the satellite vehicle to the retrieving-aircraft appears to be sufficiently large to make this technique feasible. The development of adequate automatic guidance systems for both satellite and retrieving-aircraft are required, though techniques and components exist to make this practicable at the present time.

This study, in considering the recovery of manned satellites, is essentially a compromise system. The two extreme conditions, or techniques for recovery of manned satellites being, (1) recovery by aerodynamic braking, and (2) recovery by braking rockets. Though both of these techniques may accomplish the task

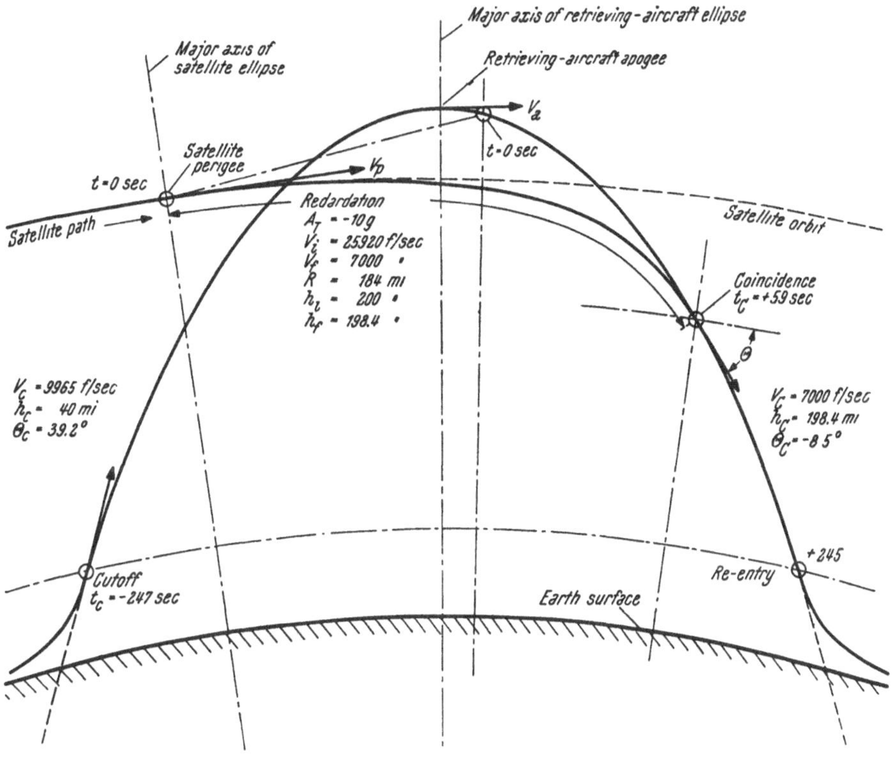

Fig. 6. Recovery system for minimum earth satellite (schematic)

of recovery, the aspects of time and economics, may not permit their use in the immediate future.

It is beyond the scope of this brief study to illustrate the advantage of this technique over that for a winged re-entry satellite utilizing combined aerodynamic and rocket braking.

The conditions for recovery, in this study, are such as to permit payload transfer from the satellite to a high-speed aircraft having only a slight performance improvement over current vehicles such as the NACA X-15. The capture velocity is arbitrarily established at 7000 f/sec at about 200 miles altitude thus requiring a cutoff velocity of 9960 f/sec at 40 miles altitude for the retrieving-aircraft. Reducing the capture velocity to lower values significantly increases the retardation propellant required on board the satellite. It is estimated that approximately 8 supporting rocket vehicles would be required to refuel the on-orbit satellite for

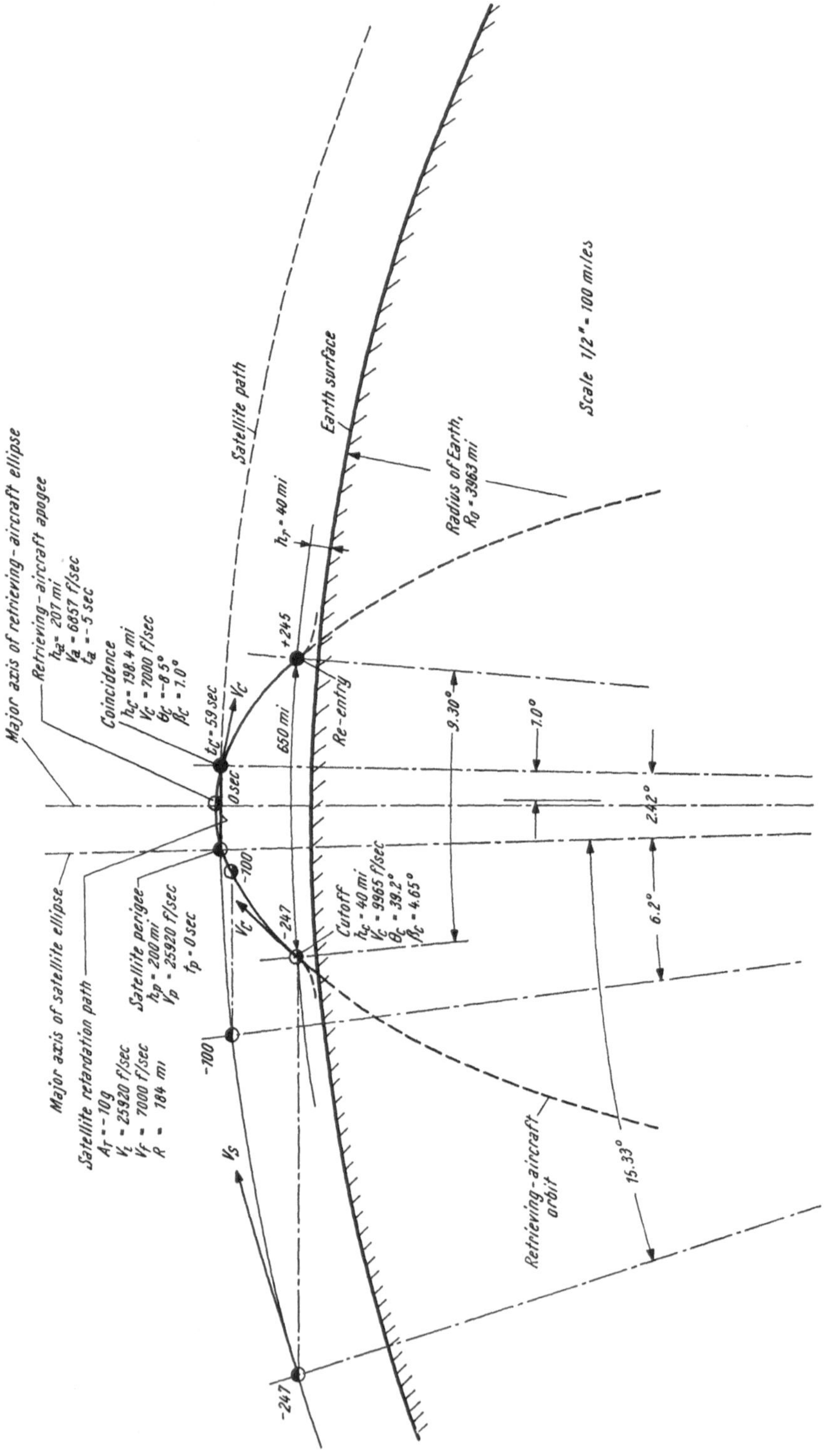

Fig. 7. Recovery system for minimum earth satellite (too scale)

the 700 lb., recovery vehicle prior to recovery. Higher speed retrieving-aircraft would significantly reduce the number of supporting rocket vehicles. Increasing the coincidence speed to about 1/2 perigee speed (13,000 f/sec) would reduce the refueling rockets to 2.

In view of the severity of the problem of satellite recovery, it is believed that this proposed technique would advance the launching date of manned earth satellites.

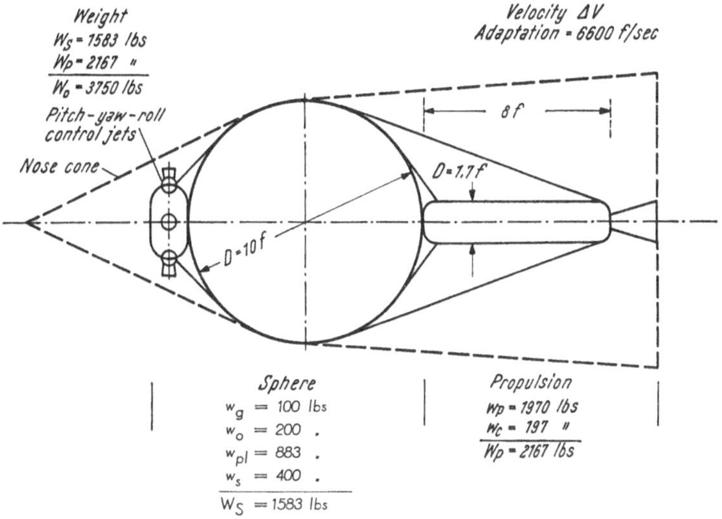

Fig. 8. Satellite configuration (ascent phase)

Fig. 5 illustrates the overall recovery system. This recovery technique assumes the following phases of operation:

(a) Launch to cutoff; (Ballistic Vehicle + Satellite), $h_c = 150$ miles
(b) Ellipse of ascent; $h_c = 150$ mi to $h_a = 600$ miles
(c) Adaptation maneuver; apogee to circular orbit, $h_a = h_{co} = 600$ miles
(d) On-orbit refueling of satellite;
(e) Adaptation maneuver; circular orbit to elliptic orbit, $h_{co} = h_a = 600$ miles, $h_p = 200$ miles
(f) Ellipse of descent; from $h_a = 600$ miles to $h_p = 200$ miles
(g) Retardation maneuver at perigee; $A_T = 10\ g$
(h) Coincidence and payload transfer; $h_c = 198.4$ miles to $h = 40$ miles.

Satellite and Retrieving-Aircraft Flight Paths

Typical flight paths for the satellite vehicle and retrieving-aircraft during the recovery operation are shown on Fig. 6 for an arbitrary retardation of -10 g. The flight path for the satellite during the retardation maneuver actually departs very little from a straight line (Fig. 6 is not to scale). In this analysis, the retardation path, for simplification uses a zero-lift, free-fall constraint during burning. The loss in altitude from perigee, at $h_p = 200$ miles, to coincidence, at $h_c = 198.4$ miles is therefore only 8500 feet and the satellite travels along its path about 184 miles. The retardation of -10 g reduces the initial satellite perigee velocity, $V_p = 25920$ f/sec, to the coincidence velocity, $V_c = 7000$ f/sec in approximately

59 sec. Retardation of -10 g's for 1 minute exposure, is assumed as an upper practical limit for manned vehicles. A lower g-level of 5 g's with a duration of 108 sec still constrains the vehicle to nearly a straight line path.

The trajectory during the free-fall condition after coincidence is simply determined by the equations defining elliptic motion, namely, (a) polar equation of ellipse, (b) direction angle function and (c) KEPLER's third law of planetary motion.

Total flight time from coincidence to re-entry at 40 miles altitude is about 186 sec. Thus, over 3 minutes time is available for payload transfer. Longer transfer times can be realized by establishing coincidence at higher altitudes. Fig. 7 illustrates the trajectories to scale.

Satellite Characteristics

a) Ascent Configuration

Within the weight restrictions of the assumed payload of 3750 lbs., it appears quite feasible to launch satellites having on-orbit payloads upwards of about 1580 lbs. Larger payloads may be placed on lower altitude orbits.

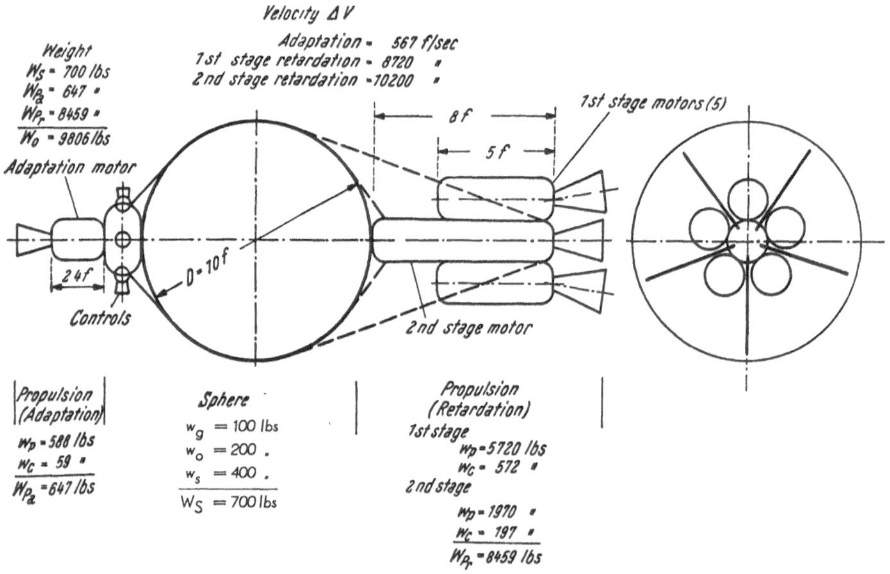

Fig. 9. Satellite configuration (retardation phase)

For an arbitrary apogee altitude of 600 miles, for a 5000 mile vehicle, the required velocity increment to establish circular motion is about a ΔV of 4500 mph. For a specific impulse of existing propulsion systems, with $I_{sp} = 275$ lbs. -sec/lb., the mass ratio, m_p/m_o, is therefore 0.525. Consequently, circularizing the 5000 mile ballistic path at apogee permits an on-orbit payload of about 1580 lbs, with

m_p = propellant weight .... = 1970 lbs.

m_s = motor structural weight .... = 200 lbs.

m_p = payload weight .... = 1580 lbs.

and m_o = vehicle gross weight .... = 3750 lbs.

It is assumed that an adequate guidance system can be designed to be included as part of this payload weight of 1580 lbs., and still permit a reasonably useful manned capsule on-orbit.

Fig. 8 illustrates the satellite ascent configuration with a 10 foot diameter, spherical, pilot compartment and propulsion system. No attempt is made to show internal arrangement.

b) Satellite Descent (Recovery) Configuration

Of the total payload (1538 lbs.,) for the ascent stage, 700 lbs., is established as the minimum weight capsule practicable for the retardation maneuver as based on the on-orbit vehicle. The remaining 883 lbs., is considered as supplementary

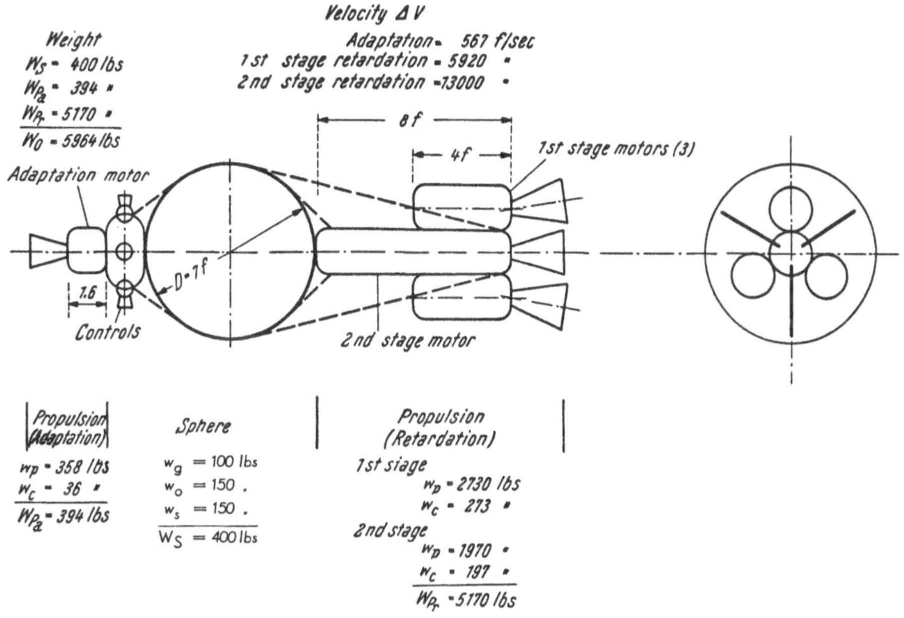

Fig. 10. Minimum recovery satellite

equipment and instrumentation and is assumed to be left on-orbit prior to the recovery phases.

Fig. 9 illustrates the descent configuration including the propulsion stages for the adaptation and retardation maneuvers. Approximately 588 lbs. of propellant are required to elliptize the initially assumed circular orbit at 600 miles. A retardation of 560 f/sec will establish an elliptic orbit having perigee at 200 miles and apogee at 600 miles. Upon adjustment to a precise 200 mile perigee altitude the adaptation motor is jettisoned, with the vehicle gross weight prior to the retardation maneuver being 9159 lbs. The propulsion system weight is 8459 lbs., for the 700 lb., manned capsule.

Approximately 8 refueling vehicles, from surface to orbit, are necessary to provide the necessary propulsion system with each refueling vehicle transporting 1250 lbs., of payload.

A minimum recovery satellite configuration is shown on Fig. 10, having a 7 foot diameter sphere. Total recovery weight of 400 lbs., is assumed with

$$\begin{aligned} w_o &= 150 \text{ lbs.} \\ w_g &= 100 \text{ lbs.} \\ w_s &= 150 \text{ lbs.} \\ \hline w_S &= 400 \text{ lbs.} \end{aligned}$$

In considering the structural weight allotted for the capsule (50 lbs.), the estimated internal pressure loads (14.7 psi) plus an arbitrary 10-g retardation requires only an 18.3 lb., skin weight for reinforced fiber glass (tensile strength of 50,000 psi).

The operating period for the minimum weight recovery satellite is assumed to be at most 2 hours. The minimum operating period would be the transit time from the 600 mile apogee to the point of perigee, or about 45 minutes. Should the satellite pilot make the decision, at, or just prior to perigee, not to commit his vehicle to the retardation maneuver, the satellite would proceed to ascend to

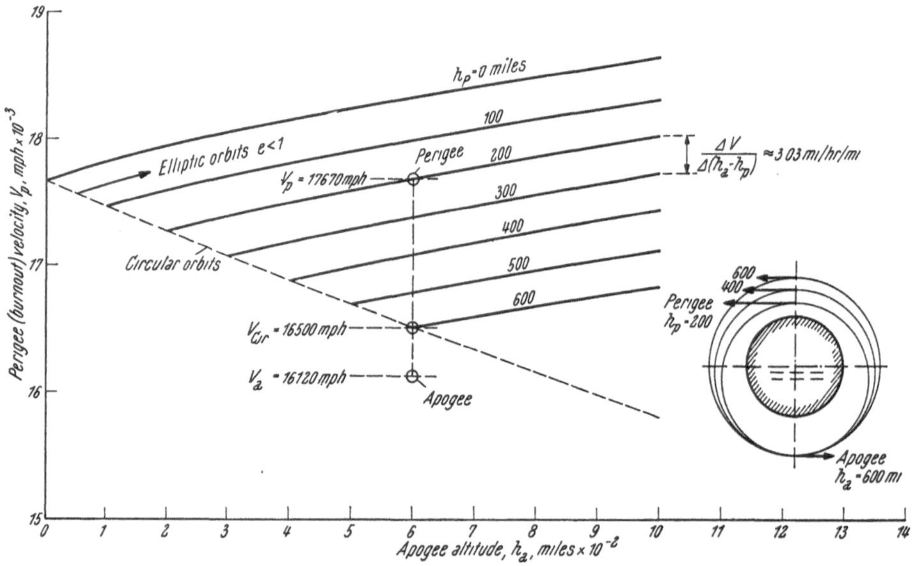

Fig. 11. Perigee (burnout) velocity vs apogee altitude for satellite on elliptic orbits

apogee, with the maximum operating time being about 90 minutes. Oxygen supply for a 2 hour stay time in the recovery satellite is then only about 0.2 lbs.

Insulation, additional structural material, and communication equipment is assumed to comprise the additional 130 lbs., of structure.

Approximately 4 refueling vehicles, are needed for the 400 lb., minimum recovery capsule.

c) *Guidance System*

Based on existing navigation systems the guidance equipment for the satellite and recovery stages may have a total of about 100 lbs. This weight would be comprised of the following:

Inertial System	75 lbs.
Control System	25 lbs.
Total Guidance Weight	100 lbs.

The guidance constraint on Vanguard represents the upper limit of errors permitted in cutoff velocity and angle of injection for close orbit operation. Guidance requirements for Project Vanguard indicate an acceptable velocity error of about $\pm$ 1 percent. Thus, at cutoff, the velocity error may be as great as $\pm$ 250 f/sec, or a factor of about 50 times more than is needed for precise orbit constraint.

The guidance system should provide sufficient accuracy, both in angle and velocity, to constrain the satellite to near circular flight, although, the pilot on board the satellite should be able to make correction maneuvers to circularize the orbit to the desired degree of accuracy based in part on earth surface optical or radio tracking station information.

The use of a relatively simple optical instrument in measuring the horizon angle could perhaps provide a reasonably accurate system to determine continuously the orbit altitude including the apogee and perigee positions.

Measurement of the horizon angle in 2 planes would establish the local vertical to a fair degree of precision and perhaps would be sufficient in monitoring adaptation maneuvers on orbit. The use of an horizon angle indicator, or a star tracker could provide the vertical reference for re-referencing of the primary guidance system for the adaptation and retardation maneuvers prior to capture. Fig. 11 illustrates the effect of a velocity variation greater than or less than that required for circularity. A velocity increment of $\pm$ 5 feet per second results in an altitude variation between apogee and perigee of about one mile.

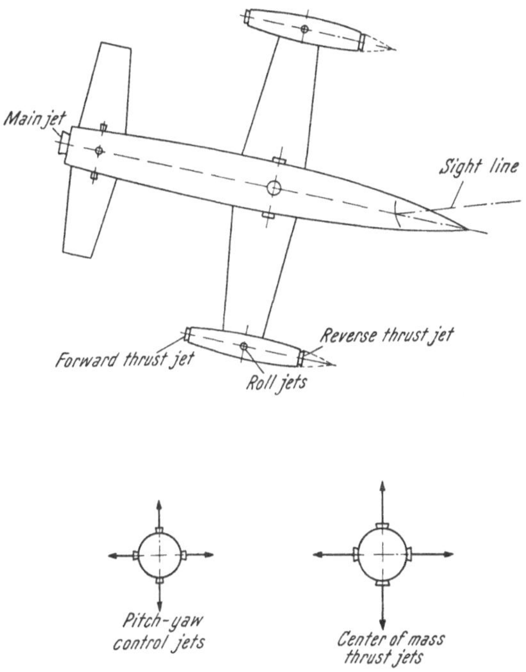

Fig. 12. Thrust and control jet configurations

Navigation for Retrieving-Aircraft

In order to permit a maximum time to effect payload transfer the point of coincidence of the two ballistic paths necessarily must occur near the retrieving-aircraft apogee point. A payload transfer time of approximately 3 minutes is realizable for the conditions selected.

The navigation problem considered here is more complicated than typical missile homing systems by reason of the added requirement that the thrust vector must be controlled in order to provide closure and coupling. The position and velocity errors of the two vehicles must be reduced to zero at coincidence. Both positive and negative thrust motors, plus the addition of control jets, must be available about all three axes. The use of the main thrust motor would be sufficient, however, if the rotation of the retrieving-aircraft about the pitch, yaw and roll axes was unrestricted. The need for a fast response plus precision control of the thrust force would dictate use of center-of-gravity thrust motors (positive and negative about the lateral and directional axes) plus forward and reverse thrust motors, on the longitudinal axis, with vernier controls. Fig. 12 illustrates the control jet arrangement.

Initial guidance of the retrieving-aircraft, prior to cutoff at 40 miles altitude is constrained by ground tracking stations up-range from the capture point. Due to the need for precise satellite orbit data, several ground tracking stations would be required in the orbit plane, perhaps as far as 1200 miles up-range from

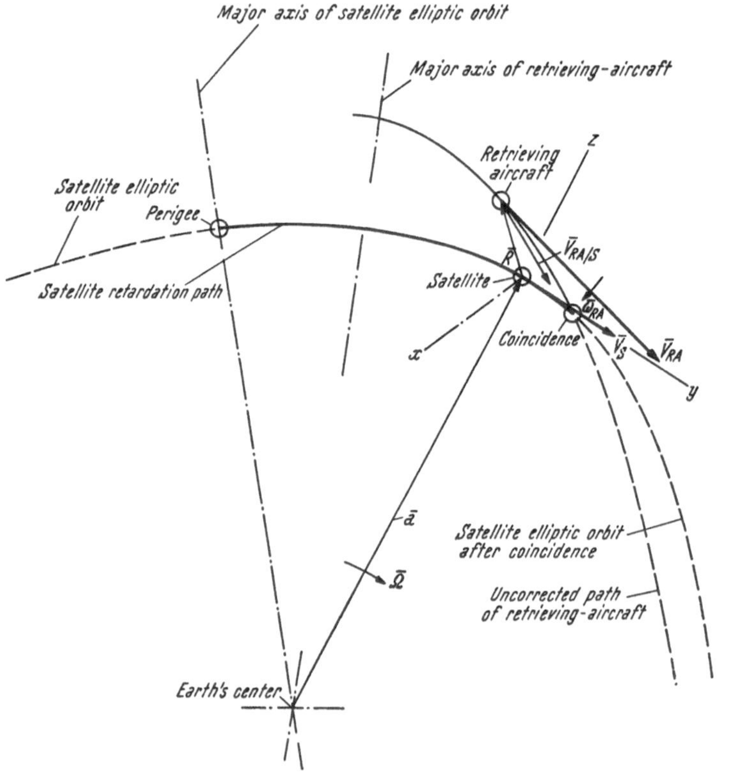

Fig. 13. Geometrical relationship between retrieving-aircraft and satellite (course predictor phase)

the capture point. The number of ground stations could be minimized by requiring the satellite to make successive approaches to perigee on its elliptic path to permit computing a precise orbit. Though, due to perturbations caused by gravity anomalies, earth's oblateness, earth's rotation and the element of time, instantaneous orbit information would be required.

Sight-line range at satellite perigee is about 100 miles. The tracking and homing system is assumed in operation out to at least 200 miles thus providing the satellite pilot approximately 20 seconds time prior to being committed to the retardation maneuver at perigee.

Since position and velocity errors will develop in the retrieving operation, from both the satellite and aircraft aspects, adequate search and track radar systems plus course-prediction and homing-with-capture (terminal) guidance must be provided. Range limitations of existing radar systems are considered adequate for the retrieving operation, that is, from acquisition at 200 miles distance down to a lower limit of about 1000 feet. Below a 1000 foot limit in range from retrieving-aircraft to satellite, other means than radar would be a necessity. Pilot monitoring throughout the terminal and capture phase is practical since time of flight is sufficient and required accelerations permit pilot control manipulation. The final approach to capture, 100 feet to contact, may best be manually executed.

Average acceleration levels for correction maneuvers are anticipated to be of the order of $\pm 1\ g$ about the free-fall condition. A position error of as great as

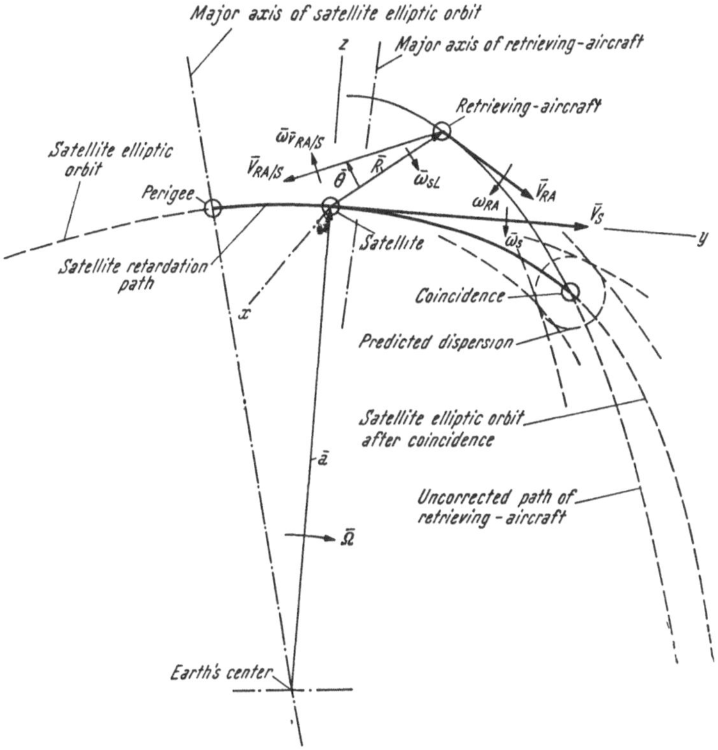

Fig. 14. Geometrical relationship between retrieving-aircraft and satellite (proportional navigation phase)

1 mile at coincidence requires only 18 seconds to eliminate. A position error of one mile simultaneously along the flight path and normal to it in two planes, will increase the required correction time, for an average 1 g acceleration, to about 22.6 seconds.

An estimated velocity error at coincidence of $\pm$ 1000 f/sec (including the velocity increase resulting from an assumed 1 mile position-error correction), for a correction acceleration of 3 g's, requires an additional 10 seconds. Thus the total correction time is about 33 seconds, and to a large extent, should occur prior to the time of coincidence.

To eliminate the need for actual vehicle contact, the use of a controlled probe from the retrieving-aircraft could assist payload transfer. Instead of placing the final constraint of requiring actual contact to be established by the guidance system. A controlled probe, capable of coupling to the satellite, could tolerate a small velocity and position differential.

Fig. 13 and 14 illustrate the navigation systems considered necessary to effect closure and capture on the satellite. These include I, a dispersion-computer and associated guidance system, and II, a proportional navigation system during the terminal phase. The operating range for System I extends from beyond perigee to the vicinity of coincidence with System II functioning throughout the terminal phase up to contact.

A single system using proportional navigation would require that the thrust vector be related to the separate retrieving-aircraft and satellite orbit errors, $\Delta\bar{a}$, $\Delta\bar{R}$, $\Delta\bar{V}_{RA}$ and $\Delta\bar{V}_S$, based on an initially established point of coincidence.

System I operation, to a large extent occurs while both vehicles follow essentially straight line paths. System I operation must be based on the measured range and velocity errors from the initially predicted flight paths for both vehicles. Though the relative range and closing velocity along the sight line changes sharply throughout the satellite's retardation path, the sight line rotational rate remains nearly zero for the major part of the flight path.

System II is assumed to function at close range with the thrust vector, $\bar{T}$, controlled by the lead angle, θ, and sight line range, R_{SL}.

References

1. Major General SCHRIEVER, Dinner Address: Astronautics Symposium, sponsored jointly by A.F.O.S.R.-Convair, February 18—20, 1957.
2. Progress on X-15: Aviation Week, p. 26, April 29, 1957.
3. C. GAZLEY, JR., and D. J. MASSON, A Recoverable Scientific Satellite. Rand Report P-958, The Rand Corporation, February 27, 1957.
4. R. W. PORTER, Recovery of Data in Physical Form. Earth Satellites as Research Vehicles. Monograph No. 2, J. Franklin Institute, June, 1956.
5. K. A. EHRICKE, Astronautical and Space-Medical Research with Automatic Satellites. Earth Satellites as Research Vehicles. Monograph No. 2, J. Franklin Institute, June, 1956.
6. J. JENSEN, Satellite Ascent Guidance Requirements. J. Astronautics **3**, 1 (1956).
7. W. WRIGLEY, Performance of a Linear Accelerometer. Orbital and Satellite Vehicles. Vol. 1, MIT, August, 1956.

Attitude Control of a Satellite Vehicle— an Outline of the Problems

By

R. E. Roberson [1]

(With 5 Figures)

Abstract — Zusammenfassung — Résumé

Attitude Control of a Satellite Vehicle — an Outline of the Problems. The attitude of a satellite vehicle must be controlled for many applications. This paper describes some of the fundamental problems associated with the design of an attitude control system. These include the choice of an attitude reference system and of reference axes within the body, and the nature of the attitude perturbation torques acting on the satellite. Attitude equations of motion are derived and a rationale for a control system synthesis is suggested. Control torque sources, the effects of vehicle configuration, and the role of attitude sensing devices are discussed.

Lagekontrolle eines künstlichen Satelliten. Die Lage eines künstlichen Satelliten im Raum muß für gewisse Anwendungszwecke reguliert werden. In der vorliegenden Arbeit werden die grundlegenden Probleme, die beim Entwurf einer solchen Lagensteuerung auftreten, beschrieben. Diese Probleme umfassen zunächst die Wahl eines geeigneten Bezugssystems und geeigneter Achsen in dem Satellitenkörper, ferner die Art der Lagenstörmomente, die an dem Körper angreifen. Bewegungsgleichungen für die Lagenänderungen werden abgeleitet und Richtlinien für den Aufbau eines Steuerungssystems werden angedeutet. Sodann werden die möglichen Kraftquellen für die Erzeugung von Steuermomenten sowie die Einflüsse der Körpergestalt und die Aufgaben der Lagenmeßgeräte besprochen.

Une esquisse des problèmes posés par le contrôle de l'orientation d'un satellite. Pour de nombreuses applications l'orientation d'un satellite doit pouvoir être gouvernée. L'article décrit quelques uns des problèmes fondamentaux associés à la conception d'un dispositif de contrôle de cette orientation. Parmi ceux-ci figurent le choix d'une référence d'orientation, le choix d'axes de référence dans le satellite et la nature des couples perturbateurs agissant sur celui-ci. Les équations du mouvement d'orientation sont établies et une approche rationnelle est suggérée pour la synthèse d'un dispositif de contrôle. Les types de couples de contrôle, les effets de la configuration du véhicule et le rôle des détecteurs d'orientation sont mis en discussion.

1. Introduction

Controlling the attitude of a satellite vehicle is important in many applications. This fact is implicit in much of the astronautical literature which discusses these applications.

On the whole, though, attitude control problems have not been discussed very extensively. They usually are mentioned in passing, further discussion being con-

[1] Autonetics, a Division of North American Aviation, Inc., 9150 E. Imperial Highway, Downey, California, U.S.A.

fined to a general description and feasibility statement concerning specific methods for control, i.e., control torque sources, or attitude sensing, i.e., detection of satellite orientation. (Some very ingenious possibilities have been proposed for both.) That attitude control has not come in for a larger share of attention in the current astronautical literature may well be that it is not such a burning question for the early satellite now contemplated, specifically Vanguard, as for the more sophisticated satellite applications one may envision for the future. The attitude of such minimum satellites may be uncontrolled or may be stabilized by imparting an initial spin to the satellite stage. It is expected that the resulting attitude behaviour will suffice for the relatively crude applications to which such satellites lend themselves.

Nevertheless, many satellite applications have been discussed which involve much more stringent attitude control requirements and more elaborate control techniques for satisfying these requirements.

When the attitude motions of a satellite are considered under the general requirement that they be kept small, or else known with high precision, a number of problems arise. The analytical framework for analyzing the motions becomes rather elaborate. Considerable pains must be taken not to discard effects which normally are ignorable but which assume potential significance under satellite conditions. Decisions must be made about attitude reference coordinate systems and reference axes in the body, some of these rather subtle.

This paper outlines the problems of this nature which arise in a study of attitude control, and indicates areas in which further work on these problems is desirable.

2. Attitude Reference Systems

The Problem

Perhaps the first question which must be faced is the choice of an attitude reference system. One must decide what reference attitude the vehicle is to maintain under ideal conditions; that is to say, the attitude with respect to which all attitude deviations are to be measured. A complete answer to this question depends, of course, on the specific application to which the satellite is to be put.

To the best of my knowledge, this problem has not been discussed in the astronautical literature, perhaps because a satellite vehicle is so often visualized in a circular orbit about a spherical earth. In this case, a natural choice for a vertical reference is the geocentric vertical at the satellite, and a natural choice for a forward reference is the forward direction in the orbital plane. However, we know that an actual orbit about the oblate earth can be neither circular nor plane. In fact, it is true that for some satellite applications the departure from the plane circular idealization can be quite significant. When the actual behaviour and the satellite application are considered, there is a question as to what other choices are reasonable and how good the above choice remains.

A brief answer is that there is a large number of "intuitively natural" choices, based on different definitions of both vertical and forward directions, and that the above choice is clearly nonoptimum in some cases. However, it will be seen that the differences among these natural choices usually are quite small.

This last statement does not mean that the problem of choosing attitude reference axes should be shrugged off. There are several reasons why the problem needs to be examined carefully. First, there is the truism that any discipline which presumed to be a science should have as sound a logical foundation as possible. This is promoted by defining the reference system with clarity and precision before launching into a discussion of physical control systems. Second, although differences among suitable reference frames may be small, if the attitude control system must constantly strive to correct a small difference it should not have to correct, its net power consumption may be prohibitive. Third, some applications inherently require extreme attitude precision. This is true, for example, when an orbit or free trajectory is corrected by a short burst of rocket power, for the direction of this applied thrust usually is critical.

The nontriviality of the question "Which reference system is best?", and the difficulty of giving a simple answer to it are illustrated by the following general remarks.

Assume that the satellite is not dynamically spherical so that it has a "long axis" (a principal axis of inertia about which the moment of inertia is smallest). It is known [4] that an attitude perturbation torque arises from the gravitational field of the earth tending to line the satellite up with its long axis normal to the local gravitational equipotential surface. Moreover, it can be shown that this can be a significant, if not major, torque source. These facts suggest that the optimum choice of vertical direction is the local gravitational vertical, for this prevents any gravitational perturbation torques from trying to pull the satellite out of its desired vertical attitude. However, one is penalized by such a choice because, even for a circular orbit (and still more for an elliptical one), this gravitational vertical moves with a nonuniform angular velocity in space as the vehicle moves around its orbit. This makes control torques necessary to impart sufficient angular acceleration to keep the satellite alined with the chosen nonuniformly rotating vertical.

The converse case also is a possibility. Suppose that the vertical reference is chosen to move at a uniform angular velocity in space[1]. Clearly, no torque is required to make the satellite attitude keep pace with the uniformly rotating reference vertical. However, in this case the reference axis is no longer normal to the gravitational equipotential surface to the vehicle and the vehicle is bound to feel a gravitational perturbation torque and excitation of attitude deviations from this source.

Any other choice for a vertical reference axis merely combines in a single case both of the two possible excitations mentioned above. It may, of course, have concomitant advantages.

The choice of a forward reference is even less definite. Although it is conventional to choose it normal to the vertical reference direction, there is no a priori requirement that this be so. Furthermore, it is not evident that it must necessarily lie in the instantaneous plane of the orbit. For example, if the satellite were being used to map the earth as suggested by [1], the optimum forward direction probably is in the plane of the local geographical vertical and the satellite velocity vector, the latter probably being the velocity relative to the earth rather than velocity relative to inertial space.

With these considerations in mind, let us turn to the explicit definition of an attitude reference system. Actually, it is convenient to define a whole class of "reasonable" systems so that the final choice can be postponed. Ultimately the choice is made to suit the purpose at hand, keeping in mind both the attitude reference needs of the particular application and the possibility of simplifying the

[1] Such a reference is the geocentric vertical at the point which moves around the orbit with the satellite's mean motion; that is, with the constant angular velocity the satellite would have if the orbit had no eccentricity but the same total energy.

attitude equations of motion by a perspicacious choice. It will be seen later that these equations contain a parametric excitation which can be shifted from place to place in the equations by various choices of reference system.

Basic Coordinate Systems

Let $\mathfrak{X}\mathfrak{Y}\mathfrak{Z}$ be a right-hand rectangular Cartesian coordinate system with its origin at the center of the earth and an orientation fixed in inertial space. Let the $\mathfrak{Y}$ axis lie in the direction of the earth's north pole. Although the orientation of $\mathfrak{X}$ and $\mathfrak{Z}$ is arbitrary, assume for definiteness that the $\mathfrak{Z}$ axis is directed toward the First Point of Aries[1]. Next, let X' Y' Z' be a similar system obtained from

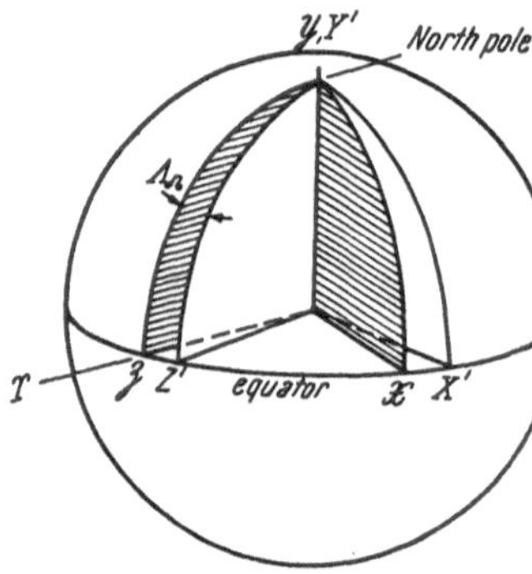

Fig. 1. Coordinate systems $\mathfrak{X}\mathfrak{Y}\mathfrak{Z}$ and X' Y' Z' in relation to the earth

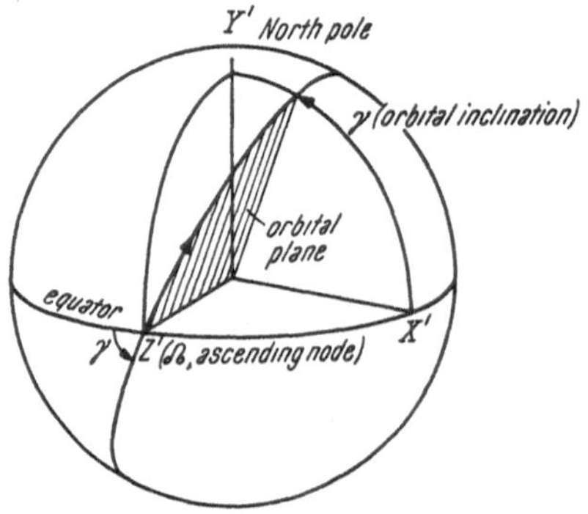

Fig. 2. Orbital plane and its relationship to X' Y' Z'

$\mathfrak{X}\mathfrak{Y}\mathfrak{Z}$ by a rotation through an angle Λ_Ω about the $\mathfrak{Y}$ axis. The significance of the notion chosen for this angle becomes apparent later. These two coordinate systems, the relation between them, and their relation to the earth are shown in Fig. 1.

Next consider a plane containing the Z' axis and inclined to the earth's equator at an angle γ (with $0 \leqq \gamma \leqq \pi$), as shown in Fig. 2. An important problem of orbit analysis is to find how the angle Λ_Ω must behave if the satellite is to remain within, or close to, this inclined plane. Thus it is natural to speak of the plane as the "orbital plane" and of the angle γ as the "orbital inclination." The convention is adopted that the satellite crosses the equator from south to north near the Z' axis, so that the intersection of this axis with the projection of the equator on the celestial sphere is the so-called "ascending node"[2] of the orbit, and the Z' axis itself is the "line of nodes."

Three generalized coordinates are required to specify completely the location of the satellite relative to the earth. One of them is chosen to be the radial distance, r, of the center of mass of the satellite from the center of the earth. The other two are angular coordinates, α and β, which are described conveniently in terms of the orbital plane.

---

[1] For the non-astronomer reader, the First Point of Aries, symbolized by ♈, is the intersection on the celestial sphere of the earth's equator and the sun's path as it passes from south to north across the equator, i.e., the vernal equinox.

[2] This ascending node is customarily designated by ☊ in astronomical literature, whence the motivation for the longitude angle notation Λ_Ω. Note that ♈ is the ascending node of the sun in its apparent motion about the earth.

Fig. 3, an edge-on view of the orbital plane from the positive Z' axis, shows the satellite at point S. Denote by y' the axis through the center of the earth normal to the orbital plane, its positive sense on the northerly side of this plane for $\gamma < \pi/2$ and on the southerly side for $\gamma > \pi/2$. The shaded plane in Fig. 3 contains both the y' axis and the satellite and intersects the orbital plane in a line defined as the z' axis. The positive sense of this axis is such that it makes an angle of less than $\pi/2$ with the ray from the center of the earth to the satellite. In these terms the dihedral angle from the $Z'y'$ plane to the $y'z'$ plane, labeled β, describes the angular advance of the satellite from its ascending node. The angle α, measured in the $y'z'$ plane positively as shown in the figure, describes the satellite departure from the orbital plane. An axis x' is adjoined to y' and z' defined above so as to form a right-hand system.

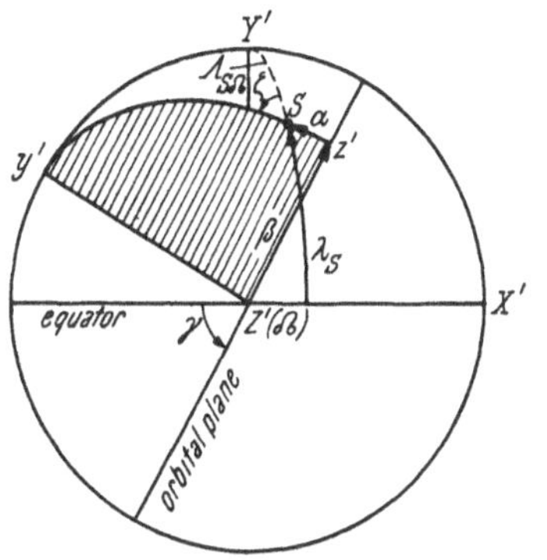

Fig. 3. Position angles of the satellite

In this paper, we consider only attitude reference systems xyz which remain "close to" $x'y'z'$. Angles Φ_1, Φ_2, Φ_3 are defined by the unit sphere diagram of Fig. 4, relating these two sets of axes. The usual "small angle" approximations are assumed to hold for the Φ-angles, this being the significance of the requirement that xyz remain "close to" $x'y'z'$. For cases where this is not true, such as those in which the attitude is desired to remain inertially fixed, a similar development is possible retaining the complete direction cosine expressions connecting xyz and $x'y'z'$ for arbitrary Φ_1, Φ_2, Φ_3. In that case, though, it may be desirable to define xyz directly in terms of $\mathfrak{X}\mathfrak{Y}\mathfrak{Z}$.

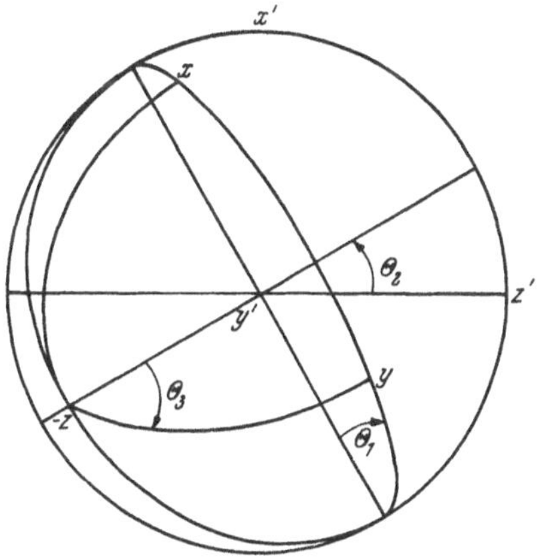

Fig. 4. Relation between $x'y'z'$ and xyz axes (Read Φ_1, Φ_2, Φ_3 for θ_1, θ_2, θ_3)

Non-Constant Angular Velocity of xyz

The fact that Φ_1, Φ_2 and Φ_3 are to remain small has an important implication; namely, that there is *no* acceptable reference frame having this characteristic which also has a constant angular velocity in inertial space. This is proved as follows. If xyz remains close to $x'y'z'$, one can represent its relative position by a small-angle vector

$$\Phi = \Phi_1 \bar{e}_{x'} + \Phi_2 \bar{e}_{y'} + \Phi_3 \bar{e}_{z'} \tag{1}$$

(where $\bar{e}_{x'}$ etc. are unit vectors along the axes indicated) such that to the first order in small quantities

$$\left.\begin{aligned} \bar{e}_x &= \bar{e}_{x'} + \Phi_3 \bar{e}_{y'} - \Phi_2 \bar{e}_{z'} \\ \bar{e}_y &= -\Phi_3 \bar{e}_{x'} + \bar{e}_{y'} + \Phi_1 \bar{e}_{z'} \\ \bar{e}_z &= \Phi_2 \bar{e}_{x'} - \Phi_1 \bar{e}_{y'} + \bar{e}_{z'} \end{aligned}\right\} . \tag{2}$$

If the angular velocity in inertial space of xyz has components ω_1, ω_2, and ω_3 along x, y, and z respectively, and $\omega_{x'}$, $\omega_{y'}$, $\omega_{z'}$ are the components relative to the primed set of the angular velocity of the primed set in inertial space, then differentiating Eq. (1) with respect to time gives (to first order in small quantities)

$$\left.\begin{aligned}\dot{\Phi}_1 &= \omega_1 - \Phi_3\,\omega_2 + \Phi_2\,\omega_3 - \omega_{x'}\\ \dot{\Phi}_2 &= \Phi_3\,\omega_1 + \omega_2 - \Phi_1\,\omega_3 - \omega_{y'}\\ \dot{\Phi}_3 &= -\Phi_2\,\omega_1 + \Phi_1\,\omega_2 + \omega_3 - \omega_{z'}\end{aligned}\right\}. \tag{3}$$

Now suppose the ω_1, ω_2, ω_3 are all constant. Then Eq. (3) can be solved by quadrature. It is easy to show that the solutions $\Phi_1(t)$, $\Phi_2(t)$ and $\Phi_3(t)$ involve integrals of the form $\int_0^t (\omega_1 - \omega_{x'})\,dt$, $\int_0^t (\omega_2 - \omega_{y'})\,dt$, $\int_0^t (\omega_3 - \omega_{z'})\,dt$ which grow unboundedly with t unless $\omega_{x'}$, $\omega_{y'}$, $\omega_{z'}$ are constant and precisely equal to ω_1, ω_2, ω_3 respectively. It is a fact, though, that they cannot be constant. For example, $\omega_{y'}$ involves principally the angular rate of the satellite about its orbit, which is not constant except for a circular orbit. However, it is known that a circular orbit cannot exist about an oblate earth. Moreover, various perturbations of the orbit cause a slow rotation of its line of nodes, which implies that oscillatory terms occur in $\omega_{x'}$, $\omega_{y'}$, $\omega_{z'}$ whose amplitudes are proportional to the angular rate of precession of the line of nodes.

It is clear that these results become invalid as the magnitude of Φ_1, Φ_2, and Φ_3 increase. However, the acceptable bounds on these quantities are well within the range of validity of the small-angle approximation used, and one infers that the reference system departs from $x'y'z'$ by an amount greater than can be tolerated if ω_1, ω_2, and ω_3 are all constant.

The principal result of these general considerations is that one is not at liberty to choose a reference coordinate system for which ω_1, ω_2, and ω_3 are all constant. Therefore there is no possibility that the general equations of attitude deviation can be reduced to equations with constant coefficients merely by a proper choice of reference axes. In short, one must be reconciled to the existence of small parametric excitations in the attitude equations, and one has only the option of shifting the locations of the excitations to various terms of the equations.

Possibilities for Choice

Although it is not exhaustive, a list of some of the most promising possibilities for both vertical and for forward directions is given in Table I[1].

Several of these cases can be discarded immediately. Those of the form (5, M) all depart from $x'y'z'$ at a significant rate because of the orbital regression effect, as shown previously above, and are unacceptable in the class of systems to which this paper is limited (even though they would be optimum in removing inertial reaction torques and parametric excitations from the attitude equations of motion). Next, if we observe that the forward velocity vector departs from the orbital plane by only a few seconds of arc, we see that in practice case (N, 1)

---

[1] The system which represents a specific combination of these possibilities is hereafter denoted by (N, M), in which N is one of the vertical choice numbers and M is one of the forward choice numbers of Table I.

is a trivial distinction from $(N, 2)$ or $(N, 3)$. Therefore the major possibilities now remaining are (N, M) with $N = 1, 2, 3, 4$ and $M = 2, 3$.

The next reduction comes from considerations of sensing feasibility. From the practical point of view, the vehicle cannot be kept in its reference attitude

Table I. *Basic attitude reference possibilities*

Vertical	
Name	**Description**
1. Geocentric	Outward-pointing geocentric radius through the vehicle
2. Geographic	Outward-pointing line through the vehicle and normal to the International Ellipsoid
3. Gravitational	Outward-pointing line through the vehicle and normal to the smoothed ellipsoidal approximation to the earth gravitational equipotential surface passing through the vehicle
4. Mean-motion geocentric	Outward-pointing geocentric radius through the point which moves with the vehicle mean motion (in the sense of celestial mechanics)
5. Constant velocity geocentric	Outward-pointing geocentric radius through the point which would move with the vehicle mean motion if there were no orbital regression
Forward	
Name	**Description***
1. None standard	Vehicle forward velocity as seen from earth or from the regressing plane
2. None standard	Forward sense of motion projected onto the instantaneous orbital plane, as seen by an earth-fixed observer directly beneath the vehicle
3. None standard	Forward sense of motion projected onto the instantaneous orbital plane, as seen by an observer in the regressing plane

* Normal to the chosen vertical direction and in the plane determined by it and by item given.

unless its deviations from the reference attitude can be sensed. Any apparent advantages from the selection of a reference system which for some reason cannot be located relative to the vehicle frame by sensing instruments are illusory. The failure to sense deviations is equivalent to introducing torque driving terms into the equations of attitude deviation which cannot be removed by an attitude control system, so that deviations are bound to result. The role of attitude sensing is discussed at greater length later, but it can be observed here that in order to use the forward reference corresponding to $M = 2$ there must be a means within the satellite of distinguishing its altitude, orbital inclination, orbital period, and the instants when it crosses the equator. In many cases, it is unlikely that all of this information can be available, and we conclude that $M = 2$ is a relatively impractical case.

These remarks can have an important practical implication. Suppose that in a specific satellite application it is desired to maintain a forward direction of the type $M = 2$. It appears generally infeasible to do this, or at least quite difficult, relative to the type $M = 3$. If one can practically mechanize only $M = 3$, must be prepared to accept a yaw error equal to the angular difference between the forward direction in inertial space and the forward direction relative to the earth, which can be about 4° at the nodes of an orbit with high inclination and an altitude of a few hundred kilometers. It follows that if $M = 3$ must be selected purely on the basis of feasibility, even though $M = 2$ would be operationally optimum, it is irrational to insist on a tight yaw control relative to the selected reference.

In summary, the choice of forward reference system which appears best for most applications is precisely the one which is intuitively "most natural," namely the forward horizontal direction in the instantaneous orbital plane.

A specific choice of vertical reference would require further analysis. It would be determined by the nature of the available sensing mechanism and by the magnitudes of the resulting inertial reaction and external torques. Four facts increase the difficulty of this choice: (1) *no* choice exists which simplifies the equations of motion to a constant coefficient system; (2) *no* choice is completely satisfactory by all criteria; (3) *all* of the remaining possibilities are characterized by ω_1, ω_2, and ω_3 of the same general form (trigonometric series of small amplitude with a fundamental equal to the orbital period), so that no obvious difference exists in the modes of parametric excitation for the four choices; and (4) all of the remaining possibilities lead to driving torques of the same general form (trigonometric series of small amplitude with a fundamental equal to the orbital period), although the amplitudes differ for the four cases, so that no one choice is obviously most favorable in avoiding quasi-resonance effects.

Forms of Φ_1, Φ_2, and Φ_3

These angles and the angular velocity components of the xyz system occur in some of the attitude deviation torque terms discussed later, so it is worthwhile to give them explicit forms for the various reference systems now suggested as major possibilities, namely $(N, 3)$ with $N = 1, 2, 3, 4$.

Suppose xyz is initially coincident with $x'y'z'$, then rotated through a small angle ε about a vector in the $x'y'$ plane whose projections on these axes are -cos δ and sin δ respectively. The four remaining cases are subsumed by

$$\left.\begin{aligned} \Phi_1 &= -\varepsilon \cos \delta \\ \Phi_2 &= \varepsilon \sin \delta \\ \Phi_3 &= 0 \end{aligned}\right\} \tag{4}$$

in which ε and δ have the explicit forms given in Table II.

The angular velocity $\bar{\omega}$ of the xyz frame in inertial space is

$$\bar{\omega} = (\omega_{x'} - \dot{\varepsilon} \cos \delta)\, \bar{e}_{x'} + (\omega_{y'} + \dot{\varepsilon} \sin \delta)\, \bar{e}_{y'} + (\omega_{z'} - \dot{\delta})\, \bar{e}_{z'}. \tag{5}$$

That is, the components ω_x, ω_y, ω_z of $\bar{\omega}$ resolved into the xyz system are

$$\left.\begin{aligned} \omega_x &= \omega_{x'} - \dot{\varepsilon} \cos \delta - (\omega_{z'} - \dot{\delta})\, \varepsilon \sin \delta + 0\,(\varepsilon^2) \\ \omega_y &= \omega_{y'} + \dot{\varepsilon} \sin \delta - (\omega_{z'} - \dot{\delta})\, \varepsilon \cos \delta + 0\,(\varepsilon^2) \\ \omega_z &= \omega_{x'} - \dot{\delta} - \omega_{y'}\, \varepsilon \cos \delta + \omega_{x'}\, \varepsilon \sin \delta + 0\,(\varepsilon^2) \end{aligned}\right\}. \tag{6}$$

Table II. *ε and δ for various vertical reference axes*

Vertical reference	ε	δ
Geocentric	0	$\arccos \frac{\cos\gamma}{\cos\lambda_S}$
Geographic	$\varepsilon_{gg} \sin 2\lambda_S$	$\arccos \frac{\cos\gamma}{\cos\lambda_S}$
Gravitational	$\varepsilon_{gr} \sin 2\lambda_S$	$\arccos \frac{\cos\gamma}{\cos\lambda_S}$
Mean-motion geocentric	$\omega_0 t - \beta$	$\frac{\pi}{2}$

ε_{gg} and ε_{gr} are constants characteristic of the oblateness of the International Ellipsoid and the gravitational equipotential surface (at the satellite) respectively. For this discussion, there is no loss in generality in supposing that the zero of time is the instant the vehicle passes through the ascending node of its orbit. The orbital frequency is ω_0 and the satellite latitude is given by $\sin\lambda_S = \sin\alpha\cos\gamma + \cos\alpha\sin\gamma\sin\beta$.

In terms of the angular velocity $\dot{\Lambda}_\Omega$ of the line of nodes,

$$\left.\begin{aligned} \omega_{x'} &= \dot{\Lambda}_\Omega \sin\gamma\cos\beta \\ \omega_{y'} &= \dot{\beta} + \dot{\Lambda}_\Omega \cos\gamma \\ \omega_{z'} &= \dot{\Lambda}_\Omega \sin\gamma\sin\beta \end{aligned}\right\} . \tag{7}$$

Using the expressions for $\dot{\Lambda}_\Omega$, $\beta(t)$ and $\dot{\beta}(t)$ given in [5] and ε, δ from Table II, one can find explicit expressions for ω_x, ω_y, ω_z as time-functions or β-functions.

3. Reference Axes in the Body

Attitude Deviation Angles

A question which is closely related to the choice of reference axes in space is the choice of reference axes in the body of the satellite vehicle. Again the answer depends upon the reason for attitude stabilization.

To see the role of the body reference axes, we first define a set of attitude deviation angles. Let XYZ be a right hand orthogonal system of axes imbedded in the body (in a sense to be made precise later), such that XYZ coincides with xyz when the vehicle is in its optimum, or desired, attitude. Then any nonoptimum attitude can be characterized by three attitude angles θ_1, θ_2, θ_3 which relate XYZ to xyz as shown in the unit sphere diagram, Fig. 5. The angles θ_1, θ_2, θ_3 are called roll, pitch, yaw respectively.

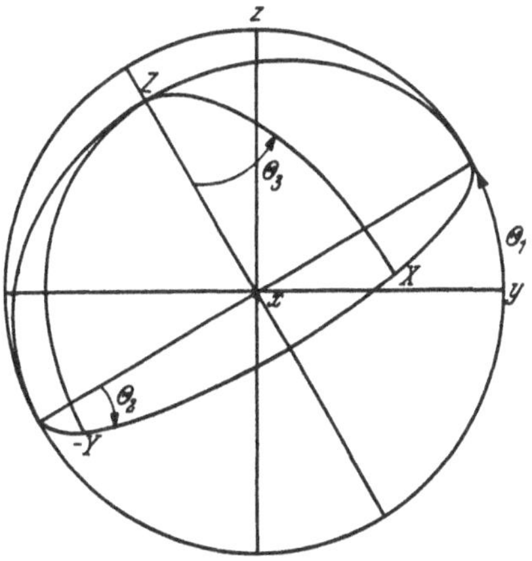

Fig. 5. Attitude deviation angles

Table III gives the direction cosine matrices among the coordinate systems xyz, $x'y'z'$ and XYZ. Some of the entries are specialized to the small-angle case, but quadratic terms in θ_1, θ_2, and θ_3 are carried even in this case because they are needed for the analysis of certain perturbation torques on the satellite.

The importance of the choice of XYZ is clear. The roll, pitch, and yaw angles are to become the basic dependent variables which describe, in some sense, the attitude motion of the satellite. We have not completed the description until we are explicit about *what* motion is being described.

Table III. *Direction cosine matrices relating coordinate systems $x'y'z'$, xyz, and XYZ*

A	x'	y'	z'
x	1	Φ_3	$-\Phi_2$
y	$-\Phi_3$	1	Φ_1
z	Φ_2	$-\Phi_1$	1

(Φ_1, Φ_2, Φ_3 all small angles)

B	x	y	z
X	$\cos\theta_2\cos\theta_3$	$\sin\theta_1\sin\theta_2\cos\theta_3+\cos\theta_1\sin\theta_3$	$-\cos\theta_1\sin\theta_2\cos\theta_3+\sin\theta_1\sin\theta_3$
Y	$-\cos\theta_2\sin\theta_3$	$\cos\theta_1\cos\theta_3-\sin\theta_1\sin\theta_2\sin\theta_3$	$\sin\theta_1\cos\theta_3+\cos\theta_1\sin\theta_2\sin\theta_3$
Z	$\sin\theta_2$	$-\sin\theta_1\cos\theta_2$	$\cos\theta_1\cos\theta_2$

(general case)

C	x'	y'	z'
X	$1-\frac{\theta_2^2+\theta_3^2}{2}-\theta_2\Phi_2-\theta_3\Phi_3$	$\theta_3+\theta_1\theta_2+\theta_2\Phi_1+\Phi_3$	$-\theta_2+\theta_1\theta_3+\theta_3\Phi_1-\Phi_2$
Y	$-\theta_3+\theta_1\Phi_2-\Phi_3$	$1-\frac{\theta_1^2+\theta_3^2}{2}-\theta_1\Phi_1-\theta_3\Phi_3$	$\theta_1+\theta_2\theta_3+\theta_3\Phi_2+\Phi_1$
Z	$\theta_2+\theta_1\Phi_3+\Phi_2$	$-\theta_1+\theta_2\Phi_3-\Phi_1$	$1-\frac{\theta_1^2+\theta_2^2}{2}-\theta_1\Phi_1-\theta_2\Phi_2$

(θ_1, θ_2, θ_3, Φ_1, Φ_2, Φ_3 all small angles)

D	x	y	z
X	$1-\frac{\theta_2^2+\theta_3^2}{2}$	$\theta_1\theta_2+\theta_3$	$-\theta_2+\theta_1\theta_3$
Y	$-\theta_3$	$1-\frac{\theta_1^2+\theta_3^2}{2}$	$\theta_1+\theta_2\theta_3$
Z	θ_2	$-\theta_1$	$1-\frac{\theta_1^2+\theta_2^2}{2}$

(θ_1, θ_2, θ_3 all small angles)

Satellite Principal Axes

A simple and obvious possibility is to choose XYZ along the principal axes of inertia of the satellite vehicle. This is a well-defined set of three axes unless the inertia ellipsoid has two or three equal axes, but an arbitrary choice can be made in such a case without loss of generality.

A major advantage of this choice is that some perturbation torques (e.g., the gravitational torque discussed in [4] and inertial reaction torques) are most naturally expressed in terms of a principal axis system. Moreover, these axes are often close to axes of physical symmetry of the vehicle, which are the natural reference frame for still other perturbation torques.

However, there is one major disadvantage to such a choice. In any but the most rudimentary satellite vehicle, there will be internal moving parts—rotating machinery, translating masses, people moving about a manned vehicle, etc. Thus the principal axes of inertia inevitably wander with respect to the vehicle frame, usually through small angles about some equilibrium orientation. One may properly question whether such wandering axes are really those which it is operationally desirable to control.

A secondary, but none the less important, point is that principal axes are conceptual rather than physical axes. They are not accessible to inspection—one cannot see or feel them—and they can be rather tedious to determine in a specific case.

Body-Fixed Axes

The axes which overcome the above objections to principal axes are a set of axes which are physically fixed in the body. They can be scribed on the body structure, so that there is no question about what is being controlled in attitude.

For maximum convenience, these body-fixed axes should be chosen so that they are close to the principal axes (assuming that the latter do not depart very far from their average position). It is evident, though, that one still must know the relationship between the two sets of axes in order to resolve the attitude perturbation torques normally expressed in terms of principal axes into the other set.

The major advantage of the body-fixed axes is not their conceptual simplicity, as important as that is, but the fact that the reason for attitude control usually arises in terms of such axes. In all of the applications mentioned previously, it is important that equipment (e.g., rocket engines) which is in a known orientation relative to these axes be attitude-controlled.

For this reason, body-fixed XYZ axes are assumed henceforth.

Case of Movable Equipment

One other case should be mentioned, that in which the equipment whose attitude it is desired to control (e.g., antennas) is not fixed relative to the satellite body, but can be rotated about one or more axes relative to it. In this case, the raison d'être of attitude control is not intrinsically related to the body-fixed axes, for one may try to control only the attitude of the equipment and let the body rotate about the controlled element as it will.

This case can be analyzed in a similar way, but is somewhat more complex because of an additional consideration which normally arises. This is the fact that the attitude of the body itself usually cannot be completely ignored, and some degree of attitude control must be enforced on it. Therefore, one has the problem of simultaneous attitude control of two or more reference frames, albeit to different degrees of accuracy, rather than merely exchanging the attitude control of the satellite body completely for the control of some of its internal parts. Although it is recognized that this case may exist, we do not treat it further in this paper.

4. Attitude Perturbation Torques

Torque Sources

In the previous sections, our attention was on logical questions—the definition of suitable attitude reference systems in space and in the satellite vehicle. We now come to the first major physical question, the nature of the torques which act to change the attitude of the satellite.

The qualitative nature of these torques is important because it indicates the vehicle design procedures which are indicated to enhance useful (stabilizing) torques and reduce undesirable (deviation-producing) torques. Their amplitudes and detailed time behavior is basic to the physical design of any attitude control system.

Some of the sources of attitude perturbation torques that can be conceived are

1. earth's magnetic field,
2. earth's electric field,
3. sun's radiation pressure,
4. pressure of electromagnetic radiation from the satellite,
5. air drag,
6. meteoroid bombardment,
7. cosmic ray bombardment,
8. gravitational fields of celestial bodies,
9. non-uniform rotation of reference coordinates,
10. moving parts within the satellite,
11. earth's gravitational field.

Most of these torques depend critically on the configuration design of the satellite, so they cannot be discussed here in the absence of a specific configuration except in generalities. However, even these general considerations can be illuminating.

Earth's Magnetic Field

The principal component of the earth's magnetic field can be represented at the surface by a dipole near the center of the earth with a magnetic moment of 8.1×10^{25} emu. It is assumed that the field at the satellite can be represented by the same dipole M. The magnetic induction in a vacuum at a field point to be associated with satellite position is, in cgs units,

$$\bar{B} = -\nabla\left(\frac{\bar{M}\cdot\bar{r}}{r^3}\right) = \frac{M}{r^3}\left(C_{Mx'}\,\bar{e}_{x'} + C_{My'}\,\bar{e}_{y'} - 2\,C_{Mz'}\,\bar{e}_{z'}\right) \tag{8}$$

where $C_{Mx'}$, $C_{My'}$, $C_{Mz'}$ are direction cosines of the magnetic dipole vector relative to the $x'y'z'$ coordinate system. [In Eq. (8), r is measured from the center of the dipole.]

It can be shown that $C_{Mx'}$ and $C_{Mz'}$ are oscillatory functions containing frequency components ω_0 and $\omega_0 \pm \Omega$ where ω_0 is the orbital frequency and Ω is the angular speed of rotation of the earth. On the other hand, $C_{My'}$ is composed of a constant term and an oscillatory term with frequency Ω. Now, for an eccentric orbit, r^{-3} contains frequency components $n\omega_0$ ($n = 1, 2, 3$). The $\sin \omega_0 t$ and $\cos \omega_0 t$ terms from this source multiply those similar terms contained in $C_{Mx'}$ and $C_{Mz'}$ to give rise to small constant components of $\overline{B}$ along the x' and z' axes.

Consider first the effect of eddy currents. Because there are varying components of magnetic field along all reference axes, eddy currents are induced in the conducting skin and other parts of the vehicle. By interacting with the earth's field, these cause a net torque on the vehicle. Exact calculations of this torque are very difficult, even when a configuration is specified, but it may be possible to find an approximate upper limit by assuming all conducting material to be compressed into a single loop of some equivalent area and resistance.

An additional torque arises for an aspheric paramagnetic body. Such a body experiences a torque tending to aline its long axis with the ambient magnetic field. It may be possible to obtain a crude estimate of torque level using the formula (7) for this effect on a simple prolate spheroid. The torque is

$$\overline{L}_M = 2\,(\overline{\mathfrak{p}} \times \overline{B})\;(\overline{\mathfrak{p}} \cdot \overline{B})/|\overline{\mathfrak{p}}| \tag{9}$$

where $\overline{\mathfrak{p}}$ is a vector whose magnitude depends on the volume, permeability and elongation ratio of the magnetic body and whose direction is that of the prolate axis.

It is easy to show that

$$\overline{L}_M = (2\,\mathfrak{p}_{y'} + \mathfrak{p}_{z'})\,(-2\,C_{My'}\,C_{Mz'})\,\bar{e}_{x'} + (2\,\mathfrak{p}_{x'} + \mathfrak{p}_{z'})\,(2\,C_{Mz'}C_{Mx'})\;\bar{e}_{y'} + \\ + (\mathfrak{p}_{x'} + \mathfrak{p}_{z'})\,(2\,C_{Mx'}C_{My'})\,\bar{e}_{z'}. \tag{10}$$

The products of direction cosines in this equation can be evaluated without difficulty. In particular, $C_{My'}C_{Mz'}$ and $C_{Mx'}C_{My'}$ can be represented by terms with orbital frequency with amplitudes (less than unity) which wax and wane at twice the earth's rotational frequency. Also $C_{Mx'}C_{Mz'}$ is oscillatory with twice orbital frequency and amplitude behavior like that of the other products. The $\overline{\mathfrak{p}}$-components are constant except for terms in the attitude deviation angles, so there is no external torque from this source which is persistent in direction. The terms involving attitude deviation angles become parametric excitation terms in the attitude equations of motion.

Finally, there is a torque from the interaction of the earth's field with any permanent magnets or current loops within the satellite. This effect is probably the easiest of the three to analyze once the nature of these internal fields is known.

Earth's Electric Field

The existence of a nominally radial electric field with an approximate intensity of 10^{-2} statvolt/cm at the surface of the earth is well known. However, no information is available on the field at satellite altitudes which may be above most of the ionized gases of the atmosphere. At this altitude, one would expect an essentially radial contribution to the field from any net charge on the earth and its ionized layers.

There are two effects to consider. First, if the satellite is aspheric there is a torque from the ambient electric field which tends to aline the body with the field (an induced electric effect analogous to the magnetic effect mentioned previously). In a radial field, this effect is qualitatively like the earth's gravitational effect to be discussed later, and introduces no new behavior. A second effect arises from the fact that the satellite is a conductor traveling at a relatively high velocity through a magnetic field. There is an equivalent electric field of the motion which then interacts with the ambient electric field.

I believe that further analysis of both electric and magnetic field effects is very desirable. However, rough preliminary calculations I have made indicate that they are not likely to produce torques of major significance unless a specific design deliberately or accidentally enhances some of the normal effects.

Sun's Radiation Pressure

The calculation of torque from radiation pressure depends first on the reflectivity of the vehicle surfaces. An upper bound on such torques provided by assuming complete reflection (zero absorption), but this may be too large by a factor which depends on surface shape and ranges up to 2 for a flat plate. The surface reflectivity will depend on its state of oxidation and pitting, and may be low enough that a realistic estimate of torque can be made on the assumption of complete absorption.

The incident radiation power (the solar constant) is about 1.93 cal/cm^2 min corresponding to a radiation pressure of 4.3×10^{-5} dyne/cm^2 with complete absorption.

The torque on the satellite is the product of the radiation pressure, the area presented to the sun, and the linear distance between the centroid of this presented area and the projection of the center of mass of the satellite onto the same area. The result, therefore, is very sensitive to the precise configuration design and may easily vary by several orders of magnitude among "reasonable" designs.

It is of some interest to note that essentially constant torques can be produced by the sun. Suppose the sun is described by angles λ_H, Λ_H relative to the inertial $\mathfrak{X}\mathfrak{Y}\mathfrak{Z}$ axes introduced previously, as follows. The unit vector which describes the position of the sun is

$$\bar{e}_H = \cos\lambda_H \cos\Lambda_H \bar{e}_{\mathfrak{x}} + \cos\lambda_H \sin\Lambda_H \bar{e}_{\mathfrak{y}} + \sin\lambda_H \bar{e}_{\mathfrak{z}}. \tag{11}$$

From this and the relation between the $\mathfrak{X}\mathfrak{Y}\mathfrak{Z}$, x' y' z' and XYZ axes, it can be shown that there is a component of $\bar{e}_H$ along the Y (pitch) axis equal to

$$\cos\lambda_H \sin\Lambda_H \cos\gamma + \sin\lambda_H \sin\Lambda_\Omega \sin\gamma - \cos\lambda_H \cos\Lambda_H \cos\Lambda_\Omega \sin\gamma + \\ + \text{terms in } \Phi,\ \theta \text{ angles}. \tag{12}$$

Because λ_H, Λ_H and Λ_Ω vary very slowly with time, the component given by Eq. (12) is essentially constant. Therefore, if the product of centroid distance and projected area mentioned previously has any constant component in the XZ-plane, the sun's radiation pressure causes an essentially constant torque about an axis normal to this latter component and contained in the XZ-plane.

Electromagnetic Radiation from the Satellite

If any such radiation, including heat radiation, occurs asymetrically with respect to the satellite's center of mass, a torque will be produced on the vehicle. Overcoming this torque is largely a matter of careful radiator design. The radiation would have to represent many kilowatts at a large lever arm to become large, although it might be likely to produce a persistent torque component which would be troublesome if not corrected.

Air Drag

PETERSEN [3] has argued that the residual atmosphere at satellite altitude is sufficiently tenuous to justify the assumption of free molecular flow. The mean free path is several orders of magnitude greater than any reasonable vehicle dimension so that reflected particles influence the momentum of the oncoming particles to a negligible extent. If one assumes that the incident particles of atomic dimensions are reflected by an ideal diffuse reflector, in accordance with LAMBERT's law the ratio of the number of particles reflected at an angle from the normal to the number reflected normally equals the cosine of the angle, whatever the angle of incidence. Under these conditions, it can be shown that the resulting torque is

$$\mathrm{L}_{AD} = \varrho\, r^2 \omega_0^2 \left(1 + \frac{c_m^2}{r^2 \omega_0^2}\right) \int\int (\bar{e}_{x'} \cdot \bar{e}_n) \left(\bar{e}_{x'} + \frac{2}{3}\bar{e}_n\right) \times \bar{r}_S \, dS \tag{13}$$

where c_m is the rms molecular velocity, $\bar{e}_n$ is the unit normal to the element of surface area dS, and rS is the vector from the center of mass to dS.

The application of this equation to vehicles with some kind of external symmetry about the XYZ axes (and therefore, to within terms of the order of Φ and θ, about the x' y' z' axes) gives both a constant aerodynamic torque, in general, and a "spring constant" representing an aerodynamic moment per unit angle of attack. The latter is effectively the pitch angle θ_2. The corresponding torque may be stabilizing or upsetting, depending on the nature of the configuration and the reference orientation chosen for it. The constant torque could be virtually eliminated by careful control of the center of mass relative to the center of pressure.

Meteoroid Bombardment

A satellite is subject to collisions with meteoroids. Of paramount importance is the question, analyzed by LANGTON [2], of whether a particular meteoroid punctures the skin and causes a system failure. However, even those which do not cause failure and those which fail to penetrate can be significant because of the torque impulses they apply to the system.

No completely satisfactory estimate exists of the occurrence frequency of various magnitudes of meteoroid torque impulse. This is because of the compounded uncertainties of meteoroid frequency, energy, and degree of momentum transfer in an impact, as well as the distributions of presented area and impact location.

It appears from the relatively meagre information which is available that the average momentum from meteoroid impacts is not very large. Even over a period of the order of a year, the net impulse seems likely to be small. However, individual impacts certainly could produce significant overturning impulses, imparting transient attitude oscillations to the satellite that might be large compared with those arising from some of the other sources I have mentioned here.

I feel that this is a problem area worthy of considerably more study.

Cosmic Ray Bombardment

Upper atmosphere data currently available [8] indicate that the energy of cosmic ray bombardment at the satellite will be of the order of (2 bev/cm^2)/sec. As 1 ev $= 1.59 \times 10^{-12}$ erg $= 1.59 \times 10^{-19}$ joule, 1 bev/sec $\approx 10^{-10}$ watts. It is apparent by comparison with the results for sun's radiation pressure that even if the cosmic rays are completely reflected, their bombardment effect will be insignificant compared to that of the radiation from the sun.

Gravitational Fields of Celestial Bodies

The equations which give the attitude torque in the field of the earth[1] can be applied to find the torque from the sun and moon. It can be shown that the latter are approximately 10^{-9} times the torque due to the earth.

Rotation of Reference Coordinates

The desired, or reference, attitude of the vehicle is not fixed in inertial space. The reference coordinate system generally is affected by the orbital regression and the orbital eccentricity and has a nonuniform angular velocity from these causes. Therefore control torques must be applied to keep the vehicle properly alined relative to this coordinate system. The tendency of the vehicle to depart from its desired attitude would appear to an observer in the satellite exactly as if it were caused by a perturbation torque.

It must be remembered that these so-called "inertial reaction torques" are actually equivalent torques in the sense that their negatives must be applied to the vehicle to prevent attitude deviations from the attitude reference, which is undergoing angular accelerations.

No special analysis is needed to derive these excitation terms, for they appear automatically when the equations of motion are written relative to body-fixed axes.

Moving Parts within the Satellite

There are many moving parts within a reasonably sophisticated satellite — rotating wheels, vibrating parts, circulating fluids, and the like. Their motion tends to produce effective torques on the satellite through inertial reaction, gyroscopic coupling, center of mass shift, axes of inertia shift, etc. Although some of these torques are very small, others can be significant sources of attitude disturbance.

Expressions for the torque from internal moving parts have been developed in a separate paper [6]. It suffices here to note that some of these torques are explicit time functions, while other small terms involve the angular velocity of the satellite's body axes, generally with variable coefficients. The latter, therefore, are actually parametric excitation terms in the equations of attitude motion.

It may be that some torques of this class are used as attitude control torques. Thus a third category, neither external driving functions nor parametric excitations, should be kept in mind. One such torque arises from constant speed rotating parts within the satellite. If the angular momentum of the part (relative to the body of the satellite) is $\overline{H}$, a constant, it produces an effective torque $-\overline{\omega}^* \times \overline{H}$ on the satellite, where $\overline{\omega}^*$ is the angular velocity of XYZ in inertial space.

Earth's Gravitational Field

An expression for the potential function of a satellite-type body has been derived elsewhere [4]. It can be used to find the torque on the satellite caused by the differential gravitational attraction on its various mass elements.

This derivation is an *exceedingly* tedious task. Although it has been done, there is nothing gained by presenting all of the terms of the result. There are twenty-seven such terms after dropping all small terms of second order and higher in the attitude deviation angles and their derivatives, and in the oblateness

---

[1] See discussion later.

of the earth, the ellipticity of the orbit, the difference angles between the $x'y'z'$ and xyz systems, the moment of inertia shifts from moving parts, and the related axes of inertia shifts.

The dominant terms of the result are easy to find, though. They are simply

$$\begin{aligned} L_G = & -3\,\omega_0^2\,(I_Y - I_Z)\,\theta_1\,\bar{e}_X \\ & -3\,\omega_0^2\,(I_X - I_Z)\,\theta_2\,\bar{e}_Y \end{aligned} \tag{14}$$

where ω_0 is the mean motion (average orbital angular velocity) of the satellite, constant except for slow air drag changes which are not considered here.

The remaining terms not given explicitly contribute both small explicit driving functions of time and small parametric excitation terms involving the angular deviation angles and their rates.

5. Attitude Control

Equations of Attitude Motion

It already has been stated that XYZ is a set of attitude reference axes imbedded rigidly in the satellite. Suppose, now, that the inertia dyadic of the composite vehicle (the body of the satellite plus all its moving parts) is divided into a constant part $\bar{\bar{\Lambda}}$ and a variable part $\bar{\bar{\gamma}}$ as described in [6]. There is no loss in generality if we suppose that XYZ are so chosen to diagonalize $\bar{\bar{\Lambda}}$, giving it the form

$$\bar{\bar{\Lambda}} = I_X\,\bar{e}_X\,\bar{e}_X + I_Y\,\bar{e}_Y\,\bar{e}_Y + I_Z\,\bar{e}_Z\,\bar{e}_Z\ . \tag{15}$$

Also, let the angular velocity $\bar{\omega}^*$ of XYZ in inertial space be resolved as

$$\bar{\omega}^* = \omega_X\,\bar{e}_X + \omega_Y\,\bar{e}_Y + \omega_Z\,\bar{e}_Z\ . \tag{16}$$

Then the attitude equations of motion take the well-known EULER form

$$\left.\begin{aligned} I_X\,\dot{\omega}_X + (I_Z - I_Y)\,\omega_Z\,\omega_Y = L_X \\ I_Y\,\dot{\omega}_Y + (I_X - I_Z)\,\omega_X\,\omega_Z = L_Y \\ I_Z\,\dot{\omega}_Z + (I_Y - I_X)\,\omega_Y\,\omega_X = L_Z \end{aligned}\right\} \tag{17}$$

where $\bar{L}$ is the totality of torques applied to the body from sources like those mentioned in the previous chapter.

In order to use these equations as a basis for an attitude control system analysis, one must express the components of $\bar{\omega}^*$ in terms of the attitude deviation angles, and must have an explicit form for the torque components.

First consider the angular velocity components. The angular velocity components ω_x, ω_y, and ω_z of the xyz coordinate system in inertial space (resolved into the xyz system) are given by Eq. (6). From Fig. 5,

$$\bar{\omega}^* = \dot{\theta}_1\,\bar{e}_x + \dot{\theta}_2\,\cos\theta_1\,\bar{e}_y + \dot{\theta}_2\,\sin\theta_1\,\bar{e}_z + \dot{\theta}_3\,\bar{e}_Z + \bar{\omega} \tag{18}$$

whence

$$\left.\begin{aligned} \omega_X = \dot{\theta}_1 + \omega_x + \omega_y\,\theta_3 - \omega_z\,\theta_2 \\ \omega_Y = \dot{\theta}_2 - \omega_x\,\theta_3 + \omega_y + \omega_z\,\theta_1 \\ \omega_Z = \dot{\theta}_3 + \omega_x\,\theta_2 - \omega_y\,\theta_1 + \omega_z \end{aligned}\right\}\,. \tag{19}$$

Here, we have used the direction cosine relations of Table III to resolve into the common basis $\bar{e}_X$, $\bar{e}_Y$, $\bar{e}_Z$, and we have confined our attention to the special case of small attitude deviation angles. The generalization to large angles is straightforward, but messy.

It is evident from Eq. (6) that ω_x and ω_z generally are small functions of time or of position in the orbit. However, ω_y contains $\omega_{y'}$ which involves the relatively large orbital angular velocity ω_0. Define an inertial reaction torque $\bar{L}_I$ as the negative of the torque which must be applied to keep the vehicle alined with the reference coordinate system because of the non-uniform motion of the latter.

$$\begin{aligned}\bar{L}_I = \ & [I_X \dot{\omega}_x + (I_Z - I_Y)\,\omega_z\,\omega_y]\,\bar{e}_X \\ & + [I_Y \dot{\omega}_y + (I_X - I_Z)\,\omega_x\,\omega_z]\,\bar{e}_Y \\ & + [I_Z \dot{\omega}_z + (I_Y - I_X)\,\omega_y\,\omega_x]\,\bar{e}_Z\end{aligned} \tag{20}$$

Then the equations of motion become, to first order terms in θ_1, θ_2, θ_3 and their derivatives:

$$\begin{aligned}& I_X\ddot{\theta}_1 + \omega_0^2\,(I_Y - I_Z)\,\theta_1 + \omega_0\,(I_X + I_Z - I_Y)\,\dot{\theta}_3 \\ & = L_X - L_{IX} - \Big\{ I_X \frac{d}{dt}\,[(\omega_y - \omega_0)\,\theta_3 - \omega_z\,\theta_2] \\ & + (I_Z - I_Y)\,[\omega_z\,\dot{\theta}_2 + (\omega_y - \omega_0)\,\dot{\theta}_3 + \omega_y\,\omega_x\theta_2 - 2\,\omega_0\,(\omega_y - \omega_0)\,\theta_1]\Big\}\end{aligned} \tag{21}$$

$$\begin{aligned}& I_Y\ddot{\theta}_2 = L_Y - L_{IY} - \Big\{ I_Y \frac{d}{dt} - \omega_x\,\theta_3 + \omega_z\,\theta_1] \\ & + (I_X - I_Z)\,[\omega_z\,\dot{\theta}_1 + \omega_x\,\dot{\theta}_3 - \omega_x\,\omega_y\,\theta_1 + \omega_y\,\omega_z\,\theta_3]\Big\}\end{aligned} \tag{22}$$

$$\begin{aligned}& I_Z\ddot{\theta}_3 + \omega_0^2\,(I_Y - I_X)\,\theta_3 - \omega_0\,(I_X + I_Z - I_Y)\,\dot{\theta}_1 \\ & = L_Z - L_{IZ} - \Big\{ I_Z \frac{d}{dt}\,[\omega_x\,\theta_2 - (\omega_y - \omega_0)\,\theta_1] \\ & + (I_Y - I_X)\,[(\omega_y - \omega_0)\,\dot{\theta}_1 + \omega_x\,\dot{\theta}_2 + 2\,\omega_0\,(\omega_y - \omega_0)\,\theta_3 - \omega_y\,\omega_z\,\theta_2]\Big\}\,.\end{aligned} \tag{23}$$

Note that the terms in curly brackets on the right-hand sides of Eqs. (21)-(23) all involve attitude deviation angles and their derivatives with small variable coefficients. These terms, therefore, represent parametric excitation of the equations. (More conventionally, they would be written on the left-hand side of the equations.) It already has been mentioned that the contribution to $\bar{L}$ from moving parts is a sum of similar parametric excitations and explicit time functions. The same is true for the contribution to $\bar{L}$ from the gravitational field of the earth, except that in this case there are the additional terms given by Eq. (14).

Let the $\bar{L}_G$ terms of Eq. (14) be carried to the left-hand sides of Eqs. (21)-(23), and let the control torques $\bar{L}_C$ applied to the satellite be split out of $\bar{L}$. Also, lump all the remaining terms on the right-hand side into two categories: the driving functions $\bar{D}$ whose coefficients involve only time, and the parametric excitation functions $\bar{P}$ whose coefficients involve θ_1, θ_2, θ_3, $\dot{\theta}_1$, $\dot{\theta}_2$, $\dot{\theta}_3$ as well as time. (Represent this set of angles and rates generically by θ.) Then the attitude equations of motion take the form, for small angles:

$$I_X\ddot{\theta}_1 + 4\,\omega_0^2\,(I_Y - I_Z)\,\theta_1 + \omega_0\,(I_X + I_Z - I_Y)\,\dot{\theta}_3 = L_{CX} + D_X(t) + P_X(\theta, t) \tag{24}$$

$$I_Y\ddot{\theta}_2 + 3\,\omega_0^2\,(I_X - I_Z)\,\theta_2 = L_{CY} + D_Y(t) + P_Y(\theta, t) \tag{25}$$

$$I_Z\ddot{\theta}_3 + \omega_0^2\,(I_Y - I_X)\,\theta_3 - \omega_0\,(I_X + I_Z - I_Y)\,\dot{\theta}_1 = L_{CZ} + D_Z(t) + P_Z(\theta, t). \tag{26}$$

The following characteristics of these equations are of particular importance.

1. Parametric excitation of the equations is inescapable. Arising from a number of separate sources, any attempt to suppress it in one place seems to make it arise in another. Thus one is necessarily forced to cope in a control system synthesis with a servo problem involving time variable system parameters. (For larger angles of attitude motion, of course, one must deal with nonlinearities as well.)

2. Aside from the parametric excitation terms, the motion in θ_2 described by Eq. (25) is decoupled from and independent of the motions in θ_1 and θ_3 described by Eqs. (24) and (26). This means that a Y axis control system represents a simpler servo problem (speaking loosely, since P_Y does not actually vanish). It also means that because of the θ_1, θ_3 mode coupling, it may suffice to exert primary control about only the X or Z axis in order to achieve effective control about both.

3. Under certain conditions on the principal moments of inertia, Eqs. (24)-(26) already are conditionally stable (again ignoring $\bar{P}$). If this were also true with $\bar{P}$ included, no information about absolute attitude would be needed to stabilize the sattelite at its desired attitude, although some attitude rate information would be required for damping. This topic is discussed later below.

One important revision to Eqs. (24)-(26) is sometimes required. If there is a net (constant) internal angular momentum within the vehicle,

$$\bar{H} = H_X \bar{e}_X + H_Y \bar{e}_Y + H_Z \bar{e}_Z,$$

then significant additional terms occur on the right-hand sides of these equations, viz. $(H_Y \omega_Z - H_Z \omega_Y)$, $(H_Z \omega_X - H_X \omega_Z)$, and $(H_X \omega_Y - H_Y w_X)$ respectively. Because of the ω_0 term in ω_Y, any components H_Z and H_X cause persistent torques in the equations for θ_1 and θ_3. The presence of such torques, when means exist for avoiding them, is a severe handicap to any attitude control system. Therefore, it is assumed that $\bar{H}$ is alined, if at all possible, so that $H_X = H_Z = 0$. When this is done, we are still left with significant terms $H_Y(\dot{\theta}_3 - \omega_0\theta_1)$ and $H_Y(\dot{\theta}_1 + \omega_0\theta_3)$ in the θ_1 and θ_3 equations respectively.

Lumping the remaining small terms with $\bar{D}$ or $\bar{P}$, we therefore rewrite Eqs. (24) and (26) as

$$I_X \ddot{\theta}_1 + \omega_0^2 [4 (I_Y - I_Z) + I_H] \theta_1 + \omega_0 (I_X + I_Z - I_Y - I_H) \dot{\theta}_3 = L_{CX} + D_X + P_X \tag{27}$$

$$I_Z \ddot{\theta}_3 + \omega_0^2 (I_Y - I_X + I_H) \theta_3 - \omega_0 (I_X + I_Z - I_Y - I_H) \dot{\theta}_1 = L_{CZ} + D_Z + P_Z \tag{28}$$

where I_H is an equivalent moment of inertia defined by $\omega_0 I_H = H_Y$.

Control Torques

In general, almost any physical effect which can cause attitude perturbation torques can be considered as a possible basis for obtaining attitude control torques. Some of these methods are unsatisfactory because the attainable torque level is too low, power requirements are too high, or they involve serious mechanization difficulties.

If one examines in detail each of the torque sources mentioned in Chapter 4, the methods which appear most attractive are inertial methods involving moving parts within the vehicle. There are a number of possibilities for such methods.

One which has been mentioned for many years in the astronautical literature is the use of flywheels which provide an inertial reaction torque when accelerated relative to the satellite frame. But other related methods might utilize the gyroscopic coupling torque between a rotating wheel and the normal angular velocity of the satellite's attitude reference system, or might be some kind of dynamic vibration absorber or gyro-stabilizer.

Another method mentioned frequently is the use of gas jets or the ejection of other particles in such a way that a pure couple is applied to the vehicle. The utility of this method probably varies with the details of the satellite objectives and design. In evaluating it one must weigh the weight penalty of orbiting the material to be ejected against the weight of the power supply for alternative methods, and must consider maximum torque and torque impulse capabilities as well as other mechanization considerations.

Regardless of the physical source of the control torques, the nature of the control problem is clear. With reference to Eqs. (25), (27), (28), one must design a control system which can provide a torque $\overline{L}_C$ as a function of θ_1, θ_2, θ_3, $\dot{\theta}_1$, $\dot{\theta}_2$, $\dot{\theta}_3$ such that the attitude motions given as the solution of these equations are kept small in some sense.

Rationale of Synthesis

The first requirement of a control system synthesis is to specify performance criteria for the system. Four conditions are reasonable to impose:

1. that the system be stable in the absence of $\overline{D}(t)$;
2. that the steady state oscillations under the influence of the parametric excitations and any periodic components of $\overline{D}$ be kept "small";
3. that the transient maxima from expected aperiodic components of $\overline{D}$ be kept "small";
4. that the decay of these transients and of non-zero initial conditions be "rapid".

The first of these would be almost a trivial question in the absence of $P(\theta, t)$, for many simple forms of $\overline{L}_C$ would suffice for stability in this case. However, the stability of a parametrically excited system is a more delicate question, and the adequacy of such simple forms in this case must be justified.

Although it is easy to state these requirements qualitatively, it is very difficult to make them quantitative. For example, in items 2 and 3, how small is "small"; in item 4, how rapid is "rapid"? Perhaps even more difficult is to decide the relative importance that should be attached to these requirements, for it may not always be possible to promote all requirements by the same choice of $\overline{L}_C$. In fact, the performance from one point of view may be worsened by the same choice that betters it from another.

I have no general answer to these difficulties. Rational design criteria are so intimately tied to the specific satellite application that elaboration of the point here is fruitless. Therefore, I content myself with mentioning the problem.

There is another problem related to the synthesis that should be discussed. It is the problem of how the synthesis should be conducted for a parametrically excited nonlinear system. Again, a general answer does not seem possible, but I recommend the following philosophy. I cannot justify it rigorously, but believe that an a posteriori check of its results is the most convincing test of validity.

The first step is to linearize the equations of attitude motion. In the present development this already has been done in deriving Eqs. (25), (27), (28). Physically, this limits the synthesis to small attitude oscillations. However, it does not imply that the resulting control system necessarily has a poorer performance for larger attitude oscillations.

The second step is to discard all parametric excitation terms, leaving an equation with constant coefficients which can be treated conveniently. This step may be expected to be permissible if one thinks of such terms as characterized by a single small-amplitude parameter and then expands the θ in powers of this small parameter as in a perturbation method. The zero*th*-order solution is just the solution of the reduced (constant coefficient) equation obtained in the present procedure. The first and higher order solutions then are obtained from identical constant coefficient equations into which the lower-order solutions enter as driving terms. It is of considerable interest that this reduction to constant coefficient equations removes all terms containing θ_1, θ_2, θ_3 which characterize the particular attitude reference system chosen. Therefore the choice need not be made at all before the initial synthesis is complete.

The principal doubt with might exist about this argument is the question of stability, because it is well known that for such generalized HILL equations as discussed here, there usually are limited regions in the parameter phase planes for which a bounded solution exists. However, it is conjectured that this is not a serious problem. The treatment of the zero*th*-order case presumably will result in a system with such damping that no resonance-like effects should be expected in obtaining the higher-order solutions. Although the perturbation method may not be the optimum practical way of treating the equations, the zero*th*-order solution may reflect the complete behavior fairly well in the case at hand.

The third step of the procedure is to analyze the control system as if the physical situation were really no more complex than is indicated by the constant coefficient equations: that is, to choose forms and gain constants for the control functions. This is a straightforward, if tedious, servo synthesis problem of conventional type.

Having found optimum control technique and parameter values by this method, the fourth and final step is to return to the nonlinear parametrically excited equations and check the adequacy of the system against the specified performance criteria. I believe that this procedure will result in a satisfactory, if not "optimum," system in most cases.

Effect of Configuration[1]

It is worth discussing briefly the effect of the satellite's configuration characteristics on the attitude control problem. The configuration enters this problem through the I_X, I_Y, and I_Z which appear in Eqs. (24)-(26).

The inherent stability in the absence of a control system is determined by the coefficients of θ_1, θ_2, and θ_3 respectively in these equations. There are six cases with stability characteristics shown in Table IV. (It is assumed for the moment that no two principal moments are equal and that $H_Y = 0$.) There is only one kind of configuration which is inherently conditionally stable. This result suggests that such a configuration is the most logical starting point for a control system synthesis.

Note, however, that if I_X and I_Y become equal, stability in θ_1 and θ_3 are lost. With this in mind it is easy to see the effect of the I_H term in Eqs. (27) and (28).

[1] In this discussion we follow the philosophy outlined above and ignore parametric excitations.

It strengthens the θ_1—θ_3 "spring" which is weakened when $I_X \approx I_Y$ and can even overcome instability in these modes occuring when $I_Y < I_X$. Thus, it may be desirable to introduce an internal angular momentum H_Y for this purpose. No choice of I_H affects the stability of the θ_2 mode. However, it cannot be increased without penalty, for it is an implicit multiplier in some of the $\overline{P}$ terms as well.

Table IV. *Stability* behavior for various configurations*

Configuration	θ_1, θ_3	θ_2
1. $I_X < I_Y < I_Z$	unstable	unstable
2. $I_Y < I_Z < I_X$	unstable	stable
3. $I_Z < I_X < I_Y$	stable	stable
4. $I_X < I_Z < I_Y$	stable	unstable
5. $I_Z < I_Y < I_X$	unstable	stable
6. $I_Y < I_X < I_Z$	unstable	unstable

*"Stability" here means conditional stability, equivalent to "not-instability".

Role of Attitude Sensing

Attitude sensing is another function which has been discussed with attitude control in the astronautical literature, usually called loosely "attitude control." By attitude sensing I mean the ability of an observer or instrumentation aboard the satellite to measure the satellite's orientation in space (and possibly one or more derivatives of orientation angles). Let us see what the role of attitude sensing is in an attitude control system.

The control torque $\overline{L}_C$ introduced into the equations of attitude motion generally must be a function of θ_1, θ_2, θ_3, $\dot{\theta}_1$, $\dot{\theta}_2$, $\dot{\theta}_3$. But in order to generate the function, some estimate of current values of these quantities must be available. It is these values which are available from an attitude sensing system.

Many possibilities have been suggested for attitude sensing systems. In general, they can be divided into devices of the following classes:

1. inertial (stabilized reference, rate gyros, etc.);
2. sightings on celestial bodies;
3. ambient fields (earth's gravitational, magnetic or atmospheric fields, cosmic rays impinging on vehicle, etc.);
4. sightings on the earth (optical, radio, etc.);
5. sightings from the earth.

Each of these methods has its own advantages and disadvantages, which must be evaluated with the satellite application in mind. A detailed survey of sensing systems is an important topic, but one beyond the scope of this paper.

There are two principal difficulties associated with attitude sensing systems. One of these is the fact that they can provide a physical measurement of attitude, but not "true values". In other words, there is inevitably some instrument error or noise which finds its way through the sensing and control systems to appear as a noise driving term exciting additional attitude motions.

The second difficulty is that sensing systems generally do not measure precisely the same attitude variables it is desired to control. For example, instead of θ_1, θ_2, θ_3, a specific instrument may measure only the departure of the Z-axis from the geographic vertical. If the θ_1, θ_2, θ_3 refer to an attitude reference system with some other vertical, such as the geocentric vertical, a conversion is necessary on the sensing information before it is used in the control system. If this problem is ignored, additional driving terms appear in the attitude equations. If the conversion is made, it often requires that the instantaneous satellite location be known and used in the conversion calculation. In general, the conversion is fraught with difficulty and potential error.

One way to reduce these problems is to dispense with as much of the attitude sensing function as possible. A way to do this is suggested by the inherent stability of the $I_Z < I_X < I_Y$ configuration, which in a sense provides its own sensing function. Because the attitude equations are already conditionally stable, no injection of θ_1, θ_2, θ_3 signals is essential for stability. The use of attitude rate signals alone, easily obtained by inertial means within the satellite, is sufficient.

There are cases, though, in which independent sensing may be important. It may be possible to give the control system improved dynamical characteristics using such a system, over those which could be obtained from the inherently stable configuration alone. In other cases, it may be desirable to have an independent measurement of attitude for future reference purposes, regardless of whether it is used as a part of the control system. For these reasons, we may regard attitude sensing mechanisms as an important class of satellite systems, worthy of detailed investigation.

References

1. P. R. Haviland, On Applications of the Satellite Vehicle. Jet Propulsion **26**, 360, 368 (1956).
2. N. H. Langton, The Thermal Dissipation of Meteorites by Bumper Screens, in: Bericht über den V. Internationalen Astronautischen Kongreß, Innsbruck (5—7 August 1954), p. 72. Wien und Innsbruck: Springer, 1955.
3. N. V. Petersen, Lifetimes of Satellites in Near-Circular and Elliptic Orbits. Jet Propulsion **26**, 341, 368 (1956).
4. R. E. Roberson and D. Tatistcheff, The Potential Energy of a Small Rigid Body in the Gravitational Field of an Oblate Spheroid. J. Franklin Institute **262**, No. 3, 209 (1956).
5. R. E. Roberson, Orbital Behavior of Earth Satellites. J. Franklin Inst. **264**, No. 3, 181, and No. 4, 269.
6. R. E. Roberson, Torques on a Satellite Vehicle from Internal Moving Parts, to appear in J. Appl. Mech.
7. W. R. Smythe, Static and Dynamic Electricity, 1st Edition. New York: McGraw-Hill Book Co., Inc., 1939.
8. J. S. Van Allen and S. F. Singer, Total Primary Cosmic Ray Energy at the Geomagnetic Equator. Physic. Rev. **79**, 206 (1950).

Meteor, Jr., a Preliminary Design Investigation of a Minimum Sized Ferry Rocket Vehicle of the Meteor Concept

By

Darrell C. Romick, Richard E. Knight and **Samuel Black**[1]

(With 15 Figures)

Abstract — Zusammenfassung — Résumé

Meteor, Jr., a Preliminary Design Investigation of a Minimum Sized Ferry Rocket Vehicle of the Meteor Concept. A previous paper outlined a concept for a large ferry-rocket vehicle and a manned earth-satellite terminal, designated *Meteor*. The present paper presents a preliminary design investigation for a minimum-sized vehicle of the same concept, called *Meteor, Jr.* While exhibiting less load-carrying efficiency than the large vehicle, the 500-ton three-stage *Meteor, Jr.* can carry approximately 1 ton of payload in addition to a crew of 2 to 4 (depending on the specific mission), and incorporates the necessary features for reasonably dependable operation as a research test vehicle, a prototype vehicle or for limited operational use. The results of the preliminary design investigation, including general arrangement, configuration, and main characteristics, are given. The resulting total vehicle is comparable in size though somewhat larger than the Atlas ICBM (Intercontinental Ballistics Missile). In a special test configuration, without appreciable payload, the regular final stage could be launched from an Atlas-sized expendable missile. The methods presented appear to be a logical approach to establishing the first manned earth-satellite, and it appears that this objective, using existing technology, could perhaps be accomplished in a period of around eight years.

Meteor Junior; vorläufige Planung einer Kleinst-Transportrakete nach Art von Meteor. Eine fruhere Arbeit befaßte sich mit dem Plan eines großen, als Transportrakete dienenden Raumfahrzeuges und einer bemannten Erdsatelliten-Station, *Meteor* genannt. Die hier vorliegende Arbeit bringt eine vorläufige Planung eines Raumfahrzeuges der gleichen Art (jedoch von Kleinstformat), *Meteor Junior* genannt. Während dieses nur eine geringere Nutzlast befördern kann als das große Fahrzeug, hat es doch — mit 3 Stufen und 500 Tonnen Startgewicht — eine Tragkraft von etwa 1 Tonne Nutzlast neben einer Bemannung von 2 bis 4 Personen (je nach dem besonderen Zweck). Das neue Raumfahrzeug hat die Fähigkeit, in verwirklichbarer Weise als Forschungsfahrzeug zu dienen, als ein Prototyp oder für einen beschränkten Wirkungskreis. Die Ergebnisse der vorlaufigen Untersuchung werden dargelegt und schließen die allgemeine Anordnung, die Gestalt und die Hauptcharakteristika ein. Das sich daraus ergebende Fahrzeug ist in seiner Größe ungefähr — wenngleich etwas größer — der Interkontinentalrakete „Atlas" vergleichbar. In einer besonderen Versuchsform, jedoch ohne nennenswerte Nutzlast, könnte die normale Endstufe von einer praktisch verwendeten Fernrakete in der Größe von „Atlas" gestartet

[1] Authors' titles: Romick, Head, Astronautics Group, Weapon Systems Dept.; Knight, Mgr., Aerodynamics Dept.; Black, Sr. Aerodynamicist, all of Goodyear Aircraft Corporation, Akron 15, Ohio, U.S.A.

werden. Die beschriebene Technik scheint eine logische Annäherung an die Errichtung eines ersten bemannten Erdsatelliten zu sein. Es scheint ferner, daß dieses Ziel unter Verwendung der heute verfugbaren Technologie vielleicht innerhalb von rund acht Jahren erreicht werden konnte.

Météor junior. Étude d'avant-projet d'un transport-fusée miniaturisé suivant la conception du Météor. Un article antérieur esquissait la conception d'un transport-fusée important et d'une station satellite terminale habitée, appelés projet *Météor*. Le présent article présente un avant-projet de fusée de transport qui est une réduction de la même conception, appelée *Météor, Jr.* Quoique moins efficiente, la fusée à trois étages *Météor, Jr.* emporte, outre un équipage de 2 à 4 hommes (suivant la mission), une charge utile d'environ une tonne pour un poids de 500 tonnes au départ. Ses caractéristiques lui permettent une sûreté raisonnable de fonctionnement comme véhicule d'expérimentation, prototype ou pour missions opérationnelles limitées. La configuration, disposition générale et caractéristiques principales sont données, telles qu'elles résultent de l'étude préliminaire. Ses dimensions sont légèrement supérieures à celles de l'ICBM Atlas. Pour les essais de configuration sans charge utile, elle pourrait être lancée par un engin non-récupérable des dimensions de l'Atlas. Les méthodes utilisées semblent constituer une approche logique à l'établissement du premier satellite habité. Cet objectif pourrait, dans l'état actuel de la technique, être atteint dans huit années.

Foreword

This paper is one of a series on the *Meteor* concept, prepared at intervals during more than five years study. Another paper, on the inherent operational reliability characteristics of this type vehicle, has not been released, but study is completed, and it is to be presented at the 12th annual meeting of the American Rocket Society in New York 2—5 December 1957. It is mentioned here because many features found necessary in the reliability study (which was being carried on concurrently) are incorporated for the first time in the vehicle presented herein.

Grateful acknowledgement is made of the very important technical assistance, participation, and consultations in various fields by FRANK A. PAKE, Manager, Thermodynamics Department; ROBERT J. COUTS, Head of General Missile Design, and CHARLES M. JAMIESON, Design Development Specialist, all of Goodyear Aircraft Corporation. Without the contributions of these and others who assisted, the authors would have been unable to accomplish this study and present it in this form at this time.

I. Introduction

When the Manned Satellite Ferry Rocket vehicle concept known as *Meteor*[1] was first formulated and studied during the early 1950's, it was realized that the first actual embodiment of any such system would have to be a relatively small version for experimental and development test work. On the other hand, it was also known that a very large version could show more favorable load-carrying efficiency and better logistic advantage.

Early consideration of these factors revealed that there was a definite relationship between ratio of payload to gross weight and vehicle size (Fig. 1). The ratio trend is one sharply improving from minimum size vehicles with practically zero payload to some value for very large vehicles whose asymptote would move toward gradual steady improvement with the advancing state of the art. The authors decided to make the first extensive study and presentation

[1] Manned Earth-satellite Terminal evolving from Earth to Orbit ferry Rockets.

based on a very large version to give a picture toward the ultimate in design, and because higher efficiency could be shown. Also investigation was desired as to whether and where fundamental size limitations might be encountered—in engine size and landing gear, for example; and it was necessary to provide an anchor point near one end, or on the asymptotic portion of the efficiency curve, as it was believed points on the lower portions could, and automatically would, be more readily established in due time.

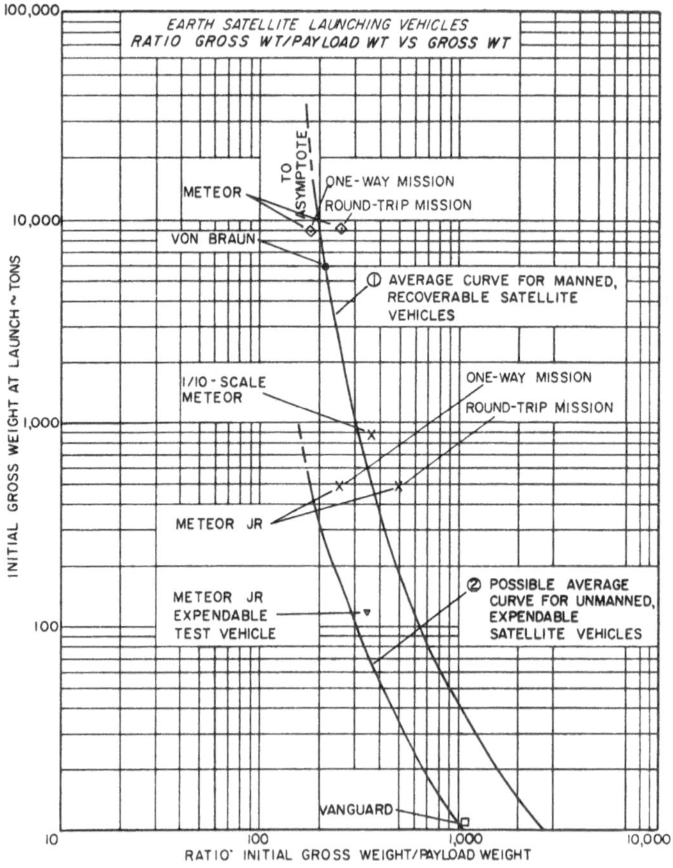

Fig. 1. Load carrying efficiency curve

Studies have since been carried out in this lower region, and the one being presented here is the result of a conviction that a most useful and important area for current consideration—the one of most practical value at the present time—would be toward the lower end of the efficiency curve, a minimum size vehicle that could adequately use the *Meteor* concept. Also, a study of a small ship, together with the previous study of a large one, will help to define generally the intermediate versions until more active attention falls in that area.

The actual scale of the minimum size is subject to differences of opinion, but *Meteor*, *Jr.* represents an initial utilization for research and development test, with possibly limited operational applications. A still smaller size was seriously considered, but providing minimum practical safety features and operational characteristics was difficult. The smaller vehicle likewise presented a minimum usefulness, which likewise tended to diminish the value of making an extensive study.

It was felt that less than a three-man crew in the final stage would seriously limit its usefulness and compromise some of the inherent advantages involved in the concept. A ton of payload was established as the minimum required for support of such a three-man crew and allowance had to be made for the weight of instruments for research and development and some safety margin allowed for additional useful weight or bulk, such as observers, equipment, and material.

In view of these and other considerations, there seemed little reason for going much smaller, particularly in view of the relationship of the vehicle size to published estimates[1] on the Atlas, and to current capabilities. Likewise, there seemed to be little reason for using larger sizes, as the 500-ton *Meteor, Jr.* is a good approximation for meeting all considerations evaluated.

Time allowance had to be made for pegging such moving parameters as engine performance and fuel development. The year 1962 was selected as the limit for these expanding parameters, not only because the elapsed time seemed reasonable as a minimum for normal program development, but also because of the difficulty in obtaining reliable predictions further into the future. To assume a higher performance of engines and fuel based on a longer time period would be less conservative and therefore a less useful approach. Fairly accurate predictable performance in 1962 likewise concides with the earliest time by which thc *Meteor, Jr.* concept could be implemented.

All the features and characteristics developed in studies of the *Meteor* concept over the past five years or more have been incorporated in the present study, including those generated in connection with the reliability paper soon to be published. In the design analysis, minutely detailed studies were generally avoided, but customary practices for achieving a valid preliminary design concept were followed. Latest techniques in use or under development for rocket vehicles, engines, guidance and control, and tracking and instrumentation, were assumed. Utilization is made of current experience in lightweight, high strength structures developed for missiles, aircraft, and airships. The study considers the latest actual hardware and the most promising ones still under development in the fields of high speed tires, brakes, landing gear, hydraulic and electric systems as well as appropriate extention of high speed aerodynamic research and development. Classified data, of course, have been avoided.

It is the opinion of the authors and others associated in this study that the results obtained represent a reliable indication of the characteristics of *Meteor, Jr.* and that taking due cognizance of other related programs under way (Vanguard, X-15, Atlas, B-58) a vehicle similar to the one described can be developed in a reasonable period of time—six or eight years to operational status.

II. Description

1. General

Meteor, Jr. is, like its big brother, a three-stage rocket designed to carry a payload to an orbit 500 mi above the surface of the earth, remain in the orbit for an indefinite period of time, or return to the earth with a similar payload, re-entering the atmosphere nearly tangentially, decelerating, and gliding to a landing. Allowance is made for required corrections and adjustments to make a rendezvous, to the return flight path or to the landing approach. In perfor-

[1] Missiles and Rockets 2, 161 (1957).

mance, the envelope is essentially the same as the large ferry-rocket described in the original paper [1] or summarized in the *Meteor* paper [3]. Configuration of *Meteor, Jr.* has undergone considerable change as the result of further study of requirements for successful operation.

2. Stages

Each part of the three-stage vehicle is recovered by re-entry glide. The two booster stages land under crew supervision, have turbojets and fairing attached,

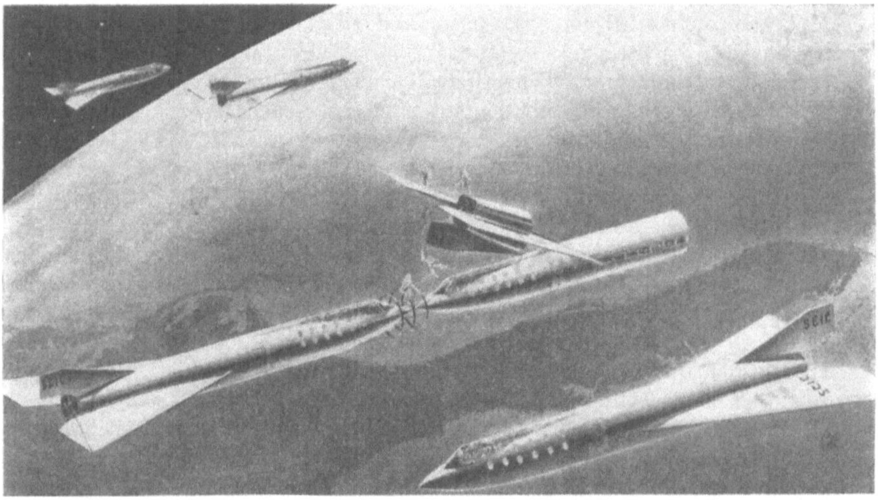

Fig. 2a. Final stages being joined to form hub of terminal

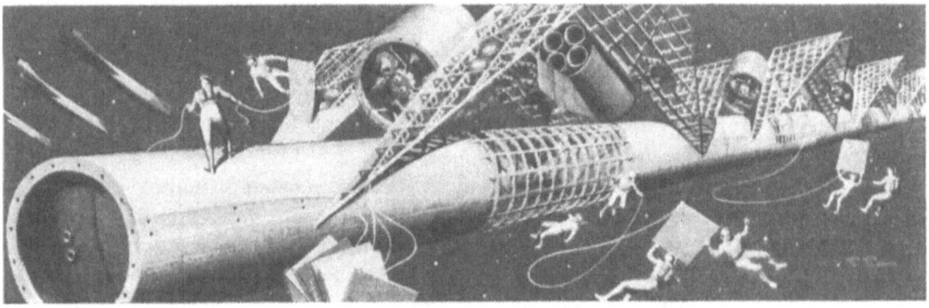

Fig. 2b. Early stages in erection of satellite terminal
These drawings (and those in Fig. 3) show design for original *Meteor* concept. *Meteor, Jr.* terminal has same configuration but smaller. Attaching framework to hub for expansion

then make an airplane-type flight back to the launching site. The third stage, after re-entry, flies direct to the base.

3. Terminal

If the third stage is not to be returned, it can become the basic unit for the satellite terminal, utilizing the cylindrical body as a tubular core and proceeding to erect around it work and living quarters in much the same manner as for the larger vehicle, only on a smaller scale [2, 3].

The tubular bodies are joined end to end and a framework erected around them to house all the terminal equipment including power generators, air conditioners, shops, storage space, service facilities, and much of the laboratory facilities.

Living quarters are provided in a large wheel structure revolving slowly around one end of the original tubular core. The rotation generates synthetic gravitational (centrifugal) forces, to some extent making living conditions normal in the otherwise weightless space state. All furnishings and equipment in the wheel must incorporate somewhat conventional supporting structure and strength, as must the associated mounting structures and floor, at least in the lower (outer) floors.

Personnel moves between the living-quarters wheel and the work-quarters portion of the terminal through cars which serve as movable air locks.

Fig. 3a. First expansion state. Wheel provides living quarters (Later stages in erection of satellite terminal)

The original terminal is, of necessity, relatively small, but the design permits expansion of all portions until the balloon-like terminal is perhaps 500 ft in diameter and 1500 ft long, and the wheel is 1000 ft in diameter. Facilities for 5000 technicians, observers, visitors, and other personnel are envisioned. One end of the tubular terminal incorporates a landing dock for future arrivals of third-stage vehicles, which can be quickly brought inside the pressure envelope for ready unloading of personnel and material.

Artist's conceptions of progressive steps in the construction of the tubular terminal and the wheel as conceived for the *Meteor* are shown in Figs. 2 and 3. The *Meteor, Jr.* terminal is similar in appearance and construction, but probably on a reduced scale.

4. Payload

Many useful combinations of payload can be obtained, despite the rather limited total weight carried. In addition to the basic crew of three men for the orbital stage (although two or four may be desirable under certain circumstances), the vehicle can carry into the orbit on round-trip missions:

Fig. 3b. Final expansion, showing landing dock
(Later stages in erection of satellite terminal)

1. Adequate supplies for a stay of two months or more;
2. Four passengers (scientists, engineers, observers) with supplies for a limited stay of about a week;
3. One-half to three-quarters of a ton of supplies, depending on other equipment or personnel carried;
4. About 1200 lb of instrumentation and scientific equipment in a wide variety of combinations, along with supplies for several days.

5. One-Way Missions

More ships need to go up than need to come back, so the payload on these one-way missions can be increased by an additional ton of supplies, equipment, or personnel. Generally speaking, the only things needed to be returned to the earth are personnel and data, and as much of the data will have previously been transmitted, it is, therefore, not necessarily a required return item.

Ships which reach the orbit and stay there can perform a number of useful functions, serving as interim or temporary stations or base facilities, standing by for emergency return of personnel to the earth or for rescue work—helping a ship that does not make orbital rendezvous because of a malfunction or aiding personnel inadvertently separated from the ship or station.

At the beginning, a number of the ships making one-way missions will be used to form the tubular core for the satellite terminal station and others will be used later as the terminal is expanded.

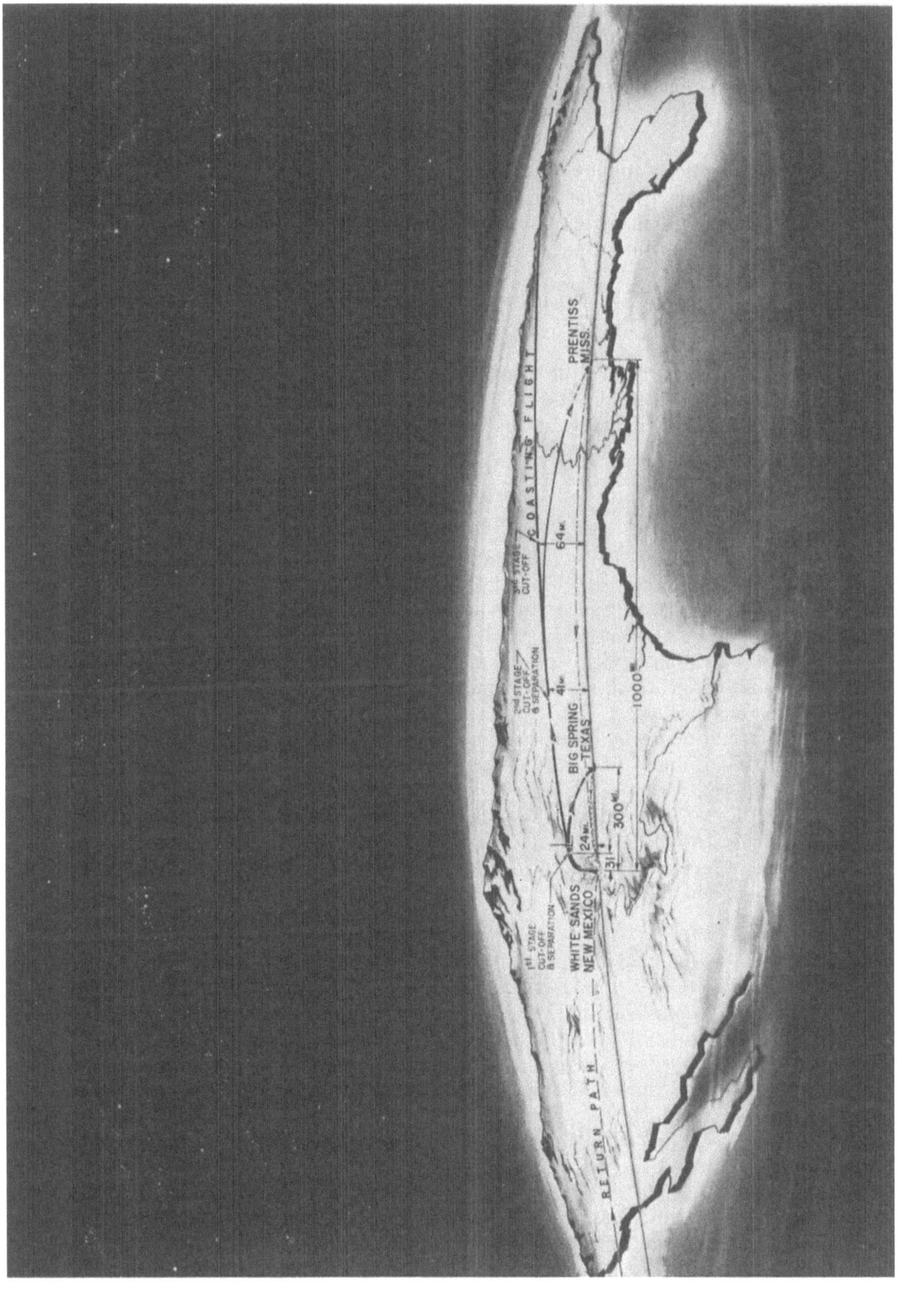

Fig. 4. Typical launch trajectory and return paths

Typical payload makeup for one-way missions, in addition to crew, can be: up to 3400 lb of cargo; or 15 passengers; or any combination of crew, equipment, and personnel not exceeding two tons. *Meteor, Jr.* offers a considerable utility potential despite its rather modest payload.

6. Exploratory Missions

Exploratory trips further into space, including, possibly, landings on the moon, can be made from the terminal by using additional rockets on the final stage. Payload for such missions is approximately the same as for a round trip from the earth to orbit. (See Appendix B.)

7. Guidance and Control

a) General

The guidance and control philosophy outlined in the first paper remains the same for *Meteor, Jr.*, as do the reliability aspects and characteristics detailed in the forthcoming reliability paper.

b) Guidance

The guidance system is essentially to determine continuously the instantaneous position and velocity vector of the vehicle and to guide it along the optimum path to its destination. Three different methods to accomplish these objectives are provided in parallel:

1. Integrators, operating on acceleration with respect to time, giving for the first integration ($1/S$) velocity; and for the second ($1/S^2$), distance or position;
2. Airborne radar devices, direction, ranging, and doppler;
3. Ground-based tracking and computing devices, relaying the results to the ship by radio link.

The information from these systems is fed to:

1. Instrument dials visible to the crew for monitoring purposes, or for manual override and control;
2. The ship's airborne guidance trajectory computers;
3. Ground stations via telemetering link.

Thus all three facilities have information from three sources for trajectory determination and control.

c) Control

Control must be provided for: (1) rocket-powered flight inside the (appreciable) atmosphere, and outside the atmosphere; (2) unpowered (ballistic) flight outside the atmosphere; and (3) aerodynamic flight for the separate vehicles.

In all cases, control (whether automatic guidance or manual) is normally through an autopilot, with provisions for manual over-ride or cutout and direct booster control. Conventional devices are used. The autopilot feeds into either one or combinations of the following:

1. Servos actuating the tiltable rocket motors (only a few of the rockets need to be gimbaled for control purposes);
2. Servos actuating the aerodynamic control surfaces, both during airplane type flight and in conjunction with the rocket motors whenever significant "q" (dynamic pressure) values are present;
3. Valves controlling reaction jets or motors, located near the wing and fin tips and in the forward part of the vehicle; this control is for use during unpowered flight outside the aerodynamically effective atmosphere.

8. Orbit

Of the many possible orbits from which to select, each offers certain advantages, each has its drawbacks. Primary consideration must be given to the end-use of the satellite, such as communication, surveillance, and mapping. An all-purpose satellite must be a compromise.

Until more data are available from the Vanguard program about atmospheric density at altitudes above 200 mi, it is difficult to place the orbit exactly, but apparently appreciable atmospheric drag disappears at 300 to 350 mi. Therefore, the 500 mi altitude chosen for *Meteor, Jr.* is free of this drawback and is the lowest practical orbit for surveillance, mapping, and similar uses. At and above 500 mi these utilizations tend to become increasingly difficult (see Fig. 4).

A higher altitude is better for communications and 1000 mi has been frequently mentioned for this purpose and other reasons. Later satellites, intended primarily for communications work, can orbit as high as 22,000 mi to permit synchronization with the earth's rotation.

Period is also a factor in orbit selection. At 500 mi the satellite will complete its orbit in 1 hr and 40 min; at 1000 mi, 2 hr.

Inclination and eccentricity must be given consideration in selecting the orbit, with the present study assuming an inclination of not over 5 to 10 deg to the ecliptic or lunar planes. Greater inclination, under some circumstances, is preferable. Eccentricity can be used to vary the orbital height to advantage, but because of many possible operational complexities, this study has assumed zero-eccentricity for *Meteor, Jr.*, and an orbital height of 500 miles—same as was used for the original *Meteor*.

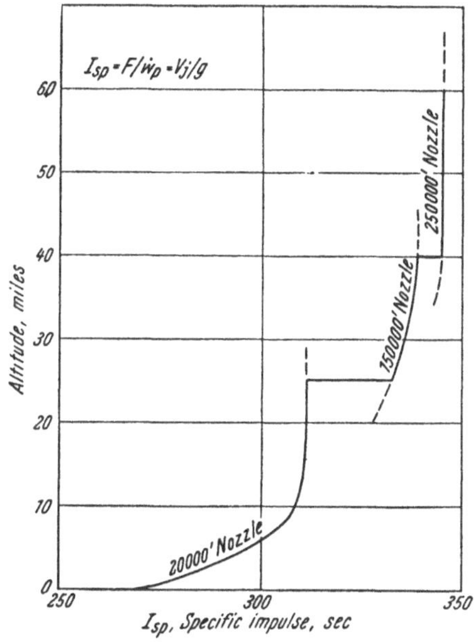

Fig. 5. Variation of specific impulse with altitude for three fixed nozzles

III. Performance Characteristics

1. General

Basic configuration, wing loadings, acceleration, and velocity gains per stage are similar for both *Meteor* and *Meteor, Jr.* Although the smaller size of *Meteor, Jr.* causes lower structural efficiency, and the drag effect is increased, recent engine performance data indicate that propellants under development will provide a somewhat compensating improvement in specific impulse within the next few years.

2. Propulsion

The propellant chosen for this study is a 1-to-3 mixture of ammonia and liquid fluorine, although fluorine (or fluorine-rich, Lox blend) with JP-X or hydrazine is also satisfactory. It is expected that by the target date of this vehicle liquid fluorine will be in relatively common use as an oxidizer, either alone or blended with Lox. Based on a chamber pressure of 500 psia, the conservative frozen equilibrium method indicates that a theoretical sea level specific impulse of 310 sec is available [4]. Technological improvements by 1963 should permit attainment of at least 90 percent of this theoretical value. An optimum sea level specific impulse of 275 sec is therefore assumed, and should be suitable for the range of fuels considered. (Specific gravity of these propellant combinations is approximately 1.0.)

By use of the design charts, on page 25 of [5], the variation of specific impulse with altitude was determined and is shown in Fig. 5. Nozzle designs for each stage were chosen to be optimum for the altitudes but were modified when necessary to stay within structural or dimensional limits.

3. Total Vehicle

Weight requirements for each of the three stages are based on the potential incremental velocities given in Table I. These velocities assume that all the

Table I. *Characteristic Velocity Calculations*

Stage	Mass ratio	C* (fps)	V/C	V (fps)
I	3.19	8,950	1.16	10,350
II	3.51	10,600	1.25	13,225
III	2.30	10,860	0.833	9,050
Total	25.8	10,120	3.25	32,920

* Note that these velocities are derived from specific impulse values that have been reduced (by 5 percent for the first and by 2 percent for the latter 2 stages) to compensate for drag, outage, and similar losses as explained in the text. Otherwise the value of C for the Final (III) Stage would be 11,110 fps and the total potential vehicle velocity would be 33,680 fps.

usable propellant is consumed in horizontal drag-free flight and indicate the capabilities of any rocket system. Mean values of jet velocity for each stage were estimated from Fig. 5 and reduced by 2 percent for outages.

A further reduction in jet velocity of 5 percent is charged to Stage I to account for its drag and compares with the 2 percent loss evaluated in [1] for the larger vehicle.

With these values, the necessary ratios of initial and final weights are 3.19, 3.51, and 2.30 for the first, second, and third stages, respectively. For the specified payload weights, the general weight breakdown and other design characteristics were determined and are given in Table II. Sufficient reserve propellant has been included for the landing operation and to permit powered approach and landing maneuver time of 5 min for the third stage and 3 min for each of the lower stages.

4. Aerodynamics

a) *Configuration*

The configuration chosen is considered to be the minimum necessary to meet the wide range of conditions encountered from hypersonic re-entry to low-speed landings. As drag is of secondary importance, the airfoil section is relatively thick (8 percent) for structural reasons, and the rounded leading edge should alleviate aerodynamic heating effects and provide good low-speed

Table II. *Basic Design Characteristics*

	Phase	Total wt (tons)	Propellant wt (tons)	Thrust (tons)	Total acceleration (g)	Max vel (mph) *	Mass ratio	Struct equip ratio	Empty wt		Payload (tons)
									Struct equip and payload (tons)	Struct and equip (tons)	
	Stage I	500	343	1112	2.23—7.1	6,667	3.18	1 : 5.82 (17.2%)	157	87**	70
	Stage II	70	50	132	1.89—6.6	15,000	3.50	1 : 5.0 (20.0 %)	20	14**	6
	Stage III	6	3.4	10	1.67—2.54	(Perigee) 18,100	2.31	1 : 3.75 (26.7 %)	2.6	1.6	1
Stage III Operations	End of initial thrust approaching apogee	3.94	1.34	10	2.54	(Apogee) 16,200	1.52	(26.7 %)	2.6	1.6	1
Stage III Operations	Start of rendezvous maneuvers after attaining orbital speed	3.73	1.13	10	2.68	(Orbital) 16,660	1.43	(26.7 %)	2.6	1.6	1
Stage III Operations	After corrections, at start of descent from orbit	3.11	1.01	10	3.22	16,660	1.48	(26.7 %)	2.1	1.6	0.5
Stage III Operations	Reserve for corrections at re-entry and landing approach	2.95	*** 0.85	*** 10/0.59	3.39/0.20	approx. 18,000	1.40	1 : 1.84 (54 %)	2.1	1.6	0.5

* Absolute velocities including earth rotational effects.

** These weights include 11.6 tons and 1.7 tons reserve fuel in Stages I and II, respectively.

*** This reserve fuel is sufficient for 4 seconds full thrust (320 fps velocity change) re-entry correction and an additional 5 minutes reserve at 0.2 g thrust at low altitude for landing approach.

characteristics. Wing loadings at landing are typical of contemporary high-speed fighter aircraft.

The selection of the wing planform, and particularly the leading edge sweep-back angle, was influenced as much or more by aerodynamic heating and structural problems as by the more usual considerations of high-speed flight.

The airfoil contour utilized was influenced by many factors: the 8-percent chord thickness by high-speed flight, structural, and low-speed maximum lift requirements; location of the maximum thickness ahead of the 50-percent chord point by low-speed lift effects; the nose radius selection considered both heating and low-speed lift. Low drag (in any flight regime) was believed to be of lesser importance for this application than is normally the case.

Static stability characteristics of the vehicle are based on [6] and [7]. The center-of-gravity locations should provide adequate static stability margins for

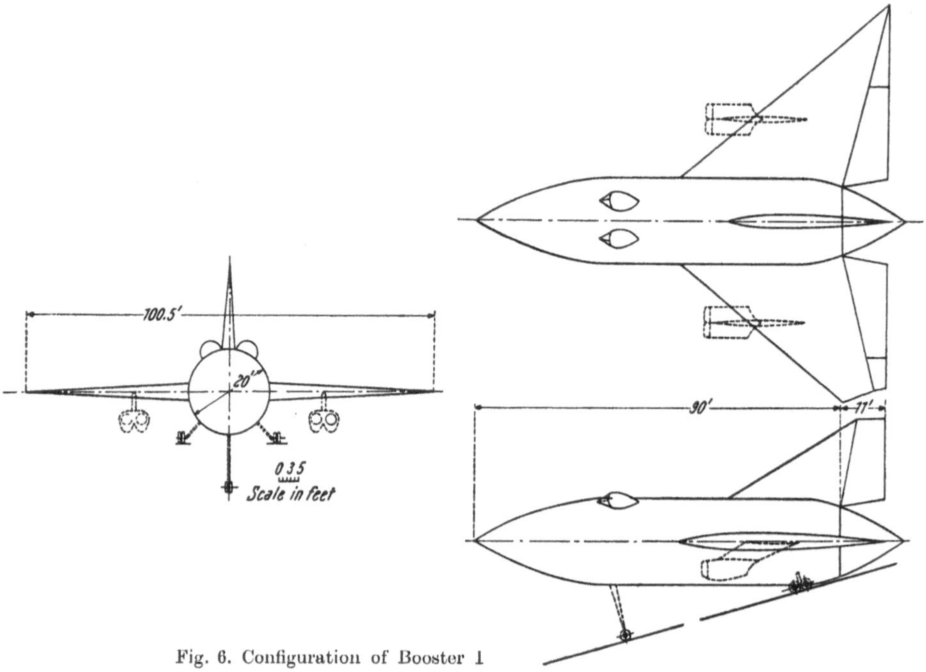

Fig. 6. Configuration of Booster 1

the individual and composite stages. The delta wings should also permit satisfactory aerodynamic control and damping characteristics. Control during rocket operation is obtained by swiveling the nozzles.

Analysis was also made to determine the thrust requirements of the airplane type configuration which is flown back to the launching site by attaching podded turbojets. As a criterion for satisfactory performance, thrust requirements should provide cruising ability plus a margin sufficient for a 1000-fpm rate of climb at 10,000-ft altitude. In terms of static sea-level thrust, it is shown that 40,000 lb, 7000 lb, and 800 lb are required for the first, second, and third stages, respectively.

b) *Test Vehicle*

A limited-performance interim test vehicle configuration was investigated from a performance standpoint, the results being given in Appendix A.

IV. Design Characteristics

1. General

This minimal size earth-to-orbit vehicle is primarily a three-stage rocket so arranged that both the first and second stages, after separation, convert

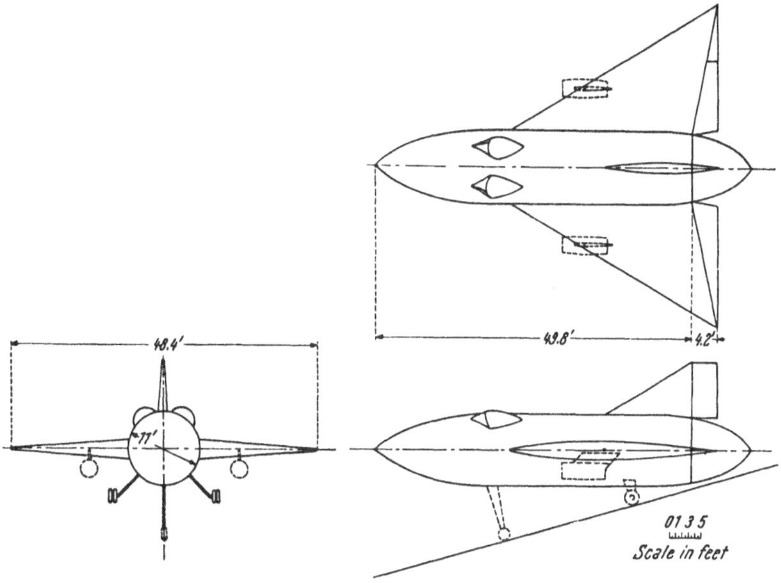

Fig. 7. Configuration of Booster II

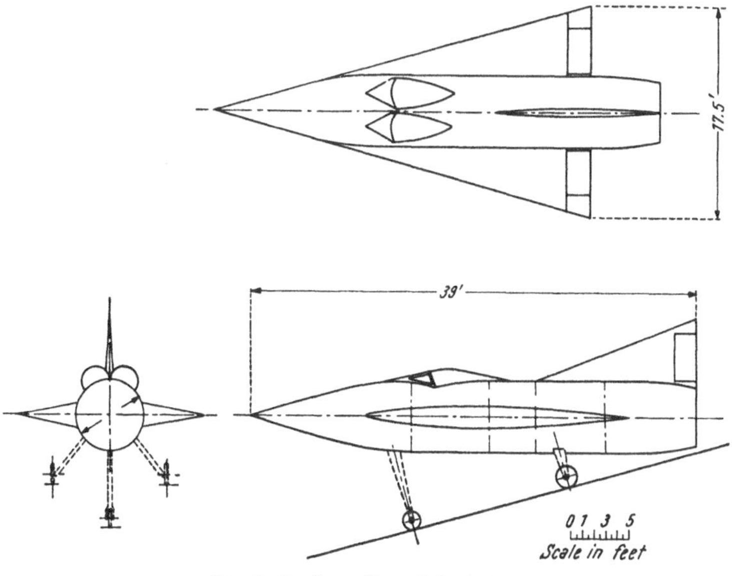

Fig. 8. Configuration of final stage

to piloted, gliding, aircraft-type vehicles that can maneuver to somewhat conventional landings. After landing, each is converted to an airplane by the addition of quick-attaching turbojet engine pods, tail fairings, and other parts for flight back to the launching site. There the added parts are removed and transported back to the landing site ready for re-use, while the vehicles themselves are

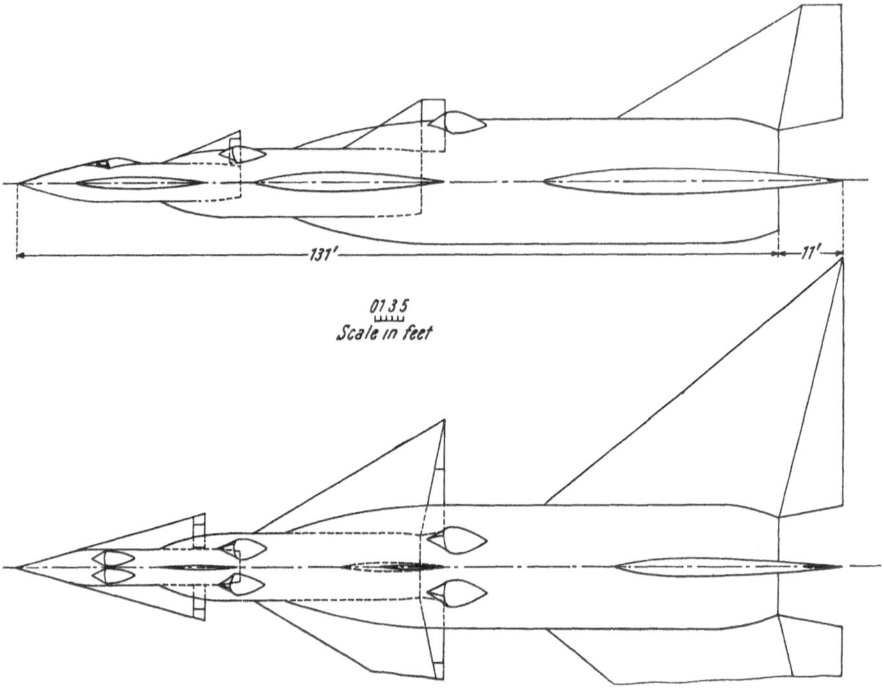

Fig. 9. The three parts of *Meteor Jr.* nested

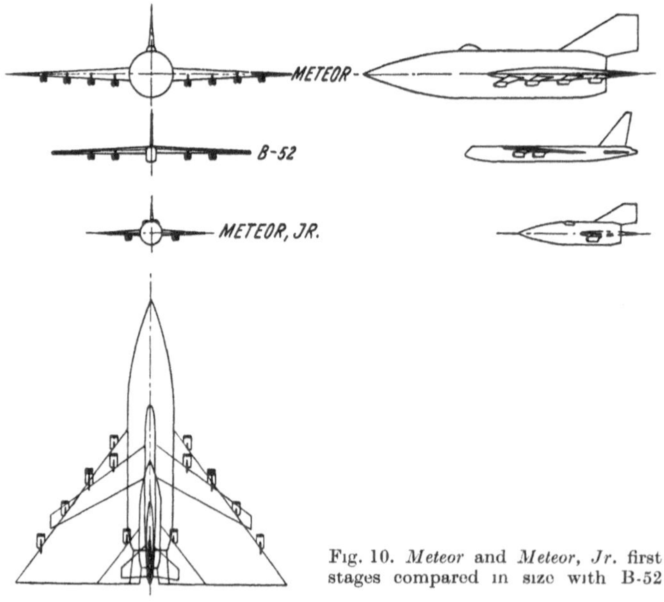

Fig. 10. *Meteor* and *Meteor, Jr.* first stages compared in size with B-52

inspected, serviced, and made ready for another launching. The final stage goes on into the orbit and later returns to the earth by aerodynamic braking, landing at the original launching site.

2. Configuration

The key features of this concept are: the nose doors and fairing that fair between the stages until after separation, when they are closed to form an ogival nose for the separated stage; the stabilizing surfaces that are designed to serve as wing and tail surfaces for the booster stages after separation; the extensionable landing gear that becomes retractable after the addition of a few parts; the crew accommodations in each stage, and the quick-attaching jet engine pods and tail fairings. These tail fairings have openings so a few of the main rockets can assist the airplane-type takeoffs, thus enabling each stage to lift off quickly and then fly on its turbojet power alone.

The third and final stage is not special to this concept, as it must be designed to make the returns from the orbit, as would presumably be the case for any such system. It does have features compatible with the concept for its utilization as a building block for a satellite terminal.

Obviously all details of the design cannot be covered in this paper. More details can be obtained from [1]. The general arrangement of *Meteor, Jr.* is outlined in Figs. 6, 7, 8, 9, and 10, which show the assembled vehicle as well as the individual stages. Booster I has a diameter of 20 ft, Booster II of 11 ft, the final stage only 6 ft.

3. Arrangement

a) Stage I

The pressurized crew compartment is located aft the nose doors on the upper side of the body and has accommodations for three men. The stage command-pilot and copilot-navigator are seated side by side with instruments and controls required for rocket-powered flight as well as those required for instrument flight in the airplane glide and ferry configurations. The flight engineer is seated behind the other crew members with instruments and controls for propulsion and auxiliary functions in both rocket-powered flight and ferry flight with jet engines. Ejection escape capsules are provided for all crew members. The nose landing gear is located below the crew compartment in the lower part of the body. The catapult for stage separation as well as the support for the second stage is in the nose ahead of the crew compartment. These components are enclosed by the nose doors after separation from the second stage. At separation the vehicle is at 24 mi altitude and is traveling at 6667 mph (including 917 for the earth's rotation). Guidance and control equipment as well as communications equipment are installed adjacent to the crew compartment.

The propellant tanks, located in the body aft the crew compartment but forward of the engines, have flexible fuel and oxidizer resistant linings to permit internal pressurization for fuel feed. For emergency jettisoning, pressure is introduced between the flexible liner and the tank wall to force the propellants out the jettison doors in the lower sides of the tanks. The fuel and oxidizer tanks are dumped in sequence to reduce hazard. All fuel is dumped except that in the reserve-restart chamber. An eight-second dump time is prescribed. This system is more fully described in the forthcoming reliability paper. A heavy frame at the juncture of the fuel and oxidizer tanks supports the front spar of the wing structure.

A firewall bulkhead is provided aft the propellant tanks to isolate the engines, support the rear spar structure, and mount the main landing gear. The multiple engine thrust chambers include peripheral units, hinge or gimbal mounted, to vary the thrust axis for pitch, yaw, and roll control in powered flight regimes where the aerodynamic controls are not effective.

The aft end of the body, tips of the wings and vertical tail surfaces have strong points for attaching to the launching carriage while in the nose-up vertical launching position.

Reaction jets are also provided at these locations for attitude control while in coasting flight.

For airplane ferry flight back to the launching site, four J-58 size turbo-jet engines mounted in a double pod on each wing are adequate. For airplane flight takeoff, several of the rocket engines are used as boosters to shorten the takeoff run. The quick demountable engine pods can be identical to those of the B-52 airplane. Jet engine fuel tanks are integral to the wing structure (or they can be attached as pods).

b) Stage II

Stage II is constructed and winged similarly to Stage I, but is smaller and has only a two-man crew, seated side-by-side with controls and instrumentation suitable arranged.

The aft end of the body has supports for mounting in the nose of Stage I, separation catapult reaction point and piston. At separation from Stage III (41 mi altitude) the vehicle is traveling 15,000 mph. (See Fig. 11.)

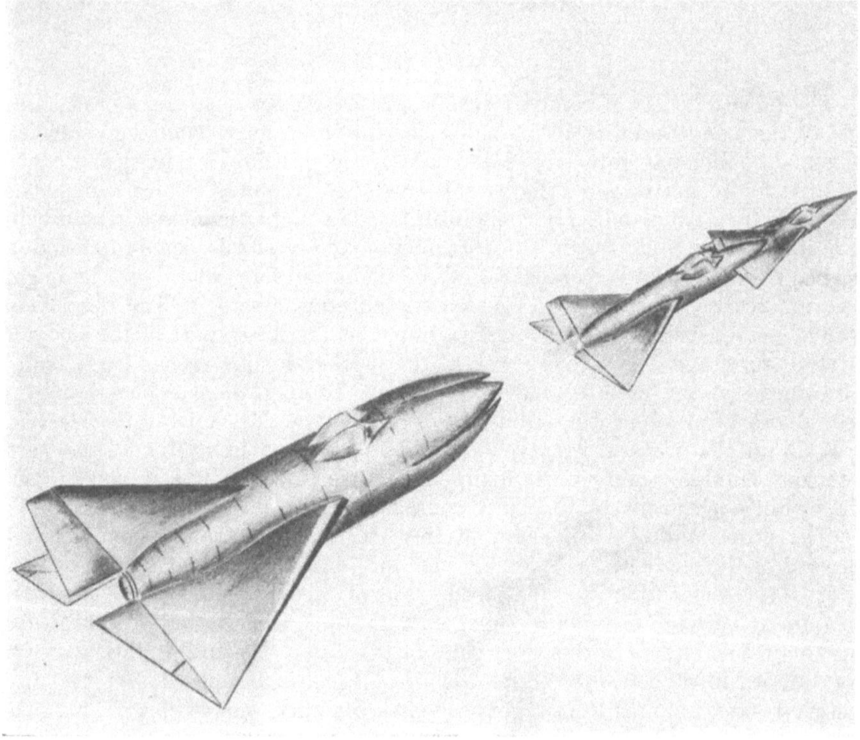

Fig. 11. Booster I separating from second stage

A single jet engine pod mounted under each wing provides sufficient thrust for ferry flight back to the base. Engines with 3500 lb thrust are sufficiently large.

c) *Stage III*

The final stage, which reaches its maximum velocity of 18,000 mph at 64 mi altitude, then drops to 16,660 mph in orbit, has different requirements, which show up in detail arrangement, construction, and equipment (see Fig. 12).

The nose is designed and arranged to withstand or counteract greater re-entry heating effects, at the same time serving as a housing and radome for equipment installed in it: radar, guidance and control, telemetering, communication, and instrumentation.

The pressurized crew compartment provides for a three-man crew, the vehicle commander and pilot sitting side by side, the flight engineer behind them. Each crewman has instruments and controls applicable to his duties. The vehicle commander has command and control of all three stages during launch, booster separation and flight of the final stage to and in the orbit, as well as return and landing.

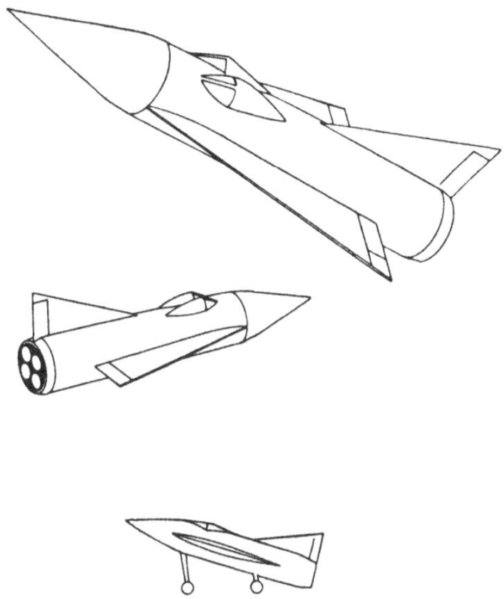

Fig. 12. *Meteor, Jr.* final stage in flight and landing

The nose landing gear is retracted into the body underneath the crew compartment. Air conditioning, pressurization, oxygen, batteries, heaters, and other auxiliary equipment are located in suitable compartments within the fuselage.

The cargo compartment is aft the crew compartment but forward of the propellant tanks.

Propellant tank arrangement is similar to the other stages, but the tanks are much smaller. The firewall bulkhead aft the tanks provides for the attachment to the second stage as well as a reaction point for the separation catapult. The heavy ring frame at the juncture of the fuel and oxidizer tanks supports the wing rear spar structure and the bulkhead forward the tanks supports the wing front spar structure and the nose gear mounting points.

To prevent depressurization due to small meteor strikes breaking the skin while in orbit, a meteor barrier skin is provided around the crew compartment and other vital areas.

The four rocket engines installed aft are hinged or gimbaled to provide control by thrust axis orientation when the aerodynamic control surfaces are not effective. There are also orientation control jets in the wing tips.

The six-foot diameter body for the final stage is probably near the minimum that can accommodate the crew in the supine or "contour chair" position, with the body axis perpendicular to the vehicle axis. This position is required to permit crew members to sustain the accelerations encountered while still retaining the necessary functional capacity for flight monitoring and other possible emergency duties.

This is less of a problem for the crew accommodations in the larger stages, where the side-by-side arrangement further alleviates the space problem and seems to enhance cockpit visibility.

To permit entering and leaving the final stage while in the orbit, an airlock entrance is provided in the crew compartment opposite the flight engineer. This airlock prevents loss of pressurization.

Although fuel reserves for the final stage vehicle are figured on a basis of normal return weight based on typical mission requirements (Table III) the vehicle structural design is based on landing with full payload and maximum fuel that could remain. This is to cover the case where, due to unforeseen contingencies, the ship does not quite attain or remain in the orbit or does not unload any payload, which would mean maximum landing weight. Fuel reserves would not be critical for this case as there would normally be excess fuel that had not been used for rendezvous maneuvers or other purposes. This might correspond to early training or shakedown flights.

Table III. *Weight and Balance Summary*

Vehicle or configuration	Weight (tons)	CG location (ft)*
Booster I, empty, glide configuration	75.4	17.2
Booster II, empty, glide configuration	12.3	13.5
Booster I, airplane configuration with fuel	97.5	18.5
Booster II, airplane configuration with fuel	25.4	15.4
Stage I (complete 3 stage vehicle) empty	89.3	26.8
Stage I (complete 3 stage vehicle) with fuel and payload	500	35.1
Stage II (last 2 stages — 2nd and final) with fuel and payload	70	21.1
Stage III (final stage only) with fuel and payload	6	13.7
Final Stage Vehicle, or return from orbit, landing configuration including payload	2.6**	17.1

* Measured from base (or rear end of body) of the vehicle or configuration.
** This is for heaviest landing condition, carrying 1 ton of return payload.

4. Propellant Volume Requirements

The propellant volume is computed on a composite density of 65 lb/ft³ to arrive at the tankage required:

Stage I	19.5 ft dia	39	ft long
Stage II	10.5 ft dia	20	ft long
Stage III	5.5 ft dia	4.5	ft long

5. Weight and Balance

The weights and sizes of the three stages are determined from the performance characteristics, with due allowance for structural, equipment, fuel, and payload weights. They are based on assumed ratios of structure and equipment to total weight (Table II), which are, in turn, based on structural and weight studies and as illustrated by weight breakdown (Tables III to VIII). These weight breakdowns are also the basis for the center of gravity determination of the individual stages. The weight and balance determination for the various combinations of all stages (Table III) are determined directly from the individual weight and balance values.

Table IV. *Final Stage Weight and Balance*

Component	Weight (lb)	Arm** (ft)	Moment (lb—ft)
Guidance	600	26.7	16,020
Pilot and co-pilot*	600	22.7	13,620
Instruments	50	24.2	1,210
Flight engineer*	300	18.4	5,520
Cargo	1,400	16	22,400
Propellant	6,800	11.1	75,480
Engine installation	200	4.8	960
Structure	1,550	14.5	22,475
Wing	500	13.5	6,750
Totals	12,000	13.7	164,435

* Weight = 200 lb for personnel, 100 lb for furnishings.
** Measured from base (or rear end of body) of the vehicle or configuration.

Notes: CG fully loaded = 13.7 ft, weight 6 tons
CG propellant out = 17.1 ft, weight 2.6 tons
CG empty = 16.8 ft, weight 1.6 tons (crew and cargo out)

Table V. *Booster II Weight and Balance*

Component	Weight (lb)	Arm** (ft)	Moment (lb—ft)
Crew	400	32	12,800
Furnishings	200	32	6,400
Structure*	18,795	15	281,925
Engine inst.	2,640	7	18,480
Wing	2,565	5.2	13,338
Totals	24,600	13.5	332,943

* Includes guidance system, air conditioning and other miscellaneous equipment.
** Measured from base (or rear end of body) of the vehicle or configuration.

Notes: CG in glide condition = 13.5 ft, total weight 12.3 tons
CG with propellant in = 18.6 ft, total weight 64 tons
Propellant added = 51.7 tons on 19.8 ft arm (includes reserve propellant of 1.7 tons)

Stage II Weight and Balance

With final stage added to nose, balance then is:

	Weight (tons)	Arm (ft)	Moment (ton-ft)
	64	18.6	1,130
	6	48.1	288.6
Totals	70	21.1	1,418.6

CG = 21.1 ft

Table VI. *Booster I Weight and Balance*

Component	Weight (tons)	Arm** (ft)	Moment (ton—ft)
Crew (3)*	0.5	56	28
Structure	58.28	20.1	1,165.6
Engine inst.	11.12	6	66.72
Wing	5.5	6	33
Totals	75.40	17.2	1,293.32

* Includes furnishings, instruments and guidance systems.
** Measured from base (or rear end of body) of the vehicle or configuration.

Empty unit	75.4	17.2	1,296.88
Propellant	354.6	29.7	10,528.12
Totals	430	27.5	11,825.00

Notes: CG in glide configuration = 17.20 ft, weight 75.4 tons
CG with propellant in (includes 11.6 tons reserve) = 27.5 ft, weight 430 tons

Table VII. *Stage I Weight and Balance (Empty)*

Component	Weight (tons)	Arm* (ft)	Moment (ton—ft)
Booster I	75.4	17.2	1,296.88
Booster II	12.3	74.7	918.81
Final stage	1.6	109	174.4
Totals	89.3	26.8	2,390.09
CG = 26.76 ft			

Table VIII. *Stage I Weight and Balance (Full)*

Component	Weight (tons)	Arm* (ft)	Moment (ton—ft)
Booster I	430	27.5	11,825.0
Booster II	64	79.8	5,107.2
Final stage	6	105.9	635.4
Totals	500	35.14	17,567.6
CG = 35.14 ft			

* Measured from base (or rear end of body) of the vehicle or configuration.

6. Engines

To lift *Meteor, Jr.* into its orbit requires engine thrust capability totaling 2,514,000 lb of thrust from 27 engines, distributed as follows:

1. Stage I — 17 engines, 130,000 lb thrust ea (at 20,000 ft)
2. Stage II — 6 engines, 44,000 lb thrust ea (at 150,000 ft)
3. Stage III — 4 engines, 5,000 lb thrust ea (at 250,000 ft)

Grouping of these engines is shown in Fig. 13 and those which are grimbaled for correction of flight path are indicated by arrows. In the final stage, each engine gimbals separately in the manner of those on the X-1 rocket research ship.

V. Operational Characteristics

1. General

Meteor, Jr. has the capability for several types of operations.

2. Unmanned Flight

Under fully automatic control, the vehicle is capable of any type of unmanned flight up to and including flight into an orbit and return to earth. Operating in this fashion, it is, therefore, capable of:

1. Testing and gradually extending its own capabilities, without risk to flight personnel in the process;
2. As an unmanned satellite, carrying a rather heavy payload of equipment, instruments and biological specimens into an orbit and, after a desired length of time, returning with its cargo.

3. Manned Flight

As a manned vehicle, *Meteor, Jr.* is capable of the same types of operations, gradually testing and extending its performance, finally going into a satellite orbit for a desired length of time, and returning to earth.

It is capable of performing any kind of manned satellite function that its payload will permit. Because the cabin includes closed system airconditioning equipment and air lock, mission times will be limited mainly by the amount of oxygen, food, water, and other supplies carried, particularly fuel or other means of supplying electric power.

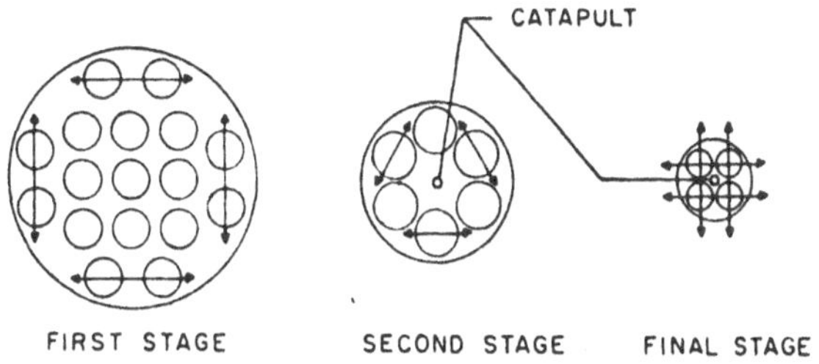

Fig. 13. Rocket nozzle arrangement. Arrows indicate directions of swing of gimbaled engines

For return flights and landing of the booster stage vehicles, landing speeds range from 160 to 200 mph, which determine the runway lengths. For airplane-type operation in returning to the landing base, runway length is not critical for takeoff, as the rockets can be used for takeoff assist. For airplane-type flight, the booster vehicles cruise at from 250 to 300 mph. This means that Booster I requires about an hour for its return flight to the launching base, Booster II up to 4 hr. By comparison the ascent into the orbit (as well as return from the orbit) requires about 1 hr, including approximately 50 min of coasting flight.

4. Handling Equipment

Handling equipment for the *Meteor, Jr.* is visualized to follow the same philosophy as that for the larger vehicle described in the original paper [1].

However, the size and weights are so much smaller that they correspond to current experience in many fields. The vehicle height is only one-half that of the larger vehicle, the heaviest weight to be handled being the first stage empty booster, which weighs about 75 tons and never has to be lifted off the ground manually. It will probably be tilted into upright position.

The Booster II vehicle is hoisted into place, but its weight is only about 12 tons before fueling. The empty weight of the final stage vehicle, which has to be lifted the highest, is only a little over 3000 lb. With these weights and sizes, ground-handling equipment experts say there should be no problems in providing good serviceable equipment for the needed requirements. Or the three stages may be nested on the ground and tilted into launch position. Actually this is the more desirable method of vehicle handling, as it involves positive motion and position at all times.

To provide maximum safety and regular operation, auxiliary ground facilities could include as many as 35 emergency fields in addition to the one main and two auxiliary landing bases for Boosters I and II. This provides fields every 25 mi on each side of the regular bases, then every 50 mi and finally every 100 mi as distances from the regular bases grow greater. The string of landing fields thus covers a distance of 2000 mi.

These could be reduced to 10 or 12 emergency fields and still give fairly good dispersion with relatively small sacrifice in operating safety — although the larger number is desirable.

These emergency fields need be little more than a single cleared runway with a shed for engine pod installation, as they would be used only infrequently. Other equipment can be flown or trucked in as needed. Each field should have lighting, instrument landing and communications facilities, however, as any field might be needed at any time. Runways should be a minimum of 12,000 ft, although at those fields used by vehicles with lower landing speeds, the runway can be shorter. Regular bases should have the 12,000 ft runway.

For airplane-type operations, thrust reversers can be utilized.

5. Maintenance

Maintenance characteristics of this vehicle are good, as it has excellent accessibility. All vehicles are essentially simple tank structures with equipment and connections at each end — engine compartment at the rear and crew compartment at the front — both offering ready accessibility to all equipment. The tank structure and system are accessible internally as well as externally without removal. Access to the inside of all tanks is gained by entering through the large dump doors. The smallest of these (on the final-stage vehicle) measures about $2 \times 2\text{-}1/2$ ft rectangular.

6. Costs and Logistics

There is nothing unique about this vehicle system that will significantly affect operating costs when compared with costs for aircraft, except for the large quantities of fuel required. By the time *Meteor, Jr.* can be implemented, the price for propellants should be between \$ 100 and \$ 500 a ton, depending on how much the price drops from the present level with increased volume requirements. The figure will probably be nearer the lower end of the price range. As 400 tons of fuel is required for each round trip (including flying back to the main base and all auxiliary flight transport associated with a flight), the fuel cost is estimated at \$ 75,000 per flight. If on a schedule of almost daily flights for a year, the annual fuel cost will be around \$ 25,000,000.

This is small compared with the vehicle development and manufacturing costs, if these vehicles approximate the costs for aircraft and missiles. An indication of total cost can be gained by comparing the empty weight of all three stages of this complete vehicle (about 180,000 lb) with the B-52 bomber, which has approximately the same weight. Complexity is comparable, so each complete *Meteor, Jr.* should cost between \$ 5,000,000 and \$ 10,000,000. Minimum requirement of 20 vehicles will cost about \$ 160,000,000; 40 vehicles will cost some \$ 275,000,000. Fuel, even for several years operation, will, therefore, be a relatively small item and the total operational cost will be dominated by the same factors as those for comparable aircraft operations — equipment inventory and amortization (including development), supporting facilities and installations, maintenance supplies, and salary and wages of operating and supporting personnel.

These are the main factors that will determine over-all operational costs and logistics, and there appears to be no reason why they should differ significantly from those for a comparable operation of modern, complex, high-performance aircraft, which costs are accepted as part of our current practice.

7. Utilization

Meteor, Jr. can be used for several purposes: as a satellite vehicle, per se, or as a transport vehicle ferrying men and material into the orbit, utilizing the payload capacity of from one to two tons. It can also be utilized as a small temporary satellite station, or as the basic building unit for a permanent orbital terminal in the manner outlined in the previous *Meteor* papers [2, 3]. The first station would be a six-foot diameter tube, which would be gradually built up to the final terminal size. The fuel tank liners can be removed and used as liquid storage tanks outside the station (providing proper insulating considerations are given to the liquids stored), and the tank cavities would then form the station body. The six-foot diameter is adequate for reasonable arrangement of equipment and work areas, with suitable free access passages to permit comfortable accommodations and reasonable operating efficiency, although this is near minimum.

The small *Meteor, Jr.* tube perhaps does not provide strength and stiffness for as long a core before expanding to larger diameter as in the case of the nine-foot tube in the original concept, but stresses are so light in the weightless pressurized station that this is probably without practical significance. In all its phases the terminal station structurally is actually a large metal balloon or airship, and has much in common with airship structures, which more nearly resemble it — in size, shape and loadings — than anything else with which we have had experience to date. The regular air conditioning and air lock installations of the component ships make adequate facilities for the initial station configuration.

Finally, the vehicle can be used for a number of specialised purposes: rescue ships, nearby space exploratory missions and deeper space probe missions of various kinds, including trips to the moon. For rescue work it would be probably adapted to push rather than tow, and for distant trips from the orbit would require fitting with proper boosters. For much of the work, these boosters can be remarkably efficient, as they can operate at greatly reduced acceleration compared with that required for getting into the orbit, with correspondingly reduced stresses and weights. The air-lock provided will be a necessary feature for many of these applications.

An intriguing potential early use might be as a research test vehicle — an extension of the X-ships, perhaps next after the X-15. By the logistic characteristics outlined in Appendix A, it appears to be demonstrable that *Meteor, Jr.* would operate more economically than any other way to fill the gap between the X-15 and satellite performance.

Meteor, Jr. so used, could take up where the X-15 program ends and finally push on upward to reach satellite velocities — generally in the way outlined in the following section. Indeed, by such methods, it may be possible to utilize successfully a single vehicle type to extend and finish the flight technology advance through the speed and altitude spectrum which was started with the X-1, and then go on to exploit the resulting operational capability.

This type vehicle appeares to encompass sufficient flexibility to give a good measure of utility in various ways — utility that may well be limited by the ingenuity of the users more than by inherent limitations imposed by the characteristics of the vehicle.

VI. Development Aspects

1. General

Meteor, Jr. embodies as one of its basic design criteria the same compatibility with a rigorous development plan that characterized the original *Meteor* concept. The final stage should be developed first, using the applicable date to come from the Vanguard, X-15, Atlas, and other pertinent programs. For getting the final stage vehicle into very high speed, high altitude trajectories and then into an orbit, development of the booster vehicles must follow closely behind on either of two paths: utilization of the regular configurations with full operational capability from the very beginning, or providing for, in addition, low-cost expendable boosters of limited capability for the first few test flights into the orbit. (See Appendix A.)

2. Flight Test Program and Methods

Prior to first flight, a realistic training and facilities testing program must be carried out, using flight simulators, small scale flight models for ground crew equipment and operational checkout, and all techniques that will train flight and ground personnel to a conditioned reflex training status. The vehicle must also be put through customary thorough ground testing.

Taxi tests are first on the vehicle test program. With turbojets installed on the experimental final stage, the taxi tests are conducted in much the same manner as for conventional aircraft. Rockets are briefly energized as the additional power source during this phase. Then come short flight tests using only the turbojets. After flight shakedown, utilization of rocket thrust in flight is begun and carried on until full thrust is realized. If desired, work may be started by remote control during any of these phases. Finally, with turbojets removed, tests are conducted with partial rocket thrust, finally full rocket thrust.

Vertical takeoff flights, using only rockets, follow a series of static firings to check the vertical starting characteristics of the vehicle. First flights may be made under automatic control, without crew. After the vehicle is shaken down in this phase, it is ready for staging.

The booster stages are taken individually through the same type of test program until all are thoroughly flight proved, their crews completely indoctrinated with them. Then staging is started with Booster II and final stage vehicle combinations, at first with automatic control. Boosters I and II are similarly combined and flight tested. Then the vehicle is ready for three-stage combination flight. All during this program, the final stage has been steadily building up to higher re-entry velocities.

It may be possible, during the earlier stages of the flight program to launch the final stage vehicle with an expendable rocket about the size of the Atlas by providing a simple nonrecoverable stage with an empty weight of about 5000 lb. Usefulness is severely limited, however, and extensive operation can be prohibitively inefficient and expensive (see Fig. 14).

The first three-stage flights will probably not go to full orbital velocity but take over where the two-stage phase ends. First flights to the orbit and return also be made by remote control of the final stage but with the two boosters manned.

Repeated flights must be made with the vehicles until their satisfactory, reliable performance is thoroughly checked and proved and the crews are completely familiar with their ships. In-flight emergencies must be practiced (after ground simulator practice) until crews are confident of the vehicles. Such a conservative approach to flight test programming usually results in lower total cost and faster over-all progress.

The experimental ship will be followed by prototypes, each given an appropriate step-by-step test program, by now well-trained personnel. Of these prototypes, some vehicle stages may be lost through accident, but with luck, proper design and training for flight contingencies and malfunctions, and a proper test programming, there should be no loss of personnel, or at least with flight hazards not exceeding those normally experienced in flight-testing new types of highperformance aircraft.

After demonstrations of the suitability of both vehicles and crews for reliable operation, the service models should be ready and flight tested, as type shakedown flights continue until the vehicle is ready for full operation in regular service. Thereafter the challenge will be with the operational directors and personnel to get the most out of the capabilities of the vehicles.

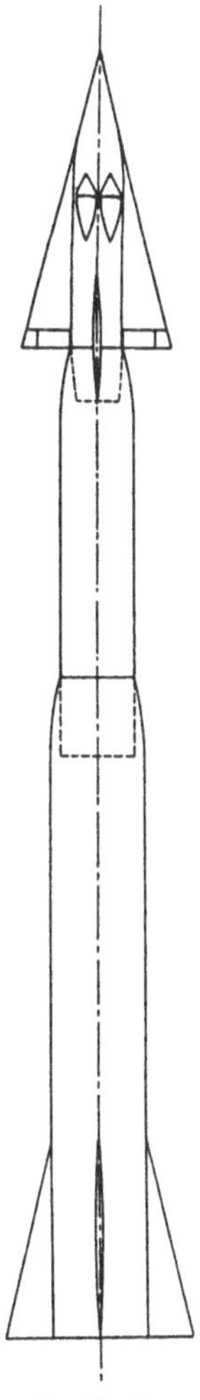

Fig. 14. Configuration of final stage *Meteor, Jr.* and Atlas-type experimental booster

3. R and D Prerequisites for Development

It must be recognized that certain *R* and *D* prerequisites are absolutely necessary prior to the development of a vehicle to be put in service in the manner described. Among the most important are:

1. Extreme upper atmospheric air data (to 200 mi alt)
— from the Vanguard program.

2. Aerodynamic drag and heating re-entry data for high speeds
— from ballistic missile and Vanguard programs.

3. Atmospheric ionization, dissociation, and other physical data and effect on communication and vehicle behavior at very high altitudes
— from various programs.

4. Data and experience on vehicle control and stability behavior beyond the usable atmosphere
— from X-15, Vanguard, and missile programs.

5. Data on man's reaction to outer space conditions, and devices for his protection under these conditions
— from X-15 and various programs of the military services, medical laboratories, and agencies.

6. Data and experience on man's reaction to flight above the atmosphere, and his ability to operate rocket aircraft under these conditions
— from X-1B and X-15 programs.

7. Continuous accumulation of cosmic ray data and effects — from various programs.
8. Continuing rocket and propellant development, facilities, and experience — from missile programs and continuation of current *R* and *D* programs.

While these programs are currently progressing satisfactorily, and seem quite likely to yield the needed data in the expected time, it must be realized that failure to obtain any of these vital data might render impractical or impossible the orderly development of the vehicle in the manner outlined. Therefore, the progress, or lack of it, in the listed programs, will be the key to (and can serve as a yardstick for measurement of) our technical ability to develop such a vehicle. The most critical aspect is probably re-entry, and progress in this area is most likely to be the factor controlling the pace of the program.

Suggested program augmentation might take the form of extended satellite work beyond the originally planned Vanguard program, and expanded very high altitude sounding rocket programs (200 to 2000 mi) along with any low-cost experience in this regime. Such programs should effectively augment already implemented programs, and yield a profitable return to many important programs currently under way.

Such activity should not significantly increase currently planned expenditures, but might even reduce costs or speed up programs through increased knowledge gained thereby.

At any rate, the effect of these activities, if not violently diverted from present trends, should permit profitable implementation of a *Meteor, Jr.*-type program properly phased, within a year or so, as data become available and effective prosecution is carried on toward initial experimental utilization by the time period assumed for the performance calculations — about 1962 or 1963. The state of the art seems to be reaching that point, and the needed concurrent programming appears to be under implementation to fit this rate of development.

VII. Summary and Conclusions

Meteor, Jr. is an attempt to meet a current need for a low-cost, efficient means to achieve a manned satellite operation by an early date, and it seems to fill quite well the objectives of this study. It exhibits a considerable usefulness and operational flexibility, in spite of its small size, and it preserves all the basic features of the original large version of the *Meteor* concept. The fast recycle time permits a large number of trips to the orbit per ship in a given time, and it provides inherently a means for testing-out progressively a stage at a time, with gradual approach to the full operating conditions. In addition, it is suitable for use as a basic building block for satellite terminal construction in the same way as originally conceived for the *Meteor*, thus greatly reducing the required number of trips into the orbit for providing a given facility.

Its operational flexibility appears to serve many purposes satisfactorily, ranging from an interim satellite to a moon-ship. And its ferrying flexibility can carry a variety of payload combinations ranging from one to two tons for different missions. Because of its modest size and because the concept inherently tends to eliminate equipment attrition, it is peculiarly attractive from an operational logistic and economy standpoint, and therefore seems to offer one good method by which to start manned satellite operation. It also establishes a natural pattern of growth for successive vehicle designs of larger size, increased utility, and load-carrying efficiency, embodying, as it does, the same or improved concepts, in accordance with the dictates of initial experience with the small vehicle.

The *Meteor, Jr.*-type vehicle offers the best opportunity yet considered by the authors for an early start toward achievement of a manned earth satellite. There seems little reason why, technically or otherwise, in view of the present state of the art, current programming in related fields, and relatively modest cost, development should not be started on a low level within a year or so, with the goal of initial flights about 1962 or 1963 and operational capability about two years thereafter.

Appendix A — A Possible Experimental Test Vehicle Configuration

A quick check has been made of the characteristics of a vehicle to launch, on a limited experimental basis, the *Meteor, Jr.* final stage developed in this study. For this purpose, the following conditions were set:

1. Payload reduced to 700 lb — enough for 3 people (personal gear), and some supplies and equipment, or for a 2-man crew with more equipment or supplies.

2. A full mission, including ascent to a 500-mi orbit (with corrections), descent, and landing.

3. Reserve fuel for landing approach maneuvers cut to 2-1/2 or 3 min at level flight thrust for this weight (instead of the 5 min allowed for regular operation with heavier load), this reduces fuel weight by 900 lb.

4. Minimum, expendable, light weight unmanned first and second booster stages to be used.

The resulting design characteristics are tabulated in Table A-1, based on the same velocity increments for each stage as the regular *Meteor, Jr.* vehicles. The empty vehicle to gross weight ratios selected by the designers (after checking that of the Viking No. 11—0.14125) for the larger first and second stages, were 0.105 and 0.115, respectively.

Table A-1. *Special Experimental Test Vehicle Configuration — Basic Design Characteristics*

Stage configuration	Total wt/lb	Propellant wt/lb	Empty wt/lb	Mass ratio	Structure and equip ratio percent	Payload (lb)
I	237,000	167,400	25,000	3.41	10.5	44,600
II	44,600	32,000	5,140	3.52	11.5	7,400
III (Final)	7,460	3,560	3,200	1.91		700

Aside from the changed fuel and payload values, the final stage vehicle is unchanged. The same specific impulse values are assumed as for the standard vehicle, and the respective required mass ratios are found to be 3.41, 3.52 and 1.91, for the first, second, and third (final) stages, respectively. This requires a total vehicle first stage weight of 118.5 tons, or about the estimated size of the Atlas ICBM[1]. This evidently means that a vehicle of the Atlas type now under development, with suitable modifications, can do the job of initial experimental test launchings of the *Meteor, Jr.* final stage vehicle into a satellite orbit — and that this can be utilized to whatever extent proves desirable, from an economic and development standpoint, as a prelude to launchings with the returnable booster vehicle stages.

---

[1] Missiles and Rockets **2,** 161 (1957).

This is entirely impractical, however, from an economic standpoint, for application beyond the initial phase. This is evident from considerations of the fuel vs vehicle costs. Using a vehicle cost of $ 40 per pound of empty vehicle, a propellant combination cost of $ 0.10 per pound, then allowing for the payload ratios (of 700 lb for this special test configuration vs that of the regular *Meteor, Jr.* — 1 ton, minimum), calculations show that for equivalent payload (1 ton) carried into the orbit, for this test configuration vehicle type (Atlas launched) approach, the vehicle cost expended is $ 3,200,000, plus $ 60,000 for fuel, for a total of $ 3,260,000 (actual costs per trip would be 1/3 this amount); whereas equivalent cost with the *Meteor, Jr.* is $ 80,000

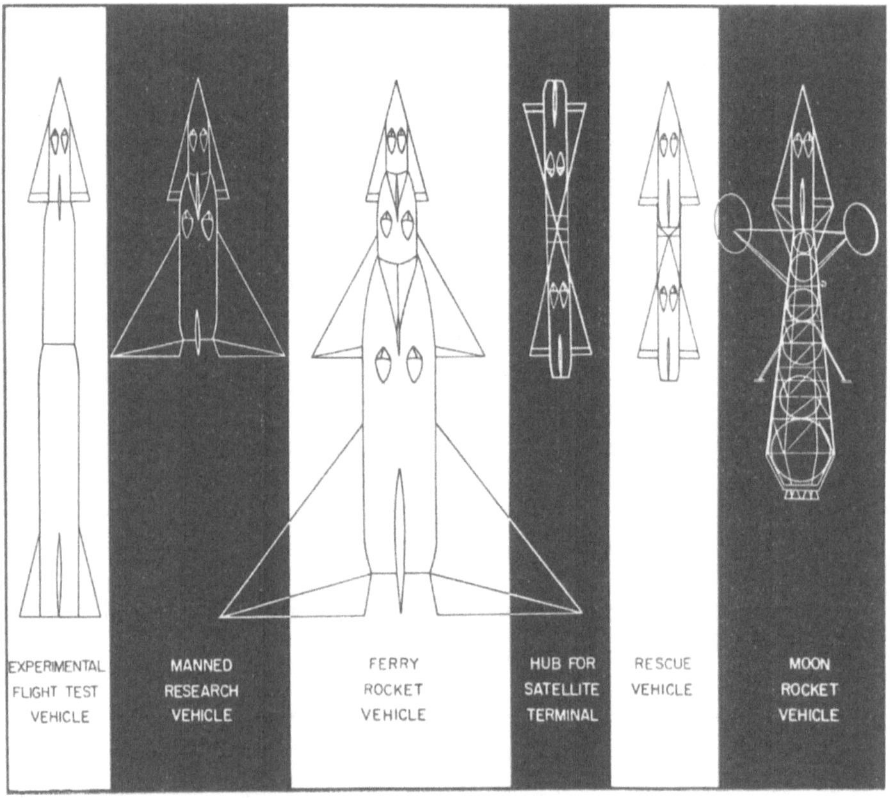

Fig. 15. Utilization versatility of *Meteor, Jr.*

for fuel, plus perhaps a nearly equivalent cost for vehicle amortization and maintenance, or a total of around $ 150,000. Other operating costs should be substantially the same, so it is seen that the direct cost per flight is about 22 times larger for the simplified test configuration as for the *Meteor, Jr.*, if use for the former is extended to long term regular use, making the cost prohibitive. This shows clearly one of the important inherent advantages of the basic *Meteor* concept.

However, the advantages offered by this early stage alternate method, making maximum use of available hardware and techniques, plus the program flexibility it can provide, affords some very attractive features for consideration for the initial program development.

Comparative uses and configurations of *Meteor, Jr.* for its several uses are shown in Fig. 15.

Appendix B — Utilization of Final Stage Vehicle for Moon Trips

A brief investigation has been made to see how the utilization of the *Meteor, Jr.* final stage vehicle for the two types of moon trips from the orbit, mentioned in Section II, would turn out from a weight, logistics, and cost standpoint.

For the exploratory (nonlanding) trips, a booster stage consisting simply of tanks and motors with associated equipment and interconnecting structure, is fitted to the vehicle. For the full moon trip, with landing, two additional booster stages are fitted to this configuration, making a 3-stage booster arrangement carrying the final stage *Meteor, Jr.* vehicle as payload. The 3 booster stages correspond to the orbit departure, moon landing, and moon take off phases, with the final stage vehicle itself providing the declaration for arrival in the orbit, or, if desired, an earth landing, which requires less than 500 mph, or about 1-1/2 percent of the total, additional. The moon-landing vehicle is, therefore, a four-stage rocket. (The upper booster stage would also need to carry a small unit for generating continuous power, probably from solar energy.)

The tanks of each stage are of a little greater capacity than that required for each stage, or phase of the operation, plus its reserves. Then some of the reserve fuel of the upper stages (never going below the minimum reserve) can be put in the lower stage tanks at the start. Then part or all of the remaining unused fuel can be pumped from the lower stage tanks to the upper stage tanks before jettisoning. This "floating" reserve increases the flexibility of utilizing the reserve, which is desirable, as reserves represent fuel to cover a reasonable amount of unforeseen developments or errors, and what proportion will occur or be needed for each stage is impossible to predict.

The velocities required for such trips have been calculated for various conditions by von Braun, Whipple, Clement, Ehricke, Lieske, and others[1]. Those of the listed references are for departures from 100 and 350 mi heights, respectively. The values calculated for this investigation for departure from a 500 mi orbit naturally range intermediate in value. For a moon trip with landing and return, the total required, allowing for suitable reserves, is approximately 8 mi per second (or about 42,000 fps). Exact values of course, depend on the amount of reserve capability allowed. Table B-1 gives the basic design characteristics assumed for the vehicle described.

Table B-1. *Basic Design Characteristics — Moon Ship Configuration*

Phase	Gross wt (lb)	Propellant (wt/lb)	Structural wt empty (lb)	Velocity increment (fps)	Mass ratio	Structural ratio (%)	Payload
I	343,000	209,000	24,000	10,500	2.55	7	110,000
II	110,000	66,800	9,900	10,500	2.55	9	33,300
III	33,300	19,800	2,500	10,000	2.46	7.5	11,000
IV	11,000	6,800	3,200	10,620	2.62	29	1,000
Totals	343,000	302,400	39,600	41,620	42.1	11.5	1,000

An exhaust velocity of 11,050 fps was assumed, or $I_{vacuum} = 345$ (same as obtained for final stage in its regular operation). This would correspond to a sea level optimum value of 1 of 275. The payload is assumed to be a 3-man crew and 400 lb of associated equipment and supplies. Actually, several hundred pounds additional carried would make very little difference (about 100 fps velocity, or a relatively small increase in gross weight), if it is left on the moon.

Whether this method of using a regular vehicle for the moon trip is a good way to do it, or whether it would be better to build a lighter-weight final stage personnel

[1] W. von Braun, F. L. Whipple, et al., Conquest of the Moon (edited by C. Ryan). New York: Viking Press, 1953; G. H. Clement, The Moon Rocket (Rand Report No. P833, revised 7 May 1956). The Rand Corporation, Santa Monica, Calif.

carrier in the orbit, cannot be determined at this time. The latter could be lighter and roomier, but the former should be quite comfortable, and has at least several advantages, one of which is that the most complex portion which must reliably provide all the necessary facilities and protection for the crew, along with navigation, guidance, and control provisions, need not be fabricated in the orbit, but can be a thoroughly ground-tested and service-proved vehicle before use for a moon trip. Only the relatively simple booster units need to be assembled and tested at the orbital terminal.

Another important advantage (especially for the nonlanding mission) is that this method eliminates the need for synchronizing with the orbital terminal station on the return, since if necessary or desirable the vehicle is capable of going right on and making its normal descent and landing on the earth at a selected spot, without the need for first making a rendezvous with the orbital terminal. Of course, a non-earth-landing vehicle could remain in the orbit and be picked up by regular ferry rockets if it missed the rendezvous (or even the proper orbit), but an independence from the need for such support could be of considerable advantage.

From the logistics and cost standpoint, utilizing the *Meteor, Jr.* vehicle should prove considerably better. It will undoubtedly be cheaper to provide, as construction work that must be done in the orbit is simplified, and the size of the total vehicle (just under 20 tons empty, 170 tons loaded) is only a little larger (40 percent to 50 percent), than the estimated gross weight of the Atlas fueled and loaded. If a fleet of three ships were to make the trip, the total weight to be ferried into the orbit (fuel and all) would be about 500 tons. With the *Meteor, Jr.* the cost of the fuel to ferry this weight to orbit would be only $ 20,000,000 or with larger ferry rockets, cost would be even less. The cost of all the booster stages for the 3 complete vehicles, at $ 100 per pound would be between $ 10,000,000 and $ 11,000,000.

For the nonlanding trip, the total ship weight is approximately 17 tons including all fuel, and is represented (approximately) by the bottom two lines (III and IV) of Table B-1. For a 3-ship fleet, the cost of fuel to ferry this weight to the orbit is a little over $ 2,000,000, and the cost of the vehicles themselves (at $ 100 per pound empty weight) would be $ 750,000.

In summing up the logistic picture, then, it appears that for moon-trips with 3-ship fleets (of which 2 would be capable of returning all personnel, in event of casualty to one of them) carrying a total of 8 to 10 people, the direct cost for all fuel and expended hardware would be around $ 2,500,000 to $ 3,000,000 for a nonlanding exploratory expedition, and for a landing expedition, around $ 30,000,000. Here, again, as with aircraft, it is obvious that these costs will be completely eclipsed by the development and operating costs, salaries, and facilities, and that the over-all cost will not differ fundamentally from that for comparable aircraft development and operating programs.

It appears then, that this concept can be utilized for initial experience and limited operations. If larger ships of the same concept are then utilized in a similar manner, costs per man and equipment carried will be reduced, and more extensive operations will become practical.

Appendix C — Rescue Missions

Rescue missions can be required for a number of reasons, the chief one being when a ship suffers severe malfunction before reaching the orbital rendezvous, and thus circles helplessly in a slightly different orbit from the satellite terminal. About the worst situation would be for the ship to be in a low orbit with a perigee and apogee at 200 and 33 mi, respectively. This involves a corresponding velocity difference from the desired orbit of a little less than 500 mph. But to this add inability to synchronize completely the rescue with the resulting faster orbit (for which approximately 20 percent should be added), and the result is a difference of about 600 mph. This is a little less than 10 percent of the final stage vehicle's total loaded performance capability.

If this velocity deficiency or difference is called V_D, then the total velocity capability required to perform a rescue mission of a ship of the same size is $2\ V_D +$

$+ V_D = 3\ V_D$, or three times the velocity of deficiency met by the ship to be rescued. This is because this deficiency has to be made up for two ships (the rescued and rescuer ships) and, in addition, the rescuing ship must first expend V_D to get into the mating "pick-up" orbit with the ship to be rescued.

As $V_D = 0.1 \times V_{III}$, where V_{III} = velocity capability of a third or final stage vehicle, then the total rescue mission velocity becomes $V_R = 3\ (0.1 \times V_{III}) = 0.3\ V_{III}$. There will be some waste or loss, so that about 1/3 of the total capabilities of these ships will be the most required for this type mission. (This estimate is conservative, it being assumed that the rescue ship is carrying a full payload, which would never actually be the case.)

Similarly, in a case where the rescued ship is heavier than the tugs (or rescue ships), by N times, then

$$\begin{aligned} V_R &= (N+1)\ V_D + V_D \\ &= (N+2)\ V_D = (N+2)\ (0.1 \times V_{III}) \\ &= \frac{N+2}{10}\ V_{III}\,. \end{aligned}$$

For example, if the ship to be rescued is five times the weight of the tug ($N = 5$), the velocity capability requirement is $0.7\ V_{III}$, or about three-quarters of total capability of the tug.

The capability of the tug to effect rescue can be exceeded under certain conditions. If the ship to be rescued is 10 times as large as the tug, then the tug's capability is exceeded by about 20 percent. It is exceeded by about 23 percent if a tug-sized ship returning from a mission to an outer orbit or to the moon misses the rendezvous by 2500 mph. Conceivable causes of missing the rendezvous by such amounts might include serious malfunction of guidance or power plant, large navigation error, loss of one ship from a group, overload, or a very large cumulative error or miscalculation.

Mathematically these conditions work out as follows:

Since $V_{III} = 6150$ mph, then

for the first case $V_R = \frac{10+2}{10}\ V_{III} = 1.2\ V_{III}$, and

for the second case $V_R = \frac{2500}{6150} \times 3\ V_{III} = 1.23\ V_{III}\,.$

In cases such as these, there are several choices for effecting rescue:

1. Take along a booster stage, such as the first booster section added for moon trips (see Appendix B).
2. If (1) is not readily available, use two rescue ships for tugs (one may take off part of the cargo load, or both may push).
3. One rescue ship can bring the rescued ship most of the way back, return to orbit to refuel, then complete the journey with the rescued ship. (This method is the least efficient, in that it takes the most fuel.)

Actual conditions for such rescue service would vary considerably, of course. Ships can work in multiples if necessary, although normally only a small portion of performance capability of one is needed. The air lock and cargo holds will often be useful, especially the former. A light outrigger framework mounted on the nose of the tug can be attached to the disabled ship and the tug becomes a pusher. This composite is maneuvered by the tug, utilizing its regular propulsion and control in the normal manner. Two or more ships acting in concert align themselves approximately with the CG of the disabled ship for best maneuverability.

If a workman or technician accidentally drifts too far from the station and is unable to return by himself, one of the tugs can pick him up, take him aboard through the air lock and stow in the tug's hold any material which has floated away with the man.

Whatever the rescue mission, *Meteor, Jr.* has all the desired characteristics and performance for easily and effectively fulfilling the requirement.

References

1. D. C. ROMICK, R. E. KNIGHT, and J. M. VAN PELT, A Preliminary Design Study of a Three-Stage Ferry Rocket Vehicle with Piloted Recoverable Stages. Presented at the 9th Annual Meeting of the American Rocket Society in New York, N.Y. American Rocket Society preprint no. 186—54. American Rocket Society, 500 Fifth Avenue, New York, N. Y., U.S.A., 1 December 1954.
2. D. C. ROMICK, Preliminary Engineering Study for a Satellite Station Affording Immediate Service with Steady Evolution and Growth. Presented at the 25th Anniversary Annual Meeting of the American Rocket Society in Chicago, Ill. American Rocket Society preprint no. 274—55. American Rocket Society, 500 Fifth Avenue, New York, N.Y., U.S.A., 21 November 1955.
3. D. C. ROMICK, Concept for *Meteor*, a Manned Earth-Satellite Terminal Evolving from Earth-to-Orbit Ferry Rockets. Proceedings of the VIIth International Astronautical Congress, Rome, 1956, p. 335. Roma: Associazione Italiana Razzi, 1957.
4. Jet Propulsion **27**, 639, 678, 679 (1957).
5. G. P. SUTTON, Rocket Propulsion Elements, 2nd Ed. New York: J. Wiley, 1956.
6. B. M. JAQUET and J. D. BREWER, Low-Speed Static-Stability and Rolling Characteristics of Low-Aspect-Ratio Wings of Triangular and Modified Triangular Platforms. National Advisory Committee for Aeronautics, Washington 25, D.C., RML8L29, 29 March 1949.
7. J. N. NEILSON, E. D. KATZEN, and K. K. TANG, Lift and Pitching-Moment Interference Between a Pointed Cylindrical Body and Triangular Wings of Various Aspects Ratios at Mach Numbers of 1.50 and 2.02. National Advisory Committee for Aeronautics, Washington 25, D.C., RMA50F06, 12 September 1950.

Spaces of Potential Visibility of Artificial Satellites for the Unaided Eye

By

Ingeborg Schmidt[1]

(With 6 Figures)

Abstract — Zusammenfassung — Résumé

Spaces of Potential Visibility of Artificial Satellites for the Unaided Eye. The visibility of satellites with unaided eyes is a problem of brightness discrimination. By computing visual ranges for the satellite in eight meridians for several positions of the sun below the horizon it is possible to determine "spaces of potential visibility". The ceilings of the spaces are given by the visual ranges, the lower limits by the perigee of the satellite and by the earth shadow. If the satellite travels through such as space it may become visible to the unaided eyes.

A satellite with a polished aluminum surface of reflectance 0.89, 20 inches (50.8 cm) in diameter, with a perigee of 200 miles (322 km) may be visible at and near the zenith as a star of 5.3rd to 5.7th magnitude, when the sun is 14° to 19° below the horizon. A diffuse reflecting satellite of reflectance 0.8, the other conditions remaining the same, may be visible with the sun 12.5° to 20° below the horizon, as a star of 5th to 5.7th magnitude. The chances for visibility of a satellite with specular reflection are to some extent better than those for a diffusely reflecting one in the meridian toward the sun, those of the diffusely reflecting are by far better in the direction opposite the sun. With a perigee of 300 miles (483 km) a satellite of 20 inches (50.8 cm) diameter will not be visible to the unaided eyes. Increasing the diameter improves the visibility markedly. A projection of the circumference of the spaces of visibility on the dome of the sky permits a demarcation of the areas of useful scanning for the satellite. In order to predict the visibility of the satellite from the spaces of potential visibility the approximate position and height of its orbit, the time of traveling through the sky and the position of the sun at this time need to be known.

The computations are made for observation from sea level at a latitude of 30 degrees and with clear air. Diverse factors making a prediction of the visibility uncertain to some extent are also discussed.

Räume potentieller Sichtbarkeit künstlicher Satelliten für das unbewaffnete Auge. Die Sichtbarkeit von Satelliten mit unbewaffnetem Auge ist ein Problem der Helligkeitsunterscheidung. Durch Berechnung visueller Bereiche für den Satelliten in 8 Meridianen für verschiedene Stellungen der Sonne unterhalb des Horizontes ist es möglich, „Räume potentieller Sichtbarkeit" zu bestimmen. Die oberen Grenzen dieser Räume werden durch die visuellen Bereiche gegeben, die unteren Begrenzungen durch das Perigäum des Satelliten und den Erdschatten. Wenn der Satellit durch einen solchen Raum wandert, kann er dem unbewaffneten Auge sichtbar werden.

Ein Satellit mit einer polierten Aluminiumoberfläche mit dem Reflexionsvermögen 0,89 und einem Durchmesser von 50,8 cm sowie einem Perigäum von 322 km kann im und nahe dem Zenit als Stern der Größe 5,3 bis 5,7 wahrgenommen werden, wenn die Sonne sich 14 bis 19° unter dem Horizont befindet. Ein diffus reflektierender Satellit mit einem Reflexionsvermögen von 0,8 kann, falls die übrigen Bedingungen gleich

[1] Division of Optometry, Indiana University, Bloomington, Indiana, U.S.A.

bleiben, als Stern 5. bis 5,7. Größe erkannt werden, wenn die Sonne 12,5 bis 20° unter dem Horizont steht. Die Möglichkeiten für die Wahrnehmbarkeit eines Satelliten mit spiegelnder Reflexion sind im Meridian gegen die Sonne einigermaßen besser als die fur einen diffus reflektierenden Satelliten; jene eines diffus reflektierenden Satelliten sind bei weitem besser in der entgegen der Sonne weisenden Richtung. Mit einem Perigäum von 483 km wird ein Satellit von 50,8 cm Durchmesser für das unbewaffnete Auge nicht sichtbar sein. Vergrößerung des Durchmessers verbessert die Sichtbarkeit beträchtlich. Eine Projektion des Umfanges der Sichtbarkeitsräume auf den Himmelsdom gestattet eine Abgrenzung der Bereiche aussichtsreicher Suche nach dem Satelliten. Um seine Sichtbarkeit aus den Räumen potentieller Sichtbarkeit voraussagen zu können, müssen die angenäherte Lage und Höhe seiner Bahn, die Zeit des Durchganges am Himmel und die Stellung der Sonne zu dieser Zeit bekannt sein.

Die Berechnungen sind für Beobachtungen in Seehöhe und einer Breite von 30° bei klarer Luft angestellt worden. Verschiedene Faktoren, welche die Vorhersage der Wahrnehmbarkeit in einem gewissen Ausmaß unsicher machen, werden ebenfalls erörtert.

Visibilité à l'œil nu des satellites artificiels de la terre. La visibilité des satellites à l'œil nu est un problème de discrimination de la brillance. Des "régions de visibilité potentielle" ont été déterminées en calculant le champ visuel de discrimination dans huit plans méridiens pour diverses positions du soleil sous l'horizon. Les limites supérieures des régions sont déterminées par le champ visuel, les limites inférieures par le périgée du satellite et l'ombre de la terre. Si le satellite traverse une de ces régions il peut devenir visible à l'œil nu.

Un satellite de 20 pouces de diamètre, ayant une surface d'aluminium poli de réflectance 0.89 et un périgée de 200 miles peut être observé aux environs du zénith comme étoile de grandeur 5.3 à 5.7, si le soleil est de 14 à 19° sous l'horizon. Si, toutes autres conditions égales, la réflexion du satellite est diffuse (réflectance 0.8), il est visible comme étoile de grandeur 5 à 5.7, le soleil étant situé de 12.5 à 20° sous l'horizon. La visibilité d'un satellite à réflexion spéculaire est un peu meilleure dans le plan méridien vers le soleil, celle d'un satellite à réflexion diffuse est bien meilleure dans la direction opposée au soleil. Si le périgée est porté à 300 miles un satellite de 20 pouces ne sera plus visible à l'œil nu. L'augmentation du diamètre améliore beaucoup la situation. Une projection des régions de visibilité sur le domé céleste est utile pour délimiter les recherches. La prédiction de la visibilité demande la connaissance de la position approximative et de la hauteur de l'orbite, et de la position du soleil pendant le temps où le satellite traversera la région visible du ciel.

Les calculs sont relatifs aux observations faites à une latitude de 30 degrés par ciel clair. Divers facteurs d'incertitude sont également discutés.

Introduction

A tracking of artificial satellites with unaided eyes is of advantage in comparison to observation through telescopes because of the greater convenience, the larger field of view and the slower apparent motion.

The following calculations about the visibility of artificial satellites for the unaided eye are based on data available in the literature about the first satellite of the project Vanguard which will be launched from the East coast of Florida sometime during the International Geophysical Year. The primary plan was to build a white diffuse reflecting plastic globe of 30 inches (76.2 cm) or 20 inches (50.8 cm) diameter. According to the newest information (Dr. J. A. VAN ALLEN) it would be most likely a 20 inch globe with a surface of polished aluminum, protected by a thin film of SiO. In the present article, both the diffusely reflecting and the specularly reflecting will be considered. The satellite will circle around the earth between the approximate latitudes of 40 degrees north and 40 degrees south of the equator. In order to confine the calculations to a geographical latitude with a certain chance for its visibility, the calculations are made for a latitude of 30 degrees. The orbit of the satellite will have a perigee no closer than 200 miles (322 km). Its orbit will appear displaced westward by 25 degrees after each revolution. Due to this displacement and due to other factors, e. g. affection of the orbit by the shape of the earth globe and seasonal variations of the day-night cycle, the satellite will be visible from different locations within the 80 degree belt of latitude, provided that it proceeds to circulate around the earth long enough.

For more details about artificial satellites see the recently published book, edited by J. A. Van Allen [1], and numerous individual papers, especially those of Tousey [2, 3].

Visibility of Satellites, a Problem of Brightness Discrimination

A satellite of 30 inches (76.2 cm) diameter seen from its nearest distance to the earth of 200 miles (321.8 km) subtends at the eye a linear angle of 0.47 seconds of arc, a satellite of 20 inches (50.8 cm) diameter subtends an angle of 0.33 seconds of arc. These dimensions are far below the resolving power of the human eye. Hence, for further considerations the satellite must be regarded as a dimensionless point physiologically, the retinal image of which is a diffraction pattern. Visual acuity being ruled out, a detection of the satellite may be possible with unaided eyes on the basis of brightness[1] differences between the satellite and the sky. When regarding the satellite as a point light source it is correct to express its visual effect in illuminance values (flux density) at the plane of the observer's entrance pupil. The magnitude which permits us to draw conclusions about the visibility of the satellite in a given direction is its *visual range*, that is the farthest distance at which it is just possible to distinguish the satellite against the sky. The visual range d can be calculated from the relationship $d = \sqrt{\frac{I \cdot T}{E_t}}$ which follows from the law of inverse squares and where E_t is the threshold incremental illuminance at the eye for the luminance of the sky in the viewing direction, I is the intensity of the light reflected from the satellite in the direction of the observer and T is the transmittance of the atmospheric layer between satellite and observer. It has been stated that a possibility to observe the satellite with unaided eyes exists only at dusk or dawn, when the observer and the atmosphere above him are already in the shadow of the earth but the satellite is still outside this shadow. By computing the visual ranges in different meridians from the position of the sun and at different zenith angles it is possible to determine "spaces of potential visibility" of the satellite. The pattern of these spaces depends on the position of the sun below the horizon. The visual ranges in different viewing directions constitute the ceiling of the space of potential visibility, the bottom is determined by the distance of the perigee from the horizon and at larger depression angles of the sun, also by the earth's shadow. The satellite should become visible for the unaided eye when it traverses such a space at the appropriate time, provided that other assumed conditions are fulfilled.

The reported computations are made under the assumption that the satellite is observed at sea level, at a latitude of 30 degrees, at a place remote from artificial

---

[1] The term "brightness" will be used as a psychological term defined as the attribute of sensation by which an observer is aware of differences in luminance. The term "visibility", will be used in its common usage to express clearness with which objects stand out from their surroundings. All photometric quantities used in this paper are defined in terms of photopic visibility functions.

illumination, in a moonless sky through clear air with no atmospheric turbulence and with the eyes perfectly adapted to the luminance of the sky surrounding the satellite, that is with a constant sensitivity of the eye.

The following symbols and abbreviations will be used in this article:

E_s — illuminance by the sun.

Ob — observer.

Z — zenith.

s.p.v. — space of potential visibility.

θ — angle of the sun below the horizon, the angle between the plane of the horizon and the tangential ray to the earth drawn from the center of the sun.

S — bearing from the sun, the meridian through the sun being $S\,0°$. A meridian plane at an angle $n°$ from the direction of the sun will be indicated by $S\,n°$.

ζ — zenith angle, subtended at *Ob* by the line of sight of the observer in direction of the satellite and by the vertical *ObZ*. $\zeta_{Sn°}$-zenith angle in the meridian plane $S\,n°$.

Zenith angle (angle of latitude) and bearing from the sun (angle of longitude) are the spherical coordinates specifying the direction of the satellite from the observer. Since $S\,0°$ is always in the direction of the sun, the coordinate system of the sky rotates with the position of the sun around the axis *ObZ* and is reversed at sundown and sunrise.

Data Required for the Computation of the Spaces of Potential Visibility

A. Height of the Earth Shadow

Although the earth shadow consists of two parts, a conical umbra (u) and a conical penumbra (p) (Fig. 1) it has been assumed cylindrical which gives an average basis of sufficient approximation for the following calculations. The boundary of this cylindrical shadow (actually, the mid-penumbral locus or the surface of half-illumination) is a tangent to the earth from the center of the sun and intersects the vertical *ObZ* at a point *H*. The effect of atmospheric refraction on the tangential ray has not been taken into consideration. The height h of the cylindrical shadow in the zenith can be calculated from $h = r_e(\sec\theta - 1)$ where r_e is the earth radius. An earth radius $r_e = 6373$ km was calculated for a latitude of 30 degrees. The diameter of the sun being 32 minutes of arc the height of the limit to the area of full sunlight can be calculated with some approximation from $u + p = re\ [\sec(\theta + 16') - 1]$. The values for h and for $u + p$ at several angles of θ are listed in Table I.

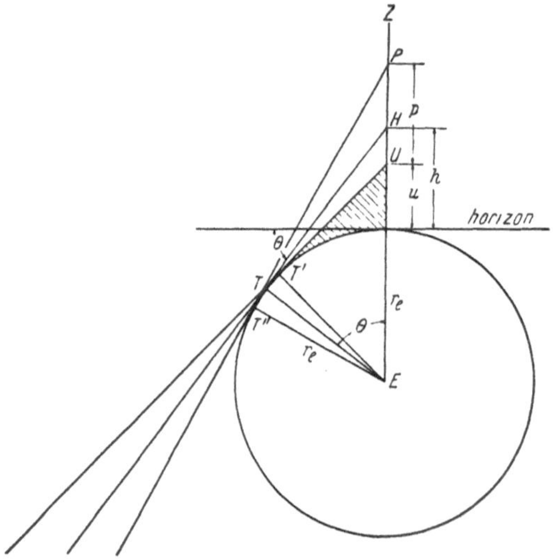

Fig. 1. Boundary of the earth shadow at the zenith. h—height of the mid-penumbral locus, p—height of the penumbra, u—height of the umbra, θ—depression angle of the sun below the horizon

Table I. *Height of the Earth Shadow at the Zenith*

θ degrees	h km	$u+p$ km
—14	195.0	202.7
—15	224.8	233.3
—19	367.1	377.9
—20	409.0	420.6
—23	550.6	564.0
—25	658.8	674.3

The upper border of the cylindrical shadow in the $S\,0° — S\,180°$ meridian can be drawn knowing the angle θ and the height of the earth shadow. In the $S\,90°$ $S\,270°$ meridian the upper border is actually a segment of a hugh ellipse but it can be assumed a parallel to the border through the point H, with sufficient

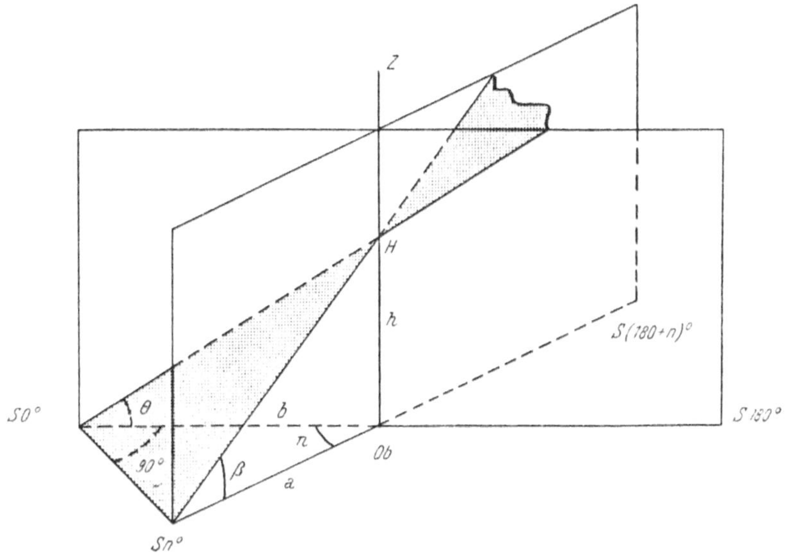

Fig. 2. Relationship of angle β, the inclination angle of the earth shadow with the horizon within an oblique meridian plane $Sn°$-S $(180 + n)°$. Angle θ is the inclination angle of the earth shadow in the meridian plane of the sun, $S\,0° — S\,180°$

approximation for the relatively small area in question. In any oblique meridian $S\,n°$ the upper limit of the earth shadow is similarly approximated by a straight line through H but inclined at an angle β with the horizon. The angle β is simply the projection of θ onto the oblique meridian plane and is determined by the formula:

$$\tan\beta = \tan\theta\ \cos n$$

the derivation of which is obvious from Fig. 2. Values at different angles θ for the oblique meridian planes considered are listed in Table II. Since the bearing from the sun can be counted either clockwise or counterclockwise, these angles are valid for both directions from the meridian through the sun.

Table II. *Angle θ of Inclination of the Earth Shadow in Oblique Meridian Planes*

Meridian Planes	Angle θ −14°	−15°	−20°
$S\ 22^1/_2°$—$S\ 202^1/_2°$	12° 58′ 18″	13° 54′ 16″	18° 35′ 11″
$S\ 45°$ —$S\ 225°$	9° 59′ 56″	10° 43′ 44″	14° 25′ 58″
$S\ 67^1/_2°$—$S\ 247^1/_2°$	5° 27′	5° 51′ 17″	7° 55′ 46″

B. Luminous Intensity of the Satellite

In order to calculate the luminous intensity of the satellite in the direction toward the observer, its *illuminance* E_s by the *sun*, its *size* and its *surface properties* need to be known.

JOHNSON [4] recently computed from a newly determined solar spectral irradiance curve an extraterrestrial solar illuminance of 12,700 ft-cd (136,703 lm/m^2) for the mean solar distance. This value was used in the present calculations.

When the satellite is well above the earth shadow, the illuminating rays will traverse no more than the upper layers of the atmosphere, which are so rarefied as to be considered optically ineffective (the "atmospheric space equivalence" of STRUGHOLD [5]). It would then be entirely correct to use the extraterrestrial E_s values. From data in the literature it may be justified to assume practically no effects by the atmosphere even for solar rays passing about 35 km to 40 km above the grazing ray to the earth. Losses of radiant flux from the sun by absorption and scattering in the lower atmosphere and refraction effects, which may partly compensate for the former, have been ignored in the following calculations. A discussion of these problems has been carried out by HULBURT [6]. Differences in distance of the satellite from the sun are negligible being insignificant in comparison to the mean distance of the sun. The illuminated atmosphere between the satellite and upper limit of the earth shadow would produce an intervening "space light" which similarly may be disregarded because of atmospheric space equivalence. An additional illuminance by the earth light has not been considered, since it is small in comparison to that of the sun.

(1) Intensity of a satellite with specular reflectance.

The formula for the intensity reflected by a sphere with specular reflection equals $I_{sp} = \frac{E_s R r^2}{4}$ where R is its reflectance and r its radius.

The visual reflectance of an aluminum mirror protected with an optimum thickness of SiO equals about 0.89 according to HASS and SCOTT [7]. This value was used in the following computations, assuming it independent of the angle of incidence and neglecting polarization. Since aluminum reflects fairly nonselectively in the visible spectrum one may assume that the color temperature of the satellite will equal that of its illuminant, the sun. The intensity of the specularly reflecting satellite is independent of the phase angle α, that is of the angle, measured from the center of the satellite, between the observer and the sun.

(2) Intensity of a satellite with diffuse reflectance.

The computation of the intensity in the direction of the observer of a dull white sphere which reflects according to LAMBERT's cosine law, is a problem similar to that of a planet. The formula as found in the Handbuch der Astrophysik [8] is following: $I_d = E_s\, R\, r^2\, 2/3\, \frac{\sin\alpha + (\pi - \alpha)\cos\alpha}{\pi}$ where E_s, R, r and α have the same meaning as above. Data about the reflectance of the "highly reflecting milky white" surface of the primarily planned satellite were not available. Therefore a value of 0.8 was assumed arbitrarily. The reflectance would very likely be non-selective so that the color temperature of the light would equal that of the sun. The phase angle decreases as the satellite moves away from the direction of the sun (from $S\,0°$ toward $S\,180°$) and the corresponding intensity increases since the illuminated portion which is visible to the observer increases. For a given depression angle θ of the sun the phase angle α will be constant within each plane which makes a constant dihedral angle with the $S\,90° — S\,270°$ meridian plane. We may call these planes of constant α and designate each by its dihedral angle, the zenith angle $\zeta_{S\,0°}$ or $\zeta_{S\,180°}$. The phase angle may be computed from the formula $\alpha = 90° + \theta - \zeta_{(S\,0°,\ S\,180°)}$, θ being negative and assuming $\zeta_{S\,0°}$ negative and $\zeta_{S\,180°}$ positive.

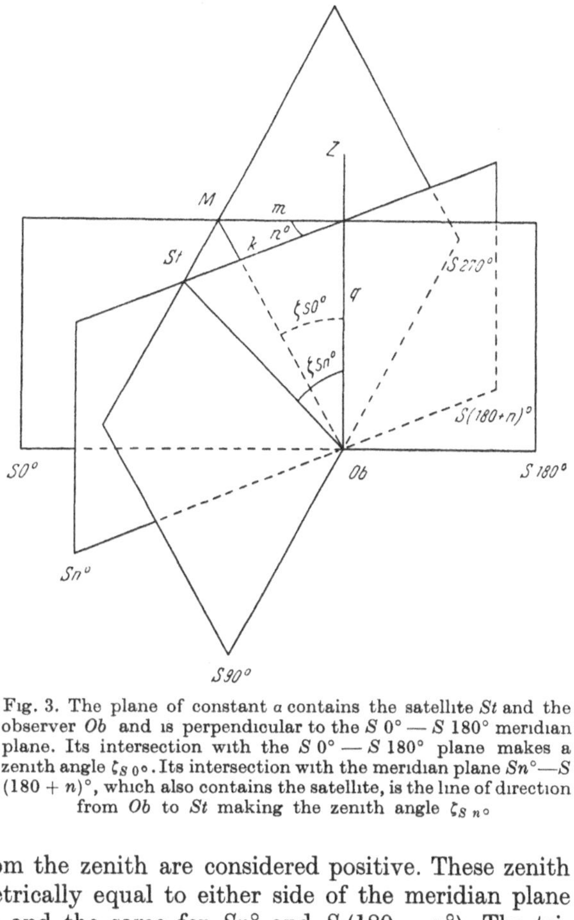

Fig. 3. The plane of constant α contains the satellite St and the observer Ob and is perpendicular to the $S\,0° — S\,180°$ meridian plane. Its intersection with the $S\,0° — S\,180°$ plane makes a zenith angle $\zeta_{S\,0°}$. Its intersection with the meridian plane $Sn°$—$S\,(180 + n)°$, which also contains the satellite, is the line of direction from Ob to St making the zenith angle $\zeta_{S\,n°}$

The phase angles were computed for 5 or for 10 degree intervals of ζ in the $S\,0°$-$S\,180°$ meridian plane. Using the same planes of constant α the corresponding zenith angles in oblique meridians were computed from the formula $\operatorname{tg}\zeta_{Sn°} = \sec n° \operatorname{tg}\zeta_{(S\,0°,\ S\,180°)}$ where $Sn°$ is the oblique meridian in question. For convenience all directions from the zenith are considered positive. These zenith angles then would be symmetrically equal to either side of the meridian plane for the directions of the sun and the same for $Sn°$ and $S\,(180 - n°)$. The trigonometry of these zenith angles in oblique meridians follows from Fig. 3. Table III contains some results of these calculations.

Table III. *Zenith Angles in Oblique Meridians Corresponding to Planes of Constant* α

Planes of constant α $\zeta_{S\,0°}$ or $\zeta_{S\,180°}$	Oblique meridians $S\,0°$ and $S\,(180—n)°$		
	22 1/2 157 1/2	45 135	67 1/2 112 1/2
5	5° 24′ 54″	7° 3′ 12″	12° 52′ 40″
10	10° 48′ 20″	14° 0′ 8″	24° 44′ 20″
20	21° 30′ 9″	27° 14′ 10″	43° 33′ 50″
30	32° 0′ 7″	39° 13′ 53″	56° 27′ 47″

The phase angle in the $S\,90° - S\,270°$ meridian plane is equal to that at the zenith.

C. Atmospheric Attenuation

On its pathway through the atmosphere toward the observer the luminous flux reflected by the satellite will be modified by absorption and scattering. This effect is known as attenuation or extinction. Besides of its dependence on the haziness of the air, it is a function of the length of the light path through the atmosphere and thus of the zenith angle. Moreover it is a function of the spectral energy distribution of the radiation. The effect will depend on the adaptation level of the eye. In case the transmitted light contains abundant short wave lengths, its luminosity will be very different in scotopic vision from that in photopic vision. In case of long wave lengths the difference is not so significant. It is evident that an all-purpose extinction factor cannot exist. From different data in the literature (Textbooks of Astronomy cite extinctions of 0.78 to 0.83 for stars observed vertically) it seemed reasonable to assume a compromise value for the visual transmittance of the atmosphere of $T = 0.800$ for clear air, when looking through a layer of dark atmosphere toward the zenith from sea level. The extinction will be practically that of the total atmosphere, so that individual differences in distance of the satellite for a given zenith angle can be neglected. We can write also $T = e^{-\tau \sec \zeta}$ where τ is an extinction coefficient for the total atmosphere and ζ is the zenith angle. The apparent displacement of the satellite toward the zenith by refraction is small even for large zenith angles, and therefore has been neglected.

D. Luminance of the Sky and Associated Threshold Incremental Illuminance at the Eye

The luminance values of the twilight sky at the depression angles of the sun in question were obtained from a paper of KOOMEN [9] and coworkers. They were measured at an altitude of 30 meters, which can be called about sea level. Since the measurements were made for every 20 degrees zenith angle and every 3 degrees of θ, between $\theta + 5°$ and $-15°$, intermediate values had to be interpolated. The sky luminance values for a sun lower than $-15°$ had to be extrapolated from a curve. The values for symmetrical directions to either side of the meridian through the sun were assumed to be symmetrically equal and the same during morning and evening twilight.

The associated threshold illuminances at the eye for a point light source were obtained from a curve published by MIDDLETON [10] (p. 97) of the so-called TIFFANY data (as reported by BLACKWELL [11]) see Fig. 4. MIDDLETON transformed the data into threshold illuminance values and into metric units. The tests were made in binocular observation with natural pupil and with no restriction to head and eye movements, which would correspond to the technique of tracking the satellite. The observers were free to use parafoveal vision at low luminances. There are several factors which make a prediction uncertain to some extent when using the TIFFANY data but there are no better available for our purpose at this time.

The observers knew the time of appearance and the approximate location of the target as a steady light. The satellite will appear at an unknown moment and at an unknown place as a moving target. A scanning for the satellite may increase the threshold by an unpredictable amount. The TIFFANY data represents an average curve of highly trained young persons. We do not consider individual

Table IV. *Visual Ranges in km of a Specularly Reflecting Satellite at Different Angles θ*

	Meridian S, Degrees	0		$22^1/_2$	45	$67^1/_2$	90		$112^1/_2$		135		$157^1/_2$		180	
	Zenith angle ζ, Degrees	0	10	10.8	14	12.9	10	20	12.9	24.7	14	27.2	10.8	21.5	10	20
Angle θ	—14°	332.9	316.2	317.3	325.7	326.6	327.2	323.1	331.5	320.8	331.7	327.4	332.3	329.7	332.0	330.5
	—15°	353.1	334.1	334.6	337.5	341.6	347.0	338.0	346.6	337.5	350.3	340.7	350.0	345.4	352.0	350.2
	—18° and more	383.9	373.5	367.8	363.7	366.9	382.7	363.4	372.8	358.3	373.5	352.8	381.6	365.4	382.7	366.8

variations by using these values. The thresholds of untrained persons and of older persons will be higher than average. The color temperature of test source and background illuminance used in the TIFFANY experiments was very likely within the range of 2500 to 3000° K, and the color temperature of the night sky with clear air is higher. In the most fortunate case the relationship in photopic

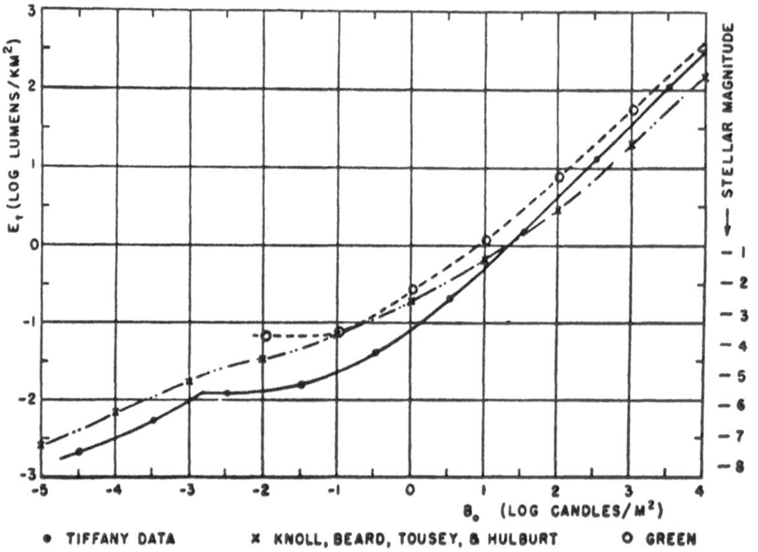

Fig. 4. Threshold illuminance E_t from a fixed, achromatic point source as a function of background illuminance B_0 (after MIDDLETON)

units may correspond. The observers in the laboratory were well adapted to the controlled luminance of the background. When scanning the sky under twilight conditions, the sensitivity of the eyes will vary.

Since the TIFFANY data were obtained with a probability of detection of 50% the values were increased by 0.3 log units to make them represent a certainty of seeing of nearly 100%.

Results

Table IV contains the visual ranges of a 20 inch specular satellite of reflectance 0.89 for different depression angles of the sun and different meridians from the direction of the sun. The values are symmetrically equal on both sides of the plane $S\,0° — S\,180°$. A depression angle of the sun smaller than — 14° has not been considered, since at about $\theta = — 13.8°$ the maximum height of the ceiling of the *s.p.v.* would be equal to the distance of the perigee from the earth; in other words, the satellite would not be visible any more. Since the sky luminance around the zenith at $\theta = — 18°$ and larger equals that of the night sky the visual ranges remain the same from θ — 18° up. The zenith angles in oblique meridians are those computed for the planes of constant a (see Table III) in order to make a comparison with the visual ranges for a diffusing satellite more convenient (e.g., in Fig. 5). The sky luminances at θ — 14° and in most cases also at θ — 15° had to be interpolated from the measurements of KOOMEN and coworkers; those of θ — 18° had to be extrapolated from a curve, therefore the results in Table IV must be regarded as an approximation only. A comparison of the visual ranges at θ — 18° and more with the height of the earth shadow at the zenith (Table I) allows one to conclude that the satellite might be visible at — 19° but not at θ — 20° or more. It follows from Table IV that the farthest distance from the observer at which the specular satellite will be perceived is about 380 km at and near the zenith. It is not difficult to recompute the results represented in Table IV in case some conditions may be varied, e. g., the size or the reflectance of the satellite.

The calculations for a 20 inch diffusing satellite of reflectance 0.8 showed that it might be visible at depression angles of the sun of — 12° to — 20.5°. The farthest distance from the observer at which this satellite will be perceived is about 495 km, at $\zeta_{S\,180°}=33°$, when $\theta = -18°$, at which time the satellite is about 415 km above the horizon. This is understandable from the fact that the intensity of the satellite increases in direction opposite the sun.

Both types of satellites have their best chance of visibility at about $\theta = -15°$ to — 18°, in agreement with TOUSEY [2, 3].

A graphical representation of the visual ranges in different meridians and at different zenith angles, together with the distance of the satellite's perigee from the horizon and the upper boundary of the earth shadow gives us an idea of the approximate shape of the *spaces of potential visibility* of satellites.

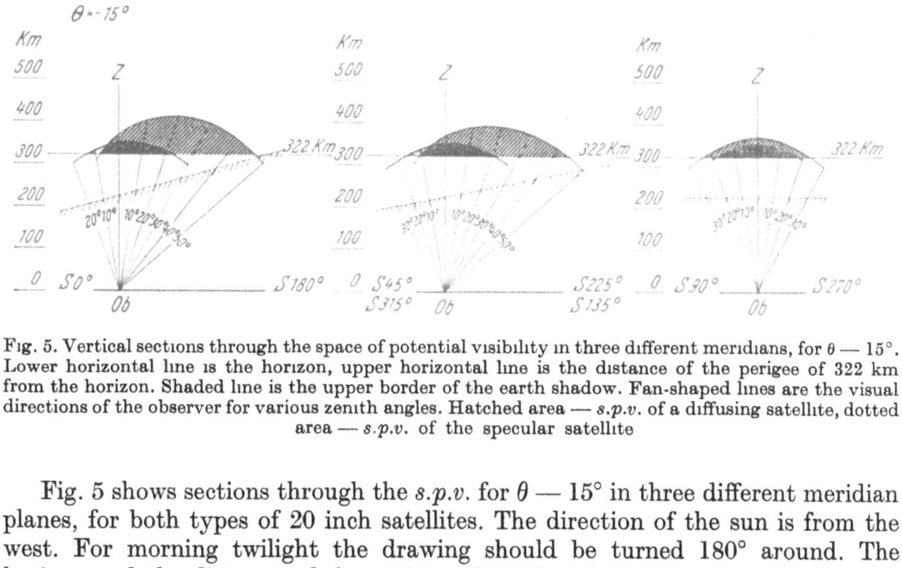

Fig. 5. Vertical sections through the space of potential visibility in three different meridians, for θ — 15°. Lower horizontal line is the horizon, upper horizontal line is the distance of the perigee of 322 km from the horizon. Shaded line is the upper border of the earth shadow. Fan-shaped lines are the visual directions of the observer for various zenith angles. Hatched area — *s.p.v.* of a diffusing satellite, dotted area — *s.p.v.* of the specular satellite

Fig. 5 shows sections through the *s.p.v.* for θ — 15° in three different meridian planes, for both types of 20 inch satellites. The direction of the sun is from the west. For morning twilight the drawing should be turned 180° around. The horizon and the distance of the perigee of 322 km (200 miles) from the horizon are drawn as straight lines neglecting the earth curvature. The earth shadow does not yet affect the shape of the bottom of the *s.p.v.* which is determined only by the perigee. These sections show that the chances for visibility of the specular satellite are worse than that of the diffusing, except in the direction toward the sun. The *s.p.v.* of the specular satellite is similar to the sector of a sphere. The maximum height of its ceiling is near the zenith, slightly decentered in the direction away from the sun, which is understandable from the distribution of the sky luminance. The *s.p.v.* of the diffusing satellite has a more nearly toric upper surface. For the $S\,0°$ meridian and up to a meridian of $S\,67^1/_2°$ to either side of the zero meridian there can be found zenith angles at which the visual ranges for both satellites are equal. From the $S\,90°$ — $S\,270°$ meridian on eastward the chances are definitely better for the diffusing satellite, which can be explained by the increase of the satellite's intensity due to decrease in phase angle and the decrease of the sky luminance in that direction. The maximum height of the ceiling is between about $\zeta_{S\,180°}=10°$ and $\zeta_{S\,180°}=20°$.

With larger depression angles of the sun the ceiling of the *s.p.v.* increases in height, though the shape remains similar. As long as there is no interference with the earth shadow the *s.p.v.* expands also in its total circumference. The bottom of the *s.p.v.* at $\theta = -15°$ and less is a plane surface. When it begins to be affected by the earth shadow it becomes inclined first in the halfdome of the sky opposite the sun. The earth shadow begins to affect the *s.p.v.* for the diffusing satellite when the sun is slightly lower than $\theta = -15°$, and for the specular satellite when the sun is slightly lower than $\theta = -16°$. From about $\theta = -19°$ up the bottom of the *s.p.v.* for both types of satellites is provided entirely by the inclined surface of the earth shadow. The specular satellite will be totally eclipsed at its visual ranges when the sun is slightly lower than 19° below the horizon; the diffusing satellite will be totally eclipsed at its visual ranges when the sun is below $\theta - 20°$.

Table V. *Visibility Limits of Various Satellites*

Diameter inches	Surface	Maximum height above horizon, km	Potential visibility limited to sun at θ, degrees	Time of potential visibility at equinox, after sunset or before sunrise, mins.	Liminal ranges of stellar magnitude of satellite
20	specular	380	-14 to -19	65 to 88	5.3 to 5.7
20	diffusing	455	-12.5 to -20	58 to 93	5 to 5.7
30	specular	575	-9 to -23	42 to 107	4.5 to 5.7
30	diffusing	710	-8 to -24.5	37 to 114	4.4 to 5.7

Table V gives a survey of the visibility limits for various satellites. Calculations of the values in this table involved many assumptions and approximations so that there is a wide range of probable error. They show the advantage of a diffusing satellite in comparison to a specular one, and the advantage of increasing the diameter. The maximum heights of the ceiling of the *s.p.v.*'s will be found under optimum conditions when the sky background is darkest. These heights permit the conclusion that it would be possible to detect a 30 inch satellite even when its perigee equals 300 miles (483 km), whereas the 20 inch satellites would then remain invisible. The stellar magnitudes at which the satellites will appear, are found from a table by G. Simpson (see Middleton [10], p. 99).

A projection of the circumferential limits of the *s.p.v.*'s on the dome of the sky outlines the areas of useful scanning for the satellite, as shown in Fig. 6, for $\theta = -14°$, $-15°$ and $-20°$. These drawings illustrate again that the chances for the specular satellite are worse than for the diffusing. The slight flattening of the area for the specular satellite toward the sun may be explained by the position of the isopter of the sky luminance at this border. The area for the diffusing satellite is of an appreciable dimension already in the direction opposite the sun. When $\theta = -15°$ the areas of useful scanning expand markedly for both types. At $\theta = -20°$ the specular satellite will be eclipsed at its visual ranges. The circumference of the *s.p.v.* for the diffusing satellite is elliptical and entirely determined by the earth shadow. The height and sidewise extent of this space will be actually smaller than shown but could not be calculated with certainty since the absorption and scattering effects of the atmosphere on the illuminating

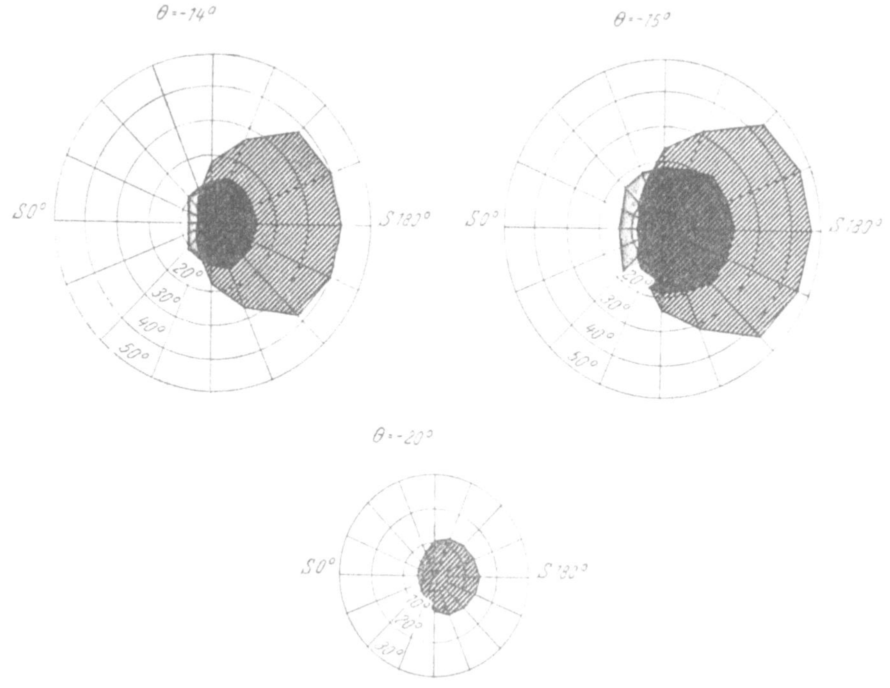

Fig. 6. Projection of the spaces of potential visibility on the dome of the sky for different depression angles of the sun. Circles are latitudes, counted from the zenith. Hatched area—*s.p.v.* of the diffusing satellite, dotted area—*s.p.v.* of the specular satellite

sunlight are unknown (though perhaps not impossible to calculate). The circumference of the space will be smaller also because it is within the penumbra of the earth shadow already, though atmospheric refraction of the sun's rays will partially compensate for this.

It should be emphasized that the predictions are valid for the most favorable conditions of observation only and that there are a number of foreseen and unforeseen variables which might affect the visibility of the satellite, e. g. changes in surface qualities by erosions from micrometeorites.

Visibility Problems in Daytime

It may be mentioned briefly that during daytime the satellite cannot be seen with unaided eyes. When observed against the sun so that no reflected light reaches the eyes, its inherent luminance, that is its luminance when observed close at hand would equal about zero. But as a very far object of zero luminance it would aquire the luminance B_s of the surrounding sky and would not be perceivable because of lack of contrast. When its position is such that it reflects sunlight toward the observer its luminance will be increased by the space light, which in the liminal case of a very far object will be equal to the luminance of the sky B_s, so that $B_a = B_i\, e^{-\tau \sec \zeta} + B_s$ where B_a is the apparent luminance of the satellite, that is its luminance when observed through the atmopshere, B_i its inherent luminance, $e^{-\tau \sec \zeta}$ the correction factor for atmospheric attenuation. The usual procedure then is to compute the contrast C from the relationship $C = \frac{B_a - B_s}{B_s}$ and to compare the result with data about contrast thresholds. However such data are not available for objects of 0.3 seconds of arc. It is also better to express the effect of small light sources, for which RICCO's law holds, as an illuminance value at the eye. Assume that the illuminated specularly reflecting satellite is observed against the zenith sky of the low luminance of 127 cd/ft² (1366.5 cd/m²) (as measured by RICHARDSON and HULBURT [12] at 670 m above sea level, on a sun 30° above the horizon). Neglecting attenuation, the illuminance E at the observers eye produced by the satellite from its nearest possible distance of 322 km equals 0.019 lm/km². The illuminance at the eye produced by the sky would be equal to $\pi \times 1366.5$ lm/m² or 4.29×10^9 lm/km². Since an additional illuminance of 0.019 lm/km² is too small to be perceived even under the assumed favorable conditions (i.e. much smaller than any reported values for the WEBER fraction) we must conclude that it is impossible to see the satellite with unaided eyes during daytime.

Time Relationship for Tracking a Satellite

The time of valitidy of the *s.p.v.*'s depends on the season and on the geographical latitude. We may assume that the 20 inch specularly reflecting satellite crosses the zenith at evening twilight time, around autumnal equinox 1957. From astronomical tables [13] where sunset is assumed at $\theta = -50'$ and the end of twilight time at $\theta = -18°$ we find a twilight duration of 80 minutes at the autumnal equinox, for a latitude of + 30°. The time of validity of the *s.p.v.*'s for $\theta = -14°$ to $\theta = -19°$ then equals from 65 to 88 minutes after sunset (see also table V). Within these 23 minutes the satellite may become visible for a few seconds only, depending on the height of its orbit. When we assume that at $\theta = -15°$ it travels through the whole 200 km diameter of the *s.p.v.* with a speed of about 7 km a second, that would last 27 seconds which is not much time for scanning.

The satellite will appear as a fast moving star. In some situations, our central and peripheral vision is more adapted to detect moving than stationary objects. The threshold for detection of motion is lower in the presence of stationary reference objects, which in this case will be the stars.

It is obvious that we should know beforehand when it is reasonable to search for the satellite. When we know the time of sunrise or sunset, the duration of twilight, the approximate height and direction of the satellite's orbit and its time of crossing the sky, we can figure out by means of the spaces of potential visibility wether it would be worthwhile to try to see the satellite.

Acknowledgment

It is pleasure to express my gratitude to Dr. HUBERTUS STRUGHOLD for the suggestion to this paper, to Dr. GORDON G. HEATH for critical advice and to Mr. FREDERICK BLACKWELL for help with the computations.

References

1. J. A. Van Allen, Scientific Uses of Earth Satellites. Ann Arbor: Univ. of Michigan Press, 1956.
2. R. Tousey, Astronaut. Acta **2**, 101 (1956).
3. R. Tousey, J. Optic. Soc. Amer. **47**, 261 (1957).
4. F. S. Johnson, J. Meteorology **11**, 431 (1954).
5. H. Strughold, J. Aviat. Med. **25**, 420 (1954).
6. E. O. Hulburt, J. Optic. Soc. Amer. **28**, 227 (1938).
7. G. Hass and N. W. Scott, J. Optic. Soc. Amer. **39**, 179 (1949).
8. Handbuch der Astrophysik **2**, 1. Grundlagen der Astrophysik II, **1**, 63. Berlin: Springer, 1929.
9. M. J. Koomen, C. Lock, D. M. Packer, R. Scolnik, R. Tousey, and E. O. Hulburt, J. Optic. Soc. Amer. **42**, 353 (1952).
10. W. E. K. Middleton, Vision through the Atmosphere. Univ. of Toronto Press, 1952.
11. H. R. Blackwell, J. Optic. Soc. Amer. **36**, 624 (1946).
12. R. A. Richardson and E. O. Hulburt, J. Geophys. Research **54**, 215 (1940).
13. American Ephemeris and Nautical Almanac for the Year 1957. Washington: US Govern. Print. Off., 1955.

Observations from the Manhigh II Balloon Capsule at 30 Kilometers[1]

By

David G. Simons[2]

(With 3 Figures)

This paper is a companion to the paper "Operation *Manhigh*," presented by Mr. Otto C. Winzen (p. 460). Mr. Winzen describes the preparation of the capsule and the flight operation as seen from the outside. This paper describes the flight as experienced by the pilot from inside the capsule.

Now that mankind has accomplished a major step toward the conquest of space by establishing an artificial earth satellite, the problems of placing man into space have become of immediate interest and of primary importance.

It was necessary to provide protection against space equivalent conditions to accomplish the *Manhigh II* flight which was used to study the environment itself and human reactions to it. Four factors were of particular interest. First the sealed cabin was studied with particular emphasis on selection and maintenance of its atmosphere. Second, the effects of isolation were of interest. Physically, the pilot observed the earth from a unique point of view. He required considerable time and planning to return from the strange looking "space world". Psychologically, he was completely alone in a truly hostile environment. Third, the effects of heavy primary cosmic rays on a human being were compared to the effects previously observed in animal experiments. Fourth, the flight provided an opportunity to learn what manner of work load and what problems arose when one attempted to conduct experiments and perform exacting tasks under conditions psychologically and physically similar to those expected in satellite flight. The only major difference from satellites was weightlessness.

Sealed Cabin

To insure that the capsule was pressure tight and would operate as a terrella, as a truly sealed cabin, the capsule was tested on the ground before flight to prove that it would lose less than 1 Kg per square cm per 24 hours when pressurized to 15 Kg/cm^2 above atmospheric. The final selection of an atmosphere represented a compromise among the conflicting interests of providing a "normal" sea level atmosphere and the hazards of hypoxia, bends, and fire. The pilot was permitted to remove the plastic face piece of the pressure suit helmet during the flight. He kept it close at hand at all times ready to clamp into position in the event of sudden loss of pressure. The total capsule atmospheric pressure of

[1] The ideas expressed are those of the author and do not constitute official Air Force statements.

[2] Major, USAF (MC), Chief, Space Biology Branch, Aero Medical Field Laboratory, Air Force Missile Development Center, Holloman Air Force Base, New Mexico, U.S.A.

6 Kg/cm^2 selected consisted of 60 % oxygen and 20 % each nitrogen and helium. This combination provided sufficient oxygen to prevent hypoxia. This also maintained the inert gas component at a sufficiently low partial pressure that there was no danger of dysbarism (bends) during ascent or in the event of sudden decompression requiring use of the partial pressure suit. Special precautions were taken to prevent a fire from starting because of the increased rate of burning observed [1].

The balloon capsule system had many safety and emergency devices engineered into it. The capsule was suspended from the balloon by an unpacked parachute that could be released from its attachment to the balloon. The release could be actuated by a switch within the capsule and by radio control from the ground, should an emergency occur. In addition to the capsule parachute, I carried a 24 foot pack parachute which was normally stowed in front of me out of the way. However, it could quickly be clipped on in an emergency situation.

Isolation—the Flight

The physical situation on this balloon flight was directly comparable to a manned satellite flight in many ways. The sky appeared more different than anything I had seen before, and should have been quite similar to what it would

Fig. 1

look like from the satellite altitude of 300 km. Undoubtedly, subconscious recognition of the fact that my existence at altitude depended upon proper functioning of the mechanical equipment or my prompt detection of any malfunction significantly influenced feelings and performance. Hours of experience as a balloon pilot had been required to initiate and execute a normal descent and landing. Likewise, in satellite flight, descent and recovery will be the most critical phases of flight when crew members are most exhausted. The exact nature and impact upon performance of these subtle stresses when combined with fatigue over long periods of time create a situation that cannot be effectively duplicated in the laboratory.

I was sealed within the capsule at 2243 hours. This was done at the Winzen Research Inc. plant so that the capsule could be fully checked for leaks and

Fig. 2

proper functioning of all components before flight. In the event of difficulty, adequate facilities were immediately available for making any necessary modifications or repairs. This was critically important because of the strict schedule limitations imposed by the exacting weather requirements of a balloon launch. After the capsule had been assembled and checked, it was loaded on a truck and moved to the open pit iron mine launch site at Crosby, Minnesota, 140 miles away.

At the mine, final adjustments and assembly procedures were completed while the balloon was being inflated. Over two years of careful planning and testing had proceeded this moment. I had complete confidence in the skill of the designer, Mr. WINZEN, and the engineers who fabricated and tested every component of the capsule. It was with an intense feeling of anticipation, and determination to do my part as well as all of the other members of the team had done theirs, that the moment of launch arrived, 0926 in the morning.

The balloon ascended rapidly at slightly more than 300 meters per minute, quickly leaving the clouds below, rising toward the ever darkening sky.

During ascent, the small bubble of gas at the top of the enormous (61 meter diameter) plastic balloon served as a knob from which the flaccid folds of balloon draped and fluttered restlessly in the air stream. Every increment of approximately 5.5 km gained in altitude caused the volume of the gas in the balloon to double. This effect was not particularly noticeable until the balloon approached ceiling altitude. While rising the last 5.5 km, the balloon expanded from half full to a complete 61 meter diameter globe. It was an impressive sight to see the balloon fully inflated, a shining ball against a dark, dark blue-purple back-

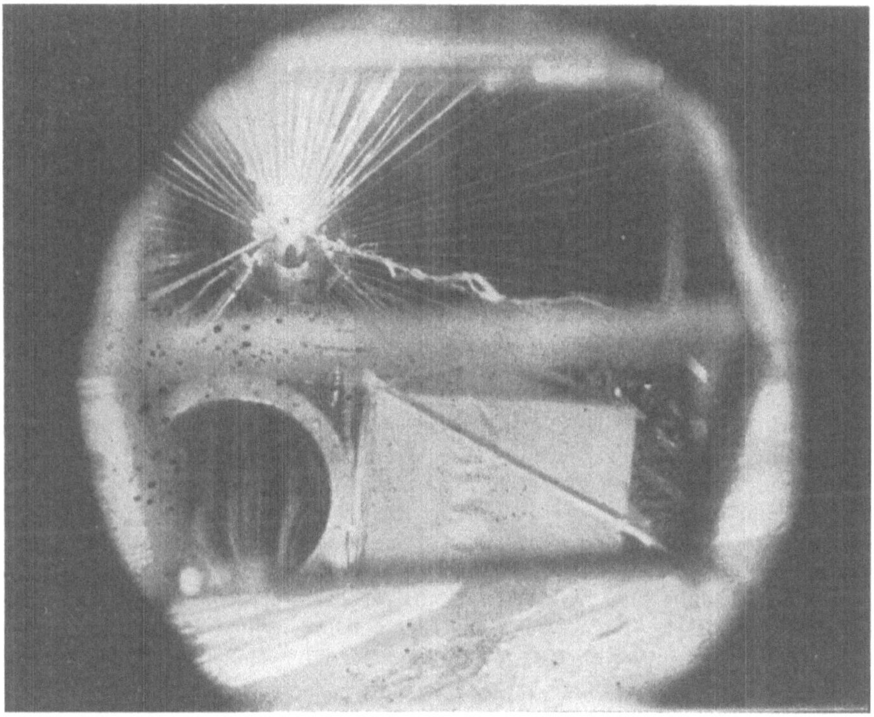

Fig. 3

ground, each seam tracing a harmonious curve that stretched upward out of view. The balloon reached ceiling two hours and fourteen minutes after launch.

The appearance of the sky at 30.5 km was startingly different. Studied carefully, the portion overhead looked deep blue-purple. When observed in a view that included the dazzling brightness of the earth and clouds below, the sky appeared completely black, or, more accurately, absent. Along the horizon, a narrow white zone separated the barely distinguishable horizon itself from the thin band of blue that represented all that was left of the blue sky to which we are accustomed. The blue color of this band shaded continuously from the white zone below to the blue-purple sky above through an arc of approximately 5°.

The brilliant earth below, limited by bands of color crowding close to the horizon leaving a blank blackness above, gave me the feeling that I was suspended high in an open dish with only a distant rim like a low fence marking its edge. This effect and the width of the white zone is illustrated in Figs. 1 and 2. The unmistakingly but faintly curved horizon made that portion of the earth which was visible from the balloon capsule look like a small part of a gigantic globe. Instead of seeming smaller and more distant from this high vantage point, the earth seemed incomprehensibly enormous compared to man's puny insignificance. It is an humbling experience to see 640 km in any direction and still just barely perceive the roundness of our planet.

One of the most memorable features of the flight was the appearance of the clouds from 30.5 km. During the middle of the afternoon, large fields of cumulus clouds appeared far below as tiny puffs of cotton dotting the earth. Their appearance is shown in Fig. 1. As the afternoon progressed, it became apparent that the cloudiness on the western horizon was the eastern border of an approaching cold front. Fig. 2 shows the broad expanse of the cloud system. Alto-stratus clouds cover the foreground with cumulonimbus buildups forming deeper into the system. Even before sunset, occasional flashes of lightning illuminated the interior of these clouds, indicating that it contained active thunderstorm areas. The frontal system was moving from northwest to southeast. The balloon and capsule, however, were drifting from east to west which is typical of winds above 24.4 km at middle latitudes during the summertime. With luck the balloon should pass over the storm area during the night. Near sunset there was a cumulo-nimbus buildup of clouds that towered above the blanket of clouds along the horizon directly to the southwest. It was later established that this correlated closely with an unusually active thunderstorm system over Denver, Colorado, 800 km away.

Fig. 3 was taken through the porthole that had a mirror system mounted just outside. The upper half of the device permitted vision vertically up toward the balloon, and the lower half down to the ground directly beneath the capsule. It had two mirrors placed at 45° to the horizontal. In the upper portion of the picture, looking through the up mirror, the segmented balloon, lighted by the orange rays of the setting sun, floats above the capsule not quite full because the gas has cooled and contracted slightly. The earth as seen through the down mirror has a slight bluish haze similar to the appearance of mountains viewed far in the distance. Above 24 km the reds and greens had a faded appearance. None of the objects on earth were distinguished by sharp, clean, bright colors from 30 km.

At the time of sunset, the balloon began to lose altitude rapidly, so it was necessary to release several exhausted batteries which had been intended for use in this manner as ballast. Approximately half an hour after sunset, what had been the white zone immediately above the horizon now became salmon pink. This pink graded smoothly into a vibrantly luminous blue. Above this crescent of blue, the sky faded quickly into the black of night, studded with a field of steady, brightly burning stars. It was such a strange and beautiful sunset that it seemed as if it must belong to another planet.

During the night the capsule settled to 24 km where it remained until sunrise. Then the sun warmed the helium, expanding it so it could again lift the balloon to nearly 30 km. The sunrise as seen from 24 km was very similar to the sunset of the night before, except that the bands of color, the pink above the horizon and the luminous blue between the pink and the dark sky above, both appeared broader. The first rays of the sun to stream through a small *v*-niche in the cloud bank formed a brilliant green flash.

After taking pictures and making observations of the sunrise, it was time to relax and enjoy breakfast. After a diet of sandwiches and candy bars, a canned Air Force in-flight meal was welcome. In fact, up to this time intense dedication to making observations or fatigue completely displaced any thought of food until the flight surgeon, Colonel STAPP, called every six hours and directed that it be eaten to maintain an adequate blood sugar level.

Throughout the early morning, the flight remained above the thunderstorm area. By 1000 hours, the western boundary of the storm area became dimly discernable. Having regained altitude, the balloon again progressed westward across the storm. By noon, it was time to release helium through the electric valve located in the apex of the balloon. By patiently continuing to valve until the desired descent rate was established, the capsule and balloon dropped quickly through strong westerly winds at 10 km. If the balloon delayed at that level, they would carry it back to the storm area before landing.

As the capsule touched the ground, the release device separated the balloon from the parachute so that the ball would not drag the capsule along the ground. In a very few minutes, helicopters and aircraft which had been faithfully following the balloon throughout the flight landed close by. The clamp release system worked perfectly. It was a delicious sensation to stretch my legs on *terra firma* beside the capsule.

Cosmic Radiation

Animal experiments [*2*] to determine the biological effects of heavy primary cosmic radiation showed that the hair of black mice frequently turned grey after exposure above 27 km for 24 hours. To compare this effect on the *Manhigh* flight, three nuclear track plates measuring 3" × 4" were taped to my body on my arms and chest. The corners of the plates were oriented over the skin by tatoo marks. Thus the track plates could be repositioned over the monitored areas after flight. No grey hairs corresponding to penetrations of heavy primaries through the nuclear track plates have been observed. Two grey hairs have appeared on the right wrist. However, the hair pattern of mice is much more dense than human skin.

In addition, experiments were conducted to detect the multi-nucleated lymphocytes response as compared to the response observed among cyclotron workers who have received measurable exposure to neutrons and other high energy particles. The results of this study are not yet available because there has not been time to read the large number of slides required.

No sensations or sensory phenomena that one might attribute to heavy cosmic ray primaries were observed.

Performance

The effectiveness of pilot performance throughout the flight is being assessed in terms of quantity and quality of observations made on the tape recorder, in terms of the accuracy of verbal reports submitted over the radio as compared to photo panel data, and in terms of the errors in logic and judgment made during the flight.

A series of many scientific observations were scheduled during the flight. They included:

a) Observations of the capsule stability throughout various phases of flight;

b) Observations of the sky brightness comparing its visual appearance to actual measured brightness using a spot brightness meter corrected for photopic vision;

c) Observation of the temperature changes in the capsule throughout the flight;

d) Observation of cloud patterns looking toward the earth;

e) Astronomical phenomena at night such as the aurora borealis and the appearance of stars.

I was expected to monitor the capsule atmosphere in terms of total pressure and oxygen partial pressure, as well as to take measurements of the carbon dioxide percentage to detect its buildup if it should occur.

With just a few exceptions, at least ten minutes of every hour were taken up preparing and sending half hourly reports of the physical conditions in the capsule and remaining supplies such as oxygen. These reports were made to the control van where Mr. WINZEN constantly monitored the progress of the flight and directed the movements of the ground and air support team.

Periodic reports of subjective reactions and impressions were scheduled throughout the flight.

In addition to learning much about the problems of living for a period of more than 24 hours under space equivalent conditions, the flight demonstrated the feasibility of using a manned balloon for making many kinds of observations from a new vantage point. The capsule proved remarkably stable while floating at altitude and should provide an excellent platform for studying planetary and solar detail and for making spectroscopic observations.

References

1. D. G. SIMONS and E. R. ARCHIBALD, Selection of a Sealed Cabin Atmosphere. (Submitted to J. Aviat. Med. for publication.)
2. D. G. SIMONS and C. H. STEINMETZ, The 1954 Aeromedical Field Laboratory Balloon Flights. J. Aviat. Med. 27, 100 (1956).

The Probability of Intelligent Life Evolving on a Planet

By

Alan E. Slater[1], BIS

Abstract — Zusammenfassung — Résumé

The Probability of Intelligent Life Evolving on a Planet. In order to estimate the likelihood of intelligent life evolving on a planet on which life has started, this paper traces briefly the line of evolution of the human species from primitive life forms, and, taking examples from animal groups at different stages along this sequence, shows that, as each group differentiated along various evolutionary paths, the particular one which eventually led to man was far from inevitable and, in fact, the chances were often against it starting at all. It therefore seems improbable that other forms of intelligent life will be found by voyagers through space.

Die Wahrscheinlichkeit der Entwicklung von intelligentem Leben auf einem Planeten. Um die Wahrscheinlichkeit der Entwicklung von *intelligentem* Leben auf einem Planeten zu beurteilen, auf dem Leben bereits entstanden ist, behandelt die vorliegende Arbeit kurz die Entwicklungslinie der Spezies „Mensch" aus primitiven Lebensformen. Durch Heranziehung von Beispielen für Tiergruppen in verschiedenen Stadien entlang dieses Entwicklungsablaufes zeigt der Verfasser, daß, da jede Gruppe sich entlang verschiedener Entwicklungswege differenzierte, die Entstehung der bevorzugten Gruppe, die vielleicht zum Menschen führen konnte, weit entfernt davon war, unabwendbar zu sein, und daß in der Tat die Chancen oft überhaupt *gegen* das Einsetzen dieser besonderen Entwicklung standen. Es scheint deshalb unwahrscheinlich, daß Raumfahrer andere Formen intelligenten Lebens finden werden.

Probabilité de l'éclosion d'une forme intelligente de vie sur une planète. Pour évaluer la probabilité d'existence d'une forme intelligente de vie sur une planète, cet article passe brièvement en revue l'évolution de l'espèce humaine depuis les formes primitives de vie. A partir d'exemples de branches animales, prises à différents stades de cette évolution, on peut voir que, durant le processus de différentiation de chaque groupe le longs de chemins évolutifs différents, la branche qui a éventuellement donné naissance à l'homme n'a pas un caractère inéluctable. En fait les circonstances mêlées à son éclosion étaient souvent adverses. Il semble par conséquent improbable que d'autres formes intelligentes de vie soient découvertes par les voyageurs de l'espace.

At the 1956 International Astronautical Congress, A. G. Haley laid down the laws which should govern human relations with sapient beings on other planets, and H. Strughold quoted an estimate by Shapley that "life of a high order" will have developed on only 100,000 planets in the entire universe. The present paper associates these two ideas, and is an attempt to estimate the likelihood of there existing on any other planet the kind of beings to whom visitors from Earth could apply Haley's rule: that the Earth space-ship must not land unless it has been invited to do so.

[1] Dell Farm, Whipsnade, Dunstable, Beds., England.

SHAPLEY's views, it must be mentioned, are not fully reported in either of the two papers quoted, though referred to in both. STRUGHOLD does not make it clear that SHAPLEY's estimate is a very conservative minimum, based on the assumption of no more than 10^{20} stars in the whole universe, of which only one in a million has planets; further, that only one planetary system in a thousand includes a planet suited for life, and on only one in a thousand of these suitable ones will higher life forms develop.

This last point does not accord with HALEY's quotation of SHAPLEY as saying that "we can no longer doubt whenever the physics, chemistry and climates are right on a planet's surface, life will emerge and persist." As to this, it must be pointed out that the continents of our Earth possess a physics, chemistry and climate eminently suited for life, as anyone can see; nevertheless, life did not start there, but had first to spread in from the sea.

The main purpose of the present paper is to assess the probability of intelligent life by tracing the evolution of man, the only intelligent species we know, from primitive animal forms, by taking examples from animal groups at various stages along the evolutionary sequence. After each stage was reached, the group differentiated along various evolutionary lines, but only one of these lines led eventually to man, and the probability of its arising from that particular group will be our particular concern.

Classification of Life Forms

For the purpose of classification, life is divided and subdivided into the following categories: Kingdom, Phylum, Class, Order, Family, Genus, Species. And, in order to cope with the diversity of life, prefixes such as "super", "sub" and "infra" can be added to these names. Man, with whom we are chiefly concerned, is classified thus:

Kingdom	Animal
Phylum	Chordate
Subphylum	Vertebrate
Class	Mammal
Order	Primate
Family	Hominid
Genus	Homo
Species	sapiens

Let us follow through from top to bottom of this list, starting with life itself.

Life

An often-used argument is: "You talk about life, but all you mean is life as we know it." How true! After all, the word "life" was invented to describe life as we know it, and for no other purpose. Life as we know it, according to the most recent investigations, appears to be based primarily on desoxyribonucleic acid (DNA for short), the molecule of which genes, the units of heredity, are composed. This molecule, it is believed, can perform two essential functions: [1] it can produce an exact replica of itself, and [2] it, and its near relation ribonucleic acid (RNA), can serve as templates for the formation of protein molecules by causing the joining up of various amino-acids, the units of which proteins are composed. These proteins include particularly protein enzymes, which regulate the growth and maintain the life of the organism. All forms of life on the Earth are based on the DNA molecule, so far as can be discovered.

Those who think there may be other forms of life than "life as we know it", elsewhere in the universe, usually assume that their biochemistry will be entirely different from ours, and often go on to suggest that their evolution could progress to more and more complicated organisms. But even in earthly life, based on nucleic acids and protein, there is an astonishing diversity of basic biochemistry in some primitive organisms; among the bacteria there are kinds which oxidize ammonia to nitrite, nitrite to nitrate, ferrous to ferric compounds, or live mainly on sulphur, or hydrogen, or hydrogen sulphide, or methane, and so on. Yet none of these has evolved into any higher form of life using the same chemical reactions. As STRUGHOLD points out, only a biochemistry based on oxygen can provide enough energy to sustain the higher life forms.

Origin of Vertebrates

The main division of the animal kingdom is into phyla, the animals in each phylum being constructed on a basic plan different from that of all other phyla. The diversity of life is reflected by a diversity of opinion on how it should be classified, and the number of phyla given by different authorities varies from about 12 to 30; SIMPSON gives 15. Some were the ancestors of others, but all, with one exception, have carried on in their own right to the present day. Only one of the 15 (or 12, or 30) gave rise to intelligent life; the remaining 14 (or 11, or 29) did not succeed in doing so, even with hundreds of millions of years at their disposal. Our own phylum is the Chordata; it consists mostly of vertebrates with jointed backbones, but includes some primitive forms with only a stiff rod to support the back. Its relation to the other phyla, in the following account, is based on the work of ROMER, YOUNG and SIMPSON.

The earliest phylum consists of the Protozoa, which are single-celled organisms; traces of them have been found in Pre-Cambrian rocks, into which their remains were incorporated more than 500 million years ago. From them evolved the earliest many-celled "higher" forms of life, which were mainly of two sorts: the sponges, with rather vaguely differentiated cells, and the Coelenterata, with two definite cell layers, inside and out (now represented by, e.g., jellyfish and corals). Only the second of these phyla gave rise eventually to intelligent life; the first has never, even to this day, developed a nervous system. So nerves, of which brains are composed, are not a necessary consequence of the evolution of many-celled organisms.

From the coelenterates several phyla with more than two cell layers evolved and were well established at the start of the Cambrian period, 510 million years ago. Many of these phyla were worm-like creatures, but two in particular gave rise to abundant descendants. One, the Annelida or segmented worms, branched into two more phyla: Mollusca (represented to-day by 80,000 species, including snails, oysters, octopus, etc.), and Arthropoda, subdivided into insects, crustaceans, spiders, etc. (the most abundant phylum, now represented by 800,000 known species, of which 625,000 are insects). The other main branch from the coelenterates gave rise first to the Echinodermata (starfish, sea urchins, etc.), and then, at last, apparently branching off from the earlier echinoderms, the Chordata (now represented by 40,000 species and therefore the third most successful phylum, if that is a measure of success).

These two main branches from the two-layered coelenterates differ fundamentally in many ways. In the first branch segmentation of the body begins at an early stage of subdivision of the fertilized ovum; in the second, no more than parts of the body ever become segmented. If only the first of these types of body structure had been tried out by the evolutionary process, instead of both, there would never have been intelligent life on the Earth.

When we try to assess the chances of the two-layered Coelenterates giving rise to a phylum which would lead to intelligent life, we are faced with the fact that, not only was the chordate phylum only one out of a dozen of those with multi-layered cells, but it was apparently the last phylum to emerge—almost as if it was an afterthought. The first sign of it in the fossil record is found in the Ordovician period, which started nearly 100 million years after the beginning of the Cambrian, when nearly all the other phyla were well enough developed to leave fossils (which means, mainly, that they had developed hard parts). If there was any sort of inevitability about the development of vertebrates, why was there no sign of them during nearly 20 % of the whole span of existence of higher life forms?

Spread to Land

Life had to spread from the sea to the land before it could develop a form from which intelligent life could later evolve. Yet, among the forms of life which stayed in the sea, those with the most highly developed brains were not evolved either among the arthropod phylum nor even the chordates (from which man's ancestry derived), but among the molluscs, and then only in one class of that phylum, the cephalopods, and in one of four orders of that class, the octopods. So, although the chordate phylum evolved the best brains on land, it did not do so in the sea; that is, no phylum has an inherent capacity for evolving a better brain than the others in every environment.

Was it inevitable that higher forms of life should spread to dry land, once land had appeared on our planet? The earliest evidence of animal life on land comes from the later part of the Silurian period, which ended 300 million years ago. Yet (according to a table given by YOUNG) the amount of land on the Earth had varied between 4,000,000 and 8,000,000 square miles throughout the preceding 200 million years during which higher life forms flourished in the sea without obtaining a foothold ashore.

There are three significant points about the way in which vertebrates came to colonize the land. Firstly, only one of the four classes then existing did so—the bony fishes. The other three—cartilage fishes, jawless fishes and placoderms—stayed in the sea. (Incidentally, the bony fishes had less well developed brains than the cartilage fishes, showing once again that evolution towards a bigger brain did not travel along a single straight line, nor was a high rate of evolution along a particular line any guarantee that it would ever lead to a brain of human standard.)

Secondly, these bony fishes did not simply spread up the beach, but, according to the available evidence, evolved in freshwater rivers; so their only use for the land, at first, was to provide fresh water for them to live in. Thirdly, when they developed primitive lungs, they used these organs, not for living entirely on land, but to survive droughts, which became more frequent during the Devonian period which followed the Silurian.

The arthropods which colonized the land, such as insects and spiders, have never managed to evolve lungs, though they have had 300 million years in which to do so. The best they could do was to develop a breathing system consisting of narrow tubes leading directly from the surface to the tissues. Since oxygen has to travel along these tubes by diffusion, and cannot do so efficiently for more than a quarter of an inch, insects have never been able to grow big enough to accommodate anything that could be called a brain.

So animals which come ashore in the course of evolution do not necessarily develop lungs. Not only that, but a large proportion of those who did so returned to the water as their evolution progressed. These bony fishes divided into two sub-classes; one of these, the ray-finned fishes, instead of improving their lungs for use on land, converted them into air bladders for regulating their depth in the water, and most of to-day's fishes are their descendants. The other sub-class divided into two orders, one of which is represented by the few "lung-fish" which still survive to-day in countries liable to drought, and the other, the lobe-finned fishes, gave to rise to amphibians by evolving legs from their fins. But the lobe-like fins only happened to be available for this purpose because, HUXLEY says, they were useful for resting on mud at bottom of stagnant water where the food supply was—a purpose quite different from walking.

Although these amphibians became our ancestors, their legs, ROMER points out, were not developed for leaving the water, but for getting back to it by travelling from a pool that had dried up to a pool that had not. He adds: "true land life seems to have been, so to speak, only the result of a happy accident."

Reptiles to Mammals

Little is known of how reptiles evolved from amphibians during the Carboniferous period (270 to 210 million years ago). But after they had done so, they became exceedingly successful, mostly on land, though some returned to the water or took to the air. From the reptile class two other classes evolved independently: the birds, which are not descended from flying reptiles; and the mammals, which evolved from only one of the 16 orders into which reptiles are classified. This order, the therapsids, arose about 200 million years ago but only lasted 50 million years, and it was lucky that mammals were able to evolve from it during that time, for they could never do so again, now that it has become extinct. In fact, the mammals were doubly lucky, for they managed to keep going in spite of having to lie low for nearly 100 million years in order to keep out of the way of the larger reptiles, which only became extinct towards the end of the Cretaceous period, 70 to 60 million years ago.

Mammals are an improvement on reptiles in many ways, one of which is warm-bloodedness—the ability to maintain a constant temperature so as to be always ready for activity, however cold the environment. Yet the big reptiles flourished because the climate was warm, so what was the advantage of warm-bloodedness to the early mammals? The writer's guess is that it enabled them to be more active in the cool of the night, to which most of their activity had to be confined until, with a change of climate to cold and dry, the reptiles' domination ceased. After that, mammals were able to expand and progress because, by the fortunate circumstance of having already developed warm-bloodedness, they were able to be just as active in the daytime in spite of the cold.

Lest anyone should think that the animal brain has an inherent tendency to advance towards a higher level of organisation, it should be pointed out that the only improvement in the basic reptile brain over the brains of bony fishes was to the extent needed for better control of its limbs; there was no development of the "higher" centres. And the earliest mammals, even after the demise of the large reptiles, still had only a reptilian type of brain; only later did they come to include the brainiest creatures on Earth; and, even then, the proportionate enlargement of different parts of the brain varied in different mammalian groups. These facts should dispose of any argument that evolution of a human type of brain from the most primitive nervous system was continuous and therefore inevitable, or that its failure to appear hitherto in other animal forms was due to a need for 500 million years of continuous progressive evolution for its development.

Evolution of Primates

Mammals are subdivided into 29 orders, of which the Primates, which include Man, form one; but nearly half the remainder are extinct. An early "infraclass" of mammals, the marsupials, became separated from the rest on reaching Australia, but nevertheless managed to evolve animals resembling the moles, wolves, squirrels and others which had evolved independently in the rest of the world; but the kangaroos were the nearest they could get to horses, and nothing resembling a primate ever arose from marsupials, showing that primates were not a necessary consequence of the appearance of mammals on the Earth.

Primate evolution was adapted to life in trees; so, if there were never any trees on the Earth, there would presumably be no primates. (The climate of Mars resembles most nearly the tundra regions of the Earth, where no trees can grow.) But trees on a planet do not necessarily bring primates into being, for there were great forests in the Devonian period, more than 200 million years before the primates appeared.

Primates are divided into four groups: prosimians, South American monkeys, Old World monkeys, and the hominoids, which include apes and men. The last three branched off independently of each other from the early prosimians before evolving a better brain, and present-day prosimians, such as the lemur, tarsier, etc., still have only a primitive mammalian brain, showing that brain improvement is not a necessary consequence of life in trees.

Prosimians show little brain development beyond that of the earliest mammals. Some got cut off in Madagascar and evolved into nothing more than further prosimian types. Others were cut off in South America and gave rise to the "New World monkeys" with larger brains, but nothing approaching the hominoid level. Only those which had the rest of the world to play with were responsible for the Old World monkeys and the hominoids, the latter having still larger brains. It looks as if brain development was proportional to the area available for its evolution; but this could not have been the only factor, as we shall see.

Arrival of Man

The last stages in the evolution of man are incompletely known because of large gaps in the fossil record, but are usually classified as follows: the hominoid "superfamily", which split off from the early primates about 40 million years ago, became divided later into two families, the apes (Pongidae) and the Hominids. The hominid line, as far as fossil evidence goes, gave rise to the genus Australopithecus (fossils 700,000 to 540,000 years old), then the genus Pithecanthropus (550,000 to 400,000 years ago), and finally the Homo, represented by two species: Homo sapiens (may have started 200,000 years ago or more, but the only certain evidence dates back about 50,000 years) and Homo neanderthalensis, which may either have branched off from Homo sapiens or started at about the same time, but became extinct 100,000 years ago.

Various fossils dating from about 25 million years ago (early Miocene) give an idea what the unspecialized common ancestors of apes and men were like. Their anatomy suggests that they were active both in the trees and on the ground; and other fossil organisms from the same time, at least in East Africa, indicate a landscape of wooded valleys with open country between. LE GROS CLARK suggests that these early hominids became adapted to moving over the ground, not in order to stay there permanently, but so that they could go on living in the trees by being able to scamper across open ground from one shrinking woodland to another—in fact, it is the story of the early spread of vertebrates from sea to land, all over again in a different setting.

No such creatures as these exist to-day—a fact which has led HUXLEY to declare: "If man were wiped out, it is in the highest degree improbable that the step to conceptual thought would again be taken, even by his nearest kin. In the ten or twenty million years since his ancestral stock branched off from the rest of the anthropoids, these relatives of his have been forced into their own lines of specialization, and have quite left behind them that more generalized stage from which a conscious thinking creature could develop."

There is no further evidence of man's ancestry from that time until the Australopithecines arrived less than a million years ago. Their skeletons, which have only been found in South Africa, bear a close resemblance to that of man, and the bones of the hip region, especially, show that they walked almost upright. But their brains, with a volume of 600 c.c., were hardly larger than the 500 c.c. of the largest present-day apes, and less than half the size of the human brain which averages 1,350 c.c. They show that the hominids walked on the ground for over 20 million years without evolving the characteristic features of the human brain, which nevertheless could, and eventually did, evolve in less than a million years.

From Australopithecus to Pithecanthropus there is another gap in the evolutionary record, but no gap in time, for they overlapped. The remains of Pithecanthropus have been found in Java and China, and while the Javan brains varied in size between 775 and 900 c.c., as estimated from skull features, those of the Chinese specimens varied from 850 to 1,300 c.c., i.e., almost up to human size. Found with these remains were the earliest evidence of primitive tools, and of the use of fire.

So it appears that hominid brains were rapidly evolving towards the human type in one part of the world, while at the same time failing to do so in another, though both groups were living on the ground and walked upright, and were nearly-related descendants from hominid stock. Evidently all these characters, though necessary for evolution towards man, could not ensure it.

Next we have two forms of Homo, with equally large brains, living on the Earth simultaneously: Neanderthal Man, whose brain was less developed than ours in front, but larger at the back; and Homo sapiens, whose brain is distinguished above all by the great size of the frontal lobes, in which his unique capacity for conceptual thought is believed to reside. The remains of both have been found in Europe, and LE GROS CLARK believes that *neanderthalensis* was an errant form of *sapiens*, showing that even the human species is capable of regressing in intelligence.

As to the frontal lobes, it is fortunate that they were available in rudimentary form, having been developed to serve the sense of smell, according to HUXLEY, who suggests that the importance of sight, rather than smell, in tree-living creatures left frontal lobes free to expand for another purpose—the purpose for which we now use them.

The foregoing story seems to give little encouragement to any belief that intelligent life on the human level is likely to be found even once among the remainder of SHAPLEY's 100,000 planets. It should also be borne in mind that, among more than a million known species of animals, man is the only intelligent one. True, he was probably responsible for exterminating his Neanderthal relation, but there is no evidence of any other animal group evolving an intelligent species for him to exterminate. Further, he has not been on Earth with his present brain for more than a quarter of a million years, if that; so if any space travellers had visited the Earth during the past 500 million years, they had less than one chance in 2,000 of finding intelligence.

Finally, at what stage in human history could extra-terrestrial visitors put across a message describing who they were, and that they wanted permission to land? Less than 200 years ago, a balloon which landed in a French field was taken for a dangerous monster and attacked with pitchforks by people far more "civilized" than many races are to-day.

It is sad that, after reading HALEY's imaginative paper, a study of the evolution of man should have led one to doubt that we shall ever have a chance to put his high principles into effect. We would all like to meet intelligences from other planets, and, still more, to find that they have discovered the secret of building a perfect civilization. If anyone can prove convincingly that the reasoning of the present paper is wrong, nobody would be better pleased than its author.

References

1. A. G. HALEY, Space Law and Metalaw. Proceedings of the VIIth International Astronautical Congress, 1956, p. 435. Rome: Associazione Italiana Razzi, 1957.
2. H. STRUGHOLD, The Ecosphere in the Solar Planetary System. Proceedings of the VIIth International Astronautical Congress, 1956, p. 277. Rome: Associazione Italiana Razzi, 1957.
3. H. SHAPLEY, On Climate and Life (Chapter 1 in: Climatic Change, edited by H. SHAPLEY). Cambridge, Mass.: Harvard University Press, 1953.
4. G. G. SIMPSON, The Meaning of Evolution. London: Oxford University Press, 1950.
5. A. S. ROMER, Vertebrate Paleontology. Chicago: University of Chicago Press, 1945.
6. A. S. ROMER, Man and the Vertebrates. London: Penguin Books, 1954.
7. J. Z. YOUNG, The Life of Vertebrates. Oxford: Oxford University Press, 1950.
8. W. E. LE GROS CLARK, The Fossil Evidence for Human Evolution. Chicago: University of Chicago Press, 1954.
9. J. S. HUXLEY, Evolution — The Modern Synthesis, p. 571. London: Allen & Unwin, 1942.

Design and Performance Data of Space Ships with Ionic Propulsion Systems[1]

By

Ernst Stuhlinger[2], ARS

(With 5 Figures)

Abstract — Zusammenfassung — Résumé

Design and Performance Data of Space Ships with Ionic Propulsion Systems. The basic relations between component masses and performance figures of electrically propelled space vehicles can be derived if a few simplifying assumptions are made. Even though many of the technical details of electric drive systems are not yet known, it appears possible to estimate the take-off mass and the initial acceleration of a space vehicle as a function of its payload, the total propulsion time, and a few design parameters which may be treated as constants. Upper limits of accelerations obtainable with present technologies can be derived. Design data and performance figures of three different space vehicles are presented. On the basis of these results, interplanetary travel with electrically propelled space ships appears principally feasible.

Konstruktions- und Leistungsdaten von Raumfahrzeugen mit Ionenantriebssystemen. Die Grundbeziehungen zwischen den Massen der Komponenten und den Leistungszahlen elektrisch angetriebener Raumfahrzeuge können abgeleitet werden, wenn einige wenige vereinfachende Annahmen gemacht werden. Obwohl viele der technischen Einzelheiten elektrischer Antriebssysteme noch nicht bekannt sind, scheint es doch möglich, die Startmasse und die Anfangsbeschleunigung eines Raumfahrzeuges als Funktion seiner Nutzlast, der gesamten Antriebszeit und einiger Konstruktionsparameter abzuschätzen, die als Konstanten behandelt werden können. Obere Grenzen der Beschleunigung, die mit Hilfe der heute verfügbaren Technologie erhältlich sind, können abgeleitet werden. Konstruktionsdaten und Leistungszahlen von drei verschiedenen Raumfahrzeugen werden angegeben. Auf der Grundlage dieser Ergebnisse scheint der interplanetarische Verkehr mit elektrisch angetriebenen Raumfahrzeugen grundsätzlich ausfuhrbar zu sein.

Conception et performances de véhicules spatiaux à propulsion ionique. Les relations fondamentales entre masses et performances de fusées à propulsion électrique peuvent être établies à l'aide de quelques hypothèses simplificatrices. Quoique plusieurs détails techniques de la propulsion électrique soient encore inconnus, il apparait possible d'estimer la masse au départ et l'accélération initiale en fonction de la charge utile, de la durée de propulsion et de quelques paramètres, qui peuvent être considérés comme constants. Les limites supérieures d'accélération qui peuvent être obtenues par les techniques actuelles sont accessibles au calcul. Les données de conception et les performances de trois engins sont présentées. Sur la base de ces résultats, les voyages interplanétaires par fusées à propulsion électrique paraissent possibles en principe.

---

[1] Statements and opinions advanced in this paper are to be understood as individual expressions of the author and do not necessarily reflect the views and opinions of ABMA or the Ordnance Corps.

[2] Army Ballistic Missile Agency, Huntsville, Alabama, U.S.A.

1. Introduction

The possibility to electrical propulsion systems for space vehicles has been discussed frequently [1], but only few papers were published with details regarding design and functioning of the various components of an electrically propelled space ship [2]. One proposal [3] describes a complete space ship with ionic drive and nuclear power source, capable of going to Mars and back. In the present study, this investigation has been extended and refined.

Ionic propulsion systems require a power source of considerable magnitude, and they produce a low thrust for a long period of time. The most influential figure in the design of an electric space vehicle is the "specific power," defined as the available electric power divided by the total mass of the power plant, and measured in kilowatts per kilogram. A second figure of great influence is the "specific mass" of the thrust chambers, defined as the total mass of the thrust chambers as a function of the ion current and the current density, and measured in kilograms times volt $^{3/2}$ per ampere. These two design constants depend on available technologies and on engineering capabilities. Once they are established, the total weight, component weights, and performance figures of a space ship can be determined as functions of payload, total propulsion time, and specific charge ε/μ of the propellant element.

A design proposal of a nuclear power plant for electrical space ships, including turbo generator and radiation cooler, was described previously [3]. The present study is based on data obtained in that investigation. The arrangement of the various components within the vehicle is assumed to be the same as in the abovementioned report.

2. Basic Relations between Space Ship Parameters

The performance of conventional rockets can be characterized by the specific impulse (in kg thrust per kg of propellant consumption per second), and by

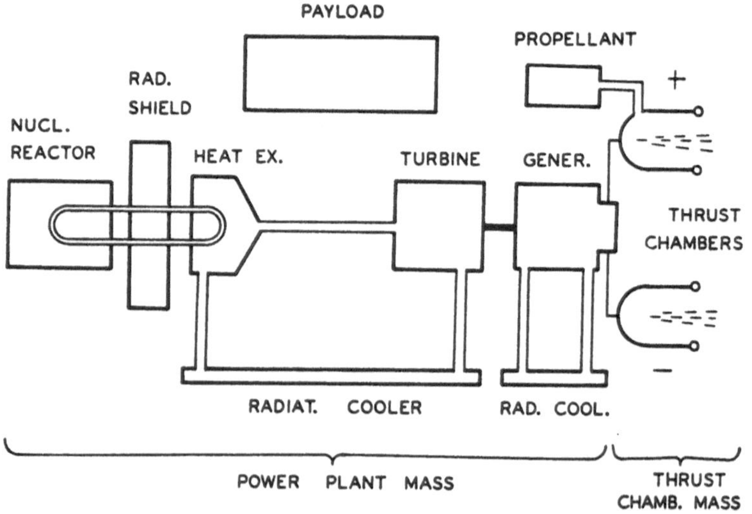

Fig. 1. Components of ionic space ship

the ratio of the initial mass of the rocket to its mass at burnout. Electrically propelled rockets have specific impulses which are 20 to 50 times higher than those of chemical rockets, but their mass ratios are considerably smaller. An electrical propulsion system requires a power plant to produce electric power, means to ionize the propellant, and a thrust unit in which the ions are accelerated by electric fields. The propulsion system will normally stay in operation for the whole duration of the flight. Thrusts obtainable from electric systems are not greater than a few kg to several hundred kg.

The most practical power source for an electric propulsion system is a nuclear reactor. It produces steam to drive a turbo-electric generator. The steam condenses in a radiation cooler from where the liquid is pumped back to the heat exchanger. The electric energy accelerates ions and electrons in a system of thrust chambers. A suitable propellant will be cesium because of its relatively large atomic mass and its low ionization potential. Cesium vapor, while passing through a gridwork of hot platinum foils, is ionized upon contact with the incandescent surfaces.

A block diagram of the electrical propulsion system is shown in Fig. 1. The various components of the system were described and discussed in detail in previous papers [3]. In these investigations, a characteristic design figure a was obtained which represents the "specific power" of the system, or the total output of electric power, L, divided by the total mass of the power plant, M_p,

$$a = \frac{L}{M_p}. \qquad (1)$$

This figure proves to be the most influential design parameter for an electric space vehicle.

The mass M_p includes the masses of reactor, shielding, heat exchanger, pumps, turbine, generator, working fluid, and radiation cooler, and also that of structural elements which are assumed to be proportional to the total power of the reactor. All of these individual masses are assumed to be proportional to the total electrical power output L. This is even true for the reactor because its mass will not be determined by the required amount of fissionable material, but by the amount of heat energy that can be removed per second from the unit reactor volume. Since this figure is practically a constant, the total mass of the reactor may be assumed proportional to its total power production. The efficiency of the power plant, as given by the ratio of electric power output to total reactor heat power, was assumed to be 20%, independent of the absolute amount of the reactor power.

The thickness of the radiation shield does not increase proportionally with the reactor power, but more slowly. Since "shadow shielding" will be applied, its area will be proportional to the reactor cross section. A further mass penalty must be paid for the cooling of the radiation shield, either in the form of radiating fins, or by including the radiation shield in the cooling system for the working fluid; the radiation shield will heat up considerably as a consequence of the absorbed gamma and neutron energies. It was assumed here that these effects together make the mass of the radiation shield roughly proportional to the total reactor power.

The mass of the thrust chambers requires special consideration[1]. Size, mass, and arrangement of the thrust chambers are governed by the space charge effects within the ion and electron jets. Unless the space charge of the ion beam is neutralized immediately behind the exit orifices of the thrust chambers by

[1] In the previous papers [3], this mass was considered proportional to the total electric power output.

combining it with the electron beams, no appreciable flow of charged particles out of the thrust chambers can be obtained.

The space charge law by Schottky and Langmuir

$$J = 1.77 \times 10^{-10} \sqrt{\varepsilon/\mu}\; U^{3/2}/x^2 = k_1\, U^{3/2} \tag{2}$$

relates the current density J of a beam of charged particles to the accelerating voltage U and the length x of the beam before neutralization. Supposing that

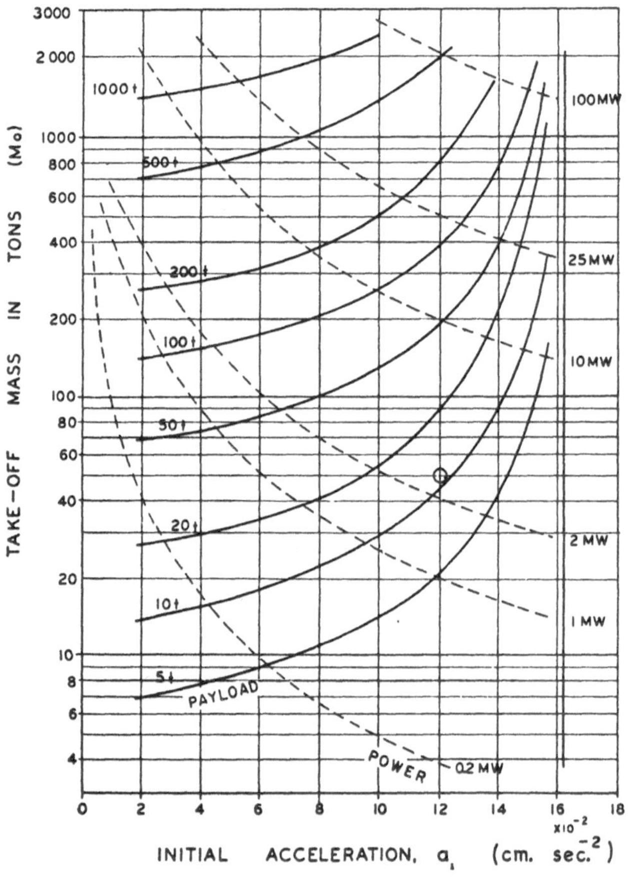

Fig. 2. Take-off mass as a function of payload and initial acceleration. Propulsion time: 10^7 sec

x is a design constant, the total mass of the thrust chambers, M_T, may be expected to be proportional to the total current I, but inversely proportional to the current density J:

$$M_T = k_2 \frac{I}{J} \tag{3}$$

where k_2 is a constant.

Since the current I is given by the total mass of propellant, M_F, the specific charge of the ions, ε/μ, and the total propulsion time τ,

$$I = \frac{M_F\, \varepsilon}{\tau\, \mu} \tag{4}$$

the mass M_T may be expressed by

$$M_T = \frac{k_2}{k_1} \frac{I}{U^{3/2}} = \beta \frac{I}{U^{3/2}}. \tag{5}$$

With $L = UI$ and Eqs. (4) and (5), the mass M_T may be expressed as

$$M_T = \beta \sqrt{(M_F \varepsilon/\mu \tau)^5 L^{-3}}. \tag{6}$$

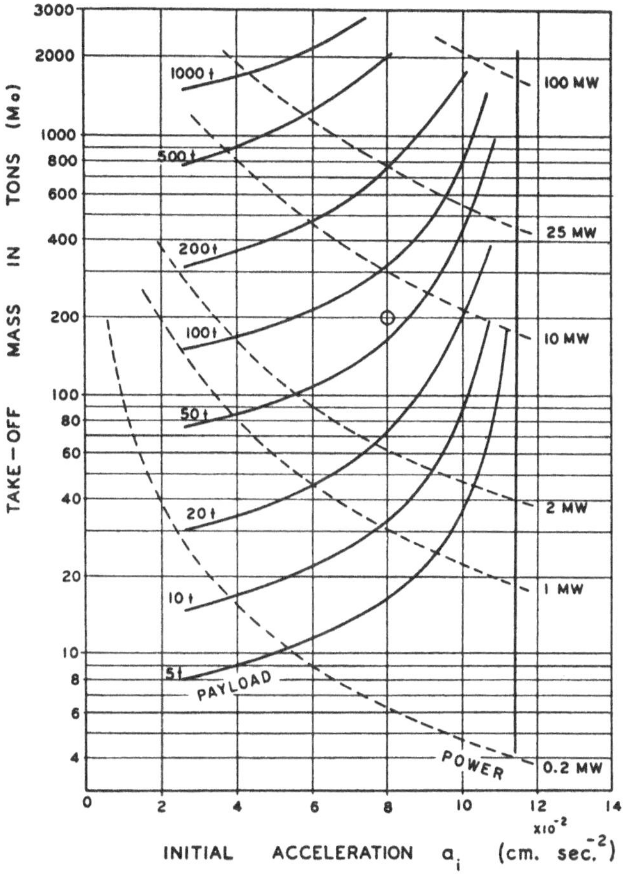

Fig. 3. Take-off mass as a function of payload and initial acceleration. Propulsion time: 3×10^7 sec

The constant β can be found by designing a representative thrust chamber for a selected current and current density, and then determining its mass from the design details.

The total initial mass of the ship, M_0, is given by

$$M_0 = M_L + M_F + M_P + M_T \tag{7}$$

where M_L is the payload; M_P includes the mass of the power plant and structural masses.

Combining Eqs. (1), (6), and (7), we obtain

$$M_0 = M_L + M_F + \frac{L}{a} + \beta \sqrt{(M_F \varepsilon/\mu \tau)^5 L^{-3}}. \tag{8}$$

Since the thrust of the ionic propulsion system, Th, may be expressed by

$$Th = \sqrt{2\, M_F\, L/\tau} \tag{9}$$

the initial acceleration a_i is

$$a_i = \sqrt{2\, M_F\, L/\tau}\, / M_0$$

or

$$a_i = \frac{\sqrt{2\, M_F\, L/\tau}}{M_F + M_L + La/\beta + \sqrt{(M_F\, \varepsilon/\mu\, \tau)^5\, L^{-3}}}\,. \tag{10}$$

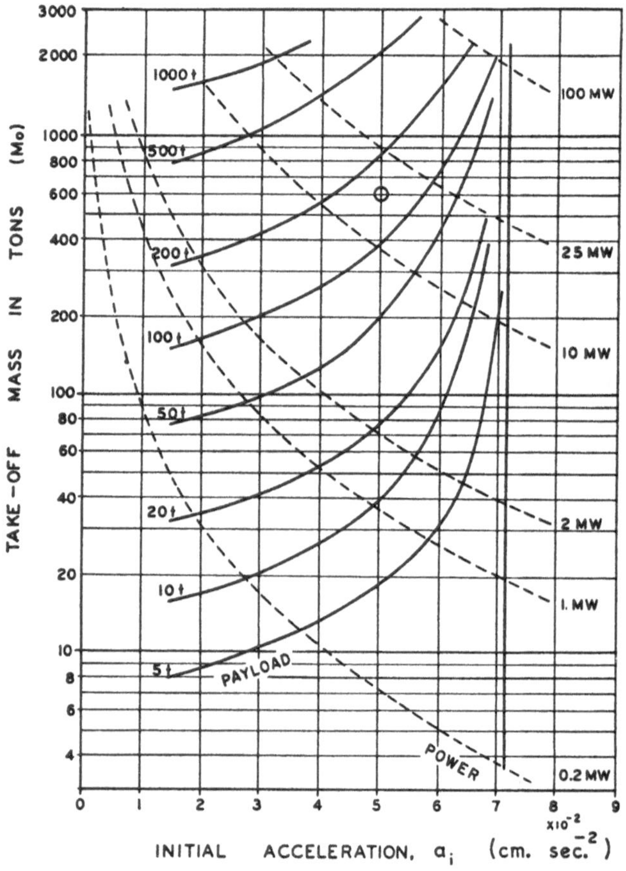

Fig. 4. Take-off mass as a function of payload and initial acceleration. Propulsion time: 10^8 sec

For selected values of travel time τ, power L, and propellant element, the optimum value for the propellant mass, M_F, can be found which provides a maximum acceleration.

Letting $\frac{da_i}{dM_F} = 0$, we obtain

$$M_F + 4\,\beta \sqrt{(M_F\, \varepsilon/\mu\, \tau)^5\, L^{-3}} - \frac{L}{a} - M_L = 0 \tag{11}$$

or

$$M_L + \frac{L}{a} - M_F - 4\, M_T = 0. \tag{12}$$

For any given set of M_L, L, τ, ε/μ, α, and β, Eq. (11) yields the amount of propellant, M_F, which makes the initial acceleration a maximum.

3. Numerical Examples

A large variety of the above 6 quantities were chosen, and the corresponding figures for M_F were obtained from Eq. (11) by graphical methods. With M_F found from Eq. (11), M_P from Eq. (1), M_T from Eq. (6), and M_0 from Eq. (7), the diagrams in Figs. 2, 3, and 4 were drawn. They show the total initial mass, M_0, of a space ship for which the initial acceleration, a_i, the payload, M_L, the travel time, τ, and the constants α, β, and ε/μ were selected. The total electric power L pertaining to any combination can be read from these diagrams.

Three representative examples were chosen for further illustration, one for a short trip (approximately 100 days), one for a medium trip (approximately 1 year), and one for a long trip (approximately 3 years). The three examples are indicated by circles in the 3 diagrams. Whereas the initial acceleration a is given by Eq. (10), the final acceleration at the end of propellant consumption, a_i, may be found from the equation

$$a_f = \frac{\sqrt{2\,M_f\,L/\tau}}{M_L + M_P + M_T}\,. \tag{13}$$

The electrical current is provided by Eq. (4), the voltage from

$$L = IU.$$

The thrust developed by the exhaust of the ions is expressed by Eq. (9). The exhaust velocity, v_{ex}, is simply

$$v_{ex} = \sqrt{2\,L\,\tau/M_F}\,. \tag{14}$$

The specific impulse, S, is related to the exhaust velocity by the equation

$$S = v_{ex}/g \tag{15}$$

where $g = 981$ cm/sec^{-2}.

The characteristic velocity v_c is found from the equation

$$v_c = v_{ex} \ln \frac{M_0}{M_0 - M_F} \tag{16}$$

where $\frac{M_0}{M_0 - M_F}$ is the mass ratio.

A number of design figures pertaining to the three selected samples are listed in Table I.

As implied by Figs. 2, 3, and 4, the initial acceleration obtainable with a given combination of propulsion time τ, specific ion charge ε/μ, specific power α, and specific thrust chamber mass β, approaches a finite value when either $M_L \to 0$ or $M_0 \to \infty$. This highest initial acceleration a_i', as a function of τ, is plotted in Fig. 5.

With decreasing τ, a_i' increases, but reaches a finite value for $\tau \to 0$. Curve I in Fig. 5 refers to Cs as propellant, and to an α and a β as computed previously (Ref. [3]). Curve II was calculated for Rb instead of Cs. For curve III, Cs was chosen for propellant, but a specific power α' which is twice as good as α,

$$\alpha' = 2\alpha.$$

Table I. *Design and Performance Figures of Three Representative Space Ships*

Specific power of powerplant	α	0.1	kw kg^{-1}		
Specific mass of thrust unit	β	3.64×10^3	tons volt$^{3/2}$ amp^{-1}		
Specific charge of propellant (Cs)	ε/μ	0.73×10^3	e.s.u. g^{-1}		
		10^7	3×10^7	10^8	sec
Propulsion time	τ	appr. 0.3	appr. 1	appr. 3	years
Payload mass	M_L	11.5	60	150	tons
Initial accel.	a_i	0.00012	0.00008	0.00005	gravity
Take-off mass	M_0	50	200	600	tons
Propell. mass	M_F	7.2	51.2	250	tons
Thrust unit mass	M_T	6.3	13.8	20	tons
Powerplant mass	M_P	25	75	180	tons
Electric power	L	2.5	7.5	18	mw
Final accel.	a_f	0.00014	0.00011	0.000086	gravity
Exhaust velocity	v_{ex}	83	94	120	km sec^{-1}
Charact. velocity	v_c	13.0	28.0	67	km sec^{-1}
Specific impulse	S	8450	9600	12200	sec
Thrust	Th	6.1	16.3	30.6	kg
Current	I	526	1250	1820	amp
Voltage	U	4750	6000	9900	volt
Payload/Take off	M_L/M_0	0.23	0.30	0.25	—

Curve IV refers to Cs and α, but it is based on a specific thrust chamber mass β' which is three times smaller than β

$$\beta' = \frac{1}{3}\beta.$$

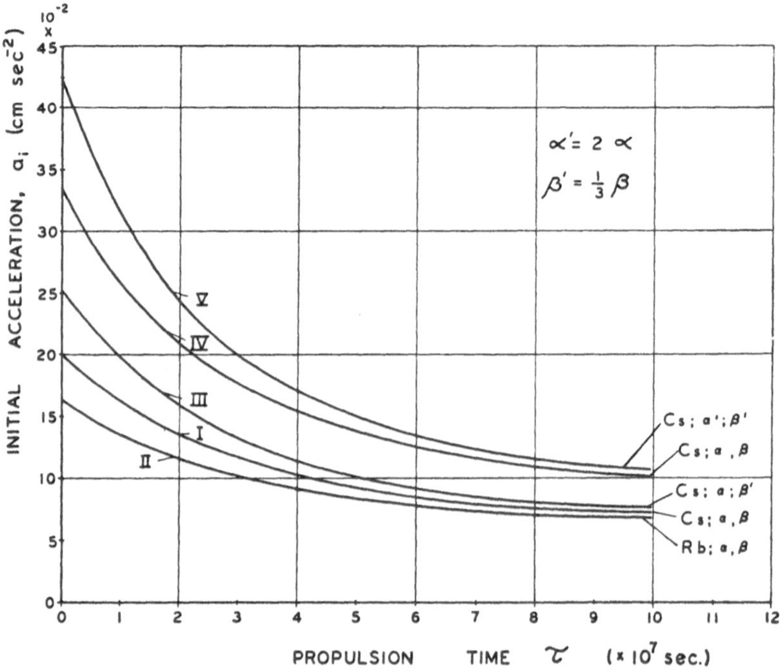

Fig. 5. Initial acceleration as a function of propulsion time and design parameters

Curve V finally shows initial accelerations a_i' for Cs combined with a specific power α' and a specific mass β'.

Fig. 5 indicates that even within a wide variation of α and β, ship accelerations do not vary too much. Practical design acceleration may be about 60 or 70 per cent of the highest possible accelerations (see Figs. 2, 3, and 4). Unless a scientific and technical breakthrough in the design of electrical power plants and ionic thrust chambers is made, it should not be excepted that the initial acceleration of an electrically propelled space ship will be greater than about 0.2 cm sec^{-2}, and probably even not greater than about 0.1 cm sec^{-2}.

4. Operational Considerations

Accelerations of this order are entirely sufficient for interplanetary travel. As shown in a previous paper [4], a space ship with an initial acceleration of only 0.067 cm sec^{-2} can make a trip to Mars in about 400 days, and the return trip in a little over 300 days. Considering the fact that a chemically propelled ship would take about 260 days for each way, but achieve a payload to take-off mass ratio of only about 1 : 70 as compared to a ratio of 1 : 3 to 1 : 4 in an electrically propelled ship, the prospects for electrical ships are most encouraging.

It may be advisable to throttle the power of the propulsion system to about 10 or 20 per cent of its full power level while the ship is traveling on the solar ellipse [4]. The increase of total travel time caused by the throttling would be only small, but the total mass would be reduced noticeably. A complete shutdown of the power plant is not advisable for the following reasons: first, to start the engines again would require considerable auxiliary power which is equivalent to additional mass; second, the power plant provides all the electric power needed for the crew, their instruments, facilities, guidance system, communication, etc., which would have to be generated by a separate power plant if the main plant were shut off; third, when the engines stop, the ship stops rotating and the crew no longer experiences the simulated gravity [3], unless the ship's rotation is maintained by auxiliary equipment like spin rockets or flywheels; fourth, a coasting ship cannot execute corrective maneuvers which will certainly become necessary during flight. Unless immediately corrected, deviations from the desired trajectory will accumulate and may become so large that they cannot be corrected at all.

It may even be desirable to keep the power plant running on a low power level during the orbiting period around a planet, not only for the reasons stated above, but also in order to compensate any atmospheric drag the ship may experience in the planetary orbit.

5. Final Remarks

It should be emphasized that the most problematic part of an electrical propulsion system is the ionic thrust chamber. Ionization of the propellant, acceleration of the ions and electrons, and neutralization of the ion beams behind the thrust chambers must work with constant efficiency over a period of years, and if possible without attendance and maintenance. Although the principle and the basic design of an ionic thrust chamber as proposed here [3] is relatively simple; early experimental studies of ionic thrust chambers are of utmost importance. Most of these experiments could be carried out on a small and inexpensive scale. Initiation of such an experimental program would be extremely desirable.

References

1. H. Oberth, Wege zur Raumschiffahrt. München-Berlin: R. Oldenbourg, 1929.
L. R. Shepherd and A. V. Cleaver, J. Brit. Interplan. Soc. **7**, 185, 234 (1948); **8**, 23, 59 (1949).
G. F. Forbes, J. Brit. Interplan. Soc. **9**, 75 (1950); J. Spaceflight **4** (1952).
L. Spitzer, Jr., J. Brit. Interplan. Soc. **10**, 249 (1951); J. Amer. Rocket Soc. **22**, 92 (1952).
H. Preston-Thomas, J. Brit. Interplan. Soc. **11**, 173 (1952).
H. S. Tsien, J. Amer. Rocket Soc. **23**, 233 (1953).
D. F. Lawden, J. Brit. Interplan. Soc. **13**, 87 (1954).
H. Michielsen, Astronaut. Acta **3**, 130 (1957). I am indebted to Mr. Michielsen for advanced information.
2. M. I. Willinsky and E. Orr, Spring Meeting of ARS, Washington, 1957.
3. E. Stuhlinger, in: Bericht über den V. Internationalen Astronautischen Kongreß, Innsbruck, 1954, S. 100. Wien-Innsbruck: Springer, 1955.
E. Stuhlinger, Electrical Propulsion System for Space Ships with Nuclear Power Source, July 1955.
4. E. Stuhlinger, Jet Propulsion **27**, Nr. 4 (1957).

Einige Optimalisationsprobleme in der Theorie der Stufenraketen und ein einfaches Verfahren zur Ermittlung der optimalen Parameter der Stufenraketen

Von

M. Subotowicz[1]

Im ersten Teil wurde ein einfaches graphisches Verfahren zur Bestimmung der optimalen Parameter der Stufenraketen vorgelegt. Es wurde der einfachste Fall der gleichen — bei optimalen Bedingungen — Parameter für alle Stufen der Großrakete diskutiert. Der volle Wortlaut dieser Arbeit ist in der Zeitschrift der Technischen Hochschule in Warschau — „Technika Rakietowa", Nr. 11, 1957, S. 173—199, Politechnika Warszawska, Warszawa — veröffentlicht.

Im zweiten Teil wurde ein analytisch begründetes Verfahren gezeigt, das schnell zum Aufdecken der optimalen Parameter einer n-Stufenrakete mit verschiedenen Brennstoff- und Konstruktions-Parametern jeder Stufe führt. Diese Arbeit wird unter dem Titel „The Optimization of the n-Step Rocket with Different Construction Parameters and Propellant Specific Impulses in Each Stage" in „Jet Propulsion" veröffentlicht.

[1] Lublin, Polen.

U. S. Army Support of Scientific Activities in Astronautics

By

H. N. Toftoy[1], ARS

Abstract — Zusammenfassung — Résumé

U.S. Army Support of Scientific Activities in Astronautics. A brief review is given of the scientific support given by the United States Army to the advancement of astronautics. The United States Army, from the beginning of rocketry in the United States, has used facilities, man power, and programs to advance the developments necessary for realization of space flight, including provisioning missiles for astrophysic experiments, long range communications and instrumentation essential to extraterrestrial flight. A description of some specific areas of support is given together with some of the results obtained.

Unterstützung wissenschaftlicher Aktivität in der Astronautik durch die U.S.-Armee. Es wird kurz über die wissenschaftliche Hilfe berichtet, die von der U.S. Army für den Fortschritt der Astronautik geleistet wurde. Die U.S. Army machte seit dem Beginn der Raketenforschung in den Vereinigten Staaten Gebrauch von Einrichtungen, menschlichen Fähigkeiten und Plänen zur Förderung der Entwicklung, die zur Verwirklichung der Weltraumfahrt führen wird. Dazu gehörten auch die Beschaffung ferngelenkter Flugkörper für astrophysikalische Versuche, ebenso das Langstrecken-Mitteilungswesen und die Instrumentierung, wie sie für den Weltraumflug unerläßlich notwendig sind. Verschiedene besondere Gebiete der wissenschaftlichen Unterstützung und einige der dabei erhaltenen Ergebnisse werden beschrieben.

L'aide apportée par la U.S. Army aux activités scientifiques dans le domaine de l'Astronautique. L'article passe brièvement en revue l'aide scientifique apportée par l'armée américaine aux progrès de l'astronautique. Depuis les débuts de la fuséonautique aux Etats-Unis, l'armée s'est servi d'équipements, a mobilisé de la main d'œuvre et a dressé des programmes pour accélérer la réalisation du vol interplanétaire. Elle a notamment fourni des engins pour les expériences d'astrophysique, des équipements de communication à grande distance et de l'instrumentation essentielle pour le vol spatial. Certains domaines dans lesquels cette aide s'est exercée sont décrits ainsi que les résultats obtenus.

It is a great honor to come before a Congress of such outstanding scientists who have the wisdom to plan upon solid foundations far-reaching progress in the conquest of our environment. It is also a particular honor to represent Lieutenant General James Gavin, Chief of Research and Development for the United States Army, who, but for pressing business, would have been personally to speak to you today.

It may seem a little surprising for a military man to take the speaker's platform during an international scientific Congress to talk about the contributions

---

[1] Major General, United States Army; Redstone Arsenal, Alabama, U.S.A.

his Army has made in a field which seems to be the exclusive prerogative of cosmopolitically-minded scientists. Most people, when thinking of their own Army, think only of protection and defense. When thinking of an actual enemy's Army, they think of destruction. These two extremes—protection and destruction—appear to be so far away from the independent and refined atmosphere of scientific work that it is not evident how pure science should draw any support from the Army. And yet, I hope to show that an Army can play an important role in scientific progress, even if we disregard supporting research completely for the moment and speak of pure science only—that type of science which appears to have no possible application to any immediate military requirement.

Many of the experiments which our scientists would like to perform for the advancement of man's fundamental knowledge are either so expensive, or would require such a large organization of men and machinery, that no scientific organization could ever hope to have the opportunity to carry out such experiments. On the other hand, military installations are sometimes in a position where it would require but little more effort and little more money to provide the means to conduct scientific experiments of the greatest scale. Such situations are a real challenge to everyone who has the power to influence military planning. Again and again, when the opportunity to lend a helping hand to pure science offered itself, it has been enthusiastically accepted by our military men because they felt that they could render a real service to their nation—a service from which all mankind can benefit.

This Congress is particularly interested in the latest developments in rockets and in space travel. These are fields of scientific endeavor to which, I believe, the United States Army was privileged to contribute a great deal. Rockets and guided missiles, are, of course, modern weapons of great importance to the military services of all nations. But as you well know, a rocket not only has the capability of carrying a warhead, it can also be equipped to carry scientific instrumentation and in this form it represents a scientific tool of almost fabulous potentialities.

As early as 1941, the United States Army Ordnance Corps became interested in the development of large supersonic rockets. At the beginning we realized that in pioneering this new field of technology the frontiers of basic science would have to be advanced. This they were at an unprecedented rate. During the period the Army was learning to walk before it ran, great technological progress was achieved. The basic information developed was made available to all agencies participating in our National program.

A carefully planned long-range research and development program was initiated and turned out to be like "casting bread upon the waters," for its returns were many fold. Significantly, some of the first efforts at development also provided vehicles for exploration of the higher atmosphere.

In the fall of 1945 the Jet Propulsion Laboratories fired the Army's *Wac-Corporal* missile to an altitude of 43 miles. From this first American liquid propelled supersonic rocket several important scientific applications evolved. The *Aerobee* upper-air sounding rocket was a scaled up version. The most historic achievement of the *Wac-Corporal* was the part it played in February 1949, when, launched from a V-2, it served as the second stage of the *Bumper* missile and established altitude and velocity records which only recently have been surpassed.

The *Bumper* itself came about from the need to prove theories and provide data on multi-stage rocket flight including (1) the separation and ignition of the second stage rocket in highly rarefied air, (2) the stability of a second stage missile launched at extremely high velocity and altitude, (3) the aerodynamic effects at high Mach numbers obtainable in no other way at that time. These research flights were strikingly successful; in reaching an altitude of 250 miles (400 kilometers) the Army was credited with being the first to send a man-made object outside the earth's atmosphere.

Shortly after the end of the war, the components of a number of German V-2 rockets were shipped to the United States, and, after certain parts which were not available had been manufactured in the United States, these missiles were assembled and fired for various purposes. Over a hundred scientists and engineers from the former rocket development center in Peenemuende voluntarily accepted government contracts to continue their work at American rocket development agencies. Approximately 80 V-2 missiles were fired under Army auspices at the White Sands Proving Ground during the years 1946 to 1949.

This test firing program also quickly developed into a scientific program of the first order. Universities and research laboratories were allotted the necessary space, power, and telemeter channels in these rocket vehicles for the purpose of conducting experiments probing into upper atmosphere. The launching pad in the White Sands desert often was a beehive of activity on firing days with dozens of scientists installing and calibrating their instruments. Dignified professors could be seen climbing around the missile with sunburnt arms and faces, but most happy with this latest scientific tool that had been placed at their disposal. The old V-2 missiles proved to be excellent work horses. They were, of course, originally designed as weapons to carry a payload of 2200 pounds of high explosive, but we modified them to carry an even bigger payload of instrumentation for scientific purposes. Their big instrument compartments became real laboratories containing such devices as cosmic ray counters, Wilson cloud chambers, spectrometers, movie cameras, ion meters, thermometers, pressure gauges, sampling bottles and the like. A living reminder of those exciting pioneering days is a little Rhesus monkey which last I heard still enjoys everybody's friendship in the White Sands area. It is probably the only living being in the Western Hemisphere whose heartbeat has been electrically timed at an altitude of 100 miles above the earth. Many scientific papers were published on the measurements taken during those early V-2 firings, and many standard tables on the high atmosphere contain the contributions of these experiments. Man-made meteorite experiments were attempted. In order to properly coordinate the experimental work proposed by the many interested agencies, an "Upper Atmosphere Panel" was established in 1946.

After termination of the V-2 launching project, high altitude experiments continued in several new projects. The best known of them is the *Aerobee* test rocket which was produced by the Navy specifically for scientific purposes and used by all three services. Though entirely separated now from military support, the *Aerobee* team, like other teams engaged today in scientific rocket experiments, obtained its first stimulation and training from the Army-sponsored firings.

Even though the merry barn-storming days of the early rocket firings are now a memory, many of the friendship between scientists and Army men, founded in those days, are still vividly alive. Even today, when the Army is busy developing some of the most modern guided missiles of our time, we still see some of our test missiles carry along with its development instrumentation an unpretentious little black box labeled with some unrelated name, such as "solar radiation chamber," "cosmic ray counter," or "artificial meteor". Our pure scientists are still enjoying, when practicable, free transportation for some scientific experiment which he otherwise could not accomplish.

The Army was privileged to pioneer in another field which may be of particular interest to members of this convention; namely, communication through interplanetary space. Communication will be as essential to space travel as it is to all forms of terrestrial travel. On 10 January 1946, at moonrise, the Signal Corps Engineering Laboratories beamed their radar signals toward the moon. Within a few seconds, the first signals ever reflected from the moon were detected upon their radarscopes. Distinct echoes were detected 22 times in 111 attempts. Some rather peculiar effects were noted; among them a rapid variation of the signal strength of the echoes, and even occasionally a total loss of signal when the equipment was apparently in normal operating condition. At other times, echoes were received as early as ten minutes before moonrise, indicating how strongly the signals were bent around the earth. In one instance, very substantial echoes were detected from an unidentified object about 120,000 miles out in space, about half the distance from the earth to the moon. This raised some rather interesting theories.

Since those early experiments, considerable progress has been made in moon relay experiments. Even more distant reflectors, like Mars and Jupiter have been included in those investigations. Only some of the anomalies in propagation and reflection have been solved so far, but appreciable gains toward interplanetary communication and navigation were achieved. Solar and galactic noise studies, investigations of radar reflection by meteor trails, or the use of artificial satellites as radar relay stations, will benefit very greatly from continued experimentation with the Army's high-powered radars.

An excellent opportunity to support scientific endeavour presented itself in 1955 when President EISENHOWER announced that the United States will launch an artificial satellite for scientific purposes during the International Geophysical Year. At that time, members of an Army Development Agency for Guided Missiles had submitted a proposal to launch a satellite with a 3-stage rocket composed of Army-developed missiles. The proposal visualized a satellite of 15 to 20 pounds, carrying instrumentation for observations of cosmic rays, solar radiations, atmospheric properties, etc. Although the Navy proposal was the one accepted for the American IGY satellite, the Army retains the potential capability to build, launch, and observe a satellite.

Members of Army development groups are greatly interested in the satellites program and will engage in the scientific evaluation work. A major part of the network of radio ground stations which is being established for the *Vanguard* Satellite Project is under the cognizance and responsibility of the Army Corps of Engineers. At the request of the Department of the Navy, the Army established and will operate six of the prime tracking stations used in the *Vanguard* program. These stations are located in Havana, Cuba; Quito, Ecuador; Ancon, Peru; Antofagasta, Chile; Santiago, Chile; and Fort Stewart, Georgia, U.S.A. Telemeter receivers located at the ground stations will record the signals of the scientific instruments carried on the satellite. Also from the tracking signals received at the ground stations, data for the radius of the earth, the oblateness of the geoid, and the density distribution of the high atmosphere will be derived with an unprecedented accuracy.

The few examples which I have outlined serve to indicate how much the United States Army is interested in advancing the frontiers of science. I trust this paper further assures you that the American Armed Services will follow the achievements of you members of the International Astronautical Federation, not only with the greatest interest and admiration, but with the real desire to render such support as they properly can in the realization of your valiant goal—the conquest of outer space.

The Meteoritic Risk to Space Vehicles

By

Fred L. Whipple[1]

Abstract — Zusammenfassung — Résumé

The Meteoritic Risk to Space Vehicles. Consideration is given to the distribution of meteoritic material and its rate of fall on the earth as functions of mass and velocity. With a simple theory, the probabilities are calculated that surfaces in space in the neighborhood of the earth may be punctured by meteoric action. A table of data and probabilities is given. It is calculated that a near-earth satellite of radius 20 inches and skin thickness 0.5 mm Al will be punctured on the average of once in five days.

Upper limits to the effects of skin erosion ón a space-exposed surface are calculated on the basis of erosion by meteoritic dust, by corpuscular radiation from the sun and by gases of the extended solar corona. The erosive effect from meteoritic dust is comparable to the combined effects from the other two causes and gives a rate of skin (Al) erosion of the order of $2 \cdot 10^{-13}$ gm/cm²/sec or less. Optical surfaces exposed to space should not be affected functionally by erosion over periods less than about a year.

Attention is given to the expected degree of accuracy of the observed data and the conclusions, particularly for the meteoritic material. The uncertainties arise from combined theoretical and observational limitations.

Die Meteoritengefahr für Raumfahrzeuge. Es werden die Verteilung von meteoritischem Material und seine Einfallsgeschwindigkeit auf der Erde als Funktionen der Masse und der Geschwindigkeit betrachtet. Mit Hilfe einer einfachen Theorie wird die Wahrscheinlichkeit berechnet, daß die Oberflächen im Weltraum in der Umgebung der Erde durch meteoritische Einwirkung durchlöchert werden. Eine Tabelle der diesbezüglichen Daten und Wahrscheinlichkeiten wird angegeben. Die Berechnung zeigt, daß ein erdnaher Satellit vom Radius 50 cm und einer Hullendicke von 0,5 mm Aluminium durchschnittlich einmal in fünf Tagen durchlöchert wird.

Obere Grenzen fur die Wirkungen der Hullenerosion auf einer dem Weltraum ausgesetzten Oberflache werden auf der Grundlage der Erosion durch meteoritischen Staub, durch Korpuskularstrahlung von der Sonne und durch die Gase der erweiterten Sonnenkorona berechnet. Die Erosionswirkung durch meteoritischen Staub ist vergleichbar der vereinigten Wirkung der beiden anderen Ursachen und gibt eine Erosionsgeschwindigkeit für eine Aluminiumhulle in der Größenordnung von 2.10^{-13}g/cm²/sec oder weniger. Dem Raum ausgesetzte optische Oberflächen dürften in ihrer Funktion nicht durch Erosionsperioden von weniger als etwa 1 Jahr angegriffen werden.

Aufmerksamkeit wird auch dem zu erwartenden Genauigkeitsgrad der Beobachtungsdaten geschenkt sowie den Schlußfolgerungen, vorzugsweise für das meteoritische Material. Die Unsicherheiten gehen auf die kombinierten theoretischen und Beobachtungsbeschränkungen zurück.

Les risques de dommage aux véhicules spatiaux causé par les météorites. La répartition des météorites et la fréquence de leur chute sur la surface de la terre est

[1] Smithsonian Institution, Astrophysical Observatory and Harvard College Observatory, Cambridge 38, Mass., U.S.A.

analysée en fonction de leur masse et de leur vitesse d'impact. Les probabilités de perforation d'une surface au voisinage de la terre sont calculées à l'aide d'une théorie simple et fournies sous forme de tables. D'après ces calculs un satellite de 50 cms. de diamètre avec un revêtement d'aluminium de 0.5 mm d'épaisseur serait perforé en moyenne une fois tous les cinq jours si son orbite s'éloigne peu de la surface terrestre.

Des limites supérieures à l'érosion du revêtement sont calculées en prenant en considération les poussières météoritiques, les radiations corpusculaires du soleil et l'action des gaz provenant de l'extension de la couronne solaire. L'érosion due aux poussières météoritiques est comparable en grandeur à celle des deux autres causes réunies et conduit à une vitesse d'érosion sur un revêtement d'aluminium de l'ordre de 2.10^{-13}gm/cm^2·sec. au plus. L'efficacité optique des surfaces polies ne serait pas affectée par l'érosion durant une période d'au moins un an.

Une attention spéciale est accordée à la précision probable des données d'observation, spécialement en ce qui concerne les météorites. Les causes d'incertitude résident aussi bien dans les limitations de la théorie que dans celles des données observables.

The Data from Meteors

Any solid surface in space, such as the skin of a space vehicle or the surface of a rocket, is subjected to the action of meteoritic particles. The effects can be conveniently separated functionally into two types: (a) punctures, and (b) surface erosion. The range of action, of course, is continuous from that by particles of only a few molecules up to objects that might even be large enough to destroy the entire vehicle or surface. Punctures are obviously of significance with regard to preserving pressure within a space vehicle, while surface erosion affects the utilization of windows and electronic equipment and the maintenance of radiative heat balance. With sufficient time, it is obvious that erosion can weaken a surface so that structural damage can occur from internal pressures or strains. Before this could happen, however, many punctures would be expected in the surface.

Atomic sputtering of the surface by corpuscular radiation from the sun is expected to be comparable in importance to meteoritic dust, at least in the neighborhood of the earth. Also, effects for the extended solar corona must be considered.

At the high relative velocities prevalent in space, a surface can be penetrated by particles whose dimensions are perhaps an order of magnitude smaller than the surface thickness. Hence for practical application of any puncture theory, one is concerned with meteoritic particles of relatively small dimensions, of radius, s, much less than 1 cm. From our knowledge of meteors we know that such particles are below the range observable by visual, photographic, or radio reflection techniques. Hence, any estimate of the numbers, velocities, masses and other fundamental characteristics must be extrapolated beyond the range of observation. On the other hand, these particles are not microscopic and are generally larger than those primarily responsible for producing the scattered sunlight in the Zodiacal Light. The meteoritic particles are relatively too rare to be well observed in terrestrial manifestations of meteoritical dust, or to appear frequently in rocket studies of surface deterioration. Probably no adequate measurements of puncturing particles can be obtained until surfaces have been exposed in space for periods of days or weeks, perhaps in a satellite vehicle.

I shall first present an extrapolation from the region of the visual, photographic and radio meteor region to the realm of the particles that would be active in puncturing and eroding structural surfaces in space. Table I is a revised version of a similar table presented for this purpose in 1952 [1]. The basic references to the general subject are available in this article and will not be presented again here. Significant changes have taken place, however, in our concepts of ordinary meteors since this earlier effort to estimate the activity of meteoritic material in space.

The Harvard Photographic Meteor Program, utilizing Super-SCHMIDT meteor cameras, has given us basic knowledge concerning the nature of meteors in the visual range. The researches on this problem by L. G. JACCHIA [2] and R. E. McCROSKY [3], coupled with those of the author [4], have led to the following conclusions:

A. Practically all photographic meteors are of cometary origin. The hyperbolic content, if present at all, is below one percent and the contributions by asteroidal material appear to be relatively small, if present. It seems relatively safe to extrapolate this conclusion to smaller bodies in space in view of arguments concerning both the observed and theoretical distributions of particle sizes.

B. The meteoric bodies, or meteoroids, manifested as ordinary meteors are extremely fragile and breakable. McCROSKY has shown that some 20 percent of them crush at pressures of the order of 1/50 atmosphere. They evidence their fragility in a number of ways, including irregular flares, appearance of fainter meteors only by flaring and, as demonstrated by JACCHIA, in more subtle phenomena involving velocity, height deceleration, luminous intensity and atmospheric density.

C. Ordinary meteoroids differ greatly from one another in physical characteristics, which may involve compressive strength, composition, density, luminous efficiency and ionizing efficiency. Meteoroids having a common origin from a given comet tend statistically to be more alike than those from different comets, although the variations are quite marked among bodies of a given cometary stream.

D. The evidence is extremely strong, although not quite conclusive, that the density of ordinary meteoroids is the order of .05 gm/cm^3. This value depends, to a limited extent, upon physical arguments from photographic meteor data and from the actual measurement of mass and density in the forward motion of the luminous train left by a meteor. Such measures, by A. F. COOK and the author [5], led to the determination of the momentum imparted to the atmosphere by a given meteoroid. Other such measurements are underway. The photographic meteor program has led to precision results in terms of meteoric velocities, luminous intensities, decelerations, and trajectories. A number of important radio meteor programs [6, 7] have added information concerning not only the diurnal and seasonal frequency distribution of meteors fainter than those that can be photographed, but also clear information concerning the production of electrons along meteoric trails. None of these techniques, however, can measure directly the density and mass of a meteor. Measurements of drag lead to the determination of a quantity, $m^{1/3}\,\varrho^{2/3}$ where m is the mass of the meteoroid and ϱ its density. Measurements of photographic intensity or radio ionization readily provide data on the luminous or ionizing efficiency of the process but none of these quantities can be evaluated directly in terms of the meteoric mass. For any of the doubly-photographed meteors, a determination of any one of the quantities—mass, density, or luminous efficiency—would immediately lead to rather precise values of the others.

The determined density of .05 gm/cm^3 appears to be an extremely low density for a solid body in space. Nevertheless the cometary origin makes such densities seem plausible while the corresponding values of luminous efficiency (the order of 10^{-3}) seem also acceptable. Note that the assignment of densities proper for stone, 3.5 gm/cm^3 or iron, 7.8 gm/cm^3, would lead to the impossible conclusion that practically all of the original kinetic energy in an ordinary visual meteor is transformed into light. The radiation processes are expected to be rather inefficient and a much lower luminous efficiency is to be expected. The conclusive evidence of fragility and the lack of structural strength, are consistent with the low density measure.

We have no direct evidence to justify our extrapolating the densities measured among the visual meteors to the extremely faint ones, or to bodies too small to be detectable by any means during their transit through the atmosphere. One might argue that smaller and smaller pieces broken from a large mass of very low density would necessarily be of higher density than the original body. This argument seems tenable but one must remember, given a cometary origin, that the particle sizes initially depend upon an unknown structure within the cometary mass before the ices sublimated due to the action of sunlight. It is quite possible initially that smaller bodies ejected from comets are indeed of lower density than larger ones. We already have rather clear-cut evidence that the meteoric material from some small comets tends to be of lower density, and generally of smaller mass and dimensions than that from larger comets. We have no sound basis upon which to assume densities for smaller particles, except for the effect of light pressure or other physical effects, discussed later in this paper.

The meanings of the various colums in Table I are described below:

Column 1—Meteor "Visual Magnitude"—The term obviously ceases to have meaning for meteors below the telescopic range. It is utilized here simply as a convenient argument, based on an adopted linearity between the luminosity and mass.

Column 2—Mass (gm).—On the basis of the Harvard Photographic Meteor Program, a meteor of visual magnitude zero is determined to have a mass of the order of 25 grams. It is assumed that the mass decreases by a factor of $10^{0.4}$ ($= 2.512\ldots$) per magnitude step. Among meteors of a given brightness, mass varies in an inverse fashion with velocity. An average velocity of 28 km/sec with respect to the earth's atmosphere has been assumed here for the brighter meteors and is reduced for fainter ones (see Column 4).

Column 3—Radius (microns).—The calculated radius is based upon the mass values given in Column 2 for spherical particles of density 0.05 gm/cm^3. Actual meteoroids will deviate markedly from sphericity and will also vary markedly in mean density.

Column 4—Assumed Velocity (km/sec).—A velocity of 28 km/sec is average for photographic meteors. Undoubtedly the velocity falls off for smaller meteoroids as we deal more and more with particles whose orbital eccentricities and dimensions have been reduced by physical effects (for discussions of some of these problems see author [8] and E. Öpik [9]). The velocity at the edge of the earth's atmosphere cannot be less than 11 km/sec because of the earth's attraction. Somewhat higher values can occur for nearly circular orbits by inclination of the orbital plane to that of the earth. A mean value of 15 km/sec has been arbitrarily chosen for the smallest meteoroids and an arbitrary gradation of velocity with magnitude adopted. Radio meteor studies may provide measures of the gradation in the near future, perhaps to the 12th magnitude.

Column 5—Kinetic Energy (ergs).—Directly calculated from the assumed values of mass and velocity.

Column 6—Penetration in Aluminum (cm).—This penetration distance, d, corresponds to the volume of material removed in a right circular cone of total apex angle 60°,

$$d = \left(\frac{9}{\pi \varrho' \zeta}\right)^{1/3} E^{1/3}, \qquad (1)$$

where ϱ' is the density of the surface struck, ζ is the heat required to remove the material from the surface (ergs/gram) and E is the kinetic energy of the meteoroid. There is little justification for this formula and the author would be delighted to revise it in terms of measurements or of a complete theory for small-particle penetration of solid materials at velocities in the range from 11 to 60 km/sec. In the present calculations, as previously, an aluminum surface is assumed to have a density $\varrho' = 2.7$ gm/cm^3 and ζ is taken as the heat to fusion, $1.18 \cdot 10^{10}$ erg/gm. The kinetic energy is given by Column 5 of Table I.

Column 7—Number Striking Earth (per day).—This number includes all meteoric bodies brighter than or of mass greater than the number indicated in the line of question. The zero value is based upon a recent determination by Dr. P. M. MILLMAN in which he concludes that $2 \cdot 10^8$ naked-eye meteors strike the earth each day. This practical limit for naked-eye meteors is here taken as 5.0 magnitude visual. The increase in number per magnitude is taken as the factor $10^{0.4}$. This relation, due to F. G. WATSON [11], is equivalent to an increase in the number of meteoroids inversely as the mass of the particle. The total mass contributed by each magnitude step is then a constant, which in this case turns out to be 54 tons per magnitude step. MILLMAN finds for brighter visual meteors that the numbers increase with magnitude more rapidly than by this law. For fainter telescopic meteors the $10^{0.4}$ increase per magnitude seems to be appropriate, although poorly determined.

Column 8—Numbers Striking 3-meter Sphere (per day).—This puncturing probability is derived directly from Column 7 on the assumption that the surface area of the earth exceeds that of a three-meter sphere by a factor of $4.51 \cdot 10^{12}$ and that the shadowing effect of the earth on a spherical body near its surfac equals 1/2. The corresponding ratio for the IGY earth satellite, a 20-inch sphere, is $2 \cdot 1.57 \cdot 10^{14}$. A correction for distance above the earth's surface is rather difficult to apply unless the precise orbital characteristics of the incoming meteoric bodies are known. Perhaps the inclusion of the factor one-half is adequate for the ordinary problem near the earth's surface. At extreme distances, greater than 10^4 km, a complete recalculation is needed with very careful attention to the orbital characteristics of meteoroids in space and of the space vehicle in question. Even after correction for the factor of one-half, the number striking the vehicle will probably fall off with increasing distance from the earth.

Punctures and Erosion on Surfaces in Space

Table I leads to the conclusion that, for example, a 20-inch diameter satellite sphere having an aluminum or magnesium surface of thickness 0.5 mm would be penetrated on the average approximately once every five days. With a penetration (Column 6) of 0.05 cm, between 18th and 19th magnitude (Column 1), we find that the earth encounters $6.2 \cdot 10^{13}$ such bodies or larger ones per day (Column 7). Correcting by the factor of $2 \cdot 1.57 \cdot 10^{14}$, from earth to satellite area, we find that the number striking the satellite per days is 0.2, the daily probability of a puncture.

For a 3-meter sphere with aluminum skin 1/8-inch thick the penetration corresponds to about a 13th visual magnitude meteor (Columns 6 and 1). The number striking the 3-meter sphere per day (Column 8) is about 0.047. The sphere should be punctured once in three weeks, on the average.

To apply Table I to other sizes and shapes of space vehicles or surfaces in space, find the skin thickness (Al or Mg) from Column 6 and read the number striking the earth per day from Column 7. Correct the area of the vehicle or

Table I. *Data Concerning Meteoroids and Their Penetrating Probabilities*

1	2	3	4	5	6	7	8
Meteor "Vis. Mag."	Mass (gm)	Radius (microns)	Ass. Vel. (km/sec)	K.E.(ergs)	Pen. in Al (cm)	No. Earth per day	No. 3-m sphere per day
0	25.0	49,200	28	$1.0 \cdot 10^{14}$	21.3	—	—
1	9.95	36,200	28	$3.98 \cdot 10^{13}$	15.7	—	—
2	3.96	26,600	28	$1.58 \cdot 10^{13}$	11.5	—	—
3	1.58	19,600	28	$6.31 \cdot 10^{12}$	8.48	—	—
4	0.628	14,400	28	$2.51 \cdot 10^{12}$	6.24	—	—
5	0.250	10,600	28	$1.00 \cdot 10^{12}$	4.59	$2 \cdot 10^{8}$	$2.22 \cdot 10^{-5}$
6	$9.95 \cdot 10^{-2}$	7,800	28	$3.98 \cdot 10^{11}$	3.38	$5.84 \cdot 10^{8}$	$6.48 \cdot 10^{-5}$
7	$3.96 \cdot 10^{-2}$	5,740	28	$1.58 \cdot 10^{11}$	2.48	$1.47 \cdot 10^{9}$	$1.63 \cdot 10^{-4}$
8	$1.58 \cdot 10^{-3}$	4,220	27	$5.87 \cdot 10^{10}$	1.79	$3.69 \cdot 10^{9}$	$4.09 \cdot 10^{-4}$
9	$6.28 \cdot 10^{-3}$	3,110	26	$2.17 \cdot 10^{10}$	1.28	$9.26 \cdot 10^{9}$	$1.03 \cdot 10^{-3}$
10	$2.50 \cdot 10^{-3}$	2,290	25	$7.97 \cdot 10^{9}$	0.917	$2.33 \cdot 10^{10}$	$2.58 \cdot 10^{-3}$
11	$9.95 \cdot 10^{-4}$	1,680	24	$2.93 \cdot 10^{9}$	0.656	$5.84 \cdot 10^{10}$	$6.48 \cdot 10^{-3}$
12	$3.96 \cdot 10^{-4}$	1,240	23	$1.07 \cdot 10^{9}$	0.469	$1.47 \cdot 10^{11}$	$1.63 \cdot 10^{-2}$
13	$1.58 \cdot 10^{-4}$	910.	22	$3.89 \cdot 10^{8}$	0.335	$3.69 \cdot 10^{11}$	$4.09 \cdot 10^{-2}$
14	$6.28 \cdot 10^{-5}$	669.	21	$1.41 \cdot 10^{8}$	0.238	$9.26 \cdot 10^{11}$	$1.03 \cdot 10^{-1}$
15	$2.50 \cdot 10^{-3}$	492.	20	$5.10 \cdot 10^{7}$	0.170	$2.33 \cdot 10^{12}$	$2.58 \cdot 10^{-1}$
16	$9.95 \cdot 10^{-6}$	362.	19	$1.83 \cdot 10^{7}$	0.121	$5.84 \cdot 10^{12}$	$6.48 \cdot 10^{-1}$
17	$3.96 \cdot 10^{-6}$	266.	18	$6.55 \cdot 10^{6}$	0.0859	$1.47 \cdot 10^{13}$	1.63
18	$1.58 \cdot 10^{-6}$	196.	17	$2.33 \cdot 10^{6}$	0.0608	$3.69 \cdot 10^{13}$	4.09
19	$6.28 \cdot 10^{-7}$	144.	16	$8.20 \cdot 10^{5}$	0.0430	$9.26 \cdot 10^{13}$	$1.03 \cdot 10$
20	$2.50 \cdot 10^{-7}$	106.	15	$2.87 \cdot 10^{5}$	0,303	$2.33 \cdot 10^{14}$	$2.58 \cdot 10$
21	$9.95 \cdot 10^{-8}$	78.0	15	$1.14 \cdot 10^{5}$	0.223	$5.84 \cdot 10^{14}$	$6.48 \cdot 10$
22	$3.96 \cdot 10^{-8}$	57.4	15	$4.55 \cdot 10^{4}$	0.0164	$1.47 \cdot 10^{15}$	$1.63 \cdot 10^{2}$
23	$1.58 \cdot 10^{-8}$	39.8*	15	$1.81 \cdot 10^{4}$	0.0121	$3.69 \cdot 10^{15}$	$4.09 \cdot 10^{2}$
24	$6.28 \cdot 10^{-9}$	25.1*	15	$7.21 \cdot 10^{3}$	0.00884	$9.26 \cdot 10^{15}$	$1.03 \cdot 10^{3}$
25	$2.50 \cdot 10^{-9}$	15.8*	15	$2.87 \cdot 10^{3}$	0.00653	$2.33 \cdot 10^{16}$	$2.58 \cdot 10^{3}$
26	$9.95 \cdot 10^{-10}$	10.0*	15	$1.14 \cdot 10^{3}$	0.00480	$5.84 \cdot 10^{16}$	$6.48 \cdot 10^{3}$
27	$3.96 \cdot 10^{-10}$	6.30*	15	$4.55 \cdot 10^{2}$	0.00353	$1.47 \cdot 10^{17}$	$1.63 \cdot 10^{4}$
28	$1.58 \cdot 10^{-10}$	3.98*	15	$1.81 \cdot 10^{2}$	0.00260	$3.69 \cdot 10^{17}$	$4.09 \cdot 10^{4}$
29	$6.28 \cdot 10^{-11}$	2.51*	15	$7.21 \cdot 10$	0.00191	$9.26 \cdot 10^{17}$	$1.03 \cdot 10^{5}$
30	$2.50 \cdot 10^{-11}$	1.58*	15	$2.87 \cdot 10$	0.00141	$2.33 \cdot 10^{18}$	$2.58 \cdot 10^{5}$
31	$9.95 \cdot 10^{-12}$	1.00*	15	$1.14 \cdot 10$	0.00103	$5.84 \cdot 10^{18}$	$6.48 \cdot 10^{5}$

* Maximum radius permitted by solar light pressure.

surface to its average area integrated over all angles: correct the result in Column 7 by 1/2 the ratio of this area to the area of the earth ($1.28 \cdot 10^{18}$ cm^2); the final number is the probability of puncture per day.

For other materials than aluminum or magnesium a correction for relative heats to fusion and for density can be made as follows:

Correction Factor to Penetration $(d) =$

$$\frac{2.7\ \text{gm/cm}^3}{\text{Density}} \cdot \frac{1.1 \cdot 10^{10}\ \text{erg/gm}^{1/3}}{\text{Heat to fusion}}. \tag{2}$$

Other penetration laws can easily be used by comparison with eq. (1).

To determine an upper limit to the erosive effect of fine meteoritic material on surfaces exposed to space we can safely estimate the total meteoritic mass available as less than 1,000 tons/day on the entire earth, with a velocity of the order of 15 km/sec. The ratio of kinetic energy to the heat to fusion is about 100 times the mass of the striking particle. Certainly the process of crater making must be less efficient for very small particles. Hence these values are all safe upper limits.

The correction factor from tons/day on the entire earth to gm/cm²/sec is $9 \cdot 10^{-18}$. Hence the maximum mass eroded from a surface will be $1/2 \cdot 1.0 \cdot 10^{3} \cdot 100 \cdot 9 \cdot 10^{-18}$ gm/cm²/sec, or $4.5 \cdot 10^{-13}$ gm/cm²/sec. For a surface density of 3 gm/cm³ the maximum depth rate of erosion will be $1.5 \cdot 10^{-13}$ cm/sec.

For optical erosion to become appreciable the surface must be roughened at least to an average depth of the order of $\lambda/2\pi$, where λ is the wavelength of the light. For $\lambda = 5000$ Å ($5 \cdot 10^{-5}$ cm), the time required will be $(5 \cdot 10^{-5}$ cm$)/(2\pi \cdot 1.5 \cdot 10^{-13}$ cm/sec), or $5 \cdot 10^{7}$ sec or 1.7 years.

Corpuscular radiation from the sun, mostly protons moving at velocities occasionally up to 3000 km/sec or more, may be quite important in eroding surfaces in space (see author [8] and F. S. SINGER [12]). Cathode sputtering at energies of the order of a thousand electron volts on solid metallic surfaces can be quite effective. The number, N, of protons/cm³ at the earth will probably average less than 600 while the average velocity, V, may be of the order of 300 km/sec. Probably the efficiency factor, E, or number of metallic atoms removed by each proton, will not exceed 1.0. For atoms of mass, m_a, the average loss of mass/cm² on the surface of a sphere is given by

$$\frac{d\,\text{mass}}{dt} = \frac{m_a E N V}{4}. \qquad (3)$$

For an aluminum surface ($m_a = 4.5 \cdot 10^{-23}$ gm) and the rather high values above for the quantities in eq. (3), the rate of corpuscular erosion is $2 \cdot 10^{-13}$ gm/cm²/sec. This value is almost identical with the value calculated for meteoritic dust. The adopted numbers for the corpuscular erosion are probably less extreme upper limits than for the dust.

Sublimation of surfaces by the extended solar corona, considered by SINGER [12], is probably comparable but smaller in erosive effect to corpuscular sputtering. The atomic energies are smaller than the 500 ev adopted for protons above, while the spacial density of coronal particles will not exceed the value of 600/cm³ adopted for the corpuscular radiation. This limit is set by electron scattering in the corona and must include both the electrons from corpuscular atoms moving radially from the sun and electrons from the extended corona, relatively at rest but with a temperature of the order of 10^{6} degrees K.

Hence we may expect an erosion rate, from all three sources, of the order of $2 \cdot 10^{-13}$ gm/cm²/sec or less. The erosion cannot become important optically over a period less than about a year.

Distance from the sun will probably not affect the rate of puncture greatly within the earth's orbit. We have no sound basis for judgment outside the earth's orbit. Probably surface erosion will decrease with solar distance following some inverse law of the distance. Possibly puncturing may become more severe in and near the asteroid belt between Mars and Jupiter.

At all solar distances within the distance of the major planets the effects of puncturing will decrease with increasing distance from the plane of the ecliptic. From the observations of the Zodiacal Light and from the radar results of G. S. HAWKINS [13], we know that meteor orbits show concentration towards the plane of the earth's orbit.

Comments on the Accuracy of Table I

The relevance of a simple distribution law for the number of meteors as a function of mass, m, over the extreme range of $2.5 \cdot 10^{10}$ in mass, even though based on reasonably good starting values, requires careful checking at the lower end. This distribution function $f(m)$ has the form

$$f(m)\, dm = \frac{C_m}{m^2}\, dm\,, \tag{4}$$

where C_m is a constant equal to $5.81 \cdot 10^7$ gm, for the assumptions on Table 1. In one magnitude interval the total mass is a constant with respect to magnitude and equal to 54 tons on the entire earth per day.

The integration of this mass over the total range of meteoric magnitudes, including a fall off in the mass distribution per magnitude interval above 5th magnitude according to MILLMAN's results, leads to a total accretion of meteoritic material of about $2 \cdot 10^3$ tons per day for the entire earth, or $2 \cdot 10^{-14}$ gm/cm²/sec. This would amount to a total accretion of about 1 gm/cm² in two million years. The total space density of solid material at the earth's distance from the sun in its orbital plane would then amount to about $4 \cdot 10^{-20}$ gm/cm³. H. PETTERSSON and H. ROTSCHI's [14] measurement of a 0.045% metallic nickel content in sea bottom clay is interpreted by E. ÖPIK [9] to indicate a total daily accretion of some 900 tons of meteoritic material by the earth. This excellent check with Table I must be accepted with considerable reserve in view of the large number of assumptions that entered into this calculation. Nevertheless, the nickel content of the deep sea oozes represents, at the present time, the only check on estimates of the *total* meteoritic infall to the earth.

From collections of magnetic spherules made over a period of a year at Table Mountain, California; Fairbanks, Alaska and other stations, P. HODGE [15] finds that magnetic spherules fall at a rate of approximately 1/cm²/day. The dimensions are in the general range of 3—5 microns; the particles are magnetic with some iron content and have some central bubbles. If we accept the mean radius of 2 microns and a density of 2.5 gm/cm³, the daily accretion rate for the entire earth is about 100 tons or approximately $1 \cdot 10^{-15}$ gm/cm²/sec. Such material would represent a rather small fraction of the total amount calculated in the table. The spherules undoubtedly represent the end-product droplets of smaller meteoroids as they enter the earth's atmosphere, and also the similar end products of larger meteoroids that break into many thousands of pieces, as observed among the brighter meteors. One would expect a correction factor of between 10 and 100 times from the total input of meteoritic spherules to the total input of meteoritic matter. If HODGE's results should be accepted as definitive, it is probable that Table I represents an underestimate rather than an overestimate of the total amount of meteoritic material falling on the earth.

The distribution function in Table I measured in radius, s, follows the law

$$f(s)\, ds = \frac{C_s}{s^4}\, ds\,, \tag{5}$$

where

$$C_s = \frac{9}{4\pi\varrho}\, C_m\,. \tag{6}$$

From measures of the distribution of light in the Zodiacal Light at great angular distance from the sun and from the FRAUNHOFER Corona (solar line spectrum) at angles near the sun H. v. DE HULST [16] finds that the distribution law for small particles is closely an inverse 2.6 power of s instead of an inverse 4 powers as in eq. (4). Similarly H. SIEDENTOPF [17] finds an inverse power of 2.85.

In the last few lines of Column 3 in Table I the radius has been calculated on the assumption that the mass distribution law still follows eq. (4) but that the density of the meteoritic material does not fall below a value such that particles will remain in the system against solar radiation pressure. For black spherical particles the ratio of solar radiation pressure to gravity is given by the following equation:

$$\frac{\text{Solar Radiation Pressure}}{\text{Gravity}} = \frac{E_s'}{4c\,G\,M\,\varrho\,s}\,, \tag{7}$$

where E_s' is the total radiation of the sun in ergs/sec, c is the velocity of light, G is the gravitational constant and M the mass of the sun.

Numerically this equation leads to the result:

$$\varrho\,s = 2.4 \cdot 10^{-4} \text{ cgs.} \tag{8}$$

Again for spherical particles eq. (7) leads to the consequence

$$m = K\,s^2\,, \tag{9}$$

where

$$K = 1.0 \cdot 10^{-3}\,. \tag{10}$$

Values of the meteoric radius, s, derived from eq. (9) are obviously upper limits since such particles would be subject to no effective gravity by the sun. Nevertheless the upper limit follows an interesting law for the distribution of particle sizes, as follows:

$$f(s)\,ds = \frac{2\,C_m}{K\,s^3}\,ds\,. \tag{11}$$

The close coincidence of this distribution law with those derived by VAN DE HULST and SIEDENTOPF suggests strongly that light pressure plays an important role in eliminating low density particles from comets. Corpuscular radiation from the sun may also play a smaller role, but will probably give the same law as that from radiation pressure. If we now assume that cometary material, the source of our small meteoritic dust in space, shows an enormous spread in density, we then see the mass distribution law may follow that for the brighter meteors while the radius distribution law for small particles can fall close to that derived for the particles that are affected strongly by light pressure.

Whether the above suggestion is relevant, the fact that a single table fits so well both the photographic and radio observations of meteors on the one hand, and our fragmentary knowledge of small particles in the solar system on the other, is highly encouraging. We may now place a certain amount of confidence in our calculations of the probable rates at which meteoric material will puncture space vehicles. The values given in Table I should be taken as fairly high limits, representing high rates of puncture probabilities, because the penetration law [eq. (1)] almost certainly overestimates the power of small particles to make holes in sheets of material. It is possible, however, that a considerable error is made for the numbers of particles in the ranges of interest and so the values in Table I represent a rough expectation of meteoritic hazards to space vehicles.

This expectation of punctures to space vehicles is much greater than earlier estimates [1]. As a consequence, the use of a "meteor bumper", as suggested by the author [1], seems even more important in planning space vehicles. The meteor bumper is simply a thin secondary layer of surface placed a few thicknesses of the major surface from it on the space side of the vehicle. The thickness of the bumper layer is probably best about 10% the thickness of the major skin. The bumper should reduce the number of punctures by a factor of approximately ten to one-hundred times by exploding the meteoritic particles far enough away from the surface of the skin that only the vapor strikes the skin and this over a large area.

The extremely valuable observations by O. E. BERG and L. H. MEREDITH [18] show that above an altitude of about 80 kilometers an exposed surface receives impacts at a rate of about 1/min/cm² from meteoritic particles. Their device has measured the radiation from meteorite impacts and has a sensitivity of the order of 0.01 ergs. The number of particles included to the bottom of Table I is not adequate to produce the number of impacts measured by BERG and MEREDITH. On the other hand, the masses permitted by Table I are quite adequate to do so for meteoritic material divided into particles with dimensions the order of 1/10 micron. It seems quite possible that some such particles may exist in adequate numbers to produce the result obtained by BERG and MEREDITH. Light pressure on particles of dimensions near that of the major solar radiation, a moderate fraction of a micron, will blow all such particles quickly from the solar system. On the other hand, it is quite possible that smaller particles, with dimensions in a minimum of the radiation pressure curve, can exist. They would be rather difficult to detect by other means than by impact measuring devices in space. Their scattering powers, of course, must be very low to prevent radiation pressure from becoming an important factor. Their densities, too, must be comparable to that of the crystalline solid material.

A discussion of this general subject can best be made when the observational data are more secure and with a considerably greater theoretical study than is possible at the present moment. But it is important to note that the result by BERG and MEREDITH does not increase the calculated rate of surface erosion by meteoritic dust as given earlier in this paper.

References

1. F. L. WHIPPLE, Meteoritic Phenomena and Meteorites. Physics and Medicine of the Upper Atmosphere, ed. by C. S. WHITE and O. O. BENSON, p. 137. Albuquerque: University of New Mexico Press, 1952.
2. L. G. JACCHIA, The Physical Theory of Meteors. VIII. Fragmentation as Cause of the Faint-Meteor Anomaly. Astrophysic. J. **121**, 521 (1955).
3. R. E. McCROSKY, Some Physical and Statistical Studies of Meteor Fragmentation. Thesis, for Doctorate. Harvard University, 1955.
4. F. L. WHIPPLE, Photographic Meteor Orbits and their Distribution in Space. Astronom. J. **59**, 201 (1954).
5. A. F. COOK and F. L. WHIPPLE, Ms. in preparation.
6. D. W. R. McKINLEY, Meteor Velocities Determined by Radio Observation. Astrophysic. J. **113**, 225 (1951).
7. Various papers in: Meteors, ed. by T. R. KAISER. London: Pergamon Press Ltd., 1955.
8. F. L. WHIPPLE, A Comet Model. III. The Zodiacal Light. Astrophysic. J. **121**, 750 (1955).
9. E. ÖPIK, Interplanetary Dust and Terrestrial Accretion of Meteoric Matter. Irish Astronom. J. **4**, 84 (1956).
10. P. M. MILLMAN and M. S. BURLAND, The Magnitude Distribution of Visual Meteors. Private communication, 1957.
11. F. G. WATSON, Between the Planets, p. 115. Philadelphia: The Blakiston Co., 1941.

12. F. S. SINGER, The Effect of Meteoric Particles on a Satellite. Presented at ARS Meeting, June, 1956.
13. G. S. HAWKINS, A Radio Echo Survey of Sporadic Meteor Radiants. Monthly Notes Roy. Astronom. Soc. **116**, 92 (1956); Variation in the Occurrence Rate of Meteors. Astronom. J. **61**, 386 (1956).
14. H. PETTERSSON and H. ROTSCHI, Nickel Content of Deep-Sea Deposits. Nature **166**, 308 (1950); The Nickel Content of Deep-Sea Deposits. Geochim. Cosmochim. Acta **2**, 81 (1952).
15. P. HODGE, Opaque Spherules in Dust Collected at Isolated Sites. Nature **178**, 1251 (1956).
16. H. v. DE HULST and J. J. REESINCK, Line Breadths and Voigt Profiles. Astronom. J. **105**, 121 (1947).
17. H. SIEDENTOPF, Zur optischen Deutung des Gegenscheins. Z. Astrophysik **36**, 240 (1955).
18. O. E. BERG and L. H. MEREDITH, Meteorite Impacts to Altitude of 103 Kilometers. J. Geophys. Res. **61**, 751 (1956).

Optical and Visual Tracking of Artificial Satellites

By

Fred L. Whipple[1] and **J. Allen Hynek**[1]

(With 1 Figure)

Abstract — Zusammenfassung — Résumé

Optical and Visual Tracking of Artificial Satellites. The visual and optical tracking of artificial satellites will be carried out by volunteer "*Moonwatch*" teams and by precision photographic stations in various parts of the world. The data when transmitted and analyzed by the computations center at Cambridge, Mass., should provide much useful geodetic, atmospheric, and astronomical information.

Optische und visuelle Verfolgung künstlicher Satelliten. Die visuelle und optische Beobachtung und Verfolgung künstlicher Satelliten wird durch freiwillige Mondbeobachter(*Moonwatch*)-Gruppen und durch Stationen fur genaue photographische Aufnahmen an verschiedenen Stellen der Erde ausgeführt werden. Die Beobachtungsdaten werden an die Rechnungszentrale in Cambridge, Mass., gesandt und dort analysiert werden. Sie werden nützliche geodätische, atmospharische und astronomische Kenntnisse vermitteln.

Repérage optique et visuel des satellites artificiels. Le repérage optique et visuel des satellites artificiels sera accompli par des équipes de volontaires „*Moonwatch*" et des stations photographiques de précision en différents endroits du globe. Les donnés transmises et analysées par le centre de Cambridge, Mass., doivent fournir d'utiles informations en géodésie, astronomie et connaissance de l'atmosphère.

The program for optical tracking of the IGY and other artificial satellites consists of three distinct parts: the visual search and acquisition program, *Moonwatch*; the photographic tracking program, of high precision; and a communications, computations and analysis center located at Cambridge, Massachusetts, at the offices of the Smithsonian Astrophysical Observatory. The assignment of this task to the Smithsonian Astrophysical Observatory has been made by the U.S. National Academy of Sciences; the financial support is being provided through the U.S. National Science Foundation. The program is under the general control of the U.S. National Committee of the International Geophysical Year.

The recent calculations of Dr. T. E. STERNE and others which indicate that the contemplated IGY satellites may remain in orbit for as long as nine years, and that therefore it is possible that several satellites may be in the skies at the same time, impose obvious conditions on the planning and organization of the three separate phases of the optical tracking program. It is, for instance, likely that for all but a small fraction of the orbital life of the satellite it will be an inert

[1] Smithsonian Institution, Astrophysical Observatory and Harvard College Observatory, Cambridge 38, Mass., U.S.A.

object, observable as are the natural satellites, by reflected sunlight. The value of the artificial satellite as a scientific vehicle over the long haul clearly depends on the precision with which the satellite can be tracked optically. The plans, preparations and organizations described below have been made with these factors in mind.

The visual search and acquisition program, *Moonwatch*, consists of volunteer groups, each with 12 or more non-professional observers equipped with small telescopes. These telescopes, of 50-mm aperture and a 12° field, will patrol a large arc of the meridian at twilight periods during the early stages of satellite launching and during the very last stages when the satellites finally plummet through the lower atmosphere. Nearly 100 *Moonwatch* groups are now active in the United States and another 50 over a number of nations in the world. The organization in each country is under the direction of an IGY representative and in the United States is monitored by the Smithsonian Astrophysical Observatory. Mr. LEON CAMPBELL, JR., is in charge of the project, and Dr. ARMAND N. SPITZ acts as special coordinator of *Moonwatch* activities.

A special "Bulletin for Visual Observers of Satellites" is published from time to time by the Smithsonian Astrophysical Observatory, and Number 8 is in preparation. The Bulletin contains basic information concerning the nature of the *Moonwatch* tracking stations, various suggestions for solutions of specific problems, and background information as to the activities of the program.

In case of radio failure in satellites or in case of satellites without radio transmitters, *Moonwatch* groups will act as a search and acquisition corps to find lost satellites and to give adequate tracking information so that orbits can be calculated. When such orbital data are sufficiently accurate, the tracking can be continued by the precision photographic stations. The computational center is now prepared to compute orbits of satellites from observations of any type as they become available.

The committee of the *Moonwatch* program in the International Astronautical Federation has been of extreme assistance to the program by assisting in the activation of non-professional observers in many countries. Particularly valuable has been the translation of the Bulletin into Spanish by Mr. T. TABANERA and his dissemination of it among Spanish reading countries.

With the potentiality of satellites launched into orbits of higher inclinations than 35° to the earth's equator, it appears that *Moonwatch* teams in latitudes higher than 35—40° north and south can contribute materially to the satellite programs of the immediate future. The Smithsonian Astrophysical Observatory will be glad to assist in any way it can in the organization and operation of any *Moonwatch* groups insofar as it is within our power to assist, but interested groups should act speedily since little time remains for organization and training before the first satellite launching.

Our experiences in the United States in the organization of *Moonwatch* teams have been most gratifying. The cooperation on the part of observers who must give freely of their time and talents has exceeded our expectations. While the mainstay of the *Moonwatch* program in the United States is the amateur astronomer, many from other technical fields have joined forces with the amateur astronomer in the establishment and operation of the typical *Moonwatch* station. In many cases, financial support for the *Moonwatch* station has come from a local civic-minded person or group of persons, but in a number of cases the individual members have constructed or purchased their own equipment. The enthusiasm exhibited by most *Moonwatch* groups in the United States, which we have been able to observe directly, has had a heartening effect on the whole optical tracking program, and we have little doubt but that similar high morale exists in *Moonwatch* teams in other countries. In Japan, for instance, we note that more than 25 *Moonwatch* stations have been established and are operating.

The value of the *Moonwatch* organization to artificial earth satellite programs can hardly be overestimated. Even apart from the basic scientific value at the initial and final stages of a satellite's life, *Moonwatch* teams will serve a useful purpose during the interim. *Moonwatch* teams will serve as the local centers of satellite information, and as the means of training new observers and of interesting technically qualified people in the space age which so clearly is at its dawn. Its great value in attracting potential young scientists into astronautics and astronomy needs no elaboration to this audience. It is interesting to muse for the moment on the biography of some future scientists that might well state that his energies were turned toward science through membership as a youth on a *Moonwatch* team!

The precision photographic program centers about the use of the BAKER-NUNN tracking camera in which the optics have been designed by JAMES G. BAKER and the mechanical system by JOSEPH NUNN. The optics are being produced by the Perkin-Elmer Corporation of Norwalk, Connecticut, and the mechanical parts by Boller and Chivens Incorporated of South Pasadena, California. A precision time standard will be maintained at each station. The crystal clock developed by ERNST NORRMAN has been improved through the efforts of ROBERT J. DAVIS and twelve of these clocks are being produced by the Norrman Company. Time will be recorded on each photograph of the satellite so that it can be read to 0.0001 seconds although generally the maintenance of absolute time at field stations cannot be expected to exceed about 0.001 seconds.

The optical system consists of a 20-inch apochromatic three-element correcting plate, with four surfaces aspherical, for a SCHMIDT system of focal length 20 inches and spherical mirror of 31 inches aperture. Good image qualities are expected over a field of diameter 30° although in practice a strip of film covering only $5 \times 30°$ will be utilized. The mounting of the camera will involve three axes so that the telescope can be pointed to any direction of the sky and the strip film oriented along an arbitrary great circle. The telescope can then follow motion along this great circle at any angular rate up to 2° per second with a precision of about 1 % of the motion.

The operation of the camera will be cyclical, consisting of motion following the expected satellite angular motion, and motion to follow the fixed stellar background as a fiduciary system. A rapidly rotating shutter will be synchronized with an exposure system so that the clock time can be photographed at a known position of the rotating shutter during each exposure. The use of film strips makes possible the quick changing of film so that short exposures of the order of 0.2 seconds can be anticipated during bright twilight conditions. With such a fast system, this exposure corresponds roughly to a sky brightness under conditions of full moonlight. The telescopes should photograph 50-cm spheres to a distance of 2500 km or more and 6-meter spheres to the moon's distance.

The photographic system, with this fiduciary star background, is expected to use a position determination of at least 2 seconds of arc normal to the trajectory of the satellite, and probably about 5 seconds of arc along the direction of motion. Since the satellite moves at a linear speed of 25,000 feet per second, the precision in time determination corresponds to about 25 feet along the trail, which would be 5 seconds of arc at 200 miles or 1 second of arc at 1,000 miles distance. Thus the expected angular precision of the photographic system is compatible with the accuracy and time. This is also true in the visual system, *Moonwatch*; here the required accuracy is a thousandfold less than in the precision photographic system, the observers being required to note the time of passage of the satellite to the nearest second of time and its position to within a degree. Since the satellite moves approximately 1 degree per second near perigee, these requirements are met.

One can expect the photographic technique to yield directions of the satellite from the tracking stations with precisions of the order of 10 meters, or conversely interpolate the positions of the stations to such an accuracy. The measures can be used according to two basic principles: (a) triangulation (effective), and (b) gravitational effects. Because of the long ranges of observation expected, it is possible that direct triangulation can be achieved between certain of the stations. Generally, however, the calculated orbit will provide a means of effective triangulation in that the position as seen at one station can be transferred to the equivalent time and position at another station. Thus triangulation will be achieved through precise orbital calculations. By this technique it is hoped to tie together the observing stations and the center of the geoid to a precision of the order of 10 meters. Much depends upon the degree to which gravitational anomalies reduce the precision of the effective transfer of position by orbital means. Until actual measurements are made on satellites, it will be difficult to estimate the degree of uncertainty or "noise" which the gravitational anomalies will introduce into the orbit and orbit calculations. Regardless of this effect, however, a number of the stations can be tied together by simultaneous or practically simultaneous observations to this order of precision. The limiting accuracy, of course, depends upon the accuracy of the basic triangulation networks separating stations in the respective networks.

It is of considerable interest that extremely high precision in the time of observation is not critical to the determination of relative positions of stations that are not located too near the northern and southern extremes of the satellite track. Since the satellite orbit will cross other stations at appreciable angles to the east-west line, the number of measurements of the position of the satellite normal to its track will determine essentially a "Sumner" plane in space. Observations made when the satellite is moving northward compared with those when it is moving southward will provide a position determined by the intersection of these planes. Each passage of the satellite will provide a determination of a plane generally inclined at a considerable angle to the other Sumner planes. The accuracy of this method in determining position will then depend upon a limited function of the orbital parameters which will be less dependent upon precise time determinations than directions obtained from a single observation.

The photographic observing stations are now under active construction, under the direction of K. G. HENIZE, at the following positions: Organ Pass, White Sands, New Mexico; West Palm Beach, Florida; atop the crater Haleakala on the island of Maui, Hawaii; at the Tokyo Astronomical Observatory, Mitaka, Japan; Naini Tal, India, in the state of Uttar Pradesh; Shiraz, Iran; Cadiz, Spain; Santa Barbara, on the island of Curaçao, Netherlands Antilles; Arequipa, Peru (at the site of the old Harvard Observatory active during the early part of this century); Villa Dolores, Argentina; Johannesburg, South Africa; and Woomera Range, Australia. These stations are being established as cooperative ventures between the respective National IGY Committee and that of the United States, and in many cases the help of local astronomers in establishing and operating these stations has been obtained. The triangulation determinations of position, as described above, will determine with high precision the shape of the geoid and the relative positions within and between geodetic networks covered by the observing stations.

Perturbations of the orbit will not only provide precise information as to the density of the high atmosphere by the effect of its drag, but will provide a very sensitive criterion of the distribution of mass in the earth. Coupled with the triangulation measurements of station position, the perturbations are expected to add appreciably to knowledge of the density distribution in the earth, particularly in the crustal volumes.

The successful achievement of the aims and objectives of the program with regard to geodetic results depends to a large extent on the ability to determine the exact orbit and orbit variations by means of high-speed computing machines. Before discussing this terminal phase of the work, the processing of the observational raw material, it is clear that the success of the program at any phase is predicated on the maintenance of an effective communications system. Such a system must operate efficiently from the time of the initial alert to *Moonwatch* stations and the rapid receipt of the first *Moonwatch* observations, through the long routine period of daily photographic observations, to the final stages of a given satellite's life when it plummets earthward, and must remain operative as long as any of the contemplated IGY satellites are aloft. High speed communications are essential for the transmittal of satellite flight data from *Moonwatch* station leaders to the computing center in Cambridge and for the transmittal from the center in Cambridge of satellite ephemerides to each of the precision stations immediately before expected satellite passage so that the BAKER-NUNN SCHMIDT telescopes can be positioned accurately and set into operation at the precise transit time.

At this time it appears that regular commercial radio or cable channels offer the best possibilities for such a communications system, with appropriate priorities for the attainment of the desired speeds of communication. In some few cases, there are no commercial services of any kind between the points involved and the United States and reliance will be placed on U.S. Governmental facilities. Examples of these latter cases are the Yap, Truk and Wake Islands in the Pacific. Wherever possible, we plan to have one prime route, and one alternative route for use in the event the prime route is inoperative or delayed for any reason.

We have assurances of the utmost cooperation from the United States owned communications systems which might be involved, and from the military services where their facilities might be employed. Both have agreed to permit the use of appropriate priorities.

It is expected that a regular routine communications schedule will be maintained between the Smithsonian Astrophysical Observatory and each of the twelve photographic stations. With respect to the much more numerous *Moonwatch* stations, it is expected that following a general alert, each *Moonwatch* team will report separately and individually to Cambridge after satellite passage. The only exception to this general plan is Japan, where arrangements contemplate that all *Moonwatch* stations in Japan will report to the photographic tracking station at the Tokyo Observatory, and that it in turn will forward the results over the same channel as that used for the exchange of data derived from that station. Similar arrangements are under discussion with respect to *Moonwatch* teams in other IGY countries.

Upon receipt of the observational data at Cambridge, which will already be in a form that can be immediately introduced into high speed computing machines, the calculations will be carried out through the courtesy of the International Business Machines Corporation by means of their No. 704 calculator at the Massachusetts Institute of Technology.

The programming and coding of this problem has been carried through the phases of preliminary orbit calculation and practical ephemeris calculations including first-order perturbations of the equatorial bulge. This work is under the direction of D. LAUTMAN. Careful studies are being made of two alternative methods for determining definitive orbits, which will include corrections to station positions and perturbational terms of the earth's distribution of mass as unknowns. Whether a general perturbations approach will be preferable to a special perturbations step-by-step integration will be determined in the near future. Possibly both methods will be attempted in order to test which is more feasible and which gives the highest precision.

The computational program will be one of successive approximations which eventually will yield refined and more accurate values of those earth's parameters

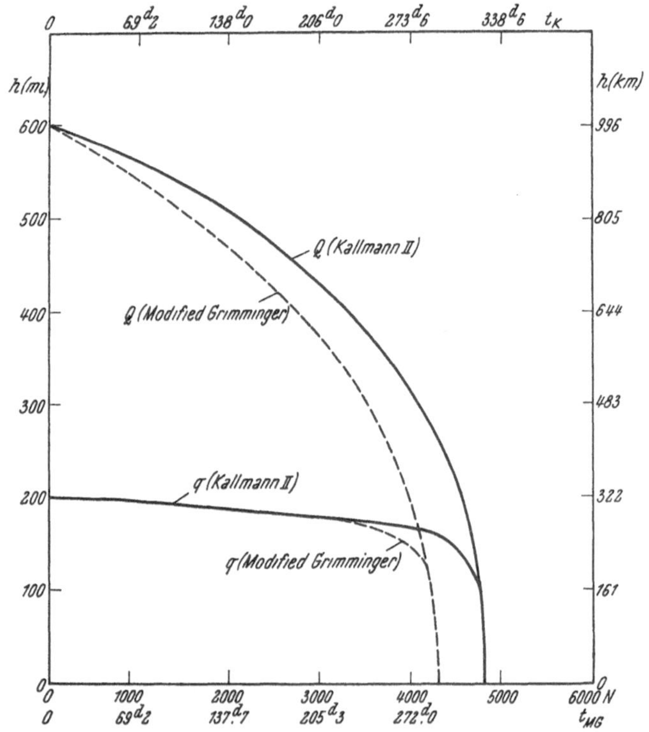

Fig. 1. Variation of perigee and apogee altitude with number of revolutions and with time

which are the basis for the theoretical aspects of geodesy. The applications of such precision measurements and calculations to the field of geodesy are unique in character and represent a fundamental step forward.

The drag of the earth's atmosphere will produce a systematic reduction in the eccentricity and in the major axis of a satellite orbit without the introduction of any errors normal to the orbital plane. The effect of the atmospheric drag occurs almost entirely in the neighborhood of perihelion because of the rapid fall of atmospheric density, roughly exponentially from the surface upward. As a consequence, aphelion distance of an eccentric orbit is mostly affected in the early stages until the eccentricity has been greatly reduced. The change in the perigee distance is rather small until it is approximately equal to the aphelion distance, at which time the decrease in center major axis is greatly accelerated and the satellite has arrived nearly at the end of its course. The nature of the effect is indicated in Fig. 1.

The rough observations and rather crude computational methods should provide good measurements of mean atmospheric densities as a function of height up to initial perigee distance down to the level at which the orbital changes become too rapid for reasonably good observation. On the other hand, a precise theory and computational effort is capable of yielding many more results concerning atmospheric density, particularly the value of the density a few kilometers above the initial perigee distance, and periodic effects or predictable cyclic effects that may occur in the earth's high atmosphere. The density variations in the high atmosphere caused by lunar tides, solar tides, diurnal effects, seasonal effects and possibly effects induced by solar activity may well be determinable from precision satellite observations coupled with a precision theory and accurate calculations. Many atmospheric effects in which the phase relationships are predictable in character are potentially amenable to statistical treatment. They will show as slight variations in the rates of change of apogee and perigee distances as a function of time, where the phases of the time variation can be predicted.

In the photographic program background starlight is refracted by practically the entire atmosphere (to one part in 10^{12}) in the same fashion as that of the satellite; hence, there is only a minor refraction correction which can be made with high-level precision. On the other hand, free electrons in the higher atmosphere produce a "refraction" of the radio waves which may amount to a few minutes of arc. A comparison of the photographic and radio measurements will, according to a suggestion by J. A. O'Keefe, make possible the determination of the total electron content of the atmosphere between the observer and the satellite. This effect will be utilized as a physical experiment in the satellite program without involving a measuring device in the satellite.

The evidence at hand, as mentioned earlier, indicates that satellites launched into the anticipated orbits in the IGY program will have lifetimes extending well beyond the period of the International Geophysical Year. Furthermore, that satellites will be launched in other programs appears to have a probability approaching inevitability. It is hoped that financial support for the continuance of essential optical observation will match this probability, even though, at the moment, the financing of the optical and visual tracking program extends only to June 30, 1959. The geodetic, atmospheric, and astronomical results from the program, however, will become accumulatively more significant with an increase in the number of observations and with extended observations of satellites in varying orbits. Note that the precision camera program can photograph 6-meter spheres to the distance of the moon.

Likewise, a continued *Moonwatch* tracking program would be of immense value in determining the fate of radio-lost satellites. Such satellites, in fact, are practically as useful for geodetic results by means of optical tracking as are their radio operating brothers. The limitations of the twilight periods of observation will be enormously relaxed for satellites in orbits at greater heights than those anticipated for the IGY satellites. We can join, therefore, with the members of this audience in anticipating future satellites and more distant satellites until indeed it may someday come to pass that the distance from the earth to our natural satellite will be amply marked by the orbits by man-made moons.

Ten Years of Plastic Balloons

By

Otto C. Winzen[1]

(With 15 Figures)

Abstract — Zusammenfassung — Résumé

Ten Years of Plastic Balloons. 1957 is the 10th anniversary year of the plastic stratosphere balloon. During this brief history the new balloons have become reliable vehicles for upper atmosphere research and operational work. From their early duties as delicate carriers of light loads, they have now been perfected to the point where they can be used for the routine launching of light loads to altitudes in excess of 120,000 feet and heavy loads to altitudes of 80,000 feet and higher. Gross weights of over 5000 pounds have been flown to date and balloons with volumes in excess of 3 million cubic feet are being built on a production basis. The proposed 5 million cubic-foot balloon can carry a gross load of 10,000 pounds to 85,000 feet, but twice that load to 70,000 feet, doubling the load at a cost of only 15,000 feet altitude. Performance capability of a family of plastic balloons is shown.

The plastic balloon is also a practical tool for the investigation of the human factors of space flight and manned space laboratories for high-altitude studies. Most spectacular applications are the recent manned high-altitude flights on USAF-Winzen Research Project *Manhigh* as well as the ONR-Winzen Research Inc. *Rockoon* launchings in which balloons are used as platforms in space-equivalent conditions for the launching of rockets, boosting their altitude capability substantially. This paper presents some details of balloon applications and takes a look into the future of balloons which stems from the development of the plastic cell. The future looks bright indeed.

Zehn Jahre Erfahrung mit plastischen Ballonen. 1957 feiert der plastische Stratospharenballon seinen 10. Geburtstag. Während ihrer kurzen Geschichte sind die neuen Ballone verläßliche Fahrzeuge zur Erforschung der Hohen Atmosphare und fur die Arbeit dort geworden. Von ihrer ersten Aufgabe als empfindliche Träger leichter Lasten sind sie bis zu dem Punkt verbessert worden, wo sie fur routinemäßigen Flug leichter Lasten bis zu Hohen von über 36 km und schwerer Lasten bis 24 km und daruber verwendet werden konnen. Gesamtgewichte von uber 2300 kg wurden bis heute emporgeflogen, und Ballone mit einem Fassungsraum von uber 85000 m³ werden auf Produktionsbasis hergestellt. Der geplante Ballon mit 142000 m³ kann eine Bruttolast von 4500 kg auf 25,5 km Höhe heben, aber die doppelte Last auf 21 km. Es tritt also bei Verdoppelung der Last ein Verlust von nur 4,5 km ein. Es wird die Leistungsfahigkeit einer Familie plastischer Ballone gezeigt.

Der plastische Ballon ist auch ein praktisches Werkzeug fur die Untersuchung der menschlichen Faktoren des Raumfluges und für bemannte Raumlaboratorien zum Studium in größeren Höhen. Die bemerkenswertesten Anwendungen sind die jüngsten bemannten Höhenflüge mit dem USAF-Winzen-Forschungsprojekt *Manhigh*, ebenso aber auch die ONR-Winzen Research Inc. *Rockoon*-Flüge, wobei Ballone als Plattformen bei weltraumäquivalenten Bedingungen fur den Start von Raketen benutzt wer-

[1] President of the Winzen Research Inc., 8401 Lyndale Avenue South, Minneapolis 20, Minnesota, U.S A.

den, wodurch die Fähigkeit, größere Höhen zu erreichen, wesentlich vermehrt wird. Die vorliegende Arbeit gibt einige Einzelheiten der Ballonanwendungen und bringt einen Blick in die Zukunft der Ballone, die sich aus der Entwicklung der plastischen Ballone ergibt. Diese Zukunft sieht in der Tat freundlich aus.

Dix années d'expérience avec les ballons en plastique. L'année 1957 marque le dixième anniversaire de l'utilisation du ballon stratosphérique en matière plastique. Durant cette période les nouveaux ballons sont devenus des instruments sûrs pour la recherche et le travail courant en haute atmosphère. D'abord porteurs fragiles de faibles charges, ils ont été perfectionnés au point de pouvoir être utilisés pour le lancement routinier de faibles charges jusqu'à 36 km et de fortes charges jusqu'à 24 km et plus. Des charges brutes de plus de 2300 kgs ont été soulevées et des ballons déplaçant plus de 85000 m³ sont en production courante. Un projet de 142000 m³ peut soulever 4500 kgs jusqu'à 25.5 km et le double de cette charge jusqu'à 21 km. Les performances d'une famille de ballons plastiques sont présentées.

Le ballon plastique est aussi un instrument pratique pour l'étude des facteurs humains du vol spatial et celle des laboratoires volants. Parmi les applications spectaculaires citons les récentes ascensions à haute altitude du projet USAF-Winzen "*Manhigh*", et les ascensions du projet ONR-Winzen-Research Inc. *Rockoon*, au cours desquelles les ballons servent de plate-forme spatiale pour le lancement de fusées, augmentant substantiellement leurs performances d'altitude. Cet article présente quelques détails sur ces applications et jette un regard sur l'avenir que promet aux ballons le développement de la cellule plastique. Cet avenir s'annonce décidément brillant.

I. History and Accomplishments

The modern plastic balloon in its very short history has become a recognized vehicle for serious upper atmosphere research operations supported by all research branches of the Armed Forces of the United States. The Navy's projects *Skyhook*, *Rockoon*, *Strato-Lab*, and the Air Force's projects *Moby Dick*, *Manhigh*, and animal cosmic-ray flights are some of the more prominent.

Fig. 1. History of balloon design, showing classical bag of spherical shape with net suspension (1784), rubberized fabric bag with catenary suspension (1930), and finally the plastic balloon with low-stress shape, load bands heat-sealed into each seam, and gondola suspension through open parachute (1950)

The original plastic stratosphere balloon developed in 1947 [1] was an extremely lightweight cell made of thin plastic film. This balloon was designed to carry a load of 70 pounds to an altitude of 100,000 feet, or above approximately 99% of the atmosphere. This original *Skyhook* balloon had a diameter of approximately 70 feet and a volume in excess of 200,000 cubic feet. Its original development was sponsored by the Office of Naval Research. Commander GEORGE W. HOOVER was Project Officer.

Until the advent of the plastic stratosphere balloon, small rubber balloons were the means of conducting a limited upper atmosphere research, although such balloons were not capable of carrying out level flight. The expensive rubberized fabric balloons were reserved for manned flights. Dr. AUGUSTE PICCARD pointed the way to the stratosphere with a large manned balloon of this type and more such flights were made in the 1930's culminating in the classical flight of the Explorer II to an altitude of 72,000 feet in 1935 [2]. This world record flight by Captains STEVENS and ANDERSON of the U.S. Army established the usefulness of a manned balloon station at high altitude and yielded valuable data on cosmic rays and other upper atmosphere phenomena. However, as a result of the tremendous cost and logistics of this type of flight, the use of balloons as upper atmosphere research vehicles faded into the twilight and no other manned flights were attempted for over 20 years. The new, inexpensive and more easily launched plastic balloon made such flights practical again. Fig. 1 shows the history of balloon design.

The development of the first plastic balloon came slowly. It was necessary to find a plastic material and then to produce the thin film which would meet the rigorous requirements of a balloon fabric. Thin polyethylene film, a household word today, is a by-product of this development.

Next came the problem of developing a production heat-sealing method to generate the miles of gas-proof seams which hold the bag together. On reflection, it is recalled that on the first plastic balloons the adhesive tape served as much to seal the many leaks as they did to support the load. And yet, the first big *Skyhook* balloon, which the author launched on 25 September 1947, reached 100,000 feet despite millions of tiny pinholes in the film, leaky seams covered with adhesive tapes, and makeshift launch method.

It is also recalled that then followed two failures, after which it was considered necessary to call the first "balloon conference". One of the next flights refused to come down for 3 days, and it became apparent that we had a long way to go to develop reliable control and radio equipment and insulated containers which would function in this new environment for extended periods. No commercial instrumentation was available. But this flight had accomplished a major break-through. It is pleasant to remember the excited phone call from Drs. HORNBOSTEL and SALANT of Brookhaven Laboratories [10] who had asked that a few cosmic ray track plates be flown on this flight. They said, "It will take years to analyze the wealth of cosmic ray events in these plates!" From that day on, cosmic ray research and plastic balloons became inseparable. Project *Skyhook* was on its way.

The plastic balloons had caused another, less scientific, sensation. One of the first flights, hanging motionless over Minneapolis and St. Paul, glowing in brilliant sunset colors 30 minutes after the sun had set on the surface, caused a near panic. The flying saucer craze became rampant. It is amusing now to recall that it was often possible to trace runaway flights by the reports of flying saucers. The balloons, clearly visible as white stars during the day, glowing fireballs when the sun set on them 30 to 40 minutes after ground sunset, had quite innocently captured the imagination of the public.

Many scientific accomplishments can be attributed to the new vehicle, and the development of balloon technology took giant steps forward in the past 10 years. The appendix shows some of the discoveries, results and improvements. After intimate association with this work since its inception, it is a list which can be reviewed with considerable satisfaction in the knowledge that this endeavor vindicated faith in the balloon by making a useful vehicle available to science.

The plastic balloon, as a result of its low cost, light weight, high load-carrying ability, high altitude capability, and relatively simple launching techniques, has started a renaissance of ballooning. The possibilities of its application challenge the imagination and make the next 10 years, now that the vehicle is generally accepted, a much more fruitful field of work. The possibility of its use not only as an aerostat but eventually as a useful tool in space work opens up a new realm of imaginative thinking. Thus, the balloon, man's oldest aerial vehicle, appears to be destined to become linked with the rocket, man's newest aerial probe, to discover the secrets of space.

II. The Modern Plastic Balloon Vehicle

The first plastic *Skyhook* balloon reached an altitude of 100,000 feet with a payload of approximately 70 pounds. It was of radically new design and construction. Instead of a fabric envelope, polyethylene film only 1/1000th of an inch thick (1 mil or 0.025 mm) formed the skin of the bag. There was no net as on Professor CHARLES' first balloon, or catenary construction as in the earlier days of the classical balloon design, which reached its highest pinnacle with the flight of the Explorer II. The Explorer II balloon represented about the ultimate advancement of the old type of balloon construction. Because of the heavy weight of the bag itself, altitude capability was limited.

The new design incorporates its stressed load-suspension members integral with the envelope. The load bands or tapes follow the meridianal seams at the edge of the gores and the load is suspended from the base of the usually sphero-conical or onion-shaped envelope. Thus, the delicate plastic skin serves primarily as a barrier material for the lifting gas and only secondarily as a load-carrying member, rather than as the main load-carrying member as in earlier balloon designs which usually employed a spherical envelope. In our latest balloon construction, the stressed load bands consisting of a laminate of polyethylene films with glass, nylon, or fortisan filaments, are integrally heat-sealed into the balloon seams during the sealing operation. These load bands are now available in strengths from 60 to 1000 pounds. Thus, it is possible to build balloons custom-tailored to the individual requirements of a particular balloon construction or performance specification. The plastic film, polyethylene, is now available in thicknesses from 1/2 to 3 mil for use as balloon material. Polyethylene even today is the optimum plastic balloon material available. It has been improved substantially in strength, cold temperature resistance and gas transmission. A new film, "Mylar" polyester, is being used experimentally but its high cost, poor tear resistance, and its inability to be heat-sealed make it as yet undesirable from a production standpoint. Laminates of these and other films are also used.

Figs. 2 and 3 show the altitude performance of a family of plastic balloons as a function of gross weight (payload plus balloon weight). Balloon weights are shown at the intersection of the appropriate curves, for polyethylene gages from 75 to 2 mil, and balloon sizes from 12-foot diameter to 233-foot diameter (volumes: 910 to 5,000,000 ft.^{3}) are given. The weight of the payload is then added to obtain gross weight and thus floating altitude.

The first curve has a range of 2000 pounds and demonstrates clearly the high-altitude capability, the second a range of 20,000 pounds of gross weight pointing out the heavy-load capability. The tables are based on the I.C.A.O. Standard Atmosphere and helium inflation. In our current work, these tables are used as the basis for the selection of a balloon to fit the performance requirements of a particular flight.

Our development in 1955 of a balloon construction with integrally heat-sealed load bands made possible the construction of a balloon which would carry heavy loads without placing undue stress upon the plastic film. As an example, the *Manhigh* balloon is made of 1.5 mil gage (.038 mm) polyethylene and incorporates 70 integrally heat-sealed load bands of 500-pound test each. Thus, strength of the balloon assembly is 35,000 pounds and its breaking strength is on the order of 12,000 pounds. The 2,000,000 ft.^{3} Winzen Research *Strato-Lab* balloon is made of 2 mil gage polyethylene and incorporates 120 integrally heat-sealed load bands of 500-pound test each. The strength of this balloon

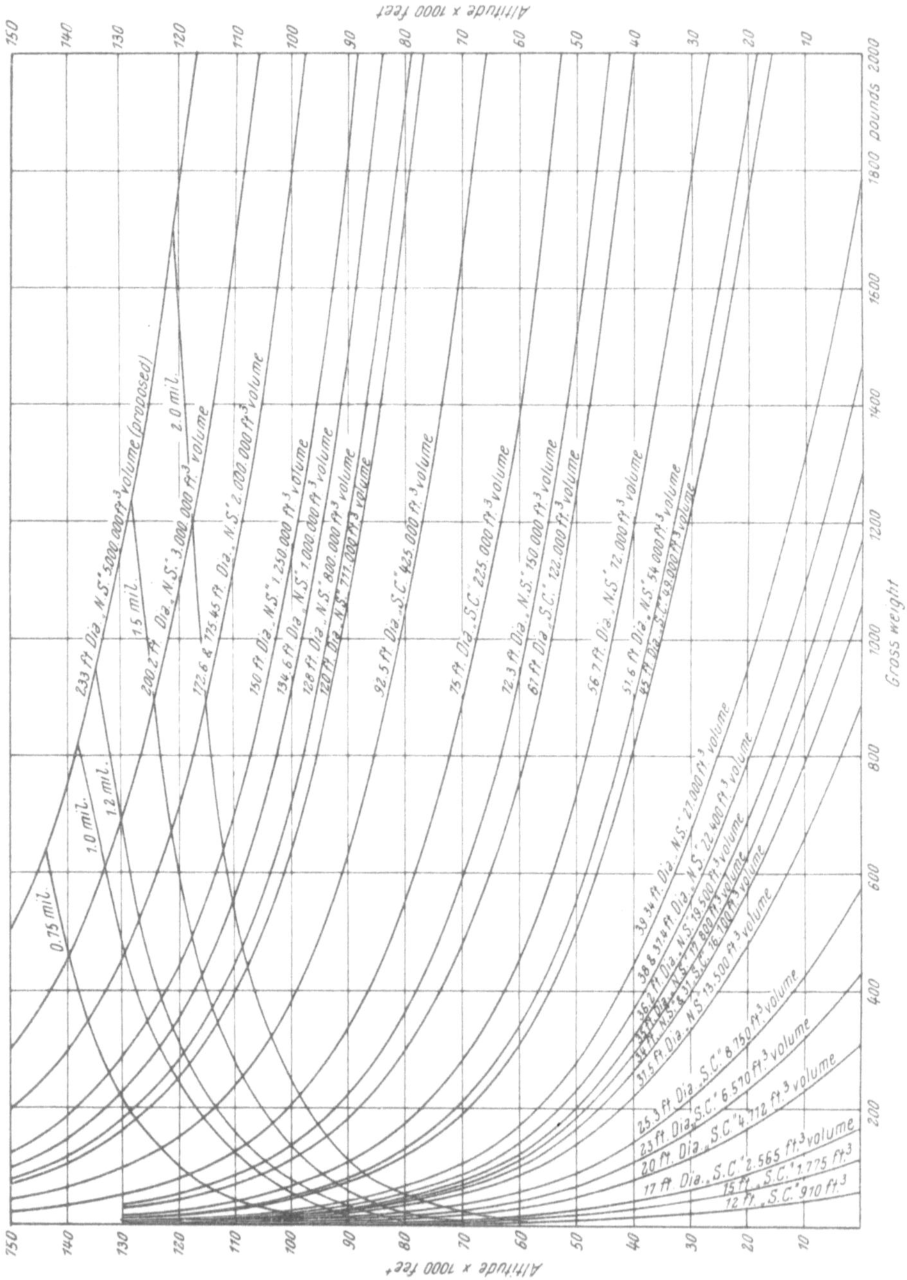

Fig. 2. Winzen Research Inc. Balloon performance chart, 2,000 lb. gross lift, date 14 December 1956, helium inflation, based on I.C.A.O. standard atmosphere, May 1954

assembly is, therefore, 60,000 pounds and its breaking strength is on the order of 20,000 pounds. Thus, a heavy duty plastic balloon is now available. Hardware, instrumentation, and launching operations have also been worked out.

Another aspect of the usefulness of the plastic balloon as a heavy-load vehicle is that the performance curve is relatively flat at the heavy-load end. As an example, the range of altitude for heavy loads is between 80,000 and 100,000 feet,

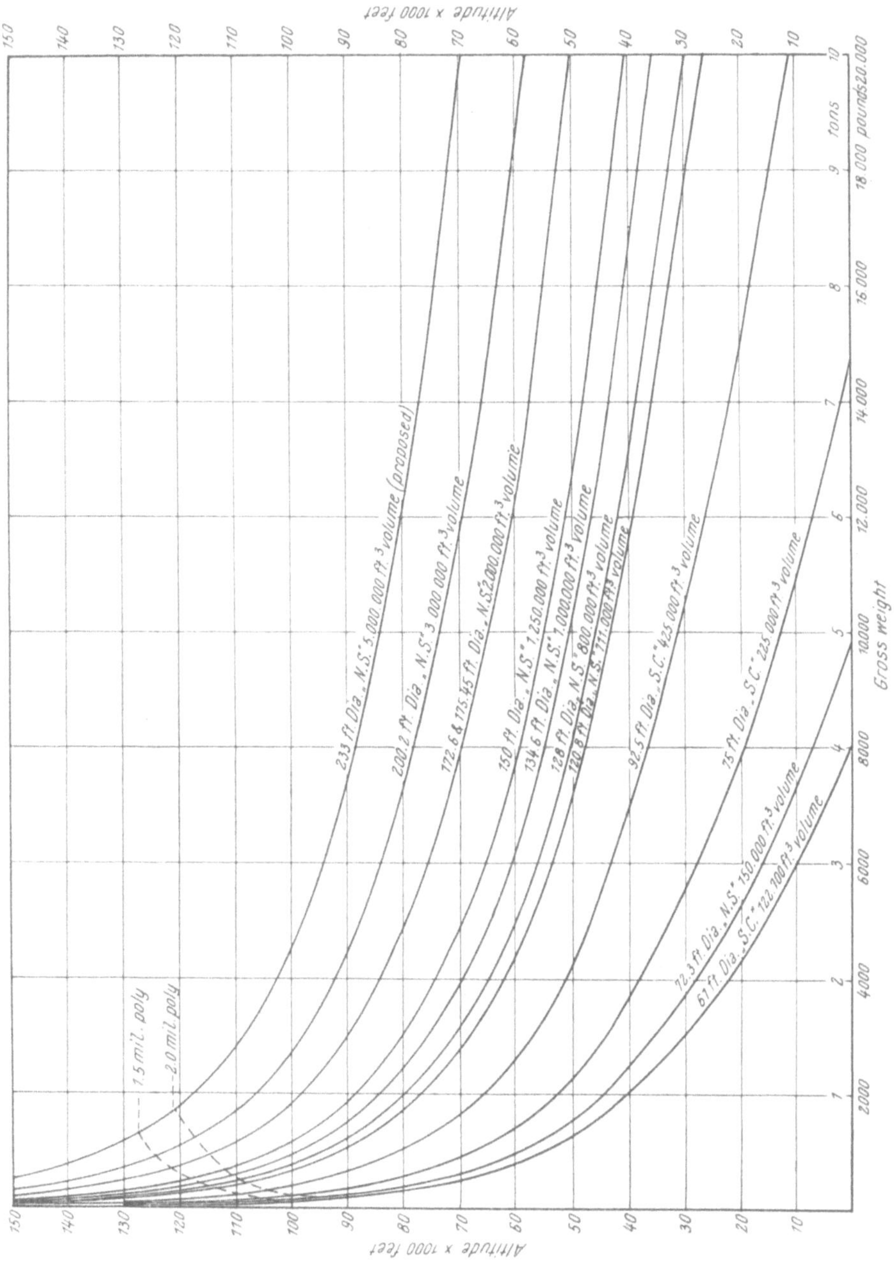

Fig. 3. Winzen Research Inc. Balloon performance chart, 20,000 lb. gross lift, date 8 October 1956, helium inflation, based on I.C.A.O. standard atmosphere, May 1954

but it is possible to carry much more weight for a small penalty of altitude. For example, the proposed 5 million cubic foot balloon can carry a gross load of 10,000 pounds to approximately 85,000 feet, but twice that load to 70,000 feet. In other words, it is possible to increase the load by 100% at a cost of only 17.5% of altitude. It almost appears that there is no limit to the size of plastic balloons which can be built or flown.

At the high altitude/low-payload end of the curves the availability of 1/2 mil polyethylene now would make it feasible to carry a 100-lb. payload to 150,000 feet pressure altitude (1.47 millibar) using the proposed 5,000,000 cubic foot balloon. (Also see enclosed conversion tables.) In this case, for each additional 15,000 feet of altitude, it is necessary to cut the gross weight (balloon and load) of the aerostat in half, and it is here also that the light-weight plastic cell really stands out. Lighter and stronger plastic materials may yet boost altitude capability substantially.

The 5 million cubic foot balloon is a realistic extrapolation into the immediate future. The success of the 3 million cubic foot balloon has prompted the immediate consideration of this larger size and production facilities are already available. Clusters of plastic balloons are considered a practical approach, but more work is needed to make them operational. When it is recalled that the plastic balloon is only 10 years old, it is difficult to foretell what its future may hold only 5 years from now.

III. Balloon Flight Operations

Modern ballooning faces a host of problems unknown to the early-day pioneer. Dense airplane traffic, telephone and high-tension lines, and high radio masts are hazards of the new era. The much higher altitudes reached by plastic balloons require basically new balloon instrumentation. These include new electrical circuity, valves, controls, transducers, and other instrumentation.

Agreements with the C.A.A. (Civil Aeronautics Administration) have been formulated for the notification of air traffic and restriction of launchings to good

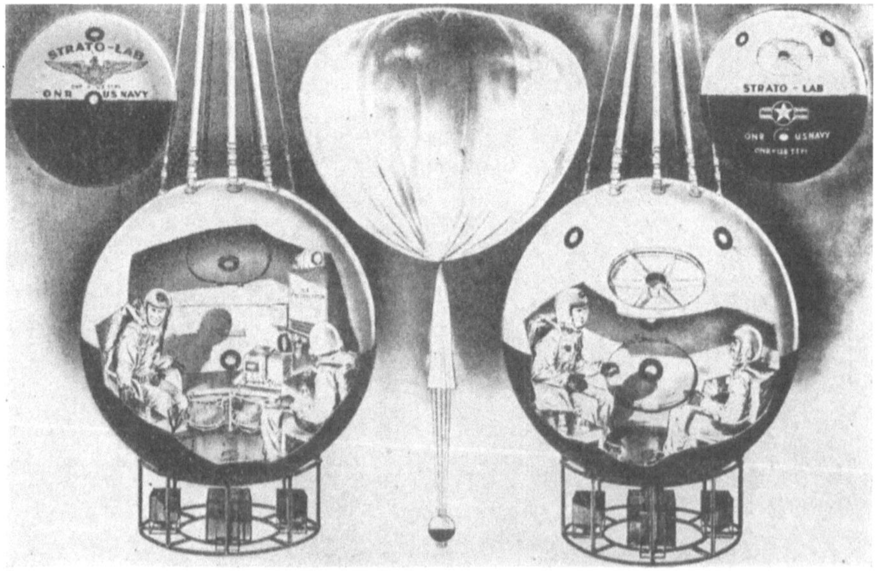

Fig. 4. Artist's view of ONR-US Navy Project *Strato-Lab*

weather conditions. A complete set of reliable balloon control instrumentation and batteries have been developed which overcomes the problems of low density, and extremes of temperatures at altitude between day and night conditions. As a result, there is now available a complete set of radio beacons, airborne telemetering of fast and slow speed (Morse code and PDM systems), special temperature and pressure transducers, multi-channel radio control instrumentation, airborne orienters, and many other types of instruments and equipment. Information telemetered from the balloon is copied by all tracking facilities and is transmitted in the Morse code or PDM system.

As an example of the modern aerostat, Fig. 4, shows an artist's view of the configuration of the Winzen Research Inc. *Strato-Lab* system at ceiling altitude. Below the balloon there is an open parachute which carries the gondola. This has become the standard method of load suspension in our flights. The gondola and its components and attachments are suspended through the shroud lines of the parachute. In case of balloon failure, or to make a parachute descent, the connection between the base of the balloon and the top of the parachute is severed by means of an electric squib cannon. The parachute opens immediately, at low opening shock, to initiate the descent of the load.

Two standard methods are used to launch the plastic balloon: First is the *Skyhook* launching method developed by the writer for the original launchings 10 years ago. It has been refined over the years and now incorporates a mobile launching platform. The system restrains only that portion of the top of the balloon which will contain, in a tight bubble, the entire volume of lifting gas required to raise the balloon. In the case of the plastic balloon, this may be a small bubble representing only 1% of the total volume of the envelope. The rest of the balloon is laid out on the ground. The load is attached to a load line, which in turn is fastened to the base of the parachute and to an anchor vehicle. At the time of launching, the launch bubble whose lift has been determined by weigh-off at the launch platform, is released and soon balloon and all items of equipment are airborne. At this time, the load line is cut at the anchor point or, by remote control, at the base of the parachute.

For heavy loads (in excess of 1 ton) this original *Skyhook* launch method is no longer practical because the dynamic stresses imposed on the plastic film at the time of launching become excessive. Vertical launchings are therefore the rule for manned balloons, for the heavy-load balloon, and for delicate light loads as telescopes. Such launchings are made from pits or from the flight deck of a carrier which can artificially create zero-wind conditions.

The vertical launch method is illustrated in Fig. 5. In the Central Plains states, launchings of this type are rarely possible from the surface, and are therefore best performed at the base of an open pit mine. The wind conditions for such a pit launching are shown in Fig. 6. The inflation proceeds in zero-wind conditions at sunrise, normally in the presence of a temperature inversion. The inflation proceeds smoothly through several plastic filling sleeves and no dynamic stresses are imposed which might damage the plastic skin.

An interesting fact in connection with the launching of plastic balloons is that even complex, manned aerostats are launched with a crew of less than 10 men. The launching of similar flights in the past required a veritable army of launching personnel.

The cost of the envelope, which is expendable after each flight, is only a small fraction of the cost of the old type of rubberized fabric cell. Either helium or hydrogen may be used for flights. The relative mobility, flexibility and ease with which launchings can be conducted by an experienced crew are further advantages of the new plastic aerostat, and have contributed substantially to its current utility.

The typical aspect of the elongated plastic balloon at launching changes during its ascent until at ceiling altitude the balloon is completely filled. Rates of ascent of plastic balloons are normally between 400 and 1400 feet per minute and this rate of ascent is maintained fairly constant during the ascent diminishing

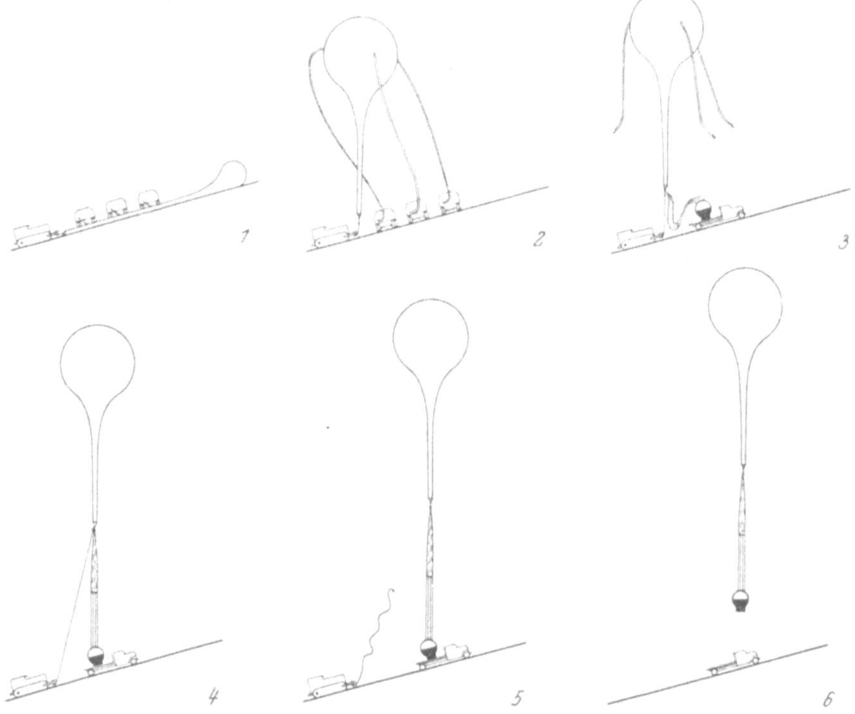

Fig. 5. Launching sequence for Winzen Research Inc. open pit flights for use on: ONR-Winzen Research Inc. Project *Strato-Lab*, ARDC-Winzen Research Inc. Project *Manhigh*

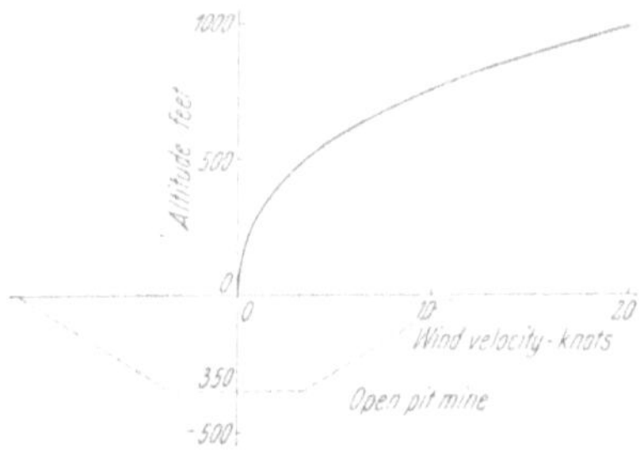

Fig. 6. Typical wind condition for open pit launch site. Diagram showing theory of pit launchings which makes it possible to extend calm conditions at the surface by the depth of the pit. In a vertical launching at the surface, the apex of the aerostat, which in a vertical launching would be 400 feet above the surface, will almost always be in a stratum of relatively high winds

slightly at the tropopause. Once the aerostat reaches ceiling altitude, that portion of the gas which constituted the free lift is valved off until the balloon is in equilibrium. It then begins a level flight in which the balloon becomes part of the surrounding air mass.

While floating at altitude, the vertical axis of the aerostat is perfectly stable. A very slow rotation is sometimes present, and this can be eliminated by the use of orienters using magnetic or other references. Floating at altitude, the balloon becomes an ideal high-altitude laboratory above the bulk of the earth's atmosphere, under space-equivalent conditions from many standpoints. The absence of accelerations, noise, vibration, and wind make it a laboratory ideally suited for high altitude observation whether manned or unmanned. In effect, the balloon becomes a low-level satellite of the earth. Flights of many days' duration have been performed on projects *Moby Dick* and *Transosonde*.

Elaborate tracking procedures have been developed for flights where the recovery of the load is mandatory. Our current tracking network consists of a single and a twin-engine aircraft, a communications van which serves as a mobile command post, and radio equipped ground vehicles for the recovery of the load. Tracking and recovery is accomplished by teamwork between aircraft and ground vehicles. The airplane is responsible for tracking the parachute or balloon-supported load to the ground and guiding the recovery crew to the landing site. Tracking is accomplished visually by our two theodolite installations, and in every case by radio direction finders homing on the balloon beacon. Radar tracking is often possible. All tracking vehicles and aircraft as well as the command post, whether it be the mobile van or the main communications center at the Minneapolis plant, have radio control capability to the balloon, and they can maintain continuous voice communication. This network virtually assures recovery of all loads flown in this area.

IV. Current Applications

Following is a partial list of current fundamental or operational research work performed with plastic stratosphere balloons:

Project Skyhook (U.S. Navy-ONR). As part of the ONR's program of basic research, the Navy offers the plastic balloon vehicle to scientists in the U.S. and abroad as a means of carrying out research at very high altitude (above 100,000 feet) in the fields of cosmic radiation, astronomy, aerial photography, and geophysics. This is the original plastic balloon project and it has done much to point out the advantages of the vehicle as a basic research tool.

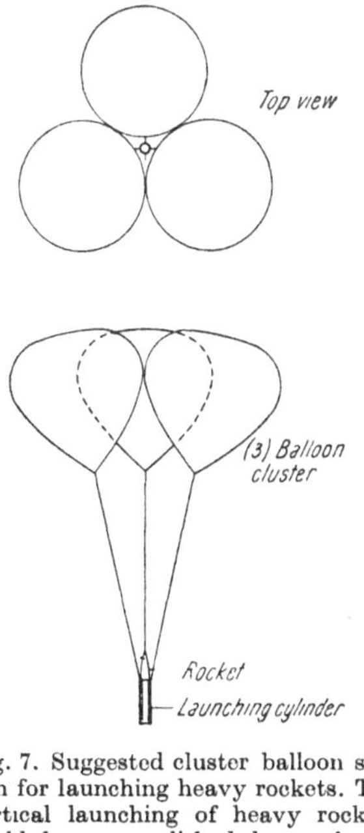

Fig. 7. Suggested cluster balloon system for launching heavy rockets. The vertical launching of heavy rockets could be accomplished by a cluster of three balloons. Three two-million cubic foot balloons could carry a three-ton rocket to 90,000 feet for firing

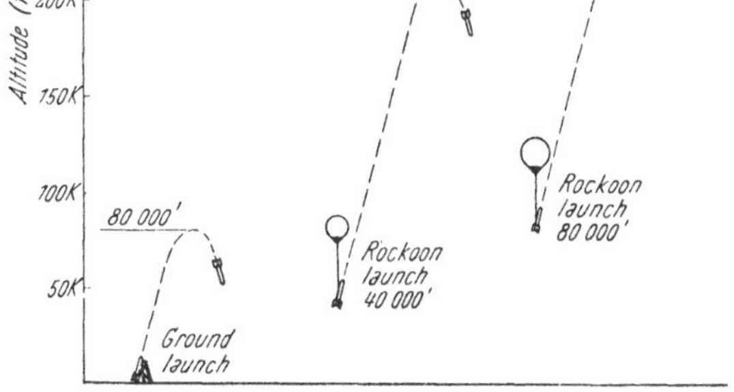

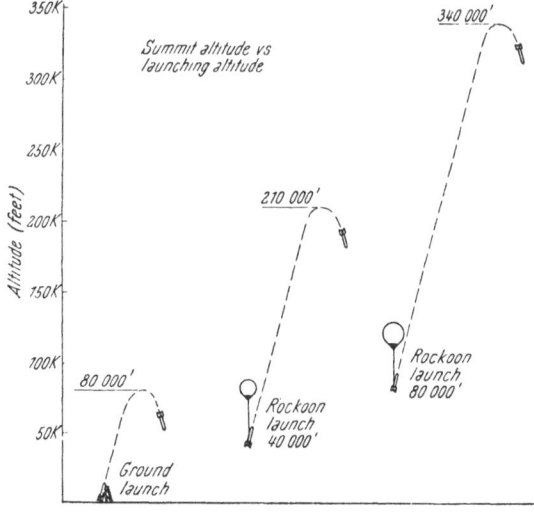

Fig. 8. Deacon rocket launching method

Moby Dick (USAF-ARDC). The Air Research and Development Command of the U.S. Air Force is engaged in a project which aims at exploring the meteorological problems including high-altitude circulation and the jet stream.

Object of this work are flights of many days' duration, therefore necessitating heavier payloads.

Project Transosonde (U.S. Navy-Bureau of Aeronautics). The project name is derived from "trans-ocean soundings". Indeed many of its flights span the great oceans. This operational project is devoted to the study of the circulation near the tropopause. Balloons fly along the isobars and directly obtain the large-scale pressure pattern currently being reconstructed on the basis of soundings in many locations. Its flights of three-day duration have in some cases spanned three-quarters of the earth's circumference. If successful in plotting high-altitude pressure patterns, it may eventually eliminate the lonely sea duty of the weather ships.

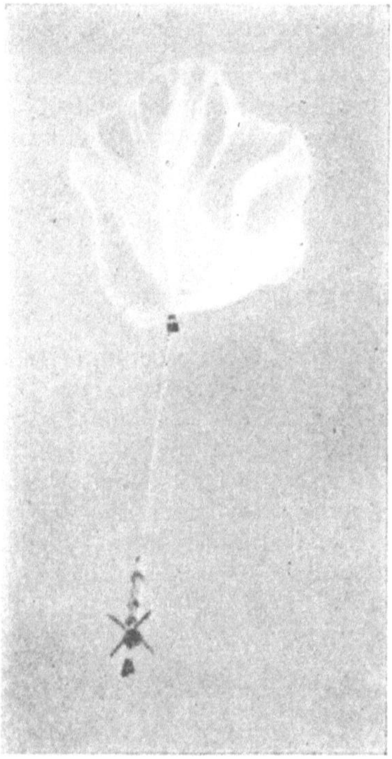

Fig. 9. ONR-Winzen Research Inc. *Rockoon* flight carrying Deacon rocket, shortly after launching. Balloon is of 72.3 foot size (150,000 cubic feet). Firing altitude is 70,000 feet

Rockoon (U.S. Navy-ONR). In 1947 the author proposed a balloon-launched sounding rocket reproduced in Fig. 7 (Reference [3]). It remained for Professor VAN ALLEN to develop the first practical and inexpensive application of this system. Project *Rockoon* explores the mysteries of cosmic radiation at altitudes much higher than the balloon vehicle is capable. The tremendous boost given a rocket when carried aloft and launched from a balloon platform is graphically and spectacularly demonstrated in Fig. 8 based on VAN ALLEN's data [4]. *Rockoon* launchings have been made by Winzen Research Inc. and General Mills Inc. from ships in the Arctic, Pacific, and Atlantic Oceans, at the equator and other latitudes. Launchings are thus made far away from inhabited land areas where such work can be accomplished in safety. The original project used a 72.3 foot diameter balloon carrying a Deacon rocket. At an altitude of 70,000 feet the rocket is fired and will normally reach a peak altitude of 300,000 feet with a payload of 30 pounds. Fig. 9 shows a *Rockoon* flight shortly after a shipboard launch.

Project *Strato-Lab* (U.S. Navy-ONR). As an extension of Project *Skyhook*, the ONR has begun a series of manned high-altitude ascents using the Winzen Research Inc. *Strato-Lab* gondola. A 2 million cubic foot balloon is capable of carrying the *Strato-Lab* gondola to an altitude in excess of 90,000 feet. The 2-man gondola carries a pilot and a scientific observer. Flights for the study of astronomical, meteorological and geophysical phenomena are now under way. Fig. 4 shows an artist's concept of the *Strato-Lab* gondola at altitude.

Animal Cosmic Ray Flights (USAF-ARDC). On this project we have carried out, over a period of 4 years, a series of animal flights designed to study the effects of cosmic radiation at altitudes up to 126,000 feet and for exposures of up to 36 hours. Mice, guinea pigs, and monkeys were flown, reminiscent of the first Montgolfière flight in which animals were carried aloft to study the hazards of flight before man would entrust himself to the atmosphere. Simultaneously, the project was instrumental in developing for the first time a true sealed cabin

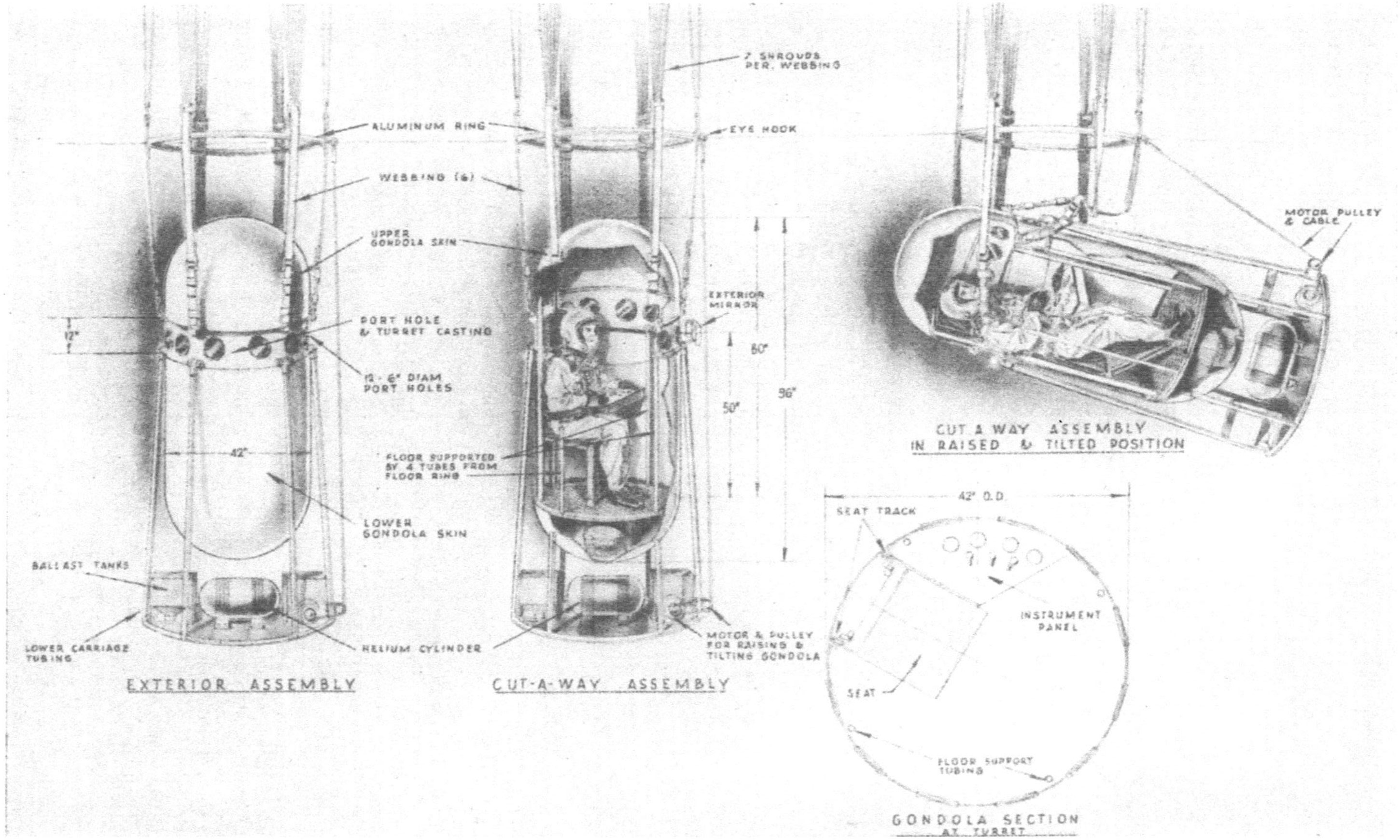

Fig. 10. Winzen Research Inc. proposal (October 1955) for Project *Manhigh*. Artist's view shows concept of one-man space capsule. Main structural member is a cast aluminum alloy turret with six portholes. All external and internal loads are attached at this turret. The upper and lower parts of the gondola consist of thin aluminum shells which are removable. The feature of inclining the gondola in flight was abandoned to permit use of a telescope on the 24-hour *Manhigh* flight. Two successful *Manhigh* flights have been completed, the first one a daylight flight to 96,000 feet, the second a 32-hour flight to 100,000 feet (August 1957)

Fig. 11. Schematic artist's drawing of the USAF-Winzen Research Inc. *Manhigh* gondola showing location of internal instruments and components. Pilot normally breathes the oxygen-helium capsule atmosphere. Equivalent cabin altitude can be selected on the panel. The *Manhigh* capsule is a minimum weight vehicle for one-man flights to altitudes in excess of 100,000 feet and for durations in excess of 24 hours permitting night and day observations and studies of human factors in space medicine. The Winzen Research *Manhigh* capsule is the first true space cabin and will serve as a research vehicle for the study of requirements for manned space flight.

1 12 Volt emergency battery; *2* Oxygen converter; *3* Emergency oxygen supply bottle; *4* Telegraph key; *5* Radio VHF; *6* Emergency oxygen control; *7* Spot photometer battery pack; *8* Oxygen control panel; *9* Chemically treated air; *10* Light; *11* Fire extinguisher; *12* Photo panel; *13* Camera; *14* Thermostat; *15* Adjustable light; *16* Electrical control panel; *17* Cold air supply; *18* Radio HF receiver; *19* Drinking water supply; *20* Floor; *21* 24 Volt communications battery; *22* Floor

which would be equally suitable for space travel. It solves the problem of extreme night and daytime temperature conditions under space-equivalent environment.

Project Manhigh (USAF-ARDC). Project *Manhigh* is the manned version of the animal-capsule flights. This sealed cabin 3 feet in diameter by 8 feet high has all the features of a true artificial environment in space. Two flights have been conducted to date, one a daylight flight to 96,000 feet using a 2 million cubic foot balloon, the other a 32-hour flight with a 3 million cubic foot balloon. Winzen Research designed and built the entire system and conducted the flight operations.

Fig. 10 shows the original proposal in an artist's conception and Fig. 11 is a diagram of the components of the interior structure of the system also shown in Fig. 12. The gondola has an artificial atmosphere of oxygen, helium and nitrogen, is insulated against heat losses at night and the intense solar radiation of the daylight hours. An air regeneration system, and a separate air cooling system form an integral part of the climate control. The gondola is designed for one-man operation, and is therefore capable of attaining altitudes in excess of 100,000 feet. Fig. 13 shows the launching of *Manhigh* Flight 2 [11, 12, 13].

Project Sky-Car (Winzen Research Inc.). This company-sponsored project uses plastic balloons from 25 to 72 ft. diameter. Currently available are two Sky-Cars or gondolae, 4.5 and 8 ft. diameter, built of aircraft tubing suitable for flights with from 1 to 6 passengers. Flights up to 12,000 feet have been conducted. Flights to higher altitudes with oxygen equipment are currently in the test stage, including a flight across the continental United States to conduct a series of experiments and, incidentally, to celebrate the 10th anniversary flight of the plastic balloon.

The manned plastic balloon serves many useful functions such as the training of military pilots prior to the conduct of stratosphere balloon flights, the testing of many components and items of instrumentation for high altitude flights, and, of course, the study of the performance of the plastic aerostat itself. Many such flights have been made and have demonstrated the safety and reliability of this operation. There are many departures from the earlier system of manned flights in addition to the use of the plastic balloon. As in all plastic balloon flights, an open gondola parachute is used. The balloon can be cut off at the moment it touches the ground so that there is no dragging, and as a result, no injury to personnel. On this project also the plastic balloon is expendable and this is practical due to its low cost. However, there have been many low wind landings in which smaller balloons were salvaged.

Fig. 12. Major DAVID G. SIMONS, pilot of *Manhigh* Flight II during final cockpit check prior to his record-breaking ascent on 19—20 August 1957. Control panel is on right, climate controls on left. The pilot is checking his portable tape recorder

V. Future Applications

This review on the eve of the 10th anniversary of the plastic stratosphere balloon has served to emphasize the phenomenal progress which we have made, particularly within the last few years. The future of plastic balloons holds many even more exciting prospects.

As an example of the modern aerostat, Operation *Manhigh II*, which carried a gondola weighing 1700 pounds loaded with more than ample electrical power and hundreds of pounds of instrumentation to an altitude in excess of 100,000 feet, accomplished a break-through by establishing a manned bridgehead in space. Thus, the plastic balloon, from its modest beginnings as an unmanned instrument carrier, now has proven its practical usefulness as a vehicle for manned research stations at high altitude. In addition to the study of hardware and the exposure of instruments at high altitude, the human problems of space medicine including its physiological and psychological aspects can go forward full speed; for after many instrumented rockets have been flown, eventually man himself will want to conquer space, to fly around the moon, and to land on other planets. Before he is able to do this, he must know his physical limitations and the effects of prolonged exposure at high altitude to establish safe tolerances for guiding design of hardware. For unless man understands the problems so he can integrate the limitations of his frail body into the flight of a rocket ship, it would be useless to start with the design of a space vehicle. The study of human factors is basic. The plastic balloon can do the pioneering work in this area, as has been demonstrated on Project *Manhigh*. Balloon stations manned by a crew of four or five are the obvious next step in the study

Fig. 13. Spectacular photo of the launching of *Manhigh* Flight II from an open pit iron mine near Crosby, Minnesota. The photo shows the 3 million cubic foot balloon rising from the pit. The height of the balloon is 280 feet. Height of the aerostat in excess of 350 feet including the open cargo parachute and the gondola. The pit is 350 feet deep

of space medicine, and such projects are already in the planning stages. The heavy-load plastic balloon made this rapid development possible [5].

Manned high altitude observatories, as a logical extension of the multiple crew balloon gondola, are envisioned as astronomical observatories above the atmosphere. These stations could also serve as weather reconnaissance posts. This capability was dramatically pointed out for the first time on the second *Manhigh* flight, which was navigated to a safe landing because of the capability of observing entire weather systems. Fig. 14 is a sketch of such a floating observatory.

A degree of navigation is possible based on the wind data we have been able to obtain through the use of the plastic balloon. Thus, there is an altitude where normally wind velocities are close to zero, making it possible to establish a fixed satellite station over a given point on the surface. Summer time circulation shows that at lower levels, the wind would carry the aerostat to the east, whereas at high levels the winds reverse and the gondola would make headway to the west. This phenomenon can be used to advantage in planning a flight trajectory which may return the aerostat to a point near the launching site or would make possible a certain degree of forward-and-reverse navigation. The plastic balloon using only small amounts of ballast to compensate for the loss of superheat at nightfall due to its low absorption of solar radiation, is ideally suited for

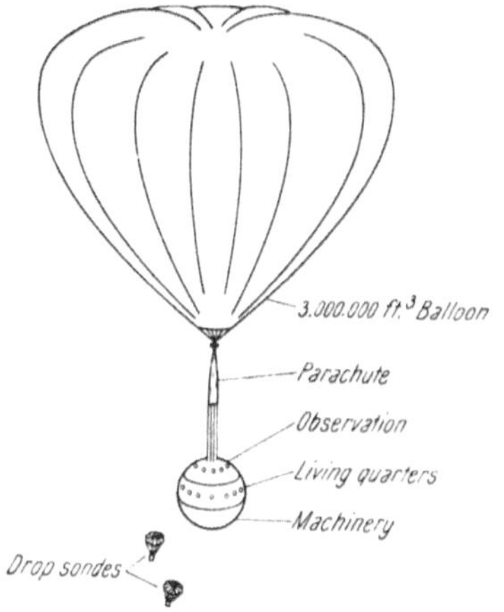

Fig. 14. Proposal for a floating observatory for meteorological and astronomical observations. This large laboratory includes three decks, the upper as an observation station, the middle section as living quarters, and the bottom for machinery and supporting equipment

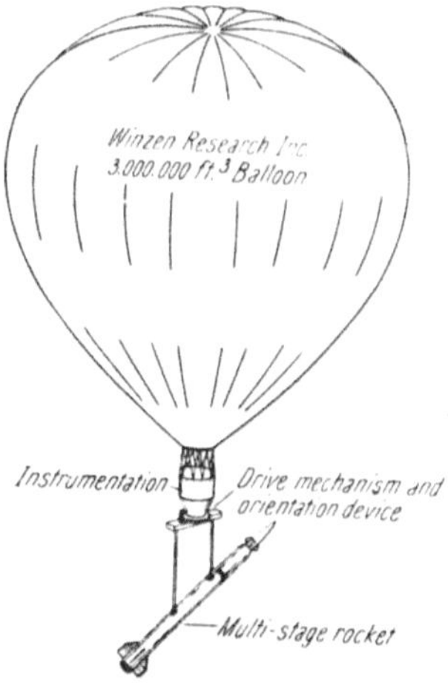

Fig. 15. Missile load suspension. The transportation of missiles to high altitudes by balloons would result in greater utilization of the available rocket power. The rocket would not have to overcome the air resistance of about 98 % of the earth's atmosphere

prolonged flights. The system developed on Project *Manhigh*, in which ballast consists of expended batteries and other items, contributes to the usefulness of the vehicle.

Fig. 15 shows an airborne rocket with an orientation device. In this system, radio control guidance is employed to orient the rocket to a given azimuth and elevation prior to firing, increasing the usefulness of research rockets by firing them inside a given range for optical and radar tracking and to insure recovery within a given area. Further, firings can be conducted with complete safety to personnel and without the need for complex firing platforms. Carriers make ideal launching sites for the balloon.

Far from being limited to the earthbound atmosphere, proposals have already been advanced to utilize balloons in connection with rockets fired to outer space. The operations outlined above point up the significant role plastic balloons are playing in space study. Thus, it is proposed to eject a plastic balloon from the third Vanguard rocket in order to enhance the possibility of visual tracking of such a satellite and, incidentally, to determine air density as a function of relative velocities between the balloon and the more dense, metallic satellite.

Winzen Research Inc. Stratosphere Balloons Conversion Tables and Data. Based on I.C.A.O. Standard Atmosphere May 1954

Altitude		Pressure			Temperature		Gas Expansion	Lift of He. lb/ft.³	Lift of H_2 lb/ft.³	$V = \frac{P}{Po}$
× 1000 Feet	Meters	Millibars	Inches of Hg.	MM of Hg.	Centigrade	Fahrenheit				
0	0	1013.25	29.92	760.01	15.0	59.0	1.00	.065953	.071050	1.000
1	304.8	981.50	28.99	736.31	13.0	55.5	1.0309	.063974	.068919	.970
2	609.6	944.50	27.894	708.55	11.0	51.8	1.0627	.062062	.066858	.941
3	914.4	909.00	26.846	681.92	9.0	49.1	1.0941	.060281	.064940	.914
4	1219.2	874.00	25.812	655.66	6.9	44.4	1.1261	.058566	.063092	.888
5	1524.0	842.00	24.867	631.66	4.8	40.7	1.1600	.056851	.061245	.862
6	1828.0	811.00	23.952	608.40	2.7	36.9	1.1947	.055203	.059469	.837
7	2133.6	780.00	23.036	585.15	+0.6	33.1	1.2300	.053620	.057764	.813
8	2438.4	751.50	22.194	563.77	—1.3	29.7	1.2690	.051971	.055987	.788
9	2743.2	724.00	21.382	543.14	—3.1	26.4	1.3106	.050322	.054211	.763
10	3048.0	696.00	20.555	522.13	—4.8	23.3	1.3550	.048673	.052435	.738
11	3352.8	669.00	19.758	501.88	—6.8	19.7	1.3966	.047222	.050872	.716
12	3657.6	644.00	19.019	483.12	—8.8	16.2	1.4409	.045771	.049309	.694
13	3962.4	620.00	18.311	465.12	—10.7	12.8	1.4881	.044320	.047746	.672
14	4267.2	596.00	17.602	447.11	—12.7	9.2	1.5385	.042869	.046183	.650
15	4572.0	572.00	16.893	429.11	—14.7	5.6	1.5898	.041484	.044690	.629
16	4876.8	549.00	16.214	411.85	—16.7	+2.0	1.6393	.040231	.043341	.610
17	5181.6	527.50	15.579	395.72	—18.8	—1.8	1.6920	.038978	.041991	.591
18	5486.4	506.50	14.959	379.97	—20.8	—5.4	1.7483	.037725	.040641	.572
19	5791.2	486.00	14.353	364.59	—22.7	—8.8	1.8083	.036472	.039291	.553
20	6096.0	466.00	13.763	349.59	—24.7	—12.4	1.8727	.035218	.037941	.534
21	6400.8	446.50	13.187	334.96	—26.7	—16.0	1.9380	.034032	.036662	.516
22	6705.6	428.00	12.640	321.08	—28.6	—19.5	2.0040	.032911	.035454	.499
23	7010.4	410.00	12.109	307.58	—30.6	—23.1	2.0704	.031855	.034317	.483
24	7315.2	393.00	11.607	294.82	—32.6	—26.7	2.1413	.030800	.033180	.467
25	7620.0	376.50	11.119	282.45	—34.5	—30.1	2.2173	.029745	.032044	.451
26	7924.8	360.00	10.632	270.07	—36.4	—33.6	2.2989	.028690	.030907	.435
27	8229.6	344.00	10.159	258.06	—38.4	—37.1	2.3866	.027634	.029770	.419
28	8534.4	329.00	9.716	246.81	—40.4	—40.7	2.4752	.026645	.028704	.404
29	8839.2	314.50	9.288	235.93	—42.3	—44.3	2.5641	.025722	.027710	.390

30	9144.0	300.50	8.875	225.43	—44.3	—47.8	2.6596	.024798	.026715	.376
31	9448.8	287.50	8.491	215.68	—46.3	—51.4	2.7624	.023875	.025720	.362
32	9753.6	275.00	8.122	206.30	—48.3	—55.0	2.8653	.023018	.024796	.349
33	10058.4	262.50	7.752	196.92	—50.3	—58.7	2.9762	.022160	.023873	.336
34	10363.2	250.00	7.383	187.55	—52.2	—62.1	3.0960	.021303	.022949	.323
35	10668.0	238.50	7.044	178.92	—54.2	—65.6	3.2258	.020445	.022026	.310
36	10972.8	227.50	6.719	170.67	—56.1	—69.2	3.3670	.019588	.021102	.297
37	11277.6	217.00	6.409	162.79	—56.3	—69.5	3.5088	.018797	.020249	.285
38	11582.4	207.00	6.113	155.29	—56.3	—69.5	3.6630	.018005	.019397	.273
39	11887.2	197.50	5.833	148.16	—56.3	—69.5	3.8168	.017280	.018615	.262
40	12192.0	188.00	5.552	141.04	—56.3	—69.5	4.0000	.016488	.017763	.250
41	12496.8	179.50	5.301	134.66	—56.3	—69.5	4.2017	.015697	.016910	.238
42	12801.6	171.00	5.050	128.28	—56.3	—69.5	4.4035	.014971	.016128	.227
43	13106.4	163.00	4.814	122.28	—56.3	—69.5	4.6296	.014246	.015347	.216
44	13411.2	155.50	4.592	116.65	—56.3	—69.5	4.8544	.013586	.014636	.206
45	13716.0	148.00	4.371	111.03	—56.3	—69.5	5.1020	.012927	.013926	.196
46	14020.8	141.00	4.164	105.78	—56.3	—69.5	5.3476	.012333	.013286	.187
47	14325.6	134.00	3.957	100.53	—56.3	—69.5	5.6180	.011740	.012647	.178
48	14630.4	128.00	3.780	96.02	—56.3	—69.5	5.9172	.011146	.012007	.169
49	14935.2	122.00	3.603	91.52	—56.3	—69.5	6.2112	.010618	.011439	.161
50	15240.0	116.50	3.441	87.40	—56.3	—69.5	6.4935	.010157	.010942	.154
51	15544.8	111.50	3.293	83.65	—56.3	—69.5	6.8027	.0096951	.010444	.147
52	15849.6	106.50	3.145	79.89	—56.3	—69.5	7.1429	.0092334	.0099470	.140
53	16154.4	101.50	2.998	76.14	—56.3	—69.5	7.5188	.0087717	.0094497	.133
54	16459.2	96.15	2.840	72.13	—56.3	—69.5	7.8740	.0083760	.0090234	.127
55	16764.0	91.80	2.711	68.87	—56.3	—69.5	8.2645	.0079803	.0085971	.121
56	17068.8	87.40	2.581	65.57	—56.3	—69.5	8.6957	.0075846	.0081708	.115
57	17373.6	83.10	2.454	62.34	—56.3	—69.5	9.1743	.0071889	.0077445	.109
58	17678.4	79.35	2.343	59.53	—56.3	—69.5	9.7087	.0067932	.0073182	.103
59	17983.2	75.75	2.237	56.83	—56.3	—69.5	10.3093	.0063974	.0068919	.097
60	18288.0	72.20	2.132	54.16	—56.3	—69.5	10.8696	.0060677	.0065366	.092
61	18592.8	69.90	2.035	51.69	—56.3	—69.5	11.3636	.0058039	.0062524	.088

Winzen Research Inc. Stratosphere Balloons Conversion Tables and Data. Based on I.C.A.O. Standard Atmosphere May 1954

Altitude		Pressure			Temperature		Gas Expansion	Lift of He. lb/ft.³	Lift of H_2 lb/ft.³	$V = \frac{P}{Po}$
× 1000 Feet	Meters	Millibars	Inches of Hg.	MM of Hg.	Centigrade	Fahrenheit				
62	18897.6	65.70	1.940	49.29	—56.3	—69.5	11.9048	.0055401	.0059682	.084
63	19202.4	62.60	1.849	46.96	—56.3	—69.5	12.3457	.0053422	.0057551	.081
64	19507.2	59.80	1.766	44.86	—56.3	—69.5	12.8370	.0051377	.0055348	.0779
65	19812.0	57.00	1.683	42.76	—56.3	—69.5	13.4228	.0049135	.0052932	.0745
66	20116.8	54.40	1.607	40.81	—56.3	—69.5	14.0845	.0046827	.0050446	.0710
67	20421.6	51.90	1.533	38.93	—56.3	—69.5	14.7059	.0044848	.0048314	.0680
68	20726.4	49.50	1.462	37.13	—56.3	—69.5	15.3846	.0042869	.0046183	.0650
69	21031.2	47.20	1.394	35.09	—56.3	—69.5	16.1290	.0040891	.0044051	.0620
70	21336.0	45.00	1.329	33.76	—56.3	—69.5	17.0068	.0038780	.0041777	.0588
71	21640.8	42.90	1.267	32.18	—56.3	—69.5	17.8891	.0036868	.0039717	.0559
72	21945.6	40.90	1.208	30.68	—56.3	—69.5	18.7266	.0035219	.0037941	.0534
73	22250.4	38.90	1.149	29.18	—56.3	—69.5	19.6464	.0033570	.0036164	.0509
74	22555.2	37.00	1.093	27.76	—56.3	—69.5	20.6186	.0031987	.0034459	.0485
75	22860.0	35.30	1.043	26.48	—56.3	—69.5	21.5983	.0030536	.0032896	.0463
76	23164.8	33.70	.995	25.28	—56.3	—69.5	22.5734	.0029217	.0031475	.0443
77	23469.6	32.20	.951	24.16	—56.3	—69.5	23.6407	.0027898	.0030054	.0423
78	23774.4	30.70	.907	23.03	—56.3	—69.5	24.8139	.0026579	.0028633	.0403
79	24079.2	29.30	.865	21.98	—56.3	—69.5	26.0417	.0025326	.0027283	.0384
80	24384.0	27.80	.821	20.86	—56.3	—69.5	27.3224	.0024139	.0026004	.0366
81	24688.8	26.60	.786	19.95	—56.3	—69.5	28.7356	.0022952	.0024725	.0348
82	24993.6	25.30	.747	18.98	—56.3	—69.5	30.1205	.0021896	.0023589	.0332
83	25298.4	24.15	.713	18.12	—55.8	—68.7	31.5457	.0020907	.0022523	.0317
84	25603.2	23.00	.679	17.25	—54.9	—67.0	33.1126	.0019918	.0021457	.0302
85	25903.0	21.95	.648	16.47	—54.1	—65.5	34.8432	.0018929	.0020391	.0287
86	26212.8	20.95	.619	15.72	—53.1	—63.7	36.7647	.0017939	.0019326	.0272
87	26517.6	19.95	.589	14.97	—52.2	—62.1	38.7597	.0017016	.0018331	.0258
88	26822.4	19.05	.563	14.29	—51.2	—60.2	40.6504	.0016224	.0017478	.0246
89	27127.2	18.20	.538	13.65	—50.4	—58.8	42.7350	.0015433	.0016626	.0234
90	27432.0	17.35	.512	13.02	—49.6	—57.4	44.8430	.0014708	.0015844	.0223
91	27736.8	16.60	.490	12.45	—48.7	—55.7	47.1698	.0013982	.0015063	.0212

92	28041.6	15.80	.467	11.85	—47.8	—54.1	49.5050	.0013322	.0014352	.0202
93	28346.4	15.15	.447	11.37	—46.8	—52.3	52.0833	.0012663	.0013642	.0192
94	28651.2	14.45	.427	10.84	—45.9	—50.7	54.6448	.0012069	.0013002	.0183
95	28956.0	13.80	.408	10.35	—45.1	—49.3	57.4713	.0011476	.0012363	.0174
96	29260.8	13.15	.388	9.86	—44.2	—47.6	60.6061	.0010882	.0011723	.0165
97	29565.6	12.60	.372	9.45	—43.3	—46.0	63.6943	.0010355	.0011155	.0157
98	29870.4	12.10	.357	9.08	—42.3	—44.3	67.1141	.00098270	.0010586	.0149
99	30175.2	11.60	.343	8.70	—41.4	—42.5	70.4225	.00093653	.0010089	.0142
100	30480.0	11.10	.328	8.33	—40.6	—41.1	73.5294	.00089696	.00096628	.0136
101	30784.8	10.60	.313	7.95	—39.7	—39.4	76.9231	.00085739	.00092365	.0130
102	31089.6	10.15	.300	7.61	—38.8	—37.8	80.6452	.00081782	.00088102	.0124
103	31394.4	9.69	.286	7.27	—37.9	—36.2	84.7458	.00077825	.00083839	.0118
104	31699.2	9.29	.274	6.97	—36.9	—34.4	89.2857	.00073867	.00079576	.0112
105	32004.0	8.89	.263	6.67	—36.0	—32.8	93.9850	.00070174	.00075600	.01064
106	32308.8	8.51	.251	6.38	—35.1	—31.2	99.5025	.00066283	.00071405	.01005
107	32613.6	8.15	.241	6.11	—34.2	—29.5	103.6269	.00063645	.00068563	.00965
108	32918.4	7.80	.230	5.85	—33.2	—27.7	108.1081	.000611007	.00065721	.00925
109	33223.2	7.48	.221	5.61	—32.3	—26.1	113.2503	.00058236	.00062737	.00883
110	33528.0	7.17	.212	5.38	—31.4	—24.5	118.7648	.00055532	.00059824	.00842
111	33832.8	6.86	.203	5.15	—30.5	—22.9	124.3781	.00053026	.00057124	.00804
112	34137.6	6.58	.194	4.94	—29.6	—21.3	130.2083	.00050652	.00054556	.00768
113	34442.4	6.30	.186	4.73	—28.7	—19.7	136.4256	.00048344	.00052080	.00733
114	34747.2	6.04	.178	4.53	—27.8	—18.1	142.8571	.00046167	.00049735	.00700
115	35052.0	5.80	.171	4.35	—26.9	—16.4	149.7006	.00044057	.00047461	.00668
116	35356.8	5.56	.164	4.17	—26.0	—14.8	156.7398	.00042078	.00045330	.00638
117	35661.6	5.32	.157	3.99	—25.1	—13.2	163.9344	.00040231	.00043341	.00610
118	35966.4	5.10	.151	3.83	—24.2	—11.5	171.8213	.00038385	.00041351	.00582
119	36271.2	4.90	.145	3.68	—23.3	—9.9	179.5332	.00036736	.00039575	.00557
120	36576.0	4.71	.139	3.53	—22.4	—8.3	187.2659	.00035219	.00037941	.00534
121	36880.8	4.52	.133	3.39	—21.5	—6.6	195.6947	.00033702	.00036307	.00511
122	37185.6	4.34	.128	3.26	—20.6	—5.1	204.9180	.00032185	.00034672	.00488
123	37490.4	4.17	.123	3.13	—19.7	—3.3	214.5923	.00030734	.00033109	.00466

Winzen Research Inc. Stratosphere Balloons Conversion Tables and Data. Based on I.C.A.O. Standard Atmosphere May 1954

Altitude		Pressure			Temperature		Gas Expansion	Lift of He. lb/ft.³	Lift of H_2 lb/ft.³	$V = \frac{P}{Po}$
× 1000 Feet	Meters	Millibars	Inches of Hg.	MM of Hg.	Centigrade	Fahrenheit				
124	37795.2	4.00	.118	3.00	—18.8	—1.8	224.2152	.00029415	.00031688	.00446
125	38100.0	3.85	.113	2.88	—18.0	—0.4	234.1920	.00028162	.00030338	.00427
126	38404.8	3.70	.109	2.77	—17.1	+1.3	244.4988	.00026975	.00029059	.00409
127	38709.6	3.55	.105	2.66	—16.2	+2.9	255.1020	.00025854	.00027852	.00392
128	39014.4	3.40	.101	2.55	—15.3	4.4	266.6667	.00024732	.00026644	.00375
129	39319.2	3.27	.097	2.45	—14.3	6.2	278.5515	.00023677	.00025507	.00359
130	39624.0	3.15	.093	2.36	—13.4	7.9	290.6977	.00022688	.00024441	.00344
131	39928.8	3.02	.089	2.27	—12.5	9.6	303.0303	.00021764	.00023447	.00330
132	40233.6	2.91	.086	2.18	—11.6	11.1	315.4574	.00020907	.00022523	.00317
133	40538.4	2.79	.083	2.09	—10.7	12.8	328.9474	.00020050	.00021599	.00304
134	40843.2	2.68	.080	2.01	—9.8	14.3	342.4658	.00019258	.00020747	.00292
135	41148.0	2.58	.077	1.93	—8.9	16.0	357.1429	.00018467	.00019894	.00280
136	41452.8	2.48	.074	1.86	—7.9	17.7	373.1342	.00017675	.00019041	.00268
137	41757.6	2.39	.071	1.79	—7.0	19.4	389.1051	.00016950	.00018260	.00257
138	42062.4	2.30	.068	1.73	—6.1	21.0	406.5041	.00016224	.00017478	.00246
139	42367.2	2.22	.065	1.67	—5.2	22.6	425.5319	.00015499	.00016697	.00235
140	42672.0	2.13	.063	1.61	—4.3	24.2	444.4444	.00014839	.00015986	.00225
141	42976.8	2.05	.060	1.55	—3.4	25.9	462.9630	.00014246	.00015347	.00216
142	43281.6	1.97	.058	1.49	—2.5	27.5	483.0918	.00013652	.00014707	.00207
143	43586.4	1.90	.056	1.43	—1.6	29.1	502.5126	.00013125	.00014139	.00199
144	43891.2	1.83	.054	1.37	—0.7	30.8	523.5602	.00012597	.00013571	.00191
145	44196.0	1.76	.052	1.32	+0.2	32.4	546.4481	.00012069	.00013002	.00183
146	44500.8	1.70	.050	1.27	1.1	33.9	568.1818	.00011608	.00012505	.00176
147	44805.6	1.63	.048	1.22	1.9	35.4	591.7160	.00011146	.00012007	.00169
148	45110.4	1.57	.046	1.18	2.7	36.8	613.4969	.00010750	.00011581	.00163
149	45415.2	1.52	.045	1.14	3.6	38.6	636.9427	.00010355	.00011155	.00157
150	45720.0	1.47	.044	1.10	4.4	39.8	662.2517	.000099589	.00010729	.00151

Ref. [7] contained an interesting article by Kurt R. Stehling and Richard Forster discussing how a four million cubic foot plastic balloon could be used to carry to 70,000 feet a solid-propellant three-stage rocket capable of carrying a four-pound payload to the moon. Stehling, as early as mid-1955, suggested a balloon launching method for small orbital satellites [9]. A 3,000,000 cubic foot plastic balloon would be used to carry the proposed three-stage vehicle weighing 13,500 pounds to 75,000 feet for launching. The Air Force Office of Scientific Research has borrowed Kurt Stehling's idea for Project *Farside*, the prime contractor for which is Ford Aeronutronics Company.

K. A. Ehricke in his latest paper, presented before this audience [6] proposes the use of balloons coated with a metallic surface for the generation of solar power in space. Such balloons would be of special designs and shapes and would be only partly coated with a metallic film. We have built balloons of this type in the past for other research applications. Even beyond the application as a generator of solar power, such balloons could be used in the study of other high altitude phenomena. The light weight, small bulk, and extremely small mass of gas required to inflate such balloons in outer space, offers exciting possibilities.

After having been connected with developments in plastic balloon flying of the last ten years, one might ask: What does the future hold? An impression that has been reconfirmed from time to time is that ballooning, far from being a thing of the past, is an important vehicle even in the rocket age and will remain an important research tool in the future [8]. We can expect developments toward better plastic films in the immediate future to improve the plastic cell itself. Electrical power sources, other than chemical means, will undoubtedly be developed. Since the balloon is predominantly a heat engine, ballast weight will be replaced by sources of heat. We may expect to see small propulsion units added to give the aerostat some degree of maneuverability at ceiling altitude. Duration of flights will be lengthened. Currently, the balloon-capsule system is the only method of maintaining a manned space-station at high altitude (above 99% of the earth's atmosphere) for an extended period of time. Such an aerial laboratory can carry out space-equivalent studies in the fields of space medicine, astronomy, geo-physics, and meteorology, with only the important factor of weightlessness missing to provide a true space environment.

The plastic balloon, which now can carry men, materials and instruments to the threshold of space, will become a significant weapon in the conquering of space.

Acknowledgments

I gratefully acknowledge the contribution of the following persons who aided in the preparation of this paper: Dr. D. G. Simons and Dr. V. E. Suomi for review and suggestions, Mr. J. C. Groth for aiding in the assembly, and Mrs. Roberta Barak for supervising the assembly.

Photos were all ably taken by Mrs. O. C. Winzen, Vice President of Winzen Research Inc. in charge of production, with the exception of Fig. 13 which is credited to Mr. Peter Marcus.

All drawings and photographs were selected from the engineering files of Winzen Research Inc. where this report was also assembled.

Appendix
Some Accomplishments of the Plastic Balloon

Scientific Results and Discoveries

1. Discovery of twice yearly wind reversal at high altitude.
2. Discovery of high energy particles of cosmic radiation.
3. Discovery of saturated layers near tropopause and haze layers in stratosphere.
4. Weather extends to 70,000' altitude with haze layers even higher.
5. Atmospheric diffusion with multiple balloon flights.
6. Cloud physics study with time-lapse movies from above.
7. High altitude visibility study.
8. Sky-brightness study in stratosphere.
9. Astronomical observations up to 100,000' altitude.
10. Temperature and density measurements.
11. Aerial photography.
12. Animal cosmic ray exposure flights of 36 hours' duration.
13. Tests of manned space cabins.
14. Parachute tests above 100,000'.
15. Optical tracking calibrations.
16. Concept and test of manned weather observation station.
17. Radio and antenna experiments.
18. Hundreds of flights to study nature of cosmic radiation.

Advances in Balloon Technology

1. New balloon construction and design.
2. Altitude capability currently to 150,000 feet.
3. Stable research platform at high altitude.
4. Capability of many days' duration.
5. Capability of traveling great distances.
6. Trajectory forecast capability (with limitations).
7. Capability of carrying heavy loads.
8. Development of complete control, communication, and sensing instrumentation.
9. Development of several new launch techniques.
10. Ease of launching with small crews.
11. Manned space cabins with daytime or day and night capability.
12. New manned balloon altitude records.
13. New open gondolae for manned flights in troposphere.
14. New inexpensive and safe low-level manned balloon vehicles.
15. Development of thin polyethylene film.
16. New production sealing methods of plastic seams, with up to 8 miles of seams per balloon.
17. New mass-production methods for plastic balloons using up to 4 acres of plastic film.
18. Self-inflating, floating hurricane ball.

References

1. O. C. WINZEN, U.S. Patent No. 2,526,719, first plastic *Skyhook* balloon.
2. National Geographic Society, book, Explorer II, 1935.
3. O. C. WINZEN, Progress Report, ONR, 10 November 1947.
4. J. A. VAN ALLEN and M. B. GOTTLIEB, The Inexpensive Attainment of High Altitudes with Balloon-Launched Rockets. State University of Iowa.
5. Winzen Research Inc. Technical Publication 5 A, The Plastic Balloon as a Heavy-Load Vehicle.
6. K. EHRICKE, Instrumented Comets—Astronautics of Solar and Planetary Probes: This volume, p. 74.
7. K. STEHLING, We Can Build a Moon Rocket Now. Missiles and Rockets **1**, 58 (1956).
8. O. C. WINZEN, Plastic Balloons in the Rocket Age. Missiles and Rockets **2**, 50 (1957).
9. K. STEHLING, Balloon-Launching an Earth Satellite Rocket. Aviation Age **24**, 16 (1955).
10. J. HORNBOSTEL and E. O. SALANT, private communication, and Physic. Rev. **79**, 184 (1950).

11. J. P. Stapp, On Top of the World. Astronautics **2**, 56 (1957).
12. J. P. Stapp, The First Space Man. Astronautics **2**, 30 (1957).
13. O. C. Winzen, Bridgehead in Space. Interavia **12**, 1040 (1957).
14. D. G. Simons, Biological Effects of Primary Cosmic Radiation. Proceedings of the VIIth International Astronautical Congress, Rome, 1956, p. 382. Roma: Associazione Italiana Razzi, 1957.

The Manhigh II Balloon Operation

By

Otto C. Winzen[1]

(With 2 Figures)

The paper I presented at this Congress entitled "10 Years of Plastic Balloons" describes the new plastic balloon vehicle and some of its research applications. Currently, the modern balloon is the only vehicle with which it is possible to carry out extended manned flights under space-equivalent conditions with a current capability of reaching altitudes above 99 % of the earth's atmosphere. It is an ideal laboratory and aerial observatory for many experiments.

I have been asked to speak in more detail of one of the applications discussed in my paper, namely the last *Manhigh* balloon flight. My other paper shows some illustrations of this project. I will speak about the part for which Winzen Research Inc. was responsible, the design and construction of the balloon and the gondola as well as the flight operation itself. Following this talk, Major SIMONS will discuss the *Manhigh* flight from the point of view of the pilot inside the gondola.[2]

In order to sustain life in the hostile environment of space, it is necessary to provide a sealed cabin for human or animal passengers. In the classical Montgolfier tradition, animals were sent up in flights which Winzen Research Inc. conducted for the U. S. Air Force, Aero Medical Field Laboratory, for the past four years. These flights had the double purpose of developing the sealed cabin for extended flights (up to 36 hours at altitudes to 126,000 feet), and also to study the effect of cosmic radiation on living tissue. The successful conclusion of this work, on which Major SIMONS reported at the I.A.F. Congress at Rome last year, led to the possibility of manned flights. Winzen Research Inc. conducted this work for the U. S. Air Force, Air Research and Development Command.

Major attention was devoted to safety. Following the animal flights, the first high altitude manned flight was a test operation, and it was successfully completed on 2 June 1957 after reaching 96,000 feet altitude in a daylight flight. The two *Manhigh* pilots were trained in the Winzen Research low-level *Sky-Car* balloon system using an open basket as the gondola. As in the stratosphere flights, the gondola is suspended through an open parachute. The expendable plastic balloon is cut off at the moment the gondola touches the ground when it returns to earth to eliminate dragging. Other pilot training and preparation included parachute indoctrination and a parachute jump, extended claustrophobia tests in the *Manhigh* gondola, ground tests in our laboratory using the climate control system of the gondola, a test of the gondola and pilot in a low pressure chamber simulating the entire flight pattern, and there were numerous blood

[1] President of the Winzen Research Inc., 8401 Lyndale Avenue South, Minneapolis 20, Minnesota, U.S.A.

[2] This volume, p. 388. Thereafter, movies and slides of this operation were shown (Editor).

tests for cosmic ray studies. Safety equipment included a pressure suit worn by the pilot for use in case of capsule failure. In addition to the gondola parachute there was a personnel parachute. Thus there were three methods of descent, normally on the balloon, in case of emergency on the gondola parachute, and, in the event of failure of the large parachute, descent by personnel chute. The pilot normally breathed the capsule atmosphere consisting of a mixture of oxygen, helium and nitrogen. It was replenished from a liquid oxygen system which had two emergency back-up systems; one, a 45-minute bottle supply, the second a 15-minute emergency supply. There were two radios. Heartbeat and respiration instruments telemetered the pilot's physiological condition. There was a radio control which permitted the ground command station to release the gondola from the balloon for parachute descent in the event the pilot should become unconscious.

The main structural member of the capsule was a central turret casting which contained the portholes and parachute as well as landing gear support points. The entire internal structure, including the seat and instrumentation, was suspended from this main casting. Artist diagrams in my other paper show the location of the various instruments and components. Upper and lower gondola shells were attached to the turret and could be released by the pilot or from the outside, either mechanically or electrically. Insulation was provided which had the double purpose of protecting the capsule from intense solar radiation during the day and extreme cold during the night hours. The landing gear constructed of aluminium tubing was designed to collapse on landing to absorb impact shock. A 5-inch refracting telescope with 4 eyepieces was fitted for the *Manhigh II* flight which featured an optically flat mirror controlled electrically by the pilot in order to give him omni-directional vision in one hemisphere. The landing gear also carried the main battery supply which was used for ballast after power was expended. An up and down mirror was installed in front of one of the portholes giving the pilot an uninterrupted view up toward the balloon and down toward the ground. He made extensive use of this device as shown in the movies and slides taken at altitude, which will be shown later.

In the final weigh-off at the company plant, the capsule was found to weigh 1700 pounds complete with pilot and all instrumentation and ballast. The gondola and balloon, packaged in a wooden crate approximately the size of a desk, were then transported to an open pit iron mine in northern Minnesota, which permitted inflating the aerostat in calm conditions for a vertical launching.

At the base of the pit, the balloon was laid out on a canvas and plastic ground cloth, and the inflation began with helium pouring into the apex of the balloon through two long plastic inflation sleeves. An electric valve located at the apex of the ballon was tested. The usual fog accompanying calm conditions and a temperature inversion prevailed in the early light of dawn. A reefing sleeve protected most of the balloon and was pulled down progressively as the inflation proceeded.

The pilot, Major DAVID G. SIMONS, had been in the capsule for several hours at this point, and the capsule had been pressure tested and its atmosphere established under laboratory conditions. As Air Force Project Officer, he was ideally suited as a subject being a flight surgeon and a scientist. He was in good spirits and took movies and still photos through the portholes of the gondola, which was prepared for the launching while the balloon was being inflated. When the lift had reached the desired value, it was transferred to the parachute and gondola and was launched with the aerostat ascending safely out of the open pit at the rate of 1000 feet per minute.

The ballon used for the flight was of 3 million cubic foot volume, had a diameter of 200.2 feet. The entire aerostat at the time of launching was over 350 feet high. Only a small bubble of gas was concentrated at the apex, the balloon

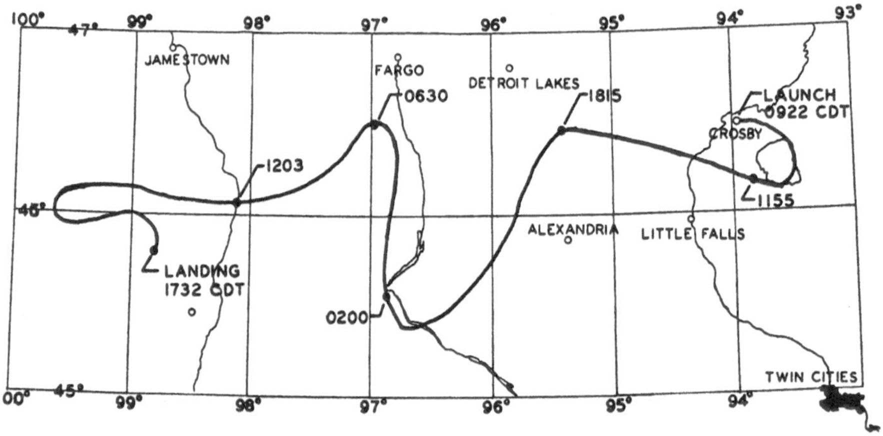

Fig. 1. Winzen Research Inc. Flight 767, USAF Project *Manhigh*, second manned flight 19—20 August 1957

attaining its full volume only at ceiling altitude in excess of 100,000 feet. The flight established a new world altitude record recognized by the F.A.I. of 101,516 feet.

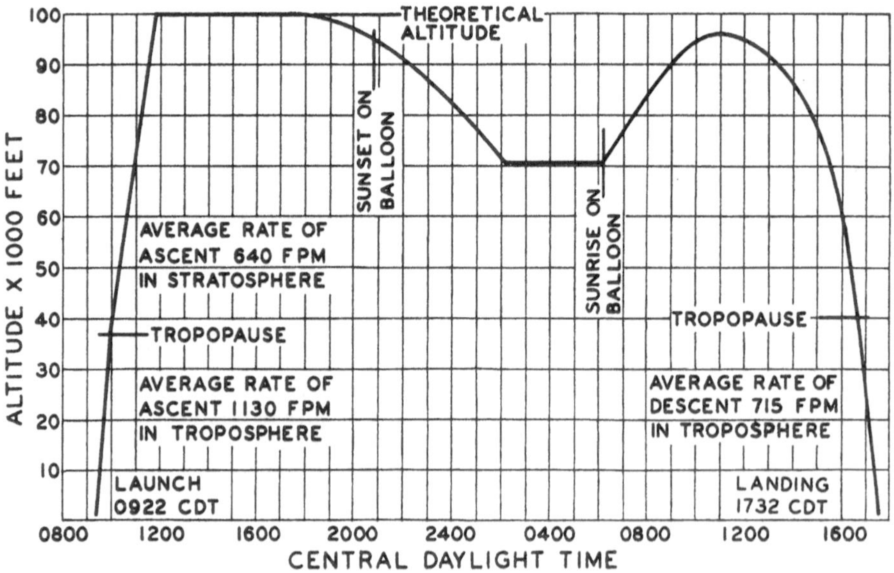

Fig. 2. Winzen Research Inc. Flight 767, USAF Project *Manhigh*, second manned flight 19—20 August 1957

The flight was tracked by several airplanes, helicopters, and ground vehicles, including a radio van which acted as the command station. Pilot heartbeat and respiration, altitude, climate of the gondola was continously telemetered and other data reported by the pilot at half-hour intervals. After a 32-hour,

10-minute flight, during which Major SIMONS became the first man to watch sunset and sunrise from above the bulk of the atmosphere, the flight landed safely near the border of North and South Dakota, approximately 200 miles from the launching site. During the night his flight had hovered over a violent thunderstorm, and he had observed an aurora borealis, a comet, the clear, un-twinkling light of the stars, the spectrum of light at the setting of Venus. During the second day he reported on the build-up of the storm and on the existence of a large opening in the storm to the west to which the aerostat was navigated for a safe landing.

As the command helicopter landed next to the gondola, the pilot had already climbed out. Except for being tired, he was in excellent physical condition. He had slept only for brief periods during the night. The trajectory and time vs. altitude curve of the flight are included as Figs. 1 and 2.

The *Manhigh II* flight has stirred the imagination of the world. Most important, in scientific circles it has dramatized the use of the balloon technique in studying the human factors of space flight, particularly their psychological and physiological aspects including the biological effect of prolonged exposure to cosmic radiation, while at the same time serving the development of the sealed space cabin.

For these flights were not conceived merely as record balloon ascensions, but as steps in the program of projecting man into space. While man waits for the rocket which will eventually be his space vehicle, he can actively study the many problems of his own adaptability for a new role in a balloon-borne sealed cabin which, except for weightlessness, permits the experience of space environment, the physiological detachment from the earth, and the establishment of criteria for space training and acclimatization in a capsule which will serve as a forerunner of the manned space vehicle.

Изучение первичного космического излучения с использованием искусственного спутника земли

С. Н. Вернов, В. Л. Гинзбург, Л. В. Курносова, Л. А. Разоренов, М. И. Фрадкин[1]

(7 рис.)

Важное место в программе научных исследований на искусственном спутнике Земли занимает изучение космических лучей. Две проблемы привлекают внимание в связи с возможностью исследований за границей атмосферы — это изучение спектра ядер по зарядам в первичном космическом излучении и изучение вариаций космических лучей.

Исследование спектра атомных ядер по зарядам в первичном космическом излучении предполагается провести, используя черенковские счетчики, а непрерывные измерения полной интенсивности космических лучей — одиночным галогенным счетчиком и ионизационной камерой.

1. Исследование ядерной компоненты первичного космического излучения

Изучение состава ядерной компоненты и энергетических спектров различных групп ядер первичного космического излучения существенно как с точки зрения теории происхождения космических лучей, так и для изучения элементарных актов взаимодействия первичных частиц высокой энергии с ядрами атомов в атмосфере.

Согласно имеющимся в настоящее время данным в состав первичной компоненты космических лучей входят протоны, α-частицы и, в значительно меньшем количестве, более тяжелые ядра.

Распределение по зарядам ядер с $Z > 2$ еще недостаточно изучено, и в отношении состава так называемой ядерной компоненты космических лучей имеется ряд вопросов, подлежащих экспериментальному исследованию. Большая часть работ по исследованию ядерной компоненты космического излучения выполнена исследователями различных стран в верхних слоях атмосферы, где остаточный слой вещества над установкой составляет примерно 15 $г/см^2$. Это приводит к тому, что истинный состав первичной компоненты за границей атмосферы может быть получен из непосредственных измерений только путем пересчета. Для такого пересчета необходимо надежно знать относительное число легких ядер, возникающих в результате разрушения более тяжелых ядер при прохождении их через остаточный слой атмосферы над установкой. Весьма существенно было бы устранить необходимость пересчета, для чего следует узмерения интенсивности потоков ядер с различным Z проводить непосредственно за границей атмосферы. Использование для этой цели ракет не может дать удовлетворительных результатов, так как интенсивность потока ядер с $Z > 2$ невелика, и за весьма ограниченное

[1] Москва, Академия наук СССР.

время пребывания ракеты вне атмосферы будет получено слишком мало данных.

Совершенно новые возможности в направлении исследования ядерной компоненты космических лучей открываются в связи с запуском искусственных спутников Земли. Установка аппаратуры на искусственном спутнике Земли позволит набрать необходимый статистический материал для определения слабых по интенсивности потоков ядер.

Одним из наиболее важных вопросов, относящихся к составу ядерной компоненты космических лучей, является вопрос о количественном соотношении между потоками легких ядер *Li*, *Be*, *B* и ядер *C*, *N*, *O*, *F*. Значение этого соотношения важно потому, что, по всей вероятности, ядра *Li*, *Be*, *B*, средняя распространенность которых во Вселенной весьма мала, испускаются источниками космических лучей лишь в ничтожном количестве, и наблюдаемые ядра *Li*, *Be*, *B* являются продуктами расщепления более тяжелых ядер при их взаимодействии с атомами межзвездной среды. Значение отношения потоков различных групп ядер у Земли можно получить путем расчета, основанного на определенных предположениях о пространственном распределении и мощности источников космических лучей и об условиях распространения частиц космического излучения в межзвездной среде. Сравнение полученного таким образом отношения с найденным в результате измерений может служить критерием правильности наших теоретических представлений. Так, например, теория, объясняющая образование космического излучения ускорением частиц за счет статистического механизма в расширяющихся турбулизированных оболочках сверхновых звезд [1], дает для соотношения между потоками ядер *Li*, *Be*, *B* и *C*, *N*, *O*, *F* значение $\gtrsim 0,1$. Вследствие неточности ряда параметров, использованных при оценке, это значение может оказаться в несколько раз больше, однако теория находится в резком противоречии с предположением, что оно много меньше 0,1.

Данные различных авторов об относительном количестве ядер *Li*, *Be*, *B* и *C*, *N*, *O*, *F* в первичном потоке нередко противоречат друг другу и, кроме того, недостаточно надежны, так как получены путем пересчета к границе земной атмосферы результатов измерений в стратосфере.

Поэтому окончательное выяснение значения отношения потоков этих двух групп ядер за границей атмосферы остается существенной проблемой.

С помощью искусственного спутника Земли может быть также внесена ясность в вопрос о наличии в первичном потоке ядер с $Z > 30$. Так как сечение взаимодействия таких ядер очень велико, то даже в случае, если такие ядра есть в первичном потоке космического излучения, на высоте, где давление равно 15—18 $г/см^2$, они не будут наблюдаться. Никаких достоверных сведений о наличии ядер с $Z > 30$ нет, но отдельные случаи их регистрации дают основания предполагать, что за границей атмосферы имеется заметный поток этих ядер. Если бы такое предположение экспериментально подтвердилось, то это имело бы весьма существенное значение для теории происхождения космических лучей. Действительно, в среднем во Вселенной ядер с $Z > 30$ очень мало [2], и вследствие большого сечения взаимодействия они должны иметь сравнительно небольшой пробег в межзвездной среде. Поэтому обнаружение заметного числа таких ядер в первичной компоненте свидетельствовало бы об аномальном богатстве источников космических лучей тяжелыми элементами.

а) Экспериментальные данные о составе первичного излучения

За последние несколько лет весьма сильно расширились сведения о первичном потоке космических лучей. Измерения потока протонов, α-частиц и ядер с $Z > 2$ проведены в большом интервале широт. Результаты измерений можно найти в работах [5—39]. Из анализа работ [5—39] видно, что измерения, проведенные на одной и той же геомагнитной широте, не всегда находятся в согласии друг с другом. По-видимому, здесь существенную роль играет различие в применяемой методике и различие в географических координатах места наблюдения. На значение последнего обстоятельства указывают работы [5—6], согласно которым в ряде случаев различие в величине потоков тяжелых ядер, измеренных в разных пунктах на одной и той же геомагнитной широте, можно устранить, если считать, что геомагнитные координаты, определяемые на основе измерения магнитного поля Земли на ее поверхности, не соответствуют магнитному полю Земли, которое воздействует на частицы космических лучей. Это предположение согласуется с результатами работы [7] по измерению потока нейтронов в космических лучах на различных широтах, согласно которой геомагнитный экватор не совпадает с „магнитным экватором для космических лучей“. К сожалению, отсутствуют систематические данные о величине потока первичных ядер в различных точках Земли, так как при измерениях применялась различная методика (фотоэмульсии, камеры Вильсона, пропорциональные, сцинтилляционные и черенковские счетчики), условия проведения экспериментов (географические и геомагнитные координаты, высота подъема, время и т.п.) были весьма разнообразны, а пересчет к границе атмосферы проводился при различных предположениях о величине используемых параметров.

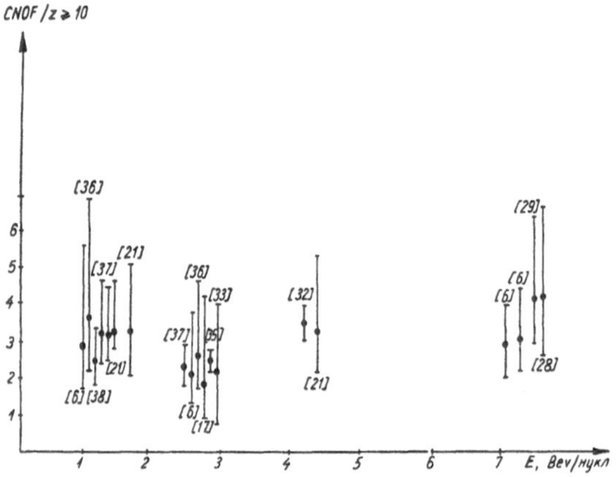

Рис. 1. Отношение потока ядер C, N, O, F к потоку ядер с $z \geq 10$

В связи с этим необходимо провести систематические измерение первичных потоков ядерной компоненты космических лучей по всей поверхности Земли, применяя для этой цели однотипную аппаратуру. Эта задача может быть разрешена, по-видимому, только на искусственном спутнике Земли. Проведение таких экспериментов не только позволит более точно узнать абсолютные значения первичных потоков различных групп ядер, но в то же время даст возможность составить представление о характере магнитного поля Земли на больших расстояниях от ее поверхности, ибо траектории частиц космических лучей определяются именно удаленной от поверхности областью земного магнитного поля.

Несмотря на заметный разнобой между приведенными в работах [5—39] абсолютными значениями потоков, оценка относительных значений потоков ядер на данной широте, т.е. ядер с данной энергией на нуклон может быть получена с удовлетворительной точностью. Если для определения относительного содержания различных ядер в космических лучах брать значения потоков, полученные в одном и том же эксперименте, то таким образом можно исключить все неопределенности, связанные с переходом от одного метода к другому. Сравнение найденных таким способом отношений проведено на рис. 1, 2 и 3, из которых видно, что в пределах ошибок (нередко веьма больших) эти отношения не зависят от энергии частиц и в среднем равны $N_{He}: N_{CNOF}: N_{Z \geq 10} \approx 53:3:1$. Доля α-частиц в первичном потоке ядер с данной энергией на нуклон составляет в среднем $11 \pm 4{,}5\%$ (для случаев, использованных на рисунках). Тот факт, что соотношение между потоками различных групп ядер в пределах ошибок не зависит от широты, указывает скорее на идентичность энергетических спектров различных групп ядер, а не на различие в них [41]. Однако точность данных, имеющихся в нашем распоряжении, слишком мала, чтобы можно было сделать вполне достоверные выводы о тождественности энергетических спектров или, напротив, об их различии, если это различие не слишком велико.

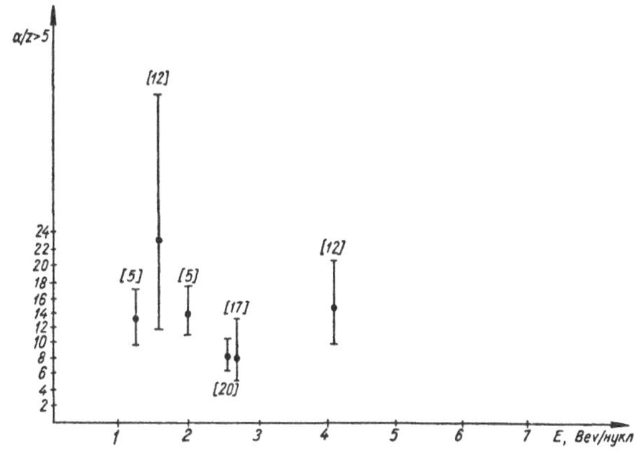

Рис. 2. Отношение потока α-частиц к потоку ядер с $z>5$

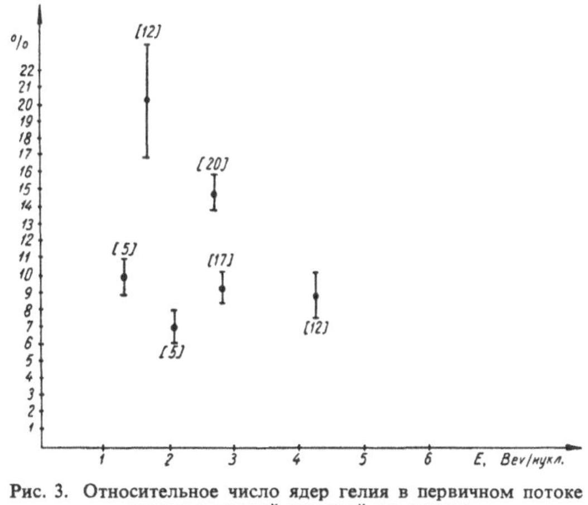

Рис. 3. Относительное число ядер гелия в первичном потоке частиц с данной энергией на нуклон

Таким образом, с точки зрения определения энергетического спектра различных групп ядер в первичном космическом излучении постановка экспериментов на искусственном спутнике Земли также сулит большие возможности, ибо будут найдены потоки частиц различной энергии (на различных широтах) при помощи одного и того же прибора, что несомненно повысит достоверность наших сведений об энергетическом спектре первичных ядер.

Одним из наиболее интересных вопросов, касающихся первичного космического излучения является, как уже указывалось выше, наличие ядер группы *Li*, *Be*, *B*. Даже в настоящее время вопрос об относительном содержании этих ядер в первичном потоке космических лучей, еще нельзя считать окончательно решенным. В отношении измерения потока этой группы ядер особенно ясно сказалась важность выбора правильной методики. В результате работы [36] было выяснено, что два различных метода идентификации следов частиц в фотоэмульсиях (определение плотности зерен с использованием эмульсий малой чувствительности [21] и измерение одновременно плотности δ-электронов и среднего угла многократного рассеяния [38]) приводят к ошибкам, если их применять для определения потока легких ядер, причем в первом случае происходит занижение, а во втором — завышение по сравнению с истинной величиной потока. Наиболее точными следует, по-видимому, считать результаты, полученные в работе [20]. Однако во всех проведенных до сих пор экспериментах для получения потока за „границей" атмосферы необходимо было делать пересчет. По этой причине остаются возможности для критики утверждения о присутствии в первичном потоке ядер *Li*, *Be*, *B* и необходимы эксперименты за границей атмосферы, чтобы окончательно решить этот вопрос.

б) Методика эксперимента по изучению спектра ядер по зарядам в первичном космическом излучении

Бóльшая часть сведений о потоках ядер с различными значениями Z получена в результате применения фотоэмульсий, поднимаемых на шарах-зондах до высот 20—30 *км*. Фотоэмульсии дают возможность при определении заряда каждого ядра досконально исследовать его поведение в эмульсии, определять одновременно с величиной потока сечение взаимодействия. При использовании этого метода почти полностью устраняется возможность спутать многозарядную частицу с другими явлениями (ядерным расщеплением, ливнем и т.п.). Однако при наличии частиц умеренной энергии метод фотоэмульсий может привести к ошибкам в измерении заряда. В некоторых случаях эффективность регистрации определенных групп ядер не достигает 100%, и по этой причине распределение ядер по зарядам может весьма сильно искажаться. С этой точки зрения предпочтительнее такие методы, в которых не происходит какой-либо дискриминации частиц по отношению к их заряду или массе. К числу таких методов относится использование счетчиков частиц, в которых электрический импульс, возникающий при прохождении через него заряженной частицы, зависит от величины заряда частицы. Применение такого рода приборов на искусственном спутнике Земли имеет еще то преимущество по сравнению с фотоэмульсиями, что позволит передавать информацию на Землю по радио. Отпадает необходимость сбрасывания со спутника и последующего нахождения на Земле приборов, регистрирующих космические лучи.

Методы, основанные на ионизации среды быстрыми заряженными частицами, обладают тем недостатком, что частицы с небольшим Z и сравнительно малой скоростью могут имитировать релятивистскую частицу с большим Z. Это связано с тем, что ионизация возрастает как с увеличением заряда, так и с уменьшением скорости. По этой причине трудно, а иногда и невозможно, измерить достаточно точно поток ядер с $Z>2$, так как протоны и α-частицы малых энергий, число которых может превышать число исследуемых ядер, будут создавать ионизацию, соответствующую пролету более тяжелого релятивистского ядра. Это замечание касается и пропорциональных счетчиков, и импульсных ионизационных камер, и сцинцилляционных счетчиков. Такого же рода ошибка создается ядерными расщеплениями в веществе счетчика, ибо при этом вылетают медленные, сильно ионизующие частицы, и появляющийся в результате на выходе счетчика сигнал будет имитировать пролет релятивистской многозарядной частицы.

Прибором, свободным от указанных выше недостатков является счетчик, в основе которого лежит использование излучения Вавилова-Черенкова [46] (черенковский счетчик). Черенковский счетчик состоит из детектора (прозрачное вещество) фотоумножителя и усилителя. В детекторе при прохождении заряженных частиц с достаточно большой скоростью (больше скорости света в этом веществе) возникает черенковское излучение. Известно, что черенковское излучение обладает тем свойством, что угол между направлением частицы и испускаемого ею излучения определяется соотношением $\cos\theta = \frac{1}{\beta n}$, где $\beta = \frac{v}{c}$, v — скорость частицы, c — скорость света, n — показатель преломления вещества детектора. Следовательно, минимальное значение скорости регистрируемой частицы равно $\beta_{min} = \frac{1}{n}$, и частицы, обладающие скоростью, меньшей β_{min}, не будут регистрироваться черенковским счетчиком. Поэтому в отличие от ионизационной камеры, пропорционального и люминесцентного счетчиков черенковский счетчик не регистрирует ядерных расщеплений и нерелятивистских частиц. Интенсивность вспышки черенковского излучения пропорциональна квадрату заряда Z^2e^2 частицы, прошедшей через детектор, зависит от скорости β, показателя преломления вещества детектора n, а также от длины пути l частицы в детекторе:

$$\Delta N = \frac{4\pi^2 Z^2 e^2}{h c^2} \Delta\nu \left(1 - \frac{1}{n^2\beta^2}\right) l,$$

где ΔN — число фотонов в интервале частот $\Delta\nu$, испущенных на пути l (*в см.*), e — элементарный заряд, с — скорость света, h — постоянная Планка. (Предпологается что показатель преломления для данного интервала частот постоянен.) Таким образом, при одной и той же скорости ядер β и одном и том же пути l величина вспышки пропорциональна Z^2, что и позволяет, регистрируя амплитуды вспышек черенковского излучения, исследовать спектр ядер по зарядам в первичном космическом излучении.

Так как магнитное поле Земли не допускает на широты ниже 40° частицы со скоростью меньше 0,94 с, то при проведении измерений в интервале широт ±40° все частицы будут иметь примерно одинаковую скорость. Постоянство же путей в детекторе можно в известных пределах обеспечить, задав направление регистрируемых частиц телескопом из счетчиков и регистрируя на выходе фотоумножителя только те импульсы от вспышек в детекторе, которые сопровождались срабатыванием счетчиков телескопа.

Основным источником разброса импульсов на выходе фотоумножителя являются флуктуации в числе фотоэлектронов, выбиваемых с фотокатода умножителя. Очевидно, что относительные флуктуации будут тем меньше, чем больше число фотоэлектронов. Поэтому весьма существенны вопросы эффективного сбора света на фотокатод, прозрачности детектора и эффективности фотокатода.

При выборе типа фотокатода умножителя необходимо наилучшим образом согласовать спектральные характеристики фотокатода со спектральным составом черенковского излучения. Наиболее подходящим с этой точки зрения является сурьмяно-цезиевый фотокатод. При использовании плексиглазового детектора и сурьмяно-цезиевого фотокатода эффективной для регистрации является область длин волн 3800—5000 Å.

Рассмотрим вопрос о флуктуациях импульсов на выходе фотоумножителя. Даже в случае отсутствия накого-либо разброса в интенсивности световых вспышек для частиц с данным Z импульсы на выходе умножителя будут получаться различными из-за статистических флуктуаций в числе фотоэлектронов, выбиваемых с фотокатода и флуктуаций в числе электронов вторичной эмиссии. В общем случае распределение в числе фотоэлектронов описывается законом Пуассона [47]. Однако при большом числе фотоэлектронов (практически при $\bar{n} \approx 10$) распределение Пуассона достаточно точно выражается зависимостью:

$$f(n) = A exp\left[-\frac{(n-\bar{n})^2}{2n}\right]$$

где n — число электронов, выбитых с катода умножителя, $\bar{n}$ — среднее значение n, A — нормировочный множитель. Полуширина P кривой распределения (т.е. ее ширина на половине максимальной высоты) равна $P = 2\sqrt{2\bar{n}\ln 2} \approx 2{,}35\sqrt{\bar{n}}$. Относительная полуширина кривой распределения, равная отношению P к среднему числу электронов $\bar{n}$, есть $\Pi = \frac{2{,}35}{\sqrt{\bar{n}}}$.

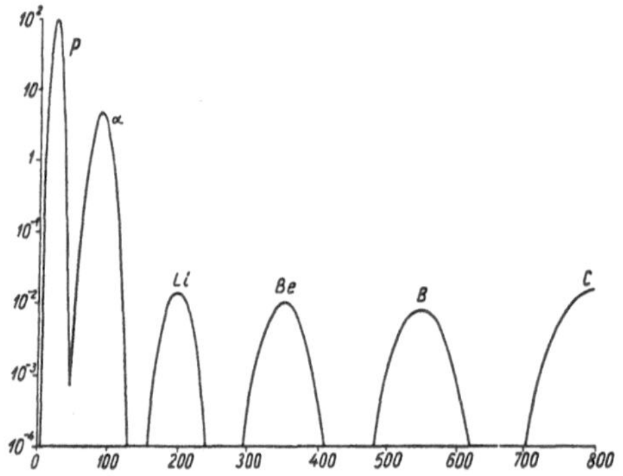

Рис. 4. Кривые распределения амплитуд импульсов на выходе черенковского счетчика (полуширина кривой распределения для однозарядных частиц принята равной 50%)

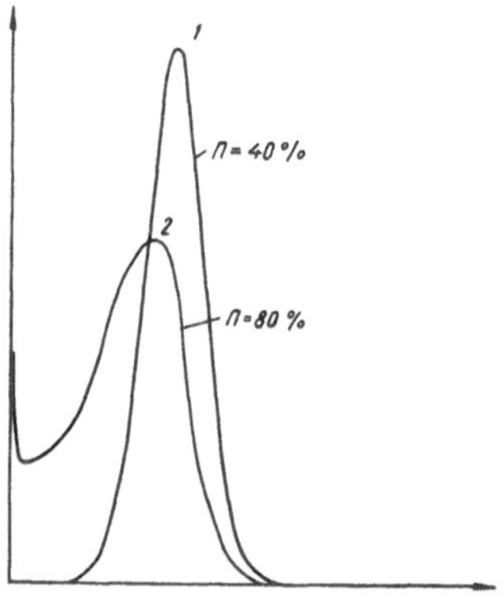

Рис. 5. Кривые распределения для $z = 1$ без учета влияния распределения по скоростям (1) и с учетом этого влияния (2)

Для того чтобы пики, принадлежащие ядрам с различными Z, разрешались, необходимо, чтобы относительная полуширина кривой распределения для однозарядной частицы составила примерно 50% (рис. 4).

Это требование накладывает ограничение на толщину плексиглазового детектора, которая должна быть не менее 2 см. В то же время толщину детектора нельзя сильно увеличивать из-за расщепления первичных ядер в веществе детектора. В связи с этим оптимальный размер детектора авторы выбрали 3,5 см.

Поперечные размеры детектора желательно иметь наибольшими, чтобы получить максимальное число зарегистрированных ядер. Помимо большой площади фотокатода, для набора статистики необходимо иметь наибольший телесный угол, в котором лежат направления регистрируемых частиц. Однако величина этого телесного угла ограничивается допуском на разброс путей частиц в детекторе. Помимо разброса в длине пути, проходимого частицами в детекторе, неравенство интенсивностей световых вспышек черенковского излучения, вызываемого прохождением частиц с данным Z, обусловлено также и другими причинами. Прежде всего, интенсивность вспышки все-таки зависит от скорости частицы и, хотя β заключено в сравнительно узких пределах, это все же приведет к значительному расплыванию пиков, соответствующих каждому значению Z. Оценка показывает, что если распределение частиц по энергиям имеет вид $N(E>E_1)=A\cdot E^{-1,2}$, то для однозарядных частиц с $\beta \geq 0,66$ (что соответствует минимальной скорости, при которой частица дает черенковское излучение в детекторе с $n=1,5$) получается увеличение полуширины кривой распределения, показанное на рис. 5. Однако для низких широт β_{min} будет определяться магнитным полем Земли, отсеивающим частицы с малой скоростью, и увеличение полуширины кривой распределения будет не столь велико.

Эффективная площадь детектора и допуск на разброс путей определяют выбор телескопического устройства, его геометрический фактор, а следовательно, и полное число частиц, регистрируемых установкой за данный промежуток времени.

Если выбрать геометрический фактор установки равным примерно 5 *см².стер.* то на основе результатов, приведенных в работах [5—39] можно получить для $Z \geq 6$ семь ядер в час, а для $Z=2$ около 50 ядер в час.

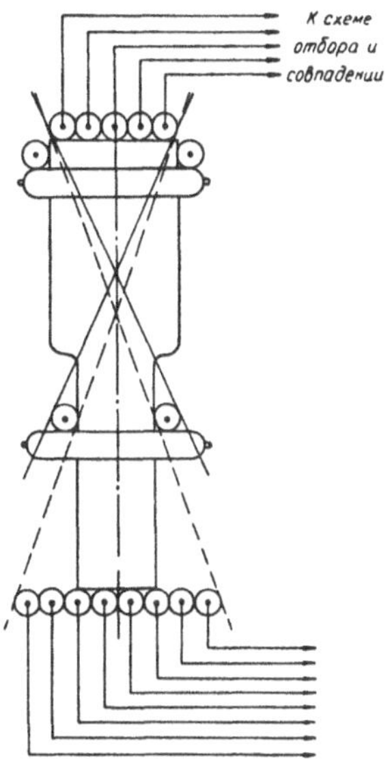

Рис. 6. Схема установки для исследования спектра ядер

Предполагаемые опыты по исследованию спектра ядер по Z предусматривают регистрацию дифференциального спектра ядер в интервале от α-частиц до кислорода. Схема установки для исследования спектра ядер, разработанной в Физическом институте им. П. Н. Лебедева Академии наук СССР, показана на рис. 6.

2. Исследование временных вариаций интенсивности космических лучей

Изучение вариаций интенсивности космических лучей имеет также существенное значение как для теории происхождения космических лучей, так и для решения ряда геофизических и астрофизических вопросов, таких как испускание Солнцем корпускулярных потоков, характер изменения магнитного поля Земли, природа магнитных бурь и т.п. Как правило, вариации интенсивности космических лучей изучаются на поверхности Земли, т.е. изучаются вариации вторичного излучения, значительная часть которых обусловлена влиянием метеорологических факторов. Кроме того, наиболее подвержена вариациям интенсивность частиц малой энергии, которые вообще почти не дают вклада в поток заряженных частиц вблизи поверхности Земли. Хотя в настоящее время теория метеорологических эффектов [3] разработана вполне удовлетворительно и влияние этих факторов можно исключить при обработке результатов, и в то же время показано [49], что вариации нейтронной компоненты на высотах гор отражают вариации интенсивности частиц малой энергии, необходимы непосредственные измерения вариаций первичного потока. Такие измерения, с одной стороны, дадут более богатый материал, а с другой, явятся проверкой существующей теории метеорологических эффектов и развиваемой Фейнбергом и Дорманом [4] теории метода коэффициентов связи, с помощью которой можно, измеряя вариации интенсивности нескольких компонент космических лучей на различных высотах, определить вариации первичного потока.

Для изучения вариаций первичного потока космических лучей необходимо поднимать аппаратуру на большие высоты, и в течение более или менее продолжительного времени аппаратура должна находиться на фиксированной высоте. Ясно, что применять для этой цели ракеты, время пребывания которых на больших высотах исчисляется минутами, невозможно. Несколько лучше обстоит дело с применением шаров-зондов. Однако в этом случае необходимо тщательно тарировать поднимаемые приборы, а для выяснения энергетических характеристик вариаций надо одновременно производить запуск установок в нескольких пунктах, различающихся геомагнитными координатами. Последнее обстоятельство весьма затрудняет и усложняет проведение экспериментов по изучению вариаций интенсивности космических лучей на больших высотах.

В связи с изложенными выше причинами трудно переоценить те возможности, которые дает установка на искусственном спутнике Земли приборов для регистрации вариаций первичного потока космических лучей. Приборы на искусственном спутнике Земли позволят зафиксировать вариации различного рода (суточные, двадцатисемидневные и др.), причем будут получены почти одновременные сведения о вариациях в различных пунктах Земли, т.е. можно будет получить синоптическую карту вариаций. Такие сведения, которых до сих пор не удавалось получить ввиду того, что станции наблюдения распределены по поверхности Земли неравномерно, позволят сделать ряд заключений о природе вариаций и их происхождении, проверить существующие теоретические представления о механизме появления больших вспышек космических лучей, о влиянии малых вспышек и т.д. Весьма существенные данные могут быть получены о влиянии солнечной активности на интенсивность космических лучей. На Солнце, по-видимому, генерируются частицы с энергией больше 10^9 эв. Ясно, что должны наблюдаться частицы и несколько меньших энергий, причем их интенсивность должна быть весьма значительной. Как известно, большие вспышки в космических лучах, коррелированные со вспышками на Солнце, наблюдаются очень редко (за последние 15 лет зарегистрировано 5 вспышек). Однако, вспышки на Солнце отмечаются значительно чаще и отсутствие коррелированных с солнечными вспышками вариаций интенсивности космических лучей объясняется, возможно, тем, что генерированные во время вспышки частицы имеют слишком малую энергию, чтобы вызвать вариации, регистрируемые на поверхности Земли. Если при малых вспышках действительно генерируются частицы малых энергий, то эффект можно будет зарегистрировать при измерениях на спутнике.

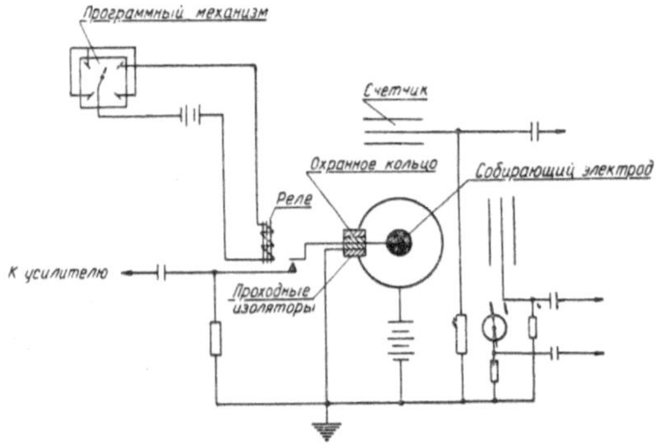

Рис. 7. Схема установки для исследования вариаций космических лучей

Вариации, период которых при регистрации на Земле составляет солнечные сутки (солнечно-суточные вариации), при измерении на спутнике дадут полуторачасовую вариацию, так как период обращения спутника вокруг Земли равен примерно 90 минутам. Такой сравнительно малый период позволит с большей достоверностью зафиксировать эти вариации и более точно определить их величину при условии, конечно, достаточно высокой статистической точности измерений.

Для изучения вариаций первичных космических лучей на спутнике намечено установить одиночный галогенный счетчик и интегрируюшую ионизационную камеру. Схема установки, разработанной в Московском государственном университете и в Якутском филиале Академии наук СССР, показана на рис. 7.

3. Некоторые другие задачи исследования первичного космического излучения

Возможность установки на искусственном спутнике Земли приборов, регистрируюших космические лучи, открывает большие перспективы в постановке других задач по исследованию первичного потока. К числу такого рода задач следует отнести измерение потока первичных протонов, выяснение роли „альбедо" земной атмосферы, определение нижнего предела для потока электронно-позитронной компоненты, энергетического спектра космических лучей, изучение взаимодействий первичных частиц с веществом и др. В связи с упомянутыми выше проблемами мы сделаем несколько замечаний.

Опыты, поставленные на искусственном спутнике Земли для измерения первичного потока ядер, дадут возможность получить еще целый ряд дополнительных сведений о свойствах первичных космических лучей, в частности о вариациях интенсивности потока различных групп ядер. Вопрос о вариациях интенсивности ядерной компоненты принадлежит к числу тех, ответ на которые не получен до сих пор.

Имеющиеся данные об изменении со временем интенсивности ядерной компоненты весьма противоречивы [23, 32, 34, 39, 42—45].

Вполне возможно, что вообще не существует регулярных суточных вариаций интенсивности тяжелых ядер, но, по-видимому, имеются определенные флуктуации интенсивности, характер и причины которых пока еще не ясны. Проведение непрерывных и длительных измерений интенсивности потоков различных групп ядер при помощи аппаратуры, установленной на спутнике, позволит наилучшим образом решить вопрос о вариациях ядерной компоненты, позволит определить, потоки каких групп ядер подвержены вариациям, как они связаны с вариациями интенсивности протонов и т.п. Регистрация интенсивности „дифференциальным" методом, т.е. каждой группы ядер в отдельности, может дать более ясные сведения о вариациях интенсивности ядерной компоненты, нежели „интегральный" метод регистрации измерений суммарной ионизации, создаваемой всем первичным космическим излучением.

Весьма важным является вопрос об энергетическом спектре космических лучей, как протонов, так и более тяжелых ядер. В области энергий до 30.10^9 эв для протонов и до 15.10^9 эв для частиц с $Z \geq 2$ распределение по энергиям можно опредилить на основе измерений широтного эффекта. Однако и в этом случае ряд обстоятельств, например отсутствие сведений о характере магнитного поля на больших расстояниях от поверхности Земли, влияние на измерение потока протонов вторичных частиц, вылетающих из атмосферы Земли, затрудняет получение достоверных данных об энергетических спектрах первичных частиц. В области высоких и сверхвысоких энергий положение еще более тяжелое: единственным источником сведений о частицах с энергией больше 10^{12} эв являются широкие атмосферные ливни. Определение спектра первичных частиц на основе изучения распределения по величине широких атмосферных ливней связано с целым рядом предположений и по этой причине недостаточно достоверно и однозначно. В связи с этим весьма перспективной представляется постановка на борту искусственного спутника Земли приборов для непосредственного измерения энергетического распределения частиц большой энергии. В качестве возможного метода решения этой задачи можно использовать измерение ионизации, создаваемой частицами электронно-фотонного ливня, образованного частицей большой энергии в слое поглотителя над ионизационной камерой [48]. Число частиц (а следовательно и создаваемая ими ионизация) в определенных пределах пропорционально энергии первичной частицы. Таким образом, используя импульсную ионизационную камеру и фиксируя величину толчка, создаваемого в ней частицами электронно-фотонного ливня, можно найти энергию первичной частицы и, собрав достаточный статистический материал, определить энергетический спектр первичных частиц в области высоких энергий. Оценка показывает, что установка площадью 300 см^2 за неделю зарегистрирует около 300 частиц с энергией 10^{13} эв.

Другой вопрос, связанный с исследованием частиц сверхвысоких энергий, состоит в том, сохраняется ли при энергиях $10^{12} \div 10^{14}$ эв соотношение между потоками протонов и более тяжелых ядер, наблюдаемое в области умеренных энергий.

Состав космических лучей в области сверхвысоких энергий в настоящее время совершенно неизвестен. Поэтому не ясно, создаются ли широкие ливни с очень большим энерговыделением протонами или ядрами. Ответить на этот важный вопрос можно будет, по-видимому, лишь поставив соответствующий эксперимент на искусственном спутнике Земли. Даже в том случае, если основная доля широких ливней создается протонами, а поток ядер с энергией $E > 10^{12}$ эв невелик, можно надеяться, что на спутнике удастся набрать достаточно богатый статистический материал для определения величины этого потока. При этом для регистрации ядер с $E > 10^{12}$ эв можно использовать комбинацию черенковского счетчика, с помощью которого будет определяться заряд частицы, и описанное выше устройство, позволяющее оценить энергию частицы.

Литература

1. В. Л. Гинзбург, Усп. Физ. Наук **62**, 37 (1957).
   В. Л. Гинзбург, С. Б. Пикельнер, И. С. Шкловский, Астроном. журнал **32**, 503 (1955).
2. H. E. Suess and H. C. Urey, Rev. Mod. Physics **28**, 53 (1956).
3. Е. Л. Фейнберг, ДАН СССР **53**, 421 (1946).
4. Л. И. Дорман, Е. Л. Фейнберг, Proceedings of the Guanajuata Conference of Cosmic Ray Physics, 1955.
5. G. J. Waddington, Nuovo Cimento **3**, 930 (1956).
6. R. E. Danielson, P. S. Freier, J. E. Naugle and E. P. Ney, Physic. Rev. **103**, 1075 (1956).
7. J. A. Simpson, K. B. Fenton, J. Katzman and D. C. Rose, Physic. Rev. **102**, 1648 (1956).
8. S. F. Singer, Physic. Rev. **80**, 47 (1950).
9. M. A. Pomerantz, J. Franklin Inst. **258**, 443 (1954).
10. G. W. McClure, Physic. Rev. **96**, 1691 (1954).

11. M. A. POMERANTZ, Physic. Rev. **95**, 1691 (1954).
12. L. GOLDFARB, H. L. BRADT and B. PETERS, Physic. Rev. **77**, 751 (1950).
13. B. PETERS, Proc. Indian Acad. Sci. Ser. A **40**, 230 (1954).
14. G. J. PERLOW, L. R. DAVIS, C. W. KISSINGER and J. D. SHIPMAN, Physic. Rev. **88**, 321 (1952).
15. A. DE MARCO, A. MILONE e M. REINHARZ, Nuovo Cimento **3**, 1150 (1956).
16. N. HORWITZ, Physic. Rev. **98**, 165 (1955).
17. J. LINSLEY, Physic. Rev. **101**, 826 (1956).
18. W. R. WEBBER and F. B. MCDONALD, Physic. Rev. **100**, 1460 (1955).
19. F. B. MCDONALD, Physic. Rev. **104**, 1723 (1956).
20. W. R. WEBBER, Nuovo Cimento **4**, 1285 (1956).
21. H. L. BRADT and B. PETERS, Physic. Rev. **77**, 54 (1950).
22. J. H. NOON, A. J. HERZ and B. J. O'BRIEN, Nature (London) **179**, 91 (1957).
23. E. P. NEY and D. M. THON, Physic. Rev. **81**, 1068 (1951).
24. L. R. DAVIS, H. M. CAULK and C. J. JOHNSON, Physic. Rev. **101**, 800 (1956).
25. J. LINSLEY, Physic. Rev. **97**, 1292 (1955).
26. C. J. WADDINGTON, Philos. Mag. **45**, 1312 (1954).
27. K. GOTTSTEIN, Philos. Mag. **45**, 347 (1954).
28. D. LAL, J. PAL, M. F. KAPLON and B. PETERS, Physic. Rev. **86**, 569 (1952).
29. H. J. TAYLOR, M. SITARAMASWAMI and P. N. KRISHNAMOORTHY, Proc. Indian Acad. Sci., Ser. A **36**, 41 (1952).
30. R. E. DANIELSON, P. S. FREIER, J. E. NAUGLE and E. P. NEY, Physic. Rev. **96**, 829 (1954).
31. R. F. HOURD, J. R. FLEMING and J. J. LORD, Physic. Rev. **95**, 647 (1954).
32. P. S. FREIER, G. W. ANDERSON, J. E. NAUGLE and E. P. NEY, Physic. Rev. **84**, 322 (1951).
33. H. FAY, Z. Naturforsch. **10a**, 572 (1955).
34. T. H. STIX, Physic. Rev. **95**, 782 (1954).
35. H. YAGODA, Physic. Rev. **99**, 1644 (1955).
36. M. F. KAPLON, J. H. NOON and G. W. RACETTE, Physic. Rev. **96**, 1408 (1954).
37. M. F. KAPLON, B. PETERS, H. L. REYNOLDS and D. M. RITSON, Physic. Rev. **85**, 295 (1952).
38. A. D. DAINTON, P. H. FOWLER and D. W. KENT, Philos. Mag. **43**, 729 (1952).
39. G. W. ANDERSON, P. S. FREIER and J. E. NAUGLE, Physic. Rev. **94**, 1317 (1954).
40. M. F. KAPLON, G. W. RACETTE and D. M. RITSON, Physic. Rev. **93**, 914 (1954).
41. S. F. SINGER, Bull. Amer. Physic. Soc. **2**, № 1, 53 (1957).
42. J. J. LORD and M. SCHEIN, Physic. Rev. **78**, 484 (1950); **80**, 304 (1950).
43. P. S. FREIER, E. P. NEY, J. E. NAUGLE and G. W. ANDERSON, Physic. Rev. **79** 206 (1950).
44. G. W. MCCLURE, M. A. POMERANTZ, Physic. Rev. **84**, 1252 (1951).
45. V. H. Yngve, Physic. Rev. **92**, 428 (1953).
46. Проблемы современной физики, сер. 5, вып. 7, изд-во иностран. лит-ры, Москва, 1953.
47. В. Л. ГРАНОВСКИЙ, Электрические флюктуации. Москва-Ленинград: ОНТИ, 1936.
48. О. Н. ВАВИЛОВ, Comptes Rendus (Doklady) **XXXIII**, № 3 (1941).
49. J. A. SIMPSON, W. FONGER, and S. B. TREIMAN, Physic. Rev. **90**, 934 (1953).

Study of the Primary Cosmic Radiation by Using Artificial Satellites of the Earth

By

S. N. Vernov, V. L. Ginzburg, L. V. Kurnosova, L. A. Razorionov and M. I. Fradkin

Cosmic-ray study is an important part of the scientific program to be conducted by the aid of the Earth satellite. Two problems of cosmic-ray physics attract attention in connection with the possibility of lifting scientific instruments to the top of the atmosphere. These are cosmic-ray time variations of different kinds and the charge-spectrum of the primary cosmic rays.

A knowledge of the primary cosmic-ray composition and of the energy spectra of different nuclei is important both from the point of view of the theory of the origin of cosmic rays as well as for investigation of the elementary interaction acts between the primary high-energy particles and nuclei in the atmosphere.

The charge-spectrum for nuclei with $Z > 2$ has not been sufficiently studied as yet. There is a number of problems concerning the so-called nuclear component which requires further experimental study. Investigators of different countries have conducted experiments with the purpose of studying the nuclear component composition, but the results obtained up to now are still insufficiently trustworthy. Most of the experiments were performed in the upper atmosphere, where the residual amount of air above the apparatus constitutes roughly 15 g/cm^2. As a result, one can determine the true composition of the primary component over the top of the atmosphere only by calculation, taking into account the diffusion equations. This calculation requires a precise knowledge of the fragmentation probability for primary nuclei interacting in the residual atmosphere above the apparatus. It would be very important to be relieved of the necessity of such calculations. To that end one should measure the fluxes of the primary nuclei with different Z directly over the top of the atmosphere. Use of rocket-borne equipment for this purpose cannot yield satisfactory results, as the flux of the nuclei with $Z > 2$ is small and very poor data would be obtained during the limited stay of the rocket outside the atmosphere.

Launching of artificial satellites opens up quite new possibilities for investigation of the primary nuclear component. Scientific instruments carried in the satellites should permit one to gather statistical data necessary for determining low intensity fluxes of the primary nuclei.

One of the most important questions bearing on the composition of the primary nuclear component is that of the ratio of the flux of light nuclei (Li, Be, B) to that of medium nuclei (C, N, O, F).

A knowledge of this ratio is important, because since the light nuclei are very rare they are probably emitted by the cosmic-rays sources in negligible quantities and the Li, Be, B nuclei observed in the upper atmosphere are fragments of heavier nuclei, which had interacted with interstellar matter.

The ratio of the fluxes of the different nuclei groups at the Earth's orbit can be computed by making some assumptions regarding the spatial distribution and intensity of the cosmic-ray sources and also the conditions of propagation of cosmic-ray particles through interstellar matter. Comparison of the ratio so obtained with that obtained experimentally may serve as a criterion of validity of our theoretical ideas. For example, according to a new theory being worked out by soviet theorists, which explains generation of cosmic rays by a statistical mechanism of acceleration in the envelopes of supernovae, this ratio is more or equal to 0.1. Because of insufficient reliable knowledge of some of the parameters used in the estimate the above mentioned value may be even higher but the theory is in striking contrast with the assumption that the ratio is much less than 0.1.

The data of various authors on the relative abundance of light (L) and medium (M) nuclei are frequently conflicting and do not seem to be sufficiently reliable since the data were obtained by calculations based on the results of measurements performed in the atmosphere. Thus, determination of the true value of this ratio remains as an essential problem.

The earth satellite may also be used to elucidate the question concerning the presence of nuclei with $Z > 30$ in the primary cosmic rays.

The present report briefly outlines a method for investigation of charge-spectra based on the VAVILOV-CHERENKOV effect. A sketch of the experimental arrangement developed in the LEBEDEV Physical Institute of the Academy of Sciences, U.S.S.R., is given.

Cosmic-ray recording apparatus installed in a satellite, may also be used to investigate a number of other problems relating to the primary component. Thus, for example it should be possible to measure the primary proton flux, to ascertain the role of the albedo of the Earth's atmosphere, to determine the lower limit of the electron-positron flux, to study the interaction between the primary particles and matter, to determine the energy spectra of the different components of the primary cosmic rays, to investigate the time variations of the primary component etc.

Instruments carried by a satellite should also directly yield information on variations of the primary cosmic-ray intensity and in particular on the variations of the low energy particle flux. It should be noted that in this case the usual diurnal variations will be manifest as one and a half hour variations. This circumstance should undoubtedly facilitate the study of the nature of these variations.

A schematic diagram of the setup for investigation of cosmic-ray variations developed jointly by the Academy of Sciences and the Moscow State University is also presented.

Некоторые задачи динамики полета к луне (тезисы к докладу)

В. А. Егоров[1]

Вопросы динамики полета к Луне в последние годы рассматривались рядом авторов [1—6], например Тюрингом, Лоуденом, Буххеймом. Однако до сих пор не были выяснены принципиально возможные формы пассивного участка траекторий полета к Луне и влияние разброса начальных данных. Объясняется это тем, что даже в простейшей постановке задача сводится к нерешенной в механике ограниченной круговой проблеме трех тел (Земля, Луна, ракета).

В настоящей работе сделана попытка дать такое решение этой задачи, которое позволяло бы для заданной цели полета к Луне выбириать траекторию, наиболее подходящую по форме и влиянию разброса начальных данных.

1. В первой части работы было проведено общее исследование траекторий полета к Луне. В системе координат, вращающейся вместе с прямой Земля — Луна, с помощью интеграла Якоби [7] были определены минимальные начальные скорости, необходимые для достижения Луны. При начале пассивного участка на высоте 200 км от поверхности Земли эти скорости равны 10,849 км/сек, независимо от направления стрельбы. Однако численное интегрирование с помощью машины показало, что если такие траектории и достигнут Луны, то только после достаточного количества оборотов (порядка сотен и более) вокруг Земли, и что минимальную начальную скорость, необходимую для достижения Луны на первом обороте, можно вычислять в невращающейся системе геоцентрических координат, не учитывая влияния Луны. При вертикальном направлении она равна 10,906 км/сек; с наклоном направления скорости эта величина несколько убывает.

Методом, аналогичным примененному В. Г. Фесенковым [8], была установлена невозможность захвата снаряда Луной в ее сфере действия для траекторий сближения (т.е. для траекторий, начинающихся у Земли и на первом обороте входящих в сферу действия Луны).

При вычислении траекторий сближения в рассматриваемой задаче, как показало численное интегрирование, можно пренебрегать влиянием Луны вне ее сферы действия и влиянием Земли — внутри этой сферы и считать, что соответствующие геоцентрические и селеноцентрические участки траекторий суть конические сечения. Это дает простую, но достаточно точную приближенную методику исследования и позволяет выявить основные закономерности и характеристики движения.

Оказалось, что селеноцентрический участок траектории сближения всегда есть гипербола, и что при начальных скоростях; лишь на 0,5 км/сек превышающих параболическую, характер закономерностей уже близок к асимптотическому, отвечающему бесконечно большим начальным скоростям.

---

[1] Москва, Академия наук СССР.

Для движения в плоскости орбиты Луны удалось приближенно построить план скоростей выхода ракеты из сферы действия и проанализировать эволюцию множества траекторий сближения с изменением вектора начальной скорости. Это позволило во второй части работы рассмотреть ряд конкретных задач о полете к Луне в плоскости ее орбиты.

2. В задаче о попадании в Луну было установлено с помощью приближенной методики, что отклонение траекторий от центра Луны является квадратичной, а не линейной функцией малых ошибок начальных данных. Были определены или оценены максимально допустимые величины ошибок по каждому из начальных данных. В частности, оказалось, что при ошибках в начальной скорости порядка 50 м/сек и в ее направлении около 0,3°, для начальных скоростей, близких к параболической, траектория еще проходит через Луну. Влияние ошибок в начальном положении при постоянном избытке начальной скорости над параболичезкой оказалось мало существенным.

В задаче об облете Луны с возвращением к Земле были установлены два класса решений. Были найдены также два класса траекторий, возвращающихся к Земле, но не отвечающих облету — так называемых долетных. Отклонение траекторий от центра Земли при возврате оказалось, как и в задаче о попаданиях, квадратичной функцией ошибок в начальных данных. При этом влияние разброса начальных данных убывает с ростом расстояния траектории сближения от Луны. При расстояниях порядка радиуса сферы действия (66000 км) точности, потребные для возвращения и Земле после облета, оказываются не выше, чем для попадания в Луну при той же начальной скорости.

Более интересная задача об облете Луны с пологим входом в атмосферу Земли имеет аналогичную классификацию. Однако вследствие относительной малости толщины атмосферы потребные точности начальных данных оказываются выше в десятки и более раз, чем в предыдущей задаче.

Задача о периодических траекториях сближения в плоскости орбиты Луны имеет одно семейство облетных решений и две последовательности семейств долетных. Облетные решения отвечают гиперболическим начальным скоростям, удалены от Земли на расстояния порядка 100000 км и неустойчивы.

Задача о наибольшем разгоне ракеты под влиянием Луны имеет решения двух классов, обходящих Луну с противоположных сторон вблизи ее поверхности. Траектория, зеркально симметричная разгонной, отвечает наибольшему торможению ракеты, возвращающейся из межпланетного полета. Вследствие малости расстояния траекторий от Луны для осуществления разгона, достаточно близкого к максимальному, требуются гораздо большие точности начальных данных, чем для попадания.

Кроме рассмотренных существуют траектории сближения, отвечающие большему или меньшему замедлению ракеты относительно Земли; других траекторий сближения не существует. Поэтому полученная классификация плоских траекторий сближения является полной.

Во всех рассмотренных задачах были прослежены решения различных классов при изменении величины и направления начальной скорости (с помощью приближенной методики). Были разработаны итерационные методы отыскания траекторий заданного назначения посредством интегрирования уравнений движения задачи трех тел на электронной цифровой машине [9]. Благодаря этому почти все результаты приближенной методики проверялись и неизменно подтверждались более точным способом. С помощью машины находились и потребные точности начальных данных. Всего было сосчитано более 1000 траекторий.

Резюмируя, можно сделать следующие выводы. Высокие требования по точности начальных данных приводят к тому, что для реализации близкого облета Луны и, особенно, близкого облета с пологим входом в атмосферу Земли, представляется необходимой коррекция пассивного участка траектории с помощью вспомогательного двигателя. Для траекторий попадания в Луну и траекторий далекого облета Луны требования по точности являются достаточно умеренными, вследствие чего полет по этим траекториям, по-видимому, может быть осуществлен без коррекции на пассивном участке, как только ракетами будут достигнуты скорости порядка параболической.

Следует отметить, что даже при необходимости коррекции для реализации полученных решений, они все же должны представлять практический интерес, так как эти решения требуют минимальной коррекции по сравнению с другими.

Заметим также, что полученные результаты могут быть использованы в качестве первого приближения при точном отыскании траекторий заданного назначения с учетом возмущений от Солнца и других второстепенных факторов. При этом метод определения траекторий сохранится.

Наконец, методика и результаты проведенной работы могут быть обобщены на случай пространственного движения и на задачи полета с Земли к внешним планетам Солнечной системы. Более подробно эти результаты рассмотрены в опубликованных работах [10, 11].

Литература

1. B. Thüring, Weltraumfahrt **3**, № 4, стр. 202 (1952); **5**, № 3, 4, 5, 69 и 103 (1954).
2. D. F. Lawden, J. Brit. Interplan. Soc. **13**, № 6, 329 (1954).
3. R. Buchheim, Artifical Satellite of the Moon, Rand Corp. P-873 (1956).
4. A. Goldstein, C. Fröberg, Kgl. Fisiogr. Sällsk. Lund Förh. **22**, № 14 (1952).
5. Г. А. Чеботарев, Бюллетень ИТА, **VI**, № 7, 487 (1957).
6. М. С. Лисовская, Бюллетень ИТА, **VI**, № 8, 550 (1957).
7. М. Ф. Субботин, Курс небесной механики, т. **II**, ОНТИ, 1937.
8. В. Г. Фесенков, Астрономический журнал, т. **23**, в. **I** (1946).
9. В. А. Егоров, труды конференции „Пути развития отечественного математического машиностроения и приборостроения", т. **II**, Москва, 1958.
10. В. А. Егоров, ДАН СССР, **113**, № 1 (1957).
11. В. А. Егоров, Успехи физических наук, т. **63**, в. **I** (1957).

Some Problems Relating to the Dynamics of the Flight to the Moon

By

V. A. Yegorov

The papers published by Br. Thüring, D. F. Lawden, R. Buchheim and others have not yet thrown light on the possible shapes of the passive segment of the trajectories of the flight to the Moon, and on the influence of errors in the initial data. This is accounted for by the fact that, even set in the simplest way, the problem amounts to a restricted circular problem of three bodies (Earth, Moon and rocket) as yet unsolved in mechanics.

The aim of the present work is to supply such a solution to the problem which would allow to select, for a given Moonflight designation a trajectory which would be most suitable from the point of view of shape and the influence of initial data scattering.

I. The first part of the work was devoted to a general investigation of the trajectories of the flight to the Moon. The minimum initial velocities required to reach the Moon have been found with the aid of JAKOBI's integral in coordinates rotating together with the straight line Earth-Moon. For an altitude of 200 km they amount to 10.849 km/sec, regardless of the direction of shooting.

Numerical integration by means of a computing machine has, however, shown that if the trajectories in question are to reach the Moon, they should make a sufficient number of revolutions (hundreds or more) around the Earth, and that the minimum speed required for reaching the Moon at the first revolution can be calculated in non-rotating geocentric coordinates, no account being taken of the influence of the Moon. This speed amounts to 10,906 km/sec for the vertical direction, decreasing somewhat with an inclination of the speed direction.

A method similar to the one used by V. G. FESENKOV was applied to establish the impossibility of capturing the missile in the sphere of action of the Moon for the rapprochement trajectories (i.e. trajectories starting at the Earth and entering the sphere of action of the Moon at the first revolution).

As proved by numerical integration, the perturbations by the Moon outside its sphere of action may be ignored as well as the perturbations by the Earth within this sphere; therefore it may be considered that the corresponding geocentric and selenocentric segments of the trajectories are conic cross sections. This provides a simple and yet sufficiently accurate approximate method of investigation and permits to disclose the basic regularities and characteristics of the motion.

It appeared that the selenocentric segment of the rapprochement trajectory is always a hyperbola and that at initial speed which exceeds the parabolic one by only 0.5 km/sec the nature of the regularities already approximates to asymptotic nature which corresponds to infinitely high initial speeds.

With regard to motion in the plane of the Moon's orbit an approximate vector-diagram of velocity at which the rocket enters and leaves the sphere of action was considered and the evolution of the whole ensemble of rapprochement trajectories with a change of the initial velocity vector was analysed. This allowed to consider, in the second part of the work, a number of specific problems of the flight to the Moon in the plane of its orbit.

II. As to the problem of hitting the Moon, it has been established by means of approximation methods that the deviation of trajectories from the centre of the Moon is a quadratic, not a linear function of the slight errors in the initial data.

The maximum permissible errors for each of the initial data were determined or evaluated. It appeared that with errors in the initial speed and in the direction of the shooting amounting to some 50 m/sec and 0·3° respectively the trajectory still passes through the Moon when the initial speeds approximate from above the parabolic speed. The effect of the errors in the initial position proved to be of little importance if difference of initial and parabolic speeds remained constant. Hitting trajectories corresponding to the various initial speeds are determined in the rotating coordinates.

Considering the problem of the flight around the Moon with a subsequent return to the Earth, two types of solutions were found. Two types of trajectories were also found, which return to the Earth but do not satisfy the requirements of envelopping the Moon; these trajectories have been termed non-envelopping ones. The deviation of the trajectories from the centre of the Earth when returning proved to be, as in the hitting problem, a quadratic function of initial data errors. In this case the effect of the initial data errors diminishes with increasing distance of the rapprochement trajectory from the Moon. With distances approximating to the sphere of action radius (66,000 km), the accuracy required for returning to the Earth after a flight around the Moon, as it appears, is not greater than in the case of hitting the Moon at the same initial speed.

The more attractive problem of flying around the Moon with a slightly oblique entrance into the Earth atmosphere is classified in a similar way. However, because of the relatively small thickness of the atmosphere, the required accuracy of the initial data is scores of times greater than in the preceding problem.

The problem of periodic rapprochement trajectories in the plane of the Moon's orbit has one series of envelopping solutions and two sequences of non-envelopping series. The envelopping solutions correspond to hyperbolic initial speeds; they pass at a distance of some 100,000 km or more from the Earth, and are unstable.

The problem of the maximum acceleration of the rocket with the help of the Moon (perturbation manoeuvre) has solutions of two types by-passing the Moon on two opposite sides near its surface. The trajectory which is mirror-symmetrical to the acceleration trajectory corresponds to the maximum slowing down of a rocket returning from an interplanetary flight. Since the distance from the trajectories to the Moon surface is too short, to obtain acceleration sufficiently near to a maximum one, considerably greater accuracy of the initial data is required than in the case of hitting.

In addition to rapprochement trajectories considered above there exist those which correspond to greater or lesser deceleration in relation to the Earth; there are no other rapprochement trajectories whatsoever. Hence the obtained classification of plane rapprochement trajectories is complete.

In all the problems analysed we traced the evolution of solutions of the various types in cases when the magnitude and the direction of the initial speed were changed; iterative methods were elaborated for finding the trajectories of a predetermined designation by integrating the motion equations of the three bodies problem on an electronic digital computer. Due to this almost all the results of the approximative method were checked and invariably confirmed by a more accurate method. The required initial data accuracy was also found by means of the computer. Altogether over 1000 trajectories were calculated.

In summing up, the following conclusions may be derived. The strict requirements concerning the accuracy of initial data mean that in order to realize a close flight around the Moon and particularly a close flight with a slightly oblique entrance into the atmosphere of the Earth, it is necessary that the passive segment of the trajectory should be corrected by means of an auxiliary engine. For trajectories hitting the Moon and those for a distant flight around the Moon, the accuracy requirements are moderate enough, hence a flight along these trajectories may probably be realized without correction at the passive segment as soon as the rockets attain a speed of the order of the parabolic speed.

It should be noted that even if a correction is necessary to effect the solutions obtained, the latter will nevertheless be of practical interest since they obviously require minimum correction as compared with other solutions.

It should also be mentioned that the results obtained may be used as the first approximation for determining accurate trajectories for a predetermined designation, taking into account perturbations from the Sun and other secondary factors. In this case the iterative methods of computing trajectories will remain the same.

Finally, it should be emphasized that the methods and the results of the work carried out may be extended to threedimensional motion and problems of the flight from the Earth to external planets of the Solar system.

Visual Observations of the Earth's Satellite in the USSR

By

A. G. Masevich[1]

Visual observations of the Earth's satellite in our country are very similar to those in the U.S.A.

The Astronomical Council of the Academy of Sciences of the USSR is in charge of these observations. We began with the designing of the equipment and ordered then at our optical industry a series of telescopes for satellite observations. Like the satellite telescope recommended for the Moonwatch Programme in the U.S.A., our 6-power telescope has a 50 mm apperture and field of vision of nearly 11°. We provided our scopes with reticles, so they could be used also for meteor observation and for students' practical training. The scope is equipped with a triped, a platform for a removable level and a vertical limb.

The USSR stations for visual observations are being organized at the Departments of Physics and Mathematics of the Universities and Pedagogical Institutes all over the country, and equipped at the expense of the USSR National Committee for the IGY.

During the observations, precise time signals will be broadcasted or telephoned to the stations. The observer sends a signal pressing a telegraph key when the satellite crosses the reticles or passes a certain stellar configuration in the field. Both time and passage signals are registered on the tape-recorder. The exact moment of the passage is determined later by a stop watch, and the position is read from a stellar chart (the Atlas by A. A. Mikhailov or Atlas by A. Becvar).

In those cases when the satellite and stars are in the field of vision simultaneously, we expect to determine the position of the satellite to within 0.5°, and the time to within 0.5^s. If the satellite is observed at early twilight and the number of stars in the field of vision is insufficient, the precision will be much less.

The ephemeris will be communicated by the Astronomical Council to the stations in advance. If the ephemeris will be uncertain, only the presumed time of observation will be communicated. Considering the great inclination of our satellites orbit we propose to observe the satellite not always in the meridian, but in the vertical that depends from the latitude of the station and crosses the orbit in a point nearest to the observer.

The observational data will be reported to the Astronomical Council.

67 stations for visual observations are organized at present in places where the probability of good weather is not too low.

An instruction for visual observations was worked out.

We intend to conduct a few test alerts to control the readiness of our stations.

Our stations situated in the zone of vision of the American satellite could participate in visual observation, provided the ephemeris are communicated to them in good time. On our part, we are ready to communicate the ephemeris of our satellite to countries that will observe it.

[1] Vice-president of the Astronomical Council of the Academy of Sciences of the USSR.

О выведении искусственного спутника земли на орбиту

Д. Е. Охоцимский, Т. М. Энеев [1]

(9 Фиг.)

В настоящей статье рассматривается вопрос о выведении искусственного спутника Земли на орбиту. Предполагается, что выведение спутника Земли осуществляется при помощи ракеты-носителя. состоящей из одной или нескольких ступеней. Исследуется, каков должен быть закон изменения во времени направления тяги реактивных двигателей, чтобы обеспечить выведение спутника на заданную орбиту с минимальным расходом топлива. Ищется также наиболее выгодный режим расходования топлива.

Решение указанных задач, проводящееся при ряде упрощающих предположений, позволяет составить определенное представление о характерных особенностях оптимального выведения на орбиту и указать пути, идя по которым, можно добиться создания ракеты-носителя с минимальным начальным весом.

В первом разделе рассматривается задача об одновременном подборе как программы для направления тяги, так и режима расходования топлива. Во втором разделе исследуется задача о подборе оптимальной программы для многоступенчатой ракеты с различным числом ступеней в предположении, что режим расходования топлива является заданным. В третьем разделе приведено обобщение задачи о выведении на случай движения в центральном поле тяготения с учетом вращения Земли.

1. Выбор оптимального режима расходования топлива и оптимальной программы для направления тяги

Решение указанной задачи будем проводить в предположении, что аэродинамические силы отсутствуют и что поле земного тяготения является плоско-параллельным. Основанием для первой гипотезы служит тот факт, что при выведении спутника на орбиту значительная часть траектории выведения будет лежать в высоких слоях атмосферы, где аэродинамические силы невелики. Замена центрального поля земли плоско-параллельным будет возможна, если предположить, что протяженность траектории выведения невелика по сравнению с радиусом Земли.

Оба указанных предположения выполняются в действительности лишь приближенно. Однако решение задачи в указанной приближенной постановке представляет тем не менее несомненный интерес, так как позволяет весьма просто получить полное решение задачи, проанализировать полученный результат и понять основные закономерности явления.

Уравнения движения спутника и ракеты-носителя в проекции на прямоугольные оси координат могут быть записаны в виде (фиг. 1).

---

[1] Москва, Академия наук СССР.

$$\frac{du}{dt} = P\cos\varphi \qquad \frac{dy}{dt} = w$$

$$\frac{dw}{dt} = P\sin\varphi - g \qquad \frac{dx}{dt} = u, \tag{1.1}$$

где u и w — горизонтальная и вертикальная проекции скорости, p — величина ускорения реактивной силы, φ — угол наклона силы тяги к горизонту. Так как тягу можно считать направленной вдоль продольной оси ракеры, то под углом φ можно понимать угол наклона оси ракеты к горизонту (угол тангажа). В уравнениях (1.1) x и y — горизонтальная и вертикальная координаты.

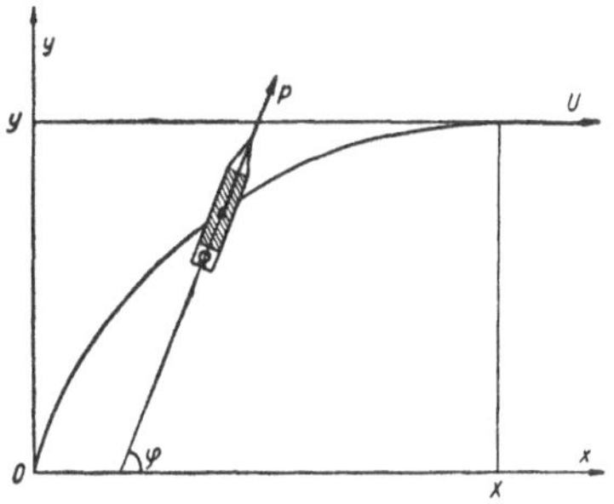

Фиг. 1

Последнее уравнение системы (1.1) служит только для определения координаты x. Если не накладывать никаких ограничений на дальность полёта, то при решении вариационной задачи это уравнение может быть отброшено, поскольку координата x в другие уравнения не входит.

Рассмотрим скорость V, которую приобрел бы спутник в идеальном случае, когда на него не действовали бы никакие другие силы кроме реактивной силы. Имеем

$$V = \int_0^t p\,dt,$$

отсюда

$$p = \frac{dV}{dt}$$

Подставляя в систему (1.1), запишем первые три уравнения в виде

$$\begin{aligned} \frac{du}{dt} &= \frac{dV}{dt}\cos\varphi \\ \frac{dw}{dt} &= \frac{dV}{dt}\sin\varphi - g \\ \frac{dy}{dt} &= w. \end{aligned} \tag{1.2}$$

Величина V представляет собой располагаемую скорость, которая может быть сообщена спутнику в процессе разгона и выведения на орбиту.

В правые части уравнений движения (1.2) входят две неопределенные функции времени $V(t)$ и $\varphi(t)$. Поставим задачу выбрать эти функции таким образом, чтобы в конце участка выведения на заданной высоте получить наибольшую горизонтальную скорость. В силу взаимости полученное решение будет обеспечивать также достижение при заданной скорости наибольшей высоты, а также достижение заданных значений высоты и скорости при минимальном расходе топлива.

Сформулируем граничные условия. Предположим, что в начале движения при $t=O$ имеем некоторую высоту y_o и некоторые значения горизонтальной и вертикальной проекций скорости u_o и w_o. Скорость V в начале движения естественно принять равной нулю. В конце движения в момент

$t=T$ высота должна быть равна $y=Y$ и скорость должна быть направлена горизонтально, т.е. $w=O$. Значение V в конце движения положим равным некоторому фиксированному значению V_k. Тем самым предполагается, что имеется некоторый запас идеальной скорости, который нужно использовать наиболее рациональным образом. Если считать, что величина скорости истечения или удельной тяги не зависит от секундного расхода топлива, то величина полной идеальной скорости V_k будет определяться отношением начальных и конечных весов ступеней и не будет зависеть от режима расходования топлива. Это означает, что задание определенного значения V_k в конце движения отвечает заданию определенного отношения начальных и конечных весов, а при заданном весе спутника означает задание определенного начального веса ракеты.

Итак, граничные условия имеют вид:

$$\text{при} \quad t=O \quad u=u_o \quad w=w_o \quad y=y_o \quad V=o, \tag{1.3}$$
$$\text{при} \quad t=T \quad w=O \quad y=Y \quad V=V_k.$$

Время движения T на участке выведения можно либо считать заданным, либо подбирать его из условия получения наибольшей скорости в конце движения. Можно при этом считать, что время движения T на участке выведения не совпадает с временем работы двигателя T_* а удовлетворяет лишь условию

$$T \geq T_*,$$

где T_* — заданная величина.

Функцию $\varphi(t)$ не будем ограничивать никакими условиями, кроме условия непрерывности. Функцию $V(t)$ как интеграл от неотрицательной величины следует считать неубывающей. Примерный график зависимости V от времени приведен на фиг. 2. Горизонтальные участки отвечают движению с неработающим двигателем.

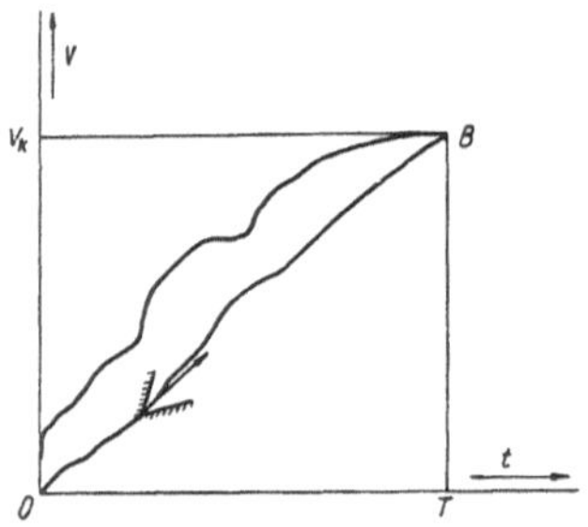

Фиг. 2

Вертикальные участки отвечают мгновенному расходованию части топлива. Если не накладывать ограничений на величину секундного расхода, то допустимой линей в плоскости (t, V) будет любая линия, соединяющая точки O и B, вдоль которой V не убывает. Если на секундный расход наложить ограничения, то допустимая линия будет удовлетворять более жесткому условию, что в каждой ее точке исходящий из этой точки элемент кривой должен быть расположен внутри или на границах некоторого угла, верхняя сторона которого отвечает движению с наибольшим, а нижняя с наименьшим секундым расходом. В предельном случае, когда наибольший допустимый расход будет бесконечно большим (мгновенное сгорание), а наименьший равен нулю (движение с выключенным двигателем), верхняя сторона угла будет идти вертикально, а нижняя горизонтально. Получим указанное выше условие неубывания величины V вдоль допустимой линии.

Скорость в конце участка разгона получим, интегрируя первое уравнение (1.2) в пределах от O до T.

$$U = \int_0^T \frac{dV}{dt} \cos\varphi \, dt + u_o \tag{1.4}$$

Кроме того, имеем два других уравнения, являющихся дифференциальными связями. Перепишем их в виде

$$\frac{dw}{dt} - \frac{dV}{dt}\sin\varphi + g = 0 \tag{1.5}$$

$$\frac{dy}{dt} - w = 0 \tag{1.6}$$

Имеем также граничные условия (1.3). Величины u_o и w_o представим в виде

$$u_o = v_o \cos\theta_o; \qquad w_o = v_o \sin\theta_o, \tag{1.7}$$

Считая v_o заданным и варьируя угол θ_o, получим из решения задачи значение оптимального угла наклона вектора начальной скорости.

Составим вспомогательной функционал

$$y = v_o\cos\theta_o + \int\limits_0^T \left\{\frac{dV}{dt}\cos\varphi + \lambda_1\left(\frac{dw}{dt} - \frac{dV}{dt}\sin\varphi + g\right) + \lambda_2\left(\frac{dy}{dt} - w\right)\right\} dt, \tag{1.8}$$

где λ_1 и λ_2 некоторые неопределенные пока функции времени. Вычислим вариацию полученного функционала, варьируя также начальный угол θ_o и время движения на участке выведения T.

Производя варьирование по всем входящим функциям и переменным параметрам и используя граничные условия, получим следующее выражение для вариации.

$$\begin{aligned}\delta y = &-v_o(\sin\theta_o + \lambda_1\cos\theta_o)\,\delta\theta_o + g\,\lambda_1\,\delta T + \\ &+ \int\limits_0^T \left\{\left(-\sin\varphi - \lambda_1\cos\varphi\right)\frac{dV}{dt}\,\delta\varphi + \left[-\frac{d}{dt}(\cos\varphi - \lambda_1\sin\varphi)\right]\delta V - \right. \\ &\left. - \frac{d\lambda_2}{dt}\,\delta y - \left(\frac{d\lambda_1}{dt} + \lambda_2\right)\delta w\right\} dt.\end{aligned} \tag{1.9}$$

Если начальный угол не фиксирован, а подбирается из условия оптимума, то имеем связь между начальным углом θ_o и значением функции $\lambda_1(t)$ в начале движения

$$\operatorname{tg}\theta_o = -\lambda_{1/t=o}, \tag{1.10}$$

получающуюся приравниванием нулю первого из выражений перед интегралом. Приравнивая нулю второе выражение, получим условие в конце участка разгона

$$\lambda_{1/t=T} = 0. \tag{1.11}$$

Если время движения фиксировано, то $\delta T = 0$ и выписанное условие для функции $\lambda_1(t)$ отпадает.

Определим теперь λ_1 и λ_2 таким образом, чтобы обратились в нуль множители при δy и δw под интегралом и тем самым эти вариации выпали из выражения для вариации функционала. Если сделать это, то выражение для вариации примет вид

$$\delta y = \int\limits_0^T \left\{(-\sin\varphi - \lambda_1\cos\varphi)\frac{dV}{dt}\,\delta\varphi + \left[-\frac{d}{dt}(\cos\varphi - \lambda_1\sin\varphi)\right]\delta V\right\} dt, \tag{1.12}$$

где функция λ_1 должна быть определена из уравнения

$$\frac{d\lambda_1}{dt} = -\lambda_2; \qquad \frac{d\lambda_2}{dt} = 0 \tag{1.13}$$

при граничном условии (1.11), которое следует учитывать в случае, когда T не фиксировано.

Из второго уравнения (1.13) имеем

$$\lambda_2 = \text{const}; \tag{1.14}$$

Подставляя это значение в первое уравнение (1.13) получим

$$\lambda_1 = c_1 + c_2 t, \tag{1.15}$$

где c_1 и c_2 некоторые постоянные. В случае, когда T не фиксировано, имеем соотношение

$$c_1 + c_2 T = 0$$

и выражение для λ_1 приведется к виду

$$\lambda_1 = -c_2 (T - t). \tag{1.16}$$

Формулы (1.15) и (1.16) показывают, что величина λ_1 является линейной функцией времени. Отметим, что эти формулы содержат одинаковое число произвольных постоянных, так как в формулу (1.16) вместо c_1 входит неизвестная величина T.

Приравнивая нулю выражение при вариации $\delta\varphi$ под интегралом в формуле (1.12), получаем

$$\operatorname{tg}\varphi = -\lambda_1 \tag{1.17}$$

и используя формулу (1.15), найдем выражение для оптимальной программы по тангажу в виде

$$\operatorname{tg}\varphi = -(c_1 + c_2 t). \tag{1.18}$$

Если T не фиксировано, то согласно (1.16) имеем

$$\operatorname{tg}\varphi = c_2 (T - t). \tag{1.19}$$

Видим, что если время не фиксировано, то в конце движения угол тангажа должен быть равен нулю и, следовательно, ось ракеты должна быть направлена горизонтально.

Формулу для оптимальной программы можно переписать еще в виде

$$\operatorname{tg}\varphi = \operatorname{tg}\varphi_0 - c_2 t, \tag{1.20}$$

где φ_0 угол тангажа в начальный момент времени.

Видим, что оптимальная программа должна быть такова, чтобы тангенс угла тангажа был линейной функцией времени. Два параметра, входящие в общее выражение для программы, надлежит выбрать таким образом, чтобы удовлетворить граничным условиям.

Сопоставление формул (1.10) и (1.17) показывает, что оптимальный начальный угол θ_0 должно удовлетворять условию

$$\operatorname{tg}\theta_0 = \operatorname{tg}\varphi_0. \tag{1.21}$$

Это означает, что направление продольной оси ракеты в начальный момент должно совпадать с направлением начальной скорости. В случае, если угол θ_0 заранее задан, это условие может не выполняться.

Формулы (1.18) — (1.20) дают закон изменения угла тангажа при любой зависимости $V(t)$. Будем теперь подбирать функцию $V(t)$ наивыгоднейшим образом, считая, что для каждой такой функции всякий раз выбран наилучший закон изменения угла тангажа.

Тогда вариация функционала примет вид

$$\delta y = \int_0^T \left\{ -\left[\frac{d}{dt} (\cos \varphi - \lambda_1 \sin \varphi) \right] \delta V \right\} dt \tag{1.22}$$

При выполнении дифференциальных условий (1.5) и (1.6) функционалы (1.4) и (1.8) совпадают. Если эти условия тождественно выполнены при варьировании, то вариации функционалов также должны совпадать. Поэтому формула (1.22) позволяет рассчитать вариацию скорости δU в конце участка разгона при изменении режима расходования топлива и связанной с ним функции $V(t)$. Необходимо отметить, что выражение для вариации (1.22) будет верно лишь в том случае, когда все изменение V заканчивается до момента T.

Формула (1.22) может быть преобразована к виду

$$\delta U = \int_0^T \Phi(t)\, \delta V \, dt, \tag{1.23}$$

где

$$\Phi = \frac{c_2 (\operatorname{tg} \varphi_0 - c_2 t)}{\sqrt{(\operatorname{tg} \varphi_0 - c_2 t)^2 + l}} \tag{1.24}$$

или

$$\Phi = c_2 \sin \varphi$$

Формулы (1.23) и (1.24) показывают, что функция $\Phi(t)$, стоящая под интегралом множителем при вариации δV, зависит только от времени. На участке (O, T) функция может либо иметь все время постоянный знак, либо менять знак, причем формула (1.24) показывает, что изменение знака Φ связано с изменением знака числителя и может поэтому иметь место только один раз.

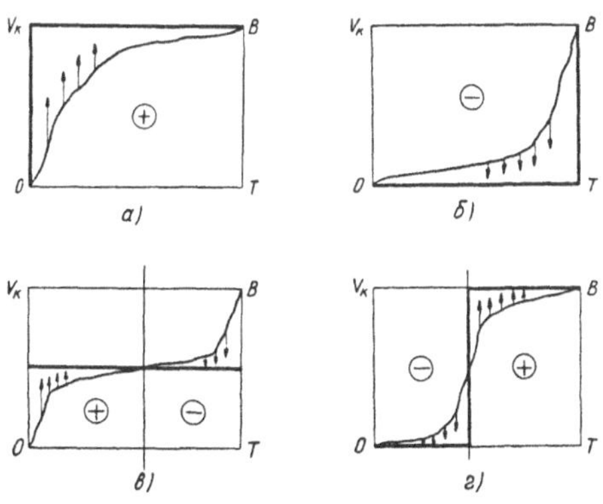

Фиг. 3

На основании полученных свойств функции $\Phi(t)$ легко получить, что эстремум функционала достигается на функциях $V(t)$ со скачкообразным изменением.

Рассмотрим отдельно различные возможные случаи.

1. Выражение Φ все время положительно (фиг. 3а). В этом случае сгорание всего запаса топлива должно происходить мгновенно в самом начале движения, так как всякую другую допустимую линию в плоскости (t, V) можно в противном случае проварьировать вверх, вследствие чего получим $\delta U > 0$, что означает возрастание скорости в конце участка выведения.

2. Выражение Φ все время отрицательно (фиг. 3б). Аналогично предыдущему получаем, что в данном случае наилучшим видом расходования

топлива будет мгновенное расходование всего запаса топлива в самом конце участка выведения при $t = T$.

3. Перемена знака Φ происходит с плюса на минус (фиг. 3в). В этом случае экстремум может достигаться лишь на ломаной, состоящей из вертикального, горизонтального и вертикального отрезков. Любую другую допустимую линию можно проварьировать в каждой из областей постоянства знака таким образом, что приращение конечной скорости будет положительным. Для того, чтобы определить положение горизонтального отрезка, напишем выражение для вариации конечной скорости при вертикальном смещении отрезка:

$$\delta U = \delta V \int_0^T \Phi(t)\, dt. \tag{1.25}$$

Если значение интеграла положительно, то этот случай аналогичен случаю 1, если отрицательно, то случаю 2. Если значение интеграла равно нулю, то это означает, что экстремум функционала может достигаться при каком-то промежуточном положении горизонтального отрезка. Следовательно, при определенных условиях может оказаться выгодным расходовать часть топлива мгновенно в самом начале движения, а оставшуюся часть расходовать мгновенно в самом конце движения.

4. Перемена знака Φ происходит с минуса на плюс (фиг. 3г). В этом случае расходование всего запаса топлива следует производить мгновенно в некоторый момент движения.

Все указанные возможности при определенных условиях могут реализоваться.

Рассмотрим случай, когда время движения не фиксировано, а выбирается из условия максимума конечной скорости. Начальный угол также не фиксирован и подбирается наилучшим. В этом случае имеем согласно (1.11), (1.17) и (1.20)

$$\operatorname{tg} \varphi_k = \operatorname{tg} \varphi_o - c_2 T = 0, \tag{1.26}$$

откуда

$$c_2 = \frac{\operatorname{tg} \varphi_0}{T}. \tag{1.27}$$

В этом случае числитель в формуле (1.24) для Φ будет иметь вид

$$\operatorname{tg}^2 \varphi_0 \left(1 - \frac{t}{T}\right) \frac{1}{T} \tag{1.28}$$

и будет положителен на всем участке (O, T). Это означает, что будет реализован случай 1). Видим, что наибольшее значение конечной скорости U в случае, когда начальный угол и время движения подбираются из условия оптимума, достигается при мгновенном израсходовании всего запаса топлива в самом начале движения. Направление начальной скорости v_0 и дополнительной скорости V, должны совпадать. Здесь и в дальнейшем значок k у величины V опускаем.

Будем в дальнейшем считать, что начальная скорость v_o включена в V, причем разбивка общего запаса скорости на две части, если это понадобится, осуществляется оптимальным образом. Можно, очевидно, ограничиться при этом рассмотрением случая $v_o = 0$. Проведем исследование указанного случая как для свободного, так и для фиксированного времени движения на участке выведения. Выясним, при каких соотношениях параметров решение возможно, и исследуем зависимость характера оптимального движения от величины задаваемого времени выведения T.

Ясно, что при $v_0 = 0$ возможны только случаи 1 и 3. Для удовлетворения граничных условий имеем соотношения:

$$v_1 \sin \varphi_1 + v_2 \sin \varphi_2 - gT = 0; \tag{1.29}$$

$$v_1 \sin \varphi_1 T - g \frac{T^2}{2} = Y, \tag{1.30}$$

где Y — заданная высота, v_1 — скорость, полученная в начальный момент движения, v_2 — скорость, полученная в конечный момент движения. Сумма

$$v_1 + v_2 = V, \tag{1.31}$$

где V — заданный запас располагаемой скорости.

Если T не фиксировано, то $v_2 = 0$. Уравнения (1.29) и (1.30) дают соотношение

$$\frac{gT^2}{2} = Y, \tag{1.32}$$

которое служит для определения времени T.

$$T = \sqrt{\frac{2Y}{g}} \tag{1.33}$$

Подставляя в любое из уравнений (1.29) или (1.30), получим

$$\sin \varphi_1 = \frac{\sqrt{2Y}}{V}. \tag{1.34}$$

Задача разрешима, если правая часть меньше единицы. Это означает, что при заданной располагаемой скорости V высота выведения Y не должна быть взята слишком большой и, наоборот, при заданной высоте запас располагаемой скорости должен быть достаточно велик, во всяком случае он должен быть больше той скорости, которая необходима, чтобы забросить тело вертикально вверх на высоту Y. При выполнении этого условия горизонтальная скорость в конце участка выведения будет

$$U = V \cos \varphi_1 \tag{1.35}$$

$$U = V \sqrt{1 - \frac{2gY}{V^2}} \tag{1.36}$$

Если T фиксировано, то условие (1.32), вообще говоря, не выполнено и не может быть найдено решения при $v_2 = 0$. Решение необходимо искать по схеме в) при промежуточном положении горизонтального отрезка.

Легко видеть, что выражение для функции Φ можно представить в виде

$$\Phi = \frac{d}{dt}\left(\frac{1}{\cos \varphi}\right). \tag{1.37}$$

Поэтому равенство нулю интеграла в формуле (1.25) дает

$$\cos \varphi_1 = \cos \varphi_2 \tag{1.38}$$

откуда

$$\varphi_2 = \pm \varphi_1. \tag{1.39}$$

Рассмотрим обе возможности и выясним, при каких соотношениях между параметрами задачи они реализуются.

1. Пусть $\varphi_2=\varphi_1$. Из формулы (1.29) имеем

$$\sin\varphi_1=\frac{gT}{V}, \tag{1.40}$$

на основании (1.30) и (1.31) получим

$$v_1=\frac{V}{2}\left(1+\frac{2Y}{gT^2}\right) \tag{1.41}$$

$$v_2=\frac{V}{2}\left(1-\frac{2Y}{gT^2}\right). \tag{1.42}$$

Так как v_2 должно быть положительно, то видим, что данный случай имеет место, если заданное время движения на участке выведения удовлетворяет условию

$$T>\sqrt{\frac{2Y}{g}} \tag{1.43}$$

т.е. больше оптимального времени выведения. Для конечной скорости имеем формулу

$$U=V\sqrt{l-\left(\frac{gT}{V}\right)^2}\,. \tag{1.44}$$

Условие, что выражение $\frac{gT}{V}$ должно быть меньше единицы является требованием на величину располагаемого запаса скорости.

2. Пусть $\varphi_2=-\varphi_1$. Тогда имеем

$$\sin\varphi_1=\frac{2Y}{VT}\,. \tag{1.45}$$

Для скоростей получим

$$v_1=\frac{V}{2}\left(1+\frac{gT^2}{2Y}\right) \tag{1.46}$$

$$v_2=\frac{V}{2}\left(1-\frac{gT^2}{2Y}\right). \tag{1.47}$$

Решение имеет место, если заданное время T меньше оптимального, т.е. если

$$T<\sqrt{\frac{2Y}{g}}\,. \tag{1.48}$$

Конечная скорость вычисляется по формуле

$$U=V\sqrt{l-\left(\frac{2Y}{VT}\right)^2}\,. \tag{1.49}$$

Условие, что выражение $\frac{2Y}{VT}$ должно быть меньше единицы, аналогично предыдущему можно рассматривать как требование, чтобы располагаемый запас скорости V был достаточно велик.

Полученные формулы показывают, что величина дополнительной скорости v_2, сообщаемой в конце движения на участке выведения, зависит от того, насколько заданное время выведения отличается от оптимального, и будет тем больше, чем больше это отличие. Если заданное время равно оптимальному, то $v_2=0$ и оба случая превращаются в предыдущий, когда время подбиралось оптимальным.

Видим также, что в обоих случаях имеем

$$v_2 < v_1,$$

т.е. вторая дополнительная скорость по величине всегда меньше первой.

Условия (1.43) и (1.48) означают, что добавка скорости должна происходить в первом случае на нисходящей, а во втором случае на восходящей ветви параболы, получающейся при движении с начальной скоростью v_1. Оба случая представлены на фиг. 4. В первом случае, когда заданное время больше оптимального, добавка скорости v_2 происходит на нисходящей ветви параболы и направлена параллельно начальной скорости v_1. Во втором случае, когда заданное время выведения меньше оптимального, добавка скорость v_2 производится на восходящей ветви параболы и направлена вниз под углом, равным по величине начальному углу φ_1.

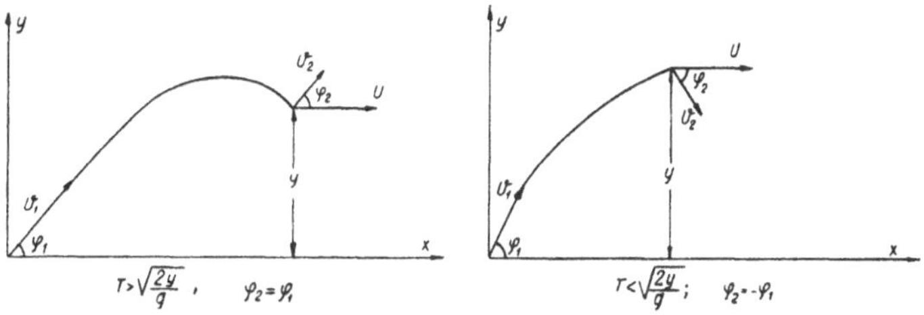

Фиг. 4

На участке выведения должны быть решены задачи подъема на заданную высоту и сообщения необходимой скорости в горизонтальном направлении. Полученные результаты показывают, что в наиболее выгодном случае, когда время выведения также выбрано оптимальным, набор скорости и набор высоты следует производить путем приложения мгновенного импульса в самом начале движения. Выведение на горизонтальное направление движения должно осуществляться за счет действия силы тяжести. Если заданное время выведения отлично от оптимального, то для выведения на горизонтальное направление необходимо использовать импульс, сообщаемый в конце участка выведения. Если заданное время излишне велико, то отклонение вектора скорости за счет силы тяжести больше необходимого и вертикальная составляющая второго импульса должна быть направлена вверх. Если заданное время мало, сила тяжести не успе вает произвести разворота вектора скорости и недостающий разворот должен быть произведен при помощи второго импульса, имеющего вертикальную составляющую, направленную вниз.

При получении решения мы исходили из предположения о допустимости мгновенного сообщения импульса и мгновенного расходования всего запаса топлива или его части. Так как в действительности это невозможно то наилучшее решение поставленной задачи о выведении получим, заменив мговенное сгорание сгоранием топлива с максимальным секундным расходом топлива, а промежуток между приложением импульсов-движением с минимальным секундным расходом или, еще лучше, движением с выключенным двигателем.

Выше было получено, что при движении в плоско-параллельном поле тяжести наилучшее выведение осуществляется за счет приложения одного импульса в начале движения. При выведении на орбиту в центральном поле сил такое решение оказывается неприемлимым, так как таким путем невозможно получить, например, круговую орбиту или орбиту не пересекающую и не касающуюся поверхности Земли. Поэтому для приложения к решению задачи о выведении в центральном поле сил, полученных выше результатов, основное значение имеет задача о выведении на орбиту в заданное время. Если заданное время не слишком велико, то размеры траектории выведения будут невелики по сравнению с радиусом Земли и гипотезы, положенные в основу приближенного рассмотрения, не будут сильно нарушены.

Рассчитаем пример, позволяющий оценить величину протяженности траектории выведения и величину потребного запаса располагаемой скорости при различных значениях задаваемого времени выведения. При рассчете примера ограничимся более важным случаем, когда задаваемое время выведения не превосходит оптимального времени выведения в плоскопараллельном поле сил.

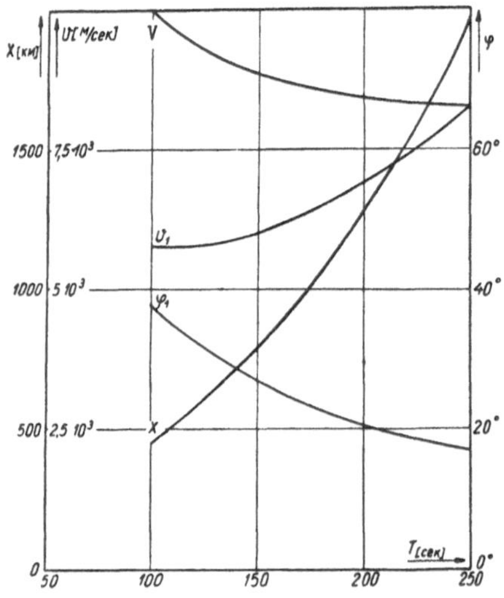

Фиг. 5

Пусть нужно вывести спутник на высоту $Y=300$ км с горизонтальной скоростью $U=7900$ м/сек.

По формуле

$$T=\sqrt{\frac{2Y}{g}}$$

получим оптимальное время выведения в плоскопараллельном поле сил

$$T\ \text{опт}=248\ \text{сек}.$$

На фиг. 5 дан для области $0.4\,T$ опт $< T < T$ опт график зависимости от T величины суммарной потребной располагаемой скорости V, начального импульса v_1, начального угла φ_1 и горизонтальной дальности X.

Видим, что по мере сокращения времени выведения суммарная потребная скорость увеличивается, величина начального импульса падает, а конечного растет. Величина начального угла с уменьшением времени выведения увеличивается, траектории становятся все более крутыми и горизонтальная дальность участка выведения уменьшается.

Видим также, что полученные величины горизонтальной дальности значительно меньше радиуса Земли. Поэтому приведенный пример имеет достаточно оснований рассматриваться как приближенный рассчет оптимального выведения спутника на орбиту.

Отметим в заключение параграфа, что, как показано выше, линейный закон изменения тангенса угла тангажа по времени не зависит от характера изменения функции $V(t)$. Он будет приложим при использовании как одноступенчатых, так и многоступенчатых ракет, а также в случае наличия перерывов между концом работы одной ступени и началом работы другой. Этот результат будет использован в следующем параграфе при рассмотрении выведения с помощью ступенчатой ракеты.

Отметим также, что при других постановках вариационной задачи закон для тангенса угла тангажа может оказаться отличным от линейного. Например, если, подбирая время движения оптимальным, фиксировать горизонтальную дальность на участке выведения, то оптимальный закон изменения по времени тангенса угла тангажа оказывается дробнолинейным. Ниже, в разделе 3, обсуждаются результаты решения вариационных задач в различных постановках с учетом отброшенных ранее факторов и приводятся получающиеся при этом оптимальные законы изменения угла тангажа по времени.

2. Движение с заданным режимом расходования топлива

Выше было получено, что наилучший закон управления по тангажу обеспечивается линейной зависимостью тангенса угла тангажа от времени

$$\operatorname{tg}\varphi = -(c_1 + c_2 t). \tag{2.1}$$

Интегрируя уравнения движения (1.1) — (1.4), получим выражения для проекций скорости и для координат в некоторый момент времени t_1:

$$u = \int_0^{t_1} p(t) \cos\varphi \, dt + u_0 \tag{2.2}$$

$$w = \int_0^{t_1} p(t) \sin\varphi \, dt + w_0 - g t_1 \tag{2.3}$$

$$x = \int_0^{t_1} (t_1 - t)\, p(t) \cos\varphi \, dt + u_0 t_1 + x_0 \tag{2.4}$$

$$y = \int_0^{t_1} (t_1 - t)\, p(t) \sin\varphi \, dt + w_0 t_1 + y_0 - \frac{g t_1^2}{2} \tag{2.5}$$

Величины $\cos\varphi$ и $\sin\varphi$ могут быть на основании (2.1) представлены как функции времени

$$\cos\varphi = \frac{1}{\sqrt{1 + (c_1 + c_2 t)^2}} \tag{2.6}$$

$$\sin\varphi = -\frac{c_1 + c_2 t}{\sqrt{1 + (c_1 + c_2 t)^2}} \tag{2.7}$$

Так как функция $p(t)$ считается заданной, то интегралы в правых частях формул (2.2) — (2.5) могут быть, вообще говоря, вычислены.

Правые части формул (2.2) — (2.5) содержат две произвольные константы, связанные с программой управления. Подбирая эти константы, можно, вообще говоря, добиться того, что при фиксированном времени движения на участке разгона будут обеспечены в конце участка разгона заданная высота y_k и равенство нулю вертикальной составляющей скорости. Подобрав таким образом константы, можно затем по формулам (2.2) и (2.4) определить величину скорости u_k и величину x_k.

Для фактического вычисления интегралов в правых частях формул (2.2 — (2.5) необходимо знать зависимость реактивного ускорения от времени, т.е. знать функцию $p(t)$. Заметим, что функция $p(t)$ является в общем случае разрывной, например, в случае составной ракеты. Поскольку составные ракеты для искусственного спутника представляют наибольший интерес, то в дальнейшем будет проведено более детальное исследование оптимального движения составной ракеты.

Для случая движения составной ракеты, используя (2.2) — (2.5), кинематические параметры в конце траектории выведения представим в виде:

$$u_k = u_o + \sum_{i=1}^{n} u_k^{(i)} \qquad x_k = x_o + \sum_{i=1}^{n} x_k^{(i)} \tag{2.10}$$

$$w_k = w_o + \sum_{i=1}^{n} w_k^{(i)} \qquad y_k = y_o + \sum_{i=1}^{n} y_k^{(i)} \tag{2.11}$$

где

$$u_k^{(i)} = \int_{t_o^{(i)}}^{t_k^{(i)}} p_i(t) \cos\varphi \, dt \tag{2.12}$$

$$w_k^{(i)} = \int_{t_o^{(i)}}^{t_k^{(i)}} p_i(t) \sin\varphi \, dt - g\,(t_k^{(i)} - t_o^{(i)}) \tag{2.13}$$

$$x_k^{(i)} = \int_{t_o^{(i)}}^{t_k^{(i)}} (t_k^{(i)} - t)\, p_i(t) \cos\varphi \, dt + (t_k^{(i)} - t_o^{(i)}) \sum_{q=o}^{i-1} u_k^{(q)} \tag{2.14}$$

$$y_k^{(i)} = \int_{t_o^{(i)}}^{t_k^{(i)}} (t_k^{(i)} - t)\, p_i(t) \sin\varphi \, dt + (t_k^{(i)} - t_o^{(i)}) \sum_{q=o}^{i-1} w_k^{(q)} - \frac{1}{2} g\,(t_k^{(i)} - t_o^{(i)})^2 \tag{2.15}$$

$$u_k^o = u_o \qquad w_k^o = w_o$$

Здесь $t_o^{(i)}$ и $t_k^{(i)}$ соответственно, начальное и конечное время движения i—ой ступени ракеты, $u_k^{(i)}$ и $w_k^{(i)}$ величины приращения, соответственно, горизонтальной и вертикальной компоненты скорости при движении i—ой ступени, $x_k^{(i)}$ и $y_k^{(i)}$, соответственно, приращения горизонтальной и вертикальной координат за время движения i—ой ступени, $P_i^{(t)}$—ускорение реактивной силы i—ой ступени.

Очевидно, что формулы (2.12)—(2.15) охватывают также и случай, когда между работой двигателей отдельных ступеней составной, ракеты имеют место паузы. В этом случае можно принять, что промежуток времени, соответствующий паузе, относится к некоей ступени ракеты, для которой

$$p_i^{(t)} = 0.$$

В дальнейшем вычисление интегралов правых частей формул (2.12) — (2.15) будет проведено для одного весьма важного частного случая, именно, случая, при котором движение каждой ступени ракеты происходит с постоянной тягой. В этом случае имеем:

$$p_i^{(t)} = \frac{gP_i}{G_i}, \tag{2.16}$$

где P_i—тяга двигателя i—ой ступени, G_i—вес i—ой ступени. Вес G_i можно представить в виде:

$$G_i = G_o^{(i)} \mu^{(i)} = G_o^{(i)} \cdot \left[1 - 1\,(-\mu_k^{(i)})\,\frac{t - t_o^{(i)}}{T_i}\right], \tag{2.17}$$

где $G_o^{(i)}$ — начальный вес i — ой ступени, $\mu_k^{(i)}$ — конечное отношение масс для i — ой ступени, T_i — время работы двигателя i — ой ступени. Вводя обозначение

$$\nu_o^{(i)} = \frac{G_o^{(i)}}{P_i} \tag{2.18}$$

получим выражение для $p_i^{(t)}$

$$p_i^{(t)} = \frac{g}{\nu_o^{(i)}\left[1 - (1 - \mu_k^{(i)})\frac{t - t_o^{(i)}}{T_i}\right]} \tag{2.19}$$

Определяя тягу двигателя i — ой ступени по формуле

$$P_i = -\frac{c_i}{g}\frac{dG_i}{dt} \tag{2.20}$$

где c_i — скорость истечения газов из сопла двигателя i — ой ступени, легко получим для времени T_i формулу

$$T_i = \frac{c_i}{g}\nu_o^{(i)}(1 - \mu_k^{(i)}). \tag{2.21}$$

Заметим, что при постоянной тяге двигателя i — ой ступени секундный расход постоянен и масса i — ой ступени ракеты является линейной функцией времени. Для i — ой ступени можно поэтому считать тангенс угла тангажа зависящим линейно не от времени, а от массы i — ой ступени или от ее относительной массы. Имеем тогда:

$$\operatorname{tg}\varphi = A_i + B_i\,\mu^{(i)}, \tag{2.22}$$

где A_i и B_i — некоторые константы, соответствующие i — ой ступени ракеты. В начале движения i — ой ступени $\mu^{(i)} = 1$ и имеем

$$\operatorname{tg}\varphi_o^{(i)} = A_i + B_i. \tag{2.23}$$

В момент окончания работы двигателя $\mu^{(i)} = \mu_k^{(i)}$ и имеем

$$\operatorname{tg}\varphi_k^{(i)} = A_i + B_i\,\mu_k^{(i)} \tag{2.24}$$

Здесь $\varphi_o^{(i)}$ и $\varphi_k^{(i)}$ значения угла тангажа, соответственно, в начале и в конце движения i — ой ступени.

Подставляя в формулы (2.8)—(2.15) выражение для $p_i(t)$ из (2.19) и выражения для $\sin\varphi$ и $\cos\varphi$ из (2.6) и (2.7), получим после интегрирования:

$$u_k = u_o + \sum_{i=I}^{n} c_i f_1^{(i)} \tag{2.25}$$

$$w_k = w_o + \sum_{i=I}^{n} c_i\left[f_2^{(i)} - \nu_o^{(i)}(1 - \mu_k^{(i)})\right] \tag{2.26}$$

$$x_k = x_o + \sum_{i=I}^{n}\frac{c_i^2}{g}\nu_o^{(i)}\left[f_3^{(i)} + (1 - \mu_k^{(i)})\sum_{q=o}^{i-I}\frac{c_q}{c_i}f_1^{(q)}\right] \tag{2.27}$$

$$y_k = y_o + \sum_{i=I}^{n}\frac{c_i^2}{g}\nu_o^{(i)}\left\{f_4^{(i)} - \frac{I}{2}\nu_o^{(i)}(1 - \mu_k^{(i)})^2 + (1 - \mu_k^{(i)})\sum_{q=o}^{i-I}\frac{c_q}{c_i}f_2^{(q)} - \right.$$
$$\left. - \nu_o^{(q)}\cdot\left(f_2^{(q)} - \nu_o^{(q)}\right)\right]\right\} \tag{2.28}$$

где безразмерные функции $f_1^{(i)} f_2^{(i)}, f_3^{(i)}$ и $f_4^{(i)}$ определяются выражениями:

$$f_1^{(o)} = \frac{u_o}{c_o} \qquad f_2^{(o)} = \frac{w_o}{c_o}$$

$$f_1^{(i)} = \frac{1}{\sqrt{1+A_i^2}} \left\{ \operatorname{Arcsh}\left(A_i + \frac{1+A_i^2}{\mu_k^{(i)} B_i} - \operatorname{Arcsh}\left(A_i + \frac{1+A_i^2}{B_i}\right)\right) \right\} \tag{2.29}$$

$$f_2^{(i)} = A_i f_1^{(i)} + \operatorname{Arcsh}(A_i + B_i) - \operatorname{Arcsh}(A_i + B_i \mu_k^{(i)}) \; (i = 1, 2, 3, \ldots n.) \tag{2.30}$$

$$f_3^{(i)} = \frac{1}{B_i} \left[\operatorname{Arcsh}(A_i + B_i) - \operatorname{Arcsh}(A_i + B_i \mu_k^{(i)})\right] - \mu_k^{(i)} f_1^{(i)} \tag{2.31}$$

$$f_4^{(i)} = \frac{1}{B_i} \left[\sqrt{1+(A_i+B_i)^2} - \sqrt{1+(A_i+B_i \mu_k^{(i)})^2}\right] - \mu_k^{(i)} f_2^{(i)} \tag{2.32}$$

Заметим кроме того, что $\nu_o^{(o)} = 0$. Наконец отметим, что константы A_i и B_i, входящие в формулы (2.29) — (2.32) в силу непрерывности программы управления по тангажу, а также в силу формул (2.22) — (2.24) связаны между собой при помощи 2 $(n-1)$ соотношений:

$$A_i + B_i \mu_k^{(i)} = A_{i+1} + B_{i+1} \qquad i = 1, 2, \ldots (n-1) \tag{2.33}$$

$$\frac{B}{B_{i+1}} = \frac{\nu_o(i)}{\nu_o(i+1)} \cdot \frac{c_i}{c_{i+1}}. \tag{2.34}$$

Последнее соотношение получается следующим образом. Согласно (2.1) $\frac{d}{dt}(\operatorname{tg}\varphi) = -c_2 = \text{const}$. С другой стороны, согласно (2.17) и (2.22)

$$\frac{d}{dt}(\operatorname{tg}\varphi) = \frac{d}{d\mu(i)}(\operatorname{tg}\varphi)\frac{d\mu(i)}{dt} = -B_i \frac{1-\mu_k(i)}{T_i} = -c_2 \tag{2.35}$$

Из (2.35), принимая во внимание (2.21), легко получить (2.34).

Формулы (2.25) — (2.21) позволяют полностью рассчитать экстремальное движение. Чтобы получить движение, удовлетворяющее заданным изопериметрическим условиям, необходимо соответствующим образом подобрать две произвольные постоянные программы управления c_1 и c_2 или, что то же самое, какие-либо две постоянные из совокупности $2n$ величин A_i и B_i например, A_1 и B_1 или A_n и B_n. Приравнивая w нулю, а y заданной величине y_k, получим для определения постоянных A_n и B_n систему двух трансцендентных уравнений, решение которых может быть найдено только подбором. Проще поэтому не определять величины A_n и B_n, а задавать их, а по ним уже рассчитывать все другие величины. Изменяя $\mu_k^{(i)}$, A_n и B_n в некотором разумном диапазоне, можно получать движения с интересным и значениями $\nu_o^{(i)}$, y_k и u_k. При этом вычисления будут гораздо более экономными, так как не надо решать уравнений, а надо лишь производить вычисления по формулам.

В дальнейшем остановимся на рассмотрении одного весьма интересного частного случая схемы составной ракеты и на нем проиллюстрируем описанную выше методику.

Рассмотрим составную ракету, у которой все ступени по ряду основных своих характеристик подобны между собой, именно, ракету, у которой

$$c_i = c \qquad \mu_k^{(i)} = \mu_k, \qquad \nu_o^{(i)} = \nu_o \tag{2.36}$$

В качестве произвольных констант программы управления примем A_n и B_n. Все остальные постоянные A_i и B_i весьма просто можно выразить через A_n и B_n. В самом деле, согласно (2.34)

$$B_i = B_n. \tag{2.37}$$

Далее, из (2.33), используя также (2.37), легко получим формулу для A_i

$$A_i = A_n + B_n (1 - \mu_k)(n - 1) \qquad i = 1, 2, \ldots n. \tag{2.38}$$

Схема расчета будет теперь такой. Задав μ_k, A_n и B_n, используя формулу (2.38), определим по формулам (2.29) — (2.32) величины $f_1^{(i)}$, $f_2^{(i)}$, $f_3^{(i)}$ и $f_4^{(i)}$. Из условия $w_k=0$ в конце движения имеем формулу

$$v_o = \frac{\sum\limits_{i=1}^{n} f_2^{(i)}}{n(1-\mu_k)}. \tag{2.39}$$

На основании этой формулы определим v_o. После этого по формулам (2.25), (2.27) и (2.28) определим величины x_k, y_k и u_k.

При выполнении подобного расчета можно задаваться не самими величинами A_n и B_n, а некоторыми величинами от них зависящими. Например, можно задаваться величиной угла тангажа в конце траектории и значением отношения масс и рассчитывать серию траекторий, меняя величину угла тангажа в начале участка разгона. Обозначая начальный угол тангажа 1—й ступени $\varphi_o^{(1)}$ через φ_o согласно (2.23), будем иметь

$$\operatorname{tg} \varphi_o = A_1 + B_1 \tag{2.40}$$

Обозначая конечный угол тангажа для n—ной ступени $\varphi_k^{(n)}$ через φ_k, согласно (2.24) будем иметь

$$\operatorname{tg} \varphi_k = A_n + B_n \mu_k \tag{2.41}$$

Наконец, принимая во внимание, что согласно формуле (2.38)

$$A_1 = A_n + B_n (1 - \mu_k)(n - 1) \tag{2.42}$$

и что $B_1 = B_n$ из соотношений (2.40), (2.41) и (2.42) получим формулы, связывающие A_n и B_n с φ_o и φ_k

$$B_n = \frac{\operatorname{tg}\varphi_o - \operatorname{tg}\varphi_k}{n(1 - \mu_k)}$$

$$A_n = \operatorname{tg} \varphi_k - \mu_k \cdot B$$

Результаты расчета представлены на фиг. 6, 7 и 8. При этом расчеты были проведены для двухступенчатой (фиг. 6), трехступенчатой (фиг. 7) и четырехступенчатой ракеты (фиг. 8). На графиках по оси абсцисс как в размерной, так и в безразмерной форме (см. ниже) отложена конечная скорость ракеты u_k, по оси ординат конечная высота выведения y_k. При выполнении расчетов значения углов тангажа в конце траектории были взяты равными $\varphi_k = 0°$; $-10°$; $-20°$; $-30°$; $-40°$; $-50°$; $-60°$. Значения μ_k были взяты

$$\mu_k = 0.1;\ 0.2;\ 0.3;\ 0.4;\ 0.5\,.$$

Для этих параметров были построены кривые, вдоль которых менялось φ_o. При этом, кривые построены для точек $\varphi_o > 60°$. Крайняя правая точка каждой кривой соответствует $\varphi_o = 60°$. Исключение составляют кривые, у которых точки, соответствующие $\varphi_o = 60°$, лежат под осью абсцисс. Значения φ_o для крайних правых точек этих кривых отдельно отмечены на графиках (фиг. 6—8). Остальные точки φ_o, отмеченные на графиках кружками, нанесены через интервал $\Delta\varphi_o = 2°$ в сторону возрастания φ_o. Оказалось, что вдоль таких кривых значения ν_o меняются монотонно. При малых значениях

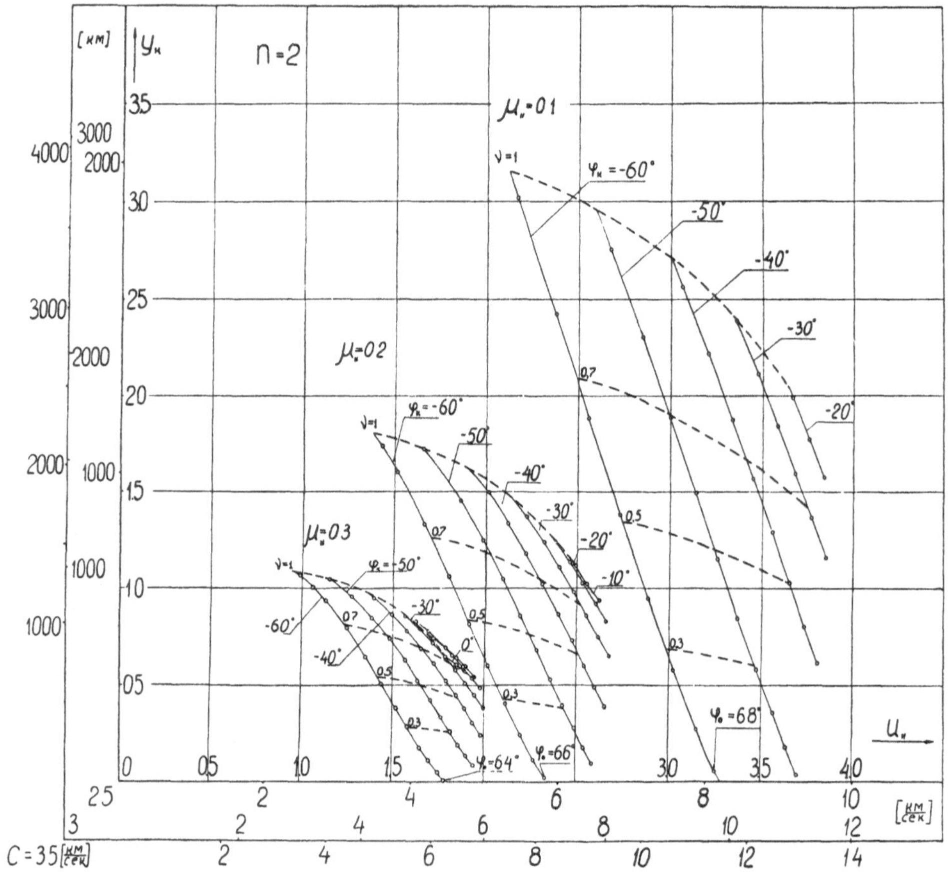

Фиг. 6

ν_o получаем точки в правой нижней части кривой, соответствующей меньшим значениям φ_o. С увеличением ν_o получаем точки, расположенные левее и выше и соответствующие большие значениям угла φ_o. Так как для практически интересных случаев ν_o не должно быть меньше единицы, то точка на кривой $\varphi_k = \text{const}$, соответствующая $\nu_o = 1$, является левой границей рассматриваемой кривой. На фиг. 6—8 нанесены линии, соединяющие точки $\nu_o = 1$, или, что одно и то же, соединяющие левые границы кривых $\varphi_k = \text{const}$ (пунктирные линии), причем нетрудно заметить, что кривые $\varphi_k = \text{const}$ могут лежать как справа, так и слева от кривой $\nu_o = 1$.

На фиг. 6—8 приведены также кривые, соответствующие другим постоянным значениям ν_o. Так, на графиках нанесены пунктирные линии, соответствующие следующим значениям ν_o

$$\nu_o = 0.7,\ 0.5,\ 0.3$$

Интересно отметить, что кривые постоянных значений ν_o могут в ряде случаев пересекаться. Так, из фиг. 6—8 легко видеть, что в случаях $n=2$, $\varphi_k=0.3$; $n=3$ $\varphi_k=0.4$; $n=4$ $\varphi_k=0.5$ имеет место пересечение кривой $\nu_o=1$ с кривыми $\nu_o=0.7$, $\nu_o=0.5$ и, по-видимому, с кривой $\nu_o=0.3$ (если продолжить кривую $\nu_o=1$ в сторону положительных значений φ_k). Это означает,

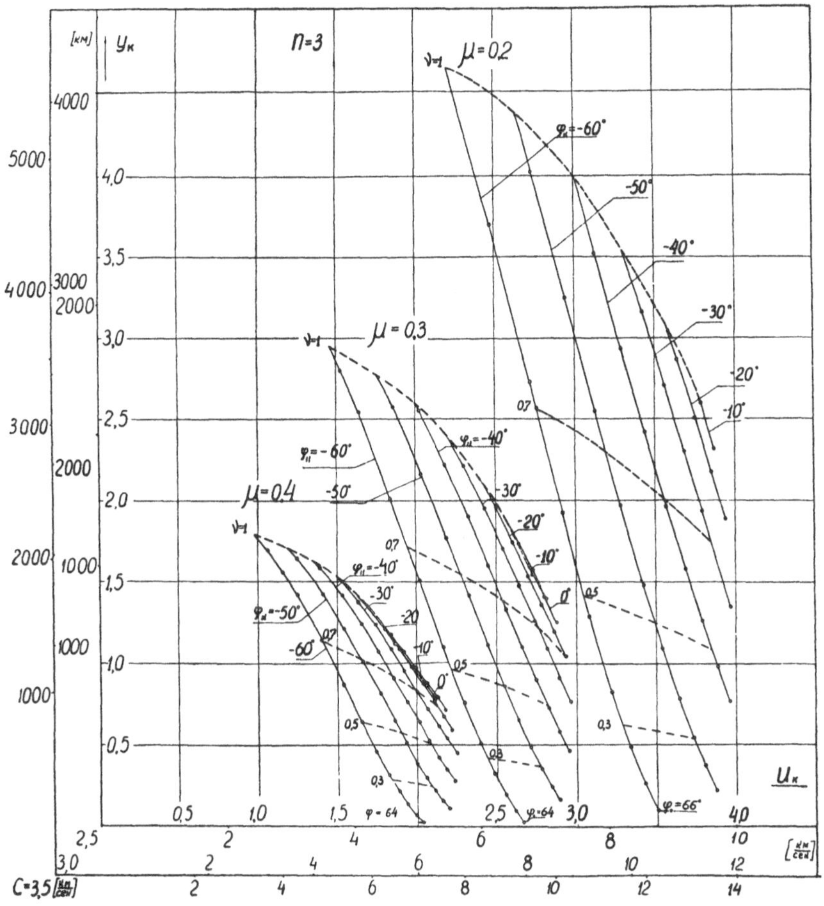

Фиг. 7

что при заданных n и φ_k и для двух различных значений ν_o, соответствующие данным значениям ν_o оптимальные программы могут обеспечить получение одной и той же комбинации $(y_k\ u_k)$.

Для того, чтобы пояснить значение указанного факта приведем пример. Допустим, что для $n=2$, $\mu_k=0.3$, $\nu_o=0.7$ и для высоты $\widetilde{y}_k=0.60$ (взята в безразмерной форме) необходимо найти оптимальную программу, обеспечивающую получение максимальной скорости $\widetilde{u}_k$.

Пользуясь фиг. 6, с помощью интерполяции, находим параметры оптимальной программы φ_o и φ_k

$$\varphi_o \approx 67^\circ \qquad \varphi_k \approx -30^\circ.$$

Найденная оптимальная программа обеспечивает получение максимальной скорости $\widetilde{u}_k$ при этом $\widetilde{u}_k=1.82$ (взята также в безразмерной форме).

Однако, из той же фиг. 6 легко видеть, что та же комбинация (y_k, u_k) может быть получена при использовании другого значения ν_0, именно, $\nu_0=1$. В этом случае параметры оптимальной программы φ_0 и φ_k будут уже другими $\varphi_0=62.5°$, $\varphi_k\approx 0°$.

С другой стороны, при сохранении стартового веса ракеты, большим значениям ν_0, согласно формуле (2.18), будут соответствовать меньшие

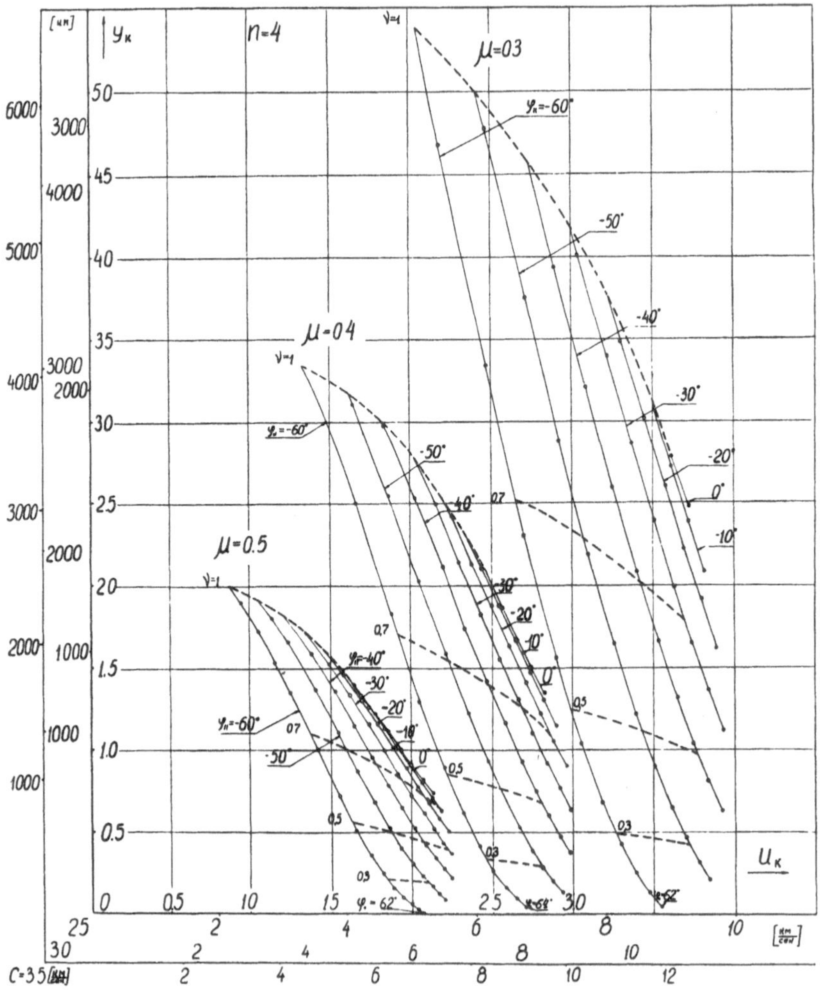

Фиг. 8

значения тяги двигателя ракеты P и, следовательно, меньший вес самого двигателя. Тогда, сохраняя n и μ_k, но обеспечивая получение для ν_0 значения $\nu_0=1$, имеем возможность поместить в ракете бо́льший полезный груз, не ухудшая при этом основных летных характеристик ракеты $-y_k$ и u_k.

Из приведенных графиков видим, что при заданном числе ступеней ракеты и при данном μ_k имеют место определенные ограничения на возможность получения различных комбинаций высоты и скорости выведения. Эти ограничения имеют двоякий характер. Чтобы показать это разобьем все кривые $\varphi_k = \text{const}$ на два класса. К I классу отнесем кривые, имеющие участки, расположенные правее и выше кривой $\nu_o = 1$ (пунктирная кривая).

К этому классу относятся, некоторые кривые $\varphi_k = \text{const}$, соответствующие случаю $n=2$, $\mu_k = 0.3$, например, кривые $\varphi_k = -20°$ и $\varphi_k = -10°$ (фиг. 6). Ко II классу отнесем кривые, целиком расположенные левее и ниже пунктирной кривой $\nu_o = 1$. К этому классу относятся, например, все кривые $\varphi_k = \text{const}$, соответствующие случаю $n=2$ $\mu_k = 0.1$ или, например, кривые $\varphi_k = -60°; -50°; -40°; -30°$; соответствующие случаю $n=2$ $\mu_k = 0.2$ (фиг. 6).

Из фиг. 6—8 легко видеть, что для семейств кривых I класса легко построить огибающие, касающиеся кривых с одинаковыми значениями μ_k (огибающие эти на фиг. не показаны). Эти огибающие ограничивают в плоскости u_k, y_k некоторые области достижимости, причем, это ограничение — энергетического характера и связано с возможностью получения при заданном запасе топлива лишь некоторых ограниченных высот и скоростей.

Другое ограничение, относящееся целиком к кривым II класса, связано с практической нецелессообразностью использования ракет, у которых тяга 1 ступени меньше ее начального веса. Пунктирная кривая $\nu_o = 1$ будет в этом случае ограничивающей кривой в плоскости (y_k, u_k), выше которой мы не можем подняться, если n и μ_k задано.

Комбинируя огибающую семейства кривых I класса и ограничивающую семейства кривых II класса, можно получить общую ограничивающую кривую, указывающую область достижимых комбинаций y_k и u_k при заданных n и μ_k. Указанная ограничивающая кривая дает нам наибольшие значения скорости, достижимой на данной высоте при данных n и μ_k или наибольшие значения высоты для заданной скорости в конце при данных n и μ_k. Кроме того, эти ограничивающие кривые указывают наибольшие значения μ_k, при котором энергетически возможно достижение данных высоты и скорости. Точки кривых $\mu_k = \text{const}$, лежащие на ограничивающей кривой, будут отвечать значениям ν_o, энергетически наиболее выгодным.

На фиг. 6—8 основной шкалой является шкала для безразмерных скорости и высоты $\widetilde{u}_k$ и $\widetilde{y}_k$, где

$$\widetilde{u}_k = \frac{u_k}{c} \qquad \widetilde{y}_k = \frac{y_k}{\frac{c^2}{g}}\,.$$

Благодаря этому приведенные графики имеют универсальный вид и пригодны для использования в случае различных скоростей истечения газов из сопла реактивного двигателя. На этих же графиках нанесены три размерные шкалы, соответствующие трем различным скоростям истечения

$$c = 2{,}5 \text{ км/сек}; \qquad 3 \text{ км/сек}; \qquad 3{,}5 \text{ км/сек}.$$

Видим, что для приведенных скоростей истечения часть кривых попадает в область таких больших высот y_k, для которых непосредственное применение развитой теории выведения в плоскопараллельном поле силы тяжести становится затруднительным. Тем не менее и в этом случае указанные кривые позволяют делать некоторые качественные суждения общего характера, относительно влияния выбора параметров ракеты и ее программы управления по тангажу на расширение области достижимых скоростей и высот.

3. Задача о выведении на орбиту с учетом переменности поля тяготения и вращения Земли

В настоящем разделе приведена более сложная постановка задачи о выведении на орбиту, учитывающая переменность поля тяготения и вращение Земли. При этом рассматривается относительное движение ракеты в системе координат, связанной с Землей.

Полагая, что траектория ракеты относительно Земли является плоской, рассмотрим декартову систему координат, начало которой помещено в точке старта, ось x направлена по касательной к Земле в сторону движения ракеты, ось у направлена вертикально ввех, ось z перпендикулярна к осям x, y и направлена так, что составляет с ними правую систему координат (фиг. 9). Уравнения движения ракеты в проекциях на оси декартовой системы координат, запишутся в виде:

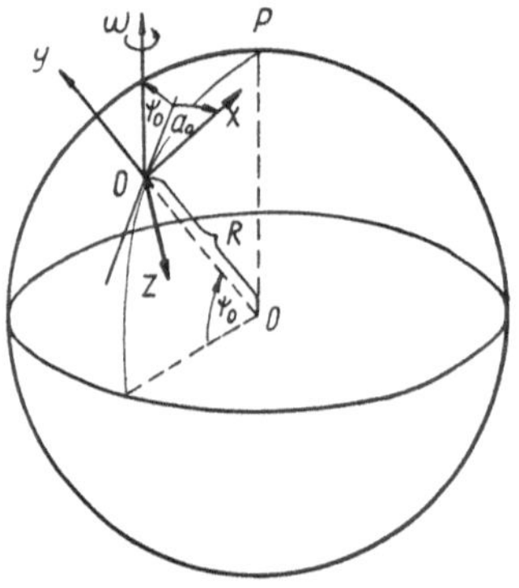

Фиг. 9

$$\left\{\begin{aligned} \ddot{x} &= -g_x + p_x - (\bar{a}_{nep})_x - (\bar{a}_{kop})_x && (3.1)\\ \ddot{y} &= -g_y + p_y - (\bar{a}_{nep})_y - (\bar{a}_{kop})_y && (3.2)\\ \ddot{z} &= -g_z + p_z - (\bar{a}_{nep})_z - (\bar{a}_{kop})_z, && (3.3)\end{aligned}\right.$$

где g_x, g_y, g_z — проекции ускорения силы тяготения и p_x, p_y, p_z — проекции ускорения реактивной силы на оси x, y, z, $(\bar{a}_{nep})_x$, $(\bar{a}_{nep})_y$, $(\bar{a}_{nep})_z$, и $(\bar{a}_{kop})_x$, $(\bar{a}_{kop})_y$, $(\bar{a}_{kop})_z$ соответственно, проекции переносного ускорения и ускорения Кориолиса на те же оси.

Проанализируем отдельные слагаемые уравнений (3.1) — (3.3). Прежде всего ясно, что в силу условия движения ракеты в плоскости x, y должно быть

$$\ddot{z} = \dot{z} = \dot{z} = 0 \qquad (3.4)$$

Далее, в силу того же условия, а также в силу центральности рассматриваемого поля тяготения ускорения силы тяготения на ось z должна быть равна нулю

$$g_z = 0 \qquad (3.5)$$

Проекция ускорения силы тяготения на оси x и $y - g_x$ и g_y можно представить в виде:

$$g_x = \frac{gR^2}{x^2 + (y+R)^2} \cdot \frac{x}{\sqrt{x^2 + (y+R)^2}} \qquad (3.6)$$

$$g_y = \frac{gR^2}{x^2 + (y+R)^2} \cdot \frac{y+R}{\sqrt{X^2 + (y+R)^2}} \qquad (3.7)$$

Проекции ускорения реактивной силы на оси x, y, z — p_x, p_y, p_z можно представить в виде:

$$\left\{\begin{aligned} p_x &= p\ \cos\varphi \cos\beta && (3.8)\\ p_y &= p\ \sin\varphi && (3.9)\\ p_z &= p\ \cos\varphi \sin\beta, && (3.10)\end{aligned}\right.$$

где φ — угол между продольной осью ракеты и плоскостью x, z (угол тангажа), β — угол между вертикальной плоскостью, проходящей через ось ракеты и плоскостью x, y (угол рысканья), p — как и ранее модуль ускорения реактивной силы — некая заданная функция времени.

Проекции ускорения Кориолиса $(\bar{a}_{kop})_x$, $(\bar{a}_{kop})_y$ и $(\bar{a}_{kop})_z$ легко выразить через кинематические параметры относительного движения, если воспользоваться формулой

$$\bar{a}_{kcp} = 2\,(\bar{\omega}\, x\, \bar{v}_{OTH}), \tag{3.11}$$

где $\bar{\omega}$ — вектор угловой скорости вращения Земли, $\bar{v}_{OTH}$ относительная скорость движения ракеты. Заметим — что для проекций угловой скорости вращения $\bar{\omega}$ на оси x, y и z имеем формулы

$$\begin{cases} \omega_x = \omega \cos\psi_o \cos a_o & (3.12)\\ \omega_y = \omega \sin\psi_o & (3.13)\\ \omega_z = -\,\omega \cos\psi_o \sin a_o & (3.14) \end{cases}$$

где ω — модуль угловой скорости вращения Земли, ψ_o — широта точки старта, a_o — угол между плоскостью x, y и меридиональной плоскостью (азимут плоскости траектории). Принимая во внимание (3.11) для проекций ускорения Кориолиса, получим известные формулы

$$\begin{cases} (\bar{a}_{kop.})_x = 2\,(\bar{\omega}\times\bar{v}_{OTH.})_x = \dot{y}\,\omega_z - \dot{z}\,\omega_y & (3.15)\\ (\bar{a}_{kop.})_y = 2\,(\bar{\omega}\times\bar{v}_{OTH.})_y = \dot{z}\,\omega_x - \dot{x}\,\omega_z & (3.16)\\ (\bar{a}_{kop.})_z = 2\,(\bar{\omega}\times\bar{v}_{OTH.})_z = x\omega_y - y\,\omega_x & (3.17) \end{cases}$$

Наконец, принимая во внимание, что в силу условий движения $z=0$, формулы для проекций ускорения Кориолиса представим в следующем окончательном виде:

$$\begin{cases} (\bar{a}_{kop.})_x = \dot{y}\,\omega_z & (3.18)\\ (\bar{a}_{kop.})_y = -\,\dot{x}\,\omega_z & (3.19)\\ (\bar{a}_{kop.})_z = \dot{x}\,\omega_y - \dot{y}\,\omega_x & (3.20) \end{cases}$$

Обращаясь теперь к проекциям переносного ускорения, заметим, что они пренебрежимо малы по сравнению с другими членами уравнений (3.1) — (3.3) и при решении рассматриваемой задачи их можно отбросить.

Принимая во внимание сделанные замечания, систему уравнений (3.1) — (3.3) запишем в виде:

$$\begin{cases} \ddot{x} = -\,g_x + p\cos\varphi\cos\beta - \dot{y}\;\omega_z & (3.21)\\ \ddot{y} = -\,g_y + p\sin\varphi + \dot{x}\;\omega_z & (3.22)\\ o = p\cos\varphi\sin\beta - (\dot{x}\;\omega_y - \dot{y}\;\omega_x) & (3.23) \end{cases}$$

Последнее уравнение системы (3.21) — (3.23) служит для определения угла β необходимого для компенсации боковых сил и обеспечивающего движение ракеты в плоскости x, y. Согласно (3.23) имеем

$$\sin\beta = \frac{\dot{x}\,\omega_y - \dot{y}\,\omega_x}{p\cos\varphi} \tag{3.24}$$

Оценка угла β, произведенная с помощью формул (3.24) показывает, что в практически интересных случаях угол β будет достаточно мал и, во всяком случае, никогда не будет превосходить несколько градусов, так что при решении рассматриваемой задачи, без большой погрешности, в уравнении (3.21) можно положить $\cos\beta = 1$.

Учитывая, что в рассматриваемой задаче протяженность траектории выведения предполагается более или менее ограниченной, с целью облегчения решения задачи, упростим выражения для компонент ускорения земного тяготения g_x и g_y, заменив формулы (3.6) и (3.7) соответствующими приближенными формулами. С этой целью функции $g_x(x, y)$ и $g_y(x, y)$ разложим в окрестности точки (0,0) в ряд по степеням x и y

$$g_x = g_x(0,0) + \left(\frac{\partial g_x}{\partial x}\right)_o \cdot x + \left(\frac{\partial g_x}{\partial y}\right)_o \cdot y + \dots \tag{3.25}$$

$$g_y = g_y(0,0) + \left(\frac{\partial g_y}{\partial x}\right)_o \cdot x + \left(\frac{\partial g_y}{\partial y}\right)_o \cdot y + \dots, \tag{3.26}$$

где для коэффициентов разложения будем иметь следующие формулы:

$$g_x(0,0) = 0 \qquad \left(\frac{\partial g_x}{\partial x}\right)_o = \frac{g}{R} \qquad \left(\frac{\partial g_x}{\partial y}\right)_o = 0$$

$$g_y(0,0) = g \qquad \left(\frac{\partial g_y}{\partial x}\right)_o = 0 \qquad \left(\frac{\partial g_y}{\partial y}\right)_o = -\frac{2g}{R}$$

В дальнейшем, в разложении (3.25), (3.26) ограничимся членами, содержащими x и y в первой степени и, следовательно, для g_x и g_y получим следующие приближенные формулы

$$g_x = g\frac{x}{R} \quad (3.27) \qquad g_y = g - 2g\frac{y}{R} \quad (3.28)$$

Принимая по внимание все сделанные замечания и упрощения, систему уравнений движения ракеты представим в следующем окончательном виде:

$$\begin{cases} \psi_1 = \dot{u} - 2\,\omega_z w + \nu^2 x - p\cos\varphi = 0 & (3.29) \\ \psi_2 = \dot{w} + 2\,\omega_z u - 2\,\nu^2 y - p\sin\varphi + g = 0 & (3.30) \\ \psi_3 = \dot{x} - u = 0 & (3.31) \\ \psi_4 = \dot{y} - w = 0 & (3.32) \end{cases}$$

где

$$\nu = \sqrt{\frac{g}{R}} \quad (3.33) \qquad \omega_z = -\,\omega\cos\psi_o \sin a_o \quad (3.34)$$

Как и в § 1 поставим вариационную задачу о достижении максимальной скорости на заданной высоте при условии, что угол между вектором конечной скорости и местным горизонтом равен нулю. Как и ранее, в силу взаимости, полученное решение будет обеспечивать достижение максимальной высоты при заданной скорости, а также достижение заданных значений высоты и скорости при минимальном расходе топлива.

Сформулируем граничные условия. Предположим, что в начале движения при $t = t_o$ имеем координаты x_o и y_o и некоторые значения горизонтальной и вертикальной проекций скорости u_o и w_o. Положим также, что кинематические параметры x_o, y_o, u_o и w_o зависят от некоторого параметра, которым может явиться в частности угол θ_o между вектором начальной скорости v_o и горизонтом в момент $t = t_o$. Вводя в последующих формулах в качестве такого параметра угол θ_o будем иметь в виду возможность замены θ_o некоторым другим параметром. Полагаем, что

$$\text{при}\, t = t_o \begin{cases} x_o = x_o\,(\theta_o) \\ y_o = y_o\,(\theta_o) \\ u_o = u_o\,(\theta_o) \\ w_o = w_o\,(\theta_o) \end{cases} \tag{3.35}$$

В конце движения в момент $t = T$ полная высота полета h фиксирована и угол между вектором скорости и местным горизонтом должен быть равен нулю. Оба указанных условия записываются в следующем виде:

$$\text{при}\, t = T \begin{cases} \sqrt{(R + y_k)^2 + x_k^2} - R = h & (3.36) \\ \dfrac{x_k}{R + y_k} = -\dfrac{w_k}{u_k} & (3.37) \end{cases}$$

В предельном случае, если x_k мало, условие (3.37) принимает вид

$$w_k = 0$$

т.е. такой же, как и в задаче, рассмотренной в разделах 1, 2.

Итак, поставим вариационную задачу. При заданной функции $p\ (t)$ найту функцию $\varphi\ (t)$, обеспечивающую максимальную горизонтальную скорость v_k на заданной высоте h, а также наивыгоднейший угол θ_o между вектором начальной скорости и горизонтом, соответствующим точке старта. Заметим, что зависимость функционала задачи v_k от функции $\varphi\ (t)$ не может быть, как это было ранее, представлена в простейшем виде.

Чтобы решить поставленную задачу необходимо, прежде всего, найти первую вариацию функционала v_k. Для нахождения первой вариации воспользуемся методом неопределенных множителей Лагранжа. С этой целью наряду с исходным функционалом

$$v_k = \sqrt{u_k^2 + w_k^2} \tag{3.38}$$

рассмотрим новый функционал y

$$y = \sqrt{u_k^2 + w_k^2} + \int_0^T \mathrm{H}\, dt \tag{3.39}$$

Для подинтегральной функции H имеем выражение

$$H = \lambda_1\psi_1 + \lambda_2\psi_2 + \lambda_3\psi_3 + \lambda_4\psi_4, \tag{3.40}$$

где λ_1, λ_2, λ_3 и λ_4 некоторые, пока неопределенные функции времени, а ψ_1, ψ_2, ψ_3 и ψ_4 определяются согласно (3.29) — (3.32). Ясно, что функционалы (3.39) и (3.38) совпадают в силу выполнения условий (3.29) — (3.32). Будут совпадать также вариации функционалов (3.38) и (3.39), если при варьировании функции φ будут удовлетворены связи (3.29) — (3.32), т.е. если при варьировании φ будут выполнены условия:

$$\delta\psi_1 = \delta\psi_2 = \delta\psi_3 = \delta\psi_4 = 0$$

Первая вариация функционала y будет иметь вид

$$\delta y = \frac{u_k}{\sqrt{u_k^2 + w_k^2}}\,\delta u_k + \frac{w_k}{\sqrt{u_k^2 + w_k^2}}\,\delta w_k + \int_0^T \delta\, H dt, \tag{3.41}$$

где

$$\delta H = H_u\,\delta u + H_{\dot u}\,\delta \dot u + H w\,\delta w + H_{\dot w}\,\delta \dot w + H_x\,\delta x + H_{\dot x}\,\delta \dot x + H_y\,\delta y + H_{\dot y}\,\delta \dot y + H_\varphi\,\delta\varphi \tag{3.42}$$

Для производных H_u, $H_{\dot{u}}$ и т.д. согласно (3.29) — (3.32) и (3.40) имеем формулы:

$$\begin{aligned} &H_u = 2\omega_z \lambda_2 - \lambda_3; \quad H_w = -2\omega_z \lambda_1 - \lambda_4; \quad H_x = \nu^2 \lambda_1; \quad H_y = -2\nu^2 \lambda_2 \\ &H_{\dot{u}} = \lambda_1; \quad H_{\dot{w}} = \lambda_2; \quad H_{\dot{x}} = \lambda_3; \quad H_{\dot{y}} = \lambda_4 \\ &H_\varphi = p(\lambda_1 \sin\varphi - \lambda_2 \cos\varphi). \end{aligned} \tag{3.43}$$

Полагая функции λ_1, λ_2, λ_3 и λ_4 непрерывными и дифференцируемыми в каждой точке, проинтегрируем по частям в подинтегральном выражении правой части (3.41) члены, содержащие производные от вариаций. В результате получим

$$\begin{aligned} \delta y = \frac{u_k}{v_k}\delta u_k + \frac{w_k}{v_k}\delta w_k + \lambda_{1k}\delta u_k + \lambda_{2k}\delta w_k + \lambda_{3k}\delta x_k + \\ + \lambda_{4k}\delta y_k - \lambda_{1o}\delta u_o - \lambda_{2o}\delta w_o - \lambda_{3o}\delta x_o - \lambda_{4o}\delta y_o + \int_o^T (\widetilde{\delta H})\, dt, \end{aligned} \tag{3.44}$$

где λ_{1o}, λ_{2o}, λ_{3o}, λ_{4o} и λ_{1k}, λ_{2k}, λ_{3k}, λ_{4k} — значения функций λ_1, λ_2, λ_3 и λ_4 соответственно в начальный и в конечный момент времени. Заметим, что вариации δu_k, δw_k, δx_k и δy_k в силу условий (3.36) и (3.37) не являются независимыми, а связаны между собой соотношениями

$$\frac{1}{R + y_k}\delta x_k - \frac{x_k}{(R + y_k)^2}\delta y_k = \frac{w_k}{u_k^2}\delta u_k - \frac{1}{u_k}\delta w_k \tag{3.45}$$

$$x_k \delta x_k + (R + y_k)\delta y_k = 0 \tag{3.46}$$

Таким образом из четырех вариаций δu_k, δw_k, δx_k и δy_k лишь две являются независимыми. В дальнейшем в качестве независимых вариаций примем вариации δu_k и δw_k. Также, согласно (3.35) вариации δu_o, δw_o, δx_o и δy_o не независимы, а связаны между собой вариацию $\delta\theta_o$ Очевидно имеем

$$\delta u_o = \frac{\partial u_o}{\partial \theta_o}\delta\theta_o, \quad \delta w_o = \frac{\partial w_o}{\partial \theta_o}\delta\theta_o, \quad \delta x_o = \frac{\partial x_o}{\partial \theta_o}\delta\theta_o, \quad \delta y_o = \frac{\partial y_o}{\partial \theta_o}\delta\theta_o \tag{3.47}$$

Пользуясь (3.45), (3.46) и (3.47), исключим из выражения, стоящего перед интегралом правой части (3.44) лишние вариации. Получим после простых преобразований

$$\begin{aligned} \delta y = \left\{\frac{u_k}{v_k} + \lambda_{1k} + \frac{w_k}{u_k^2}\left(\frac{R + y_k}{R + h}^2\right)[(R + y_k)\lambda_{3k} + x_k\lambda_{4k}]\right\}\delta u_k + \\ + \left\{\frac{w_k}{v_k} + \lambda_{2k} - \left(\frac{1}{u_k}\frac{R + y_k}{R + h}^2\right)[(R + y_k)\lambda_{3k} + x_k\lambda_{4k}]\right\}\delta w_k + \\ + \left\{\lambda_{1o}\frac{\partial u_o}{\partial\theta_o} + \lambda_{2o}\frac{\partial w_o}{\partial\theta_o} + \lambda_{3o}\frac{\partial x_o}{\partial\theta_o} + \lambda_{4o}\frac{\partial y_o}{\partial\theta_o}\right\}\delta\theta_o + \int_o^T (\widetilde{\delta H})\, dt \end{aligned} \tag{3.48}$$

Далее, для $(\widetilde{\delta H})$, стоящего под интегралом в (3.48), имеем, формулу

$$\begin{aligned} (\widetilde{\delta H}) = (2\omega_z\lambda_2 - \lambda_3 - \dot{\lambda}_1)\delta u + (-2\omega_z\lambda_1 - \lambda_4 - \dot{\lambda}_2)\delta w + \\ + (\nu^2\lambda_1 - \dot{\lambda}_3)\delta x + (-2\nu^2\lambda_2 - \dot{\lambda}_4)\delta y + p(\lambda_1\sin\varphi - \lambda_2\cos\varphi)\delta\varphi \end{aligned} \tag{3.49}$$

Подберем теперь множители λ_1, λ_2, λ_3 и λ_4 таким образом, чтобы в (3.48) под интегралом пропали члены, содержащие вариации δu, δw, δx, δy, а также члены, стоящие перед интегралом и содержащие вариации δu_k и δw_k. В результате получим систему дифференциальных уравнений для λ_1, λ_2, λ_3 и λ_4.

$$\begin{cases} \dot{\lambda}_1 = 2\,\omega_z\,\lambda_2 - \lambda_3 & (3.50) \\ \dot{\lambda}_2 = -2\,\omega_z\,\lambda_1 - \lambda_4 & (3.51) \\ \dot{\lambda}_3 = \nu^2\,\lambda_1 & (3.52) \\ \dot{\lambda}_4 = -2\,\nu^2\,\lambda_2 & (3.53) \end{cases}$$

а также два функциональных условия на правой границе, которые представим в виде:

$$\left\{ \lambda_{1k} = -\frac{u_k}{v_k} - \frac{w_k}{{u_k}^2}\,\frac{R+y_k}{R+h}^{\,2}\,[(R+y_k)\,\lambda_{3k} + x_k\,\lambda_{4k}] \right. \qquad (3.54)$$

$$\left\{ \lambda_{2k} = -\frac{w_k}{v_k} + \frac{1}{u_k}\,\frac{R+y_k}{R+h}^{\,2}\,[(R+y_k)\,\lambda_{3k} + x_k\,\lambda_{4k}] \right. \qquad (3.55)$$

Первая вариация функционала (3.39) может быть теперь представлена в виде, содержащем непосредственно связь вариации функционала с вариациями угла θ_o и функции $\varphi\ (t)$.

$$\delta 1 = \lambda_{1o}\frac{\partial u_o}{\partial \theta_o} + \lambda_{2o}\frac{\partial w_o}{\partial \theta_o} + \lambda_{3o}\frac{\partial x_o}{\partial \theta_o} + \lambda_{4o}\frac{\partial y_o}{\partial \theta_o}\ \delta\theta_o + \qquad (3.56)$$
$$+ \int_0^T p\,(\lambda_1 \sin\varphi - \lambda_2 \cos\varphi)\,\delta\varphi\,dt\,.$$

Полагая, что существует внутренний экстремум функционала (3.39) как по θ_o, так и по $\varphi\ (t)$, найдем его из условия

$$\delta y = 0 \qquad (3.57)$$

Так как вариации $\delta\theta_o$ и $\delta\varphi$ независимы, то условие (3.57) будет очевидно выполнено, если

$$\lambda_{1o}\frac{\partial u_o}{\partial \theta_o} + \lambda_{2o}\frac{\partial w_o}{\partial \theta_o} + \lambda_{3o}\frac{\partial x_o}{\partial \theta_o} + \lambda_{4o}\frac{\partial y_o}{\partial \theta_o} = 0 \qquad (3.58)$$

$$\operatorname{tg}\varphi = \frac{\lambda_2}{\lambda_1} \qquad (3.59)$$

Мы получили два соотношения, из которых первое служит для определения оптимального угла θ_o а второе для определения оптимальной программы $\varphi\ (t)$.

Система уравнений (3.50) — (3.53), (3.59), совместно с граничными условиями (3.54), (3.55), (3.58), а также совместно с уравнениями движения (3.29) — (3.32) и граничными условиями (3.35) — 3.37) дает полное решение поставленной задачи. При этом зависимость оптимальной программы управления от времени может быть получена в явном виде.

Действительно, система (3.50) — (3.53) представляет собой систему линейных дифференциальных уравнений с постоянными коэффициентами и потому может быть легко проинтегрирована. Для нахождения общего решения системы (3.50) — (3.53) необходимо найти корни характеристического уравнения этой системы. Характеристическое уравнение системы (3.50) — (3.53) имеет вид:

$$k^4 - (\nu^2 - 4\,{\omega_z}^2)\,k^2 - 2\,\nu^4 = 0 \qquad (3.60)$$

Уравнение (3.60) будет иметь два действительных и два мнимых корня

$$K_{1,2} = \pm \nu_1 \qquad k_{3,4} = \pm i\nu_2, \tag{3.61}$$

где

$$\begin{cases} \nu_1 = \dfrac{\sqrt{2}}{2} \sqrt{\sqrt{(\nu^2 - 4\omega_z^2)^2 + 8\nu^4} + (\nu^2 - 4\omega_z^2)} & (3.62) \\ \nu_2 = \dfrac{\sqrt{2}}{2} \sqrt{\sqrt{(\nu^2 - 4\omega_z^2)^2 + 8\nu^4} - (\nu^2 - 4\omega_z^2)} & (3.63) \end{cases}$$

Находя общее решение системы (3.50) — (3.53) и подставляя его в (3.59), получим общую формулу для оптимальной программы управления по тангажу, содержащую явную зависимость φ от времени t

$$\operatorname{tg} \varphi = \frac{\sigma_2 \operatorname{ch} \nu_1 (T-t) - \sigma_4 \operatorname{sh} \nu_1 (T-t) + S_2 \cos \nu_2 (T-t) - S_4 \sin \nu_2 (T-t)}{S_1 \operatorname{ch} \nu_1 (T-t) - S_3 \operatorname{sh} \nu_1 (T-t) + \sigma_1 \cos \nu_2 (T-t) - \sigma_3 \sin \nu_2 (T-t)} \tag{3.64}$$

где

$$\begin{cases} \sigma_1 = (\nu_1^2 + \nu^2 + 4\omega_z^2)\lambda_{1k} + 2\omega_z \lambda_{4k} \\ \sigma_2 = (\nu_1^2 + \nu^2 - 4\omega_z^2)\lambda_{2k} + 2\omega_z \lambda_{3k} & (3.65) \\ \sigma_3 = \dfrac{1}{\nu_2} [2\omega_z (\nu_1^2 - \nu^2 + 4\omega_z^2)\lambda_{2k} - (\nu_1^2 + \nu^2 + 4\omega_z^2)\lambda_{3k}] \\ \sigma_4 = -\dfrac{1}{\nu_1} [2\omega_z (\nu_1^2 + \nu^2 - 4\omega_z^2)\lambda_{1k} + (\nu_1^2 + 2\nu^2 - 4\omega_z^2)\lambda_{4k}] \\ \\ S_1 = (\nu_2^2 - \nu^2 - 4\omega_z^2)\lambda_{1k} - 2\omega_z \lambda_{4k} \\ S_2 = (\nu_1^2 - 2\nu^2 + 4\omega_z^2)\lambda_{2k} - 2\omega_z \lambda_{3k} \\ S_3 = \dfrac{1}{\nu_1} [2\omega_z (\nu_2^2 + \nu^2 - 4\omega_z^2)\lambda_{2k} - (\nu_2^2 - \nu^2 - 4\omega_z^2)\lambda_{3k}] & (3.66) \\ S_4 = -\dfrac{1}{\nu_2} [2\omega_z (\nu_1^2 - \nu^2 + 4\omega_z^2)\lambda_{1k} + (\nu_1^2 - 2\nu^2 + 4\omega_z^2)\lambda_{4k}]. \end{cases}$$

Чтобы довести задачи до конца, необходимо в формулах (3.65) и (3.66) определить постоянные λ_{1k}, λ_{2k}, λ_{3k} и λ_{4k}. Напомним при этом, что λ_{1k} и λ_{2k} определяются через λ_{3k}, λ_{4k} и через кинематические параметры u_k, w_k, x_k, y_k по формулам (3.54), (3.55).

Таким образом формула (3.64) определяет программу управления φ (t) как функцию времени, а также шести параметров — λ_{3k}, λ_{4k}, u_k, w_k, x_k, y_k, т.е.

$$\varphi = \varphi(t; \lambda_{3k}, \lambda_{4k}, u_k, w_k, x_k, y_k) \tag{3.67}$$

Подставим φ (t) определенное по формуле (3.64) в уравнения движения (3.29) — (3.32). Система (3.29) — (3.32) превратится после этого в неоднородную систему линейных дифференциальных уравнений с постоянными коэффициентами и с переменной правной частью, явно зависящей от времени. Решение такой системы может быть представлено в квадратурах. В результате интеграции (3.29) — (3.32) получим четыре соотношения, связывающие семь параметров u_k, w_k, x_k, y_k, λ_{3k}, λ_{4k}, и θ_o. Эти соотношения запишем в виде:

$$\begin{cases} u_k - u_o(\theta_o) + \Phi_1(\lambda_{3k}, \lambda_{4k}, u_k, w_k, x_k, y_k) = 0 & (3.68) \\ w_k - w_o(\theta_o) + \Phi_2(\lambda_{3k}, \lambda_{4k}, u_k, w_k, x_k, y_k) = 0 & (3.69) \\ x_k - x_o(\theta_o) + \Phi_3(\lambda_{3k}, \lambda_{4k}, u_k, w_k, x_k, y_k) = 0 & (3.70) \\ y_k - y_o(\theta_o) + \Phi_4(\lambda_{3k}, \lambda_{4k}, u_k, w_k, x_k, y_k) = 0 & (3.71) \end{cases}$$

где Φ_1, Φ_2, Φ_3 и Φ_4 содержат интегралы, в общем случае не берущиеся в элементарных функциях. У равнения (3.68)—(3.71) вместе с конечными соотношенями (3.36)—(3.37) и соотношением (3.58) представляют замкнутую систему семи трансцендентных уравнений с семью неизвестными — u_k, w_k, x_k, y_k, θ_o, λ_{3k}, λ_{4k}[1]. Разрешая эту систему, найдем значения кинематических параметров в конце траектории выведения u_k, w_k, x_k, y_k, оптимальное значение начального угла θ_o и значения параметров оптимальной программы управления λ_{1k}, λ_{2k}, λ_{3k}, λ_{4k}, причем, λ_{1k} и λ_{2k} вычисляются дополнительно с помощью формул (3.54) и (3.55).

Система уравнений (3.68)—(3.71), (3.36), (3.37) и (3.58) весьма сложна и громоздка, и разрешить ее в явном виде в общем случае невозможно. Поэтому при практическом использовании предлагаемого метода нахождения оптимальной программы, для решения указанной системы, целесообразно использовать какой-либо численный, итерационный медот. Не останавливаясь однако здесь на вопросе практического использования найденной оптимальной программы, проанализируем формулу (3.64) и упростим ее.

Прежде всего выясним влияние вращения Земли на выбор оптимальной программы по тангажу и покажем, что это влияние незначительно.

Рассмотрим случай, когда условия движения ракеты таковы, что

$$\omega_z = 0 \qquad (3.72)$$

Условие (3.72) выполняется, например, если $a_o = 0; \pi$ или $\psi_o = \pm \frac{\pi}{2}$

В этом случае формулы (3.62) и (3.63) для ν_1 и ν_2 сильно упрощаются и принимают вид:

$$\nu_1 = \sqrt{2\nu} \qquad \nu_2 = \nu .$$

формулы для σ_1, σ_2, σ_3 и σ_4 также упрощаются и принимают вид:

$$\begin{cases} \sigma_1 = 3\nu^2 \lambda_{1k} & \sigma_3 = -3\lambda_{3k} \\ \sigma_2 = 3\nu^2 \lambda_{2k} & \sigma_4 = -3\lambda_{4k}\frac{1}{\sqrt{2}} \end{cases} \qquad (3.73)$$

Что касается S_1, S_2, S_3 и S_4 то в этом случае из (3.66) легко получить, что

$$S_1 = S_2 = S_3 = S_4 = 0 \qquad (3.74)$$

Принимая во внимание (3.73) и (3.74) для программы $\varphi(t)$, получим значительно упрощенную формулу

$$\operatorname{tg}\varphi = \frac{\lambda_{2k}\operatorname{ch}\sqrt{2}\nu(T-t) + \dfrac{\lambda}{\nu\sqrt{2}}\operatorname{sh}\sqrt{2}\,\nu\,(T-t)}{\lambda_{1k}\cos\nu(T-t) + \dfrac{\lambda_{3k}}{\nu}\sin\nu\,(T-t)} \qquad (3.75($$

[1] Заметим, что λ_{1o}, λ_{2o}, λ_{3o} и λ_{4o}, входящие в (3.58), выражаются с помощью конечных формул через λ_{3k}, λ_{4k}, u_k, w_k, x_k и y_k.

Формула (3.75) с большой степенью точности может быть использована как формула оптимальной программы и в том случае, когда $\omega_z \neq 0$. Это легко доказать, если учесть, что вращение Земли влияет на вид оптимальной программы через величины ν_1 и ν_2.

Из (3.62) и (3.63) видим, что ω_z входит в формулы для ν_1 и ν_2 лишь в комбинации $(v^2 - 4\omega_z^2)$. Можно показать, что $4\omega_z^2$ мало по сравнению с ν^2. Пользуясь этим, с помощью (3.62) и (3.63) выведем приближенные формулы для ν_1 и ν_2

$$\nu_1 \approx \sqrt{2}\,\nu\left[1 - \frac{2}{3}\;\frac{\omega_z}{\nu}^{\,2}\right] \qquad \nu^2 \approx \nu\left[1 + \frac{2}{3}\;\frac{\omega_z}{\nu}^{\,2}\right] \tag{3.76}$$

Возьмем для ω_z максимально возможное значение, равное ω. Известно, что

$$\omega = 7.292 \cdot 10^{-5}\,\text{сек.}^{-1} \tag{3.77}$$

С другой стороны, по формуле (3.33) вычислим v

$$\nu = 1.241 \cdot 10^{-3}\,\text{сек.}^{-1} \tag{3.78}$$

Пользуясь (3.77), (3.78), вычислим максимальное значение поправочного члена в формулах (3.76)

$$\frac{2}{3}\left(\frac{\omega}{\nu}\right)^2 = 0{,}230 \cdot 10^{-2} \tag{3.79}$$

Видим, что поправка для ν_1 и ν_2 за счет вращения Земли действительно весьма незначительна и, учитывая, что рассматриваемая вариационная задача решена при ряде упрощений формул для внешних действующих сил, этой поправкой можно безусловно пренебречь.

Таким образом, формулу (3.75) можно рассматривать, как общую формулу для оптимальной программы управления по тангажу, пригодную для использования в случае различных географических условий выведения на орбиту.

Рассмотрим предельный вид формулы (3.75), соответствующий малому значению параметра νT. В этом случае, полагая $\operatorname{ch}\sqrt{2}\,\nu T$ и $\cos\,\nu T$ равными единице, а $\operatorname{sh}\sqrt{2}\,\nu T$ и $\sin\,\nu T$ равными νT, согласно (3.75), будем, очевидно, иметь

$$\operatorname{tg}\varphi = \frac{\lambda_{2k} + \lambda_{4k}\,(T-t)}{\lambda_{1k} + \lambda_{3k}\,(T-t)} \tag{3.80}$$

Формулу (3.80) можно также получить, если решать рассматриваемую задачу без учета переменности поля сил тяготения. Если при этом считать достаточно малыми величины

$$\frac{x_k}{R},\quad \frac{y_k}{R},\quad \frac{1}{\sqrt{\dfrac{R}{g}}}\quad \text{и}\quad \frac{w_k}{u_k},$$

то в результате получим формулу (2.1) предыдущего параграфа, а вместе с ней и всю теорию оптимального выведения на орбиту, развитую для плоско-параллельного поля силы тяжести.

On the Establishment of an Artificial Satellite of the Earth in Orbit [1]

By

D. E. Okhotsimsky and T. M. Eneiev

Abstract. This paper deals with the problem of placing an artificial satellite of the Earth in orbit. It is assumed that this is carried out with the help of a rocket consisting of one or several stages. The law of change in thrust with respect to time is investigated, with the object of delivering the satellite into its orbit with the minimum propellent consumption. The most economic regime of propellent consumption is sought.

The solution of these problems, carried out in a series of simplifying assumptions, allows one to form a definite concept of the characteristics of an optimum placing in orbit and also shows the way in which a vehicle can be made with the minimum initial weight.

In the first part we deal with the problem of choosing simultaneously the thrust control programme and the fuel consumption regime. In the second part we deal with the question of choosing an optimum programme for a multistage rocket with a variable number of stages assuming that the propellent consumption regime is given. In the third part is a generalisation of the problem of placing in orbit in a central gravitational field, making allowance for the Earth's rotation.

[1] An almost identical paper entitled "Some Variation Problems Connected with the Launching of Artificial Satellites of the Earth" was published in J. Brit, Interplan. Soc. **16**, **263** (1958).

Определение времени существования искусственного спутника земли и исследование вековых возмущений его орбиты

Д. Е. Охоцимский, Т. М. Энеев, Г. П. Таратынова[1]

(2 рис.)

Одним из важных вопросов, связанных с проблемой создания искусственного спутника Земли, является достаточно надежное определение времени его существования на орбите. Вследствие сопротивления атмосферы будет происходить рассеяние энергии спутника и его постепенное снижение. При движении на больших высотах в разреженных слоях атмосферы сопротивление мало, и время движения спутника может оказаться весьма значительным. При движении на сравнительно небольших высотах (порядка 100—150 км) время существования спутника невелико, и при малых поперечных нагрузках спутник может не совершить даже одного полного оборота.

В настоящее время имеется большое число работ, посвященных проблеме определения времени существования искусственного спутника. При этом достаточно точное решение дается лишь для круговой орбиты. Для оценки времени движения спутника по эллиптической орбите применяются различные приближенные методы, использующие энергетические соображения и основанные на том, что потери энергии происходят в основном в области перигея, когда спутник подходит к Земле наиболее близко. Применение этих методов не дает полного решения задачи в общем случае. Кроме того, как показывает анализ, применение приближенных методов определения времени существования в ряде случаев может приводить к существенным ошибкам.

Указанное положение вызвало необходимость разработки методики, дающей возможность достаточно быстро и надежно определять время существования спутника для общего случая его движения. Исследование выявило существование универсальных зависимостей между основными параметрами оскулирующего эллипса, такими как высота перигея и апогея или параметр и эксцентриситет. Эти зависимости справедливы для любых спутников и зависят лишь от закона распределения плотности воздуха по высоте. Указанное обстоятельство позволило свести полное решение задачи о времени жизни искусственных спутников к построению зависящей от одного параметра серии интегральных кривых уравнения первого порядка. Помещенные в статье график и таблицы дают возможность быстро определить время существования, умножив результат, взятый из таблицы или с графика, на некоторое число, просто зависящее от основных параметров спутника. Приведенные результаты позволяют не только вычислить время существования, но и определить закон изменения во времени параметров

---

[1] Москва, Академия наук СССР.

орбиты при любых заданных параметрах спутника и для достаточно широкого диапазона начальных параметров орбиты.

Интегрирование уравнений было проведено на быстродействующей электронной счетной машине (БЭСМ) Академии наук СССР. Использованный метод интегрирования позволил максимально сократить объем вычислительной работы при сохранении необходимой точности результатов. Метод является достаточно общим и может быть, по-видимому, с успехом применен во многих случаях, когда дело сводится к интегрированию уравнений, решение которых является функцией, близкой к периодической, с медленно меняющимися параметрами. Так, подобный метод был разработан и успешно использован для исследования возмущенного движения спутника в нецентральном поле сил [5].

В § 1 приведены исходные данные, принятые для расчетов. В § 2 уравнения движения спутника в оскулирующих элементах приводятся к новому независимому переменному, удобному для проводимого исследования. В § 3 дано изложение методики интегрирования дифференциальных уравнений для вековых возмущений параметра и эксцентриситета орбиты за счет действия силы сопротивления воздуха. В § 4 приводятся и обсуждаются полученные результаты. В § 5 приводится обоснование основных упрощающих допущений, принятых при расчете, и дается оценка получающейся за счет этих допущений методической ошибки. Дается также приближенное исследование вековых возмущений орбиты за счет сжатия Земли и вращения атмосферы.

Отметим, что приведенные в работе результаты числовых расчетов по определению времени существования спутника основаны на использовании определенных предположений о строении верхних слоев атмосферы. Отсутствие надежных сведений о параметрах верхней атмосферы делает численные результаты пригодными лишь для ориентировочных оценок. Следует, однако, указать, что анализ движения искусственных спутников позволит существенно пополнить данные о верхней атмосфере и даст возможность провести по изложенной методике необходимые для дальнейшего уточненные расчеты.

§ 1. Зависимость плотности атмосферы от высоты

Для расчета величины аэродинамического сопротивления необходимо знать плотность атмосферы на больших высотах. В настоящее время не имеется достаточно точных сведений относительно физических параметров верхней атмосферы. Неизвестны с достаточной степенью точности температура и состав атмосферы. В связи с различием принятых предположений о распределении температур и о составе атмосферы в различных работах приводятся существенно отличные значения плотности.

В настоящей работе использовались данные о физических параметрах верхней атмосферы, приведенные в [1].

Зависимость плотности от высоты $\varrho(y)$ аппроксимировалась формулами вида: $\varrho = \varrho_1 \Delta$,

$$\Delta = \frac{\varkappa}{\left(1 + \frac{y - y_0}{a}\right)^k} \tag{1.1}$$

где ϱ_1 — плотность воздуха для некоторой фиксированной высоты y_1. Для постоянных $\varkappa$, a, y_0 и k были приняты значения, указанные в табл. 1.

Таблица 1

Диапазон y (км)	$\varkappa$	a (км)	y_0 (км)	k
100—150	1	55	100	8
150—250	$0{,}0^2 5667$	100	150	7
250—900	$0{,}0^4 4428$	215	250	6

Отклонения значений плотности, вычисленных по формуле (1.1), от значений, рассчитанных по данным работы [1], не превышают 10—15% от величины плотности. Такая погрешность вполне допустима, принимая во внимание, что сами данные об атмосфере являются весьма приближенными.

§ 2. Уравнения движения

Движение спутника будем изучать, используя оскулирующие элементы орбиты. Уравнения движения в оскулирующих элементах имеют вид:

$$\frac{dp}{dt} = \frac{2r\sqrt{p}}{\sqrt{fM}} \cdot T$$

$$\frac{de}{dt} = \frac{\sqrt{p}}{\sqrt{fM}} \sin\theta \cdot S + \frac{\sqrt{p}}{\sqrt{fM}} \left[\left(1 + \frac{r}{p}\right)\cos\theta + e\frac{r}{p}\right] T$$

$$\frac{d\omega}{dt} = \frac{\sqrt{p}}{e\sqrt{fM}} \cos\theta \cdot S + \frac{\sqrt{p}}{e\sqrt{fM}} \left(1 + \frac{r}{p}\right)\sin\theta \cdot T - \frac{\sqrt{p}}{\sqrt{fM}} \frac{r}{p}\, ctg\, i \sin u \cdot W \qquad (2.1)$$

$$\frac{d\Omega}{dt} = \frac{\sqrt{p}}{\sqrt{fM}} \cdot \frac{r}{p} \cdot \frac{\sin u}{\sin i} W$$

$$\frac{di}{dt} = \frac{\sqrt{p}}{\sqrt{fM}} \cdot \frac{r}{p} \cos u \cdot W$$

$$\frac{d\tau}{dt} = \frac{r^2}{efM} \left[-(\cos\theta - e\sin\theta \cdot N) S + \frac{p}{r} NT\right],$$

где

$$u = \omega + \theta, \qquad r = \frac{p}{1 + e\cos\theta},$$

$$N = 2\frac{p^2}{r^2} \int\limits_0^\theta \frac{\cos\theta\, d\theta}{(1 + e\cos\theta)^3} \qquad (2.1a)$$

и θ связано с t уравнением

$$t - \tau = \frac{p^{3/2}}{\sqrt{fM}} \int\limits_0^\theta \frac{d\theta}{(1 + e\cos\theta)^2}, \qquad (2.2)$$

где p — параметр оскулирующего эллипса; e — его эксцентриситет; ω — угловое расстояние перигея от узла; Ω — долгота восходящего узла; i — наклонение орбиты; τ — время прохождения через перигей оскулирующего эллипса; θ — истинная аномалия, u — аргумент широты; S, T, W — проекции возмущающего ускорения соответственно на радиус-вектор, перпендикуляр к нему в плоскости оскулирующего эллипса и на перпендикуляр к плоскости оскулирующего эллипса; r — радиус-вектор; f — постоянная тяготения; M — масса Земли. Заметим, что значения e, входящие в подинтегральное выражение в правых частях уравнения (2.2) и формулы (2.1 а), соответствуют моменту времени t.

В случае, если правые части системы (2.1) не зависят явно от времени, то удобно ввести новое независимое переменное аргумент широты u. Чтобы перейти к этому переменному, выведем дифференциальное соотношение, связывающее u с оскулирующими элементами орбиты и с временем t. С этой целью приравняем секторную скорость возмущенного движения секторной скорости оскулирующего движения, отвечающего данному моменту времени. Справедливость такого равенства следует из определения оскулирующего движения, как эллиптического движения при отсутствии возмущений, имеющего с возмущенным движением в рассматриваемый момент времени общую точку и общий вектор скорости. Получим соотношение:

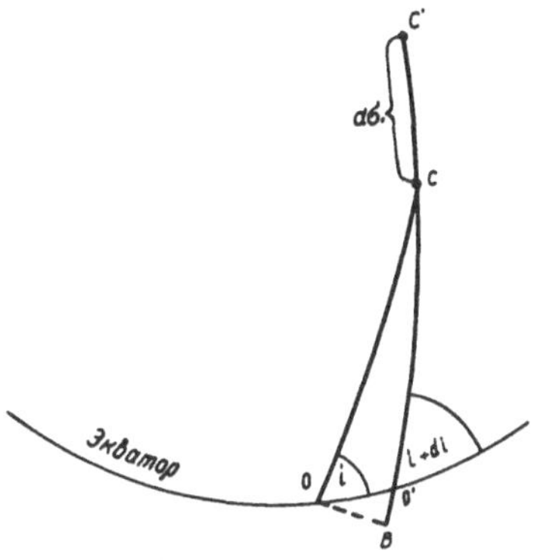

Рис. 1

$$r^2 d\sigma = \sqrt{fM \cdot p}\, dt\,, \tag{2.3}$$

где $d\sigma$ — полное угловое перемещение радиуса-вектора r за время dt. Заметим, что для невозмущенного движения соотношение (2.3) переходит в интеграл площадей, причем в этом случае очевидно, $d\sigma = d\theta$.

Найдем выражение для $d\sigma$. Из рис. 1 нетрудно видеть, что

$$d\sigma = \widehat{BC'} - \widehat{BC} = \widehat{BC'} - \widehat{OC} = \widehat{O'C'} - \widehat{OC} + \widehat{BO'}\,, \tag{2.4}$$

где $\widehat{OC}$ и $\widehat{BC'}$ — дуги оскулирующих эллипсов соответственно для моментов времени t и $t+dt$. С другой стороны, имеем

$$\widehat{OO'} = d\Omega \qquad \widehat{BO'} = \cos i\, d\Omega\,, \tag{2.5}$$

$$\widehat{OC} = u \qquad \widehat{O'C'} = u + du\,.$$

Подставляя (2.5) в (2.4), получим:

$$r^2 (du + \cos i\, d\Omega) = \sqrt{fM} \cdot \sqrt{p\, dt}$$

или

$$r^2 \left(\frac{du}{dt} + \cos i \frac{d\Omega}{dt}\right) = \sqrt{fM} \cdot \sqrt{p}\,. \tag{2.6}$$

Наконец, принимая во внимание четвертое уравнение системы (2.1), получим следующее окончательное дифференциальное соотношение для u:

$$\frac{du}{dt} = \sqrt{fM} \cdot \frac{p}{r^2} \cdot \left(1 - \frac{r^3}{fM \cdot p} \cdot \operatorname{ctg} i \sin u \cdot W\right). \tag{2.7}$$

С помощью формулы (2.7) систему (2.1) приводим к виду

$$
\begin{aligned}
\frac{dp}{du} &= \frac{2\gamma}{fM} \cdot r^3 T, \\
\frac{de}{du} &= \frac{r^2\gamma}{fM} \cdot \left[\sin\theta \cdot S + \cos\theta\left(1 + \frac{r}{p}\right) T + e\frac{r}{p} T\right], \\
\frac{d\omega}{du} &= \frac{r^2\gamma}{fMe} \left[-\cos\theta\, S + \sin\theta\left(1 + \frac{r}{p}\right) T - e\frac{r}{p} \operatorname{ctg} i \sin u \cdot W\right], \qquad (2.8)\\
\frac{d\Omega}{du} &= \frac{r^3\gamma}{fMp} \cdot \frac{\sin u}{\sin i} W, \\
\frac{di}{du} &= -\frac{r^3\gamma}{fMp} \cdot \cos u \cdot W,
\end{aligned}
$$

где

$$\theta = u - \omega, \quad \gamma = \frac{1}{1 - \frac{r^3}{fMp} \operatorname{ctg} i \sin u \cdot W} \tag{2.9}$$

Система (2.8) является замкнутой, и уравнения этой системы должны интегрироваться в общем случае совместно. Время t может быть определено после того, как система (2.8) проинтегрирована с помощью уравнения

$$\frac{dt}{du} = \frac{r^2\gamma}{\sqrt{fMp}} \tag{2.10}$$

Время прохождения через перигей τ может быть также определено после интегрирования системы (2.8) с помощью соответствующего уравнения. Уравнение это здесь не приведено, так как в дальнейшем оно не понадобится.

Вместо аргумента широты u в качестве независимого переменного может быть взят какой-либо другой угловой параметр, например, истинная аномалия θ. В этом случае система (2.1) после преобразований приведется также к виду (2.8), с той лишь разницей что, в левых частях уравнений будут стоять производные не по u, а по θ. Кроме того, величина γ при независимом переменном θ будет определяться не по формуле (2.9), а по формуле

$$\gamma = \frac{1}{1 + \frac{r^2}{fMe} \cos\theta \cdot S - \frac{r^2}{fMe}\left(1 + \frac{r}{p}\right) \sin\theta \cdot T}. \tag{2.11}$$

Полученные при этом уравнения будут отличаться от уравнений с истинной аномалией в качестве независимого переменного, приведенных в [2]. Указанные уравнения получены в [2] из уравнений (2.1) при помощи соотношения

$$r^2\, d\theta = \sqrt{fMp}\, dt, \tag{2.12}$$

справедливого для оскулирующего движения, но ошибочного для возмущенного движения, для которого вместо (2.12) должно быть использовано соотношение (2.3). Вследствие этого приведенные в [2] уравнения по истинной аномалии являются ошибочными.

Отметим, что указание на возможность перехода к истинной аномалии как независимому переменному при помощи соотношения (2.12) содержится также в работе [3].

Отметим также, что при исследовании возмущений по первому приближению, использование соотношения (2.12) допустимо, поскольку для невозмущенного эллиптического движения оно выполняется. Этим замечанием мы воспользуемся в § 5 при определении вековых возмущений орбиты.

Следует указать, что уравнения по переменному u являются более удобными для расчетов, чем уравнения по переменному θ, поскольку при малых значениях эксцентриситета вследствие имеющих место значительных изменений производная $\frac{d\theta}{dt}$ может быть весьма мала и для некоторых случаев может обращаться в нуль. Такое положение имеет место, например, при движении спутника по круговой орбите в плоскости экватора под действием силы тяготения сплюснутого земного сфероида.

§ 3. Методика определения времени существования искусственного спутника

Рассмотрим движение спутника в земной атмосфере при условии, что поле тяготения Земли является центральным. Мы пренебрегаем также вращением атмосферы вместе с Землей в ее суточном движении.. Для оценки времени существования спутника такие допущения являются вполне приемлемыми. В этом случае, очевидно, $W = O$ и система (2.8) примет вид:

$$
\begin{aligned}
\frac{dp}{du} &= \frac{2r^3}{fM} \cdot T, \\
\frac{de}{du} &= \frac{r^2}{fM}\left[\sin\theta \cdot S + \cos\theta\left(1 + \frac{r}{p}\right)T + e\frac{r}{p}T\right], \\
\frac{d\omega}{du} &= \frac{r^2}{fMe} \cdot \left[-\cos\theta \cdot S + \sin\theta\left(1 + \frac{r}{p}\right)T\right]. \\
\frac{d\Omega}{du} &= O, \qquad \frac{di}{du} = O, \qquad \theta = u - \omega.
\end{aligned}
\tag{3.1}
$$

Из (3.1) находим:

$$\Omega = \Omega_o = \text{const}, \; i = i_o = \text{const},$$

т.е. сопротивление атмосферы не вызывает вековых возмущений, долготы узла и наклонения орбиты. Оценка эффекта вращения атмосферы будет дана ниже, в § 5.

Для ускорения силы сопротивления принимаем:

$$R_x = \frac{C_x F}{m}\frac{\varrho v^2}{2},$$

где m — масса спутника, v — скорость спутника относительно воздуха, C_x — коэффициент аэродинамического сопротивления, F — площадь, к которой отнесен аэродинамический коэффициент C_x. Имеем:

$$
\begin{aligned}
S &= -\frac{C_x F}{m} \cdot \frac{\varrho \cdot v}{2} v_r, \\
T &= -\frac{C_x F}{m} \cdot \frac{\varrho \cdot v}{2} v_n.
\end{aligned}
\tag{3.2}
$$

Здесь V_r и V_n — радиальная и трансверсальная составляющие скорости:

$$
\begin{aligned}
v_r &= \sqrt{\frac{fM}{p}}\, e \cdot \sin\theta, \\
v_n &= \sqrt{\frac{fM}{p}}\,(1 + e \cdot \cos\theta).
\end{aligned}
\tag{3.3}
$$

Подставляя (1.1), (3.2) и (3.3) в (3.1) и используя формулы эллиптической теории:

$$r = \frac{p}{1 + e\cos\theta},$$

$$v = \sqrt{\frac{fM}{p}} \cdot \sqrt{1 + 2e\cos\theta + e^2},$$

получим:

$$\frac{dp}{du} = -c\,\varrho_1\,\varphi(p, e, \omega, u),$$

$$\frac{de}{du} = -c\,\varrho_1\,\psi(p, e, \omega, u), \qquad (3.4)$$

$$\frac{d\omega}{du} = -c\,\varrho_1\,\chi(p, e, \omega, u),$$

где

$$\varphi = \frac{p^2 \Delta \sqrt{1 + 2e\cos\theta + e^2}}{(1 + e\cos\theta)^2},$$

$$\psi = \frac{p \cdot \Delta \sqrt{1 + 2e\cos\theta + e^2}}{(1 + e\cos\theta)^2}(e + \cos\theta), \qquad (3.5)$$

$$\chi = \frac{p \cdot \Delta \sin\theta \sqrt{1 + 2e\cos\theta + e^2}}{e(1 + e\cos\theta)^2};$$

$\left(c\text{ — постоянная, } c = \frac{C_x F}{m} \text{ и } \theta = u - \omega\right)$.

Заметим, что непосредственное интегрирование уравнений (3.4) нецелесообразно ввиду того, что вследствие большого интервала интегрирования и ограниченного шага по u количество шагов будет весьма велико, что приведет к большому времени, потребному для интегрирования системы, а также может привести к существенному накоплению погрешности искомых функций. Чтобы избежать этого, проинтегрируем уравнения (3.5) по u от 0 до 2π и получим:

$$\Delta p = -c\,\varrho_1 \int_0^{2\pi} \varphi(p, e, \omega, u)\, du, \qquad (3.6)$$

$$\Delta e = -c\,\varrho_1 \int_0^{2\pi} \psi(p, e, \omega, u)\, du, \qquad (3.7)$$

$$\Delta\omega = -c\,\varrho_1 \int_0^{2\pi} \chi(p, e, \omega, u)\, du, \qquad (3.8)$$

где Δp, Δe, $\Delta\omega$ — изменения параметра, эксцентриситета и расстояния перигея от узла за один оборот. Заметим, что при вычислении указанных интегралов в силу малого изменения параметра p, эксцентриситета e и расстояния перигея от узла ω, в продолжение одного оборота эти элементы можно принять постоянными. Но в этом случае легко показать, что интеграл, стоящий в правой части (3.8), обращается в нуль:

$$\Delta\omega = O. \qquad (3.9)$$

Отметим, что при очень малых эксцентриситетах ($e < 0{,}0001$) принятое допущение может оказаться для ω весьма грубым, однако в этом случае для определения времени существования спутника будет существенно только изменение параметра p, которое, как это можно показать с помощью первого уравнения системы (3.1), практически не будет зависеть от ω.

Поскольку за один оборот элементы p и e изменяются весьма мало, можно с большой точностью принять, что указанные изменения равны производным от этих элементов по числу оборотов спутника $N=\frac{u}{2\pi}$. Принимая во внимание (3.9) и, полагая

$$\omega = \omega_o = \text{const},$$

получим

$$\begin{aligned} \frac{dp}{dN} &= -c\varrho_1 \cdot \int_o^{2\pi} \varphi(p,e,u)\,du, \\ \frac{de}{dN} &= -c\varrho_1 \cdot \int_o^{2\pi} \psi(p,e,u)\,du. \end{aligned} \tag{3.10}$$

Полагая $\nu = N\,c$, из уравнений (3.10) получим:

$$\begin{aligned} \frac{de}{dp} &= \frac{\int_o^{2\pi} \psi(p,e,u)\,du}{\int_o^{2\pi} \varphi(p,e,u)\,du}, \\ \frac{d\nu}{dp} &= -\frac{1}{\varrho_1 \int_o^{2\pi} \varphi(p,e,u)\,du}. \end{aligned} \tag{3.11}$$

Таким образом, задача по определению времени существования спутника свелась к интегрированию системы двух дифференциальных уравнений (3.11), правые части которых выражены через определенные интегралы от величин φ и ψ, которые согласно формулам (3.5) являются известными функциями p, e, и u. Заметим также, что правые части уравнений (3.11) не зависят от аэродинамического коэффициента C_x, а также от конструктивных параметров спутника: веса G и площади миделя F. Поэтому для заданной атмосферы уравнения (3.11) достаточно проинтегрировать один раз, а затем простым переходом от ν к N $\left(N=\frac{\nu}{c}, \text{где } c=\frac{C_x F}{G}\,g\right)$ получить количество оборотов спутника для определенных значений величин C_x. G и F. Отсюда, считая, что один оборот вокруг Земли совершается спутником примерно за 90—100 минут, можно получить достаточно точную оценку времени его существования.

При интегрировании системы (3.11) для удобства решения и экономии машинного времени интегралы, функции cos и корень квадратный в правых частях уравнений были введены с помощью дифференциальных уравнений. При решении на машине основная система уравнений (3.11) и вспомогательная система (3.13) были взяты в виде:

$$\frac{de}{dp} = \frac{\zeta_k}{\eta_k}, \quad \frac{d\nu}{dp} = \frac{1}{\varrho_1\eta_k}, \tag{3.12}$$

где

$$\begin{aligned} \frac{d\eta}{du} &= -\frac{\xi p^2 \Delta}{(1+e\lambda_2)^2}, \quad \frac{d\zeta}{du} = -\frac{\xi p\,\Delta\,(e+\lambda_2)}{(1+e\lambda_2)^2}, \\ \frac{d\xi}{du} &= -\frac{e\lambda_1}{\xi}, \quad \frac{d\lambda_1}{du} = \lambda_2, \quad \frac{d\lambda_2}{du} = -\lambda_1. \end{aligned} \tag{3.13}$$

В уравнениях (3.12) и (3.13) приняты следующие обозначения:

$$\lambda_2 = \cos u, \quad \lambda_1 = \sin u, \quad \xi = \sqrt{1 + e^2 + 2e\lambda_2}\,, \eta = -\int_0^u \frac{p^2 \Delta\xi}{(1 + e\lambda_2)^2}\,du\,,$$

$$\zeta = -\int_0^u \frac{p\,\Delta\xi\,(e + \lambda_2)}{(1 + e\lambda_2)^2}\,du, \quad \zeta_k = \zeta_{u=2\pi}, \quad \zeta_k = \zeta_{u=2\pi}\,,$$

Δ определяется формулой (1.1). В качестве ϱ_1 бралась плотность атмосферы на высоте 100 км. Кроме величин p, e, ν в каждой расчетной точке основной системы (3.12) вычислялась дополнительно скорость в перигее v_π, высота апогея h_a и высота перигея h_π по формулам

$$v_\pi = \sqrt{fM}\,\frac{1+e}{\sqrt{p}}\,, \; h_a = \frac{p}{1-e} - R, \; h_\pi = \frac{p}{1+e} - R\,,$$

где R — радиус Земли.

Интегрирование проводилось следующим образом. Для заданных значений p и e вспомогательная система (3.13) интегрировалась по u в пределах от 0 до 2π методом Рунге-Кутта с постоянным шагом. Поскольку для целей последующего интерполирования в результате расчетов требовалось получить достаточно густую сетку значений величин p и e и поскольку требуемая точность расчета была сравнительно невелика, для интегрирования основной системы (3.12) оказалось целесообразным использовать метод Эйлера с постоянным шагом. Для интегрирования вспомогательной системы (3.13) был выбран шаг $\Delta u = 6°$. При интегрировании основной системы (3.12) был выбран шаг $\Delta p = 5$ км для значений высоты апогея $h_a \leq 700$ км и $\Delta p = 10$ км для значений $h_a > 700$ км. Для принятых значений шагов методическая ошибка в определении времени движения спутника за счет приближенного интегрирования не превышала 2—5%.

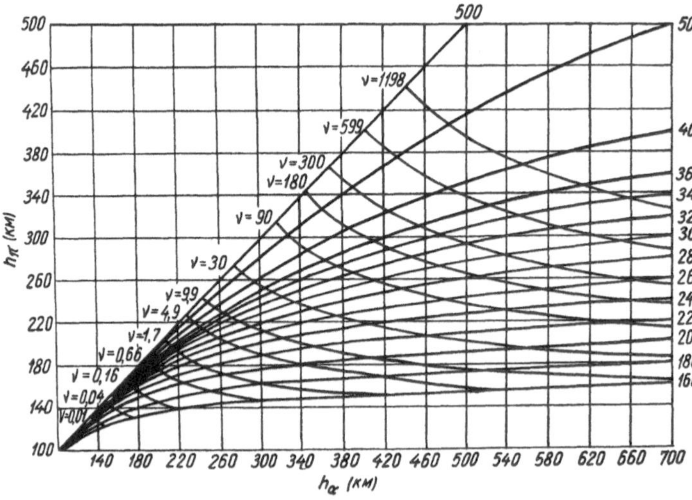

Рис. 2

В качестве исходных данных для системы (3.12) оказалось удобнее задавать начальные величины высоты апогея h_a и высоты перигея h_π, с которыми начальные значения эксцентриситета и параметра связаны соотношениями:

$$e_o = \frac{h_{ao} - h_{\pi o}}{h_{ao} + h_{\pi o} + 2R}, \quad p_o = (h_{ao} + R)(1 - e_o).$$

Поскольку каждую точку на интегральной кривой можно рассматривать как начальную для какого-то другого движения, при выборе начальных величин параметров h_a и h_π не было необходимости варьировать оба параметра. Достаточно было получить, например, серию интегральных кривых по h_π для максимального начального значения h_a. Расчеты проводились для начальной высоты апогея $h_{ao} = 1600$ км и начальных высот перигея из диапазона

$$160 \text{ км} \leq h_{\pi o} \leq 500 \text{ км}.$$

Ввиду того, что для высот больше 900 км не имелось данных относительно плотности атмосферы, принималось, что закон изменения плотности на таких высотах такой же, что и для высоты

$$250 \text{ км} \leq y \leq 900 \text{ км}.$$

Интегрирование системы уравнения (3.12) проводилось до момента достижения спутником высоты 100 км. Рассмотрение движения спутника после этого не имело смысла, так как вследствие сильного торможения об атмосферу он мог бы существовать еще лишь весьма незначительное время.

§ 4. Результаты расчетов и их обсуждение

Результаты расчетов приведены в таблице 2 и на рис. 2. В таблице 2 приведены значения величины ν [м³/кг.сек²] в зависимости от начальных значений высоты апогея h_a и высоты перигея h_π. В этой же таблице приведены значения скорости в перигее v_π [м/сек] в начале движения спутника. Используя результаты, приведенные в таблице 2, с помощью соотношения

$$N = \nu \frac{G}{F} \frac{1}{g C_x}, \tag{4.1}$$

где $g = 9{,}81$ м/сек², можно определить количество оборотов спутника и, следовательно, время его существования для любых значений аэродинамического коэффициента C_x и поперечной нагрузки $\frac{G}{F}$ [кг/м²].

На рис. 2 изображены в плоскости (h_a, h_π) два семейства кривых. Первое семейство соответствует зависимостям

$$h_\pi = f(h_a), \tag{4.2}$$

имеющим место при движении спутника вокруг Земли. Каждая точка указанных кривых соответствует своему значению ν. Линии второго семейства отвечают соотношению

$$\nu = \text{const} \tag{4.3}$$

и соединяют на кривых первого семейства точки, соответствующие одним и тем же значениям ν. Совместное рассмотрение обоих семейств кривых дает возможность не только оценить полное время существования спутника при некоторых начальных значениях высот апогея и перигея, но и оценить время, в течение которого высоты апогея и перигея меняются в некоторых определенных пределах.

На основании приведенных результатов расчетов можно сделать ряд выводов о характере изменения параметров орбиты во время движения

Таблица 2. *Значения величины v и скорости в перигее v_π*

h_π (км) \ h_a (км)	160	170	180	190	200	210	220	230	240	250
160	0.121 7811	0.167 7816	0.204 7819	0.289 7822	0.360 7825	0.426 7828	0.500 7830	0.600 7833	0.700 7836	0.792 7840
170		0.243 7806	0.315 7810	0.424 7813	0.530 7816	0.654 7818	0.800 7821	0.959 7824	1.10 7827	1.28 7829
180			0.445 7802	0.588 7807	0.750 7807	0.939 7809	1.17 7812	1.39 7816	1.61 7818	1.92 7820
190				0.763 7795	1.03 7798	1.31 7801	1.60 7805	1.95 7806	2.29 7810	2.83 7812
200					1.31 7790	1.71 7792	2.19 7795	2.69 7798	3.25 7800	3.91 7804
210						2.11 7784	2.81 7786	3.61 7788	4.31 7792	5.00 7795
220							3.43 7777	4.54 7780	5.50 7783	6.67 7786
230								5.46 7770	6.93 7775	8.46 7778
240									8.36 7766	10.4 7768
250										12.3 7758
260										
280										
300										
320										
340										
360										
400										
500										

Таблица 2 — *продолжение*

$h_{\pi\,(км)}$ \ $h_{a\,(км)}$	260	280	300	320	340	360	400	500	700	800
160	0.889 7843	1.17 7849	1.38 7854	1.67 7860	1.91 7866	2.19 7871	2.89 7884	5.42 7911	8.34 7966	10.9 7993
170	1.48 7833	1.89 7839	2.29 7845	2.66 7851	3.23 7856	3.75 7860	4.81 7876	7.50 7902	14.3 7957	20.4 7984
180	2.22 7824	2.91 7830	3.61 7836	4.28 7842	4.84 7847	5.75 7854	7.50 7865	12.7 7893	23.9 7948	31.1 7975
190	3.29 7816	4.26 7821	5.00 7827	6.20 7832	7.15 7838	8.50 7844	11.4 7856	20.0 7884	37.4 7939	50.9 7966
200	4.51 7807	5.77 7813	7.28 7818	8.75 7824	10.4 7830	12.5 7835	16.8 7847	28.3 7875	56.2 7930	73.4 7957
210	6.04 7798	7.75 7803	9.87 7810	12.3 7815	14.8 7821	17.9 7827	24.2 7838	42.0 7867	81.9 7921	110 7948
220	7.83 7789	10.3 7794	13.5 7800	16.7 7806	20.8 7812	24.5 7818	32.9 7829	57.0 7857	116 7913	152 7939
230	10.0 7780	13.4 7785	17.9 7792	22.5 7797	26.9 7803	31.6 7810	44.4 7820	77.2 7849	159 7904	211 7930
240	12.6 7772	17.1 7777	22.7 7783	28.3 7788	34.6 7795	42.1 7800	56.4 7812	101 7840	214 7895	281 7922
250	15.3 7763	21.4 7768	27.6 7774	35.6 7780	43.9 7786	52.6 7791	71.7 7803	131 7831	279 7886	367 7913
260	17.9 7753	25.7 7759	33.6 7765	43.6 7771	53.1 7777	63.9 7782	88.4 7794	164 7822	354 7877	467 7904
280		34.3 7741	47.9 7747	59.6 7753	74.7 7759	90.0 7765	127 7777	246 7805	541 7860	718 7886
300			62.2 7729	80.6 7735	101 7741	124 7748	175 7760	350 7787	789 7842	1050 7869
320				102 7718	132 7724	166 7730	238 7742	481 7769	1110 7824	1490 7851
340					162 7707	207 7713	302 7724	621 7752	1520 7807	2060 7834
360						248 7695	366 7707	832 7735	2030 7790	2760 7817
400							494 7673	1250 7702	3390 7756	4760 7782
500								3070 7617	9350 7671	14200 7698

Таблица 2 — *продолжение*

$h_{\pi\,(км)}$ \ $h_{a\,(км)}$	900	1000	1100	1200	1300	1400	1500	1600
160	13.8 8019	17.2 8045	21.0 8071	25.3 8096	30.1 8120	35.4 8144	41.2 8168	47.4 8191
170	24.8 8010	29.3 8036	33.8 8062	38.2 8087	42.5 8111	46.7 8135	50.7 8159	54.5 8182
180	38.7 8001	46.9 8027	55.4 8053	64.2 8078	73.2 8102	82.3 8126	91.6 8150	101 8173
190	62.1 7992	73.2 8018	84.3 8044	95.1 8069	105 8093	118 8117	133 8141	147 8164
200	91.9 7983	111 8009	132 8035	152 8060	173 8084	195 8108	216 8132	237 8155
210	134 7975	158 8000	182 8026	210 8051	241 8075	274 8099	308 8123	342 8146
220	191 7966	231 7991	273 8017	316 8042	358 8066	401 8090	443 8114	485 8137
230	259 7957	306 7982	364 8008	426 8033	490 8057	556 8081	623 8105	690 8128
240	353 7948	428 7974	504 7999	580 8024	657 8048	733 8072	808 8096	894 8119
250	452 7939	549 7965	656 7990	765 8015	877 8039	990 8064	1100 8087	1220 8110
260	587 7930	712 7956	840 7981	968 8006	1100 8031	1250 8055	1400 8078	1560 8102
280	907 7912	1100 7938	1300 7964	1510 7988	1740 8013	1980 8037	2220 8061	2470 8084
300	1340 7895	1630 7921	1930 7946	2270 7971	2620 7995	2980 8019	3350 8043	3710 8066
320	1900 7877	2320 7903	2400 7936	3290 7953	3420 7983	4330 8002	4850 8025	5370 8049
340	2620 7860	3240 7886	3930 7911	4630 7936	5350 7960	6080 7984	6800 8008	7520 8031
360	3540 7843	4420 7868	5370 7894	6340 7919	7320 7943	8300 7967	9380 7991	10500 8014
400	6270 7808	7850 7834	9440 7859	11100 7884	13000 7909	15000 7932	17000 7956	19000 7979
500	20200 7724	26900 7750	34400 7775	42200 7800	50300 7824	58500 7848	66600 7871	74600 7895

спутника. Высоты апогея и перигея монотонно убывают, причем для всех эллиптических орбит скорость убывания высоты апогея больше скорости убывания высоты перигея. Для сильно вытянутых орбит это различие может быть весьма существенным. Так, для орбиты с высотой перигея 300 км и высотой апогея 700 км понижение апогея на 100 км отвечает понижению перигея примерно на 6 км. Для больших значений высоты апогея эта разница будет еще более существенной. Таким образом, при большой разнице высот апогея и перигея все изменение параметров орбиты будет в течение длительного времени сводиться практически только к уменьшению высоты апогея при почти постоянной высоте перигея.

При таком изменении формы орбиты эксцентриситет орбиты будет все время убывать и стремиться к нулю. Орбита спутника будет стремиться к круговой. Из рис. 2 можно видеть, что кривые первого семейства стремятся как к ассимптоте к прямой $h_a = h_\pi$, отвечающей круговой орбите. По мере падения эксцентриситета разница между скоростями убывания апогея и перигея уменьшается.

Результаты расчетов позволяют оценить изменение времени движения спутника при изменении начальных параметров h_{ao} и $h_{\pi o}$, а также указать значения начальных параметров, позволяющих обеспечить заданное время движения наиболее простым путем.

Характер хода кривых второго семейства показывает, что время существования спутника, как и следовало ожидать, сильнее возрастает с увеличением начальной высоты перигея и слабее с увеличением начальной высоты апогея. Так, например, для орбиты с перигеем 360 км и апогеем 1500 км изменение высоты перигея на 20 км вызывает изменение времени жизни примерно на 40%, а такое же изменение апогея — примерно на 2%, т.е. в 20 раз меньше.

Ввиду того, что выведение спутника на орбиту с большой начальной высотой перигея может встретить ряд трудностей, отметим, что значительного увеличения продолжительности сиществования спутника можно достигнуть и при неизменной высоте перигея путем увеличения начальной высоты апогея, причем для этого требуется сравнительно небольшое увеличение скорости в перигее. Так, например, для орбиты с параметрами $h_\pi = 360$ км и $h_a = 700$ км, увеличение высоты апогея до 1000 км приводит к увеличению времени существования в 2,2 раза, при этом требуется увеличение скорости в перигее всего на 78 м/сек. Приведенный результат указывает на целесообразность использования вытянутых орбит, что может позволить добиться значительного увеличения продолжительности существования искусственного спутника Земли сравнительно простым путем.

Как уже указывалось, приведенные значения ν получены при определенной схеме распределения плотности воздуха по высоте. При отклонении фактических значений от принятых продолжительность существования спутника окажется иной. Уточнение данных о плотности верхних слоев атмосферы даст возможность провести по той же методике уточненный расчет значений ν для выдачи более точных прогнозов о времени существования спутника.

Так как скорость изменения параметров орбиты пропорциональна плотности атмосферы, которая быстро убывает с высотой, то изменение параметров орбиты по времени будет вначале значительно более медленным, чем впоследствии, при снижении в более плотные слои. Поэтому основное время спутника будет существовать в высоких слоях атмосферы.

Быстрое убывание плотности с высотой и медленное изменение высоты перигея указывают на то, что основное значение для времени существования спутника будет иметь величина плотности воздуха в области первоначального перигея. Этот вывод дает возможность оценить величину изменения расчетного времени существования при изменении данных о плотности атмосферы. Продолжительность существования спутника будет примерно обратно пропорциональна плотности воздуха в области первоначального перигея.

Это обстоятельство указывает на возможность оценки фактической плотности воздуха в области начального перигея по начальным значениям апогея и перигея и по фактическому времени существования спутника на орбите. Обработка результатов по ряду пусков с различными начальными высотами перигея даст возможность оценки фактического распределения плотности воздуха по высотам.

Приведем в заключение несколько примеров. Рассмотрим спутник шарообразной формы с диаметром $d=0{,}5$ м и весом $G=10$ кг. Значение коэффициента сопротивления примем равным $C_x=2$. В этом случае значение множителя, на который нужно умножить величину ν для получения времени существования спутника в оборотах, будет равно (в кг сек^{2}м$^{-3}$)

$$\frac{4G}{\pi d^2}\cdot\frac{1}{gC_x}\cong 2{,}6\,.$$

Оценку времени существования в сутках получим делением полученного значения числа оборотов на 16, что соответствует количеству оборотов в сутки для полуторачасовой орбиты:

$$n\cong\frac{N}{16}\,.$$

Примем высоту перигея и апогея равными: $h_\pi=360$ км и $h_a=800$ км, тогда из таблицы 2 находим $\nu\cong 2760$, число оборотов N будет равно 7200, а время примерно $n\cong 450$ суток, что соответствует примерно 1 году и 3 месяцам.

Для высот перигея и апогея, равных $h_\pi=500$ км и $h_a=1500$ км, имеем аналогично $\nu=66600$, $N\cong 174000$ оборотов и $n\cong 11000$ суток, что соответствует времени существования порядка 30 лет.

Таким образом, при выборе достаточно больших начальных высот перигея и апогея время существования искуственного спутника Земли может оказаться весьма значительным.

Взяв высоты перигея и апогея равными $h_\pi=200$ км и $h_a=400$ км, получим $v=16{,}8$, $N\cong 44$ и $n=2{,}7$ суток, т.е. при сравнительно малых значениях высот перигея и апогея продолжительность существования спутника на орбите оказывается небольшой.

§ 5. О вековых возмущениях параметров орбиты искусственного спутника

Проведенное исследование вопроса о времени жизни искусственного спутника было основано на изучении вековых возмущений элементов орбиты за счет сопротивления атмосферы. В настоящем параграфе будет дана оценка вековых возмущений элементов орбиты за счет влияния других возмущающих факторов. Периодические возмущения элементов орбиты не рассматриваются. Ясно, что периодические возмущения не могут оказать сколько-нибудь существенного влияния на продолжительность существования искусственного спутника.

Будем исходить из уравнений движения в параметрах оскулирующего эллипса (2.8). При учете влияния отклонения поля тяготения от центрального будем исходить из выражения для потенциала [4]

$$V = \frac{fM}{r} - \frac{\varepsilon}{3r^3}(3\sin^2\psi - 1), \tag{5.1}$$

где

$$\varepsilon = fMa^2\left(\alpha - \frac{m}{2}\right), \qquad m = \frac{\Omega_1^2 a}{g_o}, \tag{5.2}$$

ψ — широта, a — экваториальный радиус Земли, α — сжатие Земли, Ω_1 — угловая скорость суточного вращения Земли, g_o — ускорение силы земного тяготения на экваторе. Формула (4.1) дает следующие выражения для проекций возмущающего ускорения:

$$\begin{aligned} S &= \frac{\varepsilon}{r^4}[3\sin^2 i\sin^2 u - 1], \\ T &= -\frac{\varepsilon}{r^4}\sin 2i\sin 2u, \\ W &= -\frac{\varepsilon}{r^4}\sin 2i\sin u. \end{aligned} \tag{5.3}$$

При оценке дополнительных ускорений от влияния ветра за счет вращения атмосферы вместе с Землей будем исходить из приближенных формул, полученных из более точных отбрасыванием членов, содержащих параметр $\frac{\Omega_1 r}{v}$ в степени выше первой [5]:

$$\begin{aligned} S &= \frac{C_x F}{G}\frac{g}{2}\frac{v_n}{v}v_r\varrho\Omega_1 r\cos u, \\ T &= \frac{C_x F}{G}\frac{g}{2}\frac{v^2 + v_n^2}{v}\varrho\Omega_1 r\cos i, \\ W &= -\frac{C_x F}{G}\frac{g}{2}v\varrho\Omega_1 r\sin i\cos u. \end{aligned} \tag{5.4}$$

Для оценки вековых возмущений будем интегрировать правые части уравнений (2.8) по u от $u=0$ до $u=2\pi$, считая параметры эллипса постоянными и пренебрегая периодическими возмущениями параметров. Заметим также, что при интегрировании (2.8) можно принять, не допуская большой погрешности, что $\gamma=1$.

Используя такой метод, можно получить, что сжатие Земли и связанное с ним отклонение потенциала вызывает вековые возмущения долготы восходящего узла и углового расстояния перигея от узла ω. Другие характеристики орбиты, в том числе параметр и эксцентриситет, не испытывают под влиянием сжатия вековых возмущений. Это означает, что пренебрежение сжатием при расчете времени существования спутника является вполне законным. Уход узла показывает, что под влиянием сжатия плоскость орбиты вращается вокруг оси вращения Земли, сохраняя с этой осью постоянный угол.

Для расчета величины ухода восходящего узла и величины ухода расстояния перигея от восходящего узла за один оборот получим формулы:

$$\frac{d\Omega}{dN} = -\frac{2\pi\varepsilon}{p^2 fM}\cos i, \tag{5.5}$$

$$\frac{d\omega}{dN} = \frac{\pi\varepsilon}{p^2 fM}(5\cos^2 i - 1). \tag{5.6}$$

Рассмотрим для примера орбиту со средней высотой порядка 500 км. Получим для ухода узла и перигея:

$$\frac{d\Omega}{dN} = -0{,}54^\circ \cos i\,, \qquad \frac{d\omega}{dN} = 0{,}27^\circ \cdot (5\cos^2 i - 1)\,.$$

При наклонении орбиты 45° получим за оборот $\Delta\Omega \cong -0{,}38^\circ$, $\Delta\omega \cong 0{,}4^\circ$ и за сутки $\Delta\Omega \cong -6{,}1^\circ$, $\Delta\omega \cong 6{,}5^\circ$.

Формула (5.5) показывает, что быстрота ухода восходящего узла существенно зависит от наклонения и будет наибольшей для орбит, близких к экваториальным, и равной нулю для орбиты, проходящей через полюса.

Рассмотрим теперь влияние на вековые возмущения параметров орбиты спутника вращения атмосферы вместе с Землей в ее суточном движении. Используя формулы (5.4), подставляя ϱ из формулы (1.1) и интегрируя по u от 0 до 2π получим следующие оценочные формулы:

$$\left|\frac{dp}{dN}\right| < \frac{2cp^3\sqrt{p}}{\sqrt{fM}}\,\varrho_1\,\Omega_1\,I\cos i \tag{5.7}$$

$$\left|\frac{de}{dN}\right| > \frac{cp^2\sqrt{p}}{2\sqrt{fM}}\,\varrho_1\,\Omega_1\,I\cos i \tag{5.8}$$

$$\left|\frac{d\Omega}{dN}\right| < \frac{cp^2\sqrt{p}}{2\sqrt{fM}}\,\varrho_1\,\Omega_1\,I\sin 2\omega \tag{5.9}$$

$$\left|\frac{di}{dN}\right| < \frac{cp^2\sqrt{p}}{2\sqrt{fM}}\,\varrho_1\,\Omega_1\,I\sin i \tag{5.10}$$

$$\left|\frac{d\omega}{dN}\right| < \frac{cp^2\sqrt{p}}{2\sqrt{fM}}\,\varrho_1\,\Omega_1\,I\cos i\sin 2\varepsilon\,, \tag{5.11}$$

где

$$c = \frac{C_x F}{G}\,g\,. \tag{5.12}$$

Как показывает расчет, возмущения параметра и эксцентриситета за счет вращения атмосферы не превышают 10—12% соответствующих возмущений для неподвижной атмосферы. Это и понятно, так как изменение скоростного напора при переходе от абсолютного движения к относительному не должно превосходить этой величины. Наибольшее влияние будет для экваториальной орбиты, причем при движении спутника на восток сила сопротивления будет убывать, а при движении на запад возрастать по сравнению с сопротивлением при неподвижной Земле. Для орбит, проходящих вблизи полюсов, влияние вращения Земли на параметр и эксцентриситет несущественно.

Полученный результат означает, что при движении спутника на восток время существования спутника при той же величине начальных значений апогея и перигея будет больше, чем для спутника, запущенного в направлении на запад. Отличие во времени существования от случая пренебрежения вращением атмосферы, наибольшее для орбит, близких к экваториальным, не будет превосходить 10—12%.

Для других орбит отличие будет меньше. Для полярной орбиты отличие будет равно нулю. В настоящее время при весьма малой точности знания плотности верхних слоев атмосферы указанные выше величины погрешностей в определении времени жизни спутника следует признать пренебрежимо малыми.

Величины вековых возмущений долготы узла, наклонения и расстояния перигея от узла за счет вращения атмосферы оказываются весьма малыми. Для оценки этих возмущений удобно произвести оценку величины полного ухода указанных параметров за все время существования спутника. Такая оценка оказывается возможной в силу того, что как указанные возмущения, так и вековые возмущения орбиты, непосредственно влияющие на время жизни спутника, пропорциональны одним и тем же величинам — плотности атмосферы, обратной величине поперечной нагрузки и коэффициенту аэродинамического сопротивления. В силу сказанного, полученная оценка будет верна для любых спутников с различными конструктивными параметрами и различными начальными значениями параметров орбиты.

На основании расчетов получим следующие оценки.

$$\begin{aligned} |\Delta\Omega| &< 0{,}2^\circ \\ |\Delta i| &< 0{,}1^\circ \\ |\Delta\omega| &< 0{,}2^\circ . \end{aligned} \tag{5.13}$$

Приведенные результаты показывают, что отклонения получаются весьма малыми. Влияние вращения атмосферы на вековые уходы долготы узла, наклонения и углового расстояния от узла оказывается мало существенным.

Отметим, что при выполнении вычислений были использованы значения интеграла I, полученные из анализа результатов расчетов на машине.

Приведенные выше результаты показывают, что использованная в настоящей работе методика расчета времени жизни спутника является достаточно обоснованной, что подтвердилось также далее расчетами, проведенными при использовании точных уравнений движения в оскулирующих элементах [5]. Это означает, что точность изложенной методики является вполне достаточной для получения надежных прогнозов о времени существования искусственных спутников Земли.

Литература

1. С. К. Митра, Верхняя атмосфера. Москва: ИЛ, 1955 г.
2. Г. Н. Дубошин, Введение в небесную механику. Москва-Ленинград: ОНТИ, 1938 г.
3. М. Ф. Субботин, Курс небесной механики. Москва-Ленинград: ОНТИ, 1937 г.
4. Н. И. Идельсон, Теория потенциала с приложениями к теории фигуры Земли и геофизике. Москва: ОНТИ, 1936 г.
5. Г. П. Таратынова, Успехи физических наук. Гостехиздат **63**, вып. 1 (1957 г).

Determining the Time of Existence of the Artificial Earth Satellite and Studying Secular Perturbations of its Orbit

By

D. E. Okhotsimsky, T. M. Eneiev and **G. P. Taratynova**

A sufficiently reliable determination of the time of existence of the artificial Earth satellite on the orbit is one of the major points associated with the problem of producing the satellite. Due to the resistance of the atmosphere, the satellite's energy will be dissipated and it will gradually descend. When the motion takes place at great altitudes in rarefied atmosphere, the resistance is slight, and the time of satellite motion may prove to be quite considerable. When the satellite moves at comparatively small altitudes of the order of 100 to 150 kilometres, it is shortlived and having low transverse loads may be incapable of making even one complete revolution.

Lately there have appeared quite a number of papers devoted to the problem of determining the time of existence of the artificial satellite, an adequately rigorous solution being provided only for a circular orbit. To appraise the time of satellite motion along an elliptic orbit, use has been made of various approximative methods which take into consideration the energy factor and which are based on the supposition that energy loss occurs chiefly in the perigee region, when the satellite approaches the Earth most closely. The application of these methods does not provide a full solution to the problem in a general case. Furthermore as proved by analysis, the employment of approximative methods for determining the time of existence may lead to grave errors in a number of cases.

This situation has made it imperative that methods should be developed which would permit to determine rapidly and reliably enough the time of existence of the satellite in a general case of its motion. The study has revealed the existence of universal relationships between the basic parameters of an osculating ellipse such as the height of the perigee or apogee, or the parameter and eccentricity. These relationships hold good for any satellite and depend only on the law of air distribution with altitude. This fact has made it possible to reduce the full solution of the problem of the lifetime of artificial satellites to the plotting of a series of integral curves of a first order equation, depending on a single parameter. The chart and the tables so obtained allow to determine rapidly the time of existence by multiplying the result taken from the table of the chart by some number which is a simple function of the basic satellite parameters. Not only do the above results permit to calculate the time of existence; they also make it possible to establich the law of time changes in the orbit parameters for any given satellite parameters and for a sufficiently wide range of initial orbit parameters.

Integration of the equations was made on the fast computer (БЭСМ) of the USSR Academy of Sciences. The method of integration used permitted to minimize the amount of computation work while preserving the required accuracy of results. The method is general enough and can apparently be used successfully in many cases involving the integration of equations whose solution is a function approximating a periodic one, with slowly changing parameters.

The paper cites the original data used for the computations. The equations of satellite motion in osculating elements are reduced to a new independent variable convenient for the investigation in hand. This is followed by an exposition of the methods of integrating differential equations for secular perturbations of the parameter and the eccentricity of the orbit due to the effect of air resistance; then the results so obtained are discussed. The paper also provides a substantiation of the basic simplifying assumptions used in the computation, and gives an appraisal of the methodical error resulting from these assumptions. An approximative study is also made of the secular perturbations of the orbit due to Earth compression and the rotation of the atmosphere.

It should be noted that the reported results of the numerical computations of the time of satellite existence are based on the use of certain assumptions regarding the structure of the upper atmosphere. The lack of reliable data on the parameters of the upper atmosphere makes the numerical results suitable only for tentative appraisals. It should, however, be pointed out that analysis of artificial satellites motion will substantially add to the data on the upper atmosphere and will permit to use the above methods for calculations required for further refinement.

Producing the Weightless State in Jet Aircraft[1]

By

S. J. Gerathewohl[2], O. L. Ritter[3], and H. D. Stallings, Jr.[4,5]

(With 6 Figures)

Abstract — Zusammenfassung — Résumé

Producing the Weightless State in Jet Aircraft. Simple functions were used for computing the most interesting and important characteristics of the parabolic flights. The results, based upon certain flying characteristics of the T-33, F-94C, and F-104 in subsonic and transsonic flight, are in good agreement with the data obtained for the first two aircraft types during actual zero-gravity maneuvers. Certain flying safety hazards were noticed in the T-33 but remedied through appropriate measures. The F-94C Starfire proved to be superior to the T-33 with regard to safety and duration of the weightless state. If the F-104 would be made available for aeromedical research, weightlessness could be produced for more than one minute.

Herstellung des gewichtslosen Zustandes in Flugzeugen mit Strahlantrieb. Fur die Berechnung der interessantesten und wichtigsten Charakteristika parabolischer Flüge wurden einfache Funktionen benützt. Die Ergebnisse stützten sich auf bestimmte Flugcharakteristika der Flugzeugtypen T-33, F-94C und F-104 im Unterschall- und Überschallflug. Sie stimmen gut mit den Daten überein, die für die beiden erstgenannten Flugzeugtypen während der tatsächlichen Nullschwere-Manöver erhalten wurden. Gewisse Gefahren für die Flugsicherheit wurden in der T-33-Maschine bemerkt, jedoch durch geeignete Maßnahmen behoben Die F-94C-Starfire-Maschine erwies sich hinsichtlich Sicherheit und Dauer des gewichtslosen Zustandes als der T-33 überlegen. Falls die F-104 für die aeromedizinische Forschung zur Verfügung stände, könnte Gewichtslosigkeit für eine Zeitdauer von mehr als 1 Minute erzeugt werden.

Annulation du champ de la pesanteur dans les avions à réaction. Des fonctions simples ont été utilisées pour calculer les paramètres les plus importants des trajectoires paraboliques. Les résultats, basés sur certaines caractéristiques de vol des avions T-33, F-94C et F-104 en régime subsonique et transsonique, sont en bon accord avec les données expérimentales obtenues pour les deux premiers types d'avions durant des manoeuvres destinées à annuler le champ de la pesanteur. Certains dangers propres au T-33 ont été évités par l'emploi de remèdes appropriés. Le F-94C Starfire s'est révélé supérieur au T-33 du point de vue sécurité et durée de la période de vol sans pesanteur. Si le F-104 était disponible pour les recherches aéromédicales, l'absence de pesanteur pourrait être réalisée pendant plus d'une minute.

---

[1] Published in Astronaut. Acta 4, 15—24 (1958).

[2] Principal Investigator.

[3] Department of Radiobiology.

[4] Major USAF, Chief, Flight Operations and Pilot of T-33A and F-94C aircraft.

[5] All authors: School of Aviation Medicine USAF, Randolph Air Force Base, Texas, U.S.A.

In studies of human behavior under conditions of virtual weightlessness, the reduction or elimination of gravity has become an important problem of aeromedical research [1, 2, 3]. The following means were suggested or employed for producing the weightless state: Dropping a capsule from an airplane, the elevator, the "gravitron", the "sub-gravity tower", and the aircraft [4, 5, 8]. From the most practical point of view of those mentioned, only the aircraft seems to be promising for use because of its availability, safety, realism, and the longer periods and various amounts of reduced gravity that can be produced in it.

In order to appreciate the use of modern aircraft for accomplishing the weightless or "agravic" state, some of the mechanical principles of gravity and acceleration are defined below. As expressed in NEWTON's Universal Law of Gravitation, all bodies exert mutual forces of attraction upon one another. In accelerated motion, gravity is the vectorial sum of the forces of gravitation and inertia acting on a body. According to the NEWTONian relation:

$$F = m \cdot a \tag{1}$$

the force of gravity is measured most conveniently in terms of acceleration; with the acceleration due to the terrestrial gravitation 32.17 ft/sec^2 as the unit. Normally, with an object being in a state of rest, the forces of inertia are absent and the object finds itself in the "normal state of gravity" of $G = 1$.

If a body is allowed to fall freely, it is subjected to a downward acceleration of 1 g. In this case, the forces of inertia compensate the forces of gravity and the body finds itself in the state of "zero-gravity" [6]. If the body is subjected to a downward acceleration smaller than 1 g, the forces of inertia are subtracted from the gravitational force. In this case, the body is in a sub-gravity state. In both cases, a second force is superimposed upon the force of gravitation [7].

Within the gravisphere of the earth a body derives its weight from the counterforce that resists its free fall. On the other hand, a body can be defined as weightless if it is allowed to move freely under the influence of gravitation and its inertia only[1]. This is the case when the object is moving along a KEPLERian trajectory. According to the D'ALEMBERT Principle, every body — whether accelerated or not — finds itself in a state of equilibrium under the combined effects of the actual forces exerted on it, i.e., forces of gravity, forces of inertia, and external forces like propulsion, lift, drag, and support. Eq. (1) thus may be written:

$$F - m \cdot a = 0. \tag{2}$$

Now, since F represents the gravitational forces and $m \cdot a$ the "reactive effects", we write:

$$F_g - F_r = 0 \tag{3}$$

where F_r represents the reactive forces, i.e., the inertial effect associated with the acceleration of gravity. In the free fall situation in a vacuum, F_r is a quantity equal in magnitude and opposite in direction of F_g, and can be identified as the sole effect of inertia. When an airplane flies along a KEPLERian trajectory, F_r represents the inertial and external forces superimposed on F_g. In other words, the aircraft then simulates the motion of a missile cruising in a vacuum. In order to obtain lack of appression in flight through air, the craft must be guided along a parabolic arc, thereby utilizing the inertial forces for counterbalancing F_g.

---

[1] Although the terms "zero-gravity" and "weightlessness" designate the same condition, "zero-gravity" is used here to refer to the physical state, and "weightlessness" to the psychophysiological condition of the individual.

A flight maneuver of this type requires that all accelerations, except the one caused by gravitation which constantly acts downward at a magnitude of 1 g, be eliminated completely. This can be accomplished by the pilot by flying the plane through a so-called pushover, holding the needle of his accelerometer precisely between plus and minus 1 g at the zero-mark of the instrument. A push-over is a vertical-planar maneuver in which the angle of climb changes continuously from a plus to a minus value. During the maneuver the air speed decreases uniformly from an initial value v_0 to a minimum at the top of the curve, and then increases uniformly back to the initial value shortly before the pull-out. However, the horizontal component of the velocity remains constant during the entire maneuver. It is determined by the minimum speed at which the aircraft is fully controllable and stable during the push-over. Hence, the horizontal component of velocity or "minimum controlling speed" of the airplane is always somewhat higher than the stalling speed. It depends mainly upon the amount and arrangement of control-surface area built into the aircraft.

Throughout the maneuver, the plane moves at considerable air speed. Therefore, a power output of the engines ranging from an appreciable fraction to full power must be maintained to overcome drag although no lift is required. This holds for the downward as well as the upward leg of the maneuver.

The characteristics of the trajectory described above reveal that it is a parabola with vertical axis. Strictly speaking, the intended trajectory is one described by an unpropelled body in ideally frictionless space subjected to a centrally symmetric gravitational field. Generally, such a trajectory is a conic, one focal point of which always coincides with the center of attraction around which the body revolves. For sufficiently small velocities, such as are achievable by present day aircraft, the conic is a very elongated ellipse with one focal point at the center of the earth. The small section near the apex of the ellipse, emerging from the surface of the earth, can well be represented by a parabola. The condition for good approximation is that the dimensions of the section are small compared to the radius of the earth, or alternately, that the part of earth surface arched over by the section can be considered as flat instead of spherical.

It should be emphasized that the flight path is not necessarily parabolic or elliptic if a state of reduced rather than zero gravity is intended. In this case, the characteristics of the trajectory are determined by additional requirements concerning the direction of the resulting subgravity force with respect to the aircraft. If, for instance, the force is to be directed perpendicularly toward normally positioned seats regardless of the instantaneous orientation of the plane, a rather complicated functional form of the flight profile is obtained which resembles a parabola only superficially [4]. On the other hand, if the subgravity is to be directed vertically toward the surface of the earth irrespective of the orientation of the plane, the resulting trajectory is always parabolic.

For the mathematical analysis of a parabolic flight pattern, an orthogonal coordinate system can be employed having a horizontal x-axis and a vertical y-axis. The angle of climb is determined by the direction of the aircraft at the beginning of the appressionless state, i.e., at the point 0 in Fig. 1. The interesting and important information to be obtained from an analysis of the flight parabola concerns the duration of the weightless state, and its dependency upon other flight parameters.

In ballistics, the duration of the trajectory is defined as the time required for the projectile to again reach the level of its initial projection. In the flight pattern described in this paper, T is the duration of the weightless state. From Fig. 1 it can be seen that the vertical location of any point of the parabola can be expressed by:

$$y = v_0 t \sin \varepsilon - \frac{g}{2} t^2; \quad (4)$$

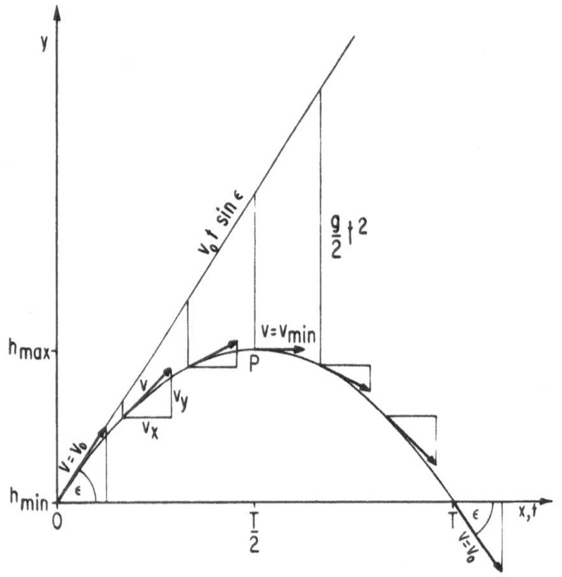

Fig. 1. Schematic of parabolic arc.
In the absence of gravity, a craft starting from point 0 at an angle ε would follow the straight line $v_0 t \sin \varepsilon$. Gravity causes the craft to fall. The vertical line segments between the straight line and the parabola represent the height of fall at the corresponding points of the trajectory. Also shown are directions and relative magnitudes of the velocity and its horizontal and vertical components. The velocity reaches its minimum at the peak P of the parabola where the vertical component vanishes. The magnitudes of the speed at begin and end of the parabolic arc are equal. The same holds for the angles ε

where v_0 = the initial velocity of the aircraft, and ε = the initial angle of climb. In order to find the duration of the flight, we determine the value t for which y again becomes 0:

$$T = \frac{2}{g} v_0 \sin \varepsilon. \quad (5)$$

Eq. (5) shows that T depends upon the value of g, the velocity v_0, and the angle of climb ε, the latter two quantities measured at the point 0. At a given v_0 the longest duration is obtained when $\varepsilon = 90^0$, i.e., when the projectile is propelled straight up. However, the airplane cannot rotate about 180^0 at zero speed on top of a straight line ascent, nor can it maneuver safely below or at its stalling speed. For reasons of good maneuverability we have to stay within the limitations of our flight pattern, i.e., the controlability on the one hand, and the maximum permissible velocity shortly before pull-out, on the other. The controlability speed is always somewhat higher than the stalling speed of the aircraft, and can be determined by flight tests. In a similar manner, the maximum permissible speed is not utilized in parabolic maneuvers but v_0 instead, which is the airspeed (IAS) available by the excess thrust for propelling the aircraft upward from the preceding pullout. If an object is hurled upward with a certain initial velocity v_0 it will, neglecting air friction, rise to its maximum height in the same time it takes to fall from that height to the ground. The same principle is true for the parabolic maneuver in which case, then, the speed at the end of the trajectory also equals v_0. Hence, our problem boils down to finding the optimal value of ε with respect to the initial velocity v_0 and the minimum maneuvering or controlability speed v_{min} of the airplane.

The optimal angle of climb can be found by considering the velocity diagram at point P in Fig. 1. Since the horizontal component of the velocity must be constant through the entire maneuver, v_x is of the same magnitude as v_{min}. Thus we write:

$$v_x = v_{min} = v_0 \cos \varepsilon, \tag{6}$$

$$\cos \varepsilon = \frac{v_{min}}{v_0}. \tag{7}$$

Eq. (7) shows that the optimal angle of climb depends upon the ratio of excess thrust and minimum controlability speed.

Finally, we want to know what relationship exists between the duration of the weightless state and the peak altitude of the maneuver. The maximum height h_{max} is defined as the greatest vertical distance reached by the aircraft as measured from the ground. In Fig. 1 the maximum height is that of the peak P of the parabola:

$$h_{max} = h_{min} + \frac{v_0^2}{2g} \sin^2 \varepsilon. \tag{8}$$

Another simple representation results if the peak of the parabola is chosen as origin of the coordinate system. In this system, the altitude at any time is

$$y = h_{max} - \frac{g}{2} t^2, \tag{9}$$

the downward vertical component of the velocity is

$$v_y = g\, t. \tag{10}$$

From the total velocity

$$v = \sqrt{v_x^2 + g^2 t^2} \tag{11}$$

the duration of one leg of the parabola is found

$$\frac{T}{2} = \frac{1}{g} \sqrt{v_0^2 - v_x^2} \tag{12}$$

where

$$v_{0y} = \sqrt{v_0^2 - v_x^2} \tag{13}$$

is the vertical component of the velocity at the endpoints of the parabola. In terms of maximal and minimal altitudes, this vertical component and the total duration can be represented by

$$v_{0y} = \sqrt{2g\,(h_{max} - h_{min})} \tag{14}$$

$$T = \sqrt{\frac{8}{g}\,(h_{max} - h_{min})}. \tag{15}$$

Finally, the optimal angle of climb is again given by formula (7) of the last section.

The flight path characteristics for producing weightlessness of maximum duration using three different types of airplanes available today in the United States are shown in Table I. The data concerning minimum controlability speed, entry and pull-out speed, and operational altitude, which served for the computation of the optimal angle of climb, peak altitude, and duration of the weightless state, were obtained through parabolic flights involving states of decreased or entirely abolished appression. Table I shows that longer periods of weightlessness can be produced when higher performance aircraft are employed.

Table I. *Characteristics of optimal flight parabola for three different aircraft*

Aircraft	Minimum controlability speed	Entry speed v_0	Starting altitude	Maximum height over ground	Angle of climb ε	Duration of virtual weightlessness
T-33A	180 knots	320 knots	18,000 feet	20,600 feet	55^0	28 seconds
F-94C	195 knots	425 knots	18,000 feet	24,400 feet	$63\frac{1}{2}^0$	40 seconds
F-104B	200 knots	800 knots	40,000 feet	66,800 feet	$75\frac{1}{2}^0$	82 seconds

In Fig. 2, 3, and 4, the total duration T of the maneuver, the height $h_{max} - h_{min}$ of the parabolic arc, and the optimal angle of climb ε_{opt} are given in the range of maximal speeds between 400 and 2000 knots for three values of the minimal speed.

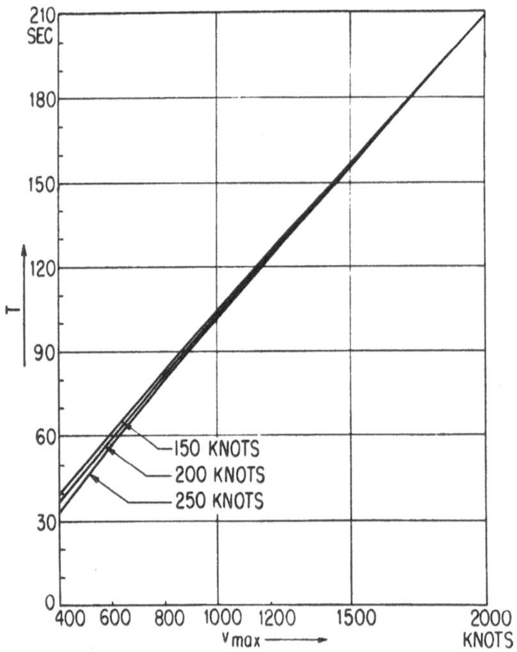

Fig. 2. Duration T of parabolic maneuver in dependence on maximal and minimal speed. The duration is approximately proportional to the maximal speed the craft can achieve. The minimal controllable speed has little influence unless it amounts to an appreciable fraction of the maximal speed

Prominent test and research pilots such as Major (CHUCK) YEAGER, Major ARTHUR MURRAY, SCOTT CROSSFIELD, BILL BRIDGEMAN, Capt. IVEN KINCHELOE, just to mention a few, have on several occasions experienced weightlessness; but very little has been published concerning the practical realization of this state. However, there is a need for the dissemination of information already available, because more and more research organizations become involved in the problem of fact-finding. Now, after we have had the opportunity to try it out ourselves, and after having accomplished more than 200 parabolic flights in the T-33 and F-94 type aircraft encompassing a total of about 2,000 individual parabolas, we feel encouraged to present our practical experiences to the interested organizations and individuals. The presentation will also show how the practical results agree with the figures given in Table I.

Since jet fighters are in great demand by many research units of the U.S. Air Force, preliminary investigation was not begun until 1955 when the School of Aviation Medicine, USAF, had assigned a Lockheed T-33A type jet aircraft powered by a J-33A-35 engine developing 4,600 lb. thrust. Inasmuch as we had no information to work with it was agreed that the exact flight patterns and working altitudes would have to be determined by a "hit and miss" method. Numerous flights were made using different altitudes, entry airspeeds, angles of attack, and power settings. From the data gathered, the most acceptable flight pattern was selected: Although attempts were made at altitudes that varied from 10,000 to 30,000 feet, the best starting altitude was found to be about 20,000 feet. Therefore, this altitude was considered the optimum working altitude because of usual lack of clouds and turbulance at this height, and because this altitude provided the requirements of safety in the event of emergency. At this altitude the parabola can be flown utilizing the maximum performance of this particular aircraft with the maximum amount of safety; i.e., we could at this level effect corrective action in case of engine trouble that would enable us either to return to the base or eject successfully.

At 20,000 feet, a sharp dive was started at 96 percent engine r.p.m. As the indicated air speed (IAS) built up to 350 knots, an altitude of 17,500 feet was usually reached. Now, a sharp pull-out was begun resulting in a "positive" $3\,G$

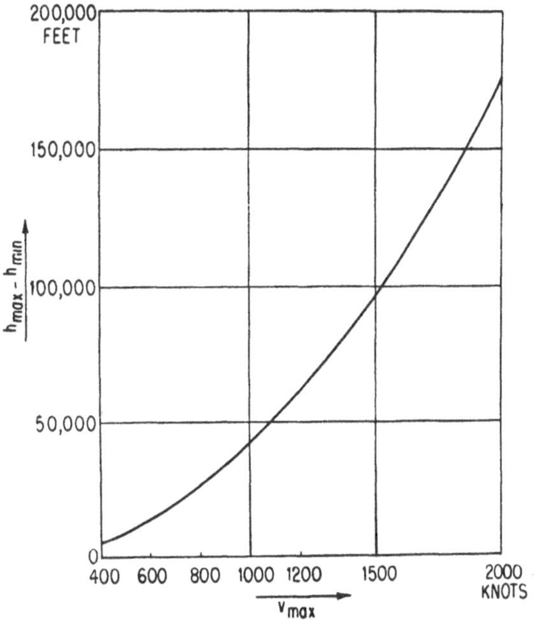

Fig. 3. Height of the parabolic arc vs. maximal speed, for a minimal speed of 200 knots. The height increases approximately with the square of the maximal speed. For minimal speeds of 150 and 250 knots, the curve is shifted up or down, respectively, by approximately 1000 feet

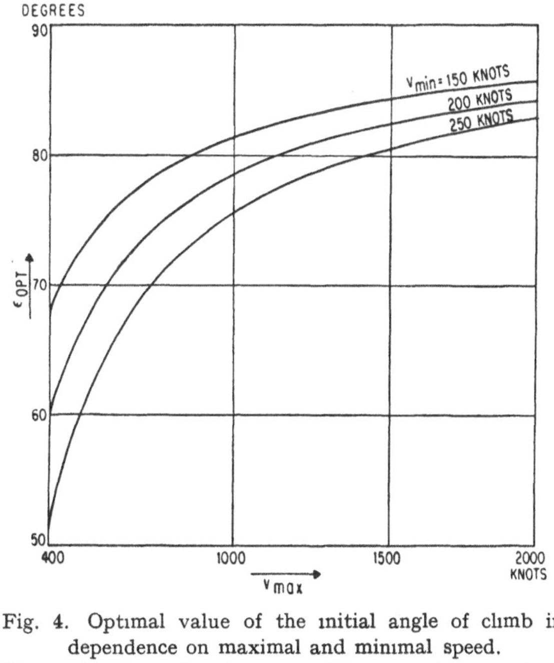

Fig. 4. Optimal value of the initial angle of climb in dependence on maximal and minimal speed. The optimal angle steepens with increasing maximal speed and approaches 90° for very large speeds

condition, and the aircraft was put into an angle of climb of about 60 degrees from the horizontal at full throttle. As the IAS dropped off to 300 knots, forward pressure was applied to the control stick and thus the push-over initiated. At this point of the maneuver, a slight and momentary yawing of the aircraft occurred but was easily controlled by aileron movements. Slight changes in forward pressure were also required to control the attitude of the plane and to maintain the zero-gravity state.

Air speed at the top of the parabola was about 180 knots, and altitude ranged from 20,000 to 20,500 feet. As the apex was reached, forward stick pressure was continued until the plane was diving at an IAS of 350 knots. When the angle of dive was approximately 75 degrees, a pull-out was started early enough to prevent the aircraft from passing its MACH limit and avoiding an excessive radial acceleration during recovery. Usually at this point, the maneuver was repeated again using the pull-out speed for the upswing into the next parabola (see Fig. 5).

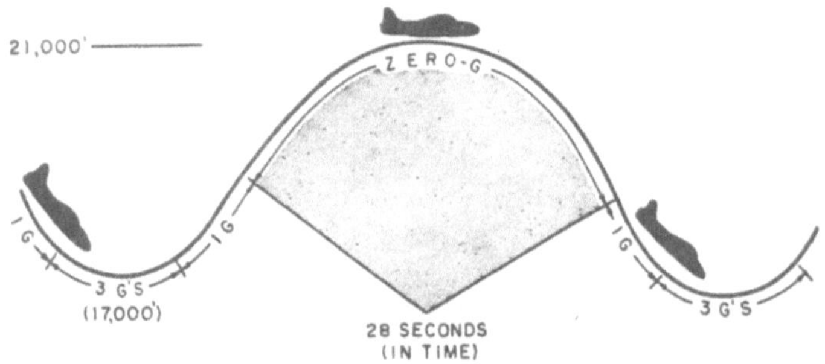

Fig. 5. T-33 "zero-*g*" flight pattern: The maneuver starts at an altitude of about 18,000 feet, reaches almost 21,000 feet at the top of the parabola, and yields about 28 seconds of virtual weightlessness

Theoretically, the IAS is not an important factor for producing the weightless state; in actual flying, however, it is of utmost importance. Yaw and roll movements of the T-33 during parabolas produced undesirable accelerations in all three axes owing to the sensitive aileron boost at low air speeds and rolling motions exaggerated by tip fuel. This was not the case in the F-94C which was used in later experiments.

The standard *g*-meter installed in all U. S. Air Force fighter aircraft was employed as a primary reference. Furthermore, the weightless state was indicated by the release of some small object, usually a cigarette lighter, a glove or in several instances a flashlight in the cockpit. When the object was stationary or floated in space before the subject's eyes, there was little doubt that zero-gravity was present. The parabolic arcs obtained in this fashion varied from 25 to 28 seconds, but the actual state of zero-gravity lasted for a few seconds only; while during the other part, the gravitational force was drastically reduced. For this reason, we call the entire condition the state of virtual weightlessness.

These T-33 flights were not without incident, however. During the sub-gravity state at altitude, fuel in the main tank had a tendency to vaporize, resulting in an overflow from the tank out through the vent drain line. This vaporization prevented the fuel pump from supplying enough fuel to the engine to maintain the required engine r.p.m. Fuel pressure was constantly observed during the run, and as soon as a reduction occurred, the parabola was discontinued and application of "positive" *g*'s usually brought engine operation back to normal.

The few power reductions that occurred during zero-gravity were corrected immediately by reducing throttle and applying "positive" *g*-forces. Complete loss of oil pressure also took place but was of little concern because of the known ability of this engine to run long periods with no lubrication whatsoever. The aircraft's hydraulic system was unaffected.

As the experiments progressed it was felt that extended parabolas were necessary in order that future research concerning man's reaction to zero-gravity could be successfully continued. Since the T-33's capabilities had been exploited to their limits, and since the power loss that had repeatedly occurred during flight was certainly not consistent with safe aircraft operation, a request was forwarded to Headquarters, United States Air Force, for assignment of a two-place jet fighter capable of high thrust and high MACH performance with a pressurized fuel system to be used in the study of weightlessness. In May 1956, an F-94C Starfire was assigned for this purpose.

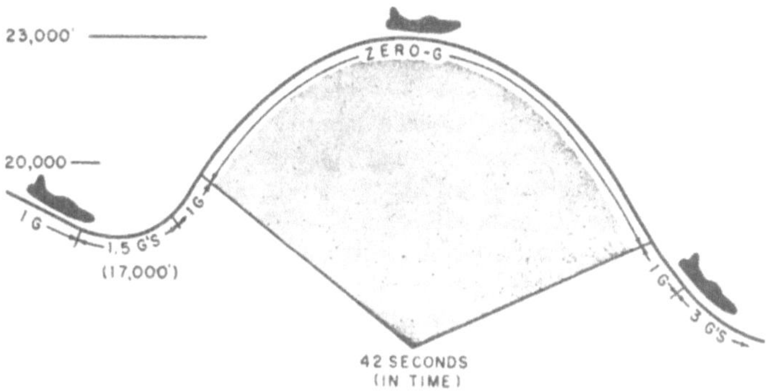

Fig. 6. F-94C "zero-*g*" flight pattern: The maneuver starts at an altitude of about 18,000 feet, reaches about 23,000 feet at the top of the parabola, and yields about 40 seconds of virtual weightlessness

Once again various patterns were attempted using different entry air speeds and altitudes. For the same reasons as given above, the most satisfying altitude was about 20,000 feet. A dive was entered, engine r.p.m. at 100 percent, and as the IAS reached 425 knots a climb of 65 degrees to 70 degrees from the horizontal was executed. With airspeed dropping off, forward stick pressure produced a stable and lengthy parabola yielding almost 40 seconds of practical weightlessness. Since the F-94C has a pressurized fuel system, the treacherous power loss was absent and the 6 degree dihedral present in the Starfire's short but able wings produced a parabola free of rock and roll that had previously presented severe control problems. The F-94C's variable elevator boost made small aft and forward control changes both easy and instantaneous, allowing small corrections to be made without varying the *G*-conditions appreciably. The high MACH rating eliminated the necessity of a too early dive recovery. In short, this ship now performed with both a high degree of efficiency and safety so necessary in our experimental flights (see Fig. 6).

The afterburner of the Starfire was also used to produce the longest period of weightlessness ever recorded, namely, 43 seconds. In this maneuver at 20,000 feet with elevator boost ratio at 11 : 1, engine, r.p.m. was increased to 100 percent and the afterburner actuated. IAS was increased to 430 knots, and the aircraft eased up into a climb of 75 degrees from the horizontal. The afterburner was used until this angle was reached in order to obtain an IAS of somewhat over 400 knots. At that time the afterburner was cut off and the parabola begun. Because of the high rate of fuel consumption, the afterburner was not used thereafter unless an extended period of weightlessness was required.

Since we still lack an accurate instrument for indicating and recording sub- and zero-gravity, and the conventional *g*-meter registers accelerations in the vertical only, an auxiliary indicator of *g*-forces in the three axes was introduced. A golf ball, painted black and white for better visual reference, was fastened to the end of a fourteen inch long nylon cord. This contraption was hung in the center top of the windshield inside the pilot's portion of the cockpit. Entry into the zero-state was determined by the conventional device. Then, as weightlessness was achieved, the pilot tapped the ball slightly from below, lifting it just enough to allow it to float. This primitive but revealing device portrayed any acceleration in the three space axes by moving in the opposite direction because of its inertia. Through practice, these movements could be kept at a minimum, and corrections were made by the pilot by simply "flying the ball" so-to-speak. Moreover, a zero-accelerograph has been installed in the F-94C registering the vertically and longitudinally acting accelerations within the range of $\pm 0.5\,g$ during the weightless phase.

Both parabola patterns, described with the T-33 and the F-94C, had their advantages and their shortcomings. The first one with the sharp dive produces longer states of weightlessness after preceding states of markedly increased acceleration. The second one with the shallow dive before being practically weightless, yields a shorter zero-state, but the preceding acceleration of $1.25\,g$ generally remains unnoticed because of the slow rate of change. Hence, the second maneuver seems to be appropriate for determining the effects of weightlessness on the body without preceding increase of weight, which also affects the human organism. The pilot has flown all zero-gravity research flights at the School of Aviation Medicine, USAF, and — assuming an accumulated weightlessness of about 3 minutes during each flight — has been weightless for almost eleven hours with no apparent untoward effects. However, this figure does not mean too much because of the short periods of exposure during each individual parabola.

References

1. F. DIXON and J. L. PATTERSON, JR., Determination of accelerative forces acting on man in flight and in the human centrifuge. U. S. Naval School of Aviation Medicine, Naval Air Station, Pensacola, Fla., July 1953.
2. S. J. GERATHEWOHL, Personal experiences during short periods of weightlessness reported by sixteen subjects. Astronaut. Acta 2, 203 (1956).
3. S. J. GERATHEWOHL, H. STRUGHOLD, and H. D. STALLINGS, JR., Sensomotor performance during weightlessness: Eye-hand coordination. J. Aviat. Med. 27, 7 (1957).
4. F. HABER, Study of subgravity states. School of Aviation Medicine, USAF, Project No. 21—34—003, Report No. 1, April 1952.
5. F. HABER and H. HABER, Possible methods of producing the gravityfree state for medical research. J. Aviat. Med. 21, 395 (1950).
6. H. HABER, Gravity, inertia and weight. In: Physics and Medicine of the Upper Atmosphere. (Edit. by C. S. WHITE and O. O. BENSON, JR.) Albuquerque, N. M.: The University of New Mexico Press, 1952.
7. H. HABER, The physical environment of the flyer. Air University, School of Aviation Medicine, USAF, 1954.
8. T. LOMONACO, M. STROLLO, and L. FABRIS, Behavior of motor coordination in subjects exposed to 3 to 0 *g* acceleration values. Proceedings of the VIIth International Astronautical Congress, Rome, Italy, 17—22 September 1956, p. 825. Roma: Associazione Italiana Razzi, 1957.

The Communication Satellite[1]

By

R. P. Haviland[2], ARS

(With 12 Figures)

Abstract — Zusammenfassung — Résumé

The Communication Satellite. The interrelations between an artificial satellite and the earth and the characteristics of satellites are reviewed to determine their effect on communication systems. The requirements for typical communication systems are studied. A set of services are proposed for integration into a large communication satellite.

Der Satellit für Übermittlung von Informationen. Die gegenseitigen Beziehungen zwischen einem künstlichen Satelliten und der Erde sowie die Charakteristika von Satelliten werden untersucht, um ihre Auswirkung auf Nachrichtenübermittlungs-Systeme zu bestimmen. Die Erfordernisse für typische Übermittlungssysteme werden studiert. Es wird vorgeschlagen, einen Satz von Einzeldiensten in einem großen Informationssatelliten zu vereinigen.

Le satellite d'information. Les caractéristiques des satellites et de leurs orbites sont passées en revue du point de vue de leurs possibilités comme transmetteurs d'information. Les performances exigées des systèmes de télécommunication sont mises à l'étude. On propose un ensemble de services d'information à intégrer dans un grand satellite artificiel.

The use of the satellite vehicle as a platform or base for communication services has been proposed previously [1, 2, 3]. This paper extends consideration of the requirements for such service in some detail, in order to arrive at the optimum conditions for design and operation of such a service.

The basic factor which makes the satellite attractive is its height. It is equivalent to a radio station having a very high antenna mast. This extends the line-of-sight coverage, and makes the use of very high radio frequencies for long distance communication feasible (as will be seen, it also makes the use of the ultra-high frequencies necessary).

The approach of this paper is to review first those characteristics of satellites, and of the satellite-earth system which affect the communication service. Next some of the aspects of communications are considered. Finally, some of the possible communication services are studied. In all cases the approach is that of the system engineer interested in established the magnitude of the problem rather than a detail solution. This is accomplished by the use of simplified models and approximations, rather than by exact solutions.

---

[1] Published in Astronaut. Acta 4, 70—89 (1958).

[2] Missile and Ordnance Systems Department, General Electric Company, Philadelphia 4, Pa., U.S.A.

A. Coverage

The problem of coverage by the satellite can be divided into three parts:

a) The instantaneous coverage of a single satellite station.

b) The total coverage of a single station.

c) The number of stations required to give continuous coverage to a given point on the earth's surface.

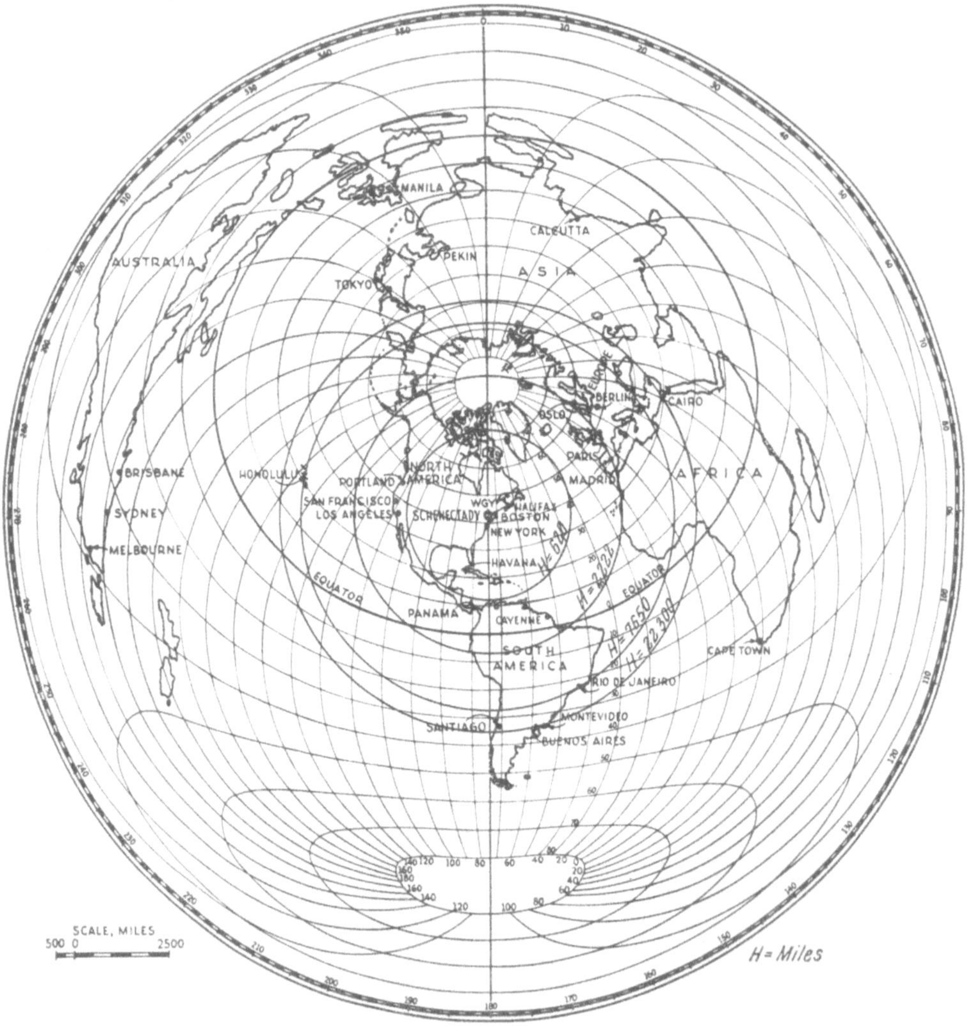

Fig. 1

The instantaneous line-of-sight coverage of a single station is readily calculated from geometrical considerations: these values are shown in Table I. The relative coverage for various heights is shown by the chart of Fig. 1. It is evident that the coverage is large even for modest heights.

The total coverage for a single station is shown in Table II for the three conditions of an equatorial, polar and 45° inclination orbits. It is evident that

Table I. *Instantaneous Coverage of a Satellite*

Height Miles	Vision Arc Miles	Subtended Angle Degrees	% of Earth
257	2,800	140	3.0
620	4,200	120	6.7
1,210	5,600	100	11.7
2,222	7,000	80	17.8
4,000	8,400	60	25.0
7,650	9,800	40	32.9
19,000	11,200	20	41.3
22,300	11,800	17	42.6

Table II. *Total Coverage of a Single Satellite*

Height	Max Latitude of Coverage, Equatorial Orbit	Time to Next Passage, Equatorial Orbit, Hours	Max Latitude of Coverage 45° Inclined Orbit	Max Interval Between Coverage, Hours	
				Equatorial	Polar
257	20°	1.7	65°	1.51	9.35
620	30°	1.9	75°	1.59	8.00
1,210	40°	2.3	85°	1.79	6.66
2,222	50°	3.0	90°	2.17	5.34
4,000	60°	4.6	90°	3.06	4.00
7,650	70°	9.7	90°	5.92	2.66
19,000	80°	97.5	90°	54.6	1.33
22,300	81.5°	(a)	90° (b)	0 (a)	1.13 (c)

(a) Appears stationary over a point on the equator.
(b) Appears to move from 45° N to 45° S.
(c) Appears to move from 90° N to 90° S.

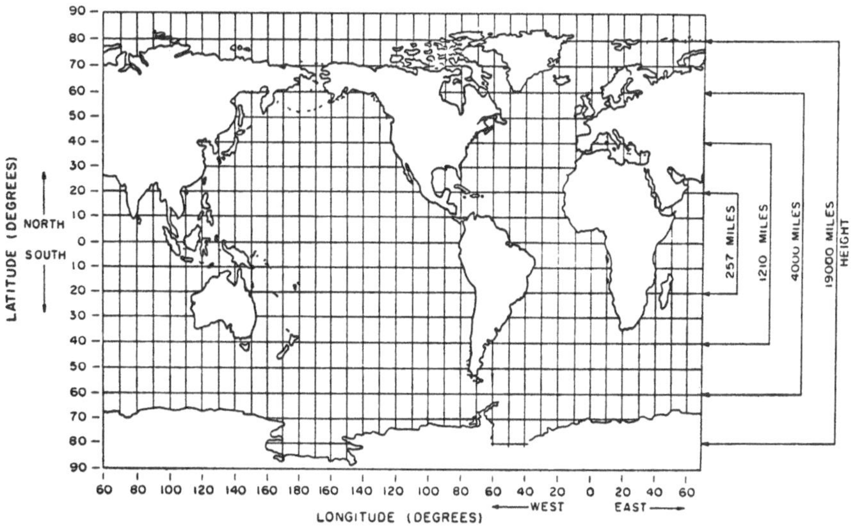

Fig. 2

the polar and inclined orbits are not desirable, since they increase the coverage of uninhabited regions at the expense of more desirable areas, and since the time to return is large compared to the orbital period. For these reasons it does not appear worthwhile to consider them further.

The relative total coverage for various heights and equatorial orbits is shown in Fig. 2. Considering the distribution of population over the earth's surface it appears that any of the orbits above 2220 miles would give satisfactory coverage.

The number of stations in equatorial orbits required to give continuous coverage to a given point is shown in Table III. From 3 to 5 stations are seen to be ample if the stations are above 2200 miles.

Table III. *Number of Satellites to Give Continuous Coverage Equatorial Orbit*

Station Height	Exact Number	Integral Number No Overlap	10% Overlap
257	8.8	9	11
620	5.9	6	8
1,210	4.4	5	6
2,222	3.6	4	5
4,000	3.0	3	4
7,650	2.6	3	4
19,000	2.2	3	3
22,300	2 2 (a)	3 (a)	3 (a)

(a) For worldwide coverage: a single station will give coverage over essentially a hemisphere.

B. The Influence of the Satellite Orbit Condition

A factor in the selection of operating condition for the communications service is the effort to be expended for various satellite orbits. This problem cannot be answered in an absolute sense at this time, but it is easy to secure a relative answer.

It is assumed first that continuity of communications is important. As has been seen, this appears to limit consideration to equatorial orbits, and requires use of several stations, except for the special case of the synchronized orbit.

Second, two "limit conditions" may be assumed for the size of the platform: at one extreme, it can be assumed that the size of the system measured by mass is independent of the number of platforms used in the service, and, at the other extreme, that the total mass of the system is independent of the number of stations and their operating conditions. The "point of equality" of mass is assumed to occur with a single station at 22,300 miles.

Finally, it may be assumed that the total effort and expense is measured adequately by the payload energy of operating (KE), which is defined by:

$$(KE) = \tfrac{1}{2} M (V_t + V_0 - V_a - V_d - V_g + V_r)^2.$$

Where

M = mass of system or, payload of ascending rockets
V_t = ideal takeoff velocity
V_0 = ideal orbital velocity
V_a = ideal velocity of arrival at orbit
V_d = loss due to drag
V_g = loss due to acceleration of gravity
V_r = gain due to rotation of earth.

Values of the first three velocity terms are derived in the Appendix. The sum of the remaining velocity terms is estimated to amount to a loss of 1,000 ft/sec.

The relative kinetic energy calculated on the basis of these assumptions is shown in Fig. 3.

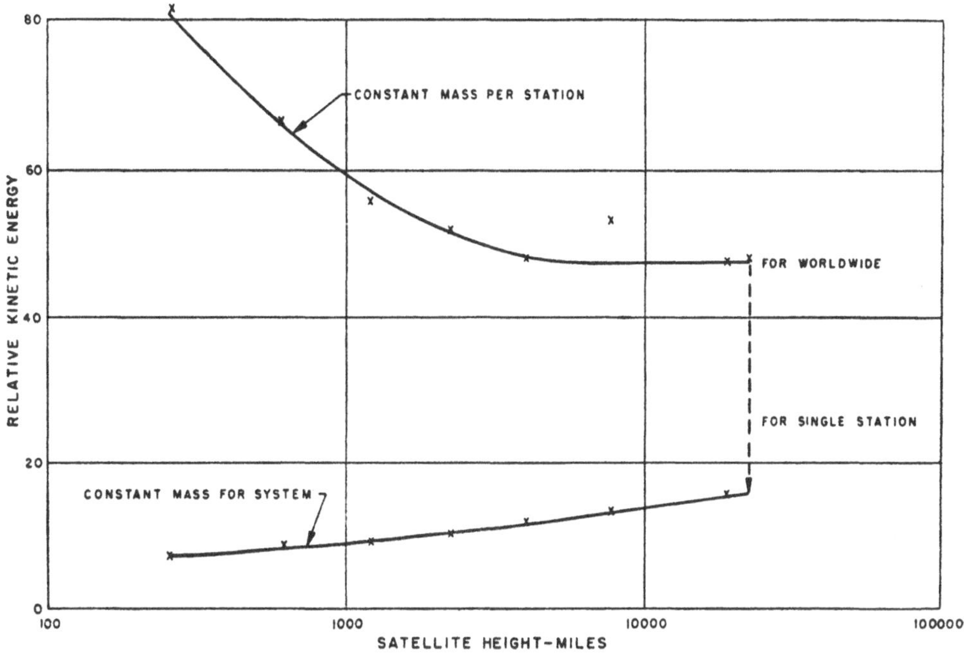

Fig. 3. Relative kinetic energy for satellite system

These curves indicate that:

a) On the basis of the pessimistic assumption of constant station mass the desired service requires less energy if the orbits are high.

b) The more optimistic assumption of constant mass to establish the service favors the lower altitudes. However, the difference between the lowest altitude for good coverage and the higher altitudes is small.

Since it is probable that the true situation is between the limit assumptions, it appears reasonable to conclude that the effort involved in establishing a satellite communication service is essentially independent of the satellite operating conditions chosen. Therefore, these may be selected to give the best grade of service.

C. The Effect of Satellite Motion

Motion of the satellite with respect to the surface of the earth introduces two problems: first, a Doppler shift is introduced into the radio frequency signals: second, if directive antennas are used on the ground they would have to track the satellite.

The Doppler shift is readily calculated for the equatorial plane and an eastward orbital motion and is found to be:

$$\frac{\Delta F}{F} = \frac{2\pi R_e}{C}\left(\frac{1}{P} - \frac{1}{24}\right)\cos\psi.$$

Where

$$\begin{aligned}\Delta F &= \text{Doppler shift, mc/s}\\ F &= \text{operating frequency, mc/s}\\ R_e &= \text{radius of earth, km}\\ C &= \text{velocity of light, km/hr}\\ \psi &= \text{elevation angle of satellite}\\ P &= \text{period of satellite, hours.}\end{aligned}$$

The value of this is not particularly large: for example, for the 1200 mile 4 hour orbit the maximum shift amounts to about 900 cycles/sec at an operating frequency of 100 mc/s. This would not be serious for a wide-band system, but would definitely require an automatic frequency control for narrow band-widths.

The amount of Doppler shift decreases with operating altitude, and becomes zero for the 24 hour orbit. This is a definite advantage.

The fact that there is, in general, relative motion of the satellite and earth indicates that any directional earth surface antenna will have to track the satellite. The tracking itself is not a serious limitation, since the maximum tracking rate cannot exceed about 4 degrees per minute, and is much less for satellite heights of 1200 miles or more. However, the acquisition of the next satellite to come into view is much more troublesome, since it will be necessary to have a rapid retrograde motion of the antenna, or to introduce two antennas and the necessary switching. These problems are also eliminated with the synchronized orbit.

It appears that the importance of these effects depends on the nature of the service. For point-to-point relays the special equipment such as AFC and tracking antennas would not be a serious problem. However, for the broadcast service, with perhaps millions of installations, such equipment would be a major system drawback.

The inverse problem, that of keeping the satellite antenna oriented with respect to earth will have to be solved for any altitude. This may be accomplished with either tracking antennas, or with fixed antennas by controlling the angular motion of the satellite. The fixed antennas appear superior. Controlling the satellite angular motion will certainly be easier with larger satellites, since the effects of personnel movement, rotating machines, etc. will tend to cancel, and in any event will give smaller angular motion for a given momentum change.

D. Propagation from the Satellite

Any practical satellite will be above at least part of the ionosphere. An estimate of influence of this on communication is made as follows:

a) The minimum useable frequency, by reciprocity, corresponds to the maximum useable frequency of normal transmission, and is

$$MUF = fc \sec \alpha.$$

Where

$$\begin{aligned}MUF &= \text{Minimum useable frequency}\\ fc &= \text{Critical frequency for vertical incidence at layer}\\ \alpha &= \text{Angle of arrival at layer.}\end{aligned}$$

b) The ionosphere layer is assumed to be very thin, to be located at the virtual height, and to be of constant intensity.

With these assumptions the arrival angle α can be determined geometrically from Fig. 4, which gives

$$\text{Cos} = \frac{R_e}{R_e + M} \cos \psi$$

$$= \cos \alpha_{max} \cos \psi$$

M = virtual height of layer

ψ = elevation angle of satellite from earth

R_e = radius of earth.

That is, the minimum useable frequency at a given elevation angle is independent of the height of the satellite above the earth.

The value of the $MUF/fc = \sec \alpha$ is given in Fig. 5 as a function of layer height for various values of elevation angle.

The values of known layer heights and critical frequency are time varying quantities, which depend on time of day, solar conditions, and other factors.

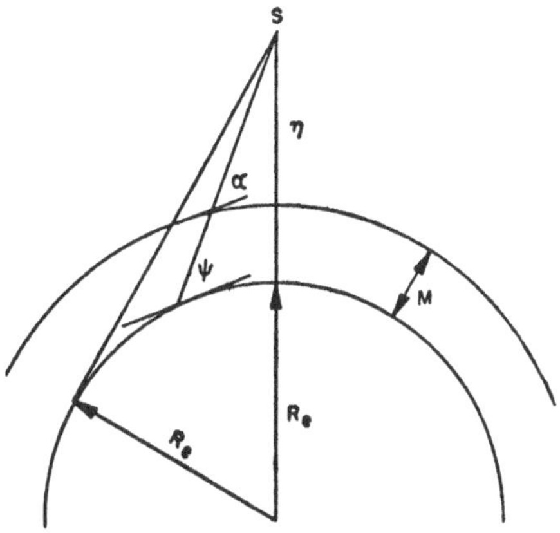

Fig. 4. Geometry of propagation. S satellite, M ionosphere layer height, η satellite height, R_e radius earth, φ elevation angle of satellite, α arrival angle at layer

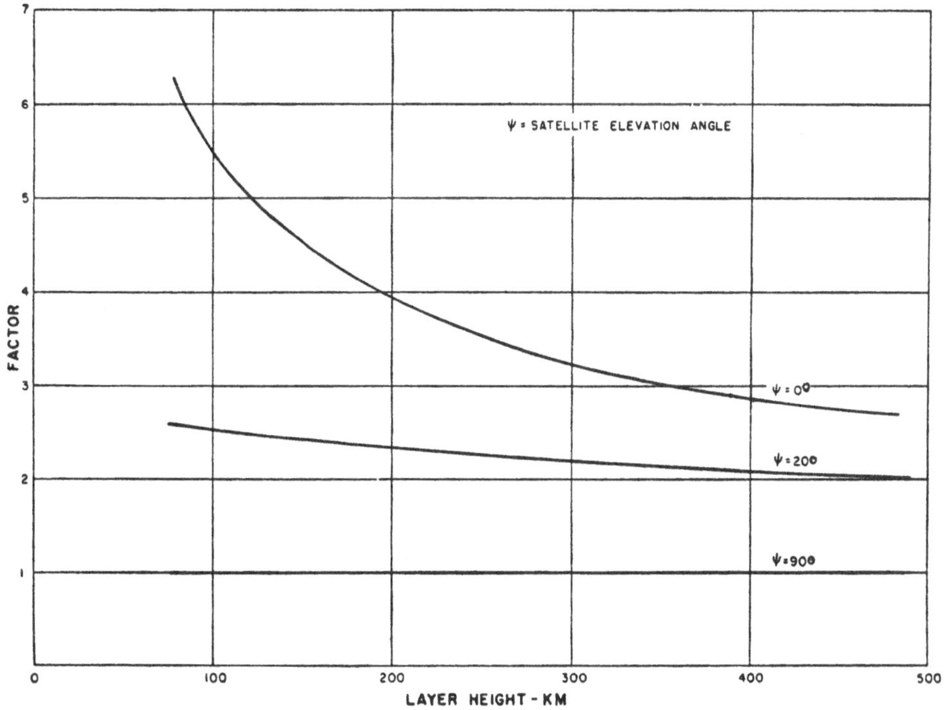

Fig. 5. MUF factor. φ satellite elevation angle

Typical values are shown in Table IV A. The corresponding minimum useable frequencies are shown in Table IV B.

Table IV

A. Typical Values of Ionosphere Critical Frequency Fc and Height

Layer Designation	Height km	Sunspot Summer	*Fc*, mc/s, Values For		
			Maximum Winter	Sunspot Summer	Minimum Winter
E	120	4	4	5	5
F_1	210	6	—	4	4
F_2	400	9	15	5	6
Es	120	12	12	12	12

B. Typical Values of MUF for Satellite Propagation

Layer Designation	Sunspot Summer	*MUF*, mc/s, Values For		
		Maximum Winter	Sunspot Summer	Minimum Winter
E	20	20	25	25
F_1	24	—	16	16
F_2	27	45	15	18
Es	60	60	60	60

Es — E sporadic.

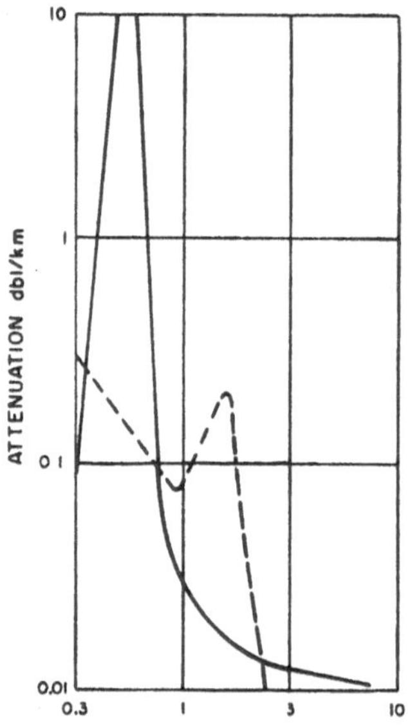

Fig. 6. Atmospheric absorption. ---- water vapor, —— oxygen

It is evident that the minimum frequency for complete coverage is around 60 mc/s. Allowing for neglected factors, such as ionospheric and atmospheric bending and absorption, it appears that it is safe to assume that frequencies above 100 mc/s are always useable.

At the other end of the frequency spectrum the maximum useable frequency is determined by atmospheric absorption. The contributions of water vapor and oxygen to this attenuation are shown in Fig. 6. These are the main components. This indicates that frequencies below 10,000 mc/s are useable.

The allowable band of frequencies for the satellite communication service thus covers the range between 100 and 10,000 mc/s. On an absolute basis this is about 10,000 times the bandwidth now available for reliable long range communication, or, on a percentage basis, just double the present spectrum. These factors indicate that there is ample room for the satellite service.

E. Basis for Communication Studies [5, 6]

Numerical calculations for the various communication systems will be on:

a) The so-called beacon equation, for one-way transmission:

$$P_t = G - A_t - A_r - K + \frac{S}{N} + F + N + B + Pr.$$

Where

P_t = transmitted power, db referred to 1 watt
G = free space attenuation between isotropic antennas, db
A_t, A_r = gain of transmitting and receiving antennas, db
K = system improvement factor, db
S/N = desired signal to noise ratio, db
F = fading factor, db
N = noise figure of receiver, db
B = line and other loss, db
Pr = theoretical receiver sensitivity, db referred to 1 watt.

b) The so-called radar equation for two-way reflection transmission, written as

$$P_t = 2G - 2A + \frac{S}{N} - K + F - \sigma + N + B + P_r - 28.$$

Where

$$\sigma = 10 \log_{10} \text{ target area in square ft.}$$

c) The energy equation, in appropriate units

$$P_t = A \frac{E^2}{120\pi}$$

A = area to be covered

E = desired signal strength.

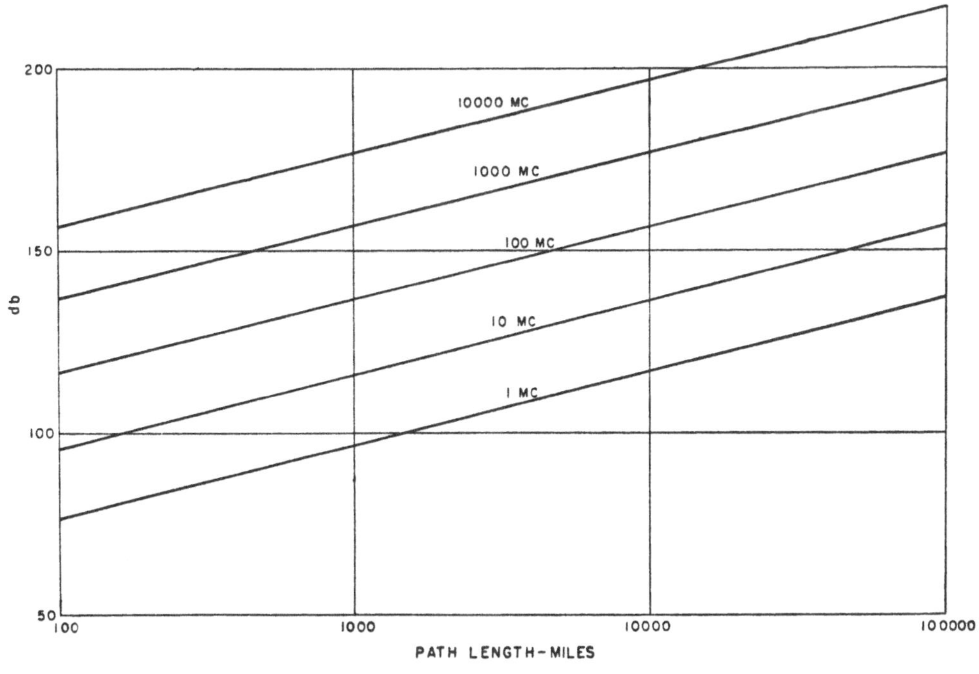

Fig. 7. Free space attenuation

The free space attenuation between isotropic antennas (path attenuation) is given by

$$G = + 36.6 + 20 \log_{10} D + 20 \log_{10} fmc.$$

Where

$$D = \text{distance of transmission, miles}$$

$$fmc = \text{operating frequency, mc/s.}$$

This factor is shown in Fig. 7.

The gain of a parabolic reflector is given by

$$G = 20 \log fmc + 20 \log_{10} d - 53$$

$$d = \text{diameter in feet.}$$

The beam width θ of the parabola is approximately

$$\theta^2 = \frac{27{,}000}{g}.$$

Where

$$g = 10^G$$

$$= \text{numerical gain.}$$

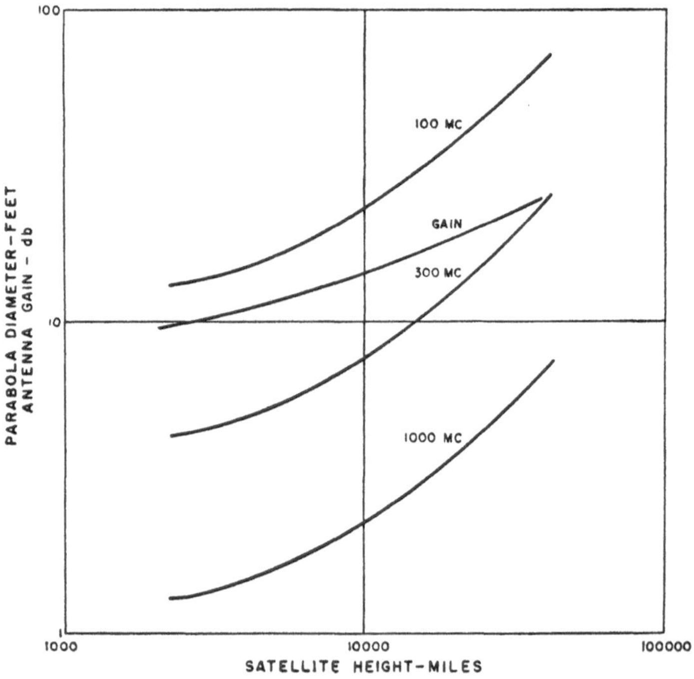

Fig. 8. Size and gain of a parabola illuminating the visible earth

For the special case of an antenna at the satellite which is to illuminate the entire visible sector of the earth's surface, the beam width is determined by geometrical considerations, and is shown in column 3 of Table I. The required antenna size as for various frequencies, and the gain of such antennas are shown in Fig. 8.

The system improvement factor is determined primarily by the type of modulation employed to convey intelligence. Typical values are:

Modulation System	Improvement Factor K
Double sideband telephony	0 db
Single sideband telephony	+3 db
FM Broadcast	+15 db
AM Telegraphy	+17 db
FSK Telegraphy	+21 db
Pulse Code Telegraphy (typical)	+28 db
Television	0 db

The theoretical receiver sensitivity is a function of bandwidth of the receiver, and is 144 db below 1 watt per megacycle of bandwidth. For typical systems the bandwidths and corresponding sensitivities are:

Modulations System	Bandwidth	Theoretical Sensitivity
Double Sideband Telephony	6 kc/s	−166 dbw
Single Sideband Telephony	3 kc/s	−169 dbw
FM Broadcast	75 kc/s	−155 dbw
AM Telegraphy	0.1 kc/s	−184 dbw
FSK Telegraphy	1.7 kc/s	−172 dbw
Pulse Code Telegraphy	3 mc/s	−139 dbw
Television Broadcast	6 mc/s	−136 dbw

The desirable signal to noise ratio depends on the system and the grade of service desired. For study purposes an average value of 20 db is satisfactory. The fading factor may be assumed to be 10 db, and the lines loss to be 3 db.

In the case of television broadcast services in the USA, the Federal Communication Commission has established the following standards for minimum signal levels, for the peak value of the signal:

Channel	Freq	Primary		Signal Levels			
				Grade A		Grade B	
	mc/s	db	uv/m	db	uv/m	db	uv/m
2−6	54−88	74	5010	68	2510	47	224
7−12	174−216	77	7080	71	3550	56	631
14−83	470−890	80	10000	74	5010	64	1585

Relative levels are in db above 1 uv/meter.

There is reason to believe that the requirements for the two higher bands are excessive when applied to the satellite service, since, in part, these were established to permit use of simple dipole antennas. From an overall system view it appears more desirable to assume antennas having an effective area which is independent of frequency. For this reason service levels will be calculated on the basis of those for the low band.

F. Preliminary Consideration of Communication Systems

Before investigating communication systems in detail it is desirable to determine the effect of height on their feasibility. This can be done by considering a limited number of system types.

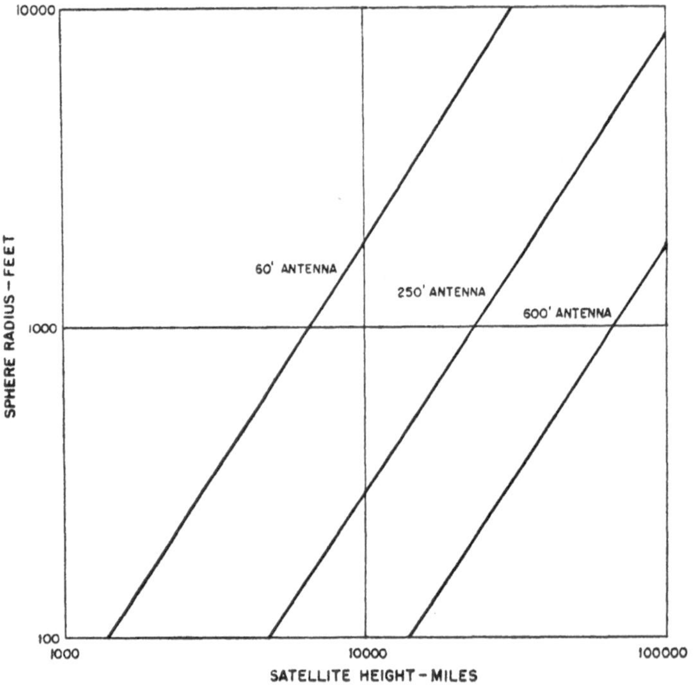

Fig. 9. Effect of height on a passive PCM relay

To secure a wide range of requirements, the systems chosen are:

a) Pulse code passive relay telegraphy

b) FM relay

c) Television area broadcast.

The following assumptions are made as to the operating conditions:

a) The PCM passive system operates on 10 cm, with a 1 megawatt ground transmitter, against a spherical reflector. The reflector size is determined as a function of height for three ground antenna sizes. The system is operative anywhere within the visible area, and provides the S/N ratios of section E.

b) The FM relay system operates on 30 cm, with a 60′ dish on the ground, and a satellite dish to completely cover the visible area, with the S/N ratios of Section E. Antenna pattern shaping to give uniform signal strength is not used.

c) The TV broadcast system is calculated to give FCC signal levels on channels 2—6 over the visible area. Antenna pattern shaping is used to give uniform power density over the illuminated area.

The effect of height on the three systems is shown in Figs. 9 to 11 inclusive. The following conclusions appear justified.

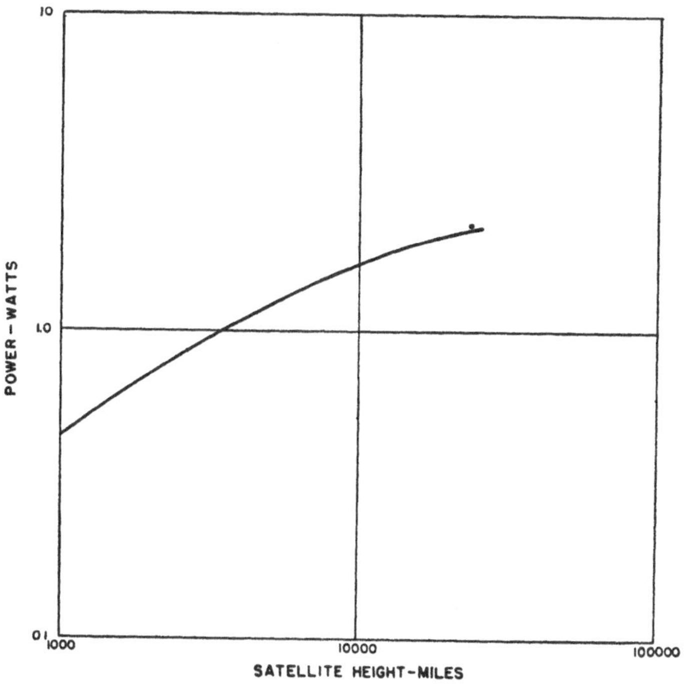

Fig. 10. Effect of height on a FM relay

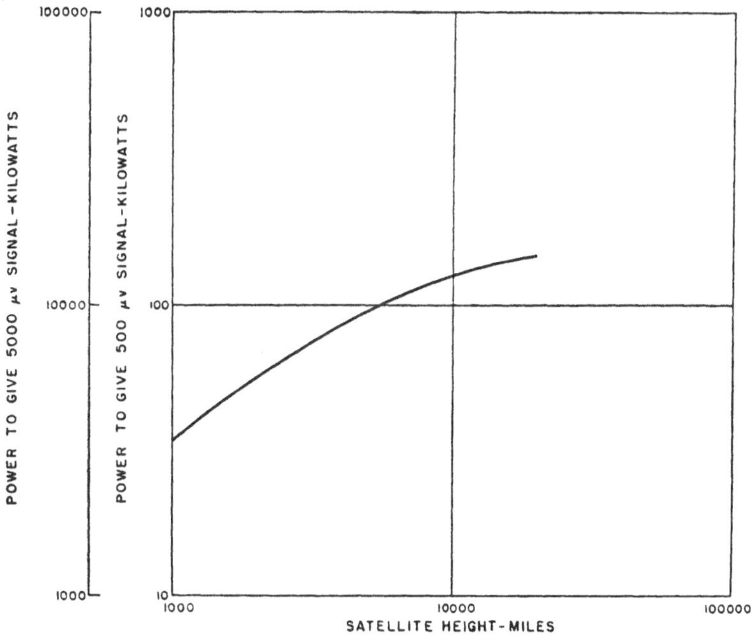

Fig. 11. Effect of height on TV broadcast

Assuming a 100′ radius sphere or equivalent as the practical limit for the satellite, the passive relay system would be useable to an altitude of about 1400 miles with a 60 foot dish, which does not appear unreasonable. At greater altitudes either the power or dish size would have to be increased. Tracking dishes of large size are probably not practical, but a stationary one might be. This indicates that the use of passive communication with the synchronized 22,300 mile satellite may be feasible, and should be studied further.

The power requirement for FM relay service is modest, and is essentially independent of satellite height. The system chosen is probably out of balance in terms of cost, since increasing the transmitted power will permit reduction in antenna size.

The power requirement in the TV broadcast service also changes slowly with altitude. Provision of the FCC Grade B signal over the visible part of the earth can be accomplished with present techniques, but the primary coverage signal level requirements appear excessive. However, it appears possible to provide the primary coverage over smaller areas of the earth surface.

Comparison of the requirements for active and passive systems shows that the power and antenna requirements are more modest for the active system. However, this may not be true if the economics of the system are considered, since equipment at the satellite will be the most expensive to install and operate.

G. The Optimum Orbit for Communication Service

The following factors, developed in the previous sections, are of importance in selection of the optimum orbital height:

a) The equatorial orbit appears to be superior in all cases.

b) The coverage of a single satellite obviously increases with altitude, but appears to be entirely adequate for any altitude of 2220 miles or greater.

c) The number of stations required to give continuous coverage to a given point decreases somewhat with increasing altitude, but not by an important factor.

d) The synchronized 22,300 mile satellite permits establishment of a "partial" service, giving continuous coverage to wide areas with a single satellite. All other cases give world wide coverage when continuous coverage is established.

e) The cost of establishing world wide service measured in terms of payload energy appears to be essentially independent of orbital altitude. The cost of continuous coverage to a point or moderate area is probably less for the synchronized orbit, since a single satellite is adequate.

f) The effects of orbital motion are smaller at great altitude, and become zero for the synchronized orbit. This is probably not important for limited communication service, but is important for the broadcast service.

g) Propagation problems are independent of orbital altitude.

h) When used as a passive reflector in the PCM relay service the system requirements are less severe at the lower altitudes. For the synchronized orbit the elimination of tracking requirements may be more important than other system factors.

i) In the FM relay service the system requirements are quite modest, and are essentially independent of altitude.

j) In the TV broadcast service the requirements appear attainable for acceptable wide area coverage, and are essentially independent of altitude. Wide area high quality coverage does not appear feasible at any altitude considered.

k) Small area high quality TV broadcast coverage problems are essentially independent of altitude.

Since only two of these factors favor the lower altitude orbits, and since the synchronized 24 hour orbit appears to be the best of the high altitude orbits, it appears reasonable to conclude that the communication satellite will be based on the use of a platform at 22,300 mile altitude.

H. Further Consideration of the Television Broadcast Service

The provision of several communication systems from the 22,300 synchronized satellite is now considered in more detail. Because of its more severe technical requirements, the Television Broadcast service is considered first.

It should be noted that the television power levels quoted previously are peak powers, measured at the synchronization level of the USA standard signal. The carrier power is smaller than this, by a factor of 0.66. However, in practice the actual radiated power must be greater than that calculated by area alone, since practical antennas do not have "ideal" patterns. Further, an aural transmitter of 50–70% of the power of the video transmitter is also required. For this study it is assumed that peak power and pattern effects compensate, so that the calculated values are the carrier level, and that the aural and video transmitters have equal power.

To review, the power requirements for illumination of the visible part of the earth are:

Service Level	Kilowatts Power Required
Primary	14,400
A	3,600
B	35

The first two of these are clearly excessive, so that it will be necessary to restrict the coverage area when the best grades of service are to be provided.

If it is assumed that a 50 kw transmitter represents a practical limit, and that the "*A*" grade coverage is desired, the total area coverage possible is 1.15×10^6 square miles, an area about equal to that part of the USA which lies east of the Mississippi River, or to about that of Western Europe. This coverage appears to be entirely ample, and does have the additional desirable feature of limiting the coverage on the earth to a span of about two time zones.

The antenna pattern needed for this limited area coverage depends on the geographical location of the area, and on the longitude of the satellite. For a satellite of 100^0 west, to give good coverage at the Americas, the beam required to cover the eastern third of the USA is approximately 2.7^0 in width and 3.3^0 in heigth. At a frequency of 100 mc/s the reflector would need to be about 270 feet in width by 200 feet in height.

Considered on a world-wide basis, it appears that the major features of the TV broadcast service would include the following:

a) A network of three satellite stations. A station for the Americas would be located at 100^0 west, one for Europe/Africa at 30^0 east, and one for Asia/Australia at 120^0 east. This nonuniform spacing is indicated by the nonuniform distribution of land masses and population.

b) Each satellite would include a general coverage channel, to furnish a "B" level signal to the entire visible part of the earth. The transmitter would have a power of 35—50 kw. The antenna, from Fig. 7, would have a beam width of 17°, and a diameter, for 100 mc/s, of 45 feet.

c) Each station would have two larger antenna systems, one for high level coverage to northern industrial areas, the other to southern areas. (It is assumed that the present acceleration of industrial development in the southern latitudes will continue.) Each of these will consist of a single large reflector, about 200 × 300 feet in size, with multiple feeds. Three transmitters per large dish appear to give ample coverage.

d) Overall, the equipment at each satellite amounts to a total of seven video transmitters, and seven aural transmitters, amounting to a total power output of 700 kw.

e) The system described lends itself to development in stages. The initial stage could consist of the single general coverage transmitter. The second stage could be achieved by adding one of the limited coverage antennas, with one or more of its transmitters, and so on.

f) The system also lends itself to expansion. This may be accomplished by increasing the number of transmitters, or by increasing the power of a transmitter. For example, megawatt transmitters are technically feasible, which would permit better service levels over wide areas.

g) It would appear that the satellite service and the more restricted coverage surface transmitters would complement each other, with the satellite service being primarily used for network-type programs, and the surface transmitters for local-interest programs.

h) Considering the demonstrated appeal of television programs, particularly for children, it is interesting to speculate on the lingual and sociological impact of the world-wide coverage possible. These speculations are not pursued here.

I. Reconsideration of Passive Relay System

It has been found that the passive relay system working against a 1000 foot radius spherical satellite of 22,300 miles requires an excessively large earth antenna.

A major factor in this lies in the nature of the assumed reflector since a sphere reflects isotropically. In the satellite application this means that all energy lying outside a 17° wide cone is wasted. A means of concentrating this energy is needed.

The first thought is a corner reflector. However, an ideal reflector returns all of its energy to the source, whereas the desired system scatters this energy across the visible segment of the earth. What is needed is a diffusing system: that is, a convex reflector. This suggests using a segment of a sphere, which is permissible in this application, since other antennas will require stabilization in any event.

The geometry of the problem requires that the reflection lie in a 17° wide cone, approximately $2\pi/100$ steradians. Since the radar cross section is defined as

$$\sigma = 4\pi\,\frac{\text{Power scattered per unit solid angle}}{\text{Power incident per unit area}}$$

the cross section in such a case is approximately $200\pi a^2$, where a is the radius of the segment base. Since the required cross section is 10^8 square feet, for the system previously assumed, the required base radius is 400 feet. This still appears large. Therefore it appears necessary to increase the size of the ground antenna.

A reasonable balance in requirements appears to be a ground antenna of 150′ diameter. This decreases the required cross section to 2.5×10^6 square feet, and the base radius of the reflecting segment to 65 feet.

This system may not require a special satellite reflector. The area of the television antenna suggested for limited coverage in the preceding section is very nearly the same as that required for the passive link. If the energy incident on it were reflected at the feed point it would be returned to earth. However, with normal feeds the energy would be concentrated into a narrow beam, so that a special design is needed.

It should be noted that the passive system is inherently broad-band, since addition channels do not require additional satellite equipment, and since the system requirements do not change as the frequency changes, providing the antenna and reflector sizes are maintained. Therefore, the system is capable of a large degree of expansion, by adding earth surface transmitter and/or receiver stations.

However, because of the power requirement, it appears that it is best limited to the relay of telegraphic signals. In this service, using PCM modulation, each transmitter is capable of handling around 100 simultaneous messages. These can be received at any point on the visible segment of the earth. In practice, it appears that the system could be made automatic, by the use of destination codes, which would allow the receiving point to select the messages directed to it.

J. Reconsideration of FM Relay Systems

The FM relay system previously discussed was based on the transmission of high fidelity music. It was found that system was not in balance, since antenna size appeared to be excessive.

By increasing the radiated power from 2 watts to 20 watts the ground antenna can be reduced to a diameter of 20 feet. This appears to be a better balance for a limited number of channels. A further increase in power is possible: if this is sufficient the system becomes attractive for the broadcast service. For example, with a 2 kw of power, the required ground antenna is a parabola slightly over 2 feet in diameter, or alternately, a corner reflector antenna of nearly the same dimension. The cost of such an antenna is sufficiently low to be attractive.

Another type of relay service in which FM techniques are useable is long distance telephone communication. By multiplexing, a system capable of carrying high fidelity music can carry 5 simultaneous speech circuits. This gives a power requirement of 0.4 watts per channel if the 60′ antenna is used.

It appears that a 5 channel link would be grossly inadequate for the relaying of telephone signals. For example, the capacity of a coaxial cable is about 600 channels. In the satellite service, since the power required is porportional to bandwidth, this would require a total power of about 250 watts.

This system is probably not the optimum for this service, since the satellite antenna covers the visible segment of the earth. By limiting the coverage to smaller areas on the earth the required power can be reduced. However, it does not appear profitable to pursue this optimization further at this time, since the requirements are modest.

K. Interstation Relays

To provide world wide service the communication system must provide for communication and relaying between the three satellite stations. Since directive antennas will be needed, this service could be on the same frequencies as are used for earth coverage. However, to avoid interference problems within the station it appears desirable to use frequencies outside the earth coverage band. This suggests use of the centimeter band.

Since the proposed satellite stations are not uniformily located about the earth, the transmission distance varies with the stations, from around 30,000 to around 40,000 miles. Assuming that 30′ dish antennas are used, that the relayed signal is 6 mc/s wide, corresponding to television signals, and that the allowance for signal/noise ratio, fading, etc. amounts to 53 db, the required power is 2.5 watts per channel. This appears to give a satisfactory system balance. For other types of relaying the same value may be used: for example such a channel could accomodate about 600 voice channels.

L. The Influence of the Communication Service on the Satellite

By now, it seems obvious that the communications satellite will have two major features: a large power supply, and a number of antenna reflectors. To arrive at the extent of these, the following assumptions are made concerning the services provided by each of the satellite stations:

a) The television service includes 2 general coverage channels, plus a total of six limited area channels, divided into a northern set and a southern set of three each.

b) There is a passive relay reflector.

c) The system includes 100 relay transmitter links, each having a capacity equivalent to 600 one way voice channels.

d) The system also includes the capacity of 100 relay receiving links.

e) Two FM broadcast transmitters are provided.

f) Interstation links are provided to each of the other two satellite stations. The total capacity of these is equivalent to 100 FM relay links.

For such a system the power requirements are:

System	Power/Transmitter	Number of Transmitters	Total
TV Broadcast	50 kw	16	800 kw
FM Relay	0.25	100	25
FM Broadcast	2	2	4
Interstation	.0025	100	.25

It is assumed that the efficiency including receiving, terminal and modulation equipment will be 50%. The total demand on the power supply is therefore about 1600 kilowatts.

The antenna requirements for this system would be, assuming maximum utilization of each antenna:

System	Antenna
TV Broadcast	1—45′ dish
	2—200′ × 270′ dish
Passive Relay	1—65′ convex reflector
FM Relay and Broadcast	1—4.5′ dish
Interstation Relay	2—30′ dishes

All antennas would be stabilized with respect to earth.

It appears evident that a system as large as this would have to be designed for manned operation on a long term basis. This, in turn, appears to require a synthetic gravity, obtained by rotation of the station. However, the antennas cannot be allowed to rotate, since they must remain fixed with respect to earth, and must also maintain the direction of antenna polarization constant.

The solution to these opposing requirements is to mount the rotating part of the station as a wheel on an axle which carries the nonrotating parts. The axle would be oriented parallel to the axis of the earth. There would be some friction, so the axle and wheel would require torque jets, to maintain constant angular velocity. Additional jets would be needed to bring the rotational period of the assembly to 24 hours, so that the antennas always point to earth, and to make corrections.

A reasonable arrangement would place the limited coverage antennas at each end of the axle, with the rotating wheel at the center. The passive reflector could bridge the ends of the axle, outside of the wheel on one side.

To avoid slip ring problems it appears desirable to keep all communication equipment in the non-rotating part, probably without air. This introduces the problem of servicing while weightless, and in a space suit, but these problems are probably less important than the possibility of perfect vacuum in tubes, the lack of dust and moisture, and so on. The communication equipment would need to be enclosed in metal, for shielding, and protection against photo-electric emission.

The power supply for such a station appears to be a much larger problem than the communication equipment itself. Solar energy does not appear feasible with the efficiencies possible at present, although a 500 kw unit has been proposed. The size and weight of a nuclear reactor is negligible, but the shielding and generation equipment would be massive. However, nuclear propulsion systems for aircraft are being seriously studied. The problems appear solvable in this application, and since the power rating is orders of magnitude larger than needed here, it appears safe to assume that a solution for the satellite is feasible.

The space requirements for the communication service have not been detailed. It appears that the 400 ton, 250 foot diameter ring of VON BRAUN [7] is too small in some respects: on the other hand the 1000′ × 3000′ station of ROMICK [8] appears oversize. Probably a station which is between VON BRAUN's and ROMICK's 1500 ton 40′ × 500′ "Second Phase" in size and mass would be satisfactory.

In any event the station would represent a major undertaking, and is clearly beyond the reach of present rocket technology. With nuclear fuels it should be straightforward: with advances in fuel energy it may become feasible. The question of desirability is separate: this depends on complex economic factors, and on sociological and political factors. These will have to be considered later.

Appendix

The velocity for establishing a satellite may be derived by assuming that the total energy required is the sum of that needed under "ideal" conditions plus that needed for "practical" considerations.

Assuming that the rocket is of the two period burning type, it can be shown that the optimum "ideal" trajectory is that shown in Fig. 12.

For this letting

$$k = \frac{R_0}{R_e},$$

if can be shown that the orbital velocity is

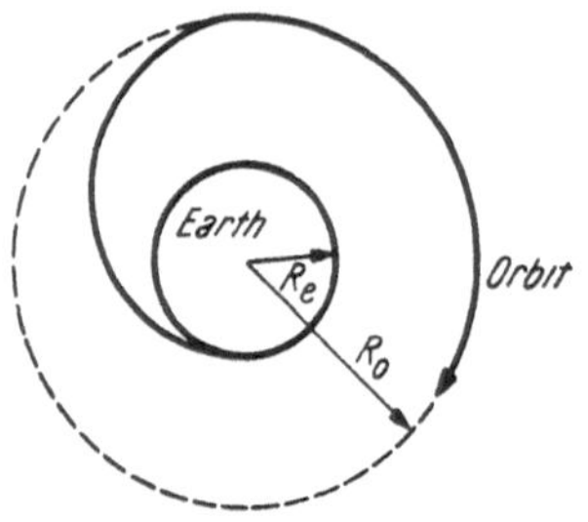

Fig. 12

$$V_0^2 = \frac{V_e^2}{2k}$$

V_e = velocity of escape.

The velocity of takeoff, which must be developed to rise to orbital height, is:

$$V_t^2 = V_e^2 \frac{k}{k+1}.$$

The rocket arrives at orbital height with an arrival velocity

$$V_a^2 = V_e^2 \frac{1}{k(k+1)}.$$

Correction terms must be added for practical effects. These are:

V_d = The equivalent velocity due to atmospheric drag.
V_g = The equivalent velocity due to gravity action during finite burning periods.
V_r = The velocity of rotation of the earth.

The total velocity required to establish the rocket in the orbit is accordingly:

$$Vi = V_t + V_0 - V_a - V_d - V_g + V_r.$$

References

1. A. C. CLARKE, Extra-Terrestrial Relays. Wireless World, October, **1945**, pp. **305–308**.
2. ARS Space Flight Committee, On the Utility of an Artificial Unmanned Earth Satellite. Jet Propulsion **25**, **71** (**1955**).
3. R. P. HAVILAND, On Applications of the Satellite Vehicle. Jet Propulsion **26**, **360–363** (**1956**).
4. F. E. TERMAN, Radio Engineers Handbook. New York: McGraw Hill.
5. J. A. WEBB, Interplanetary Communications. J. Astronautics **3**, **29** (**1956**).
6. Federal Telephone and Radio Corp., Reference Data for Radio Engineers, IT and T, New York, 4th Ed., **1956**.
7. C. RYAN, ed., Across the Space Frontier. New York: The Viking Press, **1952**.
8. D. C. ROMICK, Preliminary Study of a Satellite Station Concept, presented at ARS Annual Meeting, Chicago, November **14–18**, **1955**.

Selenoid Satellites[1]

By

W. B. Klemperer and **E. T. Benedikt**[2]

(With 2 Figures)

(*Received May 14, 1957*)

Abstract — Zusammenfassung — Résumé

Selenoid Satellites. Discussion of the LAGRANGEian solution of the restricted three-body-problem of constant configuration. The five possible positions are the inferior and superior conjunction, the opposition and two "trojan" positions or sextiles. These positions are computed for a symperiodic Moon companion of the Earth and an Earth companion of the Sun.

Selenoid-Satelliten. Es wird die LAGRANGEsche Lösung des vereinfachten Dreikörperproblems konstanter Konfiguration behandelt. Die fünf möglichen Positionen sind die untere und obere Konjunktion, die Opposition und zwei „Trojaner"-Stellungen oder Sextile. Diese Positionen werden für einen symperiodischen Begleiter des Erdmondes und einen zu Sonne und Erde „trojanischen" Erdbegleiter berechnet.

Satellites sélénoides. Discussion de la solution lagrangienne du problème simplifié des trois corps dans le cas particulier d'une configuration constante. Les cinq positions possibles sont la conjonction supérieure et inférieure, l'opposition et deux positions "trojanes" ou sextiles. Ces positions sont calculées pour un compagnon lunaire de la terre et un compagnon terrestre du soleil.

With the advent of artificial satellites traveling around the Earth, and in view of the interest evoked by several serious papers on artificial satellites circling the Moon which have been presented before previous International Astronautical Congresses and published in astronautical magazines, the question may well be of interest: Under what conditions should it be possible to establish an artificial "SELENOID" or Moon Companion on such an orbit that it remains in a constant configuration relationship with Earth and Moon? Such a space station would have interesting properties which could become useful in any lunar satellite system telecommunication network, and, conceivably, in a project of checking the Moon and Earth mass ratio.

The kinematics of such Selenoid orbits belong, of course, in the so-called Three-Body-Problem which is notorious for the difficulties with which the treatment of the general case is fraught. However, the present specific problem

---

[1] Published in Astronaut. Acta **4**, 25—30 (1958).

[2] Missiles Engineering Department, Santa Monica Division, Douglas Aircraft Comp., Inc., Santa Monica, Calif., U.S.A.

is a special case, which has been shown by the eminent mathematician, J. L. LAGRANGE, as early as 1772, to be soluble in closed form, in a celebrated treatise [1] which was awarded a prize by the then Royal French Academy of Sciences in Paris.

In that paper — with a great deal of mathematical exercise and without the benefit of a single illustrating figure — he proved that three celestial bodies can travel in constant constellation with respect to each other around their common barycenter if they are situated either on a straight line or at the corners of an equilateral triangle lying in the plane of their mutual revolution. If the Moon were traveling in a circle around the Earth-Moon barycenter, then such a Selenoid would also have to travel in a coplanar circle around it, and the three distances, Earth-Moon-Selenoid, would stay constant. However, with the Moon traveling, as it does, in an elliptical orbit, the Selenoid companion would travel in a coplanar ellipse of the same numerical eccentricity and so that the ratios of the distance between the three bodies remain constant; in other words, so that the configuration formed by them remains similar.

These conclusions were derived by LAGRANGE with great rigor and considerable apparatus of algebra. The solutions have also been derived by LAPLACE, CARATHEODORY, and BRENDEL; the latter's proof is contained in Chapter X of a textbook for students of Celestial Mechanics, by Professor H. HAPPEL, Breslau [2]. What prompted us to take another look at the problem was a desire to present the fundamentals of it in concise form and to discuss its implications for any practical establishment of a Selenoid Satellite.

We contented ourselves with simplifying the general problem by several theoretical assumptions: 1. The satellite is assumed to be a true "planetoid", that is to say, it is so small and light that it does not perceptibly influence or distort the gravity field in which it moves; also the satellite has no propulsion — it travels on a ballistic trajectory. 2. The Sun, though large and massive, is assumed too far away to influence the satellite movement perceptibly as compared to the influence of Earth and Moon. 3. Other astral bodies, such as neighboring planets, even Jupiter, are also considered as of negligible influence. 4. No tidal force nor any other energy-dissipating forces are deemed to be of sufficient influence to warrant taking them into account. 5. Only gravitational and inertial forces are taken into account, none due to electricity, magnetism, radiation, or cosmic dust. 6. We also confined our observations to the hypothetical case of a circular rather than elliptic Moon orbit.

It is at once clear that constant configurations can only exist in a plane coplanar with the Moon's orbit around the Earth. Even a small departure from the lunar revolution plane would evoke a restoring component of the lunar and terrestrial attraction of the Selenoid which would tend to bring it back to the lunar revolution plane. So long as the departure is very small it will merely cause the plane of the new orbit to appear tilted against the lunar orbit. Any large initial departure is likely to lead to an aperiodic motion and eventual loss of constellation.

The conditions of constancy of constellation in the plane of revolution are, obviously, that the forces of gravitational attraction by Moon and Earth must exactly balance the inertial force in an orbit in step with the Moon. For circular orbits traveled at constant velocity, the only inertial force present is the centrifugal force directed away from the barycenter B of the system. The equilibrium condition is therefore most logically expressed in two components, one circumferential, the other radial. The geometry of the general situation is illustrated by Fig. 1, in which the Moon-Earth distance is assigned the value of unity so that all other distances are expressed in non-dimensional terms of fractions or multiples of it, viz.:

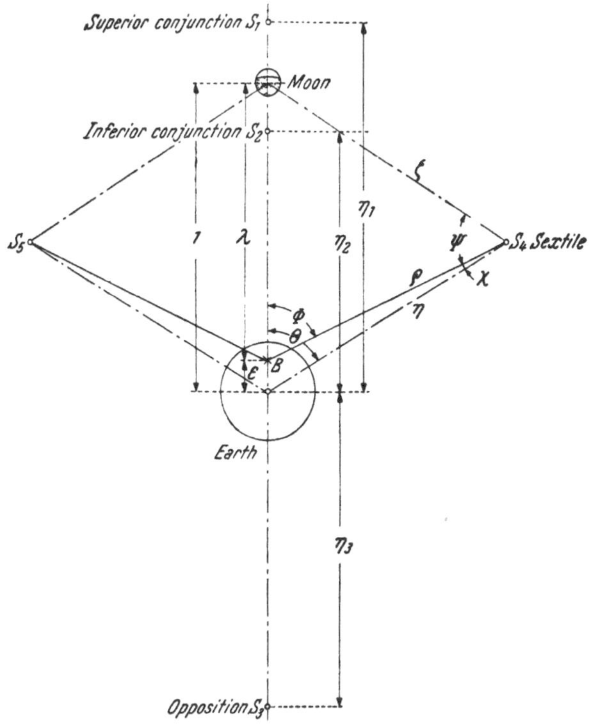

Fig. 1. Geometry

ε = the relative distance from barycenter to Earth center.
λ = the relative distance from barycenter to Moon center.
ϱ = the relative distance from barycenter to Selenoid.
η = the relative distance from Earth center to Selenoid.
ζ = the relative distance from Moon Center to Selenoid
while the angles occurring in the two triangles are:
ϕ = the angle encompassed at the barycenter between Moon and Selenoid
χ = the angle encompassed at Selenoid between Earth center and barycenter.
ψ = the angle encompassed at Selenoid between Moon center and barycenter.
θ = the angle encompassed at Earth center between Moon and Selenoid.

It will be noted that $\lambda + \varepsilon = 1$ and that the ratio of the two sections, viz., ε/λ, is the inverse of the mass ratio of Earth and Moon. (This is readily verified by considering that the centrifugal forces of the two bodies, which are balanced by their mutual attraction, must be equal; these centrifugal forces are proportional to their respective masses and their respective radii of revolution, the angular velocity being common.)

Considering first the circumferential equilibrium of forces (per unit mass) at the Selenoid Satellite, it can be expressed, in non-dimensional terms, by merely equating the sine components of the two NEWTONian attractions, cancelling all constants common to both sides, viz.:

$$\frac{\sin\chi}{\eta^2} = \frac{\varepsilon}{\lambda}\,\frac{\sin\psi}{\zeta^2}. \tag{1}$$

However, as can be gleaned from Fig. 1,

$$\sin\chi = (\varepsilon \sin\phi)/\eta$$

and

$$\sin\psi = (\lambda \sin\phi)/\zeta$$

so that eq. (1) becomes

$$\frac{\varepsilon \sin\phi}{\eta^3} = \frac{\varepsilon \sin\phi}{\zeta^3}$$

which can be fulfilled by $\eta = \zeta$. Hence one locus of points of circumferential equilibrium is the bisecant of the Earth-Moon distance, independently of the Earth-Moon mass ratio. Another such locus is the Earth-Moon axis itself, because for it $\sin\chi = \sin\psi = 0$ in eq. (1). These two straight lines are indicated dot-dashed in Fig. 2.

Now the radial equilibrium of the forces (per unit mass) is expressed by equating the sum of the cosine components of the two attractions to the centrifugal force on the Selenoid, cancelling identical factors which express, on the left side of the equation, the mutual Earth-Moon attraction and, on the right side, the centrifugal force of either body; viz.:

$$\lambda \frac{\cos\chi}{\eta^2} + \varepsilon \frac{\cos\psi}{\zeta^2} = \varrho\,. \tag{2}$$

As can also readily be gathered from Fig. 1:

$$\cos\chi = (\varrho^2 + \eta^2 - \varepsilon^2)/2\,\varrho\,\eta$$
$$\cos\psi = (\varrho^2 + \zeta^2 - \lambda^2)/2\,\varrho\,\zeta$$

and

$$\varrho^2 = \lambda\,\eta^2 + \varepsilon\,\zeta^2 - \varepsilon\lambda.$$

Inserting these in (2) yields

$$\frac{1 - \eta^2\,(1+\lambda)/\varepsilon - \zeta^2}{\eta^3} + \frac{1 - \zeta^2\,(1+\varepsilon)/\lambda - \eta^2}{\zeta^3} = 2\left(1 - \frac{\eta^2}{\varepsilon} - \frac{\zeta^2}{\lambda}\right) \tag{3 a}$$

or the quintic between η and ζ:

$$\frac{2}{\varepsilon}\eta^5\,\zeta^3 + \frac{2}{\lambda}\eta^3\,\zeta^5 - 2\,\eta^3\,\zeta^3 - \eta^5 - \zeta^5 - \frac{1+\varepsilon}{\lambda}\eta^3\,\zeta^2 - \frac{1+\lambda}{\varepsilon}\eta^2\,\zeta^3 + \eta^3 + \zeta^3 = 0. \tag{3 b}$$

The locus of all points compatible with this equation is a curve whose main branch departs but little from the lunar orbit path except where it bulges out to loop around the Moon, while another short loop swings inside from the Moon center, as shown on Fig. 2 [1].

The only places where the satellite could be in omnidirectional equilibrium, and hence capable of following the constellation constantly, are those where both eqs. (1) and (2) are fulfilled, i.e. where the curves cross the dot-dashed axes on Fig. 2. These are the five points marked S_1, S_2, S_3, S_4, and S_5. Their astronomical constellation names, their coordinates as distances from Earth (η), Moon (ζ) and barycenter (ϱ) (all referred to unit Moon-Earth distance), and their bearing angle from the Earth against the Moon are listed in the following table:

[1] If the barycenter were not inside the Earth, then a third branch of the curve would exist; it would very closely follow the Thales circle constructed between the Earth center and the barycenter, as far as it lies outside of the Earth, because at these points the Earth's attraction is normal to the radius vector and the Moon's influence is comparatively weak.

		η	ζ	ϱ	θ
S_1	Superior Conjunction	1.1676	.1676	1.1555	0
S_2	Inferior Conjunction	.8590	.1510	.8369	0
S_3	Opposition	.993	1.993	1.005	180^0
S_4 or S_5	Sextile[1]	1.000	1.000	.994	$\pm 60^0$

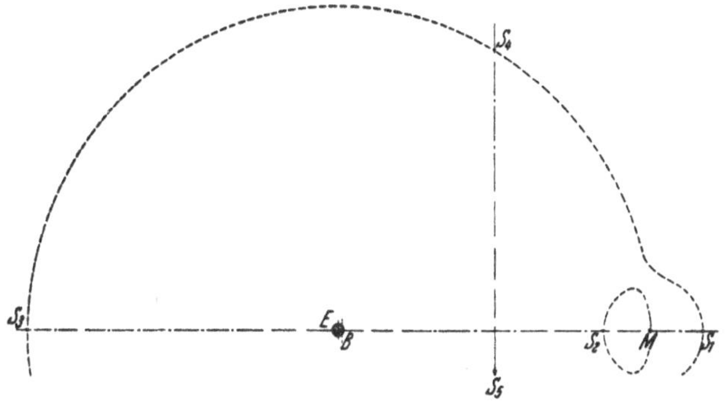

Fig. 2. Loci of peripheral and radial component equilibrium

The common quintic equation for the collinear points reads

$$\overset{+}{\pm}\eta^5 \overset{-}{\mp} (2+\varepsilon)\eta^4 \overset{+}{\pm} (1+2\varepsilon)\eta^3 - (1 \overset{+}{\pm} \varepsilon)\eta^2 + 2(1-\varepsilon)\eta = 1-\varepsilon. \qquad (4)$$

This follows from eq. (2) by letting $\overset{+}{\pm}\eta = \overset{\mp}{+}\zeta = 1$ and observing that $\varepsilon + \lambda = 1$, also $\eta - \varrho = \overset{+}{\pm} 1$, and $\cos\chi = 1$, $\cos\psi = \overset{\pm}{+} 1$. That the sextile point is also a solution is readily verified by trying $\eta = \zeta = 1$ in eq. (3 a) or (3 b), with $\cos\theta = 1/2$.

In general, all five possible constant constellations are kinematically unstable, except that, as ROUTH has shown, the sextile is stable when $\varepsilon/\lambda < 1/25$; small orbits around the libration points are then possible. In the conjunction constellations (the only ones where the Moon is close enough to exert a restoring influence which would be perceptible as a small periodic oscillation superimposed upon the Selenoid orbit motion), the frequency ω of such oscillation would bear a relationship to the lunar orbit frequency Ω, viz., $\omega/\Omega = \sqrt{\lambda/\eta^3 + \varepsilon/\zeta^3}$ for perturbations normal to the orbital plane and slightly different for perturbations tangential to the orbit. The numerical values of those frequency ratios for the mass ratio of the real Moon would be:

	Normal to Orbital Plane	Tangential to the Orbit
Superior conjunction	1.79	1.87
Inferior conjunction	2.27	2.33

Against radial perturbations the conjunction positions are unstable. In the opposition case, the Moon's influence is so weak that the effect of a perturbation cannot be interpreted as manifesting itself in an oscillation about the old orbit; it simply establishes a new one which is not in permanence of constellation. Any perturbation of the Selenoid in the orbital plane would entail a change in orbital period. Any perturbation out of the plane would cause an ecliptic inclination of the new

[1] Also called Trojan constellation after some planetoids named after Trojan heroes discovered to be in apparently permanent nearly equilateral triangle arrangement with respect to Sun and Jupiter.

orbit element against the old, which implies a continued oscillation at lunar frequency.

If any real selenoid vehicle were to be placed in space, it would presumably first have to be brought into one of the five libration points and then given a reactive impulse to match its velocity vector to that required to stay in permanence in this constellation. To be sure, the orbit of the Selenoid would have to be not circular but elliptical, geometrically similar to that of the real Moon. This ellipticity is not great so that the behavior would not differ very much from that described under the hypothetical simplifying assumptions. The real path and time table of the artificial Selenoid could readily be computed by reference to the Moon. From a practical viewpoint the constellation of inferior conjunction would probably be the most interesting, inasmuch as the Selenoid should be directly visible in front of the Moon and distinguishable either against the lighted or the dark part of it.

Perhaps it is worth mentioning that in the sextile and, particularly, in the opposition constellations the equilibrium of the Selenoid is much less precarious in peripheral than in radial direction. In fact, anywhere in the arc of the locus of radial equilibrium between these positions the Moon's influence is so weak that an error in bearing would create only a very slow departure from the initial constellation. For instance, in a position of quadrature, i.e., at 90^0 from the Moon, the peripheral pull component of the Moon would be only of the order of a millionth of a g. Hence it may be possible to counteract it for an appreciable period by the thrust of a very weak gas jet after the manner of a Satelloid.

Appendix

Symperiodic Earth Companions

Some day it might also become of interest to navigate a space vehicle into a position where it can remain in a fixed constellation with respect to the Earth and Sun. Again there will be — aside from perturbations — five equilibrium positions. Three of these, the sextiles and the opposition, would be two and six months, respectively, ahead or behind the Earth in its annual orbit around the Sun, practically at the same distance from Sun as Earth. The other two are the superior and inferior conjunction which are both located very close to 1% of the Sun's mean distance from the center of the Earth, approximately 1.5 million km, assuming a mass ratio of Earth to Sun of 3.037×10^{-6}.

This mass ratio is so small that it becomes permissible in eq. (4) to anticipate that $\eta = 1 + \delta$ is very close to 1, so that when the powers of η are developed, all terms higher than δ^3 or $\varepsilon\,\delta^2$ can be dropped for a first approximation. Hence eq. (4) reduces to $3\,\delta^3 - \varepsilon\,(1 + \delta)^2 = 0$ and within 2/3% accuracy: $\delta = \sqrt[3]{\varepsilon/3}$.

This approximation gives a good idea of the distances involved, viz., almost 4 times that of the Moon from the Earth. For any practical navigation purposes the ellipticity of the Earth's orbit would have to be taken into account and also the perturbations due to occasional passages of the Moon, and perhaps Venus and Jupiter. The perturbation caused by the Moon's motion is of an amplitude of the order 1/140 of the Earth's attraction; that of Venus in closest constellation comes to about 1/900, that of Jupiter to 1/570.

References

1. J. L. LAGRANGE, Les Oeuvres de LAGRANGE (The Works of LAGRANGE). J. A. SERRET, Editor. Paris: Gauthier-Villars, 1873.
2. H. HAPPEL, Das Dreikörperproblem. Leipzig: K. F. Koehler, 1941.

On Relativistic Rocket Mechanics[1]

By

J. M. J. Kooy[2], NVR

(With 1 Figure)

Abstract — Zusammenfassung — Résumé

On Relativistic Rocket Mechanics. Most authors, treating rectilinear rocket flight in non gravitational space, consider the rocket plus exhausted masses as *one* mechanical system, and apply the relativistic law of conservation of momentum. In this article, only the rocket plus still included fuel is considered as the mechanical system and the differential equation of rectilinear motion, and also of curvilinear motion in non gravitational space, is directly derived from the basic equation of relativistic mechanics. Further the astronautical aspect of the relativistic time dilatation is discussed. The acceleration of the vehicle with respect to a general metric field is of paramount importance. In order to define this general metric field, the space time structure of the universe as a whole may play a fundamental part.

Über relativistische Raketenmechanik. Die meisten Autoren gehen bei der Behandlung der geradlinigen relativistischen Raketenbewegung im schwerefreien Raum von dem relativistischen Impulssatz aus, und betrachten dabei die Rakete und die bereits ausgestoßene Masse als *ein* mechanisches System. In dieser Arbeit wird nur die Rakete und die noch darin befindliche Stoßmasse als das mechanische System betrachtet und die Bewegungsgleichung der geradlinigen und auch der allgemeinen Raketenbewegung im schwerefreien Raum direkt aus der Grundgleichung der relativistischen Mechanik hergeleitet.

Weiter wird der astronautische Aspekt der Zeitdilatation diskutiert. Die fundamentale Bedeutung der Beschleunigung des Fahrzeuges in Beziehung auf ein allgemeines metrisches Feld wird hervorgehoben. Zur näheren Bestimmung dieses allgemeinen metrischen Feldes wird der weltumfassende Raum-Zeit-Zusammenhang herangezogen.

Sur la mécanique relativiste des fusées. Pour appliquer la loi relativiste de la conservation de la quantité de mouvement, la plupart des auteurs conçoivent l'ensemble de la fusée et des gaz expulsés comme *le* système mécanique. Dans le présent article le système mécanique ne comporte plus que la masse de la fusée à l'instant considéré et l'équation du mouvement, même non rectiligne, est dérivée directement de l'équation fondamentale de la mécanique relativiste. La dilatation relative du temps est alors discutée du point de vue de l'astronautique. L'accélération du véhicule par rapport à un champ métrique général joue un rôle particulièrement important. La structure spatio-temporelle de l'univers est essentielle dans la définition de ce champ.

---

[1] Published in Astronaut. Acta 4, 31–58 (1958).

[2] Lector K. M. A., St. Ignatiusstraat 99 a, Breda, Holland.

In classic mechanics we have as fundamental equation:

$$\overline{K} = m\frac{d\overline{v}}{dt}. \tag{1}$$

In (1) $\overline{K}$ is the force by which the particle with mass m is acted upon, whereas v denotes the speed of the particle and t the time coordinate. Now it is very essential to observe that the force $\overline{K}$ does not formally follow from (1), but must be *defined* in each physical case. Without such an independent definition of force, it has no sense to speak about the mass m, for no method could be indicated in order to measure m. In relativistic mechanics we have as fundamental equation:

$$\overline{K} = \frac{d}{dt}\frac{m_0\,\overline{v}}{\sqrt{1-v^2/c^2}} = m_0\frac{d}{dt}\left(\frac{\overline{v}}{\sqrt{1-v^2/c^2}}\right), \tag{2}$$

in which m_0 is the rest mass of the particle, acted upon by the force $\overline{K}$. Again we can observe that it is very essential that the force $\overline{K}$ does not follow formally from (2), but must again be *defined* in each physical case. For otherwise no method can be indicated to measure m_0.

Following this natural line of thought, let us try to derive in direct way the differential equation of rectilinear motion of a rocket in non gravitational space, assuming constant fuel consumption and constant exhaust speed, controlled by the crew. In that case $\overline{K}//\overline{v}$, so that then the corresponding general eq. (2) reduces to:

$$K = m_0\frac{d}{dt}\left(\frac{v}{\sqrt{1-v^2/c^2}}\right) = m_0\frac{\dot{v}}{(\sqrt{1-v^2/c^2})^3}. \tag{3}$$

In order to apply the general eq. (3) in the case of a rectilinear rocket motion, we can consider the rocket + still included fuel as the mechanical system coming into play. The rest mass of this system then varies with time, so that then in the general eq. (3) m_0 represents the *instantaneous rest mass*. The force K by which the system is acted upon must now be defined in this special case, independent on eq. (3). Therefore let us firstly imagine that the rocket is *at rest* on a proofstand, while the rocket motor works (see Fig. 1).

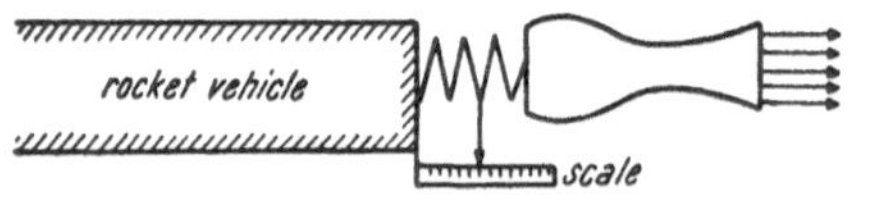

Fig. 1

Let the force by which the rocket vehicle is acted upon be directly measured by the compression of a spring (dynamometer), indicated by a pointer sliding along a scale.

By definition, the force by which the rocket vehicle is acted upon *is* then this pointer reading. Now in this case of the rocket at rest on the proofstand, we may be quite sure of it that the force, as defined in this way, will be equal to the momentum generated per second by expansion of the gases in the nozzle. If a_0 be the exhaust speed, Δm_0^* the rest mass of the expelled matter per second and c the speed of light, this momentum generated per second becomes:

$$\frac{\Delta m_0^*\,a_0}{\sqrt{1-a_0^2/c^2}} = \mu_0\,a_0.$$

If the corresponding pointer reading of the dynamometer be P, we may then write:

$$P = \mu_0\,a_0.$$

Let us now indicate the time coordinate by t_0 and assume that at $t_0 = 0$ the rest mass of the rocket vehicle + included fuel be M_0. The total quantity of energy expelled per second, *energy of rest mass included*, becomes:

$$\frac{\Delta m_0^* c^2}{\sqrt{1 - a_0^2/c^2}},$$

and the rest mass of this total energy (= rest mass energy + kinetic energy) becomes:

$$\frac{\Delta m_0^*}{\sqrt{1 - a_0^2/c^2}} = \mu_0.$$

Hence the rest mass of the rocket vehicle + still included fuel at any time t_0 becomes:

$$M_0 - \mu_0 t_0.$$

Let us now, as next stage in our line of thought, consider our rocket vehicle in rectilinear flight in non gravitational space. Let us thereby assume — for sake of simplicity — that the mass of the rocket motor + content is still very small in comparison to the instantaneous mass of rocket vehicle + still included fuel. We can then again *define* as magnitude of the thrust the *pointer reading* of the dynamometer. (This pointer reading may be conveyed to the terrestrial observer by telemetering.) Then we have to substitute in (3) for K the quantity:

$$\frac{\Delta m_0^* a_0}{\sqrt{1 - a_0^2/c^2}} = \mu_0 a_0,$$

in which now Δm_0^* and a_0 are the corresponding quantities *as measured by the crew of the rocket vehicle*, that is as to the system of reference at rest as to the rocket vehicle, and with a corresponding clock, at present in the rocket vehicle. The time coordinate, as measured by this astronautical clock, will now be indicated by t_0, whereas the terrestrial time will be denoted by t. Hence, applying the general eq. (3) we obtain in this way as differential equation of motion:

$$(M_0 - \mu_0 t_0) \frac{dv/dt}{(1 - v^2/c^2)^{3/2}} = \mu_0 a_0. \tag{4}$$

Eq. (4) is quite in accordance with the result of KRAUSE [1], who considered rocket + exhausted fuel as the mechanical system coming into play and applied accordingly the relativistic law of conservation of momentum. Further following KRAUSE, we can multiply (4) by $dt_0 = \sqrt{1 - v^2/c^2}\, dt$, so that then:

$$\frac{dv}{(1 - v^2/c^2)} = \frac{\mu_0 a_0 dt_0}{M_0 - \mu_0 t_0}.$$

If the times $t = 0$ and $t_0 = 0$ coincide, and at $t = 0$, $v = 0$, we obtain by integration [1] the well known formula of ACKERET [2]:

$$\left(\frac{1 + v/c}{1 - v/c}\right)^{c/2a_0} = \frac{M_0}{M_0 - \mu_0 t_0}, \tag{5}$$

by which the mass ratio can be computed required for obtaining a speed v. From (5) we find:

$$M_0 - \mu_0 t_0 = M_0 \left(\frac{1 - v/c}{1 + v/c}\right)^{c/2a_0}. \tag{6}$$

Substituting (6) in (4) we obtain:

$$M_0 \left(\frac{1 - v/c}{1 + v/c}\right)^{c/2a_0} \frac{dv/dt}{(1 - v^2/c^2)^{3/2}} = \mu_0 a_0,$$

whence follows:

$$\frac{M_0}{\mu_0\, a_0}\int\limits_0^v \left(\frac{1 - v/c}{1 + v/c}\right)^{c/2a_0} \frac{dv}{(1 - v^2/c^2)^{3/2}} = t. \tag{7}$$

In (7) the integral can always be evaluated numerically. By (7) we can compute the time required for obtaining the final speed v or any speed v during the powered rectilinear flight.

If in this way $v = v(t)$ is found numerically, the way passed through at time t becomes:

$$s = \int\limits_0^t v(t)\, dt. \tag{8}$$

In case of a photon rocket no matter, but photons are expelled. If E be the energy expelled per second astronautical time, the mass of this energy is equal E/c^2, and the momentum E/c, so that then: $P = E/c$.

In that case the differential equation of motion for rectilinear flight and constant thrust in non gravitational space becomes:

$$\left(M_0 - \frac{E}{c^2}\, t_0\right) \frac{dv/dt}{(1 - v^2/c^2)^{3/2}} = \frac{E}{c}. \tag{9}$$

Hence in this case eq. (5) becomes:

$$\left(\frac{1 + v/c}{1 - v/c}\right)^{1/2} = \frac{M_0}{M_0 - (E/c^2)\, t_0}, \tag{10}$$

and eq. (7) reduces to:

$$\frac{c\, M_0}{E}\int\limits_0^v \left(\frac{1 - v/c}{1 + v/c}\right)^{1/2} \frac{dv}{(1 - v^2/c^2)^{3/2}} = t. \tag{11}$$

Now the expression under the integral sign can be transformed to:

$$\sqrt{\frac{1 - v/c}{(1 + v/c)\,(1 - v/c)^3\,(1 + v/c)^3}}\, dv =$$

$$= \frac{dv}{\sqrt{(1 + v/c)^4\,(1 - v/c)^2}} = \frac{dv}{(1 - v/c)\,(1 + v/c)^2}.$$

Hence we may write for (11):

$$\frac{c\, M_0}{E}\int\limits_0^v \frac{dv}{(1 - v/c)\,(1 + v/c)^2} = t. \tag{12}$$

Now we can split up the integral in (12) in partial fractions, by writing

$$\frac{1}{(1 - v/c)\,(1 + v/c)^2} = \frac{A}{1 - v/c} + \frac{B}{1 + v/c} + \frac{C}{(1 + v/c)^2}.$$

Hence:

$$1 = A\,(1 + v/c)^2 + B\,(1 - v/c)\,(1 + v/c) + C\,(1 - v/c).$$

Or:

$$1 = A\,(1 + 2\,v/c + v^2/c^2) + B\,(1 - v^2/c^2) + C\,(1 - v/c).$$

Thus:

$$1 = A + B + C + v/c\,(2\,A - C) + v^2/c^2\,(A - B).$$

This relation must avail for all possible values of v, from which follows:

$$\left.\begin{aligned} A+B+C&=1\\ 2A-C&=0\\ A-B&=0\end{aligned}\right\}\quad \text{thus } A=1/4,\quad B=1/4,\quad C=1/2.$$

Hence by (12):

$$\frac{c\,M_0}{E}\int\limits_0^v\left(\frac{1/4}{1-v/c}+\frac{1/4}{1+v/c}+\frac{1/2}{(1+v/c)^2}\right)dv=t,$$

so that:

$$\frac{c\,M_0}{E}\left\{\frac{c}{4}\ln\frac{1+v/c}{1-v/c}+\frac{v}{2\,(1+v/c)}\right\}=t. \tag{13}$$

By means of (13) we can find for any v the corresponding value for t, so that also $v=v(t)$ can be determined numerically. The way passed through at any time t then again follows from (8).

Further by (10) and (13) we can find the rest mass of photon rocket and still included fuel at any terrestrial time t during the powered flight. For by (13) we can find for any speed v the corresponding terrestrial time t, and then the corresponding value of $M_0-(E/c^2)\,t_0$ follows from (10). In case of the rocket expelling matter, we can find in similar way by (6) and (7) the rest mass of rocket vehicle + still included fuel at any time t of the powered flight.

Now by (5) we can already conclude that the exhaust speed must be at least of the order of magnitude of the speed of light, if we like to obtain a speed v comparable with c. Of course the most advantageous case is the photon rocket, but we have to bear in mind that we have today no idea how to obtain a concentrated photonbeam giving a thrust of some technical importance. As well known a large exhaust speed can be obtained by acceleration of charged particles (ions) in an electric field. If V be the electric potential passed through, e the charge and m_0 the restmass of each particle, the exhaust speed a_0 follows from:

$$e\,V=m_0\,c^2\left(\frac{1}{\sqrt{1-a_0^2/c^2}}-1\right).$$

If n be the number of particles jettisoned per second, the corresponding jet power becomes

$$n\,e\,V=n\,m_0\,c^2\left(\frac{1}{\sqrt{1-a_0^2/c^2}}-1\right),\qquad\text{in which}\qquad n\,m_0=\Delta\,m_0{}^*.$$

Of course an equal number of particles of opposite charge must be expelled, in order to keep the vehicle neutral.

It would be very advantageous if this acceleration of electric particles could be combined in some way by applying atomic energy obtained by fusion of nucleons, being the most powerful source known today.

Let us imagine we have a gaseous mixture of ${}^1_2\mathrm{H}$ and ${}^1_3\mathrm{H}$ nucleons, through which an electric discharge occurs. Let us indicate the particle charge by e and the masses of the ${}^1_2\mathrm{H}$ and ${}^1_3\mathrm{H}$ nucleons by $2\,m_0$ and $3\,m_0$. If then v_1 be the speed of a ${}^1_2\mathrm{H}$ nucleon after having passed through a potential drop V, we obtain:

$$e\,V=2\,m_0\,c^2\left(\frac{1}{\sqrt{1-v_1^2/c^2}}-1\right),$$

whence follows:

$$v_1 = c\sqrt{1 - \frac{4\,m_0^2\,c^4}{(e\,V + 2\,m_0\,c^2)^2}} = c\sqrt{\frac{e^2\,V^2 + 4\,e\,V\,m_0\,c^2}{e^2\,V^2 + 4\,e\,V\,m_0\,c^2 + 4\,m_0^2\,c^4}} \tag{14}$$

and similar as to ${}_3^1\mathrm{H}$:

$$v_2 = c\sqrt{\frac{e^2\,V^2 + 6\,e\,V\,m_0\,c^2}{e^2\,V^2 + 6\,e\,V\,m_0\,c^2 + 9\,m_0^2\,c^4}}\,. \tag{15}$$

Hence after having passed through a potential drop V, the relative speed of the accelerated ${}_2^1\mathrm{H}$ nucleon as to the accelerated ${}_3^1\mathrm{H}$ nucleon becomes:

$$v_{\text{rel}} = \frac{v_1 - v_2}{1 - v_1\,v_2/c^2}\,. \tag{16}$$

Substituting (14) and (15) in (16) we obtain:

$$v_{\text{rel}} = F(V). \tag{17}$$

By means of (17) we can determine the potential drop $V_{\min}$ in order to obtain the required fusion:

$${}_2^1\mathrm{H} + {}_3^1\mathrm{H} = {}_5^2\mathrm{He} \to {}_4^2\mathrm{He} + {}_1^0n. \tag{18}$$

From (16) it follows that v_{rel} tends to the speed of light if V increases, for if we write $v_1 \approx c$, we obtain by (16):

$$\lim_{v_1 \to c} v_{\text{rel}} = \frac{c - v_2}{1 - v_2/c} = c\,.$$

Of course only fusion occurs in case of an encounter. If the density of the mixture is small, such encounters will be rare and the fusion energy which will be gained will only be a small fraction of the energy required for maintaining the current. By increasing this density a value will be attained at which the gain of fusion energy will cancel the loss of energy required for current maintenance. On the other hand, the density of the mixture may only be increased to such an extent, that the fusion heat development remains below a certain value. In this way it seems possible to come to a controlled employ of fusion energy, which could be used for driving the turbo generators maintaining the current. The ${}_2^1\mathrm{H}$ and ${}_3^1\mathrm{H}$ nucleons and also the formed ${}_4^2\mathrm{He}$ particles can then be used as propulsion material. The quantity of expelled ${}_4^2\mathrm{He}$ particles however must necessarily be small and their contribution to the jet can be disregarded. Thereby, as a consequence of the special theory of relativity, the scattering influence of the Coulomb repulsion tends to zero, if the exhaust speed tends to c. Also in the gaseous mixture, by the magnetic field of the convection current, the charged particles will be acted upon by a Lorentz force, causing a tendency to keep these particles in stream direction, thereby protecting the walls of the vessel. On the other hand, the neutron radiation due to the disintegration

$${}_5^2\mathrm{He} \to {}_4^2\mathrm{He} + {}_1^0n$$

will give rise to serious difficulties. For the walls a material must be used which can endure this neutron bombardment.

Further, as already stated above, always an equivalent number of electrons must be emitted in order to keep the space ship neutral and maintain the thrust. The ejection of these electrons can further contribute to the thrust.

If Δm_0^* be the total rest-mass of ${}^1_2H$ and ${}^1_3H$ nucleons expelled per second astronautical time, the total thrust due to these nucleons becomes:

$$S = \frac{2/5 \, \Delta m_0^* v_1}{\sqrt{1 - v_1^2/c^2}} + \frac{3/5 \, \Delta m_0^* v_2}{\sqrt{1 - v_2^2/c^2}}, \tag{19}$$

in which v_1 and v_2 follow from (14) and (15). For simplicity let us neglect the contribution of the expelled electrons to the thrust, as well as their mass. If we then assume S [formula (19)] as constant, and write:

$$\frac{2/5 \, \Delta m_0^*}{\sqrt{1 - v_1^2/c^2}} + \frac{3/5 \, \Delta m_0^*}{\sqrt{1 - v_2^2/c^2}} = \mu_0' \tag{20}$$

we obtain instead of (4):

$$(M_0 - \mu_0' t_0) \frac{dv/dt}{(1 - v^2/c^2)^{3/2}} = S \, . \tag{21}$$

Hence again multiplying by $dt_0 = dt \sqrt{1 - v^2/c^2}$ we obtain:

$$\frac{dv}{1 - v^2/c^2} = \frac{S \, dt_0}{M_0 - \mu_0' t_0} = - \frac{S}{\mu_0'} \frac{d(M_0 - \mu_0' t_0)}{M_0 - \mu_0' t_0},$$

or after integration:

$$\frac{c}{2} \ln \frac{1 + v/c}{1 - v/c} = \frac{S}{\mu_0'} \ln \frac{M_0}{M_0 - \mu_0' t_0} \, .$$

Or:

$$\left(\frac{1 + v/c}{1 - v/c}\right)^{c\mu'_0/2S} = \frac{M_0}{M_0 - \mu_0' t_0} \, .$$

Hence:

$$M_0 - \mu_0' t_0 = \left(\frac{1 - v/c}{1 + v/c}\right)^{c\mu_0'/2S} M_0 .$$

This introduced in (21) gives:

$$M_0 \left(\frac{1 - v/c}{1 + v/c}\right)^{c\mu_0'/2S} \frac{dv/dt}{(1 - v^2/c^2)^{3/2}} = S.$$

Or:

$$\frac{M_0}{S} \int_0^v \left(\frac{1 - v/c}{1 + v/c}\right)^{c\mu_0'/2S} \frac{dv}{(1 - v^2/c^2)^{3/2}} = t,$$

whence follows again $v = v(t)$ and $s = \int_0^t v(t) \, dt$. For the rest it may be observed that in the discharge vessel also neutral hydrogen or another neutral medium could be brought, which is heated up by the fusion heat and exhausts through a nozzle. In this way however, assuming a temperature of about 3000° C in the discharge vessel, only exhaust speeds of the order of 10 km/sec. can be obtained.

Now let us generalize (4) by assuming that μ_0 and a_0 are not constant, but prescribed functions of t_0. We then obtain:

$$\left(M_0 - \int_0^{t_0} \mu_0(t_0)\, dt_0\right) \frac{\dot{v}}{(1 - v^2/c^2)^{3/2}} = \mu_0(t_0)\, a_0(t_0).$$

Then again multiplying by $dt_0 = \sqrt{1 - v^2/c^2}\; dt$ we find:

$$\frac{dv}{1 - v^2/c^2} = \frac{\mu_0(t_0)\, a_0(t_0)\, dt_0}{M_0 - \int_0^{t_0} \mu_0(t_0)\, dt_0},$$

so that

$$c \int_0^v \frac{d(v/c)}{1 - v^2/c^2} = \int_0^{t_0} \frac{\mu_0(t_0)\, a_0(t_0)\, dt_0}{M_0 - \int_0^{t_0} \mu_0(t_0)\, dt_0}. \tag{22}$$

Or:

$$\frac{c}{2} \ln \frac{1 + v/c}{1 - v/c} = \varphi(t_0). \tag{23}$$

In (22) the right member $\varphi(t_0)$ can be numerically evaluated for any value of t_0, so that $v = v(t_0)$ becomes known. Then further:

$$\frac{dt_0}{dt} = \sqrt{1 - 1/c^2\, [v(t_0)]^2},$$

so that:

$$\int_0^{t_0} \frac{dt_0}{\sqrt{1 - 1/c^2\, [v(t_0)]^2}} = t. \tag{24}$$

By numerical integration of (24) we find $t_0 = t_0(t)$, so that then also $v(t_0) = v(t_0(t)) = v(t)$ becomes known. Then further the way passed through at any time t during the powered rectilinear flight again follows from:

$$s = \int_0^t v(t)\, dt.$$

Let us now consider the more restricted case that $a_0(t_0) = a_0 = \text{const}$, whereas $\Delta m_0^* = \Delta m_0^*(t_0)$ is variable and prescribed. In that special case (22) simplifies to:

$$\frac{c}{2} \ln \frac{1 + v/c}{1 - v/c} = a_0 \int_0^{t_0} \frac{\mu_0(t_0)\, dt_0}{M_0 - \int_0^{t_0} \mu_0(t_0)\, dt_0}.$$

Or writing

$$M_0 - \int_0^{t_0} \mu_0(t_0)\, dt_0 = M(t_0),$$

we obtain:

$$\frac{c}{2}\ln\frac{1+v/c}{1-v/c} = -a_0 \int\limits_{M_0}^{M(t^0)} \frac{dM(t_0)}{M(t_0)},$$

so that:

$$\frac{c}{2\,a_0}\ln\frac{1+v/c}{1-v/c} = \ln\frac{M_0}{M(t_0)},$$

or:

$$\left(\frac{1+v/c}{1-v/c}\right)^{c/2a_0} = \frac{M_0}{M(t_0)} = \frac{M_0}{M_0 - \int\limits_0^{t_0} \mu_0(t_0)\,dt_0}. \tag{25}$$

(25) is again the well known formula of Ackeret which avails for all cases of rectilinear flight in non gravitational space, as long as a_0 is kept constant. (For the rest, as soon as we think in terms of the speed of light, ordinary gravitational forces practically do not count.) The formulae above not essentially alter if we consider the more general case that different types of particles are simultaneously emitted with different exhaust speeds.

As example let us consider a space ship emitting positive and negative particles (the normal case of ion propulsion).

Let $\Delta\, m_{0_1}{}^*$ be the rest mass of the positive particles and $\Delta\, m_{0_2}{}^*$ the rest mass of the negative particles emitted per second astronautical time t_0 and let a_{0_1} and a_{0_2} be the corresponding exhaust speeds as to the rocket system of reference. Let $\Delta\, m_{0_1}{}^*$, and thus also $\Delta\, m_{0_2}{}^*$, and further a_{0_1} and a_{0_2} be prescribed as functions of t_0. Then for rectilinear motion we obtain:

$$\left.\begin{aligned} &\frac{dv/dt}{(1-v^2/c^2)^{3/2}} = \frac{\mu_{0_1}(t_0)\,a_{0_1}(t_0) + \mu_{0_2}(t_0)\,a_{0_2}(t_0)}{M_0 - \int\limits_0^{t_0}(\mu_{0_1}(t_0)+\mu_{0_2}(t_0))\,dt_0} \\ &\text{in which} \\ &\mu_{0_1} = \frac{\Delta\, m_{0_1}{}^*(t_0)}{\sqrt{1-\frac{[a_{0_1}(t_0)]^2}{c^2}}}, \qquad \mu_{0_2} = \frac{\Delta\, m_{0_2}{}^*(t_0)}{\sqrt{1-\frac{[a_{0_2}(t_0)]^2}{c^2}}}. \end{aligned}\right\} \tag{26}$$

Again by multiplying with $\sqrt{1-v^2/c^2}\,dt = dt_0$ and subsequent integration we obtain:

$$c\int\limits_0^v \frac{d(v/c)}{1-v^2/c^2} = \frac{c}{2}\ln\frac{1+v/c}{1-v/c} = \int\limits_0^{t_0} \frac{\mu_{0_1}(t_0)\,a_{0_1}(t_0) + \mu_{0_2}(t_0)\,a_{0_2}(t_0)}{M_0 - \int\limits_0^{t_0}(\mu_{0_1}(t_0)+\mu_{0_2}(t_0))\,dt_0}\,dt_0. \tag{27}$$

The right member of (27) can be integrated numerically, so that $v = v(t_0)$ can be determined. Then further we obtain:

$$\int\limits_0^{t_0} \frac{dt_0}{\sqrt{1-\frac{1}{c^2}\,[v(t_0)]^2}} = t \rightarrow t_0 = t_0(t),$$

so that:

$$v(t_0) = v(t_0(t)) = v(t), \; s = \int_0^t v(t)\, dt.$$

If a_{01} and a_{02} are maintained at constant value, and for the rest $\Delta m_{01}{}^* = \Delta m_{01}{}^*(t_0)$ and $\Delta m_{02}{}^* = \Delta m_{02}{}^*(t_0)$ be prescribed, we may write (27):

$$\frac{c}{2} \ln \frac{1 + v/c}{1 - v/c} = a_{01} \int_0^{t_0} \frac{\mu_{01}(t_0)\, dt_0}{M(t_0)} + a_{02} \int_0^{t_0} \frac{\mu_{02}(t_0)\, dt_0}{M(t_0)}.$$

Now let us write:

$$\frac{1}{\sqrt{1 - \frac{a_{01}{}^2}{c^2}}} = \beta_1, \qquad \frac{1}{\sqrt{1 - \frac{a_{02}{}^2}{c^2}}} = \beta_2.$$

We then obtain:

$$\frac{c}{2} \ln \frac{1 + v/c}{1 - v/c} = a_{01}\, \beta_1 \int_0^{t_0} \frac{\Delta m_{01}{}^*(t_0)\, dt_0}{M(t_0)} + a_{02}\, \beta_2 \int_0^{t_0} \frac{\Delta m_{02}{}^*(t_0)\, dt_0}{M(t_0)}.$$

Let us further write:

$$a_{02} = k_1\, a_{01}, \qquad \Delta m_{02}{}^*(t_0) = k_2 \Delta m_{01}{}^*(t_0).$$

Then:

$$\frac{c}{2} \ln \frac{1 + v/c}{1 - v/c} = a_{01}\, (\beta_1 + k_1\, k_2\, \beta_2) \int_0^{t_0} \frac{\Delta m_{01}{}^*(t_0)\, dt_0}{M(t_0)}.$$

Now:

$$- dM(t_0) = [\mu_{01}(t_0) + \mu_{02}(t_0)]\, dt_0 =$$
$$= [\beta_1 \Delta m_{01}{}^*(t_0) + \beta_2 \Delta m_{02}{}^*(t_0)]\, dt_0 = (\beta_1 + \beta_2\, k_2)\, \Delta m_{01}{}^*(t_0)\, dt_0,$$

so that we may write:

$$\frac{c}{2} \ln \frac{1 + v/c}{1 - v/c} = a_{01} \frac{\beta_1 + k_1 k_2 \beta_2}{\beta_1 + k_2 \beta_2} \int_0^{t_0} \frac{(\beta_1 + k_2\, \beta_2)\, \Delta m_{01}{}^*(t_0)\, dt_0}{M(t_0)}.$$

Hence if we write:

$$\frac{\beta_1 + k_1 k_2 \beta_2}{\beta_1 + k_2 \beta_2} = \alpha,$$

we get:

$$\frac{c}{2} \ln \frac{1 + v/c}{1 - v/c} = - a_{01} \alpha \int_{M_0}^{M(t_0)} \frac{dM(t_0)}{M(t_0)}.$$

Hence:

$$\frac{c}{2\, \alpha\, a_{01}} \ln \frac{1 + v/c}{1 - v/c} = \ln \frac{M_0}{M(t_0)}. \qquad \text{Or:} \qquad \left(\frac{1 + v/c}{1 - v/c}\right)^{c/2\, \alpha\, a_{01}} = \frac{M_0}{M(t_0)}. \tag{28}$$

(28) is then again the formula of Ackeret for this special case. Let us now proceed to the general case of relativistic curvilinear motion of the same rocketship, emitting positive and negative particles, moving in non gravitational

space. In this case the directions of speed and thrust will be different. The rest masses of the positive and negative particles jettisoned per second astronautical time and the corresponding exhaust speeds are again indicated as above. Further the magnitude of the thrust (= the force by which rocket vehicle + still included fuel is acted upon) is again equal to the dynamometer pointer reading as imagined before, and therefore becomes

$$\mu_{01}\, a_{01} + \mu_{02}\, a_{02} \qquad \text{[compare eq. (26)]}.$$

The *direction* of this thrust now depends on the instantaneous position of the axis of the rocket, and will be different for terrestrial observer and occupant of the vehicle, on account of the LORENTZ contraction of the space of the latter. Therefore in order to avoid mathematical complications, let us prescribe $\Delta\, m_{01}{}^*$ and $\Delta\, m_{02}{}^*$, being the rest masses of the positive and negative particles jettisoned per second astronautical time as well as the magnitudes a_{01} and a_{02} of the corresponding astronautical exhaust speeds, as functions of t_0, whereas the common direction (thrust direction) of the vector quantities $\overline{a}_{01}$ and $\overline{a}_{02}$, coinciding with the instantaneous axial direction of the rocket, at any astronautical time t_0, *be prescribed with respect to the terrestrial system of reference.* (The numerical values of the vector quantities $\overline{a}_{01}$ and $\overline{a}_{02}$ are a_{01} and a_{02}.)

Hence in accordance with the fundamental eq. (2) we then obtain as differential equations of motion:

$$\left.\begin{aligned} \left\{M_0 - \int_0^{t_0} [\mu_{01}(t_0) + \mu_{02}(t_0)]\, dt_0\right\} \frac{d}{dt}\left(\frac{\overline{v}}{\sqrt{1 - v^2/c^2}}\right) &= \mu_{01}(t_0)\, \overline{a}_{01}(t_0) + \mu_{02}(t_0)\, \overline{a}_{02}(t_0) \\ \frac{dt_0}{dt} &= \sqrt{1 - v^2/c^2}. \end{aligned}\right\} \quad (29)$$

Then in (29) the force on the right of the vector equation, as prescribed function of astronautical time t_0, refers to the terrestrial system of reference xyz, as to magnitude *and* direction. (This terrestrial system of reference may be chosen with the earth center as origin, and does not rotate with respect to the celestial sky.) (29) is a system of 4 simultaneous differential equations of first order, with t as independent variable, and v_x, v_y, v_z and t_0 as dependent variables. (The vector equation stands for 3 component equations corresponding with x, y and z direction.) Starting from the initial state:

$$t = 0 \rightarrow t_0 = 0 \rightarrow v_x = 0, \qquad v_y = 0, \qquad v_z = 0,$$

this system can be solved numerically, whether by the method of RUNGE-KUTTA, or by the method of successive approximations. Let us write:

$$\left.\begin{aligned} &\frac{\mu_{01}(t_0) a_{01_x}(t_0) + \mu_{02}(t_0)\, a_{02_x}(t_0)}{M_0 - \int_0^{t_0} [\mu_{01}(t_0) + \mu_{02}(t_0)]\, dt_0} = p_x(t_0) \\ &\text{and similar } p_y(t_0) \text{ and } p_z(t_0). \end{aligned}\right\} \quad (30)$$

Carrying out the differentiations

$$\frac{d}{dt}\left(\frac{v_x}{\sqrt{1 - (v_x{}^2 + v_y{}^2 + v_z{}^2)/c^2}}\right), \text{ etc.,}$$

we then obtain the following set of equations:

$$\left.\begin{aligned}
&\frac{1-(v_y^2+v_z^2)/c^2}{(1-(v_x^2+v_y^2+v_z^2)/c^2)^{3/2}}\frac{dv_x}{dt}+\frac{v_x v_y/c^2}{(1-(v_x^2+v_y^2+v_z^2)/c^2)^{3/2}}\frac{dv_y}{dt}+\\
&\qquad+\frac{v_x v_z/c^2}{(1-(v_x^2+v_y^2+v_z^2)/c^2)^{3/2}}\frac{dv_z}{dt}=p_x(t_0)\\
&\frac{v_y v_x/c^2}{(1-(v_x^2+v_y^2+v_z^2)/c^2)^{3/2}}\frac{dv_x}{dt}+\frac{1-(v_x^2+v_z^2)/c^2}{(1-(v_x^2+v_y^2+v_z^2)/c^2)^{3/2}}\frac{dv_y}{dt}+\\
&\qquad+\frac{v_y v_z/c^2}{(1-(v_x^2+v_y^2+v_z^2)/c^2)^{3/2}}\frac{dv_z}{dt}=p_y(t_0)\\
&\frac{v_z v_x/c^2}{(1-(v_x^2+v_y^2+v_z^2)/c^2)^{3/2}}\frac{dv_x}{dt}+\frac{v_z v_y/c^2}{(1-(v_x^2+v_y^2+v_z^2)/c^2)^{3/2}}\frac{dv_y}{dt}+\\
&\qquad+\frac{1-(v_x^2+v_y^2)/c^2}{(1-(v_x^2+v_y^2+v_z^2)/c^2)^{3/2}}\frac{dv_z}{dt}=p_z(t_0)\\
&\qquad\frac{dt_0}{dt}=\sqrt{1-\frac{v_x^2+v_y^2+v_z^2}{c^2}}.
\end{aligned}\right\}\tag{31}$$

The first three eqs. (31) are linear in the differential quotients dv_x/dt, dv_y/dt and dv_z/dt, so that we can solve these quantities from these equations. Then we can write (31):

$$\left.\begin{aligned}
&\frac{dt_0}{dt}=\sqrt{1-\frac{v_x^2+v_y^2+v_z^2}{c^2}}=F_1(v_x,v_y,v_z)\\
&\frac{dv_x}{dt}=F_2(t_0,v_x,v_y,v_z)\\
&\frac{dv_y}{dt}=F_3(t_0,v_x,v_y,v_z)\\
&\frac{dv_z}{dt}=F_4(t_0,v_x,v_y,v_z).
\end{aligned}\right\}\tag{32}$$

Starting from the initial state $t=0\to t_0=0,\ v_x=0,\ v_y=0,\ v_z=0$, let us integrate (32) by the method of RUNGE-KUTTA. We then obtain the scheme:

$$\left.\begin{aligned}
&t\to h,\ t_0\to k,\ v_x\to l,\ v_y\to n,\ v_z\to q.\\
&k=\frac{1}{6}k_1+\frac{1}{3}k_2+\frac{1}{3}k_3+\frac{1}{6}k_4 \qquad n=\frac{1}{6}n_1+\frac{1}{3}n_2+\frac{1}{3}n_3+\frac{1}{6}n_4\\
&l=\frac{1}{6}l_1+\frac{1}{3}l_2+\frac{1}{3}l_3+\frac{1}{6}l_4 \qquad q=\frac{1}{6}q_1+\frac{1}{3}q_2+\frac{1}{3}q_3+\frac{1}{6}q_4\\
&k_1=F_1(v_x,v_y,v_z)\,h \qquad l_1=F_2(t_0,v_x,v_y,v_z)\,h\\
&k_2=F_1\left(v_x+\frac{l_1}{2},v_y+\frac{n_1}{2},v_z+\frac{q_1}{2}\right)h \qquad l_2=F_2\left(t_0+\frac{k_1}{2},\right.\\
&\qquad\qquad\left.v_x+\frac{l_1}{2},v_y+\frac{n_1}{2},v_z+\frac{q_1}{2}\right)h
\end{aligned}\right\}\tag{33}$$

$$\left.\begin{aligned}
&k_3 = F_1\left(v_x + \frac{l_1}{2}, v_y + \frac{n_1}{2}, v_z + \frac{q_1}{2}\right) h \qquad l_3 = F_2\left(t_0 + \frac{k_2}{2},\right. \\
&\qquad\qquad\qquad\qquad\qquad\qquad\qquad \left. v_x + \frac{l_2}{2}, v_y + \frac{n_2}{2}, v_z + \frac{q_2}{2}\right) h \\
&k_4 = F_1\left(v_x + l_3, v_y + n_3, v_z + q_3\right) h \qquad l_4 = F_2\left(t_0 + k_3, v_x + l_3,\right. \\
&\qquad\qquad\qquad\qquad\qquad\qquad\qquad \left. v_y + n_3, v_z + q_3\right) h
\end{aligned}\right\} \tag{33}$$

and similar n and q.

The computation according to the scheme (**33**) must be carried out in the sequence $k_1\, l_1\, n_1\, q_1\, k_2\, l_2\, n_2\, q_2\, k_3\, l_3\, n_3\, q_3\, k_4\, l_4\, n_4\, q_4$, and is suitable for a digital electronic computer. Repeating the procedure with half the number of steps, the difference in result is about 15 $\times$ the order of magnitude of the remaining inaccuracy of the result of the first computation.

By this step by step integration, we then find:

$$t_0 = t_0(t), \qquad v_x = v_x(t), \qquad v_y = v_y(t), \qquad v_z = v_z(t). \tag{34}$$

Then further the coordinates of the rocket vehicle at any terrestrial time t become:

$$x = \int_0^t v_x(t)\, dt + x_0, \qquad y = \int_0^t v_y(t)\, dt + y_0, \qquad z = \int_0^t v_z(t)\, dt + z_0. \tag{35}$$

The integrals in (35) must again be computed numerically. In connection with the first eq. (**34**) we can then further determine:

$$\begin{aligned}
v_x &= v_x(t_0), & v_y &= v_y(t_0), & v_z &= v_z(t_0) \\
x &= x(t_0), & y &= y(t_0), & z &= z(t_0).
\end{aligned} \tag{36}$$

After having carried out the computation, we can arrange the program of navigation for the crew, in terms of astronautical time t_0. The required consecutive directions of navigation (directions of rocket axis) can be indicated by remote celestial objects (spiral nebulae).

We can also proceed in different way by *prescribing* the orbit in space and time, and determine the required program of navigation. In that case the functions

$$x = x(t), \qquad y = y(t), \qquad z = z(t) \tag{37}$$

are given, so that the functions:

$$v_x = v_x(t), \qquad v_y = v_y(t), \qquad v_z = v_z(t) \tag{38}$$

are known.

We have again to start from the system (29). Let us write:

$$\left.\begin{aligned}
\frac{d}{dt}\left(\frac{\overline{v}}{1 - v^2/c^2}\right) &= \frac{\mu_{01}(t_0)\, \overline{a}_{01}(t_0) + \mu_{02}(t_0)\, \overline{a}_{02}(t_0)}{M_0 - \int_0^{t_0} [\mu_{01}(t_0) + \mu_{02}(t_0)]\, dt_0} \\
\frac{dt_0}{dt} &= \sqrt{1 - \frac{v^2}{c_2}}.
\end{aligned}\right\} \tag{29'}$$

If $\overline{v} = \overline{v}(t)$ (hence (38)) is prescribed, the left member of the first eq. (29′) as vector function of t is known. Let us indicate this left member by $\overline{p}(t)$. Further by the last eq. (29′) we obtain:

$$t_0 = \int_0^t \sqrt{1 - \frac{[v_x(t)]^2 + [v_y(t)]^2 + [v_z(t)]^2}{c^2}}\, dt.$$

The integral must be computed numerically. In this way we find $t = t(t_0)$. Then $\overline{v} = \overline{v}(t) = \overline{v}(t(t_0)) = \overline{v}(t_0)$, so that also $\overline{p} = \overline{p}(t_0)$ as function of t_0 becomes known.

Then the first eq. (29) is equivalent with the three component equations:

$$\left.\begin{array}{l} p_x(t_0) = \dfrac{\mu_{01}(t_0)\, a_{01_x}(t_0) + \mu_{02}(t_0)\, a_{02_x}(t_0)}{M_0 - \int_0^{t_0} (\mu_{01}(t_0) + \mu_{02}(t_0))\, dt_0} \\ \text{and similar for } y \text{ and } z. \end{array}\right\} \quad (39)$$

We may now write the first eq. (39):

$$M_0 - \frac{\mu_{01}(t_0)\, a_{01_x}(t_0) + \mu_{02}(t_0)\, a_{02_x}(t_0)}{p_x(t_0)} = \int_0^{t_0} [\mu_{01}(t_0) + \mu_{02}(t_0)]\, dt_0.$$

Hence:

$$\left.\begin{array}{l} -\dfrac{d}{dt_0}\left(\dfrac{\mu_{01}(t_0)\, a_{01_x}(t_0) + \mu_{02}(t_0)\, a_{02_x}(t_0)}{p_x(t_0)}\right) = \mu_{01}(t_0) + \mu_{02}(t_0) \\ \text{and similar for } y \text{ and } z. \end{array}\right\} \quad (40)$$

Now let us assume $\Delta m_{01}{}^*$ and $\Delta m_{02}{}^*$ constant with time, and let us write again: $a_{02} = k_1 a_{01}$, $\Delta m_{02}{}^* = k_2 \Delta m_{01}{}^*$. We then obtain:

$$-\frac{d}{dt_0}\left(\frac{\Delta m_{01}{}^* \dfrac{a_{01_x}}{\sqrt{1 - (a_{01_x}{}^2 + a_{01_y}{}^2 + a_{01_z}{}^2)/c^2}} + k_2 \Delta m_{01}{}^* \dfrac{k_1 a_{01_x}}{\sqrt{1 - k_1{}^2 (a_{01_x}{}^2 + a_{01_y}{}^2 + a_{01_z}{}^2)/c^2}}}{p_x(t_0)}\right) =$$

$$= \frac{\Delta m_{01}{}^*}{\sqrt{1 - (a_{01_x}{}^2 + a_{01_y}{}^2 + a_{01_z}{}^2)/c^2}} + \frac{k_2 \Delta m_{01}{}^*}{\sqrt{1 - k_1{}^2 (a_{01_x}{}^2 + a_{01_y}{}^2 + a_{01_z}{}^2)/c^2}}$$

Hence carrying out a stage further, the system (40) becomes:

$$\left.\begin{array}{l} \left(\Delta m_{01}{}^* \dfrac{a_{01_x}}{\sqrt{1 - (a_{01_x}{}^2 + a_{01_y}{}^2 + a_{01_z}{}^2)/c^2}} + \right. \\ \left. + k_2 \Delta m_{01}{}^* \dfrac{k_1 a_{01_x}}{\sqrt{1 - k_1{}^2 (a_{01_x}{}^2 + a_{01_y}{}^2 + a_{01_z}{}^2)/c^2}}\right) p_x'(t_0) + \\ - p_x(t_0) \left\{\Delta m_{01}{}^* \dfrac{d}{dt_0}\left(\dfrac{a_{01_x}}{\sqrt{1 - (a_{01_x}{}^2 + a_{01_y}{}^2 + a_{01_z}{}^2)/c^2}}\right) + \right. \end{array}\right\} \quad (41)$$

$$\left.\begin{aligned} &\frac{+\, k_1 k_2 \Delta m_{01}{}^* \dfrac{d}{dt_0}\left(\dfrac{a_{01_x}{}^{\cdot}}{\sqrt{1 - k_1{}^2 (a_{01_x}{}^2 + a_{01_y}{}^2 + a_{01_z}{}^2)/c^2}}\right)\Big\}}{[p_x(t_0)]^2} = \\ &= \frac{\Delta m_{01}{}^*}{\sqrt{1 - (a_{01_x}{}^2 + a_{01_y}{}^2 + a_{01_z}{}^2)/c^2}} + \\ &+ \frac{k_2 \Delta m_{01}{}^*}{\sqrt{1 - k_1{}^2 (a_{01_x}{}^2 + a_{01_y}{}^2 + a_{01_z}{}^2)/c^2}} \\ &\text{and similar for } y \text{ and } z. \end{aligned}\right\} \quad (41)$$

In (41),

$$p_x'(t_0) = \frac{dp_x(t_0)}{dt_0} = \frac{dp_x(t)}{dt}\frac{dt}{dt_0} = p_x'(t)\cdot\sqrt{1 - \frac{[v(t)]^2}{c^2}},$$

in which again $t = t(t_0)$ must be substituted, and similar $p_y'(t_0)$ and $p_z'(t_0)$. The quantities $p_x'(t)$, $p_y'(t)$ and $p_z'(t)$ are obtained by differentiation of the left members of (31), whereas $\dot{v}_x$, $\dot{v}_y$, $\dot{v}_z$ and $\ddot{v}_x$, $\ddot{v}_y$, and $\ddot{v}_z$, appearing in the expressions, are again known functions following from (38). Further carrying out the differentiation in (41), we obtain a system of 3 simultaneous differential equations, which are linear in da_{01_x}/dt_0, da_{01_y}/dt_0 and da_{01_z}/dt_0, so that the equations can be solved with respect to these three differential quotients. We then obtain the system

$$\left.\begin{aligned} \frac{da_{01_x}}{dt_0} &= F_1(a_{01_x}, a_{01_y}, a_{01_z}, t_0) \\ \frac{da_{01_y}}{dt_0} &= F_2(a_{01_x}, a_{01_y}, a_{01_z}, t_0) \\ \frac{da_{01_z}}{dt_0} &= F_3(a_{01_x}, a_{01_y}, a_{01_z}, t_0). \end{aligned}\right\} \quad (42)$$

Let us again assume that at $t = 0$, $t_0 = 0$, $v_x = 0$, $v_y = 0$, $v_z = 0$. Then at the time of start $t_0 = 0$ (coinciding with $t = 0$), we obtain:

$$M_0 \sqrt{(\dot{v}_x)^2_{t_0=0} + (\dot{v}_y)^2_{t_0=0} + (\dot{v}_z)^2_{t_0=0}} =$$
$$= \frac{\Delta m_{01}{}^* (a_{01})_{t_0=0}}{\sqrt{1 - (a_{01})^2_{t_0=0}/c^2}} + \frac{k_1 k_2 \Delta m_{01}{}^* (a_{01})_{t_0=0}}{\sqrt{1 - k_1{}^2 (a_{01})^2_{t_0=0}/c^2}}$$

from which $(a_{01})_{t_0=0}$ can be solved numerically [because the functions (38) are prescribed, also the functions $\dot{v}_x = \dot{v}_x(t)$, $\dot{v}_y = \dot{v}_y(t)$, $\dot{v}_z = \dot{v}_z(t)$ are determined, so that also the quantities $(\dot{v}_x)_{t_0=0} = 0$, $(\dot{v}_y)_{t_0=0} = 0$, $(\dot{v}_z)_{t_0=0} = 0$ are known]. We then further obtain:

$$[(a_{01})_{t_0=0}]_x = (a_{01})_{t_0=0} \frac{(\dot{v}_x)_{t_0=0}}{\sqrt{(\dot{v}_x)^2_{t_0=0} + (\dot{v}_y)^2_{t_0=0} + (\dot{v}_z)^2_{t_0=0}}}$$

and similar

$$[(a_{01})_{t_0=0}]_y \quad \text{and} \quad [(a_{01})_{t_0=0}]_z.$$

[At $t_0 = 0$ (coinciding with $t = 0$) the directions of $\bar{a}_{01}$ with respect to the terrestrial system and rocket system of reference are identical.]

Now, starting from the initial state

$$t_0 = 0 \rightarrow [(a_{01})_{t_0=0}]_x, \ [(a_{01})_{t_0=0}]_y, \ [(a_{01})_{t_0=0}]_z,$$

we can integrate the system (42) by the method of successive approximations, or by the method of RUNGE-KUTTA. Again choosing the last one, we obtain the scheme:

$$t_0 \rightarrow h \qquad a_{01_x} \rightarrow k, \qquad a_{01_y} \rightarrow l, \qquad a_{01_z} \rightarrow n$$

$$\left.\begin{aligned} k &= \frac{1}{6} k_1 + \frac{1}{3} k_2 + \frac{1}{3} k_3 + \frac{1}{6} k_4 \\ l &= \frac{1}{6} l_1 + \frac{1}{3} l_2 + \frac{1}{3} l_3 + \frac{1}{6} l_4 \\ n &= \frac{1}{6} n_1 + \frac{1}{3} n_2 + \frac{1}{3} n_3 + \frac{1}{6} n_4 \end{aligned}\right\}$$

Then:

$$\left.\begin{aligned} k_1 &= F_1 (a_{01_x}, a_{01_y}, a_{01_z}, t_0)\, h \\ k_2 &= F_1 \left(a_{01_x} + \frac{k_1}{2}, a_{01_y} + \frac{l_1}{2}, a_{01_z} + \frac{n_1}{2}, t_0 + \frac{h}{2}\right) h \\ k_3 &= F_1 \left(a_{01_x} + \frac{k_2}{2}, a_{01_y} + \frac{l_2}{2}, a_{01_z} + \frac{n_2}{2}, t_0 + \frac{h}{2}\right) h \\ k_4 &= F_1 (a_{01_x} + k_3, a_{01_y} + l_3, a_{01_z} + n_3, t_0 + h)\, h \end{aligned}\right\}$$

and similar n and l.

The computation must be carried out in the sequence $k_1\, l_1\, n_1\, k_2\, l_2\, n_2\, k_3\, l_3\, n_3\, k_4\, l_4\, n_4$. The accuracy can again be estimated as above.

[After having determined:

$$a_{01_x} = a_{01_x}(t_0), \qquad a_{01_y} = a_{01_y}(t_0), \qquad a_{01_z} = a_{01_z}(t_0)$$

in this way, we obtain:

$$a_{01_x} = a_{01_x}(t_0(t)) = a_{01_x}(t) \text{ and similar } a_{01_y}(t) \text{ and } a_{01_z}(t).]$$

These quantities a_{01_x}, a_{01_y}, a_{01_z}, determined as functions of t_0, refer to the terrestrial system of reference.

Further the coordinates of the spaceship with respect to the terrestrial system of reference, as functions of t_0 become:

$$x(t) = x(t(t_0)) = x(t_0) \text{ and similar } y(t_0) \text{ and } z(t_0),$$

so that by these results of computation a program of navigation can be indicated, using different remote celestial objects as consecutive directive points. The required numerical value a_{01}, at any astronautical time t_0, then becomes:

$$a_{01} = \sqrt{[a_{01_x}(t_0)]^2 + [a_{01_y}(t_0)]^2 + [a_{01_z}(t_0)]^2}.$$

After these pure theoretical considerations, let us turn to the question what the formulae have to tell us about the prospects of interstellar space flight in a remote future. Firstly let us consider the power required. Let us call the quantity of kinetic energy per second astronautical time, for maintaining the jet, J (jetpower).

Then:

$$J = \Delta m_0^* c^2 \left(\frac{1}{\sqrt{1 - a_0^2/c^2}} - 1 \right). \tag{43}$$

The corresponding thrust then becomes:

$$P = \frac{\Delta m_0^* a_0}{\sqrt{1 - a_0^2/c^2}}. \tag{44}$$

Expanding in (43) $(1 - a_0^2/c^2)^{1/2}$, we obtain:

$$J = \Delta m_0^* c^2 \left[\frac{1}{2} \frac{a_0^2}{c^2} + \frac{3}{8} \frac{a_0^4}{c^4} + \text{etc.} \right].$$

Hence if $a_0 \ll c$, (43) and (44) reduce to:

$$J = \frac{1}{2} \Delta m_0^* a_0^2 \qquad \text{and} \qquad P = \Delta m_0^* a_0,$$

so that then:

$$J = \frac{1}{2} P a_0. \tag{45}$$

If we take as example $a_0 = 10{,}000$ km/sec $= 10^7$ m/sec, $(a_0/c)^2 = 1/900$, so that we can use the last formula (45). If we further assume for P the very small value of 15 kg we then obtain:

$$J = 7.5 \times 10^7 \text{ kg/sec} = 10^6 \text{ metric H.P.}$$

Hence for $a_0 = 10{,}000$ km/sec, we require for any 15 kg thrust a million H.P. If our interstellar space ship has a weight of 150,000 kg, we should obtain an acceleration of 0.0001 g if g = acceleration due to gravity at the terrestrial surface. This acceleration may be sufficient for an interplanetary electric space ship, for an interstellar space ship that must reach a final speed comparable with c, it is uttermost inapplicable. By the very long acceleration period all profit of relativistic time dilatation would be excluded. Further with this exhaust speed, we obtain according to (5) a mass ratio of astronomical magnitude. The space ship would have to jettisone itself completely. On the other hand, if we increase the acceleration to g, the jet power becomes 10^{10} metric H. P. Even if our propulsion instalment would have a thermal efficiency of 99% the space ship would at once be destroyed by the waste heat. And meanwhile the required mass ratio still remains the same: If we increase the exhaust speed a_0, the energy situation becomes still more severe. If we increase the initial mass of the space ship, maintaining the same acceleration and exhaust speed, the situation remains the same. We must confess that we are confronted with insurmountable difficulties.

Only if mankind succeeds in generating the enormous jet powers required practically without any waste heat development, interstellar space flight with a speed comparable with c will have a chance. Further fusion of nucleons will be insufficient. A complete conversion of matter in jet energy, without intermediate heat stage, will be necessary.

The solution of this formidable problem is perhaps not for ever excluded, but in every case it seems to be far beyond our technical range for the first ages to come.

And even if this problem will have been solved in a remote future, still enormous mass ratios will be required, as soon as we leave the stellar neighbourhood of the solar system.

In order to pass *in free flight* through a distance of about 100 lightyears — as measured by the terrestrial observer — in an *astronautical* time lapse of about 10 years, we require a time dilatation:

$$\frac{dt_0}{dt} = \sqrt{1 - \frac{v^2}{c^2}} \approx \frac{1}{10},$$

whence follows a travel speed

$$v = c\sqrt{1 - \frac{1}{100}} \approx 0{,}99\, c.$$

Hence with this speed a distance of 99 lightyears (as measured by the terrestrial observer) is passed through in an *astronautical* time lapse of 10 years. If $a_0 = c$ (photon rocket), the required mass ratio in order to obtain this speed becomes by (5):

$$\sqrt{\frac{1 + 0.99}{1 - 0.99}} \approx \sqrt{200} \approx 14.$$

If we like again to reduce this speed to zero at the end of the voyage, the required mass ratio becomes $(\sqrt{200})^2 = 200$. This simple consideration already shows that the nearest stars can then be reached within the range of a human life.

Now let us contemplate which mass ratio will be required in order to pass through *in free flight*, after a period of acceleration, a distance of 1.5×10^6 lightyears (= distance of nearest galaxy) in about 15 years astronautical time. Then the required time dilatation becomes:

$$\frac{dt_0}{dt} = \sqrt{1 - \frac{v^2}{c^2}} = \frac{15}{1.5 \times 10^6} = 10^{-5}.$$

Hence

$$\frac{v}{c} = \sqrt{1 - 10^{-10}} \approx 1 - \frac{1}{2} \cdot 10^{-10}.$$

By (5) the required mass ratio then becomes, if $a_0 = c$:

$$\sqrt{\frac{2 - \frac{1}{2} \cdot 10^{-10}}{\frac{1}{2} \cdot 10^{-10}}} \approx \sqrt{\frac{2}{\frac{1}{2} \cdot 10^{-10}}} = 2 \cdot 10^5.$$

If we should like not only to attain the travel speed $v = c\,(1 - \frac{1}{2} \cdot 10^{-10})$, but also again to reduce this speed to zero in order to visit the remote object, a mass ratio of $4 \cdot 10^{10}$ will be required:

From these figures it already follows that whatever may be our technical achievements in the remote future, still intergalactic distances will be invincible — at least if we like to pass through the way in one human life of today and base on the rocket as principle of locomotion. In fact Sänger comes to the same conclusion [3].

But why only to base our hope on the rocket principle? The mighty astronautical barrier: *gravitation*, is in itself still a complete mystery. Why is gravitation always attractive? Is there any interdependence between gravitation and electrodynamic forces? Is perhaps "gravitation" some weak mean effect of an internuclear state of motion? If so, it seems in principle probable to attack gravitation directly on the atomic level, by influencing this internal state, perhaps converting gravitational attraction into gravitational repulsion, with simultaneous neutralisation of all stresses due to inertia. Here we are scarcely at the very beginning of scientific research. But whatever may be the ultimate way according to which interstellar spaceflight will once be realized despite all obstructions (including the obstruction of interstellar matter), the application of the *time dilatation* will remain an essential independent feature. Therefore let us now turn to this point.

The time dilatation, as well as the LORENTZ contraction, which are fundamental features of the special theory of relativity, will be well known to the reader. The special theory of relativity is strongly confirmed by experiment, so that there can be no doubt about the real appearance of the time dilatation when the travel speed tends to the speed of light. However, by the reciprocity of these phenomena, as a consequence of the formulae describing the space and time interdependence, it seems at first view that the astronaute cannot have, *at the return*, any real advantage of the time dilatation as to the terrestrial inhabitant who was still staying at home. For by this reciprocity, from the astronaute's point of view the physical processes on earth slow down in the same ratio. However, these reciprocal relations are only valid for observers in free flight, moving with respect to each other with constant speed. As soon as the space ship accelerates with respect to the terrestrial observer, this reciprocity is disturbed. At the end of this paper we return to this point.

Further we have to bear in mind that the paradoxical reciprocity is only a consequence of our restricted outlook in terms of separated space and time. The physical world however is not a world of instantaneous objects, but a world of events. Events, which are simultaneous for one observer may have different time coordinates for another observer, so that the point events, constituting for one observer an object at a given time, will not constitute at all an object at a given time for another observer.

Hence an "instantaneous object" has no absolute existence, it is a product of pure mindspinning corresponding with our conventional restricted outlook on the world in terms of separated space and time, a result of our preference of permanence in the structure of events. The *events* are the basic constituents of the physical world. This world is neither static nor dynamic, but only *is*, in non temporal sense. Hence comparing the worldlines of the astronaute and the terrestrial inhabitant between the intersection points corresponding with the start and the return of the space ship, we have only to count the number of clock beats (point events) along each world line. (The clocks concerned are supposed to be physically quite similar.) We have then all reason to expect that the number of clock beats along the terrestrial world line will be the larger one, and that the gain of time for the astronaute will be practically in accordance with the result obtained by using the formulae of the special theory of relativity, the system of the terrestrial observer thereby considering as "the system at rest". Of course we now surpass the special theory of relativity, by adding — also independent on the general theory of relativity — the hypothesis that there must be a general metric field, defined by the total space time structure of the universe by which "the system at rest" can be defined in each special region. Is there any reason for such a supposition? In this connection it must be emphasized that in the special theory of relativity no explanation is given of the space and time interdependences, as described by the formulae. Now in cosmical respect, there is an enormous difference between the state of the terrestrial inhabitant and the astronaute, travelling with a speed nearly to c. The last one has not only this enormous speed with respect to the earth, but also with respect to all other masses, at least, of the cosmical neighbourhood. This last fact may have a predominant influence on the rate of all physical processes going on in the space ship and may ultimately be also responsible for the time dilatation and LORENTZ contraction as described by the formulae of the special theory of relativity.

According to the conceptions of MACH and EINSTEIN, the mass of any particle can be conceived as a function of its interdependence upon all other masses throughout the universe. This conception is strongly emphasized by modern quantum statistics.

However, we have already advanced above that the objective world of physics is not a world of material objects, but a world of events, not only extending in space, but also in time. In connection with this insight the principle of MACH must not be applied to a spatial conception "mass", but to the more general four-dimensional conception "action" (energy $\times$ time). From this a kind of four-dimensional causality emerges, which can be expressed as follows in mathematical terms.

The action density $F(x, y, z, t)$ at any point $xyzt$ of space time, in which x, y and z are space coordinates and t a cosmical time coordinate, must be a function of the distribution of action throughout all space time. In symbols, we then obtain:

$$F(x, y, z, t) = \int \psi(x, y, z, t, x', y', z', t')\, F(x', y', z', t')\, \sqrt{g'}\, dx'\, dy'\, dz'\, dt' \tag{46}$$

in which ψ represents the type of (unknown) interconnection, while the integral must be extended throughout space time, which may be curved and closed. $\sqrt{g'}$ is a function of x', y', z' and t', depending on the metric structure of space time.

For an intellect knowing the function ψ in specialized form, it is theoretically possible to solve this integral equation, by which all events throughout space time would be known.

For the rest, eq. (46) can be conceived in such a way that the action density $F(x, y, z, t)$ at any point $xyzt$ of space time can be considered, to some extent, as the superposition of the action densities at all other points $x'y'z't'$, any term of the summation being multiplied by a certain factor:

$$\psi(x, y, z, t, x', y', z', t')\, \sqrt{g'}\, dx'\, dy'\, dz'\, dt'$$

dependent on the nature of this interconnection. Hence, to some extent, we can say that at any point of space time, all other events throughout space time are represented.

An integral equation of the type (46) has in general no solution, unless the function ψ is of special form. If we write more generally $\psi = \lambda^{-1}\,\psi'$, in which λ is a parameter, we can find an approximate solution by putting:

$$F(x,y,z,t) = \sum_{i=1}^{n} c_i\, \gamma_i(x,y,z,t), \tag{47}$$

the γ_i being functions chosen in a suitable way, and the c_i unknown constants. Substituting (47) in the integral eq. (46) we find, after arranging:

$$\sum_{i=1}^{n} c_i\, [\lambda\, \gamma_i(x,y,z,t) - \xi_i(x,y,z,t)] = 0, \tag{48}$$

in which:

$$\xi_i = \int \psi'(x,y,z,t,x',y',z',t')\, \gamma_i(x',y',z',t')\, \sqrt{g'}\, dx'\, dy'\, dz'\, dt'. \tag{49}$$

The integral in (49) must be extended throughout space time. Determining the constants c_i in such a way that (47) satisfies (46) at n points scattered throughout space time, we obtain by (48) n homogeneous equations in the c_i, from which the unknown quantities c_i/c_j can be solved, after determination of λ from the equation:

$$D = 0, \tag{50}$$

D being the determinant of the system.

For the rest, the method of least squares can also be applied. Writing (48) in abbreviated form:

$$\sum_{i=1}^{n} c_i \eta_i = 0,$$

we then obtain for the c_i the set of equations:

$$\frac{d}{dc_j} \int \left[\sum_{i=1}^{n} c_i \eta_i \right]^2 \sqrt{g}\, dx\, dy\, dz\, dt = 0, \qquad (j = 1, 2 \ldots n),$$

in which $\sqrt{g}$ is a function of $x,y,z,$ and t.

Or:

$$\sum_{i=1}^{n} c_i \int \eta_i \eta_j \sqrt{g}\, dx\, dy\, dz\, dt = 0, \qquad j = 1, 2 \ldots n, \tag{51}$$

the integral being extended throughout space time. λ then follows from the equation:

$$D' = 0, \tag{52}$$

D' being the determinant of the system (51).

If the metric structure of space time is conceived as dependent on $F(x,y,z,t)$, the integral eq. (46) becomes an equation of non linear type. Further it may be of some interest to observe that, if space time were flat and infinite, and the action density finite throughout space time, and if further the function ψ were only dependent on the interval, the convergence of this function would necessarily be stronger than the convergence of the reciprocal of the fourth power of the interval.

Further the reversed problem, to determine the function ψ, if $F(x,y,z,t)$ throughout space time would be known, is of interest. We can then assume as approximation of ψ:

$$\psi = \sum_{i=1}^{n} c_i \gamma_i'(x,y,z,t,x',y',z',t') \tag{53}$$

in which the γ_i' are functions, chosen in a suitable way.

Substituting (53) in (46), we find:

$$F(x,y,z,t) = \sum_{i=1}^{n} c_i\, \alpha_i(x,y,z,t), \tag{54}$$

in which:

$$\alpha_i(x,y,z,t) = \int F(x',y',z',t')\, \gamma_i'(x,y,z,t,x',y',z',t') \sqrt{g'}\, dx'\, dy'\, dz'\, dt', \tag{55}$$

in which the integral must be extended throughout space time. Applying the method of least squares, the c_i can be computed from the equations:

$$0 = \frac{d}{dc_j}\int\left[F(x,y,z,t) - \sum_{i=1}^{n} c_i\,\alpha_i\right]^2 \sqrt{g}\,dx,\,dy,\,dz,\,dt \Bigg\}, \quad j = 1, 2 \ldots n \tag{56}$$

in which $\sqrt{g}$ again is a function of $x,y,z,$ and $t,$ the integral being extended throughout space time. Carrying out (56), we then obtain the system:

$$\int F(x,y,z,t)\,\alpha_j(x,y,z,t)\,\sqrt{g}\,dx\,dy\,dz\,dt = \sum_{i=1}^{n} c_i \int \alpha_i(x,y,z,t)\,\alpha_j(x,y,z,t)\,\sqrt{g}\,dx\,dy\,dz\,dt \Bigg\}, \quad j = 1, 2 \ldots n \tag{57}$$

from which the c_i can be solved.

For the rest it may be observed that, if we confine our attention to the distribution of action on a cosmical scale, disregarding the fluctuations of smaller order of magnitude, it is very well possible that the corresponding function ψ obtains a less complicated form, suitable for mathematical treatment.

From the foregoing it follows that no event exists on itself; an event is only real by its relations with all other events. Now we have to emphasize that the cosmical time coordinate t in eq. (46) may not be identified with the time coordinate in the special theory of relativity. It will be obvious that the system of coordinates $xyzt$ in (46), embracing the whole universe in space and time, must be chosen in such a way that thereby the "function of universal interdependence" ψ becomes as simple as possible. In order more nearly to determine this most suitable system $xyzt$, we must know something more of the space time structure of the universe as a whole.

Now the conception "universe" implies in the very first place the feature of *completeness*, so that the geometrical representation of the universe ought to express this feature. Now the plain idea of a time endless in past and future and of space of infinite extent, that is the conception of a universe infinite in space and time, has essentially the characteristic of non completeness. For whatever extension it may be thought to have, it always has to surpass its own limits in order to embrace more without ever becoming all embracing. But just the conception "universe" is equivalent with "all embracing existence". Hence if we really believe in the universe (= "all embracing existence"), we have to accept that it is of finite extent. Or better said: we must try to represent it in geometrical terms by an image of finite extent, because only in doing so an infinite regression of "all embracing" existence is avoided.

Further there must be with respect to any event, past and future, and with respect to any body, a surroundings of other bodies, so that the image may not be bounded, neither in time, nor in space. Hence space time must be not bounded and nevertheless finite, in the same way as a closed surface is not bounded in itself, although finite in extent. Another general point of interest is, that the physical properties of material systems vary if all dimensions are increased or decreased in the same ratio. Hence the geometry of space time has also to reflect this general feature, and must therefore be *not* invariant as to conformal transformation.

Going on, it seems a reasonable hypothesis that everywhere in the universe, throughout a region of sufficient large dimensions the general (over all) structure of interdependence of the events, that is the general structure of space time, is the same. In this way we come to the idea of a hyperspherical space time model, in which space time is represented by a four dimensional spherical surface.

Let us now turn — after these reflections — to the physical world as it appears to our senses. This visible world then appears to us as an enormous system of clusters of galactic systems (spiral nebulae) which are on a large scale evenly distributed in space.

Now the most remarkable feature of this gigantic system of clusters of galaxies is that it expands, in such a way that all the mutual distances increase with time, and every galactic cluster can be considered as the center of expansion. It is this universal tendency of scattering, by which we can really speak of a universe, instead of a mere haphazard aggregate of masses. According to HUBBLE, taking into account the newest estimates, the speed of recession increases at a rate of about 55 km/sec per million light years.

Hence assuming in first approximation this expansion as linear with time "since the beginning" (in which now time is a cosmical time coordinate), this beginning must be some 5500 million years ago. In our spherical space time, this "beginning" is no absolute birth of the universe, but only a radiant of worldlines of galactic clusters. In other words on the cosmical scale we have to do with a system of galactic worldlines, radiating from a point. These world lines will be geodesics (large circles on the four dimensional spherical surface). Then, in order to account for an approximate linear expansion "since the beginning", we must assume that the radius of space time is very large in comparison to the geodesic distance of the radiant. If we accept in this model for space and time the same natural units as in the special theory of relativity (which seem to express a still mysterious feature of the absolute world), this geodesic distance (in four dimensional sense) would be 5500 million light years.

The way in which this hyperspherical model of space time can be tested by an astronomical program of observation has already been indicated in a B.I.S. article [4]. The interesting question, if the galacted radiant observed be the only radiant of space time or not, will not be discussed here. If there were only one radiant, we should obtain an elliptical space time model, a slightly modified kind of spherical continuum in which opposite points are considered as identical.

Accepting the cosmical space time structure indicated above as a rough approximation of the truth, we are now enabled to specify our natural reference system x,y,z,t in eq. (46).

As expression of the squared line element of spherical space time we obtain:

$$ds^2 = r^2 \left[dT^2 + \sin^2 T \left\{dX^2 + \sin^2 X \left(d\theta^2 + \sin^2 \theta \, d\varphi^2\right)\right\}\right] \qquad (58)$$

in which r denotes the radius of space time and X, θ and φ are the coordinates of instantaneous spherical space. The time T (expressed in arc measure) is *cosmical time*; the cosmical time direction in any small and therefore practically flat region of hyperspherical space time is prescribed; it is the direction of the worldline of a point with respect to which the distribution of the speeds of the spiral nebulae is radial symmetric, the radial speeds increasing with the distance. For the rest the choice of the space coordinates X, θ and φ is not prescribed.

The center of gravity of the Milky Way system can practically be considered as such a point, and in connection with this, also the earth. (Thereby we assume that the speeds of the spiral nebulae are only due to the expansion of space; small superimposed motions are disregarded.)

From the above it follows that in the model advanced, *only by orientation with respect to the radiant of galactic cluster worldlines, the cosmical time coordinate can be defined.* This is in accordance with the remark advanced above, that the "cosmical expansion" is the first physical evidence that we can really speak of a universe and not only of a haphazard aggregate of events. If there would be more galactic radiants in space time, cosmical time comparable with our ordinary conception of time can only have physical significance in the neighbourhood of a radiant. Of course we can then mathematically extend our cosmical time (with "our" radiant as base of orientation) throughout spherical space time, in the same way as we can imagine meridians subdividing the terrestrial surface. But this extrapolated cosmical time will then only correspond with our sense of time in the neighbourhood of our radiant. If we split up the four dimensional world in space and time as defined by (58) we obtain a harmonically oscillating spatial world of which the intergalactic world space is spherical. The radius of space curvature then becomes in connection with (58):

$$R = r \sin T. \tag{58'}$$

Then in connection with the natural units of space and time as they appear to give the most simple description in the range of the special theory of relativity, we may expect that the present radius of curvature will be of the order of magnitude of 5500 million light years.

However, we have to bear in mind that the actual world is the hyperspherical world of events, and that the conception of an oscillating spatial world is only a restricted view, approaching our ordinary life conceptions. Further this oscillating spatial world is only the complete counterpart of the whole if the structure of the visible universe (in space and time) can be extrapolated over the whole. In that case of elliptical space time — which in itself only *is*, being neither static, nor dynamic, T only varies from 0 to $\pi/2$. Returning to the integral eq. (46) this equation becomes in case of spherical or elliptical space time, as defined by (58):

$$F(X, \theta, \varphi, T) = r^4 \int \psi(X, \theta, \varphi, T, X', \theta', \varphi', T')$$

$$\cdot F(X', \theta', \varphi', T') \sin^3 T' \sin^2 X' \sin \theta' \, dX' \, d\theta' \, d\varphi' \, dT', \tag{59}$$

in which the integral must be extended throughout space time.

Now returning to relativistic rocket mechanics, it seems logical to introduce the general metric field defined everywhere (or at least in our cosmical space time neighbourhood) by the observer for which the distribution of the recession speeds of the clusters of galaxies is radial symmetric. In connection with the foregoing considerations it seems very probable that the speed of any vehicle with respect to this general metric field will be of predominant importance as to the processes, going on in the vehicle. The coordinates in (59) also refer to this general metric field.

It will be obvious that a terrestrial system of reference including terrestrial time, with the earth as origin, not rotating with respect to the celestial sky (or better the sky of galaxies) practically coincides with the general metric field as defined above, in the small and therefore flat region of our cosmical neighbourhood.

The function ψ in (59) must be such that the solution $F(X,\theta,\varphi,T)$ of the eq. (59) gives a distribution of action density, defining the events throughout space time in such a way that in any small and therefore flat region also the reciprocal Lorentz contraction and the reciprocal time dilatation is accounted for, as described by the formulae of the special theory of relativity, if we compare the measurements of an observer at rest with an observer in uniform motion, with the special addition that "at rest" refers to the general metric field. By this addition we only *seem* to return to the old conception of absolute motion. For there are no instantaneous objects in the four dimensional world "which move", but only *events*.

In our considerations above we have still assumed that the red shift of spectral lines of the distant nebulae must be explained in convential way as DOPPLER-effect, leading to the conception of an expanding universe. However, we have still to bear in mind that in cosmology we have to deal with values of time, of distances and of large scale physical phenomena which far exceed our terrestrial experience.

Considering the latest preliminary results of the investigation of the large scale distribution of matter, it appears that the stellar systems are arranged in *clusters of galaxies*. These clusters appear to be space fillers, so that only by way of exception a galaxy belongs to no definite cluster at all. In case of globular clusters of galaxies, the density distribution appears to be similar as in an isothermic gassphere, satisfying the well-known differential equation of EMDEN [5]. Hence the tendency of clustering is produced by gravitation. Further the average speed of galaxies in a cluster relative to their neighbours is of the order of magnitude of 100 km/sec. to 5000 km/sec.; on the other hand, no lateral motions of galactic clusters with respect to one another have been observed. This whole dynamical image of the visible world points into the direction of a *static* universe. Moreover, investigating the frequency of clusters of galaxies as a function of their angular diameters, the preliminary results of observation also favour the hypothesis of a *non*-expanding universe, whereas the radius of space curvature (assuming spherical space) obviously appears still to be too large for detection [6]. Therefore ZWICKY suggests that the universal red shift in the spectra of very remote objects may be caused by a transfer of energy from the traveling light quanta to a possible immense background of zero energy of the whole universe.

Hence, in connection with these newest observational results we have — for sake of completeness — also to consider the possibility that the spatial universe is non-expanding. Thereby also the discordance between the age of the spatial universe and the age of the celestial objects, as required in astrophysical theories, would disappear.

Now in case of a static universe, if intergalactic space is spherical — which again may be assumed in connection with the same general reasons as advanced before — the radius of space curvature must be constant with "time", in which "time" refers to any observer at rest with respect to the galactic cluster centers.

Then as to space time as a whole, assuming homogeneity of metric structure, two main possibilities can be advanced: space time may be either cylindrical, being infinite in time direction, or space time may be spherical.

It seems not to be very probable that the image of a cylindrical space time is in accordance with the actual universe. For if there were an infinite past time, it seems inconceivable that the universe is still in a state of energy far from complete dissipation.

If space time is spherical and the radius of space curvature is constant with time, instantaneous space can only be represented by a largest three dimensional spherical manifold in space time, so that the radius of space curvature and the radius of space time must coincide and be equal every-where. This largest threedimensional spherical manifold is then at the *here-now* in question, perpendicular to the local time direction, which may vary with the local entropy gradient or direction of worldlines. Hence this instantaneous space must be only considered as an extrapolation of the flat spatial region of this here-now.

The radius of space curvature (identical with the radius of space time) could then be determined in principle by counts of centers of clusters of galaxies in different concentric spherical shells of definite thickness, as follows: let n be the number of centers of galactic clusters per unit of volume in our flat neighbourhood. Then the total number of galactic clusters at distances between RX and $R(X + dX)$ (in which X denotes the geodesic distance in arc measure) becomes:

$$4\pi R^2 \sin^2 X \cdot R\,dX \cdot n = 4\pi R^3 \sin^2 X\,dX \cdot n = A.$$

The distance (between RX and $R(X + dX)$ can be measured by using the constant of Hubble for the red shift, determined within our flat neighbourhood, assuming in first approximation that this quantity is independent on the distance of the clusters. Hence we have to count the number of galactic clusters of which the galaxies have red shifts between the corresponding limits, so that A can be determined observationally. If we only investigate in this way a solid angle ω of the sky, and denote the corresponding counted number by A', we obtain:

$$A = \frac{4\pi}{\omega} A'.$$

On the other hand, if intergalactic space were flat, the number of galactic clusters at distances RX and $R(X + dX)$ would be:

$$n \cdot 4\pi R^2 X^2 R\,dX = n \cdot 4\pi R^3 X^2\,dX = B,$$

so that B can be computed. We then further obtain:

$$\frac{A}{B} = \frac{\sin^2 X}{X^2}, \qquad \text{or} \qquad \sqrt{\frac{A}{B}} = \frac{\sin X}{X}. \tag{60}$$

Further the distance $D = RX$, hence

$$R = \frac{D}{X}. \tag{61}$$

Now in (60):

$$\sin X = X - \frac{X^3}{3!} + \frac{X^5}{5!} - \text{etc.},$$

so that:

$$\sqrt{\frac{A}{B}} = 1 - \frac{X^2}{3!} + \frac{X^4}{5!} - \text{etc.}$$

Neglecting X^4, etc., we obtain:

$$\frac{X^2}{3!} = 1 - \sqrt{\frac{A}{B}}, \qquad \text{or} \qquad X = \sqrt{6\left(1 - \sqrt{\frac{A}{B}}\right)}. \tag{62}$$

Hence the radius of space curvature the follows from:

$$R = \frac{D}{\sqrt{6(1 - \sqrt{A/B})}}. \tag{63}$$

Of course R may then turn out to be much too large for detection with the telescopic power available for the first time to come.

Concluding, it will be obvious that in our considerations of relativistic rocket motion above, the general metric field will now (in case of a static universe) be determined by all observers everywhere at rest with respect to the general mass of galactic clusters.

If S_1, S_2 and S_3 are the centers of three galactic clusters, we can define S_1 as the origin of a rectangular system xyz, whereas S_2 is situated on the x axis and S_3 in the xz plane. We thereby assume that the spatial region embracing S_1, S_2 and S_3 is sufficiently small, so that it can be considered as a flat region. *In case of an expanding universe,* S_2 and S_3 have only radial speeds with respect to S_1, so that as to the system xyz there will be no centrifugal field due to rotation. The system xyz then locally represents the general metric as defined above. Only with respect to this system xyz, or as to a system $x'y'z'$ having a *uniform motion* with respect to system xyz, the fundamental dynamical eq. (2) may be applied — disregarding gravitational fields.

Hence from the dynamical point of view, only the *acceleration* as to the general metric field (being locally the system xyz mentioned above), or as to any other inertial system $x'y'z'$ as defined above, is of fundamental importance. By this acceleration in general the mass

$$\frac{m_0}{\sqrt{1-(v^2/c^2)}}$$

increases, and remains constant as long as v remains constant. The time dilatation can then be dynamically "explained" as due to this increase of mass.

Summarizing the interdependence of general metric field and time dilatation we can say: only if the space ship accelerates or decelerates with respect to the general metric field there will be as to the space ship system of reference an inertia field. Also, if the acceleration of the space ship with respect to the general metric field is unless zero, there will be a rate of variation of the rate of the space ship clock as to any system of reference having a uniform motion with respect to the general metric field. This rate of variation will be a deceleration of the rate of the space ship clock if the relative speed of the space ship clock increases and an acceleration of the rate of the space ship clock, if the relative speed of the space ship decreases.

(In mathematical symbols we have:

$$\frac{dt_0}{dt} = \sqrt{1-\frac{v^2}{c^2}} = \left(1-\frac{v^2}{c^2}\right)^{1/2}$$

$$\frac{d^2t_0}{dt^2} = -\frac{(v/c^2)\,(dv/dt)}{\left(1-(v^2/c^2)\right)^{3/2}}$$

in which v denotes the numerical value of the relative speed.
Hence if

$$\frac{dv}{dt} > 0, \frac{d^2t_0}{dt^2} < 0,$$

$$\frac{dv}{dt} < 0, \frac{d^2t_0}{dt^2} > 0.$$

As soon as

$$\frac{dv}{dt} = 0, \frac{d^2t_0}{dt^2} = 0,$$

so that then the rate of the space ship clock with respect to the used inertial system of reference will be constant. In accordance with the rate of the space ship clock, the rate of all physical processes going on in the space ship will be varied.)

References

1. H. G. L. KRAUSE, Relativistische Raketenmechanik. Astronaut. Acta **2**, 30 (1956).
2. J. ACKERET, Zur Theorie der Raketen. Helvet. Physica Acta **19**, 103 (1946).
3. E. SÄNGER, Die Erreichbarkeit der Fixsterne, Abbildung 2. Kongreßvortrag Rom, 1956.
4. J. M. J. KOOY, Space Travel and Future Research into the Structure of the Universe. J. Brit. Interplan. Soc. **15**, 248 (1956).
5. F. ZWICKY, Morphological Astronomy, p. 135. Berlin-Göttingen-Heidelberg: Springer, 1957.
6. Ibidem, pp. 166–170.

Interplanetary Ballistic Missiles
A New Astrophysical Research Tool[1]

By

S. F. Singer[2], ARS

(*Received April 4, 1957*)

Abstract — Zusammenfassung — Résumé

Interplanetary Ballistic Missiles, a New Astrophysical Research Tool. The idea of exploding an H-bomb on the moon must have occurred to many people and no originality can be claimed for the concept. However, no proposal seems to have been put forward which is at all realistic or indicates a good reason for such an undertaking. Yet an agreement on atomic disarmament may raise the question of how to dispose of the bombs usefully.

The purpose of the present paper is to show that the explosion of an H-bomb carried to the moon by means of a missile is completely feasible and approximately equal in difficulty to the firing of an ICBM (Intercontinental Ballistic Missile) with an H-bomb warhead; to show that it is desirable from the point of view of testing, and worthwhile scientifically. These three factors combined, in the author's opinion, make such a project highly desirable. Most of the paper is devoted to the many astrophysical applications of this new research tool; the legal and geopolitical aspects of this operation are beyond the scope of this paper.

Interplanetarische ballistische Geschosse, ein neues astrophysikalisches Forschungsgerät. Der Gedanke, eine Wasserstoffbombe auf dem Mond explodieren zu lassen, muß vielen Leuten gekommen sein; für diesen Plan kann keinerlei Originalität beansprucht werden. Immerhin scheint kein Vorschlag gemacht worden zu sein, der überhaupt realistisch ist oder eine gute Begründung für ein solches Unternehmen gibt. Ein Abkommen über Atomwaffenabrüstung muß aber die Frage aufwerfen, wie man die Bomben nützlich verwenden kann.

Der Zweck der vorliegenden Arbeit ist es, zu zeigen, daß die Explosion einer Wasserstoffbombe, die mittels eines ferngelenkten Flugkörpers auf den Mond gebracht wird, durchaus ausführbar und in ihrer Schwierigkeit annähernd gleich dem Abfeuern eines interkontinentalen ballistischen Flugkörpers mit einem Wasserstoffbomben-Kriegskopf ist. Es soll ferner gezeigt werden, daß dies einerseits vom Standpunkt der Erprobung wünschenswert, anderseits wissenschaftlich lohnend ist. Die Zusammenfassung dieser drei Gesichtspunkte macht nach Ansicht des Verfassers ein solches Projekt äußerst anstrebenswert. Der größte Teil der Untersuchung ist den vielerlei astrophysikalischen Anwendungen dieses neuen Forschungsgerätes gewidmet; die gesetzliche und geopolitische Seite des Unternehmens liegt außerhalb des Gesichtskreises dieser Arbeit.

---

[1] Published in Astronaut. Acta **4**, 59—69 (1958).

[2] Professor of Physics, Physics Department, University of Maryland, College Park, Md., U.S.A.

Engins balistiques interplanétaires, instrument nouvel de recherche de l'astrophysique. L'idée de faire exploser une bombe H sur la lune doit avoir germé dans plus d'un esprıt et ne peut être considéré comme originale. Cependant aucune proposition réaliste ou bien motivée n'a été avancée pour la réalisation d'un tel projet. Et pourtant un accord en matière de désarmement atomıque soulèverait le problème d'utıliser à bon escient les bombes existantes.

Le but du présent article est de montrer que l'explosıon d'une bombe H, portée jusqu'à la lune par un engin, est une entreprise réalisable, d'une difficulté équivalente au lancement d'une fusée balistique intercontinentale avec tête atomique. De plus cette entreprise, interéssante en tant qu' essai, serait riche d' enseignements scientifiques. Dans l'opinion de l'auteur, ces facteurs combinés rendent la réalisation de ce projet hautement désirable.

L'article est surtout consacré aux nombreuses applications astrophysiques de ce nouvel instrument de recherche. Les aspects légaux et géopolitiques de l'opération sortent du cadre de ces considérations.

Introduction

It is safe to say that there are probably not very many peaceful applications for thermonuclear explosions. Yet if a general atomic disarmament is agreed upon, some applications must be found for existing bombs.

Certainly exploding such devices far away from the earth is a desirable step. An explosion on the surface of the moon is not only peaceful but also useful and, therefore, worthwhile. In particular, such explosions can provide a new astrophysical tool which can tell us a great deal about the environment of the earth, the properties of the interplanetary medium, and in general deepen our understanding about the nature of the universe.

The transportation of the thermonuclear device to the moon involves rockets and guidance techniques which are of importance for astronautics. It is, therefore, appropriate that this paper be presented at the VIIIth Congress of the International Astronautical Federation. In fact, we as astronautics enthusiasts should take special pride in the fact that astronautics provides at least one peaceful application of thermonuclear weapons.

A project to explode thermonuclear bombs on the surface of the moon raises many problems and questions. Some of them at least will be answered in this paper, e.g.: (1) How difficult is such a program compared, for example, with firing an ICBM; (2) What are the hazards involved in such an operation; (3) Will the flash from the explosion blind us; (4) Will the surface of the moon be destroyed; (5) Will the orbit of the moon be affected; (6) What can we observe about the explosion, in addition to the light flash; (7) What after-effects are likely to be encountered and how will they affect life on the earth; (8) What precisely can we learn about the nature of the interplanetary medium, about the environment of the earth, about the surface of the moon? (9) How could we start, and how soon?

The explosion of H-bombs on the moon is certainly an intermediate goal. A less ambitious proposal, and a necessary first step, would be to explode bombs beyond our atmosphere at varying distances. Such experiments can give us important information about astrophysical quantities, e.g., the density of hydrogen gas in the vicinity of the earth. But there has been something about the moon which has intrigued man for centuries and the idea of creating a permanent crater as a mark of man's work is an appealing one. Also the scientific pay-offs are considerably higher.

The growth potentials of this set of experiments are in two directions. (1) One would be to explode thermonuclear devices on other planets as well, in order to study the composition of their atmospheres, perhaps by studying the selective absorption of the light flash from the explosion or an induced photoluminescence. (2) Once we have learned to explode safely thermonuclear devices away from the earth, we can construct truly cosmic thermonuclear devices resembling stars and we can, therefore, initiate a new experimental field of astrophysics: star making. Much may be learned from such pursuits, perhaps even the secret of carrying on a safe continuous thermonuclear reaction on the earth for the purpose of producing electric power. Certainly this is one of the most challenging of all problems facing mankind at the present time.

The discussion in this paper will be on a non-mathematical level, although it will be made quite evident how the conclusions are calculated. For those who wish to do so, an appendix is available. The accuracy of calculation is not particularly high and can be improved mainly by using data on H-bomb explosions which are at the present time not openly available. However, order of magnitude results are quite satisfactory for the purpose of the present paper.

The purpose of this paper is to propose the establishment of IPBM's (Interplanetary Ballistic Missiles) which, armed with a hydrogen bomb warhead, could be used to blast craters on the surface of the moon. In this paper we (A) first develop a justification and discuss the many advantages of such a scheme and try to show how this type of weapon testing may promote world peace. We next take up (B) the interesting technical aspects of this operation to establish feasibility, and finally discuss (C) scientific benefits of this kind of operation.

A. Advantages of IPBM's

Sending IPBM's to the moon in order to explode H-bomb warheads seems to have several very attractive features, not the least of which is the large distance of the moon from the earth (384,400 km or 240,000 miles). As H-bombs become more powerful, the point will soon be reached where tests on the earth may present a health hazard to the world's population as a whole, including the nation conducting the tests. Our proposal would do away with this hazard by exploding H-bombs on the moon, and measuring their performance from the earth: (1) by means of gamma-ray detectors in the high atmosphere, (2) the light flash from the ground, (3) by carefully scaling the lunar crater produced by the explosion. By these methods (discussed later) one can judge quite easily the effectiveness and power of the H-bomb device. The H-bomb race between the big powers would then be reduced to the much more tractable problem of seeing who could make a bigger crater on the moon.

This scheme has many advantages: (1) First of all it can give a good record of the power of the H-bomb for all to see. By conducting the explosion on the moon and in full view of all nations, any competent scientist will be able to deduce the effectiveness of a similar explosion on the earth.

(2) H-bomb testing on the earth is difficult and expensive, particularly for the larger weapons[1]. It must be difficult also to measure the efficiency of the H-bomb explosion on earth because of the influence of the atmosphere, and terrain. For example, most of the radiant energy of the bomb is absorbed by the atmosphere which takes up momentum and produces a strong pressure wave. A test conducted on the moon in a condition of vacuum would permit a better calculation of the energy output and therefore of the efficiency of the bomb.

(3) The testing of long-range space vehicles is also difficult. There are no test ranges in existence which extend over half the globe, and in any case, it would be difficult to find someone to check the accuracy of the impact point at the other end. This problem does not exist with a moon rocket since there the impact point is clearly in view. However, to make it visible one needs an indicator and an H-bomb explosion does provide such an indicator. It is therefore apparent

[1] Cost of Operation Redwing is quoted as more than $ 150 million (Pegasus, Aug. 1957).

that the combination of an IPBM plus H-bomb makes a nice package for testing both the missile and the warhead. That this test is realistic both from the point of view of propulsion and guidance is discussed later in Section B.

(4) An additional point is that after the test is completed, and the data from the explosion have been reported, one is left with a nice small crater on the moon which is unnamed and therefore provides unique opportunities for perpetuating the names of presidents, primeministers, and party secretaries.

(5) There need be no limit to the size of such a lunar bomb. Since its use as a weapon on the earth is clearly impractical, it had better be termed a "device" instead of a "bomb". From an astrophysical point of view it would be interesting to construct and test truly cosmic fusion devices, to simulate processes in the interior of stars. With larger devices and higher temperatures one may test new types of fusion reactions; or investigate the effects of rapid rotation leading to mixing; or construct such fusion devices in concentric layers of varying composition to investigate the burning of stars where mixing is inhibited.

(6) Finally, we should mention the actual scientific benefits of such tests. The results of an explosion on the moon's surface can be used scientifically in various ways: a) it gives a means of checking the prediction of the theory of meteor craters; by knowing the energy output of the bomb and measuring afterwards the size of the crater. b) The explosion results in the ejection of a great deal of vaporized rock, of fine dust, and of rock in the form of rubble. A large fraction of the material escapes from the gravitational field of the moon and is captured by the earth's gravitational field. In entering the earth's atmosphere it provides a unique opportunity for meteor observations since the fragments will produce light trails and ionization in the upper atmosphere. The very smallest dust particles probably will not be visible in the atmosphere but their optical effects can be studied by means of the sunlight which they scatter. These problems are discussed in Section C. It will be shown there that studies of this artificially produced "cosmic dust" simulates an enhancement of interplanetary dust and can be used to study the motion of the interplanetary gas with respect to the earth, as well as the effect of interplanetary magnetic fields and of the earth's magnetic field on the motion of the interplanetary gas. We can also study the effects of the dust on the earth's weather and climate.

We have here a new type of probe for studying the conditions in the outermost regions of the earth's atmosphere. The prospects for astrophysics and geophysics from such tests are therefore quite exciting. In Section C we calculate the expected effects, their observability, and indicate the type of results which one may obtain and how they will illuminate for us the conditions existing in the earth's environment and in interplanetary space, as well as the properties of our own atmosphere.

B. Technical Aspects

We want here to compare the technical problems involved in constructing and testing an IPBM as against the conventional ICBM. The three points considered are propulsion, guidance, and the actual testing of the fusion device.

1. Propulsion

For a ballistic rocket one requires a burn-out velocity of orbital magnitude, approximately 25,000 feet per second; for a moon rocket about 35,000 feet per second or an increase of 40%. Offhand, therefore, the IPBM would seem to require a great deal more in the way of propulsion, perhaps one or two more stages. But this is not quite correct, mainly because the IPBM payload can be made quite light. For the IPBM the problem of re-entry does not exist since the moon has no atmosphere. Therefore, with no need to protect it from the atmospheric heating, all of the weight of the warhead can be given over to the actual H-bomb. Neither is it necessary to include fuzing mechanisms and other complicated bomb devices. For the IPBM, we need only a small clock, which activates the bomb for a short period after a time interval given by the travel time from the earth to the moon along a predetermined trajectory, plus a simple impact or proximity fuse. The resulting saving in weight of the warhead can, of course, be realized in a simplification of the propulsion system. This would seem to make the problem of an IPBM not much more difficult than an ICBM from a propulsion point of view[1].

2. *Guidance*

As can be seen from the comparison in Table I, the guidance problem for a moon rocket is simpler than the corresponding problem for an ICBM designed to hit a target at a distance of an earth radius. In order to make this comparison we have used data published by the RAND Corporation on sample trajectories from the earth to the moon; even if a different regime is used, the data in Table I give an approximate idea of the accuracy required. The lower accuracy required in the IPBM guiding system reflects itself, of course, in a reduction in the weight of the last stage and therefore a reduction in the size and complexity of the overall propulsion system.

Table I. *Comparison of guidance accuracy for IPBM and ICBM*

	Nominal Velocity	Velocity Error	Angular Error
IPBM[a]	35 000 ft/sec	+ 35[b] − 25	± 0.1°
ICBM[c]	20 800 ft/sec	± 2[d]	± 0.02° horizontal plane ± 0.3°[d] vertical plane

a) GEORGE H. CLEMENT, The Moon Rocket, in "Earth Satellites as Research Vehicles". Philadelphia: Franklin Institute, 1956.
b) To hit surface of moon facing the earth.
c) For a range of 1 earth radius (4000 miles) and a range accuracy of ± 1 mile (along a minimum energy trajectory).
d) R. C. WENTWORTH, private communication.

3. *Bomb Testing*

If we take a nominal 20 megaton H-bomb[2], we can calculate the size crater produced on the moon. Within the accuracy of available theories and assuming a detonation below the surface layer of the moon, this turns out to be a crater of angular diameter one second of arc, which is just visible. For a comparison of crater sizes see Table II. The release of this energy will produce a light flash corresponding to a luminous intensity as a star of magnitude −16, or 1/10,000 that of the sun. If we assume a visual acuity of one minute of arc, then to a dark adapted eye the point of light from the H-bomb flash will be twice as bright as a point on the sun. The gamma-rays and neutrons from the explosion will not reach the lower atmosphere and present no hazard; measurement

[1] It is unfortunately not possible to supply actual numbers here.
[2] 1 MT (megaton TNT equivalent) = 4.2×10^{22} ergs.

of their radiation flux will, however, give considerable information on the operational efficiency of the bomb and on the absorption properties of the interplanetary gas between the moon and the earth (see discussion in Section C). As can be seen, therefore, exploding an H-bomb on the moon allows an unexcelled measurement of its power, efficiency and radiation properties.

Table II. *Observable effects of lunar H-bomb explosions*

	Diameter of Crater (ft)	Depth (ft)	Angular Diameter at Distance of Moon	Crater Volume (cu mi)
(Arizona meteorite)	4150	700	0.67″	
10 MT bomb[1]	5500	950	0.9″	3×10^{-2}
(Ungava, Quebec meteorite)	11,500	1,330	1.9″	
100 MT bomb[1]	12,500	1,800	2.0″	3×10^{-1}
1000 MT bomb[1]	25,000	3,000	4.0″	2.4
Piazzi Smyth (lunar craters)	6 mi.	3,500	5.1″	
Mädler (lunar craters)	20 mi.	7,500	17.2″	
Tycho (lunar craters)	54 mi.	12,000	46″	
Copernicus (lunar craters)	56 mi.	11,000	48″	
Clavius (lunar craters)	146 mi.	16,100	125″	

[1] Assuming an explosion below the surface of the moon.

C. Scientific Aspects

The scientific applications of the lunar H-bomb explosion as a new astrophysical research tool are so numerous and of such great variety that it might be best to organize them under various headings. We will discuss, therefore: (1) the visible phenomena seen from the ground, (2) the X-ray and gamma-ray phenomena seen from the top of the atmosphere, and (3) the radio phenomena observable at the earth. All of these constitute electromagnetic radiations coming at the instant of the explosion. Shortly after the explosion, there will arrive on the earth corpuscular radiation consisting mainly of neutrons. Their measurement is discussed under *4*. Finally there will arrive on the earth particles from the lunar crater, ranging from tiny dust particles which filter slowly through the earth's atmosphere, to larger drops which become solidified, up to fair-sized boulders which can be recovered and analyzed chemically; they lose little mass in traversing the earth's atmosphere. The rock gas and the extremely fine dust ("smoke") from the explosion will not reach the earth or at least will not be easily detectable as it reaches the earth; but because of its light scattering power it can be seen by its optical effects, as discussed under point *1*.

1. Visible Effects[1]

We can describe in a non-mathematical way what an observer on earth may see. Let us assume that the impact occurs on the dark side of the moon, just beyond the terminator line; the observer is located on the night side of the earth. Now a nuclear fusion explosion releases a lot of energy per gram of material, 10^7 times higher than TNT, which releases 4.2×10^{22} erg/10^{12} gm

[1] The use of bomb explosions to produce light flashes has been suggested by K. A. Ehricke and others, to aid in the acquisition and tracking of space vehicles.

while in the transformation $4\,H \rightarrow He$ about 1/1000 of the mass appears as energy according to the EINSTEIN relation $E = MC^2$. Therefore 1 gm of bomb material produces $10^{-3} \times (3 \times 10^{10})^2 \sim 10^{18}$ erg/gm, assuming 100% efficiency. This energy appears mainly in the form of radiation, although a small fraction will undoubtedly go to the kinetic energy of the bomb material and container. One-half of this energy goes outward from the moon and will be visible on the earth as a brilliant flash of light. The ultraviolet portion of the radiation is absorbed in the high atmosphere, but the gamma-rays will be able to penetrate quite deep into the atmosphere and should be detectable at balloon altitudes. After this initial flash is over, we will begin to see the effects of the radiation directed towards the moon. Again the visible radiation will be absorbed by the adjacent rock which will be immediately vaporized. The gamma-rays and neutrons will penetrate deeper and extend the volume of nearly instantaneous vaporization. The rock gas in this deeper layer will be under high pressure which is released in an explosion. It is this explosion which gouges out a large crater. In the immediate vicinity of the explosion the material will be vaporized; at some distance the rock will be molten, and further still it will be pulverized and broken up into rubble. All of this material will be ejected with high velocity. Most of the volume is ejected in the form of rock dust and rock rubble, very similar to the case of a crater produced by the impact of a meteorite.

The hot rock gas (with a temperature of at least 2200 °K) will be visible as a point of light, of orange-red color; it slowly expands as the hot gas diffuses from the point of explosion. Its rate of cooling is very slow so that the glow will remain visible for many minutes.

Some seconds after the explosion a new phenomenon comes into view. The escape velocity of the moon is only 2.3 km/sec, while the mean thermal velocity of the rock gas is ~ 1 km/sec and the rock dust, too, is ejected with high velocities. As the gas and dust rise above the lunar surface, they will enter the region of solar illumination, with spectacular consequences. The gas will scatter sunlight and start to give off a bluish color, almost as bright as the daylight sky, i.e. not much different from the brightness of the illuminated surface of the moon. The dust on the other hand will diffract sunlight and appear as a white dot which slowly expands.

It now becomes possible to study the diffusion and motion of the rock gas and rock smoke, the latter consisting of very fine particles of about optical size 0·5 micron to a few microns in diameter. Its rate of diffusion will indicate the density of the interplanetary gas between the moon and the earth, while its motion will indicate the motion of the interplanetary gas with respect to the earth. This question of the existence of a streaming effect or "interplanetary wind" is one of the most fascinating aspects of the whole investigation and could shed a great deal of light on the properties of the earth's outermost atmosphere, i.e. the region between 20 to 60 earth radii, the latter being the distance to the moon. In particular, we would like to know about the control of magnetic fields over the motion of the interplanetary gas by studying the motion of the rock gas and the rock smoke. Does the earth's magnetic field still predominate at these distances or has the interplanetary field become strong enough to take over? These studies can be closely tied in with such geomagnetic problems as the origin of the *aurora*, the existence of a ring current to explain *magnetic storms*, and the deflection of low energy *cosmic rays* by magnetic fields near the earth. A solution of the geomagnetic question by means of this new technique would therefore illuminate many of these most urgent astrophysical problems and would tie in with a whole range of other astrophysical phenomena.

2. *Ultraviolet, X-rays and Gamma-rays*

It may be expected that a good percentage of the electromagnetic radiation from the explosion will be in the ultraviolet region of the spectrum and even in the X-ray and gamma-ray region because of the high temperatures produced in the explosion and because of various nuclear reactions which go on. Some of the latter are a result of the absorption of neutrons by materials surrounding the point of explosion. In any case these radiations would come to the earth, having been absorbed more or less depending on their wavelength and depending on the nature and amount of interplanetary gas between the moon and the earth. Herein lies the most interesting application of the measurements, namely to find out what types of gas exists between the moon and the earth, i.e. in the earth's outermost atmosphere, what its state of ionization is, and how much of it there is. Probably the best way to tackle this problem is by measuring the spectrum of the ultraviolet and X-ray radiation at various levels of the ionosphere with rocket-borne spectrographs. A good fraction of the radiation will, of course, be absorbed only in the earth's atmosphere itself and will produce an enhanced ionization there which might be directly measurable from the ground by noting increased radio reflections, i.e. by ionospheric sounding methods. Once we reach the region of gamma-rays the absorption coefficient becomes small; they will be able to reach the atmosphere down to balloon altitudes so that measurements will be facilitated.

3. *Corpuscular Radiation*

A large number of neutrons are emitted in the fundamental fusion reactions of an H-bomb. The energies are as high as several Mev and may range down to very low energies depending on the nature of the materials surrounding the bomb. The fastest neutrons will arrive near the earth about 10 seconds after the explosion and those of lower velocity and lower energy will arrive later. The very slow neutrons will not arrive at the earth for two reasons: (1) They will decay along the way since their halflife is only 12 minutes: Then they become protons and will be affected by the earth's magnetic field and deflected away from the atmosphere. (2) Also, low energy neutrons are likely to be scattered very effectively by the hydrogen nuclei between the moon and the earth. Here again we have a means of measuring the amount of material between the moon and the earth provided our detection methods can be refined sufficiently.

4. *Radiosignal from the Explosion*

A rather new point which should be considered quantitatively is the production of a large radio burst similar to that produced by a stroke of lightning in the atmosphere. This burst of radio energy comes about because there is created suddenly a large amount of ionized material and the differential expansion of positive and negative charges will undoubtedly carry with it some electromagnetic effects. It now remains to be determined which of these can be detected on the earth. Effects of very high frequency, more than 50 megacycles, propagate through the ionosphere and could therefore be viewed from sea level by means of radio telescopes pointed towards the moon. In the interval between 50 down to perhaps 1 megacycle the ionosphere acts as an effective barrier and if observations are desired they would have to be carried out from rocket-borne receivers at altitudes of at least 200 km. Lower frequency radio waves, however, may penetrate through the ionosphere in a manner very similar to that of the so-called "radio whistlers". If so, they will follow the magnetic lines of force. It would be tremendously important to find out where the radio energy ends up on the earth: in other words, which magnetic line of force does it follow? There exist now many radio whistler receiving stations. They are quite simple to set up and require only a long wire which acts as an antenna and a high gain audio amplifier. Since large geographic coverage is necessary, amateurs could participate in the observations very effectively. The purpose would be mainly to find the impact zones of the radio power on the earth's surface and thereby establish the probable path of the radio signal and, therefore, the trajectory of the magnetic line of force.

5. *Lunar Material Captured by the Earth*

The lunar material captured by the earth must be massive enough so that the earth's gravitational field overcomes the forces of the interplanetary wind which tend to blow away the very fine particles, the rock smoke. Here we are considering particles of about 100 micron diameter and larger. These micrometeorites will enter the earth's atmosphere some considerable time after the explosion. Depending on their initial velocity, which must be larger than the escape velocity from the moon (about 2.3 km per sec), they will reach the earth with times as short as a few hours up to several days. This means that they will be distributed rather uniformly over the earth's surface. If they are charged by the solar ultraviolet radiation as recent studies indicate, and if the frictional effect on the gas between the moon and the earth is still appreciable, then they will enter the earth's atmosphere with velocities much smaller than the escape velocity of 11.2 km per sec. But long before they approach the earth's atmosphere itself they will be deflected by the earth's magnetic field because of their charge and low momentum. In fact, it is fairly sure that they will be deflected towards the poles and probably be observed only in the auroral zones. Because of their low velocity they will not produce any spectacular effects in the earth's atmosphere but filter down slowly into the atmosphere and be detected by collection methods from balloons or mountain tops. One interesting detection method which is not usually applicable to micrometeorites could be applied here. Since we can be fairly certain that these particles would be radioactive because of the intensity of neutron bombardment from the H-bomb explosion, radioactive or radiochemical techniques can be used to make the detectors much more sensitive.

One of the most interesting effects which these dust particles could produce are changes of the earth's weather and associated climatic changes. An important school of thought holds that climatic changes are caused by dust particles, either accreted by the earth as it moves through interstellar dust clouds, or due to dust produced in planetary or asteroidal collisions, or dust produced in volcanic eruptions on the earth itself. Here we can produce a controlled influx of dust in large quantities which can be used to verify some of these hypotheses, e.g. will the incoming dust provide additional nucleation centers for water vapor and thus cause increased rain fall as has been suggested by Bowen. Will this increased rain fall remove water vapor from the atmosphere, decrease the percentage of cloud cover, therefore lower the albedo and therefore increase the earth's mean temperature? Or will the incoming dust act chiefly as a reflector or diffractor of solar visible radiation, its size being too small to affect the infrared radiation emanating from the earth, with the end result that the albedo will increase and the earth's climate will get colder? It is possible that both effects will happen one after the other. It would be fascinating to be able to study this problem in a controlled laboratory type of investigation.

In a number of places on the earth there have been found little glassy objects called tectites which are thought to be of extraterrestrial origin. Urey, the noted cosmochemist, has suggested that they are lunar material which has been splashed out when meteorites impacted on the moon, that they solidified on the way to the earth and then landed at sea level. This represents another hypothesis which could be verified by a lunar H-bomb explosion.

Particular interest attaches to the larger or faster objects coming from the moon. The faster objects will act as meteors and produce luminous and ionization effects in the upper atmosphere which will provide us with a controlled means of studying the upper atmosphere, almost like artificial meteors. The larger objects, particularly those which enter with very low velocities, corresponding to the escape velocity from the earth, may land without having lost much of their mass and without having been greatly heated[1]. These objects might impact directly on the earth, or could be captured by perturbations, chiefly by the earth's atmosphere. Such braking ellipses will cause eventual entry velocities of less than 11 km/sec. With these samples in hand we have an unequalled opportunity of studying the nature of the lunar surface material long before we have an opportunity of visiting the moon in person. Even a simple measurement of the density of the rock sample will tell us whether there is stratification of rocks (as in the earth) indicating that the moon has been liquid at one time. Very fascinating chemical analyses can be carried out which undoubtedly will tell us more about the origin of the moon by comparing chemical ratios with those of the earth.

Of particular interest to the writer is the question of how long this surface material has formed the surface of the moon. This may be determined by a method which I have termed the "cosmic ray age" method. As cosmic rays bombard the surface of the moon, they will produce many nuclear disintegrations. In these nuclear disintegrations two interesting nuclides are produced in large quantities, tritium and helium-3. Tritium is radioactive with a halflife of about 12 years while helium-3 is stable. In the samples of lunar surface we could measure both the tritium and the helium-3. The tritium concentration would tell us immediately the present level of cosmic ray intensity hitting the moon. By comparing it with the cosmic ray intensity which we know to exist in free space we could, therefore, tell whether the moon possesses a magnetic field which keeps away some of the cosmic radiation. From the helium-3 content of the sample we can tell immediately how long the sample has been exposed to cosmic rays, i.e. how long it has formed part of the surface of the moon, because the cosmic rays will not produce helium-3 in any material which is more than a few centimeters below the surface. We would expect that the helium-3 content of the sample corresponds to a time interval between the time when the particular portion of the moon was hit by meteorites and the present. This method would then allow us to take various points on the lunar surface and establish a time history for the lunar surface and, therefore, for the bombardment of the lunar surface by meteorites. In this manner we might be able to state when various large meteorites hit the moon and we will be able to tell something about the prehistoric frequency of meteorite impacts. One might expect that it is much higher than the present rate of meteorite impacts but by how much? The answer to this question will tell us much about the origin of meteorites and asteroids.

In summary, we see that experiments carried on by H-bombs shot to the moon (and eventually to other planets) by means of IPBM's will touch on many

[1] The possibility of blasting loose portions of the moon and recovering them on the earth for chemical study has also been suggested by UREY in a recent issue of the journal "Observatory".

problems of current interest, only some of which have been mentioned here. Such experiments will open up fascinating prospects for astrophysics, for cosmochemistry and for studies of the origin of the solar system.

We hope to have demonstrated that the project is both feasible and worthwhile, and can therefore look forward to seeing it become reality in the near future.

Acknowledgements

The suggestion to think about the effects of H-bomb explosions in space came from FRED C. DURANT III in a conversation about two years ago. I am especially indebted to Dr. E. J. ÖPIK for discussing critically the contents of the present paper, for contributing many important ideas and for making available to me many of his calculations, some of them unpublished, on meteor craters and related topics.